Student Study Guide and Solutions Manual
for Atkins and Jones's

CHEMICAL PRINCIPLES
The Quest for Insight

Fifth Edition

John Krenos
Rutgers, The State University of New Jersey

Joseph Potenza
Rutgers, The State University of New Jersey

Carl Hoeger
University of California, San Diego

Laurence Lavelle
University of California, Los Angeles

Yinfa Ma
Missouri University of Science and Technology

W.H. Freeman and Company
New York

ISBN-13: 978-1-4292-3135-0
ISBN-10: 1-4292-3135-1

Printed in the United States of America

Third printing

W. H. Freeman and Company
41 Madison Avenue
New York, NY 10010
Houndmills, Basingstoke RG21 6XS, England

www.whfreeman.com

CONTENTS

PREFACE

This Study Guide and Solutions Manual accompanies the textbook *Chemical Principles: The Quest for Insight,* Fourth Edition, by Peter Atkins and Loretta Jones—an authoritative and thorough introduction to chemistry for students anticipating careers in science or engineering disciplines. We have followed the order of topics in the textbook chapter for chapter. The parallel between the symbols, concepts, and style of this supplement and the textbook enable the reader to easily move back and forth between the two.

Much of the Study Guide is presented in outline style with material highlighted by bullets, arrows, and tables. In general, bullets offset major items of importance for each section and arrows provide explanatory material, descriptive material, or some examples to reinforce a concept. This telegraphic style should help the student obtain a broad perspective of large blocks of material in a relatively short period of time, and may prove particularly useful in preparing for examinations. We believe that the Study Guide will be most useful following a careful reading of the text. Chapter **Sections** follow those in the text and contain descriptive material (bullets and arrows) as well as numerous examples. Many examples are worked out in detail. While we have covered most of the material in the text, we have not attempted to be encyclopedic. To help obtain a broad overview of the material, important equations are highlighted in boxes. Students may find this aspect of the guide particularly useful before exams. Where appropriate, we have introduced supplementary tables either to amplify material in the text, to clarify it further, or to summarize a body of material. Sprinkled throughout the guide are **Notes**; in the main, these are designed to point out common pitfalls.

The Solutions Manual section contains solutions and answers to the odd-numbered exercises in the textbook, including the Fundamentals sections. To display intermediate results we have disregarded rules concerning significant figures, but have adhered to them in reporting the final answers. In exercises with multiple parts, the properly rounded values are used for subsequent calculations. Student answers may differ slightly if only the final values are rounded, a method recommended by some instructors and the text, but not readily implemented in a printed manual.

ACKNOWLEDGEMENTS

The Study Guide authors, John Krenos and Joseph Potenza, are indebted to many individuals who have made significant contributions to it.

First and foremost, we wish to thank Beth Van Assen for a thorough and incisive reading of the drafts. Her critical suggestions led to improved readability and scientific accuracy of the Study Guide. Many examples and important sections of descriptive text were clarified and expanded with her help. Her encouragement, patience, and persistence were essential to the completion of this project.

We also thank two of our colleagues at Rutgers for critically reading three chapters. Harvey Schugar carefully reviewed Chapter 16. Our knowledge and appreciation of inorganic chemistry were broadened greatly by his comments and analysis. Spencer Knapp critiqued Chapters 18 and 19 and gave us a quick tutorial on modern organic chemistry. Many of the chemical structures were prepared with his help.

We also wish to gratefully acknowledge the help and encouragement of the staff at W. H. Freeman and Company. We especially thank Jessica Fiorillo (chemistry editor) for asking us to author this guide, Jodi Isman (project editor) for guiding us through the laborious process of developing the original manuscript from scratch and also for helping with the editing of this edition, and Dave Quinn (supplements editor) for encouraging us to keep to a tight schedule.

Finally, we extol the efforts of the authors of the textbook, Peter Atkins and Loretta Jones, in creating a new approach to teaching general chemistry. Beginning an introductory chemistry textbook with quantum theory is a logical, but challenging approach. The authors succeed by treating the fundamentals of chemistry (an "as needed" review) in a separate section and by minimizing the coverage of a great deal of material (mostly historical) presented in other books. In our opinion, this approach leads to the development of all the major topics in chemistry enlightened and enlivened in the first instance by the molecular viewpoint.

The Solutions Manual authors, Laurence Lavelle, Yinfa Ma, and Carl Hoeger, would like to thank Loretta Jones, and Danielle S. Sherwood for their invaluable assistance in ensuring the accuracy of this text. To Terry Kennair, who assisted with two plots that have been included in this manual. And finally, to Jenness Crawford for project managing this text with attentiveness.

STUDY GUIDE

Chapter 1 ATOMS: THE QUANTUM WORLD
INVESTIGATING ATOMS (Sections 1.1–1.3)

1.1 The Nuclear Model of the Atom

- **Subatomic particles**
    - → Electron: mass ($m = 9.1094 \times 10^{-31}$ kg)   charge ($-e = -1.602\,177 \times 10^{-19}$ C)
    - → Proton: mass ($m = 1.6726 \times 10^{-27}$ kg)   charge ($e = 1.602\,177 \times 10^{-19}$ C)
    - → Neutron: mass ($m = 1.6749 \times 10^{-27}$ kg)   charge $= 0$

- **Nucleus**
    - → *Nucleons* (protons and neutrons) occupy a small volume at the center of the atom. The binding energy of the nucleus is attributed to a strong force (nuclear) acting over a very short distance.
    - → The radius of the nucleus (assumed to be spherical) is given roughly by $r_{nuc} = r_0\, A^{1/3}$, where $r_0 \approx 1.3 \times 10^{-15}$ m $= 1.3$ fm.

- **Atom**
    - → Atomic number: $Z = N_p =$ number of protons in the nucleus
    - → Atomic mass number: $A = N_p + N_n =$ number of protons and neutrons in the nucleus
    - → Uncharged atom: $N_p = N_e$ (number of protons equals the number of electrons)
    - → *Electrons* occupy a much larger volume than the nucleus and define the "size" of the atom. The binding energy of the electrons is attributed to a weak force (coulomb) acting over a much longer distance than the strong force.

1.2 The Characteristics of Electromagnetic Radiation

- **Oscillating amplitude of electric and magnetic field**
    - → Wave characterized by *wavelength* and *frequency*
    - → Wave always travels at the speed of light (constant for a given medium)
    - → No known upper or lower limit of frequency and wavelength

Distance behavior (fixed time *t*)   **Time behavior (fixed position *x*)**

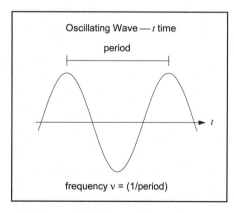

Speed of light (distance/time) = wavelength / period = wavelength × frequency

$$c = \lambda \nu$$ speed of light = wavelength × frequency

SI units: (m·s^{-1})   (m)   (s^{-1})   [1 Hz (hertz) = 1 s^{-1}]

Note: The speed of light c (c_0 in vacuum $\approx 3.00 \times 10^8$ m·s^{-1}) depends on the medium it travels in. Medium effects on wavelength in the visible region are small (beyond three significant figures).

1.3 Atomic Spectra

- **Spectral lines** $\rightarrow$ Discharge lamp of hydrogen

    H_2 + electrical energy $\rightarrow$ H + H* [$* \equiv$ asterisk denotes excited atom]

    H* $\rightarrow$ H$^{(*)}$ + $h\nu$ [$^{(*)} \equiv$ denotes a less excited atom]

- **Lines form a discrete pattern** $\rightarrow$ Discrete energy levels

- **Hydrogen atom spectral lines**
    - $\rightarrow$ Spectral lines imply discrete energy level for electrons in atoms
    - $\rightarrow$ Johann Rydberg's general equation
    - $\rightarrow$ $n_2 = n_{\text{upper}}$ and $n_1 = n_{\text{lower}}$

$$\nu = \mathscr{R}\left(\frac{1}{n_1^2} - \frac{1}{n_2^2}\right) \quad n_1 = 1, 2, \ldots \quad n_2 = n_1 + 1, n_1 + 2, \ldots \qquad \mathscr{R} = 3.29 \times 10^{15} \text{ Hz} = \text{Rydberg constant}$$

Rydberg expression reproduces pattern of lines in H atom emission spectrum. The value of $\mathscr{R}$ is obtained empirically.

Note: Lines with a common n_1 can be grouped into a series and some have special names:

$n_1 = 1$ (Lyman), 2 (Balmer), 3 (Paschen), 4 (Brackett), 5 (Pfund).

QUANTUM THEORY (Sections 1.4–1.7)

1.4 Radiation, Quanta, and Photons

- **Black body**
    - $\rightarrow$ Perfect absorber and emitter of radiation
    - $\rightarrow$ Intensity of radiation for a series of temperatures leads to two laws (Stefan-Boltzmann and Wien's)

    - $\rightarrow$ Stefan–Boltzmann law:

$$\frac{\text{Power emitted (watts)}}{\text{Surface area (meter}^2)} = \text{constant} \times T^4$$

    Wavelength corresponding to maximum intensity = λ_{max}

    - $\rightarrow$ Wien's law: $\boxed{T\lambda_{\text{max}} = \text{constant}}$ where constant = 2.88 K·mm

    At higher temperature, maximum intensity of radiation shifts to lower wavelength.

- **Energy of a quantum (packet) of light (generally called a photon)**
    - $\rightarrow$ Postulated by Max Planck to explain black body radiation
    - $\rightarrow$ Resolved the "ultraviolet catastrophe" of classical physics, which predicted intense ultraviolet radiation for all heated objects ($T > 0$)
    - $\rightarrow$ Quantization of electromagnetic radiation

$$\boxed{E = h\nu}$$ photon energy = Planck constant × photon frequency

SI units: (J) $(h = 6.6261 \times 10^{-34}\ \text{J·s})$ (s^{-1})

- **Photoelectric effect**
  - → Ejection of electrons from a metal surface exposed to photons of sufficient energy
  - → Indicates that light behaves as a particle

$$\boxed{E_K = h\nu - \Phi}$$ E_K = kinetic energy of the ejected electron, Φ = threshold energy (work function) required for electron ejection from the metal surface, and $h\nu$ = photon energy

  - → $h\nu \geq \Phi$ required for electron ejection

- **Bohr frequency condition**

$$\boxed{h\nu = E_{\text{upper}} - E_{\text{lower}}}$$ Relates photon energy to energy difference between two energy levels in an atom

1.5 The Wave–Particle Duality of Matter

- **Wave behavior of light** → Diffraction and interference effects of superimposed waves (*constructive* and *destructive*)
- **Matter has wave properties**
  - → Proposed by Louis de Broglie
  - → Consider matter with mass m and velocity v
  - → Such matter behaves as a wave with a characteristic wavelength

$$\boxed{\lambda = \frac{h}{mv}}$$ de Broglie wavelength for a particle with linear momentum $p = mv$

  - → Wave character of electrons verified by electron diffraction

1.6 The Uncertainty Principle

- **Complementarity of location (x) and momentum (p)**
  - → Uncertainty in x is Δx; uncertainty in p is Δp
  - → Limitation of knowledge (p and x cannot be determined simultaneously)

  $\boxed{\Delta p \Delta x \geq \hbar/2}$ Heisenberg uncertainty principle, where $\hbar = h/2\pi$

  - → $\boxed{\hbar \text{ is called "h bar"}}$ $\hbar = 1.0546 \times 10^{-34}\ \text{J·s}$

  - → Refutes classical physics on the atomic scale

1.7 Wavefunctions and Energy Levels

- **Classical trajectories** → Precisely defined paths

- **Wavefunction ψ** → Gives *probable* position of particle with mass m

- **Born interpretation** $\rightarrow$ Probability of finding particle in a region is proportional to ψ^2

- **Schrödinger equation**

    $\rightarrow$ Allows calculation of ψ by solving a differential equation

    $\rightarrow$ $H\Psi = E\Psi$; H is called the Hamiltonian

    $\rightarrow$ H represents the sum of potential energy and kinetic energy in a system

- **Particle in a box**

    $\rightarrow$ Mass m confined between two rigid walls a distance L apart

    $\rightarrow$ $\psi = 0$ outside the box and at the walls (boundary condition)

$$\Psi_n(x) = \left(\frac{2}{L}\right)^{1/2} \sin\left(\frac{n\pi x}{L}\right) \qquad n = 1,\ 2,\ \ldots$$

$\Psi_n(x) =$ wavefunction that satisfies the Schrödinger equation between the box limits. n is a *quantum number*.

Note: A node is a point in the box where $\psi = 0$ and ψ *changes* sign. $\boxed{\psi^2 \geq 0,\ \text{always}}$

$$E_n = \frac{n^2 h^2}{8mL^2}$$

$E_n =$ allowed energy values of a particle in a box

Note: $n = 1$ gives the zero-point energy E_1.

$E_1 \neq 0$ implies residual motion.

$$\Delta E = E_{n+1} - E_n = \frac{(2n+1)h^2}{8mL^2}$$

Energy difference between two neighboring levels

- **Probability as a function of position in the box**

    $\rightarrow$ Plot is shown for the first three levels

    $\rightarrow$ ψ^2 as a function of the dimensionless variable x/L

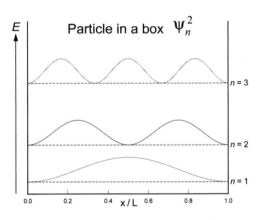

THE HYDROGEN ATOM   (Sections 1.8–1.11)

1.8 The Principal Quantum Number

- **Charge on an electron** $\rightarrow$ $q = -e = -1.602\,177\,33 \times 10^{-19}$ C

- **Vacuum permittivity** $\rightarrow$ $\varepsilon_0 = 8.854\,187\,817 \times 10^{-12}$ $C^2 \cdot N^{-1} \cdot m^{-2}$

- **Coulomb potential energy** $\boxed{V(r) = \dfrac{q_1 q_2}{4\pi\varepsilon_0 r}}$

$$V(r) = \frac{\text{product of charges on particles}}{4\pi \times \text{permittivity of free space} \times \text{distance between charges}} \qquad \text{(SI system)}$$

Units of $V(r)$: $J = \dfrac{C^2}{(C^2 \cdot N^{-1} \cdot m^{-2})\, m} = N \cdot m = (kg \cdot m \cdot s^{-2})\, m = kg \cdot m^2 \cdot s^{-2}$

→ For the H atom, $V(r) = \dfrac{(-e)(+e)}{4\pi\varepsilon_0 r} = -\dfrac{e^2}{4\pi\varepsilon_0 r}$

→ Solution of the Schrödinger equation using $V(r)$ above leads to energy levels E_n given below.

- **H atom and one-electron ion energy levels (He$^+$, Li^{2+}, Be^{3+}, etc.)**

  → Solutions to the Schrödinger equation

$$E_n = -h\,\mathcal{R}\left(\frac{Z^2}{n^2}\right) \qquad Z=1 \ \& \ n = 1, 2, 3, \ldots \qquad (n \text{ is dimensionless})$$

$$\mathcal{R} = \frac{m_e e^4}{8 h^3 \varepsilon_0^2} = 3.289\,842 \times 10^{15} \text{ Hz} \qquad \text{Units:} \quad \frac{kg \cdot C^4}{(J \cdot s)^3 (C^2 \cdot N^{-1} \cdot m^{-2})^2} = s^{-1} \equiv Hz$$

→ All quantum numbers are dimensionless. With $\mathcal{R}$ in units of frequency, the H atom energy-level equation has the same form as the Planck equation, $E = h\nu$.

→ E_n = energy levels (states) of the H atom. Note the *negative* sign.

→ $\mathcal{R}$ = Rydberg constant, calculated exactly using Bohr theory *or* the Schrödinger equation

→ Z = atomic number, equal to 1 for hydrogen

→ n = principal quantum number

→ As n increases, energy increases, the atom becomes less stable, and energy states become more closely spaced (more dense).

→ Integer n varies from 1 (ground state) to higher integers (excited states) to ∞ (ionization).

→ Energies of H atom states vary from $-h\mathcal{R}$ ($n = 1$) to 0 ($n = \infty$). States with $E > 0$ are possible and correspond to an ionized atom in which the energy > 0 equals the kinetic energy of the electron.

1.9 Atomic Orbitals (AOs)

- **Definition of AO**

  → Wavefunction (Ψ, psi) describes an electron in an atom.

  → Orbital (Ψ^2) holds 0, 1, or 2 electrons.

  → Orbital can be viewed as a *cloud* within which the point density represents the *probability* of finding the electron at that point.

  → Orbital is specified by *three* quantum numbers (n, ℓ, m_ℓ; see next page).

- **Wavefunction**
  - → Fills all space
  - → Depends on the *three* spherical coordinates: r, θ, ϕ
  - → Written as a product of a radial $[R(r)]$ and an angular $[Y(\theta,\phi)]$ wavefunction; mathematically, $\psi(r,\theta,\phi) = R(r)Y(\theta,\phi)$

- **Wavefunction for the H atom 2s orbital** $\rightarrow$ $n = 2$, $\ell = 0$, and $m_\ell = 0$

$$R(r) = \frac{\left(2 - \dfrac{r}{a_0}\right)e^{-r/2a_0}}{(2a_0)^{3/2}} \qquad Y(\theta,\phi) = (4\pi)^{-1/2} \qquad a_0 = \frac{4\pi\varepsilon_0\hbar^2}{m_e e^2} = \left\{ \begin{array}{l} 5.29177\times10^{-11}\ \text{m} \\ \text{(Bohr radius)} \end{array} \right\}$$

$$\text{Units: } R(r) = \text{m}^{-3/2} \qquad Y(\theta,\phi) = \text{none} \qquad a_0 = \frac{(\text{C}^2\cdot\text{N}^{-1}\cdot\text{m}^{-2})(\text{J}\cdot\text{s})^2}{\text{kg}\cdot\text{C}^2} = \text{m}$$

Three Quantum Numbers [n, ℓ, m_ℓ] Specify an Atomic Orbital

Symbol	Name	Allowed Values	Constraints
n	Principal quantum number	$= 1, 2, 3, \ldots$	Positive integer
ℓ	Orbital angular momentum quantum number	$= 0, 1, 2, \ldots, n{-}1$	Each value of n corresponds to n allowed values of ℓ.
m_ℓ	Magnetic quantum number	$= \ell, \ell{-}1, \ell{-}2, \ldots, -\ell$ $= 0, \pm1, \pm2, \ldots, \pm\ell$	Each value of ℓ corresponds to $(2\ell+1)$ allowed values of m_ℓ.

- **Terminology (nomenclature)**

  shell: AOs with the same n value

  subshell: AOs with the same n and ℓ values;

           $\ell = 0, 1, 2, 3$ equivalent to s-, p-, d-, f-subshells, respectively, or

            s-orbital $\Rightarrow$ $\ell = 0$ $m_\ell = 0$

            p-orbital $\Rightarrow$ $\ell = 1$ $m_\ell = -1, 0,$ or $+1$

            d-orbital $\Rightarrow$ $\ell = 2$ $m_\ell = -2, -1, 0, +1,$ or $+2$

            f-orbital $\Rightarrow$ $\ell = 3$ $m_\ell = -3, -2, -1, 0, +1, +2,$ or $+3$

- **Physical significance of the wavefunction $\psi(r,\theta,\phi)$**

  For all atoms and molecules, $\psi^2(r,\theta,\phi)$ is proportional to the probability of finding the electron at a point r, θ, ϕ. We can also regard $\psi^2(r,\theta,\phi)$ as the electron density at point r, θ, ϕ.

- **Plot of electron density for the 2s-orbital of hydrogen**

Computer-generated electron density dot diagram for the hydrogen atom 2s-orbital. The nucleus is at the center of the square and the density of dots is proportional to the probability of finding the electron. Notice the location of the spherical node.

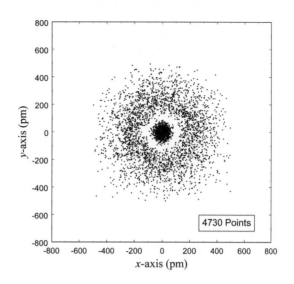

- **Concept of AOs** $\rightarrow$ Two interpretations are useful:

    1 Visualize AO as a *cloud of points,* with the density of points in a given volume proportional to probability of finding the electron in that volume.

    2 Visualize AO as a *surface* (boundary surface) within which there is a given probability of finding the electron. For example, the 95% boundary surface, within which the probability of finding an electron is 95%.

- **Boundary surfaces** $\rightarrow$ *Shapes* of atomic orbitals

s-orbital	*spherical*
p-orbital	*dumbbell* or *peanut*
d-orbital	*four-leaf clover* or *dumbbell* with equatorial *torus* (doughnut)

- **Number and type of orbitals** $\rightarrow$ Follow from allowed values of quantum numbers

    s-orbitals: s means that $\ell = 0$; if $\ell = 0$, then $m_l = 0$. So, for each value of n, there is *one s*-orbital.

    > *one* for each n : $1s, 2s, 3s, 4s, \ldots$

    p-orbitals: p means that $\ell = 1$; if $\ell = 1$, then $m_\ell = -1, 0, +1$, and $n > 1$. So, for each value of $n > 1$, there are *three p*-orbitals, p_x, p_y, p_z.

    > *three* for each $n > 1$: $2p_x, 2p_y, 2p_z$; $3p_x, 3p_y, 3p_z$; $\ldots$

    The *np*-orbitals are referred to collectively as the 2*p*-orbitals, 3*p*-orbitals, and so on.

    d-orbitals: d means that $\ell = 2$; if $\ell = 2$, then $m_\ell = -2, -1, 0, +1, +2$, and $n > 2$.

    So, for each value of $n > 2$, there are *five d*-orbitals, $d_{xy}, d_{xz}, d_{yz}, d_{x^2-y^2}$, and d_{z^2}.

    > *five* for each $n > 2$: $3d_{xy}, 3d_{xz}, 3d_{yz}, 3d_{x^2-y^2}$, and $3d_{z^2}$;
    >
    > $4d_{xy}, 4d_{xz}, 4d_{yz}, 4d_{x^2-y^2}$, and $4d_{z^2}$; $\ldots$

    The *nd*-orbitals are referred to collectively as the 3*d*-orbitals, 4*d*-orbitals, and so on.

- **Wavefunction Properties**
  - → Positive and negative regions designated as + and − on orbital diagrams
    [$n > 1$ (radial node), $\ell > 0$ (angular node)]
  - → Possess nodal surfaces defined by $\psi = 0$ (wavefunction changes sign).
    Example: All $2p$ orbitals have positive and negative lobes separated by a nodal plane.

1.10 Electron Spin

- **Spin states**
  - → An electron has *two* spin states, represented as ↑ and ↓ **or** α and β.
  - → Are described by the spin magnetic quantum number, m_s
  - → For an electron, only two values of m_s are allowed: $+\frac{1}{2}, -\frac{1}{2}$.

1.11 The Electronic Structure of Hydrogen

- **Degeneracy of orbitals**

    In the H atom, orbitals in a given shell are *degenerate* (have the same energy).
    For many-electron atoms, this is not true and the energy of a given orbital depends on n and ℓ.

- **Ground and excited states of H**

    In the ground state, $n = 1$, and the electron is in the $1s$-orbital. The first excited state corresponds to $n = 2$, and the electron occupies one of the *four* possible orbitals with $n = 2$ ($2s$, $2p_x$, $2p_y$, or $2p_z$). Similar considerations hold for higher excited states.

- **Ionization of H**

    Absorption of a photon with energy $\geq h\mathcal{R}$ ionizes the atom, creating a free electron and a free proton. Energy in excess of $h\mathcal{R}$ is utilized as kinetic energy of the system.

- **Summary**

    The state of an electron in a H atom is defined by four quantum numbers $\{n, \ell, m_\ell, m_s\}$. As the value of n increases, the size of the H atom increases.

MANY-ELECTRON ATOMS
(Sections 1.12–1.14)

1.12 Orbital Energies

- **Many-electron atoms**
  - → Atoms with more than one electron
  - → Coulomb potential energy equals the sum of *nucleus-electron* **attractions** and *electron-electron* **repulsions**.
  - → Schrödinger equation cannot be solved exactly.
  - → Accurate wavefunctions obtained numerically by using computers

- **Variation of energy of orbitals** → For orbitals in the same shell but in different subshells, a combination of *nucleus-electron* attraction and *electron-electron* repulsion influences the orbital energies.

- **Shielding and penetration** → Qualitative understanding of orbital energies in atoms

- **Shielding** → Each electron in an atom is *attracted* by the nucleus and *repelled* by all the other electrons. In effect, each electron feels a *reduced* nuclear charge ($Z_{eff}\,e$ = effective nuclear charge). The electron in an occupied orbital is *shielded* to some extent from the nuclear charge and its energy is raised accordingly.

- **Penetration** → The *s-*, *p-*, *d-*, ... **orbitals** have different shapes and different electron density distributions. For a given *shell* (same value of the principal quantum number *n*), **s-electrons** tend to be closer to the nucleus than **p-electrons**, which are closer than **d-electrons**. We say that, other things equal, **s-electrons** are more *penetrating* than **p-electrons**, and **p-electrons** are more *penetrating* than **d-electrons**.

Note: Shielding and penetration can be understood qualitatively on the basis of the nucleus-electron potential energy term: $-(Ze)e/4\pi\varepsilon_0 r$, where Ze is the nuclear charge and r the nucleus-electron distance. *Shielded* electrons have the equivalent of a reduced Z value ($Z_{eff} < Z$) and therefore *higher* energy; *penetrating* electrons have the equivalent of a reduced value of r and therefore *lower* energy.

- **Review** → Z has three equivalent meanings:
  Nuclear charge (actually, Ze)
  Atomic number
  Number of protons in the nucleus

- **Consequences of *shielding* and *penetration* in many-electron atoms**
  → Orbitals with the same *n* and different ℓ values have *different* energies.

  → For a given *shell* (*n*), subshell energies *increase* in the order: $ns < np < nd < nf$.
  Example: A 3*s*-electron is lower in energy than a 3*p*-electron, which is lower than a 3*d*-electron.
  → Orbitals within a given *subshell* have the *same* energy.
  Example: The five 3*d*-orbitals are degenerate for a given atom.
  → *Penetrating* orbitals of higher shells may be lower in energy than less *penetrating* orbitals of lower shells.
  Example: A penetrating 4*s*-electron may be lower in energy than a less penetrating 3*d*-electron.

1.13 The Building-Up Principle

- **Pauli exclusion principle**
  → No more than *two* electrons per orbital
  → *Two* electrons occupying a single orbital must have paired spins: $\boxed{\uparrow\downarrow}$
  → *Two* electrons in an atom may *not* have the same *four* quantum numbers.

- **Terminology**

  closed shell: Shell with maximum number of electrons allowed by the exclusion principle

  valence electrons: Electrons in the outermost occupied shell of an atom; they occupy the shell with the largest value of *n* and are used to form chemical bonds.

  electron configuration: List of all occupied *subshells* or *orbitals*, with the number of electrons in each indicated as a numerical superscript. Example: Li: $1s^2 2s^1$.

- **Building-Up (*Aufbau*) Principle** → Order in which electrons are added to *subshells* and *orbitals* to yield the *electron configuration* of atoms

    The ($n + \ell$) rule → Order of filling subshells in *neutral atoms* is determined by filling those with the *lowest* values of ($n + \ell$) first. Subshells in a group with the same value of ($n + \ell$) are filled in the order of increasing n (*topic not covered in the text*).

- **Usual filling order of subshells ($n + \ell$ rule):** $1s < 2s < 2p < 3s < 3p < 4s < 3d < 4p < 5s < 4d < ...$

    Note: Ionization or subtle differences in shielding and penetration can change the order of the energy of subshells. *Thus, energy ordering of subshells is not fixed absolutely, but may vary from atom to atom or from an atom to its ion.*

- **Orbitals within a subshell** → Hund's rule:

    Electrons add to *different* orbitals of a subshell with spins *parallel* until the subshell is half full.

- **Applicability** → The building-up principle in combination with the *Pauli exclusion principle* and *Hund's rule* accounts for the *ground-state* electron configurations of atoms. The principle is generally valid, **but there are exceptions.**

1.14 Electronic Structure and the Periodic Table

- **Order of filling subshells**
    - → Understanding the organization of the periodic table
    - → Straightforward determination of (most) electron configurations

- **Terminology** → Given in the following two tables

Groups:	Columns in the periodic table, labeled 1–18 horizontally
***s*-block elements:**	Groups 1, 2; *s*-subshell fills
***p*-block elements:**	Groups 13–18; *p*-subshell fills
***d*-block elements:**	Groups 3–12; *d*-subshell fills
transition elements:	Groups 3–11; *d*-subshell fills
Main-group elements:	Groups 1, 2 and 13–18
Lanthanides:	4*f*-subshell fills
Actinides:	5*f*-subshell fills

Note: A transition element has a partially filled *d*-subshell either as the element or in any commonly occurring oxidation state. Thus, Zn, Cd, and Hg are not transition elements. The ($n - 1$)*d* electrons of the *d*-block elements are considered to be valence electrons. Groups 1, 2, and 13–18 are alternatively labeled with Roman numerals I–VIII, which correspond to the number of valence electrons in the element.

Valence shell:	Outermost occupied shell (highest n)
Period:	Row of the periodic table
Period number:	Principal quantum number of valence shell
Period 1:	H, He; $1s$-subshell fills
Period 2:	Li through Ne; $2s$-, $2p$-subshells fill
Period 3:	Na through Ar; $3s$-, $3p$-subshells fill
Period 4: (first long period)	K through Kr; $4s$-, $3d$-, $4p$-subshells fill

- **Periodic table** → Two forms displayed:
  One shows the elements, the other the *final* subshell filled.
  Look at the tables to understand the terminology.

Periodicity of Elements in the Periodic Table

Period
↓
Group (1–18) →

Period	1	2	3	4	5	6	7	8	9	10	11	12	13	14	15	16	17	18
1	H																	He
2	Li	Be											B	C	N	O	F	Ne
3	Na	Mg											Al	Si	P	S	Cl	Ar
4	K	Ca	Sc	Ti	V	*Cr*	Mn	Fe	Co	Ni	*Cu*	Zn	Ga	Ge	As	Se	Br	Kr
5	Rb	Sr	Y	Zr	*Nb*	*Mo*	Tc	*Ru*	*Rh*	***Pd***	*Ag*	Cd	In	Sn	Sb	Te	I	Xe
6	Cs	Ba	Lu	Hf	Ta	W	Re	Os	Ir	*Pt*	*Au*	Hg	Tl	Pb	Bi	Po	At	Rn
7	Fr	Ra	Lr	Rf	Db	Sg	Bh	Hs	Mt	Ds	Rg	Cp	Uut	Uuq	Uup	Uuh		Uuo

Lanthanides	*La*	*Ce*	Pr	Nd	Pm	Sm	Eu	*Gd*	Tb	Dy	Ho	Er	Tm	Yb
Actinides	*Ac*	***Th***	*Pa*	*U*	*Np*	Pu	Am	*Cm*	Bk	Cf	Es	Fm	Md	No

- **Electron configurations of the elements** → See the Appendix section on *Ground-State Electron Configurations* in the text.
- **17 italicized elements**
  - → Exceptions to the $(n + \ell)$ rule
  - → Differ by the placement of *one* electron
    Example: Cr: $1s^2 2s^2 2p^6 3s^2 3p^6 4s^1 3d^5$ = $[Ar]\, 4s^1 3d^5$ (not $\ldots 4s^2 3d^4$ as might be expected)
- **2 bold, italicized elements** → Differ by the placement of *two* electrons (Pd, Th)
  - **Pd**   We expect $[Kr]\, 4d^8 5s^2$, but *actually* find $[Kr]\, 4d^{10}$.   *No 5s-subshell electrons*
  - **Th**   We expect $[Rn]\, 5f^2 7s^2$, but *actually* find $[Rn]\, 6d^2 7s^2$.   *No 5f-subshell electrons*
- **Recently named elements**
  - → **Rg** = Roentgenium ($Z = 111$) *official* [in honor of Wilhelm Röntgen (Roentgen)]
  - → **Cp** = Copernicium ($Z = 112$) *proposed* [in honor of Nicolaus Copernicus]

Order of Filling Subshells in the Periodic Table

Period ↓	Group (1–18) →																	

```
Period                        Group (1–18)  →
  ↓
        1                                                                      18
  1   | 1s | 2                                          13  14  15  16  17  | 1s |
  2   | 2s |                                                            2p
  3   | 3s |      3   4   5   6   7   8   9   10  11  12                3p
  4   | 4s |                         3d                                 4p
  5   | 5s |                         4d                                 5p
  6   | 6s |                         5d                                 6p
  7   | 7s |                         6d                                 7p
            ↑
                                     4f
                                     5f
```

THE PERIODICITY OF ATOMIC PROPERTIES
(Sections 1.15–1.21)

1.15 Atomic Radius (r)

- **Definition of atomic radius**
  - → Half the distance between the centers of neighboring atoms (nuclei)
  - → For metallic elements, r is determined for the solid.

    Example: For solid Zn, 274 pm between nuclei, so r (atomic radius) = 137 pm

  - → For nometallic elements, r is determined for diatomic molecules (covalent bond).

    Example: For I_2 molecules, 266 pm between nuclei, so r (covalent radius) = 133 pm

- **General trends in radius with atomic number**
  - → r decreases from *left* to *right* across a period   (*effective* nuclear charge increases)
  - → r increases from *top* to *bottom* down a group   (change in valence electron principal quantum number and size of valence shell)

1.16 Ionic Radius

- **An ion's share of the distance between neighbors in an ionic solid (cation to anion)**
  - → Distance between the nuclei of a neighboring cation and anion is the sum of two ionic radii
  - → Radius of the oxide anion (O^{2-}) is 140 pm.

    Example: Distance between Zn and O nuclei in zinc oxide is 223 pm; therefore, the ionic radius of Zn^{2+} is 83 pm [$r(Zn^{2+})$ = 223 pm − $r(O^{2-})$].

  - → Cations are **smaller** than parent atoms; for example, Zn (133 pm) and Zn^{2+}(83 pm).
  - → Anions are **larger** than parent atoms; for example, O (66 pm) and O^{2-}(140 pm).

- **Isoelectronic atoms and ions** → Atoms and ions with the same number of electrons

    Example: Cl^-, Ar, and K^+, and Ca^{2+}

1.17 Ionization Energy (I)

- I → Energy needed to remove an electron from a gas-phase atom in its lowest energy state

- **Symbol** → Number subscripts (*e.g.*, I_1, I_2) denote removal of successive electrons.
  For a given species, $I_2 > I_1$ always.

- **General periodic table trends in I_1 for the main-group elements**
  - → Increases from *left* to *right* across a period   (Z_{eff} increases)
  - → Decreases from *top* to *bottom* down a group   (change in principal quantum number n of valence electron)

- **Exceptions** → For Groups 2 and 15 in Periods 2–4, I_1 is *larger* than the neighboring *main group element* in Groups 13 and 16, respectively. Repulsions between electrons in the same orbital and/or extra stability of completed and half completed subshells are responsible for these exceptions to the general trends.

- **Metals toward the lower left of the periodic table (*e.g.*, Cs, Ba)**
  - → Have low ionization energies
  - → Readily lose electrons to form cations

- **Nonmetals toward the upper right of the periodic table (*e.g.*, F, O)**
  - → Have high ionization energies
  - → Do not readily lose electrons

1.18 Electron Affinity (E_{ea})

- **Energy *released* when an electron is *added* to a gas-phase atom**

- **Periodic table trends in E_{ea} for the main-group elements**
  - → Increases from *left* to *right* across a period   (Z_{eff} increases)
  - → Decreases from *top* to *bottom* down a group   (change in principal quantum number n of valence electron)
  - → Generally, the same as for I_1 with the major exceptions displayed in Fig. 1.54

- **Summary of trends in r, I_1, and E_{ea}**
  - → All depend on Z_{eff} and n of outer subshell electrons.
  - → Recall that there are *many* exceptions to the general trends, as noted earlier.

1.19 The Inert-Pair Effect

- **Tendency to form ions two units lower in charge than expected from the group number relation**

  - → Most pronounced in *p*-block elements near the bottom of the groups

  - → Due in part to the different energies of the valence *p*- and *s*-electrons

  - → Important for the *lower* two members of Groups 13, 14, and 15; for example, Pb(IV) & Pb (II) or Sb(V) & Sb(III)

  - → Valence *s*-electrons are called a "lazy pair."

1.20 Diagonal Relationships

- **Diagonally related pairs of elements often show similar chemical properties.**
  - $\rightarrow$ Diagonal band of metalloids dividing metals from nonmetals
  - $\rightarrow$ Similarity of Li and Mg (react directly with N_2 to form nitrides)
  - $\rightarrow$ Similarity of Be and Al (both react with acids and bases)

1.21 The General Properties of the Elements

- ***s*-block elements (Groups 1 and 2)** $\rightarrow$ All are reactive metals (except H); they form *basic* oxides (O^{2-}), peroxides (O_2^{2-}), or superoxides (O_2^-).

- **Compounds of *s*-block elements are ionic, except for beryllium.**

  Notes: *Hydrogen* (a nonmetal) is placed by itself, or *more usually* in Group 1, because its electronic configuration ($1s^1$) is similar to those of the alkali metals: [noble gas] ns^1.

  Helium, with electronic configuration $1s^2$, is placed in Group 18 because its properties are similar to those of neon, argon, krypton, and xenon.

- ***p*-block elements (Groups 13–18)** $\rightarrow$ Members are metals, metalloids, and nonmetals.

  Metals: Group 13 (Al, Ga, In, Tl); Group 14 (Sn, Pb); Group 15 (Bi)

  Note: These metals have relatively low I_1, but *larger* than those of the *s* block and *d* block.

  Metalloids: Group 13 (B); Group 14 (Si, Ge); Group 15 (As, Sb); Group 16 (Te, Po)

  Note: Metalloids have the *physical* appearance and properties of *metals* but behave *chemically* as *nonmetals*.

  Nonmetals: Group 14 (C); Group 15 (N, P); Group 16 (O, S, Se); Groups 17 & 18 (All)

  Notes: High E_{ea} in Groups 13–17; these atoms tend to *gain* electrons to complete their subshells. Group 18 noble-gas atoms are generally *nonreactive*, except for Kr and Xe, which form a few compounds.

- ***d*-block elements (Groups 3–12)** $\rightarrow$ All are metals (*most* are transition metals), with properties intermediate to those of *s*-block and *p*-block metals.

  Note: *All* Group 12 cations retain the filled *d*-subshell. For this reason, these elements (Zn, Cd, Hg) are *not* classified as transition metals. Recall that in this text *d*-orbital electrons are considered to be valence electrons for *all* the *d*-block elements.

- **Characteristically, many transition metals**
  - $\rightarrow$ Form compounds with a variety of oxidation states (oxidation numbers)
  - $\rightarrow$ Form alloys
  - $\rightarrow$ Facilitate subtle changes in organisms

- ***f*-block elements (lanthanides and actinides)**
  - $\rightarrow$ Rare on Earth
  - $\rightarrow$ Lanthanides are incorporated in *superconducting* materials.
  - $\rightarrow$ Actinides are all *radioactive* elements.
  - $\rightarrow$ Most actinides do not occur naturally on Earth.

Chapter 2 CHEMICAL BONDS

IONIC BONDS (Sections 2.1–2.4)

2.1 The Ions That Elements Form

- **Cations** → Remove outermost electrons in the order np, ns, $(n-1)d$

- **Metallic s-block elements and metallic p-block elements in periods 2 and 3**
  - → Form cations by losing electrons down to the noble-gas core
    - **Examples:** Mg, [Ne]$3s^2$ → magnesium(II), Mg^{2+}, [Ne];
      - Al, [Ne]$3s^2 3p^1$ → aluminum(III), Al^{3+}, [Ne]

- **Metallic p-block elements in periods 4 and higher**
  - → Form cations with complete, typically unreactive d-subshells
    - **Example:** Ga, [Ar]$4s^2 3d^{10} 4p^1$ → gallium(III), Ga^{3+}, [Ar]$3d^{10}$

- **Metallic d-block elements**
  - → Lose s-electrons and often a variable number of d-electrons
    - **Examples:** iron(II), Fe^{2+}, [Ar]$3d^6$ and iron(III), Fe^{3+}, [Ar]$3d^5$

- **Many metallic p-block elements**
  - → May lose either their p-electrons or all their s- and p-electrons
    - **Example:** Sn, [Kr]$5s^2 4d^{10} 5p^2$ → tin(II), Sn^{2+}, [Kr]$5s^2 4d^{10}$ *or* tin(IV), Sn^{4+}, [Kr]$4d^{10}$

- **Anions** → Add electrons until the next noble-gas configuration is reached.
    - **Examples:** Carbide (methanide), C^{4-}, [He]$2s^2 2p^6$ or [Ne]   {octet}
      - Hydride, H^-, $1s^2$ or [He]   {duplet}

2.2 Lewis Symbols (Atoms and Ions)

- **Valence electrons**
  - → Depicted as dots; a pair of dots represents two paired electrons
  - → Lewis symbols for neutral atoms in the first two periods of the Periodic Table:

    H•   He••   Li•   Be••   •B•   •C•   :N•   :O•   :F•   :Ne:

    - Note: Ground-state structures for B and C are :B• and :C• . B and C have one and two unpaired electrons, respectively.

- **Variable valence**
  - → Ability of an element to form two or more ions with different oxidation numbers
  - → Displayed by many d-block and p-block elements
    - **Examples:** Lead in lead(II) oxide, PbO, and in lead(IV) oxide, PbO_2
      - Iron in iron(II) oxide, FeO, and in iron(III) oxide, Fe_2O_3

  - → Lewis symbols for ions: use brackets where necessary to show charge associated with ion as a whole.
    - **Example:** Ca^{2+}, 2 $\left[:\overset{..}{\underset{..}{Cl}}: \right]^-$

2.3 The Energetics of Ionic Bond Formation

- **Ionic bond** → Electrostatic attraction (coulombic) of *oppositely* charged ions
- **Ionic model** → Energy for the *formation* of ionic bonds is supplied *mainly* by coulombic attraction of *oppositely* charged ions. This model gives a good description of bonding between the ions of metals (particularly *s*-block) and those of nonmetals.

- **Ionic solids /ionic crystals**
  - → Are assemblies of cations and anions arranged in a regular array
  - → The ionic bond is *nondirectional*; each ion is bound to *all* its neighbors.
  - → Typically have *high* melting and boiling points and are brittle; NaCl, for example
  - → Form electrolyte solutions if they dissolve in water

- **Formulas of compounds composed of monatomic ions**
  - → Formulas are predicted by assuming that atoms forming cations lose all valence electrons and those forming anions gain electrons in the valence subshell(s) until each ion has an octet of electrons or a duplet in the case of H, He, and Be.
  - → For cations with variable valence, the oxidation number is used.
    Example: Iron(III) chloride, $FeCl_3$
  - → Relative numbers of cations and anions are chosen to achieve electrical neutrality, using the smallest possible integers as subscripts.

2.4 Interactions Between Ions

- **In an ionic solid**
  - → All the cations repel each other, and all the anions repel each other.
  - → Each cation is attracted to all the anions to a greater or lesser extent.
  - → In this view, an ionic bond is a "global" characteristic of the entire crystal.

- **Coulomb potential energy between two ions, E_P**

$$E_{P,12} = \frac{(z_1 e) \times (z_2 e)}{4\pi\varepsilon_0 r_{12}} = \frac{2.307 \, z_1 z_2}{r_{12}} \times 10^{-19} \, \text{J·nm}$$

e is the elementary charge; z_1 and z_2 are the charges on the ions; r_{12} is the distance in nanometers (nm) between the ions; ε_0 is the vacuum permittivity.

- **Lattice energy** → Potential energy difference: ions in solid *minus* ions infinitely far apart

- **Madelung constant (A)** $$E_P = -A \times \frac{|z_1 z_2| N_A e^2}{4\pi\varepsilon_0 d}$$

  A is a positive number; d is the distance between the centers of nearest neighbors in the crystal.

  - → A depends on the *arrangement* of ions in the solid.
  - → In the expression above, E_P is always negative and A is positive.
  - → So, ionic crystals with **small ions** (short interionic distances), **large values of A**, and **highly charged ions** tend to have large lattice energies.

- **Values of A** → 1.747 56 (NaCl structure) [$d = 0.2798$ nm for NaCl (actual value near 0 K)]
  1.762 67 (CsCl structure) [$d = 0.351$ nm for CsCl (sum of ionic radii)]

- **Potential energy of an ionic solid**
  - → Takes attractive and repulsive interionic interactions into account

- → Is given by the Born-Meyer equation below

- **Born-Meyer equation**
$$E_{P,min} = -\frac{N_A |z_A z_B| e^2}{4\pi\varepsilon_0 d}\left(1-\frac{d^*}{d}\right)A$$

The constant d^* is commonly taken to be 34.5 pm. It derives from the repulsive effects of overlapping electron charge clouds.

- **Refractory material** → Substance that can withstand high temperatures; MgO, for example

---

COVALENT BONDS (Sections 2.5–2.8)

2.5 Lewis Structures

- **Covalent bond**
  - → Pair of electrons *shared* between two atoms
  - → Located between two neighboring atoms and *binds* them together

  Examples: Nonmetallic elements such as H_2, N_2, O_2, F_2, Cl_2, Br_2, I_2, P_4, and S_8

- **Rules**
  - → Atoms attempt to complete duplets or octets by sharing pairs of valence electrons.
  - → Valence of an atom is the number of bonds it can form.
  - → A line (–) represents a shared pair of electrons.
  - → A lone pair of nonbonding electrons is represented by two dots (:).

  Example: H–H, (single bond) duplet on each atom (valence of hydrogen = 1) and :N≡N:, (triple bond and lone pair) octet on each atom (valence of nitrogen = 3)

- **Lone pairs of electrons** → Electron pairs not involved in bonding

  Example: The electrons indicated as dots in the Lewis structure of N_2 above

2.6 Lewis Structures of Polyatomic Species

- **Lewis structures**
  - → Show which atoms are bonded (atom connectivity) and which contain lone pairs of electrons
  - → Do not portray the *shape* of a molecule or ion

- **Rules**
  - → Count total number of valence electrons in the species
  - → Arrange atoms next to bonded neighbors
  - → Use minimum number of electrons to make all single bonds
  - → Count the number of nonbonding electrons required to satisfy octets
  - → Compare to the actual number of electrons left
  - → If lacking a sufficient number of electrons to satisfy octets, make *one extra bond* for each deficit pair of electrons.
  - → Sharing pairs of electrons with a neighbor completes the octet or duplet.
  - → Each shared pair of electrons counts as one *covalent bond* (line).

    | One shared pair | **single bond** | (–) | Bond order = 1 |
    | Two shared pairs | **double bond** | (=) | Bond order = 2 (multiple bond) |
    | Three shared pairs | **triple bond** | (≡) | Bond order = 3 (multiple bond) |

- **Bond order** → Number of bonds that link a specific pair of atoms
- **Terminal atom** → Bonded to only one other atom
- **Central atom** → Bonded to at least two other atoms
- **Molecular ions** → Contain *covalently* bonded atoms: NH_4^+, Hg_2^{2+}, SO_4^{2-}
- **Rules of thumb**
    → Usually, the element with the lowest I_1 is a central atom, but electronegativity is a better indicator (see Section 2.12). For example, in HCN, carbon has the lowest I_1 and is the central atom. It is also less electronegative than nitrogen.

    → Usually, there is a symmetrical arrangement about the central atom. For example, in SO_2, OSO is symmetrical, with S as the central atom and the two O atoms terminal.

    → Oxoacids have H atoms bonded to O atoms; H_2SO_4 is actually $(HO)_2SO_2$, with two O atoms and two OH groups bonded to S.

    Examples: Ethyne (acetylene), C_2H_2 (10 valence electrons): H–C≡C–H

    Hydrogen cyanide, HCN (10 valence electrons): H–C≡N:

    Ammonium ion, NH_4^+ (8 valence electrons): $\begin{bmatrix} H \\ | \\ H-N-H \\ | \\ H \end{bmatrix}^+$

2.7 Resonance

- **Multiple Lewis structures**
    → Some molecules can be represented by different Lewis structures (*contributing structures*) in which the *locations* of the electrons, but not the nuclei, vary.
    → Multiple Lewis structures used to represent a given species are called *resonance structures*.

- **Electron delocalization** → Species requiring multiple Lewis structures often involve *electron delocalization* with some electron pairs distributed over more than two atoms.

- **Blending of structures** → Double-headed arrows (⟷) are used to relate contributing Lewis structures, indicating that a blend of the contributing structures is a better representation of the bonding than any one.

- **Resonance hybrid** → Blended structure of the contributing Lewis structures

    Example: N_2O, nitrous oxide, has $2(5) + 6 = 16$ valence electrons or eight pairs.

    :N̈=N=Ö: ⟷   :N≡N–Ö:

    Blending of two structures, both of which obey the octet rule.

    Note: The central N atom is the least electronegative atom (Section 2.12).

    Note: A triple bond to an O atom is found in species such as BO^-, CO, NO^+, and O_2^{2+}.

Example: C_6H_6, benzene, has $6(4) + 6(1) = 30$ valence electrons, or 15 pairs.

Kekulé structures   or   Resonance hybrid

Note: Carbon atoms lie at the vertices of the hexagon (the six C–H bonds radiating from each corner are not shown in the stick structures). The last structure depicts six valence electrons *delocalized* around the ring. The resonance hybrid suggests that all CC bonds in benzene are equivalent, which, in contrast to the Kekulé structures, agrees with experiment.

2.8 Formal Charge

- **Formal charge** → An atom's number of valence electrons (V) minus the number of electrons assigned to it in a Lewis structure

- **Electron assignment in a Lewis structure**
  - → Atom possesses all of its lone pair electrons (L) and half of its bonding electrons (B) [B means shared bonding electrons]
  - → Formal charge $= V - (L + \frac{1}{2}B)$

- **Contribution of individual** → Structures with individual formal charges
  Lewis structures to a   closest to **zero** usually have the lowest energy and
  resonance hybrid   are the major contributors.

  Example: N_2O: :N̈=N=Ö: ⟷ :N≡N–Ö: ⟷ :N̈–N≡O:
                 –1 +1   0   0 +1 –1   –2 +1 +1
                                                                        (high formal charges)

- **Plausibility of isomers** → The isomer with lowest formal charges is *usually* preferred.

  Examples: N_2O: :N̈=N=Ö: and   :N̈=O=N̈:
                   –1 +1   0   –1 +2 –1
                  (low formal charges)   (high formal charges)

          HCN: H–C≡N: and   H–N≡C: (first structure with lower formal charges preferred)
                    0   0   0   0 +1 –1

Summary

- **Formal charge**
  - → Indicates the extent to which atoms have gained or lost electrons in a Lewis structure (covalent bonding)
  - → Exaggerates the *covalent* character of bonds by assuming that electrons are shared equally
  - → Structures with the lowest formal charges usually have the lowest energy (major contributors to the resonance hybrid).
  - → Isomers with lower formal charges are usually favored.

- **Oxidation number** → Exaggerates the *ionic* character of bonds

  Example: Carbon disulfide, $:\ddot{S}=C=\ddot{S}:$ Here, C has an oxidation number of +4 (all the electrons in the double bonds are assigned to the more electronegative S atoms) and a formal charge of zero (the electrons in the double bonds are shared equally between the C and S atoms).

EXCEPTIONS TO THE OCTET RULE (Sections 2.9–2.11)

2.9 Radicals and Biradicals

- **Radicals**
  - → Species with an unpaired electron
  - → All species with an odd number of electrons are radicals.
  - → Are highly reactive, important for reactions in the upper atmosphere, cause rancidity in foods, degradation of plastics in sunlight, and perhaps contribute to human aging

    Examples: CH_3 (methyl radical), OH (hydroxyl), OOH (hydrogenperoxyl),
    NO (nitric oxide), NO_2 (nitrogen dioxide), O_3^- (ozonide ion)

    $$\left[:\ddot{O}-\ddot{O}-\ddot{O}: \right]^- \longleftrightarrow \left[:\ddot{O}-\ddot{O}-\ddot{O}: \right]^- \longleftrightarrow \left[:\ddot{O}-\ddot{O}-\ddot{O}: \right]^-$$

    Note: Ozonide ion does not follow the octet rule.

- **Biradicals** → Species containing *two* unpaired electrons

    Examples: O (oxygen atom) and CH_2 (methylene) [unpaired electrons on a single atom]
    O_2 (oxygen molecule) and larger organic molecules [on different atoms]

- **Antioxidant** → Species that reacts rapidly with radicals before they have a chance to do damage

    Examples: Vitamins A, C, and E; coenzyme Q; substances in coffee, orange juice, chocolate

2.10 Expanded Valence Shells

- **Expanded valence shells**
  - → More than eight electrons associated with an atom in a Lewis structure (expanded octet)

    A *hypervalent compound* contains an atom with more atoms attached to it than is permitted by the octet rule. Empty *d*-orbitals are utilized to permit octet expansion.

  - → Electrons may be present as bonding pairs or lone pairs.

  - → Are characteristic of nonmetal atoms in *Period 3 or higher*
    First- and second-period elements do not utilize expanded valence shells.

- **Variable covalence**
  - → Ability to form different numbers of covalent bonds
  - → Elements showing variable covalence include (next page):

Valence Shell Occupancy

Elements	8 electrons	10 electrons	12 electrons
P	PCl_3 / PCl_4^+	PCl_5	PCl_6^-
S	SF_2	SF_4	SF_6
I	IF	IF_3	IF_5

Example: Major Lewis structures for sulfuric acid, H_2SO_4 (32 valence electrons)

In the structure on the right, all atoms have zero formal charge and the S atom is surrounded by 12 electrons (expanded octet). *Because it has the lowest formal charges, this structure is expected to be the most favored one energetically and the one to make the greatest contribution to the resonance hybrid.* An additional Lewis structure with one double bond is not shown. Similar examples include SO_4^{2-}, SO_2, and S_3O (S is the central atom).

2.11 The Unusual Structures of Some Group 13/III Compounds

- **Incomplete octet**
  - → Fewer than eight valence electrons on an atom in a Lewis structure

    Example: BF_3,

    The single-bonded structure with an incomplete octet makes the major contribution.

  - → Other compounds with incomplete octets are BCl_3 and $AlCl_3$ vapor at high temperature

  - → At room temperature, aluminum chloride exists as the dimer, Al_2Cl_6, which follows the octet rule (see structure 33 in the text).

- **Coordinate covalent bond** → One in which both electrons come from one atom

    Example: Donation of a lone pair by one atom to create a bond that completes the octet of another atom with an incomplete octet. In the Lewis structure for the reaction of BF_3 with NH_3, both electrons in the B–N bond are derived from the N lone pair.

IONIC VERSUS COVALENT BONDS (Sections 2.12–2.13)

2.12 Correcting the Covalent Model: Electronegativity

- **Bonds** → All molecules may be viewed as resonance hybrids of pure covalent and ionic structures.

  Example: H_2 The structures are $H-H \longleftrightarrow H^+[H\!:\!]^- \longleftrightarrow [\:\!H]^- H^+$

  > Here, the two ionic structures make equal contributions to the resonance hybrid, but the single covalent structure is of major importance.

- **Partial charges**
  - → When the bonded atoms are different, the ionic structures are not energetically equivalent.
  - → Unequal sharing of electrons results in a **polar covalent bond.**

    Example: HF The structures are $H-F \longleftrightarrow H^+[\:\!\ddot{\underset{..}{F}}\:\!]^- \longleftrightarrow [\:\!H]^-[\:\!\ddot{\underset{..}{F}}\:\!]^+$

  Note: The electron affinity of F is greater than that of H, resulting in a small negative charge on F and a corresponding small positive charge on H:
  $$E_{ea}(F) > E_{ea}(H) \implies {}^{\delta+}H\!-\!F^{\delta-} \text{ (partial charges, } |\delta+| = |\delta-|)$$
  $$E(H^+F^-) \ll E(H^-F^+) \implies H^-F^+ \text{ is a very } minor \text{ contributor}$$

- **Electric dipole** → A partial *positive* charge separated from an equal but *negative* partial charge

- **Electric dipole moment (μ)**
  - → Size of an electric dipole: partial charge times distance between charges
  - → Units: debye (D)
  - → 4.80 D $\equiv$ an electron $(-)$ separated by 100 pm from a proton $(+)$
  - → $\boxed{\mu = (4.80\ \text{D}) \times \delta \times (\text{distance in pm}/100\ \text{pm})}$

- **Electronegativity (χ)**
  - → Electron-attracting power of an atom when it is part of a bond
  - → Mulliken scale: $\chi = \frac{1}{2}(I_1 + E_{ea})$
  - → Follows same periodic table trends as I_1 and E_{ea}
  - → *Increases from left to right and from bottom to top*
  - → Pauling numerical scale based on bond energies [$\chi(F) = 4.0$] is used in text (qualitatively similar to Mulliken's scale).
  - → See Fig. 2.12 in the text for a tabulation of values for selected elements.

- **Rough rules of thumb**

$$
\begin{aligned}
(\chi_A - \chi_B) &\geq 2 & &\text{bond is } essentially \text{ ionic} \\
0.5 \leq (\chi_A - \chi_B) &\leq 1.5 & &\text{bond is } polar \text{ covalent} \\
(\chi_A - \chi_B) &\leq 0.5 & &\text{bond is } essentially \text{ covalent}
\end{aligned}
$$

2.13 Correcting the Ionic Model: Polarizability

- **Ionic bonds** → All have *some* covalent character.
  A cation's positive charge attracts the electrons of an anion or atom in the direction of the cation (*distortion* of spherical electron cloud).

- **Highly *polarizable* atoms and ions** → Readily undergo a *large* distortion of their electron cloud

  Examples of polarizable species: large anions and atoms such as I^-, Br^-, Cl^-, I, Br, and Cl

- **Polarizing power**
  - → Property of ions (and atoms) that cause large distortions of electron clouds
  - → Increases with decreasing size and increasing charge of a cation
    Examples of species significant polarizing power: the small and/or highly charged cations Li^+, Be^{2+}, Mg^{2+}, and Al^{3+}

- **Significant covalent bonding character**
  - → Bonds between highly polarizing cations and highly polarizable anions have significant covalent character.
  - → The Be^{2+} cation is highly polarizing and the Be–Cl bond has significant covalent character (even though there is an electronegativity difference of 1.6).
  - → In the series AgCl to AgI, the bonds become more covalent as the polarizability (and size) of the anion increases $(Cl^- < Br^- < I^-)$.

<hr>

THE STRENGTHS AND LENGTHS OF COVALENT BONDS
(Sections 2.14–2.16)

2.14 Bond Strengths

- **Dissociation energy (D)**
  - → Energy required to separate bonded atoms in neutral molecules
  - → Bond breaking is *homolytic,* which means that each atom retains half of the bonding electrons.
  - → Determines the *strength* of a chemical bond
    The *greater* the dissociation energy, the *stronger* the bond.
  - → D is defined exactly for diatomic (two-atom) molecules.
    For polyatomic (greater than two-atom) molecules, D also depends on the other bonds in the molecule. However, for many molecules, this dependence is slight.

- **Values of D (Table 2.3)**
  - → Vary from about 139 kj·mol^{-1} (I–I single bond) to 1062 kj·mol^{-1} (C≡O triple bond)

- *D* **for several diatomic molecules**

$$H–H(g) \rightarrow H(g) + H(g) \qquad D = 424 \text{ kJ·mol}^{-1} \text{ (strong)}$$
$$H–F(g) \rightarrow H(g) + F(g) \qquad D = 543 \text{ kJ·mol}^{-1} \text{ (strong)}$$
$$F–F(g) \rightarrow F(g) + F(g) \qquad D = 146 \text{ kJ·mol}^{-1} \text{ (weak)}$$
$$I–I(g) \rightarrow I(g) + I(g) \qquad D = 139 \text{ kJ·mol}^{-1} \text{ (weak)}$$
$$O=O(g) \rightarrow O(g) + O(g) \qquad D = 484 \text{ kJ·mol}^{-1} \text{ (strong)}$$
$$N≡N(g) \rightarrow N(g) + N(g) \qquad D = 932 \text{ kJ·mol}^{-1} \text{ (very strong)}$$

2.15 Variation in Bond Strength (Table 2.4)

- **Bond strength** → For polyatomic molecules, bond strength is defined as the *average* dissociation energy for one type of bond found in different molecules.

  For example, the tabulated C–H single bond value is the *average* strength of such bonds in a selection of organic molecules, such as methane (CH_4), ethane (C_2H_6), and ethene (C_2H_4).

Note: Values of average dissociation energies in text Table 2.4 are actually those for the property of *bond enthalpy* (see Chapter 7), measured at 298.15 K. The bond *dissociation energies* of diatomic molecules given in Table 2.3 apply at 0 K (absolute zero).

- **Factors influencing bond strength**
  - → Bond multiplicity   (C≡C > C=C > C–C)
  - → Resonance   (C=C > C⋯C (benzene) > C–C)
  - → Lone pairs on neighboring atoms   (F–F < H–H)
  - → Atomic radii   (HF > HCl > HBr > HI)

  (The smaller the radius, the stronger the bond.)

2.16 Bond Lengths

- **Bond length**
  - → Internuclear distance, at the potential energy minimum, of two atoms linked by a covalent bond
  - → Helps determine the overall size and shape of a molecule
  - → Evaluated by using spectroscopic or x-ray diffraction (for solids) methods
  - → For bonds between the same elements, length is *inversely* proportional to strength.

- **Factors influencing bond length**
  - → Bond multiplicity   (C≡C < C=C < C–C)
  - → Resonance   (C=C < C⋯C (benzene) < C–C)
  - → Lone pairs on neighboring atoms   (F–F > H–H)
  - → Atomic radii   (HF < HCl < HBr < HI)

  (The smaller the radius, the shorter the bond.)

  Note: These trends are the *opposite* of the ones for bond strength.

- **Covalent radius**
  - → Contribution an atom makes to the length of a covalent bond
  - → *Half* the distance between the centers (nuclei) of neighboring atoms joined by a covalent bond (*for like atoms*)
  - → Covalent radii may be added together to estimate bond lengths in molecules.
  - → Tabulated values are *averages* of radii in polyatomic molecules.
  - → *Decreases* from left to right in the periodic table
  - → *Increases* in going down a group in the periodic table
  - → Decreases for a given atom with increasing multiple bond character

  Example: Use the covalent radii given in Fig. 2.21 in the text to estimate the several bond lengths in the acetic acid molecule, CH_3COOH.

Lewis Structure:

$$\begin{array}{c} H \quad \ddot{O} \\ | \quad\quad || \\ H-C-C-\ddot{O}-H \\ | \\ H \end{array}$$

Bond type	Bond length estimate	Actual
C–H	77 pm + 37 pm = 114 pm	≈109 pm
C–C	77 pm + 77 pm = 154 pm	≈150 pm
C–O	77 pm + 74 pm = 151 pm	134 pm
C=O	67 pm + 60 pm = 127 pm	120 pm
O–H	74 pm + 37 pm = 111 pm	97 pm

Chapter 3   MOLECULAR SHAPE AND STRUCTURE

THE VSEPR MODEL (Sections 3.1–3.3)

3.1 The Basic VSEPR (Valence-Shell Electron-Pair Repulsion) Model

- **Valence electrons about central atom(s)** → Controls shape of a molecule

- **Lewis structure** → Shows distribution of *valence* electrons in bonding pairs (bonds) and as lone pairs

- **Bonds, lone pairs** → Regions of high electron density that repel each other (Coulomb's law) by rotating about a central atom, thereby maximizing their separation

- **Bond angle(s)** → Angle(s) between bonds joining atom centers

- **Multiple bonds** → Treated as a *single* region of high electron concentration in VSEPR

- **Electron arrangement** → Ideal locations of bond pairs and lone pairs about a central atom (angles between electron pairs define the geometry)

Typical Electron Arrangements in Molecules with *One* Central Atom

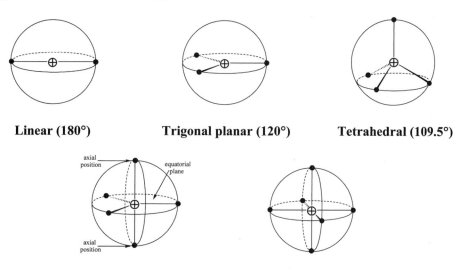

Linear (180°)   **Trigonal planar (120°)**   **Tetrahedral (109.5°)**

Trigonal bipyramidal   **Octahedral (90° and 180°)**
(120° equatorial, 90° axial-equatorial)

- **VSEPR formula, AX_nE_m**   (See Section 3.2 in the text)
  - → A = central atom
  - → X_n = n atoms (*same* or *different*) bonded to central atom
  - → E_m = m lone pairs on central atom
  - → Useful for generalizing types of structures

  Note: Molecular shape is defined by the location of the *atoms alone.*
  The bond angles are ∠ X–A–X.

  Examples: Linear, BeF_2 (AX_2); trigonal planar, BF_3 (AX_3); tetrahedral, CH_4 (AX_4); trigonal bipyramidal, PCl_5 (AX_5); octahedral, SF_6 (AX_6)

Typical Shapes of Molecules with *One* Central Atom*

Linear Angular Trigonal planar Trigonal pyramidal T-shaped Tetrahedral

Seesaw Square planar Trigonal bipyramidal Square pyramidal Octahedral Pentagonal bipyramidal

* These figures depict the locations of the atoms only, *not* lone pairs.

- **VSEPR (sometimes pronounced "vesper") method** (See Toolbox 3.1 in the text.)
  - → Write the Lewis structure(s). If there are resonance structures, pick *any* one.
  - → Count the number of electron pairs (bonding and nonbonding) around the central atom(s). Treat a multiple bond as a *single* unit of high electron density.
  - → Identify the *electron arrangement*. Place electron pairs as far apart as possible.
  - → Locate the *atoms* and classify the *shape* of the molecule.
  - → Optimize *bond angles* for molecules with *lone pairs* on the central atom(s) with the concept in mind that repulsions are in the order

    lone pair-lone pair > lone pair-bonding pair > bonding pair-bonding pair.

    Example: Use the VSEPR model to predict the *shape* of tetrafluoroethene, C_2F_4.

    Lewis structure:

    There are three concentrations of electron density around each C atom, so the shape is *trigonal planar* about each C atom. C_2F_4 is an AX_3AX_3 species.

3.2 Molecules with Lone Pairs on the Central Atom

- **VSEPR method for molecules with lone pairs ($m \neq 0$)**
  - → Lone pair electrons have preferred positions in certain electron arrangements.
  - → In the *trigonal bipyramidal* arrangement, lone pairs prefer the *equatorial* positions, in which electron repulsions are minimized.
  - → In the *octahedral* arrangement, all positions are *identical*. The preferred electron arrangement for *two lone pairs* is the occupation of *opposite* corners of the octahedron.

→ Lone pairs contribute to the electron arrangement about a central atom and influence the shape of molecules, which is determined by the location of the nuclei.

Example: Use the VSEPR model to predict the shape of trichloroamine, NCl_3.

Lewis structure: :Cl—N—Cl: The four electron pairs around N yield a *tetrahedral*
                      |
                    :Cl:

electron arrangement, while the shape is *trigonal pyramidal* (like a badminton birdie). Because lone pair-bonding pair repulsion is greater than bonding pair-bonding pair repulsion, the bond angles [$\angle$ Cl–N–Cl] will be less than 109.5°.

3.3 Polar Molecules

* **Polar molecule** → Molecule with a *nonzero* dipole moment

* **Polar bond** → Bond with a *nonzero* dipole moment

* **Molecular dipole moment** → *Vector sum* of bond dipole moments

● **Representation of dipole** → Single arrow with head pointed toward the *positive* end of the dipole

  Example: Lewis structure: H—Cl: and polar (dipole) representation: H←Cl

● **Predicting molecular polarity**
  → Determine molecular *shape* using VSEPR theory.
  → Estimate electric dipole (bond) moments (Section 2.12 in the text) or use Fig. 3.7 in the text to decide whether the molecular *symmetry* leads to a *cancellation* of bond moments (*nonpolar molecule*, no dipole moment) or not (*polar molecule*, nonzero dipole moment).

  Example: Is XeF_2 a polar or nonpolar molecule?

    Lewis structure: :F—Xe—F: The electron arrangement is *trigonal bipyramidal;* the shape is *linear* (lone pairs prefer the equatorial positions). The bond angle is 180°, the bond dipoles cancel, and the molecule is *nonpolar:* F→Xe←F.

VALENCE-BOND (VB) THEORY (Sections 3.4–3.7)

3.4 Sigma (σ) and Pi (π) Bonds

● **Two major types of *bonding* orbitals** → σ and π

● **σ-orbital**
  → Has no *nodal surface* containing the interatomic (bond) axis
  → Is cylindrical or "sausage" shaped
  → Formed from overlap of two *s*-orbitals, an *s*-orbital and a *p*-orbital end-to-end, two *p*-orbitals end-to-end, a certain hybrid orbital and an *s*-orbital, a *p*-orbital end-to-end with a certain hybrid orbital, or two certain hybrid orbitals end-to-end

- **π-orbital**
  - → *Nodal plane* containing the interatomic (bond) axis
  - → Two cylindrical shapes (lobes), one above and the other below the nodal plane
  - → Formed from side-by-side overlap of two *p*-orbitals

- **Examples of σ- and π-orbital formation**
  - → Two 1*s*-orbitals of H atoms combine to form a σ-orbital of H_2 (left figure below)
  - → Side-by-side overlap of two *p*-orbitals to form a π-orbital (right figure below)

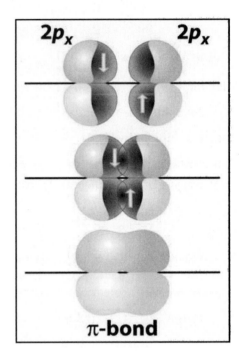

- **Electron occupancy**
  - → σ- and π-orbitals can hold 0, 1, or 2 electrons, corresponding to no bond, a half-strength bond, or a full covalent bond, respectively.
  - → In any orbital, spins of two electrons must be *paired* (opposite direction of arrows).

- **Types of bonds according to valence-bond theory**
  - → **Single bond** is a σ-bond.   [H_2, for example]
  - → **Double bond** is a σ-bond plus one π-bond.   [O_2, for example]
  - → **Triple bond** is a σ-bond plus two π-bonds.   [N_2, for example]

3.5 Electron Promotion and the Hybridization of Orbitals *and*
3.6 Other Common Types of Hybridization

- **Hybrid orbitals**
  - → Produced by mixing (hybridizing) orbitals of a central atom
  - → A construct consistent with observed shapes and bonding in molecules

Hybridization Schemes

Electron arrangement around the central atom	Hybrid orbitals (number)	Angle(s) between σ-bonds	Example(s) [underlined atom]
Linear	sp (two)	180°	$\underline{Be}Cl_2$, $\underline{C}O_2$
Trigonal planar	sp^2 (three)	120°	$\underline{B}F_3$, $\underline{C}H_3^+$
Tetrahedral	sp^3 (four)	109.5°	$\underline{C}Cl_4$, $\underline{N}H_4^+$
Trigonal bipyramidal	sp^3d (five)	120°, 90°, and 180°	$\underline{P}Cl_5$
Octahedral	sp^3d^2 (six)	90° and 180°	$\underline{S}F_6$

- **Determination of hybridization schemes**
  - → Write the Lewis structure(s). If there are resonance structures, pick *any* one.
  - → Determine the number of lone pairs and σ-bonds on the central atom.
  - → Determine the number of orbitals required for hybridization on the central atom to accommodate the lone pairs and σ-bonds.
  - → Determine the hybridization scheme from the preceding table.
  - → Use VSEPR rules to determine the molecular *shape* and *bond angle(s)*. (π-bonds are formed from the overlap of atomic orbitals [unhybridized] in a side-by-side arrangement.)

    Example: Describe the bonding in BF_3 using VB theory and hybridization. The shape is trigonal planar (VSEPR) and the hybridization at B is sp^2 (see table above) and the three σ-bonds are formed from overlap of $B(sp^2)$ orbitals with F(2p) orbitals ($B(sp^2)$–F(2p) bonds).

Note: Detailed studies indicate that both *s*- and *p*-orbitals take part in the bond formation of terminal atoms, such as F above. For simplicity, we use the approach above.

- **Central atoms** → Each central atom can be represented using a hybridization scheme, as in the example below.

  Example: Use VB theory to describe the bonding and shape of acetaldehyde, a molecule with two central atoms.

  Lewis structure:
  $$H-\overset{\overset{\displaystyle H}{|}}{\underset{\underset{\displaystyle H}{|}}{C}}-\overset{\overset{\displaystyle \cdot\cdot\atop O}{\|}}{C}-H$$

  3-D representation:

  Bonds: C–C σ-bonds from overlap of methyl C sp^3 with aldehyde C sp^2 hybrid orbital [Csp^3–Csp^2]; C=O σ-bond from overlap of aldehyde sp^2 hybrid orbital with methyl C sp^3 orbital; C=O π-bond from side-by-side overlap of C *2p* and O *2p* atomic orbitals; C–H single bonds from overlap of H1s AOs and Csp^3 hybrid orbitals (methyl C atom) and H1s AO and Csp^2 hybrid orbital

  Angles: Three $\angle$ H–C–H ~ 109.5°, three $\angle$ H–C–C ~ 109.5°, one $\angle$ C–C=O ~ 120°, one $\angle$ O=C–H ~ 120°, and one $\angle$ C–C–H ~ 120°

  Molecular shape: Tetrahedral at methyl carbon atom, trigonal planar at aldehyde C atom

3.7 Characteristics of Multiple Bonds

- **Alkanes (C_nH_{2n+2}, n = 1, 2, 3, ...)** → Characteristics: tetrahedral geometry and sp^3 hybridization at C atoms; all C–C and C–H single (σ) bonds; rotation allowed about C–C single bonds

  Examples: Methane, CH_4; and propane, $CH_3CH_2CH_3$

- **Alkenes (C_nH_{2n}, n = 2, 3, 4, ...)** → Characteristics: one C=C double bond (σ plus π); other C–C bonds and all C–H bonds single (σ); trigonal-planar geometry and sp^2 hybridization at double-bonded C atoms; tetrahedral geometry at other C (sp^3) atoms; rotation *not* allowed about double C=C bond; rotation allowed about C–C single bonds

  Examples: Ethene (ethylene), $H_2C=CH_2$; and propylene, $CH_3CH=CH_2$

- **Alkynes (C_nH_{2n-2}, n = 2, 3, 4, ...)** → Characteristics: one C≡C triple bond (σ plus two π); other C–C bonds and all C–H bonds (σ); linear geometry and sp hybridization at triple-bonded C atoms; tetrahedral geometry at other C (sp^3) atoms

  Examples: Ethyne (acetylene), HC≡CH; 2-butyne (dimethylacetylene), $H_3CC≡CCH_3$

- **Benzene (C_6H_6)**

  → **Characteristics:** *Planar* molecule with hexagonal C framework; trigonal-planar geometry and sp^2 hybridization at C atoms; σ framework has C–C and C–H single bonds; $2p$-orbitals (one from each C atom) overlap side-by-side form *three π-bonds (six π-electrons) delocalized over the entire six C atom ring*

  → See Lewis resonance structures in Section 2.7.

- **Double bonds**

  → Consist of *one* σ- and *one* π-bond; are always *stronger* than a single σ-bond:
  C=C is *weaker* than two single C–C σ-bonds
  N=N is *stronger* than two single N–N σ-bonds
  O=O is *stronger* than two single O–O σ-bonds

  → Formed readily by Period 2 elements (C, N, O)
  → Rarely found in Period 3 and higher period elements
  → Impart rigidity to molecules and influence molecular shape

  Example: Describe the structure of the carbonate anion in terms of hybrid orbitals, bond angles, and σ- and π-bonds.

  Lewis resonance structures:

  Trigonal-planar geometry with sp^2 hybridization at C, $120°$ $\angle$ O–C–O angles, and Csp^3-$O2p$ single bonds. Remaining C $2p$-orbital can overlap with O $2p$-orbitals to form a π-bond in each of three ways corresponding to the resonance structures. Therefore, each carbon-oxygen bond can be viewed as one σ-bond and one-third of a π-bond.

MOLECULAR ORBITAL (MO) THEORY (Sections 3.8–3.12)

3.8 The Limitations of Lewis's Theory

- **Valence Bond (VB) deficiencies**
  - → Cannot explain *paramagnetism* of O_2 (see text Box 3.2 for explanation of *paramagnetism* and *diamagnetism*)
  - → Difficulty treating *electron-deficient* compounds such as diborane, B_2H_6
  - → No simple explanation for *spectroscopic properties* of compounds such as color

- **Molecular Orbital (MO) advantages**
  - → Addresses *all* of the above shortcomings of VB theory
  - → Provides a *deeper* understanding of electron-pair bonds
  - → Accounts for the *structure and properties* of metals and semiconductors
  - → More facile for *computer calculations* than VB theory

3.9 Molecular Orbitals (MOs)

- **MO theory**
  - → In MO theory, electrons occupy MOs that are *delocalized* over the *entire* molecule.
  - → In VB theory, bonding electrons are *localized* between the two atoms.

- **Molecular Orbitals (MO)**
  - → Formed by superposition (**linear combination**) of atomic orbitals (**LCAO-MO**)
  - → **Bonding orbital** (constructive interference):
    Increased amplitude or electron density between atoms
  - → **Antibonding orbital** (destructive interference):
    Decreased amplitude or electron density between atoms

- **Types of Molecular Orbitals**
  - → **Bonding MO:** energy *lower* than that of the constituent AOs
  - → **Nonbonding MO:** energy *equal* to that of the constituent AOs
  - → **Antibonding MO:** energy *greater* than that of the constituent AOs

- **Rules for combining AOs to obtain MOs**
  - → *N* atomic orbitals (AOs) yield *N* molecular orbitals (MOs)
  - → Conservation of orbitals
  - → Little overlap of inner shell AOs, and these MOs are usually *nonbonding.*

- **MO energy-level diagrams**
  - → Relative energies of original AOs and resulting MOs are shown schematically in energy-level diagrams.
  - → Arrows used to show electron spin and to indicate location of the electrons in the separated atoms and the molecule

Example: Formation of MOs by LCAO method and MO energy-level diagram for the formation of Li_2 from Li atoms. Formation of bonding and antibonding orbitals:

Energy-level diagram

Energy

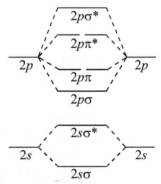

3.10 The Electron Configurations of Diatomic Molecules

- **Procedure for determining the electronic configuration of diatomic molecules**
  - → Construct all possible MOs from *valence-shell* AOs.
  - → Place *valence electrons* in the lowest energy, unoccupied MOs.
  - → Follow the **Pauli exclusion principle** and **Hund's rule** as for AOs: Electrons have *spins paired* in a fully occupied molecular orbital and enter unoccupied *degenerate* orbitals with *parallel spins*.

- **Valence-shell MOs for Period 1 and Period 2 *homonuclear* diatomic molecules**

Correlation Diagrams

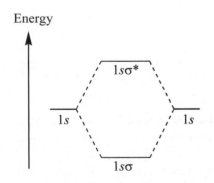

Schematic order of MO energy levels for H_2, He_2

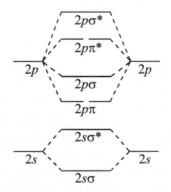

Schematic order of MO energy levels for Li_2, Be_2, B_2, C_2, N_2

Schematic order of MO energy levels for O_2, F_2, Ne_2

- **Bond order (b)** → $$b = \frac{1}{2}(N_e - N_e^*)$$

  N_e = total number of electrons in *bonding* MOs

  N_e^* = total number of electrons in *antibonding* MOs

Examples: b in Li_2 = 1 (two valence electrons in bonding $2s\sigma$ MO). b in H_2^- = $\frac{1}{2}$ (two electrons in $1s\sigma$ bonding MO, one electron in $1s\sigma^*$-antibonding MO). O_2^+ has 11 valence electrons, which are placed in the $2s\sigma$, $2s\sigma^*$, $2p\sigma$, two *degenerate* $2p\pi$, and one of the *degenerate* $2p\pi^*$ MOs. The valence electron configuration is $(2s\sigma)^2 (2s\sigma^*)^2 (2p\sigma)^2 (2p\pi)^4 (2p\pi^*)^1$. There is one unpaired electron. N_e = 8 and N_e^* = 3. Bond order = $\frac{1}{2}(8 - 3)$ = 5/2.

3.11 Bonding in Heteronuclear Diatomic Molecules

- **Bond properties**

  → Polar in nature

  → Atomic orbital from the *more* electronegative atom is *lower* in energy.

  → *Bonding* orbital has a *greater* contribution from the *more* electronegative atom.

  → A*ntibonding* orbital has a *greater* contribution from the *less* electronegative atom.

- **Result of bond properties is modified, unsymmetrical correlation diagrams.**

    Example:

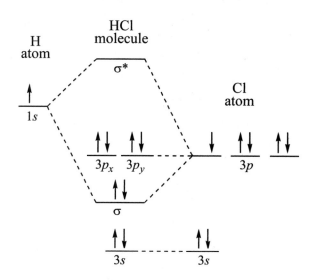

Bonding in HCl using the MO theory. The correlation diagram shows occupied valence orbitals for H and Cl atoms, resulting MOs from overlap of $1s$ H and $3p_z$ Cl orbitals, lone-pair-nonbonding orbitals $3p_x$ and $3p_y$ of Cl, and the nonbonding Cl $3s$ orbital. The electronic configuration is $(Cl\ 3s)^2 \sigma^2 (Cl\ 3p_x)^2 (Cl\ 3p_y)^2$.

3.12 Orbitals in Polyatomic Molecules

- **Bond properties**
  → All atoms in the molecule are encompassed by the MOs.
  → Electrons in MOs bond all the atoms in a molecule.
  → Electrons are *delocalized* over all the atoms.
  → Energies of the MOs are obtained by spectroscopy.

- **Effectively localized electrons** → While MO theory requires MOs to be *delocalized* over all atoms, electrons sometimes are *effectively localized* on an atom or group of atoms.

  Example: See the HCl example above with nonbonding $Cl\,3p_x$ and $Cl\,3p_y$ orbitals effectively localized on Cl.

- **Combining VB and MO theories** → VB and MO theory are sometimes jointly invoked to describe the bonding in complex molecules.

  Example: In benzene, *localized sp^2* hybridization (hybrids of $C\,2s$-, $C\,2p_x$-, and $C\,2p_y$-orbitals on each C) is used to describe the σ-bonding framework of the ring, and *delocalized* MOs (LCAOs of six $C\,2p_z$-orbitals) to describe the π-bonding.

- **Triumphs of MO theory**
  - → Accounts for paramagnetism of O_2
  - → Accounts for bonding in electron-deficient molecules

    Example: In diborane (B_2H_6, with 12 valence electrons), *six* electron pairs bond *eight* atoms using *six* MOs. VB theory with only localized covalent bonds requires *eight* electron pairs and fails to describe the bonding. A combination of MO and VB theories has also been used to describe the bonding in B_2H_6 as four B–H single bonds (8 electrons) and two, two-electron, 3-center B–H–B delocalized bonds (4 electrons).

    Diborane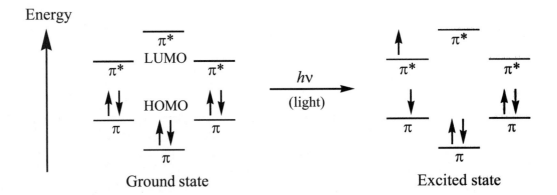

  - → Provides a framework for interpreting visible and ultraviolet (UV) spectroscopy

    Example:

Energy

| Ground state | $h\nu$ (light) → | Excited state |

For benzene (schematic energy levels shown above), the UV spectrum is interpreted as the excitation of an electron from the *highest occupied molecular orbital* (**HOMO**, *bonding* π MO) to the *lowest unoccupied molecular orbital* (**LUMO**, *antibonding* π* MO). Excitation changes the bond order of the π-system from 3 to 2.

Chapter 4 THE PROPERTIES OF GASES

THE NATURE OF GASES (Sections 4.1–4.3)

4.1 Observing Gases

- **Gases** → Examples of *bulk matter*, forms of matter consisting of large numbers of molecules

- **Atmosphere** → Not uniform in composition, temperature (T), or density (d)

- **Physical properties** → Most gases behave similarly at low pressure and high temperature.

- **Compressibility**
  - → Ease with which a gas undergoes a volume (V) decrease
  - → Gas molecules are widely spaced and gases are highly compressible. Gas molecules are in ceaseless chaotic motion.

- **Expansivity**
  - → The ability of a gas to fill the space available to it rapidly
  - → Atomic and molecular gases move rapidly and respond quickly to changes in the available volume.

4.2 Pressure

$$\text{Pressure} \equiv \frac{\text{force}}{\text{area}} \quad or \quad P = \frac{F}{A}$$

Opposing force from a confined gas

- **Barometer**
  - → A glass tube, sealed at one end, filled with liquid mercury, and inverted into a beaker also containing liquid mercury (Torricelli)
  - → The height (h) of the liquid column is a measure of the atmospheric pressure (P) expressed as the length of the column for a particular liquid.
  Common pressure units: mmHg, cmHg, cmH$_2$O

$$P = \frac{F}{A} = \frac{mg}{A} = \frac{(dV)g}{A} = \frac{(dhA)g}{A} = dhg$$

  where m = mass of liquid, g = acceleration of gravity
  ($9.806\,65$ m·s^{-2} at the Earth's surface), and d = density of liquid

  - → The atmospheric pressure P when the height h of a column of mercury in a barometer is *exactly* 0.76 m at $0°C$ is $101\,325$ kg·m^{-1}·s^{-2}, a result which is obtained by using the value of the density of mercury = $13\,595.1$ kg·m^{-3} and the gravitational constant at the surface of the Earth.
  Note: 1 Pa (pascal) = 1 kg·m^{-1}·s^{-2} = 1 N·m^{-2} (SI unit of pressure)

- **Manometer (see Fig. 4.5 in the text)**
  - → U-shaped tube filled with liquid connected to an experimental system
  - → "open-tube" pressure of system = atmospheric pressure when levels are equal
  - → "open-tube" pressure of system = pressure difference between system and atmospheric pressure when levels are unequal
  - → "closed-tube" pressure (system) = difference in level heights

- **Gauge pressure** $\rightarrow$ Difference between the pressure inside a vessel containing a compressed gas (*e.g.*, a tire) and the atmospheric pressure

4.3 Alternative Units of Pressure

- **Atmosphere** $\rightarrow$ 1 atm = **760** Torr (**bold type = exact**)
  1 atm = **101 325** Pa = **101.325** kPa
  1 bar = 10^5 Pa = **100** kPa
  1 Torr = 1 mmHg (Not exact, but correct to better than 2×10^{-7} Torr)

THE GAS LAWS (Sections 4.4–4.8)

4.4 The Experimental Observations

- **Boyle's law** (fixed amount of gas n at constant temperature T)

- **Gas behavior** $\rightarrow$ Boyle's law describes an isothermal system with a fixed amount of gas: $\Delta T = 0$, $\Delta n = 0$ (constant T, n).

$$\text{Pressure} \propto \frac{1}{\text{volume}} \quad \text{or} \quad P \propto \frac{1}{V} \quad \text{or} \quad PV = \text{constant}$$

  The functional form is a *hyperbola* (plot of P versus V), but a plot of P versus V^{-1} gives a straight line.

- **Changes between two states (1 and 2) at constant temperature**

$$P_1 V_1 = \text{constant} = P_2 V_2 \quad \text{or} \quad \frac{P_2}{P_1} = \frac{1/V_2}{1/V_1} = \frac{V_1}{V_2}$$

- **Deviations** $\rightarrow$ At low temperature and/or high pressure

- **Charles's law (fixed amount of gas n at constant pressure P)**

- **Gas behavior** $\rightarrow$ Charles's law describes an isobaric system with a fixed amount of gas: $\Delta P = 0$, $\Delta n = 0$ (constant P, n).

$$\text{Volume} \propto \text{temperature} \quad \text{or} \quad V \propto T \quad \text{or} \quad V/T = \text{constant}$$

  The functional form is *linear*.

- **Kelvin temperature scale** $\rightarrow$ Two *fixed* points define the absolute temperature scale:

  $T \equiv 273.16$ **K**   The triple point of water, where ice, liquid, and vapor are in equilibrium (see Chapter 9).

  $T \equiv 0$ **K**   Defined from the limiting behavior of Charles's law. V approaches 0 at fixed *low* pressure (extrapolation).

- **Celsius temperature scale** $\rightarrow$ $t(^\circ\text{C}) \equiv T(\text{K}) - 273.15$ (exactly)

  Notes: The freezing point of water at 1 atm external pressure is 0°C or 273.15 K. The symbol t is also used for time.

- **Changes between two states (1 and 2)**

$$V_1/T_1 = \text{constant} = V_2/T_2 \quad \text{or} \quad \frac{V_2}{V_1} = \frac{T_2}{T_1}$$

- **Deviations** $\rightarrow$ At high pressure and/or low temperature
- **Another aspect of gas behavior (fixed amount of gas *n* at constant volume *V*)**

  $\rightarrow$ $\boxed{\text{Pressure} \propto \text{temperature} \quad \text{or} \quad P \propto T \quad \text{or} \quad P/T = \text{constant}}$

  $\rightarrow$ This relation describes an *isochoric* system: $\Delta V = 0$ (constant V). The functional form is *linear*.

- **Changes between two conditions (1 and 2)**

$$P_1/T_1 = \text{constant} = P_2/T_2 \quad \text{or} \quad \frac{P_2}{P_1} = \frac{T_2}{T_1}$$

- **Deviations** $\rightarrow$ For small volume

- **Avogadro's principle** (gas at constant pressure *P* and temperature *T*)

  $\rightarrow$ A given number of gas molecules at a particular P and T occupies the same volume regardless of chemical identity.

- **Gas behavior** $\boxed{\text{Volume} \propto \text{amount} \quad \text{or} \quad V \propto n \quad \text{or} \quad V/n = \text{constant}}$

- **Molar volume** $\boxed{V_\text{m} = \dfrac{V}{n} = \text{constant} \qquad \text{for } \Delta P = 0 \text{ and } \Delta T = 0}$

- **Changes between two conditions (1 and 2)**

$$V_1/n_1 = \text{constant} = V_2/n_2 \quad \text{or} \quad \frac{V_2}{V_1} = \frac{n_2}{n_1}$$

- **Deviations** $\rightarrow$ At high pressure and/or low temperature

- **Another aspect of gas behavior (gas at constant volume *V* and temperature *T*)**

  $\boxed{\text{Pressure} \propto \text{amount} \quad \text{or} \quad P \propto n \quad \text{or} \quad P/n = \text{constant}}$

- **Changes between two conditions (1 and 2)**

$$P_1/n_1 = \text{constant} = P_2/n_2 \quad \text{or} \quad \frac{P_2}{P_1} = \frac{n_2}{n_1}$$

- **Deviations** $\rightarrow$ For small volumes
- **The ideal gas law, *PV = nRT* (*R* = constant)**
  $\rightarrow$ Incorporates Boyle's law, Charles's law, and Avogadro's principle
  $\rightarrow$ R is the universal gas constant described below.
- **Gas behavior** $\rightarrow$ As P approaches 0, all gases behave *ideally* (*limiting law*).

- **Derivation**

Boyle	**Charles**	**Avogadro**

  $$PV = \text{constant}_1 \qquad V = \text{constant}_2 \times T \qquad P = \text{constant}_3 \times n$$

  $$PV = (\text{constant}_3 \times n)(\text{constant}_2 \times T) = nRT \qquad \textit{equation of state}$$

- **Gas constant** $\rightarrow R = 8.205\,74 \times 10^{-2} \; \text{L·atm·K}^{-1}\text{·mol}^{-1}$

  $\rightarrow R = 8.314\,47 \; \text{L·kPa·K}^{-1}\text{·mol}^{-1} = 8.314\,47 \; \text{J·K}^{-1}\text{·mol}^{-1}$

4.5 Applications of the Ideal Gas Law

- **Changes between two conditions (1 and 2)**

  $$\frac{P_1 V_1}{n_1 T_1} = \frac{P_2 V_2}{n_2 T_2}$$

  Combined gas law

  Note: The more general form of the combined gas law allows the amount of gas to change.

  Note: When using this relationship, remember that the actual units used in manipulating ratios of quantities are *not* important. Consistency in units is essential. *Absolute* temperature, however, *must* be used in all the equations.

- **Molar volume** $V_m = \dfrac{RT}{P}$

- **Standard ambient temperature and pressure (SATP)** $\rightarrow$ 298.15 K and 1 bar

- **Standard temperature and pressure (STP)** $\rightarrow$ 0°C and 1 atm

  [or 273.150 K and 1.013 25 bar]

- **Molar volume at STP**

  $$\rightarrow \; V_m = \frac{RT}{P} = \frac{(8.205\,8 \times 10^{-2} \; \text{L·atm·K}^{-1}\text{·mol}^{-1})(273.150 \; \text{K})}{(1 \; \text{atm})} = 22.4141 \; \text{L·mol}^{-1}$$

4.6 Gas Density

- **Molar concentration** $\dfrac{n}{V} = \dfrac{1}{V_m} = \dfrac{P}{RT}$

- **Molar concentration at STP**

  $$\rightarrow \; \frac{n}{V} = \frac{1}{V_m} = \frac{1}{22.4141 \; \text{L·mol}^{-1}} = 4.461\,48 \times 10^{-2} \; \text{mol·L}^{-1}$$

- **Density of a gas** $\boxed{d = \dfrac{m}{V} = \dfrac{nM}{V} = \dfrac{MP}{RT}}$

- **Density of a gas at STP** $\rightarrow \quad d = \dfrac{m}{V} = \dfrac{nM}{V} = (4.461\,48 \times 10^{-2}\ \text{mol·L}^{-1}) \times M$

4.7   The Stoichiometry of Reacting Gases

- **Accounting for reaction volumes** $\rightarrow$ Use mole-to-mole calculations as described in Fundamentals Sections L and M. Follow the flow chart given below and then convert to gas volume by using the ideal gas law.

$$Mass\ of\ A \rightarrow molar\ mass \rightarrow moles\ of\ A \rightarrow stoichiometry$$
$$\rightarrow moles\ of\ B \rightarrow molar\ volume \rightarrow volume\ of\ B$$

- **Balanced chemical equation** $\rightarrow$ See Example 4.6 and the following self-test exercises in the text.

  Example: $NH_4NO_3\,(s) \rightarrow N_2O\,(g) + 2\,H_2O\,(g)$ A total of three moles of gas products is produced from one mole of solid.

- **Molar volumes of gases**
  - $\rightarrow$ Generally more than 1000 times those of liquids and solids (condensed phases)
  - $\rightarrow$ Reactions that produce gases from condensed phases can be explosive.

4.8   Mixtures of Gases

- **Behavior** $\rightarrow$ Mixtures of nonreacting gases behave as if only a single component were present.

- **Partial pressure** $\rightarrow$ Pressure a gas would exert if it occupied the container alone

- **Law of partial pressures** $\rightarrow$ Total pressure of a gas mixture is the sum of the partial pressures of its components [John Dalton]
  $$P = P_A + P_B + \ldots \ \text{ for mixture containing A, B, } \ldots$$

- **Components at the same T and V** $\rightarrow$ $n = n_A + n_B + \ldots = $ total amount of mixture

$$P = \frac{nRT}{V} \qquad \text{and} \qquad P_A = \frac{n_A RT}{V}$$

$$x_A = \frac{n_A}{n} = \frac{P_A}{P} = \text{mole fraction of A in the mixture}$$

Thus, $P_A = x_A P$

- **Humid gas** $\rightarrow$ For air, $P = P_{\text{dry air}} + P_{\text{water vapor}}$, and $P_{\text{water vapor}} = 47$ Torr at body temperature (37°C).
  Note: See Section 9.3

MOLECULAR MOTION (Section 4.9–4.11)

4.9   Diffusion and Effusion

- **Diffusion** $\rightarrow$ Gradual dispersal of one substance through another substance

- **Effusion**
    - $\rightarrow$ Escape of a gas through a small hole (or assembly of microscopic holes) into a vacuum
    - $\rightarrow$ Depends on molar mass and temperature of gas
- **Graham's law**
    - $\rightarrow$ Rate of effusion of a gas *at constant temperature* is *inversely* proportional to the square root of its molar mass M.

    For two gases A and B

    $$\frac{\text{Rate of effusion of A}}{\text{Rate of effusion of B}} = \sqrt{\frac{M_B}{M_A}}$$

    - $\rightarrow$ Rate of effusion is proportional to the average speed of the molecules in a gas *at constant temperature*.

    $$\frac{\text{Average speed of A molecules}}{\text{Average speed of B molecules}} = \sqrt{\frac{M_B}{M_A}}$$

    - $\rightarrow$ Rate of effusion is *inversely* proportional to the time taken for given volumes or amounts of gas to effuse *at constant temperature*.

    $$\frac{\text{Rate of effusion of A}}{\text{Rate of effusion of B}} = \frac{\text{Time for B to effuse}}{\text{Time for A to effuse}} = \sqrt{\frac{M_B}{M_A}}$$

- **Effusion at different temperatures** $\rightarrow$ Rate of effusion and average speed increase as the square root of the temperature.

    $$\frac{\text{Rate of effusion at } T_2}{\text{Rate of effusion at } T_1} = \frac{\text{Average speed of molecules at } T_2}{\text{Average speed of molecules at } T_1} = \sqrt{\frac{T_2}{T_1}}$$

- **Combined relationship** $\rightarrow$ Average speed of molecules is *directly* proportional to the square root of the temperature and *inversely* proportional to the square root of the molar mass.

    $$\text{Average speed of molecules in a gas} \propto \sqrt{\frac{T}{M}}$$

4.10 The Kinetic Model of Gases

- **Assumptions**
    1. Gas molecules are in continuous random motion.
    2. Gas molecules are infinitesimally small particles.
    3. Particles move in straight lines until they collide.
    4. Molecules do not influence one another except during collisions.

- **Collisions with walls** → Consider molecules traveling only in one dimension x with an average of square velocity of $\langle v_x^2 \rangle$.

- **Pressure on wall** → $P = \dfrac{Nm\langle v_x^2 \rangle}{V}$, where N is the number of molecules, m is the mass of each molecule, and V is the volume of the container

- **Mean square speed** → $v_{rms}^2 = \langle v^2 \rangle = \langle v_x^2 + v_y^2 + v_z^2 \rangle = \langle v_x^2 \rangle + \langle v_y^2 \rangle + \langle v_z^2 \rangle = 3\langle v_x^2 \rangle$

   rms = root mean square

- **Pressure on wall** → $P = \dfrac{Nmv_{rms}^2}{3V}$ and $N = nN_A$, where N_A is the Avogadro constant

   $P = \dfrac{nN_A m v_{rms}^2}{3V} = \dfrac{nM v_{rms}^2}{3V}$, where $M = mN_A$ is the molar mass

$$\boxed{PV = \dfrac{nM v_{rms}^2}{3} = nRT}$$ The term nRT is from the ideal gas law.

- **Solve for v_{rms}** $\boxed{v_{rms} = \sqrt{\dfrac{3RT}{M}}}$ v_{rms} is the **root mean square** speed of the particle.

   Note: *Beware of units.* Use R in units of J·K^{-1}·mol^{-1} and convert M to kg·mol^{-1}. The resulting units of speed will then be m·s^{-1}.

$$\boxed{\text{Molar kinetic energy} = \left(\dfrac{1}{2}mv_{rms}^2\right)N_A = \left(\dfrac{1}{2}m\right)\left(\dfrac{3RT}{M}\right)N_A = \dfrac{3}{2}RT, \text{ because } mN_A = M}$$

- **Kinetic model of gases**
  - → Consistent with the ideal gas law
  - → Suggests that the root mean square (rms) speed of gas molecules is proportional to the square root of the gas temperature

4.11 The Maxwell Distribution of Speeds

- **Symbols** → Let v represent a particle's speed, N the total number of particles, $f(v)$ the Maxwell distribution of speeds, Δ a *finite* change, and d an *infinitesimal* change (*not* density) in the following relationships.

- **For a *finite* range of speeds**

$$\boxed{\Delta N = Nf(v)\Delta v, \text{ where } f(v) = 4\pi\left(\dfrac{M}{2\pi RT}\right)^{3/2}v^2 e^{-Mv^2/2RT}}$$

- **Temperature dependence of $f(v)$**
  - $\rightarrow$ As T increases, the fraction of molecules with speeds greater than a specific speed increases. See Figure 4.27 (hot molecules move faster than cool ones on average). We will use this concept again in Chapter 14, when we study the rates of chemical reactions.

- **Dependence of $f(v)$ on molar mass M**
  - $\rightarrow$ As M decreases, the fraction of molecules with speeds greater than a specific speed increases. See Figure 4.26 (light molecules move faster than heavy ones on average).

- **For an *infinitesimal* range of speeds**

$$\boxed{\frac{\mathrm{d}N}{N} = f(v)\,\mathrm{d}v}$$

Note: Calculus can be used with the Maxwell distribution of speeds to obtain the following properties that are only mentioned in a homework exercise in the text.

Average speed

$$\boxed{\langle v \rangle = \int_0^\infty v f(v)\,\mathrm{d}v = \sqrt{\frac{8RT}{\pi M}}}$$

Most probable speed

$$\boxed{v_{\mathrm{mp}} = \sqrt{\frac{2RT}{M}}\ , \ \text{where}\ \frac{\mathrm{d}f(v)}{\mathrm{d}v} = 0 \quad \text{(maximum in distribution)}}$$

THE IMPACT ON MATERIALS: REAL GASES
(Sections 4.12–4.14)

4.12 Deviations from Ideality

- **Definition** $\rightarrow$ Attractions and repulsions between atoms and molecules cause deviations from ideality in gases.
  - $\rightarrow$ Attractions have a longer range than repulsions.

- **Evidence** $\rightarrow$ Gases condense to liquids when cooled or compressed (attraction).
  Liquids are difficult to compress (repulsion).

- **Compression factor (Z)** $\rightarrow$ A measure of the effect of intermolecular forces
  - $\rightarrow$ Is the ratio of the observed molar volume of a gas to that calculated for an ideal gas

$$\boxed{Z \equiv \frac{PV_{\mathrm{m}}}{RT} = \frac{V_{\mathrm{m}}}{V_{\mathrm{m}}^{\text{ideal}}}}$$

For an ideal gas, $Z = 1$
For H_2, $Z > 1$ at all P (repulsions dominate)
For NH_3, $Z < 1$ at low P (attractions dominate)

- **For many gases, attractions dominate at low pressures ($Z < 1$), while repulsive interactions dominate at high pressures ($Z > 1$) [methane is an example, see Fig. 4.28 in the text].**

4.13 The Liquefaction of Gases

- **Joule–Thomson effect**
  - → When attractive forces dominate, a real gas *cools* as it expands.
  - → In this case expansion requires energy, which comes from the kinetic energy of the gas, lowering the temperature. (Compression and expansion through a small hole lowers temperature further.) Exceptions are He and H_2, for which repulsion dominates.
  - → The effect is used in some refrigerators and to effect the condensation of gases such as oxygen, nitrogen, and argon.

4.14 Equations of State of Real Gases

- **Virial equation** →
$$PV = nRT \left(1 + \frac{B}{V_m} + \frac{C}{V_m^2} + \cdots \right)$$

B = second virial coefficient
C = third virial coefficient, etc.

> **Note:** **Virial coefficients** depend on temperature and are found by fitting experimental data to the virial equation. This equation is more general than the van der Waals equation but is more difficult to use to make predictions.

- **van der Waals equation** →
$$\left(P + a\frac{n^2}{V^2} \right)(V - nb) = nRT$$

where a and b are the **van der Waals parameters** (determined experimentally)

- **Rearranged form** →
$$P = \frac{nRT}{V - nb} - a\frac{n^2}{V^2}$$

where a accounts for the attractive effects and b accounts for the repulsive effects

- **van der Waals Z** →
$$Z \equiv \frac{P_{vdW}V_m}{RT} = \frac{V}{V - nb} - \frac{an}{RTV} = \frac{1}{1 - (nb/V)} - \frac{an}{RTV}$$

- **Repulsion**
  - → Repulsive forces imply that molecules cannot overlap.
  - → Other molecules are *excluded* from the volume they occupy.
  - → The *effective* volume is then $V - nb$, not V.
  - → Thus, b is a constant corresponding to the volume excluded per mole of molecules.
  - → The potential energy describing this situation is that of an *impenetrable* hard sphere.
  - → The units of b are $L \cdot mol^{-1}$.

- **Attraction**
  - → Attractive forces lead to clustering of molecules in the gas phase.
  - → Clustering reduces the total number of gas phase species.
  - → The rate of collisions with the wall (pressure) is thereby reduced.
  - → Because this effect arises from attractions between *pairs* of molecules, it should be proportional to the *square* of the number of molecules per unit volume $(N/V)^2$ or equivalently, to the *square* of the molar concentration $(n/V)^2$.
  - → The pressure is predicted to be reduced by an amount $a(n/V)^2$, where a is a positive constant that depends on the strength of the attractive forces.
  - → The potential energy describing this situation is that of long-range attraction with the form $-(1/r^6)$, where r is the distance between a pair of gas phase molecules. This type of attraction primarily arises from *dispersion* or *London* forces (see Chapter 5).
  - → The units of a are $atm \cdot L^2 \cdot mol^{-2}$.

Summary

- **Virial equation** → Describes real gases

- **van der Waal's equation**
  - → An *approximate* equation of state
  - → Takes molecular volume and intermolecular forces into account

- **Ideal gas equation**
  - → Least accurate of the gas phase equations of state
  - → Ignores molecular volume and intermolecular forces
  - → Easiest to use

Chapter 5   LIQUIDS AND SOLIDS

INTERMOLECULAR FORCES   (Sections 5.1–5.6)

5.1   The Origin of Intermolecular Forces

- **Physical states** → Solid, liquid, and gas

- **Phase** → Form of matter uniform in both chemical composition and physical state

- **Condensed phase**

  → Solid or liquid phase

  Examples: $Ag(s)$; $Sn/Pb(s)$ alloy, composition variable; $H_2O(s)$; $H_2O(l)$; 1% $NaCl(s)$ in $H_2O(l)$

  → Molecules (or atoms or ions) are close to each other all the time and intermolecular forces are of major importance.

- **Coulomb potential energy E_P**

  → The interaction between two charges, q_1 and q_2, separated by distance r is (see Fundamentals Section A in the text) $\Rightarrow$ $\boxed{E_P \propto \dfrac{q_1 q_2}{r}}$

  → Almost all intermolecular interactions can be traced back to this fundamental expression.

  Note: The term *intermolecular* is used in a general way to include atoms and ions.

5.2   Ion-Dipole Forces

- **Hydration**

  → Attachment of water molecules to ions (cations and anions)

  → Water molecules are polar and have an electric dipole moment μ.

  → A small positive charge on each H atom attracts anions and a small negative charge on the O atom attracts cations.

- **Ion-dipole interaction**

  → The potential energy (interaction) between an ion with charge $|z|$ and a polar molecule with dipole moment μ at a distance r is $\Rightarrow$ $\boxed{E_P \propto -\dfrac{|z|\,\mu}{r^2}}$

  → For proper *alignment* of the ion and dipole, the interaction is *attractive*: Cations attract the *partial* negative charges and anions attract the *partial* positive charges on the polar molecule.

  → Shorter range (r^{-2}) interaction than the Coulomb potential (r^{-1})

  → Polar molecule needs to be *almost* in contact with ion for substantial ion-dipole interaction.

- **Hydrated compounds**

  → Ion-dipole interactions are much *weaker* than ion-ion interactions, but are relatively strong for *small, highly charged* cations (large $|Z|$, small r).

  → Accounts for the formation of salt hydrates such as $CuSO_4 \cdot 5\,H_2O$ and $CrCl_3 \cdot 6\,H_2O$

- **Size effects**

  → Li^+ and Na^+ (small) tend to form hydrated compounds.

  → K^+, Rb^+, and Cs^+ (larger) tend not to form hydrates.

  → NH_4^+ (143 pm) is similar in size to Rb^+ (149 pm) and forms *anhydrous* compounds.

- **Charge effects** → Ba^{2+} and K^+ are similar in size, yet Ba^{2+} (larger charge) forms *hydrates*.

5.3 Dipole-Dipole Forces

- **Dipole alignment**   → In solids, molecules with dipole moments tend to *align* with partial positive charge on one molecule near the partial negative charge on another.

- **Dipole-dipole interaction in solids**
  - → The interaction between two polar molecules with dipole moments μ_1 and μ_2 *aligned* and separated by a distance r is $\quad\Rightarrow\quad$ $\boxed{E_P \propto -\dfrac{\mu_1\,\mu_2}{r^3}}$
  - → *Attractive* interaction in solids for head-to-tail *alignment* of dipoles
  - → Shorter range (r^{-3}) interaction than ion-dipole (r^{-2}) or Coulomb potential (r^{-1})
  - → For significant interaction, polar molecules need to be *almost* in contact with each other.

- **Dipole-dipole interactions in gas-phase molecules**
  - → Dipole-dipole interactions are much *weaker* in gases than in solids. The potential energy of interaction is $\quad\Rightarrow\quad \boxed{E_P \propto -\dfrac{\mu_1^{\,2}\,\mu_2^{\,2}}{r^6}}$
  - → Because gas molecules are in motion (rotating as well), they experience only a *weak* net attraction because of *occasional* alignment.

- **Dipole-dipole interactions in liquid-phase molecules**
  - → Same potential energy relationship as in the gas phase, but the interaction is slightly *stronger* because the molecules are closer.
  - → The liquid *boiling point* is a measure of the *strength* of the intermolecular forces in a liquid. The boiling point of isomers (see the Glossary in the text for definition) is *often* related to the strength of their dipole-dipole interactions.
  - → Typically, the *larger* the dipole moment, the *higher* the boiling point.

    Example: *Cis*-dibromoethene (dipole moment ≈ 2.4 D, bp 112.5°C) vs. *trans*-dibromoethene (zero dipole moment, bp 108°C)

5.4 London Forces

- **Nonpolar molecules**   → Condensation of *nonpolar* molecules to form liquids implies the existence of a type of intermolecular interaction other than those described above.

- **London force**
  - → Occurs in addition to any dipole-dipole interactions
  - → Is universal and applies to all molecules
  - → Accounts for the attraction between any pair of ground-state molecules (*polar* or *nonpolar*)
  - → Arises from *instantaneous* partial charges (*instantaneous* dipole moment) in one molecule *inducing* partial charges (dipole moment) in a neighboring one
  - → Exists between atoms and *rotating* molecules as well
  - → Strength depends on polarizability and shape of molecule.

    Example: Other things equal, rod-shaped molecules tend to have stronger London forces than spherical molecules because they can approach each other more closely.

- **Polarizability, α**
  - → Of a molecule is related to the ease of deformation of its electron cloud
  - → Is proportional to the total number of electrons in the molecule
  - → Because the number of electrons correlates with molar mass (generally increases), polarizability does as well.

→ The potential energy between molecules (polar or nonpolar) with polarizability α_1 and α_2, separated by a distance r, is $\Rightarrow$ $\boxed{E_P \propto -\dfrac{\alpha_1 \alpha_2}{r^6}}$

Note: Same (r^{-6}) interaction as dipole-dipole (r^{-6}) with *rotating* molecules, but the London interaction is *usually* of greater strength at normal temperatures.

- **Liquid boiling point**
  → A measure of the strength of intermolecular forces in a liquid
  → In comparing the interactions discussed in Sections 5.3 and 5.4, the *major* influence on boiling point in both *polar* and *nonpolar* molecules is the London force.

    Example: We can predict the relative boiling points of the nonpolar molecules: F_2, Cl_2, Br_2, and I_2. Boiling point correlates with polarizability, which depends on the number of electrons in the molecule. The boiling points are expected to increase in the order of F_2 (18 electrons), Cl_2 (34), Br_2 (70), and I_2 (106). The experimental boiling temperatures are: F_2 (−188°C), Cl_2 (−34°C), Br_2 (59°C), and I_2 (184°C).

- **Dipole-induced dipole interaction**
  → The potential energy between a polar molecule with dipole moment μ_1 and a nonpolar molecule with polarizability α_2 at a distance r is $\Rightarrow$ $\boxed{E_P \propto -\dfrac{\mu_1^2 \alpha_2}{r^6}}$
  → Also applies to molecules that are both polar. Each one can induce a dipole in the other.
  → Weaker interaction than dipole-dipole
    Example: Carbon dioxide $(\mu = 0)$ dissolved in water $(\mu > 0)$

5.5 Hydrogen Bonding

- **Hydrogen bonds**
  → Interaction *specific* to certain types of molecules (strong attractive forces)
  → Account for unusually high boiling points in ammonia (NH_3, −33°C), water (H_2O, 100°C), and hydrogen fluoride (HF, 20°C)
  → Arise from a H atom, *covalently* bonded to either an N, O, or F atom in *one* molecule, strongly attracted to a *lone pair* of electrons on an N, O, or F atom in *another* molecule
  → Are *strongest* when the three atoms are in a *straight line* and the distance between terminal atoms is within a particular range
  → Common symbol for the H bond is three dots: $\cdots$
  → *Strongest* intermolecular interaction between *neutral* molecules
    Examples: N–H$\cdots$:N hydrogen bonds between NH_3 molecules in pure NH_3
    O–H$\cdots$:O hydrogen bonds H_2O molecules in pure H_2O

- **Hydrogen bonding in gas-phase molecules**
  → Aggregation of some molecules persists in the vapor phase.
  → In HF, fragments of *zigzag* chains and $(HF)_6$ rings are formed. In CH_3COOH (acetic acid), *dimers* are formed. The abbreviated (no C–H bonds shown) Lewis structure for the dimer $(CH_3COOH)_2$ is shown on the right.

- **Importance of hydrogen bonding**
  → Accounts for the open structure of solid water
  → Maintains the shape of biological molecules

→ Binds the two strands of DNA together

5.6 Repulsions

- **Nature of repulsions**
  - → Molecules (or atoms that do not form bonds) that are very close together repel one another.
  - → The Pauli exclusion principle forms the basis of understanding this repulsion.

    Example: As two He atoms approach each other, at short separations, their $1s$ atomic orbitals overlap and form a bonding and an antibonding molecular orbital.

    The bonding molecular orbital is filled with two electrons and the remaining two electrons are required by the exclusion principle to fill the antibonding orbital.

    An antibonding orbital is more antibonding in character than a bonding orbital is bonding. The result is an increase in energy as the two atoms merge into each other.

    The same effect occurs for all molecules in which the atoms have filled shells (for example, H_2), even though the details of the bonding and antibonding orbitals they form may be much more complicated.

    The result is that all molecules repel each other when they come into contact and their orbitals overlap.

- **Electron density**
  - → Electron density in all atomic orbitals (AOs) and the molecular orbitals (MOs) they form decreases exponentially toward zero at large distances from the nucleus.
  - → The overlap between orbitals on neighboring molecules will also depend exponentially on their separation.
  - → As a result, repulsions between molecules usually depend exponentially on separation. Repulsions are effective only when the two molecules are very close together, but once they are close, increase sharply.
  - → The strong dependence of repulsion on separation is the underlying reason why objects around us have definite, well defined shapes.

LIQUID STRUCTURE (Sections 5.7–5.8)

5.7 Order in Liquids

- **Liquid phase**
  - → Mobile molecules with restricted motion
  - → Between the extremes of gas and solid phases
- **Long-range order**
  - → Characteristic of a *crystalline solid*
  - → Atoms or molecules are arranged in an orderly pattern that is repeated over long distances.
- **Short-range order**
  - → Characteristic of the *liquid phase*
  - → Atoms or molecules are positioned in an orderly pattern at nearest-neighbor distances only.
  - → Local order is maintained by a continual process of forming and breaking nearest-neighbor interactions.

5.8 Viscosity and Surface Tension

- **Viscosity**
    - $\rightarrow$ Resistance of a substance to flow: *Greater* viscosity yields *slower* flow.
    - $\rightarrow$ Viscous liquids include those with hydrogen bonding between molecules.

        Examples: H_3PO_4 (phosphoric acid) and $C_3H_8O_3$ (glycerol), each forms many H bonds; liquid phases of metals; and long chain molecules that can be entangled such as hydrocarbon oil and greases

- **Viscosity and temperature** $\rightarrow$ *Usually* viscosity *decreases* with *increasing T.* An exception is the unusual behavior of sulfur, for which, viscosity initially increases with increasing T because the S_8 rings break into chains that tangle together. Viscosity falls again at higher temperature as the chains break into smaller fragments.

- **Surface tension (several definitions)**
    - $\rightarrow$ Tendency of surface molecules in a liquid to be pulled into the interior of the liquid by an imbalance in the intermolecular forces
    - $\rightarrow$ *Inward pull* that determines the resistance of a liquid to an increase in surface area
    - $\rightarrow$ A measure of the *force* that must be applied to *surface molecules* so that they experience the same *force* as molecules in the *interior* of the liquid.
    - $\rightarrow$ A measure of the tightness of the surface layer   [Symbol: γ (gamma)   Units: $N\cdot m^{-1}$ or $J\cdot m^{-2}$]

- **Capillary action**
    - $\rightarrow$ Rise of liquids up narrow tubes when the *adhesion* forces are greater than *cohesion* forces
    - $\rightarrow$ **adhesion:**   Forces that bind a *substance* to a *surface*
    - $\rightarrow$ **cohesion:**   Forces that bind *molecules* of a substance together to form a *bulk material*

- **Meniscus (curved surface that a liquid forms in a tube)**
    - $\rightarrow$ *Adhesive* forces greater than *cohesive* forces (forms a $\smile$ shape)
    - $\rightarrow$ *Cohesive* forces greater than *adhesive* forces (forms a $\frown$ shape)
    - $\rightarrow$ Glass surfaces have exposed O atoms and O–H groups to which hydrogen-bonded liquids like H_2O can bind. In this case, the *adhesive* forces are greater than the *cohesive* ones. Water *wets* glass, forms a $\smile$ shape at the surface, and undergoes a capillary rise in a glass tube.
    - $\rightarrow$ Mercury liquid does not bind to glass surfaces. The *cohesive* forces are greater than the *adhesive* ones. Mercury does *not* wet glass, forms a $\frown$ shape at the surface, and undergoes a capillary *lowering* in a glass tube.

SOLID STRUCTURES (Sections 5.9–5.14)

5.9 Classification of Solids

- **Amorphous solid** $\rightarrow$ Atoms or molecules (neutral or charged) lie in random positions.
- **Crystalline solid**
    - $\rightarrow$ Atoms, ions, or molecules are associated with points in a **lattice,** which is an *orderly* array of equivalent points in three dimensions.
    - $\rightarrow$ Structure has long-range order.
- **Crystal faces** $\rightarrow$ Flat, well-defined planar surfaces with definite interplanar angles

- **Classification of crystalline solids**

    metallic solids: Cations in a sea of electrons   **Examples:** Fe(s), Li(s)

    ionic solids: Mutual attractions of cations and anions   **Examples:** NaCl(s), Ca(SO$_4$)$_2$(s)

    molecular solids: Discrete molecules held together by the *intermolecular forces* discussed earlier   **Examples:** sucrose, C$_{12}$H$_{22}$O$_{11}$(s); ice, H$_2$O(s)

    network solids: Atoms bonded *covalently* to their neighbors throughout the entire solid   **Examples:** diamond, C(d); graphite planes, C(gr)

5.10   Molecular Solids

→ Solid structures that reflect the nonspherical nature of their molecules and the relatively weak intermolecular forces that hold them together

→ Characterized by low melting temperatures and less hardness than ionic solids

→ May be amorphous (wax) or crystalline (sucrose)

5.11 Network Solids

- **Crystals**

    → Atoms joined to neighbors by *strong* covalent bonds that form a network extending throughout the solid

    → Characterized by high melting and boiling temperatures

    → Tend to be hard and rigid structures (diamond)

- **Elemental network solids**

    → Network solids formed from one element only, such as the allotropes graphite and diamond in which the carbon atoms are connected differently

    → *Allotropes* are forms of an element with different solid structures.

- **Ceramics**

    → Usually oxides with a network structure having great strength and stability because covalent bonds must break to deform the crystal

    → Tend to shatter rather than bend under stress
    Examples: Quartz, silicates, and high-temperature superconductors

5.12 Metallic Solids

- **Close-packed structures**

    → Atoms occupy smallest total *volume* with the *least* amount of empty space.

    → Metal atoms are treated as *spheres* with radii *r*.

    → Type of metal determines whether a close-packed structure is assumed.

- **Close-packed layer**

    → Atoms (*spheres*) in a planar arrangement with the *least* amount of empty space — contour of the layer has *dips* or *depressions*.

    → Each atom in a layer has *six* nearest neighbors (hexagonal pattern).

    → Layers stack such that bottoms of spheres in one layer fit in *dips* of the layer below.

- **Stacked layers**   → In a close-packed structure, each atom has *three* nearest neighbors in the layer above, *six* in its original layer, and *three* in the layer below for a total of 12 nearest neighbors.

- **Coordination number (CN)**

    → Number of nearest neighbors of each atom in the solid

→ Impossible to pack identical spheres (metal atoms) with CN > 12

- **Two arrangements of stacking close-packed layers**

    Pattern: ABABABAB…   hcp ≡ *hexagonal close-packing*

    ABCABCABC…   ccp ≡ *cubic close-packing*

- **Occupied space in hcp/ccp structure** → 74% of space is occupied by the spheres and 26% is *empty*.

- **Tetrahedral hole**
    → Formed when a *dip* between *three* atoms in one layer is *covered* by *another* atom in an adjacent layer
    → Two tetrahedral holes per atom in a close-packed structure (hcp or ccp)

- **Octahedral hole**
    → Space between *six* atoms, *four* of which are at the corners of a square plane, forms an octahedral hole. The square plane is oriented 45° with respect to two close-packed layers. *Two* atoms in the square plane are in layer **A** and the other *two* in layer **B**. There is *one* additional atom in layer **A** that is *above* the square plane and *another* in layer **B** that is *below* the square plane. The result is an octahedral arrangement of 6 atoms with a hole in the center (see the diagram on the right side below).
    → One octahedral hole per atom in a close-packed structure (hcp or ccp)

Figure of an octahedron and an octahedron rotated to show the relation to two close-packed layers

- **Body-centered cubic (bcc) structure**
    → Coordination number of 8 (*not* close-packed)
    → Spheres touch along the *body* diagonal of a cube.
    → Can be converted into close-packed structures under high pressure

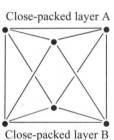

- **Primitive cubic (pc) structure**
    → Coordination number of 6 (*not* close-packed)
    → Spheres touch along the *edge* of the cube.
      Only one known example: Po (polonium)

Crystal structure	Coordination number	Occupied space	Examples
ccp (fcc)	12	74%	Ca, Sr, Ni, Pd, Pt, Cu, Ag, Au, Al, Pb Group 18: noble gases at low *T*, except He
hcp	12	74%	Be, Mg, Ti, Co, Zn, Cd, Tl
bcc	8	68%	Group 1: alkali metals, Ba, Cr, Mo, W, Fe
pc	6	52%	Po (covalent character)

5.13 Unit Cells

- **Lattice** → A regular array of equivalent points in three dimensions
- **Unit cell** → Smallest repeating unit that generates the full array of points
- **Crystal**
    → Constructed by associating atoms, ions, or molecules with each lattice point

→ **Metallic crystals:** One metal atom per lattice point [ccp (fcc), bcc, pc]
Two metal atoms per lattice point [hcp]

- **Unit cells for metals**
  - → Cubic system: pc, bcc, and fcc (ccp)
  - → Hexagonal system: *primitive unit cell* with *two* atoms per lattice point (hcp)

- **Unit cells in general**
  - → Edge lengths: a, b, c
  - → Angles between two edges: α (between edges b and c); β (between a and c); γ (between a and b)

- **Seven crystal systems (unit cell shapes)** → Cubic, hexagonal, tetragonal, orthorhombic, rhombohedral, monoclinic, and triclinic

- **Bravais lattices**
  - → 14 basic patterns of arranging points in three dimensions
  - → Each pattern has a different unit cell (see figure below right)

The 14 Bravais Lattices ⇒

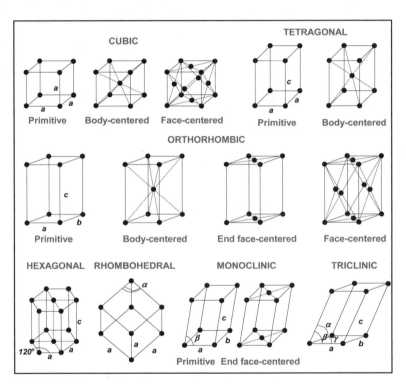

- **Volume of several unit cells**

  Cubic: $V = a \times a \times a = a^3$
  $$[\alpha = \beta = \gamma = 90°]$$

  Tetragonal: $V = a \times a \times c = a^2 c$
  $$[\alpha = \beta = \gamma = 90°]$$

  Orthorhombic: $V = a \times b \times c = abc$
  $$[\alpha = \beta = \gamma = 90°]$$

- **Properties of unit cells**
  - → Each *corner point* shared among 8 cells [1/8 per cell]
    8 corner points × 1/8 per cell = 1 corner point per cell [*all cells*]
  - → Each *face-centered point* shared between 2 cells [1/2 per cell]
    6 face points × 1/2 per cell = 3 face points per cell [*face-centered cells*]
    2 face points × 1/2 per cell = 1 face point per cell [*end face-centered cells*]
  - → Each *body-centered point* unshared [1 per cell]
    1 body point × 1 per cell = 1 body point per cell [*body-centered cells*]

- **Properties of *cubic system* unit cells** → *One* metal atom occupies each *lattice point*

  Primitive

  One metal atom (*sphere*) at each *corner point* in a unit cell
  [*one* atom (radius r) per cell (edge length a)]
  Spheres *touch* along an *edge:* $a = 2r$
  Volume of the unit cell: $V = a^3 = (2r)^3 = 8r^3$
  Density of a pc metal: (M = molar mass):

  $$d = \frac{(1\text{ atom per cell})(\text{mass of one atom})}{(\text{volume of one unit cell})} = \frac{M/N_A}{a^3} = \frac{M/N_A}{8r^3}$$

  Body-centered

  One metal atom (*sphere*) at each *corner point* in a unit cell
  One metal atom (*sphere*) at the *center* of the unit cell
  Total atoms in unit cell = 8 (1/8) + 1 (1) = 2
  [*two* atoms (radius r) per unit cell (edge length a)]

  Spheres *touch* along the *body diagonal:* $\sqrt{3}a = 4r$ or $a = \dfrac{4}{\sqrt{3}}r$

  Volume of the unit cell: $V = a^3 = \dfrac{4^3}{3^{3/2}}r^3 \approx 12.3168r^3$

  Density of a bcc metal:

  $$d = \frac{(2\text{ atoms per cell})(\text{mass of one atom})}{(\text{volume of one unit cell})} = \frac{(2)(M/N_A)}{a^3} = \frac{(2)(M/N_A)}{\left(4^3/3^{3/2}\right)r^3}$$

  Face-centered

  One metal atom (*sphere*) at each *corner point* in a unit cell
  One metal atom (*sphere*) at the *center* of each *face* in the unit cell
  Total atoms in unit cell = 8 (1/8) + 6 (1/2) = 4
  [*four* atoms (radius r) per unit cell (edge length a)]

  Spheres *touch* along *face diagonal:* $\sqrt{2}a = 4r$ or $a = \dfrac{4}{\sqrt{2}}r = \sqrt{8}r$

  Volume of the unit cell: $V = a^3 = 8^{3/2}r^3$
  Density of a fcc (ccp) metal:

  $$d = \frac{(4\text{ atoms per cell})(\text{mass of one atom})}{(\text{volume of one unit cell})} = \frac{(4)(M/N_A)}{a^3} = \frac{(4)(M/N_A)}{\left(8^{3/2}\right)r^3}$$

- **Calculations** → Determining the *unit cell* type from the *measured density* and *atomic radius* of metals in the *cubic* system; the *atomic radius* of a metal atom is determined from the crystal type and the edge length of a unit cell, which are both obtained by x-ray diffraction techniques (see Major Technique 3 in text).

Example: The *density* of lead is 11.34 $g \cdot cm^{-3}$ and the *atomic radius* is 175 pm. Assuming the unit cell is *cubic,* we can determine the type of Bravais lattice. Using the equations given above, calculate the density of Pb assuming pc, bcc, or fcc Bravais lattices. Results are: pc, 8.02 $g \cdot cm^{-3}$; bcc, 10.4 $g \cdot cm^{-3}$; fcc, 11.3 $g \cdot cm^{-3}$. The Bravais lattice is most likely fcc.

5.14 Ionic Structures

- **Model [r (tetrahedral hole) < r (octahedral hole) < r (cubic hole)]**
  - → Spheres of *different* radii and *opposite* charges representing cations and anions
  - → Larger spheres (*usually* anions) *customarily* occupy the unit cell lattice points. Smaller spheres (*usually* cations) fill *holes* in the unit cell.
- **Radius ratio [Use rule with caution, there are many exceptions.]**
  - → Ratio of radius of the *smaller* sphere to the radius of the *larger* sphere, symbol ρ
  - → Defined for ions of the *same* charge number [$z_{cation} = |z_{anion}|$]

Radius Ratio–Crystal Structure Correlations

radius ratio ≤ 0.414:	tetrahedral *holes* fill (fcc for *larger* spheres) [*zinc-blende structure* (one form of ZnS)]
0.414 < radius ratio < 0.732:	octahedral *holes* fill (fcc for *larger* spheres) [*rock-salt structure* (NaCl)]
radius ratio ≥ 0.732:	cubic *holes* fill (pc for *larger* spheres) [*cesium-chloride structure* (CsCl)]

- **Coordination number** → Number of nearest-neighbor ions of *opposite* charge
- **Coordination of ionic solid** → Represented as (cation coordination number, anion coordination number)
- **Properties of the *ionic structure* of unit cells**

Zinc blende
  - → Radius ratio (ZnS) = $r(Zn^{2+})/r(S^{2-})$ = (74 pm)/(184 pm) = 0.40
  - → S^{2-} ions at the *corners* and *faces* of a fcc unit cell
    [*four* anions (atomic radius r_{anion}) per unit cell (edge length a)]
  - → Zn^{2+} ions in *four* of the *eight* tetrahedral holes in the unit cell
    [*four* cations (atomic radius r_{cation}) per unit cell (edge length a)]
  - → Cation and anion spheres *touch* in the tetrahedral locations.
    Each cation has *four* nearest-neighbor anions, and each anion has *four* nearest-neighbor cations: (4,4)-coordination.

Rock salt
  - → Radius ratio (NaCl) = $r(Na^+)/r(Cl^-)$ = (102 pm)/(181 pm) = 0.564
  - → Cl^- ions at the *corners* and *faces* of a fcc unit cell
    [*four* anions (atomic radius r_{anion}) per cell (edge length a)]
  - → Na^+ ions in all *four* octahedral holes in the unit cell
    [*four* cations (atomic radius r_{cation}) per cell (edge length a)]

→ Cation and anion spheres *touch* along the cell edges.

Each cation has *six* nearest-neighbor anions, and each anion has *six* nearest-neighbor cations: (6,6)-coordination.

→ Spheres *touch* along the *edge:* $a = 2\,r_{cation} + 2\,r_{anion} = 566$ pm

→ Volume of the unit cell: $V = a^3 = 1.81 \times 10^8$ pm^3 = 1.81×10^{-22} cm^3

→ Density of NaCl: $M(58.44$ g·mol$^{-1})$ and d(experimental) = 2.17 g·cm^{-3}

$$d = \frac{(4 \text{ NaCl per cell})(M/N_A)}{a^3} = \frac{4(58.44/N_A)}{(1.81 \times 10^{-22})} = 2.14 \text{ g·cm}^{-3}$$

Cesium chloride

→ Radius ratio (CsCl) = $r(Cs^+)/r(Cl^-)$ = (170 pm)/(181 pm) = 0.939

→ Cl$^-$ ions at the *corners* of a pc (*primitive cubic*) unit cell
[*one* anion (radius r_{anion}) per cell (edge length a)]

→ Cs$^+$ ion at the *center* of the *cube* in the unit cell
[*one* cation (radius r_{cation}) per cell (edge length a)]

→ Cation and anion spheres *touch* along the *body diagonal* of the cell.

Each cation has *eight* nearest-neighbor anions and each anion has *eight* nearest-neighbor cations: (8,8)-coordination.

→ Spheres *touch* along the *body diagonal:* $\sqrt{3}\,a = 2\,r_{cation} + 2\,r_{anion} = 702$ pm   and $a = 405$ pm

→ Volume of the unit cell: $V = a^3 = 6.64 \times 10^7$ pm^3 = 6.64×10^{-23} cm^3

→ Density of CsCl: $M(168.36$ g·mol$^{-1})$ and d(experimental) = 3.99 g·cm^{-3}

$$d = \frac{(1 \text{ CsCl per cell})(M/N_A)}{a^3} = \frac{1(168.36/N_A)}{(6.64 \times 10^{-23})} = 4.21 \text{ g·cm}^{-3}$$

- **Calculations**
  → Determining the *ionic structure* in the *cubic* system from *ionic radii* using the *radius-ratio rule*
  → Determining the *density* of the *ionic solid* from the *density* of the *unit cell*
  Note: Usually within 10% of the experimental value

THE IMPACT ON MATERIALS (Sections 5.15–5.16)

5.15 Liquid Crystals

→ Materials that cannot be characterized as solid, liquid, or gas

→ Substances that flow like viscous liquids, but molecules form a moderately ordered array similar to that in a crystal.

→ Examples of a *mesophase,* an intermediate state of matter with the fluid properties of a liquid and some molecular ordering similar to a crystal

- **Isotropic material** → Properties independent of the direction of measurement.
Ordinary liquids are isotropic with viscosity values equal in every direction.

- **Anisotropic material** $\rightarrow$ Properties depend on the direction of measurement
  Certain rod-shaped molecules form liquid crystals, in which molecules are free to slide past one another along their axes but resist motion perpendicular to that direction.

- **Classes of liquid crystals** $\rightarrow$ Differ in the arrangement of the molecules
  (see Figs. 5.47, 5.48, and 5.49 in the text)

  nematic phase: Molecules lie together in the same direction but are *staggered*.

  smectic phase: Molecules lie together in the same direction in *layers*.

  cholesteric phase: Molecules form *nematiclike layers*, but the molecules of neighboring layers are *rotated* with respect to each other. The resulting liquid crystal has a *helical* arrangement of molecules.

- **Thermotropic liquid crystals**
  - $\rightarrow$ Made by melting solid phase
  - $\rightarrow$ Exist over small temperature range between solid and liquid
    Example: *p*-azoxyanisole

- **Lyotropic liquid crystals**
  - $\rightarrow$ Layered structures produced by the action of a solvent on a solid or a liquid.
  - $\rightarrow$ Examples include cell membranes and aqueous solutions of detergents and lipids (fats).

5.16 Ionic Liquids

- $\rightarrow$ Molecular substances tend to have low melting points.
- $\rightarrow$ Ionic, network, and metallic substances tend to have high melting points.

- **Ionic liquids**
  - $\rightarrow$ Exceptions to the rules above
  - $\rightarrow$ Ionic substances with low melting points
  - $\rightarrow$ Can be liquid at room temperature and below
  - $\rightarrow$ Constitute a new class of solvents developed to have low vapor pressures but dissolve organic compounds
  - $\rightarrow$ Compounds in which one of the ions (usually the cation) is a large, organic ion that prevents the liquid from crystallizing at ordinary temperatures

    Example: Cation: 1-Butyl-3-methylimidazolium (Margin Fig. 16 in the text)

    Anion: Tetrafluoroborate: BF_4^-

  - $\rightarrow$ One formulation is used to dissolve rubber in old tires for recycling.
  - $\rightarrow$ Another is used to extract radioactive waste from groundwater.

Chapter 6   INORGANIC MATERIALS

METALLIC MATERIALS   (Sections 6.1–6.4)

6.1 The Properties of Metals

- **Mobility of electrons**
    - → A metal can be defined as an array of cations bonded by a sea of electrons.
    - → Responsible for characteristic luster and light reflectivity properties of a metal
    - → Accounts for *malleability, ductility,* and *electrical conductivity* of metals
    - → Produces a relatively strong metallic bond that leads to high melting points for most metals

- **Slip planes**
    - → Plane of atoms that, under stress, may slip or slide along an adjacent plane
    - → Characteristic of *close-packed* structures
    - → *Cubic close-packed* structures (ccp) have eight sets of slip planes in different directions.
    - → *Hexagonal close-packed* structures (hcp) have only one set of slip planes.
    - → Large number of slip planes accounts for *malleability* of ccp metallic crystals (may be bent, flattened, or pounded into different shapes). **Examples:** Coinage metals Cu, Ag, and Au
    - → Metallic crystals with the hcp structure tend to be brittle (only one slip plane). **Examples:** Zn, Cd

6.2 Alloys

- **Properties**
    - → Alloys are metallic materials made by mixing two or more molten metals.
    - → May be homogeneous or heterogeneous
    - → Alloys are often softer than the component metals and melt at a lower temperature.

- **Heterogeneous** → Mixture of crystalline phases; various samples have different compositions; for example, solders and mercury amalgams
    - → Melt and freeze over a range of temperatures

- **Homogeneous** → Atoms of different elements are distributed uniformly; for example, brass, bronze, and coinage metals. See Table 6.1 in the text for the composition of typical homogeneous alloys.

- **Substitutional**
    - → Atoms nearly the same size (< 15% difference) can substitute for each other more or less freely (mainly *d*-block elements)
    - → Lattice is distorted and electron flow is hindered.
    - → Alloy has lower thermal and electrical conductivity than pure element, but is harder and stronger.
    - → Homogeneous examples include the Cu-Zn alloy, brass (up to 40% Zn in Cu), and bronze (metal other than Zn or Ni in copper: casting bronze is 10% Sn and 5% Pb).

- **Interstitial**
    - → Small atoms (> 60% smaller than host atoms) can occupy holes or *interstices* in a lattice.
    - → Interstitial atoms interfere with electrical conductivity and the movement of atoms.
    - → Restricted motion makes the alloy harder and stronger than the host metal.
    - → Examples include low-carbon steel (C in Fe, which is soft enough to be stamped), high-carbon steel (hard and brittle unless subject to heat treatment), and stainless steel (a mixture of Fe with metals such as Cr and Ni, which aid in its resistance to corrosion).

6.3 Steel

- **Iron**
    - → *Pig iron*, which contains 3–5% C, ~2% Si, and lesser amounts of other impurities, is produced by reduction of iron ores in a blast furnace.
    - → *Cast iron* is similar to pig iron, but with some impurities removed.
    - → Both forms are hard and brittle, whereas *pure iron* is malleable and more flexible.
    - → In the blast furnace, limestone is used to remove impurities, such as SiO_2, Al_2O_3, and P_4O_{10}.
- **Steels**
    - → Steels are homogeneous interstitial alloys (solid solutions) made from *pig iron*. First, *pig iron* is purified to *lower* the carbon content and to remove other impurities. Then, appropriate metals are *added* to form the desired steel.
    - → Steels typically contain 2% or less carbon; their hardness, tensile strength, and ductility depend on the carbon content. Higher carbon content produces harder steel. The strength of steel can be increased greatly by heat treatment. See Table 6.2 in the text for the effect of carbon content on steel.
    - → The corrosion resistance of steel is significantly improved by alloying with other metals, for example, highly corrosion-resistant *stainless steel* is an alloy containing about 15% Cr by mass.

6.4 Nonferrous Alloys

- **Properties**
    - → Nonferrous alloys are prepared by mixing molten elements, followed by molding or by *sintering*, mixing powdered metals, followed by molding the mixture by pressing it while hot.
    - → Can be *homogeneous* or *heterogeneous* solutions
    - → *Homogeneous alloys* include brass, bronze, coins containing gold or silver, and alnico, a magnetic alloy of aluminum, nickel, and cobalt used in speakers and mobile phones.
    - → *Heterogeneous alloys* include tin-lead solder, mercury amalgams for dental fillings, and Bi/Cd, which melts over a range of temperatures, except for one composition (eutectic), which melts at a fixed temperature like a pure compound.
- **Substitutional alloys (see also Section 6.2 above)**
    - → Homogeneous alloys with metals of similar radius. **Example:** Brass, consisting of Cu and Zn
    - → Typically harder than the pure metals but with lower electrical and thermal conductivity

HARD MATERIALS (Sections 6.5–6.11)

6.5 Diamond and Graphite

- **Graphite**
    - → The thermodynamically stable allotrope of carbon (Soot contains small crystals of graphite.)
    - → Produced pure commercially by heating C rods in an electric furnace for several days
    - → Contains sp^2-hybridized C atoms
    - → Consists structurally of large sheets of fused benzene-like hexagonal units. A π-bonding network of delocalized electrons accounts for the high electrical conductivity of graphite.
    - → When certain impurities are present, graphitic sheets can slip past one another, and graphite becomes an excellent dry lubricant.

→ Only soluble in a few liquid metals

→ Soot and carbon black have commercial applications in rubber and inks. Activated charcoal is an important, versatile purifier. Unwanted compounds are adsorbed onto its microcrystalline surface.

→ A chicken-wire-like sheet of carbon atoms in graphite is called a *graphene* sheet. Graphene, a single sheet of graphite, is a new material with exceptional promise in the electronics industry.

→ Very pure sheets of graphene can be prepared and stacked together with water molecules that act as a kind of glue between them.

→ The result is a very strong, flexible, but very thin, paperlike material that conducts electricity but is tougher than diamond.

- **Diamond**

    → Found naturally embedded in *kimberlite*, a soft rock

    → Hardest substance known and an excellent conductor of heat

    → An excellent abrasive, and the heat generated by friction is rapidly conducted away

    → Natural diamonds are uncommon. Synthetic diamonds are produced from graphite at high pressures (> 80 kbar) and temperatures (> 1500°C) and by thermal decomposition of methane (the preferred method).

    → Diamond is soluble in liquid metals, such as Cr and Fe, but less soluble than graphite. This solubility difference is utilized in synthesizing diamond at high pressure and high temperature.

    → Carbon in diamond is sp^3-hybridized, and each C atom in a diamond crystal is bonded directly to four other C atoms and indirectly interconnected to all of the other C atoms in the crystal through C–C single σ-bonds.

    → The σ-bonding network of localized electrons accounts for the electrical insulating properties of diamond.

6.6 Calcium Carbonate

- **Calcium compounds**

    → Used as structural materials in organisms, building, and engineering applications

    → Small, doubly charged Ca^{2+} cation and CO_3^{2-} anion give $CaCO_3(s)$ a high lattice energy, consistent with the applications noted above.

    → $CaCO_3$ occurs naturally as chalk, limestone, and marble, a very dense form with colored impurities, most commonly, Fe cations.

    → The most common forms of pure calcium carbonate are *calcite* and *aragonite* (hard, more dense, less abundant that *calcite*).

    → All carbonates are the fossilized remains of marine life and are soluble in acids.

    → Acid rain erodes limestone buildings and statuary.

    → In nature, carbonates are used as structural materials in seashells ($CaCO_3$) and in bones $[Ca_3(PO_4)_2]$.

6.7 Silicates

- **Silica or silicon dioxide, SiO_2**

    → Derives its strength from its covalently bonded network structure

    → Occurs naturally in pure form only as quartz and sand

    → Quartz contains helical chains of SiO_4 units which share O atoms.

    → Above 1500°C, quartz transforms to the mineral *cristobalite*, which has a diamond-like structure with C–C bonds replaced by Si–O–Si units.

    → Sandstone and granite are materials based on silica and silicates

- **Silicates, compounds containing SiO_4^{4-} units**
    - → View as tetrahedral silicon oxoanions with considerably covalent Si–O bonds
    - → Properties depend on charge on each tetrahedron, number of corner-shared O atoms, and the way in which linked tetrahedra lie together.
    - → Materials with silicates include gemstones, fibers, and glasses (see below)
    - → *Orthosilicates,* the simplest silicates, contain SiO_4^{4-} ions. **Examples:** Na_4SiO_4 and $ZrSiO_4$, *zircon,* an inexpensive diamond substitute.
    - → *Pyroxenes* contain chains of tetrahedral SiO_4^{4-} *units* in which one O atom bridges two Si atoms (Si–O–Si) from adjacent tetrahedra, such that the "average" unit is SiO_3^{2-}. Cations placed along the chains provide electrical neutrality. **Examples:** $Al(SiO_3)_2$ and $CaMg(SiO_3)_2$
    - → *Amphiboles* contain chains of silicate ions that form a cross-linked ladder-like structure containing $Si_4O_{11}^{6-}$ units. One example is the fibrous mineral *tremolite,* $Ca_2Mg_5(Si_4O_{11})_2(OH)_2$, one form of *asbestos,* a material that withstands extreme heat but is a known carcinogen. Almost all amphiboles contain hydroxide ions attached to the metal.
    - → Other types of silicates have sheets containing $Si_2O_5^{2-}$ units, with cations lying between the sheets and linking them together. One example is *talc,* $Mg_3(Si_2O_5)_2(OH)_2$. Additional structural types of silicon oxides exist.
- **Aluminosilicates**
    - → *Aluminosilicates* form when a Si^{4+} ion is replaced by Al^{3+} plus additional cations for charge balance. One example is *mica,* $KMg_3(Si_3AlO_{10})_2(OH)_2$.
    - → The extra cations hold the sheets of tetrahedra together, accounting for the hardness of these materials. *Feldspar* is a silicate material in which more than half the silicon is replaced by aluminum. **Example:** $KAlSi_3O_8$, a typical feldspar.

6.8 Cements and Concrete

- **Cements**
    - → Building blocks in construction are usually held together by binders called *cements.*
    - → Most common type is *Portland cement,* which is made by heating a mixture of crushed limestone, clay or shale, sand, and oxides such as iron ore in a kiln (see Table 6.3 in the text).
    - → Different types of Portland cement are used to meet different requirements.
    - → When heated, cements lose water and harden.
    - → See the text for chemical reactions describing the hardening process.
- **Concrete**
    - → Consists of a binder and a filler
    - → Filler is usually gravel, but sometimes polymer or vermiculite pellets are added to lower the density. *Vermiculite* is a low-density, clay-like silicate material with a structure that has been expanded by heating.
    - → Binder is cement, usually Portland cement. *Mortar* is a cement-like substance used to hold bricks and other building materials together.

6.9 Borides, Carbides, and Nitrides

- **Borides**
    - → Many borides of the metals and nonmetals are known.
    - → Formulas of borides are often unrelated to the position of the two elements in the periodic table. **Examples**: AlB_2, CaB_6, $B_{13}C_2$, $B_{12}S_2$, Ti_3B_4, TiB, and TiB_2.

→ B atoms in borides commonly form extended structures such as zigzag chains, branched chains, or networks of hexagonal rings.

→ Some borides are used as rocket nozzles and turbine blades because of their hardness, heat stability, and network structures.

- **Carbides**

  → Silicon carbide, SiC (*carborundum*), an extremely hard covalent carbide and an excellent abrasive, is produced by the reaction of silicon dioxide with carbon at 2000°C:

  $$SiO_2(s) + 3C(s) \rightarrow SiC(s) + 2CO(g)$$

  → Silicon carbide has a diamond-like structure and fractures easily.

  → *Interstitial carbides* are formed by direct reaction of a *d*-block metal with C at temperatures above 2000°C. The C atoms lie in holes in close-packed arrays of metal atoms. Bonds between the C and metal atoms stabilize the lattice. **Examples:** WC, Cr_3C, and Fe_3C, a component of steel

- **Nitrides**

  → When B is heated in NH_3, it forms boron nitride: $2B(s) + 2NH_3(g) \rightarrow 2BN(s) + 3H_2(g)$. The structure BN is similar to that of graphite, but with the planes of hexagons of C atoms replaced by hexagons of alternating B and N atoms.

  → BN is a fluffy, slippery powder that can be pressed into a solid rod that is easy to machine. Unlike graphite, BN is a good insulator.

  → At high pressure, BN is converted to a diamond-like crystalline material called *borazon*.

  → BN nanotubes, similar to those formed by C, are semiconductors.

6.10 Glasses

- **Preparation and properties**

  → A *glass* is an ionic solid with an amorphous structure similar to a liquid.

  → Glasses have short-range order but lack long-range order.

  → Glass is characterized by a network structure based on a nonmetal oxide, commonly silica, SiO_2, melted with metal oxides that function as *network modifiers*.

  → The development of optical fibers led to considerable recent advances in glassmaking.

  → Heating SiO_2 ruptures Si–O bonds, enabling metal ions to form ionic bonds with some of the O atoms. The nature of the glass depends on the metal added.

  → *Soda-lime glass* (about 90% of all glass), used for windows and bottles, is made by adding Na^+ (12% Na_2O) and Ca^{2+} (12% CaO). Heating Na_2CO_3 (soda) produces the Na_2O; heating $CaCO_3$ (lime) produces the CaO.

  → *Borosilicate glass,* such as Pyrex, is produced using less soda and lime and adding 16% B_2O_3. These glasses expand very little on heating and are used for ovenware and laboratory glassware.

  → In glass, tiny crystallites do not scatter light, so glass is transparent.

- **Optical fibers**

  → Made by drawing fiber from heated glass

  → Coated with plastic

  → Transmit information much faster than wires

  → Vary in composition in cross-section to avoid loss of light

- **Reaction with acids and bases**

  → Glass is resistant to attack by most chemicals.

→ But, the silica in glass can react with HF:

$$SiO_2(s) + 6\,HF(aq) \rightarrow SiF_6{}^{2-}(aq) + 2\,H_3O^+(aq)$$

→ Removal of silica from glass by the ions F^- (from HF), OH^-, and $CO_3{}^{2-}$ is called *etching*.

→ Silica also reacts with the Lewis base OH^- in molten NaOH and the carbonate anion in molten Na_2CO_3 at 1400°C:

$$SiO_2(s) + 2\,Na_2CO_3(l) \rightarrow Na_4SiO_4(l) + 2\,CO_2(g)$$

6.11 Ceramics

- **Preparation and properties**

Ceramics

→ Inorganic materials, such as clays, hardened by heating to a high temperature

→ Typically very hard, insoluble in water, and stable to corrosion and high temperatures

→ Can be used at high temperatures without failing, and resist deformation

→ Tend to be brittle, however

→ Often are oxides of elements on the border between metals and nonmetals

→ Used in many automobile parts, including spark plugs, pressure and vibration sensors, brake linings, and catalytic converters

→ Some *d*-metal oxides and compounds of B and Si with C and N are also ceramic materials.

→ Most ceramics are electrical insulators, but some are semiconductors and superconductors.

Aluminosilicate ceramics

→ Formed in many cases by heating clays to expel water trapped between sheets of tetrahedral aluminosilicate units

→ The result is a rigid heterogeneous mass of small interlocking crystals bound together by glassy silica.

→ One form is white *china clay,* used to make porcelain and china. It is free of iron impurities that tend to make clays reddish brown.

→ Methods developed to make ceramics less brittle are the *sol-gel process* and the *composite material technique.* These methods have produced *sol-gels* and *aerogels.*

Aluminum Oxide

→ A ceramic found in several different forms

→ *Corundum*, α-alumina, a hard, stable, crystalline substance

→ *Emery*, impure microcrystalline corundum, an abrasive

→ *Sapphires*, large, single crystals of alumina doped with iron and titanium impurities

MATERIALS FOR NEW TECHNOLOGIES (Sections 6.12–6.17)

6.12 Bonding in the Solid State *and*
6.13 Semiconductors

- **Ability to conduct electricity**
  - $\rightarrow$ Desirable property of many solids
  - $\rightarrow$ There are two types of such conduction.
  - $\rightarrow$ Electronic conduction – charge carried by electrons as in metals and graphite
  - $\rightarrow$ Ionic conduction – charge carried by ions as in molten salts and electrolyte solutions

- **Molecular Orbital (MO) theory**
  - $\rightarrow$ Accounts for electrical properties of metals and semiconductors
  - $\rightarrow$ Treats solid as a large molecule with delocalized elections spread over the entire solid

- **Bands**
  - $\rightarrow$ Groups of MOs having closely spaced (essentially continuous) energy levels (see Figs. 3.40 and 6.26 in the text)
  - $\rightarrow$ Formed by the spatial overlap of a very large number of AOs
  - $\rightarrow$ Separate into two groups (one mostly *bonding,* the other mostly *antibonding*)

- **Conduction band** $\rightarrow$ Empty or partially filled band

  Example: Formation of a conduction band in Na by the overlap of 3*s*-orbitals

- **Valence band** $\rightarrow$ Completely filled band of MOs (insulators such as molecular solids)

- **Band gap** $\rightarrow$ An energy range between bands in which there are *no* orbitals

- **Electronic conductors** $\rightarrow$ Metals, semiconductors, and superconductors

Insulator:	Does not conduct electricity: substance with a full *valence band* far in energy from an empty *conduction band* (large *band gap*)
Metallic conductor:	Current carried by *delocalized* electrons in *bands* Conductivity *decreases* with *increasing* temperature Substance with a partially filled *conduction band* (gap *not* relevant)
Semiconductor:	Current carried by *delocalized* electrons in *bands* Conductivity *increases* with *increasing* temperature Substance with a full *valence band* close in energy to an empty *conduction band* (small *band gap*): Excitation of some electrons from the *valence* to the *conduction band* yields conductivity.
Superconductor:	*Zero* resistance (infinite conductivity) to an electric current below a definite transition temperature

- **Electrical conductivity**
  - $\rightarrow$ Symbol is κ (kappa). Electrical conductivity of a sample of length l and cross-sectional area A is defined in terms of its resistance R.
  - $\rightarrow$ The equation is $\kappa = \dfrac{l}{A \times R}$
  - $\rightarrow$ The units are Siemens per meter (S m^{-1}, where 1 S = 1 Ω^{-1}, with Ω, omega, the symbol for ohm).

→ The conductivity of a metal is usually about 10^7 S m^{-1}; that of a semiconductor about 10^{-4} S m^{-1}.

→ A metallic conductor has a much higher electrical conductivity than a semiconductor, but it is the temperature dependence of the conductivity that distinguishes the two types of conductors.

- **Enhanced (extrinsic) semiconductors** → Prepared by *doping*, that is, adding a small amount of impurity

- **n-type semiconductor** → Adding some *valence electrons* to the *conduction band* (n = electrons added)

  Example: A small amount of As (Group 15) added to Si (Group 14): Extra electrons are transferred from As into the previously empty *conduction band* of Si.

- **p-type semiconductor** → Removing some *valence electrons* from the *valence band* (p = positive "holes" in the *valence band*)

  Example: A small amount of In (Group 13) added to Si (Group 14): Valence electron deficit is created in the previously full *valence band* of Si, thereby becoming a *conduction band*.

- **p-n junctions**
  → p-type semiconductor in contact with an n-type semiconductor
  → Solid-state electronic devices

  Example: One type of transistor is a p-n sandwich with two separate wires (*source* and *emitter*) connected to the n-side. It acts as a *switch* in the following way: When a positive charge is applied to a polysilicon contact (*base* wire) positioned between the *source* and *emitter* wires, a current flows from the *source* to the *emitter*. Other examples include diodes and integrated circuits.

6.14 Superconductors

- **Superconductor**
  → Loss of electrical resistance when a substance is cooled below a characteristic transition temperature (T_s)

  → Conventional superconductivity arises from the ability of electrons to pair by making use of lattice vibrations.

  → The electron pairs, which are called **Cooper pairs** after the scientist who first proposed the mechanism, can travel almost freely through the lattice, rather like oxen yoked together are less deflected by obstacles than individual oxen.

  → A Cooper pair forms when one electron distorts cations in its vicinity, and those cations attract a second electron. Both electrons are attracted together weakly in this way; the Cooper pair (and the resulting superconductivity) survives only if the temperature is low enough to avoid having the pair eliminated by lattice vibrations.

  → Superconductivity was first observed in 1911 in mercury, for which $T_s = 4$ K. Over the years, many other metallic superconductors were identified, some having transition temperatures as high as 23 K.

  → Such **low-temperature superconductors** must be cooled with liquid helium, which is very expensive. As a result, they have not found widespread use.

- **High-temperature superconductors (HTSCs)**
  → If T_s is > 77 K, liquid nitrogen can be used as the coolant (0.2% of the cost of liquid helium).

→ Almost all high-temperature superconductors (HTSCs) are hard, brittle ceramic oxides that have sheets of copper and oxygen atoms sandwiched between layers of either cations or a combination of cations and oxide ions, and all are derived from their respective parent insulators by doping.

→ The layered structures result in strongly anisotropic electrical and magnetic properties. Electric current flows easily along the planes of the copper–oxygen sheets, but only weakly perpendicular to them. Thus, the materials must be positioned with the copper–oxygen planes in a favorable orientation.

→ From 1986 to 1988, the record-high transition temperature rose from 35 K for a ceramic lanthanum–copper oxide material doped with barium to 93 K for an yttrium–barium–copper oxide material (1987), to 125 K for a thallium–barium–calcium–copper material (1988).

→ By 2006, the highest transition temperature attained was 130 K. Brittle ceramic materials are problematic for use as electrical wires. One possible solution has been to deposit the superconducting material on the surface of wire or tape made of a metal such as silver.

→ Twenty-five kilograms of superconducting wire can carry as much current as 1800 kg of today's copper-based electrical cable.

→ Potential applications of superconductors include electrical transmission (little or no heat loss in wires) and energy-efficient transportation (magnetically levitated trains, *etc.*).

6.15 Luminescent Materials

- **Light emission from materials**
  → *Incandescence* – light emission from a heated object, such as a lamp filament or hot soot in a candle flame. See also *black-body radiation* in Section 1.4 of the text.
  → *Luminescence* – light emission caused by other processes including light absorption, chemical reaction, impact with electrons, radioactivity, or mechanical shock.
  → *Chemiluminescence* – light emission caused by chemical reaction. The reaction products are formed in energetically excited states that decay by light emission.
    Example: The reaction of hydrogen atoms and fluorine molecules provides an example of *infrared chemiluminescence*. HF is produced with considerable *vibrational* energy, and decays by *emitting infrared* radiation (see Major Technique 1 in text Chapter 2 for a discussion of *infrared absorption*).
  → *Visible chemiluminescence* is usually produced by *electronic* excitation of product molecules.
  → *Bioluminescence* – a form of chemiluminescence produced by living organisms, such as fireflies and certain bacteria.

- **Light emission from molecules**
  → *Fluorescence* – *prompt* emission of light from molecules excited by radiation of higher frequency. Normally, absorption of ultraviolet radiation by a molecule leads to emission in the visible region; higher frequency *absorption* leads to lower frequency *emission*.
  [**Note:** With high-power lasers, simultaneous absorption of two or more photons by a molecule may lead to emission of a single photon with a *greater* frequency than that of the individual photons absorbed.]
  → *Phosphorescence* – *slow* or *delayed* emission of light from molecules excited by radiation of higher frequency. In this case, the initially excited molecule in a singlet state (spins paired, ↑↓) undergoes a transition to a triplet state (spins parallel, ↑↑) that decays to a singlet state slowly because spin pairing is a relatively slow process.

→ *Triboluminescence* (from the Greek word, *tribos,* a rubbing) – luminescences produced by a mechanical shock to a crystal. It is readily observed in striking or grinding sugar crystals. Trapped nitrogen gas escapes while in an excited state and produces the radiation.

- **Phosphors**
  → *Phosphorescent materials* that glow when activated by the impact of fast electrons, as well as high frequency radiation such as ultraviolet or x-ray

  → Clusters of three *phosphors* are used for each dot in a color TV or computer screen. Commonly used phosphors are europium-activated yttrium orthovanadate, YVO_4 for the red color, silver-activated zinc sulfide for blue, and copper-activated zinc sulfide for green.

  → *Fluorescent lamps* utilize fluorescent materials activated by ultraviolet light. Mercury atoms in the lamp are excited by electrons in a discharge and emit light at 254 and 185 nm, which is absorbed by a phosphor coated on the lamp's surface. A commonly used phosphor is *calcium halophosphate,* $Ca_5(PO_4)_3F_{1-x}Cl_x$, doped with manganese(II) and antimony(III) ions. An Sb(III)-activated phosphor emits blue light and a Mn(II)-activated phosphor emits yellow light. The net result is light with an approximately white spectral range.

  → *Fluorescent materials* have important applications in medical research. Dyes such as *fluorescein* are attached to protein molecules to probe biological reactions. *Fluorescent materials,* such as sodium iodide and zinc sulfide, can be activated by radioactivity and are used in scintillation counters to measure radiation (see Chapter 17). Light-emitting diode (LED) displays also use luminescent materials.

6.16 Magnetic Materials

- **Paramagnetism**
  → Tendency of a substance to move into a magnetic field It arises when an atom or molecule has at least one unpaired electron, which aligns with the applied field.

  → Because spins on neighboring atoms or molecules are aligned almost randomly, *paramagnetism* is very weak and alignment of electron spins is lost when the magnetic field is removed.

- **Ferromagnetism**
  → In some *d*-metals, when the unpaired electrons of many neighboring atoms align with one another in an applied magnetic field, the much stronger effect of **ferromagnetism** arises. The regions of aligned spins are called **domains**, and they survive even after the applied field is turned off.

  → Much stronger than paramagnetism, ferromagnetic materials are used to make permanent magnets.

  → Used in the coatings of cassette tapes and computer disks. Electromagnetic recording heads align large numbers of spins as the tape passes underneath, and the spin alignment in the domains remains for years.

  → **Antiferromagnetic** material – neighboring spins are locked into an *antiparallel* arrangement, so the magnetic moments cancel. Manganese is antiferromagnetic.

  → **Ferrimagnetic** material – the spins on neighboring atoms are different, and although they are locked together in an antiparallel arrangement, the two magnetic moments do not completely cancel.

  Examples: *Ferromagnetism* occurs in alloys such as alnico and some *d*-metal compounds, such as the oxides of iron and chromium. Ceramic magnets are used in refrigerator magnets, and are made of barium ferrite ($BaO \cdot nFe_2O_3$) or strontium ferrite ($SrO \cdot nFe_2O_3$) by compressing the powdered ferrite in a magnetic field and heating it until it hardens. Because these magnets are ceramic, they are hard and brittle, with low densities. However, they are the most widely used magnets because of their low cost.

- **Ferrofluids**
  - → Ferromagnetic liquids, suspensions of finely powdered magnetite, Fe_3O_4, in a viscous, oily liquid (such as mineral oil) that contains a detergent (such as oleic acid, a long-chain carboxylic acid).
  - → The iron oxide particles do not settle out, because they are attracted to the polar ends of the detergent molecules, which form clusters called *micelles* around the particles. The nonpolar ends of the detergent molecules point outward, allowing the micelles to form a colloidal suspension in the oil.
  - → When a magnet is brought near a ferrofluid, the particles in the liquid try to align with the magnetic field, but are kept in place by the oil. As a result, it is possible to control the flow and position of the ferrofluid by means of an applied magnetic field.
  - → One application of ferrofluids is in the braking systems of exercise machines. The stronger the magnetic field, the greater the resistance to motion.

- **Molecular magnets**
  - → Single molecule magnets, contains several *d*-block metal atoms bonded to groups of nonmetal atoms such as carbon, hydrogen, and oxygen.
  - → Respond to a magnetic field as they were nano-size compass needles and have a great potential for miniaturizing electronic storage media such as computer disks.
  - → Typically the *d*-block metal atoms are imbedded in cage-like structures and some of the nano-size molecular magnets have interesting shapes, such as rings, tubes, or spheres.

6.17 Composite Materials

- **Composite materials**
  - → Two or more substances combined into a heterogeneous material, while retaining their individual characteristics
  - → Combine the advantages of the component materials
  - → Can exhibit properties superior to those of the component materials
  - → Fiberglass is strong and flexible; it consists of inorganic materials in a polymer matrix.
  - → Seashells owe their strength to a tough organic matrix and their hardness to the calcium carbonate crystals imbedded in the matrix.
  - → Bones have low density, but are strong because of their composite nature. In bone tissue, crystals of phosphate salts are embedded in fibers of a natural polymer called *collagen* that hold them in place but allow a small amount of flexibility.
  - → Artificial bone and cartilage mimic bone in that they have inorganic solids mixed into a crack-resistant polymer matrix. The result is composite material with flexibility and great strength.
  - → Some lightweight composites can have three times the strength-to-density ratio of steel.
    - **Examples:** 1) The graphite composite used for tennis rackets composed of graphite fibers imbedded in a polymer matrix
      2) The fibrous refractory insulation tiles used on the body of the space shuttle containing alumina-borosilicate fibers in a silica ceramic matrix

- **Ceramic materials**
  - $\rightarrow$ Hard, but tend to be brittle
  - $\rightarrow$ Tough materials such as flakes of titanium diboride, TiB_2, can be distributed throughout ceramics to strengthen them. Even if a crack did form, it would be unable to propagate past the first flake that it encountered.
  - $\rightarrow$ A composite *polylactic acid polymer* containing ceramic flakes is used to provide rapid mends for broken bones. The material is injected into a cracked bone, sets at body temperature into a scaffold that acts like bone tissue, and promotes healing as new bone cells form around it.

NANOMATERIALS (Sections 6.18–6.20)

6.18 The Nature and Uses of Nanomaterials

- **Nanoparticles**
  - $\rightarrow$ Particles ranging in size from 1 to 100 nm, larger than molecules but too small to exhibit bulk properties
  - $\rightarrow$ Can be manufactured and manipulated at the *molecular level* using nanotechnology

- **Nanomaterials**
  - $\rightarrow$ Materials composed of *nanoparticles* or *regular arrays* of molecules or atoms such as *nanotubes*
  - $\rightarrow$ Properties differ from the properties of atoms and from those of bulk materials
  - $\rightarrow$ Used to create miniature circuits and drug delivery systems

- **Quantum dots**
  - $\rightarrow$ Three-dimensional clusters of semiconducting materials, such as cadmium selenide (CdSe), containing 10 to 10^5 atoms
  - $\rightarrow$ Made in solution or by depositing atoms on a surface, with the size of the nanocrystal being determined by the details of the synthesis
  - $\rightarrow$ Color varies with the radius of the quantum dot and it is readily observed in suspensions of CdSe quantum dots of different sizes.
  - $\rightarrow$ Some quantum dots emit light when an excited electron drops to a lower level in the dot. An application of this phenomenon is the monitoring of processes taking place in biological cells.

    Example: A CdSe quantum dot can be attached to the surface of a cell by an organic linking molecule. The spatial distribution of emission intensity (labeled molecule) can then be viewed with a microscope.

- **Understanding quantum dots**
  - $\rightarrow$ The particle-in-a-box energy levels can be used to predict the qualitative behavior of an electron trapped in a spherical cavity of radius r. The relevant equation from Section 1.7 is now

$$E_n = \frac{n^2 h^2}{8 m_e r^2}$$

  - $\rightarrow$ As the radius r of the quantum dot increases, the energy level separations decrease and radiation of lower frequency (lower energy) is needed to effect transitions. Or, as noted above, color varies with the radius of the quantum dot.

6.19 Nanotubes

- **Carbon nanotubes**
  - → Cylinders a few nanometers in diameter made from rolled sheets of graphite-like arrays of atoms
  - → Carbon and boron nitride are the most common nanotubes.
- **Carbon nanotubes**
  - → Graphene sheets of millions of carbon atoms rolled into a cylinder only 1–3 nm in diameter
  - → First identified in 1991 when two graphite rods sealed inside a container of He gas were subjected to an electric discharge
  - → Tensile strength parallel to the axis of the tube is the greatest of any material measured.
  - → Strength-to-mass ratio is 40 times that of steel because of the very low density of this material.
  - → Carbon nanotubes conduct electricity because of an extended network of delocalized π-bonds that extends from one end of the tube to the other.
  - → The conductivity of nanotubes is a function of the manner in which the tubes are formed.
  - → With hexagon points aligned along the long axis of the tube, conductivity is similar to that of metals – electrons travel through the π-orbitals with little resistance.
  - → With points aligned perpendicular to the long axis, nanotubes behave as semiconductors.
  - → Like diamonds, carbon nanotubes conduct heat well.
  - → Carbon nanotubes are good candidates for the developing miniature integrated circuits because of their electrical and thermal conductivity properties.
  - → Tiny 60-atom carbon spheres were also seen, previously discovered, and named buckminsterfullerene in 1985.
- **Other nanotubes**
  - → Some 50 different types of nanotubes and nanospheres have been synthesized from substances.
  - → MoS_2 and WS_2 nanotubes and nanospheres serve as solid lubricants.
  - → TiO_2 nanotubes show promise as a means of storing hydrogen gas.
  - → Nanospheres of ruthenium coated with platinum serve as a catalyst for the purification of hydrogen fuel.
  - → Boron nitride nanotubes are also of great interest. Unlike carbon nanotubes, boron nitride nanotubes have electrical characteristics that do not depend on the diameter of the tube or the way the sheets are rolled. They have a moderately large band gap, and so they are considered to be only weakly semiconducting.
  - → Carbon nanotubes grow until they are capped by bonds to another element or by five-membered rings, which pucker the structure, forming bends along the length of tubes and domes at their ends.
  - → For BN nanotubes, however, five-membered rings cannot form, because they would require boron–boron or nitrogen–nitrogen bonds, which are too weak to maintain the structure.
  - → Instead, BN nanotubes are capped by the atom of a metal such as tungsten.
  - → Nanotubes can be used as templates or tiny molds for nanostructures of other elements. As very small test tubes, nanotubes filled with biomolecules such as cytochrome c may act as nanosensors for medical applications.
  - → Nanotubes carrying hydrogen molecules could become a medium for hydrogen-powered vehicles.

6.20 Preparation of Nanomaterials

- **Nanotechnology**
    - → Two major methods are used to produce nanomaterials, physical and chemical.
    - → **Physical** methods such as lithography, "top-down" approaches (materials are molded into shape by use of physical nanotechnology methods – see Fig. 6.46 in the text)
    - → **Chemical** methods such as the solution and vapor-phase synthesis, "bottom-up" approaches (molecules are allowed to assemble themselves into specific patterns by use of intermolecular interactions)

        Example: In template synthesis, metal ions are plated electrochemically on porous polymer membranes, which are then dissolved to leave metal nanotubes.

Chapter 7 THERMODYNAMICS: THE FIRST LAW

SYSTEMS, STATES, AND ENERGY (Sections 7.1–7.7)

Overview

- **Thermodynamics** → Branch of science concerned with the relationship between heat and other forms of energy

- **Energy** → The capacity to do work

- **Laws of thermodynamics**
  - → Generalizations based on experience with *bulk* matter
  - → Not derivable, but understandable without knowledge of the behavior of atoms or molecules

- **First law of thermodynamics**
  - → Consequence of the **law of conservation of energy**
  - → Quantitative description of energy changes in physical (*e.g.*, phase changes) and chemical (*e.g.*, reactions) processes
  - → Energy changes in relation to heat and work:
    A paddlewheel stirrer immersed in a liquid produces heat from mechanical work (temperature rise in the liquid).

    A hot gas expanding in an insulated cylinder coupled to a flywheel produces mechanical work from heat (temperature drop in the gas).

- **Second law of thermodynamics**
  - → Criterion for spontaneity (see Chapter 8)
  - → Explains why some chemical reactions and other processes occur spontaneously while others do not

- **Statistical thermodynamics**
  - → Laws of thermodynamics reflected in the behavior of large numbers of atoms or molecules in a sample
  - → Link between the atomic level and the behavior of bulk matter

- **Chapter goals** → Gain insight into heat and work and their relationship to energy changes in physical and chemical processes

7.1 Systems

- **The "universe" or "world"** → Composed of a system and its surroundings

- **System** → Portion of the universe of interest
  Examples: A beaker of ethanol, a frozen pond, 20 g of $CaCl_2$(s), and the Earth itself.

- **Surroundings**
  - → Remainder of the universe (everything that is *not* in the system)
  - → Where observations and measurements of the system are made
    Example: The observation of heat transferred to or from a system is made in the surroundings.

- **Boundary**
  - → Dividing surface between the system and the surroundings
  - → Can be fixed (a closed, rigid container) or variable (a piston containing an expanding gas)

- **Types of systems**

  Open system
  - → Both *matter* and *energy* can be exchanged between the system and surroundings.
    - **Example:** An example is a half-liter of water in an open beaker. Matter can cross the boundary (water evaporates). Energy can cross the boundary (heat from the surroundings may enter the system).

  Closed system
  - → *Energy* can be exchanged between the system and surroundings, but *matter* cannot.
    - **Example:** The refrigerant in an air-conditioning system. As the refrigerant expands, energy (heat) is withdrawn from the space to be cooled. As the refrigerant is compressed, energy (heat) is supplied to another space that is warmed. The mass of the refrigerant is unchanged.

  Isolated system
  - → Neither *matter* nor *energy* can be exchanged between the system and surroundings.
    - **Example:** An *approximately* isolated system is a substance in a well-insulated container, such as ice in a Styrofoam ice-chest or coffee in a covered Styrofoam cup.

7.2 Work and Energy

- **Work, *w***
  - → Process of achieving motion against an opposing force
    - **Examples:** Raising a weight, compressing a gas, winding a spring, charging a battery
  - → Fundamental thermodynamic property (see Table 7.1 in the text *Varieties of Work*)
  - → Definition: work = force × distance
  - → Units: (force) *newton*, N; $1 \text{ N} = 1 \text{ kg·m·s}^{-2}$
    (work) *joule*, J; $1 \text{ J} = 1 \text{ N·m} = 1 \text{ kg·m}^2 \cdot \text{s}^{-2}$

- **Internal energy, *U***
  - → Internal energy is the *total* energy stored in a system.
  - → Includes energy of all atoms, electrons, and nuclei in a system
  - → U cannot be measured.
  - → *Only* a change in U ($\Delta U = U_{\text{final}} - U_{\text{initial}}$) is measurable.
  - → An extensive property, varies with the size of the system (see Fundamentals Section A in the text)
  - → For energy transferred to a system by doing work on the system, $\Delta U = w$.
  - → Work done on a system is positive ($\Delta U > 0$), work done by a system is negative ($\Delta U < 0$).

7.3 Expansion Work

- **Nonexpansion work** → Does not involve a change in volume
  - **Examples:** Change the position of a mass, discharging a battery
- **Expansion work** → Change in volume of the system against an external pressure
  - **Example:** Inflating a tire or balloon

- **Calculating expansion work with a constant external pressure, P_{ex}**
  - → Consider a gas confined in a volume, V, by a piston with surface area, A.
  - → Work = force × distance = $F \times d = (P_{ex} \times A) \times d = P_{ex} \Delta V$ (see Section 4.2), where d is the distance the piston is displaced (*not* the density).
  - → $\boxed{w = -P_{ex} \Delta V}$ By convention, work done *on* the system is positive (+) in sign.

    Units: $1\ Pa \cdot m^3 = 1\ kg \cdot m^{-1} \cdot s^{-2} \times 1\ m^3 = 1\ kg \cdot m^2 \cdot s^{-2} = 1\ J$
             $1\ L \cdot atm = 10^{-3}\ m^3 \times 101\ 325\ Pa = 101.325\ Pa \cdot m^3 = 101.325\ J$ (exactly)

- **Free expansion** → Expansion into a vacuum ($P_{ex} = 0$ and $w = 0$). No work is done because there is no opposing force.

- **Reversible process**
  - → A process that can be reversed by an *infinitesimal* change in a variable
  - → For a change in pressure, the external pressure must always be infinitesimally different from the pressure of the system. A net change is effected by a series of infinitesimal changes in the external pressure followed by infinitesimal adjustments of the system.
  - → For a change in temperature, the temperature of the system must change by a series of infinitesimal amounts of either heat flow or of work done.
  - → Processes in which change occurs by *finite* amounts are *irreversible* in nature.

- **Reversible, isothermal expansion or compression of an ideal gas**

  - → $\boxed{w = -nRT \ln \dfrac{V_{final}}{V_{initial}} = -nRT \ln \dfrac{P_{initial}}{P_{final}}}$

  - → The change in pressure is carried out in infinitesimal steps (text derivation in Section 7.3).

7.4 Heat

- **Heat, q**
  - → Fundamental thermodynamic property
  - → Energy transferred as a result of a temperature difference, $\Delta T = T_2 - T_1 > 0$
  - → Energy flows as *heat* from a region of high temperature, T_2, to one of low temperature, T_1.
  - → SI unit: *joule*, $1\ J = 1\ N \cdot m = 1\ kg \cdot m^2 \cdot s^{-2}$

    Common unit: The *calorie*, $1\ cal \equiv 4.184\ J$, is widely used in biochemistry, organic chemistry, and related fields. The *nutritional calorie*, Cal, is 1 kcal.

- **Sign of q defined in terms of the system**
  - → If heat flows from the surroundings into the system, $q > 0$ (*endothermic* process).
  - → If heat flows from the system into the surroundings, $q < 0$ (*exothermic* process).
  - → If no heat flows between the system and surroundings, $q = 0$ (*adiabatic* process).
    - **Example:** A thermally insulating (adiabatic) wall (Styrofoam or vacuum flask) prevents the passage of energy as heat.
  - → *Diathermic* (nonadiabatic) walls permit the transfer of energy as heat.

- **Internal energy, U**
  - → For energy transferred to a system by the flow of heat, $\Delta U = q$ ($w = 0$).
  - → If the system is initially at temperature T_1 and $\Delta T = T_2 - T_1 > 0$, then $q > 0$ and $\Delta U > 0$.
  - → **Thermal energy:** Sum of the potential and kinetic energies of all atoms, molecules, and ions in a system.

7.5 The Measurement of Heat

- **Heat capacity, C, of a pure substance**
  - → Is an *extensive* property.
  - → Is the amount of heat absorbed *by a sample of the substance* per degree Celsius rise in temperature.

- **Specific heat capacity, C_s, of a pure substance**
  - → Is an *intensive* property.
  - → Is the amount of heat absorbed *by one gram* of a substance per degree Celsius rise in temperature (values for common materials are given in Table 7.2 in the text)
  - → $C_s = C/m$ Units: $J \cdot (°C)^{-1} \cdot g^{-1}$ or $J \cdot K^{-1} \cdot g^{-1}$ (commonly, the *former*)
  - → $\boxed{q = C\Delta T = mC_s\Delta T}$

- **Molar heat capacity, C_m, of a pure substance**
  - → Is an *intensive* property.
  - → Is the amount of heat absorbed *by one mole* of a substance per degree Celsius rise in temperature (values for some substances are given in Table 7.2 in the text)
  - → $C_m = C/n$ Units: $J \cdot (°C)^{-1} \cdot mol^{-1}$ or $J \cdot K^{-1} \cdot mol^{-1}$ (commonly, the *latter*)
  - → $\boxed{q = C\Delta T = nC_m\Delta T}$
  - → Values of molar heat capacities increase with increasing molecular complexity. They depend on the temperature and the state of the substance. $C_m(\text{liquid}) > C_m(\text{solid})$

- **Calorimeter**
  - → Device used to monitor or measure heat transfer in a system by the temperature changes that take place in the surroundings
  - → In calorimetry, $q_{cal} = q_{surr}$
  - → For an *exothermic* process, $q_{cal} > 0$; the temperature of the calorimeter increases, $\Delta T > 0$.
  - → For an *endothermic* process, $q_{cal} < 0$; the temperature of the calorimeter decreases, $\Delta T < 0$.
  - → The heat capacity of a calorimeter, C_{cal}, is the amount of heat absorbed *by the calorimeter* per degree Celsius rise in temperature.
  - → $\boxed{q_{cal} = C_{cal}\Delta T}$
  - → A calorimeter's heat capacity is determined by adding a known amount of heat, supplied electrically, and measuring the temperature rise in the calorimeter (see Example 7.4 in the text).
    Units of heat capacity: $J \cdot (°C)^{-1}$ or $J \cdot K^{-1}$

7.6 The First Law

- **Combining heat and work**
  - → $\boxed{\Delta U = q + w}$
  - → Change in internal energy of a system is the *sum* of the heat *added* ($q > 0$) *to* or *removed* ($q < 0$) *from* the system *plus* the work done *on* ($w > 0$) or *by* ($w < 0$) the system.

- **First law of thermodynamics**
  - → States that the internal energy of an *isolated* system is constant.
  - → Is an extension of the law of conservation of energy.
  - → Is a generalization based on experience and cannot be proven.

Note: If the first law were false, one could construct a "perpetual motion machine" by starting with an isolated system, removing it from isolation, and letting it do work on the surroundings. It could then be placed in isolation again, and its internal energy allowed to return to the initial value. This feat has never been accomplished despite many attempts.

- **Constant volume process, $\Delta V = 0$**
  - $\rightarrow$ No expansion work is done
  - $\rightarrow$ $\Delta U = q$ [no work of any kind (expansion, electrical, *etc.*) is done]
  - $\rightarrow$ Heat absorbed or released by a system equals the change in internal energy for a constant volume process if no other forms of work are done: $\Delta U = q$.

- **State of a system** $\rightarrow$ Defined when all of its properties are fixed

- **State function**
  - $\rightarrow$ Property that depends only on the current state of the system
  - $\rightarrow$ Independent of the manner in which the state was prepared
    Examples of state functions include temperature, T; pressure, P; volume, V; and internal energy, U. (Heat, q, and work, w, are *not* state functions.)

- **Path**
  - $\rightarrow$ Sequence of intermediate steps linking an initial and a final state
  - $\rightarrow$ Consider the transition from state 1 to state 2. The changes in the state functions are independent of path. The quantities ΔU, ΔP, ΔV, and ΔT are uniquely determined. The amount of heat transferred or the work done depends on the sequence of steps followed.

7.7 A Molecular Interlude: The Origin of Internal Energy

- **Internal energy, U** $\rightarrow$ Energy stored as potential and kinetic energy of molecules

- **Potential energy** $\rightarrow$ Energy an object has by virtue of its position

- **Kinetic energy**
  - $\rightarrow$ Energy an object has by virtue of its motion
  - $\rightarrow$ For atoms, kinetic energy is *translational* in nature.
  - $\rightarrow$ For molecules, kinetic energy has *translational, rotational,* and *vibrational* components.
  - $\rightarrow$ *Translation* is the motion of the molecule as a *whole*.
  - $\rightarrow$ See Section 1.7 in the text for the particle-in-a-box translational energy levels.
  - $\rightarrow$ *Rotation* and *vibration* are motions *within* the molecule and do not change the center of gravity (mass) of the molecule.
  - $\rightarrow$ See Box 2.2 in the text for the rigid-rotor rotational energy levels.
  - $\rightarrow$ See Major Technique 1 in the text for a description of vibrational motion.

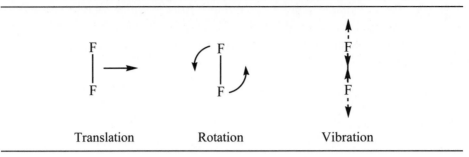

| Translation | Rotation | Vibration |

- **Degrees of freedom**
  - $\rightarrow$ Modes of motion (translational, rotational, and vibrational)
  - $\rightarrow$ *Mainly* translational and rotational degrees of freedom store internal energy at room temperature.
  - $\rightarrow$ Vibrational modes are largely inactive at room temperature. We will ignore them.

- **Equipartition theorem**
  - $\rightarrow$ *Average* kinetic energy of each degree of freedom of a molecule in a sample at temperature T is equal to $\frac{1}{2}kT$. The Boltzmann constant $k = 1.380\,658 \times 10^{-23}$ J·K^{-1}.
  - $\rightarrow$ A molecule moving in *three* dimensions has *three* translational degrees of freedom.

    $U(\text{translation}) = 3 \times \frac{1}{2}kT = \frac{3}{2}kT$

    For 1 mol of molecules, $U_m(\text{translation}) = N_A(\frac{3}{2}kT) = \frac{3}{2}RT$ $(R = N_A k)$.

  - $\rightarrow$ A linear molecule has *two* rotational degrees of freedom.

    $U(\text{rotation, linear}) = 2 \times \frac{1}{2}kT = kT$

    For 1 mol of molecules, $U_m(\text{rotation, linear}) = N_A(kT) = RT$.

  - $\rightarrow$ A nonlinear molecule has *three* rotational degrees of freedom.

    $U(\text{rotation, nonlinear}) = 3 \times \frac{1}{2}kT = \frac{3}{2}kT$

    For 1 mol of molecules, $U_m(\text{rotation, nonlinear}) = N_A(\frac{3}{2}kT) = \frac{3}{2}RT$.

  - $\rightarrow$ A molecule of an *ideal gas* does *not* interact with its neighbors and the potential energy is 0. For an ideal gas, the internal energy is independent of volume; it depends *only* on temperature, $U_m(T)$.

  - $\rightarrow$ A molecule of a liquid, solid, or real gas does interact with its neighbors. The potential energy is an important component of the internal energy, which has a significant dependence on volume. The internal energy depends on both temperature and volume, $U_m(T,V)$.

ENTHALPY (Sections 7.8–7.12)

Overview

- **Heat transfer at constant volume, q_V (notation not used in the text)**
  - $\rightarrow$ $\boxed{\Delta U = q_V}$ for a system in which only expansion work is possible
  - $\rightarrow$ Combustion reactions are studied in a bomb calorimeter of constant volume.

- **Heat transfer at constant pressure, q_P (notation not used in the text)**
  - $\rightarrow$ $\boxed{\Delta H = q_P}$ a new state function (*enthalpy*) for a system at constant pressure
  - $\rightarrow$ Most chemical reactions occur at constant pressure largely because reaction vessels are usually open to the atmosphere.

7.8 Heat Transfers at Constant Pressure

- **Definition of enthalpy, H**
  - $\rightarrow$ $\boxed{H = U + PV}$ a state function because U, P, and V are state functions
  - $\rightarrow$ Change in enthalpy of a system, $\Delta H = \Delta U + \Delta(PV)$
    At constant pressure, $\Delta H = \Delta U + P\Delta V = q_P + w + P\Delta V = q_P - P_{ex}\Delta V + P\Delta V = q_P$
    for a system open to the atmosphere and for which only expansion work may occur.
    Examples: System open to the atmosphere; gas or liquid in a piston with constant opposing pressure (see Fig. 9.11 in the text for an example of a system open to the outside)

- **Heat change for a constant pressure process**
  - $\rightarrow$ *Exothermic* process: one in which heat is released by the system into the surroundings, $\Delta H = q_P < 0$
  - $\rightarrow$ *Endothermic* process: one in which heat is absorbed by the system from the surroundings, $\Delta H = q_P > 0$
  - $\rightarrow$ Because most chemical reactions take place at constant pressure, enthalpy is of great importance in chemistry.

7.9 Heat Capacities at Constant Volume and Constant Pressure

- **General definition of heat capacity** $\rightarrow$ $\boxed{C = \dfrac{q}{\Delta T}}$

  At constant volume $\rightarrow$ $\boxed{C_V = \dfrac{q_V}{\Delta T} = \dfrac{\Delta U}{\Delta T}}$ **At constant pressure** $\rightarrow$ $\boxed{C_P = \dfrac{q_P}{\Delta T} = \dfrac{\Delta H}{\Delta T}}$

- **Molar heat capacities**
  - $\rightarrow$ $C_{V,m} = C_V/n$ and $C_{P,m} = C_P/n$
  - $\rightarrow$ $C_{V,m} \approx C_{P,m}$ *liquids and solids*, but $C_{P,m} > C_{V,m}$ *gases*

- **Molar heat capacity relationship for an ideal gas (see Section 7.10 in the text)**
  - $\rightarrow$ $C_P = C_V + nR$ *any amount of an ideal gas*
  - $\rightarrow$ $\boxed{C_{P,m} = C_{V,m} + R}$ *one mole of an ideal gas*
  - $\rightarrow$ For gases, $C_{P,m} > C_{V,m}$ because at constant pressure only some of the heat added raises the temperature; the rest is used to expand the gas (expansion work).

7.10 A Molecular Interlude: The Origin of the Heat Capacities of Gases

- **Contributions to the molar heat capacity for a monatomic ideal gas**
  - $\rightarrow$ Consider only translational motion.
  - $\rightarrow$ $U_m = \frac{3}{2}RT$ and $\Delta U_m = \frac{3}{2}R\Delta T$ (molar internal energy depends only on T)

$$\rightarrow \quad \boxed{C_{V,\mathrm{m}} = \frac{\Delta U_{\mathrm{m}}}{\Delta T} = \frac{\frac{3}{2}R\Delta T}{\Delta T} = \frac{3}{2}R} = 12.471\,77 \ \mathrm{J \cdot K^{-1} \cdot mol^{-1}} \quad \text{and} \quad \boxed{C_{P,\mathrm{m}} = \frac{\Delta H_{\mathrm{m}}}{\Delta T} = \frac{5}{2}R}$$

- **Contributions to the molar heat capacity for a linear-molecule ideal gas**
  - $\rightarrow$ Consider translational motion and 2 degrees of freedom in rotational motion.
  - $\rightarrow$ $U_{\mathrm{m}} = \frac{5}{2}RT$ and $\Delta U_{\mathrm{m}} = \frac{5}{2}R\Delta T$ (molar internal energy depends only on T)

$$\rightarrow \quad \boxed{C_{V,\mathrm{m}} = \frac{\Delta U_{\mathrm{m}}}{\Delta T} = \frac{\frac{5}{2}R\Delta T}{\Delta T} = \frac{5}{2}R} = 20.786\,28 \ \mathrm{J \cdot K^{-1} \cdot mol^{-1}} \quad \text{and} \quad \boxed{C_{P,\mathrm{m}} = \frac{\Delta H_{\mathrm{m}}}{\Delta T} = \frac{7}{2}R}$$

- **Contributions to the molar heat capacity for a nonlinear-molecule ideal gas**
  - $\rightarrow$ Consider translational motion and 3 degrees of freedom in rotational motion.
  - $\rightarrow$ $U_{\mathrm{m}} = 3RT$ and $\Delta U_{\mathrm{m}} = 3R\Delta T$ (molar internal energy depends only on T)

$$\rightarrow \quad \boxed{C_{V,\mathrm{m}} = \frac{\Delta U_{\mathrm{m}}}{\Delta T} = \frac{3R\Delta T}{\Delta T} = 3R} = 24.943\,53 \ \mathrm{J \cdot K^{-1} \cdot mol^{-1}} \quad \text{and} \quad \boxed{C_{P,\mathrm{m}} = \frac{\Delta H_{\mathrm{m}}}{\Delta T} = 4R}$$

Note: The contribution from molecular rotation is for room temperature. At lower temperatures, this contribution diminishes and the heat capacity approaches the value for a monatomic gas. At room temperature, a small contribution to the heat capacity from molecular vibrational motion leads to a value slightly larger than the ones given above.

7.11 The Enthalpy of Physical Change

- **Physical changes (phase transitions) at constant temperature and pressure**
  - $\rightarrow$ Molecules in liquids and solids are held together by intermolecular attractions.
  - $\rightarrow$ Phase changes that increase the separation between molecules (vaporization) require energy and are endothermic.
  - $\rightarrow$ Phase changes that decrease the separation between molecules (condensation) give off energy and are exothermic.

  - $\rightarrow$ Vaporization: liquid $\rightarrow$ vapor $\Delta H_{\mathrm{vap}} = H_{\mathrm{vapor,m}} - H_{\mathrm{liquid,m}}$ > 0 (*endothermic*)

        Condensation: vapor $\rightarrow$ liquid $\Delta H_{\mathrm{cond}} = H_{\mathrm{liquid,m}} - H_{\mathrm{vapor,m}}$ < 0 (*exothermic*)

  - $\rightarrow$ Fusion (melting):   solid $\rightarrow$ liquid $\Delta H_{\mathrm{fus}} = H_{\mathrm{liquid,m}} - H_{\mathrm{solid,m}}$ > 0 (*endothermic*)

        Freezing: liquid $\rightarrow$ solid $\Delta H_{\mathrm{freez}} = H_{\mathrm{solid,m}} - H_{\mathrm{liquid,m}}$ < 0 (*exothermic*)

  - $\rightarrow$ Sublimation: solid $\rightarrow$ vapor $\Delta H_{\mathrm{sub}} = H_{\mathrm{vapor,m}} - H_{\mathrm{solid,m}}$ > 0 (*endothermic*)

        Deposition: vapor $\rightarrow$ solid $\Delta H_{\mathrm{dep}} = H_{\mathrm{solid,m}} - H_{\mathrm{vapor,m}}$ < 0 (*exothermic*)

  - $\rightarrow$ $\Delta H_{\mathrm{vap}} = -\Delta H_{\mathrm{cond}}$ $\Delta H_{\mathrm{fus}} = -\Delta H_{\mathrm{freez}}$ $\Delta H_{\mathrm{sub}} = -\Delta H_{\mathrm{dep}}$

  - $\rightarrow$ $\Delta H_{\mathrm{sub}} = \Delta H_{\mathrm{fus}} + \Delta H_{\mathrm{vap}}$ (for the same T and P)

       Values of $\Delta H_{\mathrm{fus}}°$ and $\Delta H_{\mathrm{vap}}°$ for selected substances are given in Table 7.3 in the text.

  - $\rightarrow$ Liquids may evaporate at any temperature, but they boil only at the temperature at which the vapor pressure of the liquid is equal to the external pressure of the atmosphere.

7.12 Heating Curves

- **Heating (cooling) curve**
  - → Is a graph showing the variation of the temperature, T, of a sample as heat is added (removed) at a constant rate.
  - → Two types of behavior are seen in a heating curve.
  - → The temperature increase of a single phase to which heat is supplied has a positive slope that depends on the value of the heat capacity. The larger the heat capacity is, the more gentle the slope.
  - → At the temperature of a phase transition, two phases are present and the slope is 0 (no temperature change). The *greater* the heat associated with the phase change, the *longer* the length of the flat line.
  - → As more heat is added to the substance, the relative amounts of the two phases change. The temperature does not rise again until only one phase remains.

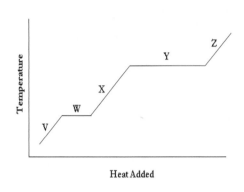

In this schematic heating curve, line V with positive slope represents the heating of a solid phase. The flat line W shows the phase transition between solid and liquid (fusion).

Line X represents the heating of the liquid after the solid has completely melted.

Line Y shows the phase transition between liquid and vapor (boiling at constant external pressure).

Finally, line Z represents the heating of the vapor after the liquid disappears. Because the enthalpy of vaporization is greater than the enthalpy of fusion, the length of line Y is longer than line W.

THE ENTHALPY OF CHEMICAL CHANGE (Sections 7.13–7.21)

7.13 Reaction Enthalpies

- **Thermochemical equation**
  - → A chemical equation with a corresponding enthalpy change, ΔH, expressed in kJ for the *stoichiometric* number of moles of each reactant and product
  - → Applying the principles of thermodynamics to chemical change

- **Enthalpy of reaction, ΔH_r**
  - → The enthalpy change in a thermochemical reaction expressed in kJ·mol^{-1}, where "mol" refers to the number of moles of each substance as determined by the stoichiometric coefficient in the balanced equation
  - → Allows calculation of the amount of heat released or required for any mass of reactant consumed or product formed in a chemical reaction
  - → Combustion is reaction with oxygen, $O_2(g)$.

7.14 The Relation Between ΔH and ΔU

- **Reactions with liquids and solids only**
  - $\rightarrow$ Recall that $\Delta H = \Delta U + \Delta(PV)$
  - $\rightarrow$ $\Delta H \approx \Delta U \implies q_P \approx q_V$
  - $\rightarrow$ Only a small change in the volume of reactants compared to products at constant P

- **Reactions with ideal gases only**
  - $\rightarrow$ $\Delta H = \Delta U + \Delta(PV) = \Delta U + \Delta(nRT) = \Delta U + (\Delta n)RT$ *constant temperature*
  - $\rightarrow$ The equation is exact for ideal gases only.

- **Reactions with liquids, solids, and gases**
  - $\rightarrow$ $\Delta H \approx \Delta U + (\Delta n_{gas})RT$ *constant temperature*
  - $\rightarrow$ $\Delta n_{gas} = n_{final} - n_{initial} =$ change in the number of moles of gas between products and reactants
  - $\rightarrow$ The equation is a very good approximation. The value of $(\Delta n_{gas})RT$ is usually much smaller than ΔU and ΔH.

7.15 Standard Reaction Enthalpies

- **Standard conditions**
  - $\rightarrow$ Values of reaction enthalpy, ΔH_r, depend on pressure, temperature, and the physical state of each reactant and product.
  - $\rightarrow$ A standard set of conditions is used to report reaction enthalpies.

- **Standard state**
  - $\rightarrow$ Standard state of a substance is its pure form at a pressure of *exactly* 1 bar.
  - $\rightarrow$ For a solute in a liquid solution, the standard state is a solute concentration of 1 mol·L^{-1} at a pressure of *exactly* 1 bar.
  - $\rightarrow$ Standard state of liquid water is pure water at 1 bar. Standard state of ice is pure ice at 1 bar. Standard state of water vapor is pure water vapor at 1 bar.

  Note: Compare the definition of standard state in thermodynamics to the definition of standard temperature and pressure, STP, for gases (Section 4.5). For gases, standard temperature is 0°C (273.15 K) and standard pressure is 1 atm (1.013 25 bar).

- **Standard reaction enthalpy, $\Delta H_r°$**
  - $\rightarrow$ Reaction enthalpy when reactants in their standard states form products in their standard states
  - $\rightarrow$ Degree symbol, °, is added to the reaction enthalpy.
  - $\rightarrow$ Reaction enthalpy has a *weak* pressure dependence, so it is a very good approximation to use standard enthalpy values even when the pressure is *not* exactly 1 bar.

- **Temperature convention**
  - $\rightarrow$ Most thermochemical data are reported for 298.15 K (25°C).
  - $\rightarrow$ Temperature convention is *not* part of the definition of the standard state.

7.16 Combining Reaction Enthalpies: Hess's Law

- **Hess's law**
  - $\rightarrow$ Overall reaction enthalpy is the sum of the reaction enthalpies of the steps into which a reaction can be divided.

    Example: Add two reactions (1 and 2) to get a third (3), add their reaction enthalpies.
    $$\Delta H_r(1) + \Delta H_r(2) = \Delta H_r(3)$$

→ Consequence of the fact that the reaction enthalpy is path independent and depends only on the initial reactants and final products

→ Used to calculate enthalpy changes for reactions difficult to carry out in the laboratory

→ Used to calculate reaction enthalpies for unknown reactions

- **Using Hess's law**
  → Write the thermochemical equation for the reaction whose reaction enthalpy is unknown.
  → Write thermochemical equations for intermediate reaction steps whose reaction enthalpies are known.
  → These steps must sum up to the overall reaction whose enthalpy is unknown.
  → Sum the reaction enthalpies of the intermediate reaction steps to obtain the unknown reaction enthalpy (see Toolbox 7.1 in the text).

7.17 The Heat Output of Reactions

- **Heat output in a reaction**
  → Heat is treated as a reactant or product in a stoichiometric relation.
  → In an endothermic reaction, heat is treated as a reactant.
  → In an exothermic reaction, heat is treated as a product.

- **Standard enthalpy of combustion, $\Delta H_c°$**
  → Change in enthalpy per mole of a substance that is burned in a combustion reaction under standard conditions (see values in Table 7.4 in the text)
  → Combustion of hydrocarbons is exothermic, and heat is treated as a product.

- **Specific enthalpy**
  → Enthalpy of combustion per *gram* ($kJ.g^{-1}$)
  → Practical measure of a fuel's value (see values in Table 7.4 in the text).

- **Energy conservation and biomass fuels**
  → Biomass energy is being used 5 million times faster that it is being stored
  → Consequently, the importance of energy conservation and the use of alternative energy fuels have grown (see Box 7.2 in the text).

7.18 Standard Enthalpies of Formation

- **Treatment of chemical reactions**
  → Possible thermochemical equations are nearly innumerable.
  → Method to handle this problem is the use of a common reference for each substance. The reference is the *most stable* form of the elements.
  → Standard enthalpies of formation of substances from their elements are tabulated and used to calculate standard reaction enthalpies for other types of reactions.

- **Standard enthalpy of formation, $\Delta H_f°$**
  → Enthalpy change for the formation of one mole of a substance from the most stable form of its elements under standard conditions
  → Elements (most stable form) → substance (one mole) $\Delta H° = \Delta H_f°$

  For example, $C(s) + 2 H_2(g) + \frac{1}{2}O_2(g) \rightarrow CH_3OH(l)$ $\Delta H_f° = -238.86 \, kJ.mol^{-1}$

  → Values of $\Delta H_f°$ at 298.15 K for many substances are listed in Appendix 2A.

→ Most stable form of the elements at 1 bar (standard state) and 298.15 K
(thermodynamic temperature convention):

Metals and semimetals are solids with atoms arranged in a lattice (see Chapter 5). The
exception is $Hg(l)$.

Nonmetal solids are $B(s)$, $C(s, graphite)$, $P(s, white)$, $S(s, rhombic)$, $Se(s, black)$, and $I_2(s)$.

Monatomic gases are $He(g)$, $Ne(g)$, $Ar(g)$, $Kr(g)$, $Xe(g)$, and $Rn(g)$. *Group 18*

Diatomic gases are $H_2(g)$, $N_2(g)$, $O_2(g)$, $F_2(g)$, and $Cl_2(g)$.

Nonmetal liquid is $Br_2(l)$.

Note: $P(s, red)$ is actually more stable than $P(s, white)$, but the white form is chosen because it is
easier to obtain pure. Also, some of the nonmetal solids have molecular forms, for example, P_4
and S_8.

- **Standard enthalpy of reaction, $\Delta H_r°$**

  → Use the most stable form of the elements as a reference state to obtain enthalpy changes for any
  reaction if $\Delta H_f°$ values are known for each reactant and product.

  → $\boxed{\Delta H_r° = \sum n\Delta H_f°(\text{products}) - \sum n\Delta H_f°(\text{reactants})}$

  In this expression, values of n are the stoichiometric coefficients and the symbol $\sum$ (sigma)
  means a summation.

  → Procedure to obtain $\Delta H_r°$: 1) convert the reactants into the most stable form of their elements
  $[-\sum n\Delta H_f°(\text{reactants})]$; 2) recombine the elements into products $[+\sum n\Delta H_f°(\text{products})]$.

  This is an application of Hess's law. Use $\Delta H_f°$ values in Appendix 2A to obtain $\Delta H_r°$.

7.19 The Born–Haber Cycle

- **Ionic solids and lattice enthalpy**

  → Cations and anions are arranged in a three-dimensional lattice (see Sections 5.13 and 5.14) and
  held in place mainly by coulombic interactions (see Section 2.4).

  → Lattice Enthalpy: enthalpy required to separate the ions in the solid into gas phase ions,
  $\Delta H_L = H_m(\text{ions, g}) - H_m(\text{ions, s}) > 0$. See values in Table 7.6 in the text.

  → Lattice enthalpy cannot be measured directly.

  → Calculate lattice enthalpy using a Born–Haber cycle.

- **Born–Haber cycle**

  → Thermodynamic cycle constructed to evaluate the lattice enthalpy from other reactions in which
  the enthalpy changes are known

  → Steps in the Born–Haber cycle that start and end with the most stable form of the elements in
  amounts appropriate to form one mole of the ionic compound

  Example: Determine the lattice enthalpy of $CaF_2(s)$. Start with one mole of $Ca(s)$
  and one mole of $F_2(g)$. Two moles of F atoms are then required.

  (1) *Atomize:* Atomize the metal element to form cations and the nonmetal element to form
  the anions. **In all cases, ΔH (atomization) > 0.**

  $Ca(s) \rightarrow Ca(g)$ $\Delta H_r° = \Delta H_f°[Ca(g)] = 178 \text{ kJ·mol}^{-1}$

  $F_2(g) \rightarrow 2\,F(g)$ $\Delta H_r° = 2\Delta H_f°[F(g)] = 2(79 \text{ kJ·mol}^{-1}) = 158 \text{ kJ·mol}^{-1}$

  ΔH (atomization) = (178 + 158) kJ·mol^{-1} = +336 kJ·mol^{-1}

  (2) *Ionize (cation):* Ionize the gaseous metal atom to form the gaseous cation. Several steps
  may be required. Because electrons created in the formation of cations in the ionization
  steps are consumed in the formation of anions in the electron-gain steps (3), energy and

enthalpy are considered to be interchangeable when treating ionization and electron gain.
In all cases, ΔH(ionization, cation) > 0.

$Ca(g) \rightarrow Ca^+(g) + e^-(g) \qquad \Delta H_r° = I_1[Ca] = 590 \text{ kJ·mol}^{-1}$

$Ca^+(g) \rightarrow Ca^{2+}(g) + e^-(g) \qquad \Delta H_r° = I_2[Ca] = 1145 \text{ kJ·mol}^{-1}$

$Ca(g) \rightarrow Ca^{2+}(g) + 2e^-(g) \qquad \Delta H_r° = I_1 + I_2 = (590 + 1145) \text{ kJ·mol}^{-1} = 1735 \text{ kJ·mol}^{-1}$

ΔH(ionization, cation) = +1735 kJ·mol^{-1}

(3) *Ionize (anion):* Attach electron(s) to the gaseous nonmetal atom to form the gaseous anion. Several electron affinity values may be required. Recall that for electron gain, $\Delta H = -E_{ea}$ for each electron added.

ΔH(ionization, anion) < or > 0

$F(g) + e^-(g) \rightarrow F^-(g) \qquad \Delta H_r° = -E_{ea}[F] = -328 \text{ kJ·mol}^{-1}$

$2F(g) + 2e^-(g) \rightarrow 2F^-(g) \qquad \Delta H_r° = -2E_{ea} = -656 \text{ kJ·mol}^{-1}$

ΔH(ionization, anion) = −656 kJ·mol^{-1}

(4) *Latticize:* Form the lattice of ions in the solid from the gaseous ions. The enthalpy of lattice formation is the negative of the lattice enthalpy, ΔH_L.

ΔH(lattice formation) < 0

$Ca^{2+}(g) + 2F^-(g) \rightarrow CaF_2(s) \qquad \Delta H_r° = -\Delta H_L \qquad value\ unknown$

ΔH(lattice formation) = −ΔH_L

(5) *Elementize:* Form the most stable form of the elements from the ionic solid.

ΔH(element formation) < 0

$CaF_2(s) \rightarrow Ca(s) + F_2(g) \qquad \Delta H_r° = -\Delta H_f°[CaF_2(s)] = 1220 \text{ kJ·mol}^{-1}$

ΔH(element formation) = +1220 kJ·mol^{-1}

(6) *Lattice enthalpy:* The sum of all the enthalpy changes for the complete cycle is 0.

$0 = \Delta H$(atomization) + ΔH(ionization, cation) + ΔH(ionization, anion) + ΔH(lattice formation) + ΔH(element formation)

$0 = 336 + 1735 + (-656) + (-\Delta H_L) + 1220 \text{ kJ·mol}^{-1} = 2635 \text{ kJ·mol}^{-1} - \Delta H_L$

ΔH_L = +2635 kJ·mol^{-1} for CaF$_2$(s) $\rightarrow$ Ca^{2+}(g) + 2F$^-$(g)

- **Summary** → Strength of interactions between ions in a solid is determined by the lattice enthalpy, which is obtained from a Born–Haber cycle.

7.20 Bond Enthalpies

- **Bond enthalpy, ΔH_B**
    - → Enthalpy change accompanying the breaking of a chemical bond in the gas phase
    - → Difference between the standard molar enthalpy of the fragments of a molecule and the molecule itself in the gas phase
    - → Reactants and products in their standard states (pure substance at 1 bar) at 298.15 K
    - → Bond enthalpies for diatomic molecules are given in Table 7.7 in the text.

- **Mean (average) bond enthalpies**
  - → In polyatomic molecules, the bond strength between two atoms varies from molecule to molecule.
  - → Variations are not very large, and the average values of bond enthalpies are a guide to the strength of a bond in any molecule containing the bond (see Table 7.8 in the text).

- **Using mean bond enthalpies**
  - → Atoms in the gas phase are the reference states used to estimate enthalpy changes for any gaseous reaction. A value of ΔH_B is required for each bond in the reactant and product molecules.
  - →
  $$\Delta H_r^\circ \approx \sum_{\text{reactants}} n\Delta H_B(\text{bonds broken}) - \sum_{\text{products}} n\Delta H_B(\text{bonds formed})$$
  - → In this expression (not in the text), values of n are the stoichiometric coefficients and the symbol $\sum$ (sigma) means a summation.
  - → The procedure is to convert the reactants into gaseous atoms $[+\sum n\Delta H_B(\text{bonds broken})]$ and then recombine the atoms into products $[-\sum n\Delta H_B(\text{bonds formed})]$. This is an application of Hess's law.
  - → Compare bond enthalpy to dissociation energy in Sections 2.14 and 2.15.

7.21 The Variation of Reaction Enthalpy with Temperature

- **Temperature dependence of reaction enthalpy**
  - → ΔH_r° needs to be measured at the temperature of interest.
  - → Approximation method is possible from the value of ΔH_r° measured at one temperature if heat capacity data on the reactants and products are available.

- **Kirchhoff's law**
  - → The difference in molar heat capacities of the products and reactants in a chemical reaction is
  $$\Delta C_P = \sum n C_{P,\text{m}}(\text{products}) - \sum n C_{P,\text{m}}(\text{reactants})$$
  - → An estimation of the reaction enthalpy at a temperature of T_2 if the value at a temperature of T_1 and the difference in molar heat capacities are known is
  $$\Delta H_{r,2}^\circ = \Delta H_{r,1}^\circ + \Delta C_P(T_2 - T_1)$$
  Kirchoff's law
  - → A major assumption is that the molar heat capacities are independent of temperature.
  - → The law in this form does not account for phase changes, for which separate steps must be added.

Chapter 8 THERMODYNAMICS:
THE SECOND AND THIRD LAWS

ENTROPY (Sections 8.1–8.8)

8.1 Spontaneous Change

- **The big question:** What is the *cause* of *spontaneous change?*

- **Spontaneous (or *natural*) change** → Occurs *without* an external influence, can be fast or slow

   Examples: Diamond converts to graphite (*infinitesimally slow rate*). Iron rusts in air (*very slow, but noticeable change*). $2H_2(g) + O_2(g) \rightarrow 2H_2O(g)$ (*very fast reaction or explosion*).

- **Nonspontaneous change** → Can be effected by using an external influence (using energy from the surroundings to do work on the system)

   Examples: Liquid water can be *electrolyzed* to form hydrogen and oxygen gases. Graphite can be converted to diamond under extremely high pressures. Gases can be compressed by applying an external pressure.

8.2 Entropy and Disorder

- **Spontaneous changes** → Any spontaneous change is accompanied by an *increase* in the *disorder* of the universe (*system* plus the *surroundings*).

- **Entropy (S)** → A measure of disorder; *increase* in disorder leads to an *increase* in S.

- **Second law of thermodynamics**
   - → For a spontaneous change, the entropy of an *isolated* system *increases.*
   - → For a spontaneous change, the entropy of the *universe increases.*
     [The universe is considered to be a (somewhat large) isolated system.]

- **Isolated system** → No exchange of energy or matter with the surroundings

- **Macroscopic definition of entropy, infinitesimal change** $\boxed{dS = \dfrac{dq_{rev}}{T}}$ (See Section 8.3)

- **Finite change, isothermal process** $\boxed{\Delta S = \dfrac{q_{rev}}{T}, \text{ for constant } T}$

   - → Heat transfer processes are carried out *reversibly* to evaluate ΔS. (Temperature of surroundings *equals* that of the system at all points along the path.)
   - → Reversible path: The system is only infinitesimally removed from equilibrium at all points along the path.

- **Entropy is an extensive property** → Proportional to the amount of sample

- **Entropy is a state function** → Changing the path does not change ΔS.

   Note: According to the definition of ΔS, we must calculate ΔS by using a *reversible path.* The result then applies to *any* path because ΔS is *independent of path.*

8.3 Changes in Entropy

- **Entropy increases**
  - → If a substance is heated (increase in thermal disorder)
  - → If the volume of a given amount of matter increases (increase in positional disorder)

- **Temperature dependence of entropy** (*any substance*)
  - → Let n = number of moles of a substance, C_P = heat capacity ($J \cdot K^{-1}$) at constant P, $C_{P,m}$ = molar heat capacity ($J \cdot K^{-1} \cdot mol^{-1}$) at constant P, C_V = heat capacity at constant V, $C_{V,m}$ = molar heat capacity at constant V, T_1 = initial temperature, and T_2 = final temperature.

  - → Isobaric heating of a substance (constant P)
  $$\Delta S = C_P \ln\frac{T_2}{T_1} = nC_{P,m} \ln\frac{T_2}{T_1}$$

  - → Isochoric heating of a substance (constant V)
  $$\Delta S = C_V \ln\frac{T_2}{T_1} = nC_{V,m} \ln\frac{T_2}{T_1}$$

  - → Above relationships assume that the heat capacity is *constant* over the range of T_1 to T_2.

    Example: If two moles of Fe are cooled from 300 K to 200 K at constant P, the entropy change is
    $$\Delta S = nC_{P,m} \ln\frac{T_2}{T_1} = (2 \text{ mol})(25.08 \text{ J} \cdot K^{-1} \cdot mol^{-1}) \ln\frac{200 \text{ K}}{300 \text{ K}} = -20.3 \text{ J} \cdot K^{-1}; \ \Delta S < 0.$$

    ΔS of the system *decreases* as expected from the *decrease* in temperature.

- **Volume dependence of entropy** (*ideal gas*)
  - → n = number of moles, R = gas constant ($8.314\,51 \text{ J} \cdot K^{-1} \cdot mol^{-1}$), V_1 = initial volume, and V_2 = final volume

  - → Isothermal volume change of an *ideal gas* (constant T)
  $$\Delta S = \frac{q_{rev}}{T} = -\frac{w_{rev}}{T} = nR \ln\frac{V_2}{V_1}$$

  - → Recall: $\Delta U = 0$ because the energy of an *ideal* gas depends only on its temperature.

- **Pressure dependence of entropy** (*ideal gas*)
  - → n = number of moles, P_1 = initial pressure, and P_2 = final pressure

  - → Isothermal pressure change (*ideal gas*, constant T), $\Delta U = 0$, (Boyle's Law) $\dfrac{V_2}{V_1} = \dfrac{P_1}{P_2}$

  and
  $$\Delta S = nR \ln\frac{V_2}{V_1} = nR \ln\frac{P_1}{P_2}$$

    Example: Three moles of an ideal gas expand from 20 L to 80 L at constant T.
    $$\Delta S = nR \ln\frac{V_2}{V_1} = (3 \text{ mol})(8.314\,51 \text{ J} \cdot K^{-1} \cdot mol^{-1}) \ln\frac{80 \text{ L}}{20 \text{ L}} = 34.6 \text{ J} \cdot K^{-1}$$

    ΔS of the system increases as expected from the volume increase.

Example: Four moles of an ideal gas undergo a pressure increase from 0.500 atm to 1.75 atm at constant T.

$$\Delta S = nR \ln \frac{P_1}{P_2} = (4 \text{ mol})(8.314\,51 \text{ J}\cdot\text{K}^{-1}\cdot\text{mol}^{-1}) \ln \frac{0.500 \text{ atm}}{1.75 \text{ atm}} = -41.7 \text{ J}\cdot\text{K}^{-1}$$

ΔS of the system decreases as expected from the volume decrease.

8.4 Entropy Changes Accompanying Changes in Physical State

- **Changes of physical state**
  - → Fusion ≡ fus (solid to liquid); vaporization ≡ vap (liquid to vapor); sublimation ≡ sub (solid to vapor)
  - → Solid-to-solid phase changes; for example, Sn(gray) → Sn(white)
  - → Normal boiling point, T_b: temperature at which a liquid boils when $P = 1$ atm.

- **For the phase changes above, at the transition temperatures**
  - → Temperature of substance remains *constant* during the phase change.
  - → Transfer of heat is *reversible.*
  - → Heat supplied is identified with *enthalpy change* because pressure is constant (for these reversible processes at constant P, $q_{rev} = q_P = \Delta H$).

- **Entropy of vaporization**
  - → Normal boiling point, T_b: T at which liquid boils when $P = 1$ atm

  - → $\boxed{\Delta S_{vap} = \dfrac{\Delta H_{vap}}{T_b}}$ ΔS_{vap} is the *entropy of vaporization* (units: J·K^{-1}·mol^{-1} or kJ·K^{-1}·mol^{-1}).

  - → $\Delta S_{vap}°$ = *standard entropy of vaporization* (liquid and vapor both pure and both at 1 bar).

- **Trouton's rule**
  - → For many liquids, $\Delta S_{vap}° \text{ (}liq\text{)} \approx 85$ J·K^{-1}·mol^{-1}
  - → Rationale: Approximately the same increase in *positional disorder* occurs when *any* liquid is vaporized (gas molecules are far apart and moving rapidly).
  - → Exceptions: Liquids with very weak or very strong intermolecular interactions (examples below)

    Examples: Liquid helium, He (20 J·K^{-1}·mol^{-1}); *hydrogen-bonded* liquids such as water, H_2O (109 J·K^{-1}·mol^{-1}), and methanol, CH_3OH (105 J·K^{-1}·mol^{-1})

- **Entropy of fusion (melting)**
  - → $\boxed{\Delta S_{fus}° = \dfrac{\Delta H_{fus}°}{T_f}}$, where T_f is the *melting* point under standard conditions.

  - → $\Delta S_{fus}°$ is the *standard entropy of fusion* (solid and liquid both pure and both at 1 bar).
  - → $\Delta S_{fus}°$ is smaller than $\Delta S_{vap}°$ because a liquid is only slightly more *disordered* than its solid, whereas a gas is considerably more disordered than its liquid.

    Example: Methane melts *reversibly* at 1 bar and $-182.5°C$: $CH_4\text{ (s)} \rightarrow CH_4\text{ (l)}$. Under these conditions, the enthalpy of fusion is $+0.936$ kJ·mol^{-1}. The standard (°) molar entropy of fusion is given by $\Delta S_{fus}° = \dfrac{\Delta H_{fus}°}{T_f} = \dfrac{936 \text{ J}\cdot\text{mol}^{-1}}{90.65 \text{ K}} = 10.3$ J·K^{-1}·mol^{-1}.

    Because $\Delta S > 0$ for *melting*, the final state is more disordered than the initial state, as expected.

8.5 A Molecular Interpretation of Entropy

- **Absolute value of the entropy** $\rightarrow$ If entropy S is a measure of disorder, then a perfectly ordered state of matter (perfect crystal) *should* have *zero* entropy.

- **Third law of thermodynamics** $\rightarrow$ Entropies of perfect crystals are the same at $T = 0$ K.

  (Thermal motion *almost* ceases at $T = 0$ K, and by convention $S = 0$ for perfect crystals at 0 K.)

- **Boltzmann approach** $\rightarrow$ Entropy increases as the number of ways that molecules or atoms can be arranged in a sample *at the same total energy* increases.

- **Boltzmann formula** $\rightarrow$ $\boxed{S = k \ln W}$; k = Boltzmann constant = $1.380\,658 \times 10^{-23}$ J·K^{-1}

  [**Note:** $R = N_A k$ J·K^{-1}·mol^{-1} = the gas constant.]

  W = number of *microstates* available to the system at a certain energy

- **Boltzmann entropy** $\rightarrow$ Also called statistical entropy

- **Microstate**
  - $\rightarrow$ *One* permissible arrangement of atoms or molecules in a sample with a given total energy
  - $\rightarrow$ W = total number of permissible arrangements corresponding to the same total energy, also called an **ensemble**
  - $\rightarrow$ If only one arrangement is possible, $W = 1$ and $S = 0$

- **Boltzmann interpretation** $\rightarrow$ Spontaneous change occurs toward more probable states.

- **Scaling S**
  - $\rightarrow$ For N molecules or atoms, the number of *microstates* is related to the number of molecular *orientations* permissible: $W = (orientations)^N$ and

  $$\boxed{S = k \ln W = k \ln(orientations)^N = N\, k \ln(orientations)}$$

  - $\rightarrow$ For N_A molecules or atoms, $W = (orientations)^{N_A}$ and

  $$\boxed{S = k \ln(orientations)^{N_A} = N_A k \ln(orientations) = R \ln(orientations)}$$

- **Residual entropy** $\rightarrow$ For some solids, $S > 0$ at $T = 0$. (Such solids retain some *disorder*.)

  Note: Units of S are usually J·K^{-1}·mol^{-1}.

  Example: The residual entropy of CO(s) at 0 K is 4.6 J·K^{-1}·mol^{-1}. The number of *orientations* of CO molecules in the crystal is calculated as

  $$S = R \ln(orientations); \quad (orientations) = e^{S/R} = e^{(4.6/8.314\,51)} = e^{(0.553\,25)} = 1.7 .$$

  This is less than 2, the number expected for a random orientation (*disorder*) of CO molecules at 0 K (see Fig. 8.7 in the text), suggesting some ordering at 0 K, possibly caused by alignment of the small permanent dipoles of neighboring CO molecules.

8.6 The Equivalence of Statistical and Thermodynamic Entropies

- Boltzmann's *molecular interpretation* compared with the *thermodynamic approach*

- Qualitative comparison

  Volume increase in an ideal gas:

  thermodynamic approach $\rightarrow$ $\Delta S = nR \ln \dfrac{V_2}{V_1}$; if $V_2 > V_1$, then $\Delta S > 0$

  molecular approach $\rightarrow$ Consider the gas container to be a box. From particle-in-a-box theory, as V increases, energy levels accessible to gas molecules pack more closely together and W increases (see Fig. 8.9 in the text).

  So, $\Delta S = S_{V_2} - S_{V_1} = k \ln \dfrac{W_2}{W_1} > 0$.

  Temperature increase in an ideal gas:

  thermodynamic approach $\rightarrow$ $\Delta S = nC_{(V \text{ or } P,\mathrm{m})} \ln \dfrac{T_2}{T_1}$; if $T_2 > T_1$, then $\Delta S > 0$

  molecular approach $\rightarrow$ From particle-in-a-box theory, at low T, gas molecules occupy a small number of energy levels (W = small); at higher T, more energy levels are available (W = larger) (see Fig. 8.10 in the text).

  So, $\Delta S = S_{T_2} - S_{T_1} = k \ln \dfrac{W_2}{W_1} > 0$.

  Note: For liquids and solids, similar reasoning for the T dependence of S applies.

- Quantitative comparison
  - $\rightarrow$ Assume that the number of microstates available to any molecule is proportional to the volume available to it: $W = \text{constant} \times V$.
  - $\rightarrow$ For N molecules, $W = (\text{constant} \times V)^N$
  - $\rightarrow$ For an expansion of N_A molecules from V_1 to V_2, $\boxed{\Delta S = N_A k \ln\left(\dfrac{\text{constant} \times V_2}{\text{constant} \times V_1} \right) = R \ln\left(\dfrac{V_2}{V_1} \right)}$

    which is *identical* to the thermodynamic expression.

  Summary $\rightarrow$ Macroscopic and microscopic approaches give the same predictions for ΔS. The Boltzmann approach provides deep insight into entropy on the molecular level in terms of the energy states available to a system.

8.7 Standard Molar Entropies, $S_\mathrm{m}°(T)$

- Use of different formalisms
  - $\rightarrow$ Use Boltzmann formalism to *calculate* entropy (sometimes difficult to do).
  - $\rightarrow$ Use thermodynamic formalism to *measure* entropy.

- Standard ($P = 1$ bar) molar entropies of *pure* substances
  - $\rightarrow$ Determined experimentally from *heat capacity* data ($C_{P,\mathrm{m}}$), *enthalpy* data (ΔH) for phase changes, and the third law of thermodynamics

→ For a gas at final temperature T (assuming only one solid phase and heat capacities independent of temperature)

$$S_m^\circ(T) = S_m^\circ(0) + \int_0^{T_f} \frac{C_{P,m}(\text{solid})dT}{T} + \frac{\Delta H_{\text{fus}}^\circ}{T_f} + \int_{T_f}^{T_b} \frac{C_{P,m}(\text{liquid})dT}{T} + \frac{\Delta H_{\text{vap}}^\circ}{T_b} + \int_{T_b}^{T} \frac{C_{P,m}(\text{gas})dT}{T}$$

→ If we assume that $S_m(0) = 0$ according to the third law (no residual entropy), then

$$S_m^\circ(T) = \int_0^{T_f} \frac{C_{P,m}(\text{solid})dT}{T} + \frac{\Delta H_{\text{fus}}^\circ}{T_f} + \int_{T_f}^{T_b} \frac{C_{P,m}(\text{liquid})dT}{T} + \frac{\Delta H_{\text{vap}}^\circ}{T_b} + \int_{T_b}^{T} \frac{C_{P,m}(\text{gas})dT}{T}$$

Note: For solids, only the first term in the equation is used if only *one* solid phase exists. For liquids, the first three terms are used. For gases, all five terms are required.

Example: Plots of $C_{P,m}$ as a function of T and $C_{P,m}/T$ as a function of T for solid copper, Cu(s), are shown below. In the second graph, the area under the curve ($T = 0$ to $T = 298.15$ K) yields the experimental value of the entropy for Cu(s): For the pure substance at 1 bar, S_m° (298.15 K) = 33.15 J·K^{-1}·mol^{-1}.

 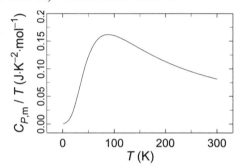

- **Appendix 2A**

  → Lists of values of *experimental* standard molar entropies at 25°C for many substances (*standard* ≡ *pure* substance at 1 bar)

  → Entropies of gases tend to be larger than those of solids or liquids ($S_{\text{gas}} > S_{\text{liquid}}$ *or* S_{solid}), as expected from the association of entropy with *randomness* and *disorder*.

  → Other things equal, molar entropy increases with molar mass. Heavier species have more energy levels available to them than lighter ones, so W and S are larger. (See Fig. 8.13 in the text for use of particle-in-a-box model to understand this better.)

8.8 Standard Reaction Entropies, ΔS_r°

- **Standard reaction entropies, ΔS_r°** → Determine similarly to calculation of standard enthalpy of reaction, ΔH_r°, for a chemical reaction.

- **For any chemical reaction** → $\boxed{\Delta S_r^\circ = \sum n S_m^\circ(\text{products}) - \sum n S_m^\circ(\text{reactants})}$

  Example: Calculate ΔS_r° for the reaction $2\,\text{Li(s)} + \text{Cl}_2(\text{g}) \rightarrow 2\,\text{LiCl(s)}$.
  Using S_m° values from Appendix 2A, we obtain

  $$\Delta S_r^\circ = \sum n S_m^\circ(\text{products}) - \sum n S_m^\circ(\text{reactants}) = 2\,S_m^\circ(\text{LiCl}) - [2\,S_m^\circ(\text{Li}) + S_m^\circ(\text{Cl}_2)]$$
  $$= 2(59.33) - [2(29.12) + 222.96] = -162.54 \text{ J·K}^{-1}\text{·mol}^{-1}.$$

$\Delta S_r° < 0$, as expected for a net decrease in the number of moles of gas when reactants form products.

- **Generalizations**
  - → In the surroundings, for an exothermic reaction, $\Delta S_{surr} > 0$; for an endothermic reaction, $\Delta S_{surr} > 0$.
  - → For reaction entropy, the entropy of gases dominates changes.
  - → $\Delta S_r°$ is *positive* if there is a net *production* of gas in a reaction.
  - → $\Delta S_r°$ is *negative* if there is a net *consumption* of gas in a reaction.
- **Properties of S and ΔS** → S is an *extensive* property of state, like enthalpy (H) and ΔH.
- **For any process or change in state**
  - → *Reverse* the process, *change* the sign of ΔS.
  - → *Change* the amounts of all materials in a process; make a *proportional change* in the value of ΔS.
  - → Add *two* reactions together to get a *third* one; *add* the ΔS values for the first *two* reactions to get ΔS for the *third* reaction.

GLOBAL CHANGES IN ENTROPY (Sections 8.9–8.11)

8.9 The Surroundings

- **Second law** → For *spontaneous* change, the entropy of an *isolated* system *increases*.
- *Any* **spontaneous change**
  - → Entropy of system *plus* surroundings *increases*.
  - → Criterion for spontaneity *includes* the *surroundings;* $\Delta S_{tot} = \Delta S + \Delta S_{surr}$.
- **Criterion**
  - → A process is *spontaneous* as written if $\Delta S_{tot} > 0$.
  - → A process is *nonspontaneous* as written if $\Delta S_{tot} < 0$.
    In this instance, the *reverse* process is spontaneous.
  - → For a system at *equilibrium*, $\Delta S_{tot} = 0$ (see Section 8.11).

- **Calculating ΔS_{surr} for a process at constant T and P** → $\boxed{\Delta S_{surr} = \dfrac{q_{surr}}{T} = -\dfrac{\Delta H}{T}}$

  - → The heat capacity of the surroundings is vast, so its temperature remains constant
  - → We assume that, at constant pressure, the *heat* associated with the process is transferred *reversibly* to the surroundings, so $\Delta S_{surr} = q_{surr,rev}/T$.
  - → For a system at constant P, $q_P = \Delta H_{sys}$
  - → So, $\Delta S_{surr} = -\Delta H_{sys}/T$ as in the box above.

    Note: The process occurring in the system itself may be *reversible* or *irreversible*.

    Example: The entropy change of the surroundings when 1 mol of $H_2O(l)$ vaporizes at 25°C is

    $$\Delta S_{surr} = \frac{q_{surr}}{T} = -\frac{\Delta H_{vap}}{T} = -\frac{40700 \text{ J} \cdot \text{mol}^{-1}}{(273.15 + 25) \text{ K}} = -137 \text{ J} \cdot \text{mol}^{-1} \cdot \text{K}^{-1} .$$

    The entropy of the surroundings decreases while that of the system increases.

8.10 Overall Change in Entropy

- **Spontaneity depends on ΔS *and* ΔS_{surr}** $\rightarrow$ Four cases arise because ΔS and ΔS_{surr} can *each* be either positive (+) or negative (−).

Case	[ΔS_{tot}	=	ΔS	+	ΔS_{surr}]	Spontaneity
1	+		+		+	*always* spontaneous
2	?		−		+	spontaneous, if $\|\Delta S_{surr}\| > \|\Delta S\|$
3	?		+		−	spontaneous, if $\|\Delta S\| > \|\Delta S_{surr}\|$
4	−		−		−	*never* spontaneous

- **Application to chemical reactions**
    - $\rightarrow$ *Exothermic* reactions ($\Delta S_{surr} > 0$) correspond to Cases 1 and 2, that is, *spontaneous if $\Delta S > 0$ or if $\|\Delta S_{surr}\| > \|\Delta S\|$ when $\Delta S < 0$.*
    - $\rightarrow$ *Endothermic* reactions ($\Delta S_{surr} < 0$) correspond to Cases 3 and 4, that is, *spontaneous* only if $\Delta S > 0$ *and* $\|\Delta S\| > \|\Delta S_{surr}\|$.

- **For a given change in state (same ΔS), the *path* can affect ΔS_{surr}, ΔS_{tot}, and *spontaneity.***
    - $\rightarrow$ Comparison of a *reversible* and an *irreversible* isothermal expansion of an *ideal gas* for the same initial and final states (see Example 8.12 in the text):
        - (1) ΔS (*irreversible*) = ΔS (*reversible*) (entropy is a state function)
        - (2) $\Delta U = q + w = 0$ and $\Delta H = 0$ because energy and enthalpy of an ideal gas depend only on T, and temperature is constant for this process ($\Delta T = 0$).
        - (3) *q and w* are different for the two processes (q and w are *path* functions). Heat given off to the surroundings ($-q = q_{surr}$) is different for the two processes.
        - (4) A *reversible* process does the *maximum* work w (see Fig. 7.6 in the text and Chapter 7). Because $q + w = 0$, a *reversible* process delivers *less* heat to the surroundings.
        - (5) Because $\Delta S_{surr} = q_{surr} / T$, ΔS_{surr} (*irreversible*) > ΔS_{surr} (*reversible*).
        - (6) Thus, ΔS_{tot} (*irreversible*) > ΔS_{tot} (*reversible*) = 0.
        - (7) The *reversible* process corresponds to one in which the system is only *infinitesimally* removed from equilibrium as the process proceeds, whereas the *irreversible* process [ΔS_{tot}(*irreversible*) > 0] is *spontaneous*.

- **Clausius inequality and the second law**
    - $\rightarrow$ From point (6) above, we can write in general that $\Delta S \geq q/T$. This is the Clausius inequality.
    - $\rightarrow$ For an isolated system, $q = 0$ and $\Delta S \geq 0$.
        Summary: **For an isolated system, ΔS cannot decrease (another statement of the second law).**

8.11 Equilibrium

- **System at equilibrium** $\rightarrow$ No tendency to change in forward or reverse direction without an external influence
- **Types of equilibrium**
    thermal: No tendency for heat to flow in or out of the system
    Example: An aluminum rod at room temperature

mechanical: No tendency for any part of a system to move

Example: An undeformed spring with no tendency to stretch or compress

physical: Two phases of a substance at the transition temperature with no tendency for either phase to increase in mass

Example: Steam and water at the normal boiling point (100°C)

chemical: A mixture of reactants and products with no *net* tendency for the formation of either more reactants or of more products

Example: A saturated solution of sucrose in water in contact with solid sucrose

- **Universal thermodynamic criterion for equilibrium**

  → $\Delta S_{tot} = 0$ for any system at equilibrium.

  If this were not true, ΔS_{tot} would be greater than 0 in either the forward or the reverse direction and *spontaneous change* would occur until $\Delta S_{tot} = 0$.

  → Total entropy change ($\Delta S_{tot} = \Delta S + \Delta S_{surr}$) may be calculated to determine whether a system is at equilibrium.

GIBBS FREE ENERGY (Sections 8.12–8.16)

8.12 Focusing on the System

- **ΔG, an alternative criterion for *spontaneity***

  (1) $\Delta S_{tot} = \Delta S + \Delta S_{surr}$ (always true)

  (2) If P and T are constant, $\Delta S_{surr} = -\dfrac{\Delta H}{T}$.

  (3) If P and T are constant, $\Delta S_{tot} = \Delta S - \dfrac{\Delta H}{T}$ (function of *system only*).

  (4) Definition of Gibbs free energy, $G \equiv H - TS$

  (5) If T is constant, $\Delta G = \Delta H - \Delta(TS) = \Delta H - T\Delta S$.

  (6) According to (5) and (3), $\Delta G = -T\Delta S_{tot}$.

- **Summary**
- If a process at constant T and P is

  → *spontaneous* ($\Delta S_{tot} > 0$), then $\Delta G < 0$.

  → *nonspontaneous* ($\Delta S_{tot} < 0$), then $\Delta G > 0$.

  → at *equilibrium* ($\Delta S_{tot} = 0$), then $\Delta G = 0$ as well.

- **Dependence of spontaneity on ΔS and ΔH of a system at constant T and P**

  → Four cases arise because *both* ΔS and ΔH can be *either* positive (+) *or* negative (−).

  → Spontaneity depends on the sign and magnitude of ΔH and ΔS, as well as the temperature T.

Case	[ΔG	=	ΔH	−	$T\Delta S$]	Spontaneity
1	−		−		+	*always* spontaneous
2	?		−		−	spontaneous, if $|\Delta H| > T|\Delta S|$
3	?		+		+	spontaneous, if $T|\Delta S| > |\Delta H|$
4	+		+		−	*never* spontaneous

Case 1

→ *Exothermic* process or reaction ($\Delta H < 0$) with $\Delta S > 0$

→ *Enthalpy* and *entropy* favor *spontaneity.*
   Enthalpy- and *entropy*-driven process or reaction

→ *Temperature* dependence: If ΔH and ΔS are T independent, the reaction is *spontaneous* at all T.

Case 2

→ *Exothermic* process or reaction ($\Delta H < 0$) with $\Delta S < 0$

→ *Only enthalpy* favors *spontaneity.*
   If the process is spontaneous, it is *enthalpy*-driven.

→ *Temperature* dependence: If ΔH and ΔS are T independent, the reaction is *spontaneous* at low T (*enthalpy* "wins") but *nonspontaneous* at high T. A crossover temperature exists at which $\Delta H = T\Delta S$ and the system is at equilibrium.

Case 3

→ *Endothermic* process or reaction ($\Delta H > 0$) with $\Delta S > 0$

→ *Only entropy* favors *spontaneity.*
   If the process is spontaneous, it is driven by *entropy (*entropy-driven process).

→ *Temperature* dependence: If ΔH and ΔS are independent of T, the reaction is *nonspontaneous* at low T and *spontaneous* at high T (*entropy* "wins"). A crossover temperature exists at which $\Delta H = T\Delta S$ and the system is at equilibrium.

Case 4

→ *Endothermic* process or reaction ($\Delta H > 0$) with $\Delta S < 0$

→ *Nonspontaneous* process or reaction ($\Delta G > 0$)

→ *Temperature* dependence: If ΔH and ΔS are independent of T, then the reaction is *nonspontaneous* at all T.

- **Temperature dependence of *G* for a *pure substance***

  → $G = H - TS$

  → The free energy of a substance decreases with increasing T.

  → But, for a given substance, $S_{m,solid} < S_{m,liq} < S_{m,gas}$.

  → So $G_{m,solid}$ decreases more slowly than $G_{m,liq}$ which decreases more slowly than $G_{m,gas}$.

  → The above forms the basis for a thermodynamic understanding of melting, vaporization, and sublimation (see Figs. 8.25 and 8.26 in the text).

8.13 Gibbs Free Energy of Reaction, ΔG_r

- **Chemical reactions** → ΔG_r and ΔG_r° are defined in a manner similar to the reaction enthalpy, ΔH_r, and standard reaction enthalpy, ΔH_r°.

- **Definitions (n = stoichiometric coefficient)**
  → If G_m is the molar Gibbs free energy of a reactant or product, then the Gibbs free energy of reaction is

$$\Delta G_r = \sum n G_m (\text{products}) - \sum n G_m (\text{reactants})$$

  → If G_m° is the *standard* molar Gibbs free energy, the standard Gibbs free energy of reaction is

$$\Delta G_r^\circ = \sum n G_m^\circ (\text{products}) - \sum n G_m^\circ (\text{reactants})$$

  → Values of G_m or G_m° cannot be determined directly. So the equations given above *cannot* be used to determine ΔG_r and ΔG_r°.

  → Values of ΔG_r and ΔG_r° are determined from the Gibbs free energies of *formation*, ΔG_f and ΔG_f°.

- **Standard Gibbs free energy of formation, ΔG_f°**
  → Standard Gibbs free energy of *formation* of a compound or element
  → Gibbs free energy change for the formation of one mole of a compound from the most stable form of its elements under standard conditions ($P = 1$ bar for each reactant and product)
  → $\Delta G_f^\circ \equiv 0$ for *all elements* in their most stable form (same convention as for enthalpy).

- **Calculating ΔG_f° for a given compound**

  1. *Write and balance the formation reaction.*
     One mole of compound on the product side and the most stable form of the elements with appropriate coefficients on the reactant side

  2. *Calculate ΔH_f° and ΔS_f° for the reaction by using the data in Appendix 2A.*
     The value of ΔS_f° is obtained from the standard molar entropy values as follows:
     $\Delta S_f^\circ = S_m^\circ (\text{compound}) - \sum n S_m^\circ (\text{reactants})$.

  3. *Solve for the formation of one mole of compound, using the expression*

$$\begin{aligned} \Delta G_f^\circ &= \Delta H_f^\circ - T \Delta S_f^\circ \\ &= \Delta H_f^\circ (\text{compound}) - T \left[S_m^\circ (\text{compound}) - \sum n S_m^\circ (\text{reactants}) \right] \end{aligned}$$

- **Thermodynamically *stable* compound is characterized by**
  → A *negative* standard free energy of formation ($\Delta G_f^\circ < 0$).
  → A *thermodynamic* tendency to *form* from its elements.
  Examples: Ores such as $Al_2O_3(s)$ and $Fe_2O_3(s)$

- **Thermodynamically *unstable* compound is characterized by**
  → A *positive* standard free energy of formation ($\Delta G_f^\circ > 0$).
  → A *thermodynamic* tendency to *decompose* into its elements.
  Examples: Some hydrocarbons such as acetylene (C_2H_2) and benzene (C_6H_6)

- **Properties of thermodynamically *unstable* compounds**
  - → *Many* decompose into their elements over a *long* time span.
  - → *Thermodynamically unstable* substances (like liquid octane and diamond) that decompose into their elements *slowly* are said to be *kinetically stable*.

- **Classification of thermodynamically *unstable* compounds**

  labile: Decompose *or* react readily, for example, TNT and NO

  nonlabile: Decompose *or* react slowly, for example, liquid octane

  inert: Exhibit virtually no reactivity *or* decomposition, for example, diamond

- **Calculating the standard Gibbs free energy of reaction $\Delta G_r°$ for a given chemical reaction**
  1. *Write the balanced chemical equation for the reaction.*
  2. *Calculate $\Delta H_r°$ and $\Delta S_r°$ for the reaction by using the data in Appendix 2A.*
  3. *Solve for $\Delta G_r°$, using the expression*

$$\Delta G_r° = \Delta H_r° - T\Delta S_r°$$
$$= \left[\sum n\Delta H_f°(\text{products}) - \sum n\Delta H_f°(\text{reactants})\right] - T\left[\sum nS_m°(\text{products}) - \sum nS_m°(\text{reactants})\right]$$

  4. *Alternatively, calculate $\Delta G_r°$ from the $\Delta G_f°$ values listed in Appendix 2A.*

$$\Delta G_r° = \sum n\Delta G_f°(\text{products}) - \sum n\Delta G_f°(\text{reactants})$$

  The equation directly above yields a faster result, but does *not* allow for an estimation of the *temperature* dependence of $\Delta G_r°$.

  Example: Calculate $\Delta G_r°$ at 25°C for the reaction
  $$2\,Al_2O_3(s) + 3\,C(\text{graphite,s}) \rightarrow 3\,CO_2(g) + 4\,Al(s).$$

  $\Delta G_r° = \sum n\Delta G_f°(\text{products}) - \sum n\Delta G_f°(\text{reactants})$
  $= 3\,\Delta G_f°[CO_2(g)] + 4\,\Delta G_f°[Al(s)] - \{2\,\Delta G_f°[Al_2O_3(s)] + 3\,\Delta G_f°[C(\text{graphite,s})]\}$
  $= 3(-394.36 \text{ kJ·mol}^{-1}) - 2(-1582.3 \text{ kJ·mol}^{-1}) = 1981.5 \text{ kJ·mol}^{-1}$

  Since $\Delta G_r°$ is positive, the reaction is not expected to be *spontaneous* at 25°C under standard conditions.

  Note: $\Delta G_f°[Al(s)]$ and $\Delta G_f°[C(\text{graphite,s})]$ are zero (elements in their standard states).

8.14 The Gibbs Free Energy and Nonexpansion Work

- **Definition of free energy** → The maximum nonexpansion work that can be obtained from a process at constant P and T

- **Processes at constant T and P**
  Expansion work (performed against an opposing pressure)
  - → $dw = -PdV$ (*infinitesimal* change in volume, see Chapter 7)
  - → $w = -P\Delta V$ (*finite* change in volume, constant opposing P)
  - → Expansion work is performed against an opposing pressure.
  - → The sign of w is defined in terms of the *system*:

    If work is done *on* the system *by* the surroundings, $w > 0$ (the system gains energy).
    If work is done *by* the system *on* the surroundings, $w < 0$ (the system loses energy).

Nonexpansion work

→ dw_e (*infinitesimal* change, subscript "e" stands for extra)

→ w_e (*finite* change)

→ Any other type of work, including electrical work, mechanical work, work of muscular contraction, work involved in neuronal signaling, and that of chemical synthesis (making chemical bonds)

- **Relationship between nonexpansion work and (finite) Gibbs free energy changes**

  → For a reversible process, $\Delta G = w_{rev,e}$ (constant P and T)

  → $w_{rev,e}$ = *maximum* nonexpansion work obtainable from a constant P, T process

  → *Reversible* process: maximum amount of work is done by system on the surroundings ($w_{rev,e}$ is negative).

  → All real systems are *irreversible* and the maximum nonexpansion work is never obtained.

  Example: The maximum electrical work obtainable at 1 bar and 25°C by burning one mole of propane in a fuel cell with *excess* oxygen according to the equation

  $$C_3H_8(g) + 5\,O_2(g) \rightarrow 3\,CO_2(g) + 4\,H_2O(l) \text{ is given by}$$

  $$w_{rev,e} = \Delta G_r^\circ = \sum n\Delta G_f^\circ(\text{products}) - \sum n\Delta G_f^\circ(\text{reactants})$$

  $$= 3\Delta G_f^\circ[CO_2(g)] + 4\Delta G_f^\circ[H_2O(l)] - \{\Delta G_f^\circ[C_3H_8(g)] + 5\Delta G_f^\circ[O_2(g)]\}$$

  $$= 3(-394.36 \text{ kJ·mol}^{-1}) + 4(-237.13 \text{ kJ·mol}^{-1}) - (-23.49 \text{ kJ·mol}^{-1})$$

  $$= -2108.11 \text{ kJ for 1 mol of } C_3H_8(g)$$

  The negative value for $w_{rev,e}$ indicates work done on the surroundings.

8.15 The Effect of Temperature

- **For any chemical reaction at constant P and T**

$$\boxed{\Delta G_r^\circ = \Delta H_r^\circ - T\Delta S_r^\circ = \left[\sum n\Delta H_f^\circ(\text{prod}) - \sum n\Delta H_f^\circ(\text{react})\right] - T\left[\sum nS_m^\circ(\text{prod}) - \sum nS_m^\circ(\text{react})\right]}$$

  → Cases 1–4 from the table in Section 8.12 of this *Study Guide* apply to any chemical reaction, assuming that ΔH_r° and ΔS_r° are independent of temperature.

  → In this case, the temperature dependence of ΔG_r° arises from the $(-T\Delta S_r^\circ)$ term.

  Example: To calculate the temperature range over which the reaction $H_2(g) + \frac{1}{2}O_2(g) \rightarrow H_2O(g)$ is spontaneous, use $\Delta G_r^\circ = \Delta H_r^\circ - T\Delta S_r^\circ$. Set $\Delta G_r^\circ = 0$ to determine the crossover temperature, where neither the forward nor reverse reaction is favored.

  $$T_{crossover} = \frac{\Delta H_r^\circ}{\Delta S_r^\circ}$$

  $$= \frac{\Delta H_f^\circ[H_2O(g)]}{S_m^\circ[H_2O(g)] - \{S_m^\circ[H_2(g)] + (1/2)S_m^\circ[O_2(g)]\}}$$

  $$= \frac{-241.82 \text{ kJ·mol}^{-1}}{0.188\,83 \text{ kJ·K}^{-1}\cdot\text{mol}^{-1} - \{0.130\,68 + (1/2)(0.205\,14)\}\text{kJ·K}^{-1}\cdot\text{mol}^{-1}}$$

  $$= \frac{-241.82 \text{ kJ·mol}^{-1}}{-0.044\,42 \text{ kJ·K}^{-1}\cdot\text{mol}^{-1}} = 5444 \text{ K}$$

  In this case, enthalpy favors *spontaneity*; entropy does not. At 298.15 K, the reaction is spontaneous $[\Delta G_r^\circ = \Delta H_r^\circ - T\Delta S_r^\circ = -241.82 - (298.15)(-0.044\,42) =$

-228.58 kJ·mol^{-1}]. It is spontaneous until $T = 5444$ K. Above 5444 K, $H_2O(g)$ is expected to decompose *spontaneously* into its elements.

8.16 Impact on Biology: Gibbs Free Energy Changes in Biological Systems

- **Metabolism** → In metabolic processes, reaction steps may have a *positive* (unfavorable) free energy of reaction. They can be coupled to spontaneous reactions to make the overall (or net) reaction spontaneous.
- **Concept of coupled chemical reactions**
    → A way to make *nonspontaneous* reactions occur without changing temperature or pressure
    → A prominent process in biological systems, but also found in nonbiological ones
    → If reaction (1) has a *positive* reaction free energy [$\Delta G_r°(1) > 0$] (*nonspontaneous*), it can be coupled to reaction (2) with a *more negative* reaction free energy [$\Delta G_r°(2) < 0$], such that reaction (3), the sum of the two reactions, has a *negative* reaction free energy [$\Delta G_r°(3) < 0$]:
    $\Delta G_r°(3) = \Delta G_r°(1) + \Delta G_r°(2) < 0$ (overall reaction is *spontaneous*).

    Example: The sugar glucose (a food) is converted to pyruvate in a series of steps. The overall process may be represented as:

    glucose + other reactants → 2 pyruvate + other products; $\Delta G_r° = -80.6$ kJ·mol^{-1}.

    The relatively large *negative* value of $\Delta G_r°$ indicates that the overall reaction occurs spontaneously under *biochemical* standard conditions. The *biochemical standard state* corresponds more nearly to typical conditions in a cellular environment. The first step in the metabolic process is the conversion of glucose into glucose-6-phosphate:

    glucose + phosphate → glucose-6-phosphate + H_2O;   (1) $\Delta G_r°(1) = +14.3$ kJ·mol^{-1}.

Glucose   Glucose-6-phosphate

The *positive* standard free energy implies *nonspontaneity* for reaction (1). To generate glucose-6-phosphate, reaction (1) is coupled with (2), the hydrolysis of adenosine triphosphate (ATP) to yield adenosine diphosphate (ADP). Reaction (2) has a favorable $\Delta G_r° < 0$ value (*spontaneous*):

ATP + H_2O → ADP + inorganic phosphate;   (2) $\Delta G_r°(2) = -31.0$ kJ·mol^{-1}.

Adding (coupling) reactions (1) and (2) gives a net spontaneous reaction (3):

glucose + ATP → glucose-6-phosphate + ADP (3)

$\Delta G_r°(3) = \Delta G_r°(1) + \Delta G_r°(2) = -16.7$ kJ·mol^{-1}

Note: Reactions in metabolic pathways are *catalyzed* by *enzymes*. Step (1) is catalyzed by the enzyme hexokinase. Biological catalysts are discussed in Chapter 14 and sugars, a class of carbohydrates, are discussed in Chapter 19.

Chapter 9 PHYSICAL EQUILIBRIA

PHASES AND PHASE TRANSITIONS (Sections 9.1–9.7)

9.1 Vapor Pressure

- **Phase** $\rightarrow$ A specific physical state of matter such as a solid, liquid, or gas.

- **Liquids** $\rightarrow$ Evaporate to form a gas or vapor

   Puddle of rainwater evaporates. The reaction $H_2O(l) \rightarrow H_2O(g)$ goes to completion and the puddle disappears. Phase equilibrium is *not* attained here.

- **Gases** $\rightarrow$ Condense to form a liquid or solid at sufficiently low temperature

   Water vapor condenses to form rain. Steam condenses on cold surfaces.

- **Equilibrium** $\rightarrow$ In a closed system, both liquid and gas phases exist in *equilibrium*.

   Represent equilibrium system using double-headed arrow: $H_2O(l) \rightleftharpoons H_2O(g)$

- **Liquid-gas phase equilibrium** $\rightarrow$ *Dynamic equilibrium*

   Rate of evaporation equals rate of condensation.

   Example: In a *covered* jar half-filled with water at room temperature, the liquid and gas phases are in equilibrium, $H_2O(l) \rightleftharpoons H_2O(g)$. Water vapor exhibits a characteristic equilibrium pressure, its *vapor pressure*.

- **Vapor pressure, P**

   $\rightarrow$ Characteristic pressure of a vapor above a *confined* liquid or solid in dynamic equilibrium (*closed system*)

   $\rightarrow$ Depends on temperature, increasing rapidly with increasing temperature, T

   $\rightarrow$ Different vapor pressure for different liquids and solids

   Examples: Several solids with measurable vapor pressures are camphor, naphthalene, *para*-dichlorobenzene (mothballs), and dry ice [$CO_2(s)$].

   Virtually all liquids have measurable vapor pressures, but some [*e.g.*, $Hg(l)$] may be very low at room temperature.

- **Sublimation** $\rightarrow$ Evaporation of solid to form a gas **Example:** [$CO_2(s) \rightarrow CO_2(g)$]

- **Volatility**

   $\rightarrow$ Related to the ability to evaporate

   $\rightarrow$ Liquids with high vapor pressures at a given temperature are *volatile* substances.

- **Characteristics of equilibrium between phases**

   $\rightarrow$ *Dynamic equilibrium* occurs when molecules enter and leave the individual phases at the same rate, and the total amount in each phase remains unchanged.

   $\rightarrow$ Molar free energies of the individual phases are equal.

   substance (phase 1) $\rightleftharpoons$ substance (phase 2), $\Delta G_m = 0$

9.2 Volatility and Intermolecular Forces

- *Increasing* the strength of intermolecular forces in a liquid

   $\rightarrow$ *Decreases* the volatility

   $\rightarrow$ *Decreases* the vapor pressure at a given temperature

   $\rightarrow$ *Increases* the normal boiling point, T_b (defined in Fig. 9.3 in the text and in Section 9.4)

- **London forces**
    - $\rightarrow$ More massive molecules are less volatile.
    - $\rightarrow$ More electrons in a molecule produce stronger forces.

- **Dipolar forces**
    - $\rightarrow$ For molecules with the same number of electrons, dipolar molecules tend to be less volatile substances with only London forces.
    - $\rightarrow$ A *greater* dipole moment yields a *lower* volatility.

- **Hydrogen bonding** $\rightarrow$ Molecules that form H–bonds may produce even less volatile substances than dipolar molecules.

- **Ionic forces** $\rightarrow$ Salts are essentially nonvolatile.

    Note: As with many rules in chemistry, the above qualitative guidelines must be applied cautiously.

- **Quantitative approach to vapor pressure** (see derivation in Section 9.3 in the text)
    - $\rightarrow$ $\ln(P/P^\circ) = \dfrac{-\Delta G_{vap}{}^\circ}{RT}$ *and* $\Delta G_{vap}{}^\circ = \Delta H_{vap}{}^\circ - T\Delta S_{vap}{}^\circ$ *lead to*

$$\ln(P/P^\circ) = \frac{-\Delta H_{vap}{}^\circ}{RT} + \frac{\Delta S_{vap}{}^\circ}{R}$$

- **Insight from the above equation**
    - $\rightarrow$ Because $\Delta S_{vap}{}^\circ$ is about the same for *all* liquids (Trouton's rule), the vapor pressure of a liquid depends mainly on $\Delta H_{vap}{}^\circ$, which is always a positive quantity.
    - $\rightarrow$ *Stronger* intermolecular forces result in a *larger* $\Delta H_{vap}{}^\circ$ and a *decrease* in vapor pressure.
    - $\rightarrow$ For a given liquid, an *increase* in temperature results in an *increase* in vapor pressure.

9.3 The Variation of Vapor Pressure with Temperature

- **Solve the vapor pressure equation for two temperatures** (P_2 at T_2 and P_1 at T_1)

    - $\rightarrow$
$$\ln\frac{P_2}{P_1} = \frac{\Delta H_{vap}{}^\circ}{R}\left(\frac{1}{T_1} - \frac{1}{T_2}\right)$$
    Clausius–Clapeyron equation

- **Uses of the Clausius-Clapeyron equation**
    - $\rightarrow$ Measure the vapor pressure of a liquid at two temperatures (P_2 at T_2 and P_1 at T_1) to *estimate* the standard enthalpy of vaporization $\Delta H_{vap}{}^\circ$.

$$\Delta H_{vap}{}^\circ \approx \left(\frac{R}{\left(T_1^{-1} - T_2^{-1}\right)}\right)\ln\frac{P_2}{P_1}$$

    - $\rightarrow$ Measure $\Delta H_{vap}{}^\circ$ and the vapor pressure of a liquid at one temperature (P_1 at T_1) to *estimate* the vapor pressure at any different temperature (P_2 at T_2).
    - $\rightarrow$ Measure $\Delta H_{vap}{}^\circ$ and the vapor pressure at one temperature (P_1 at T_1). *Estimate* the normal boiling point of the liquid ($T_b = T_2$, when $P_2 = 1$ atm) or the standard boiling point of the liquid (T_2, when $P_2 = 1$ bar).

→ Determine $\Delta H_{vap}°$ and the vapor pressure at the normal boiling temperature (P_1 at T_b). *Estimate* the vapor pressure at a different temperature (P_2 at T_2).

Note: In all cases, the term *estimate* is used because the assumption that $\Delta H_{vap}°$ is independent of temperature is not strictly true; it is an approximation.

9.4 Boiling

- **Liquid in an open container** → Rate of vaporization is greater than the rate of condensation, so the liquid evaporates.
- **Boiling** → Vapor pressure of liquid equals atmospheric pressure.
  Rapid vaporization occurs throughout the entire liquid.
- **Boiling point** → Temperature at which the liquid begins to boil
- **Normal boiling point** → Boiling point at 1 atm, T_b
  Within experimental error, $T_b = 99.974°C \approx 100°C$ for water.
- **Standard boiling point** → Boiling point at 1 bar
- **Effect of pressure on the boiling point**

  → An *increase* of pressure on a liquid leads to an *increase* in the boiling point. Pressure cookers make use of this fact.

  → A *decrease* of pressure on a liquid leads to a *decrease* in the boiling point. This helps explain why water boils at a lower temperature on a mountaintop. It also accounts for vacuum distillation at low temperatures to purify liquids that decompose at higher temperatures.

9.5 Freezing and Melting

- **Freezing and melting** → Two common phase transitions, one the reverse of the other

- **Freezing (melting)** → Solidification of a liquid (liquefaction of a solid)

- **Freezing (melting) point** → Temperature at which liquid freezes (solid melts)
  For a given substance, freezing and melting points are identical.

- **Normal freezing point, T_f** → Temperature at which solid begins to freeze at 1 atm
  Within experimental error, $T_f = 0°C$ and 1 atm for water.

- **Standard freezing point** → Temperature at which solid begins to freeze at 1 bar
  Difference between T_f and the standard freezing point is very small.

- **Supercooled liquid**

  → A pure liquid may exist below its freezing point if it is cooled extremely slowly.

  → Thermodynamically unstable with respect to formation of solid

  → If heat is withdrawn slowly from pure water, the temperature may drop below 0°C, indicating supercooling. At some point, a tiny crystal (nucleus) of ice forms, and the entire sample crystallizes rapidly to form ice whose temperature rises from the heat released by sudden freezing.

- **Pressure dependence of the freezing point:**

  Most substances → Solid phase is denser than the liquid phase (smaller molar volume), so the solid *sinks* as it forms. Increasing pressure favors the phase with the smallest density; thus, the freezing point *increases* with *increasing* pressure.

Few substances → Solid phase is less dense than the liquid phase (larger molar volume) and the solid *floats* as it forms. An increase in pressure favors the phase with the greatest density; thus, the freezing point *decreases* with *increasing* pressure. **Examples:** Water and bismuth

Note: The *anomalous* behavior of water is critical for the survival of certain species. If ice sank, bodies of water would freeze solid in winter, killing aquatic life.

9.6 Phase Diagrams

- **Component**
  - → A single substance **Examples:** Aluminum, octane, water, sodium chloride
  - → A chemically independent species

- **Single-component phase diagram** → Map showing the most stable phase of a substance at different pressures and temperatures

- **Phase boundary**
  - → Lines or curves separating regions on a phase diagram
  - → Represents a set of P and T values for which *two* phases coexist in dynamic equilibrium

- **Triple point**
  - → Point where *three* phase boundaries intersect
  - → Corresponds to a single value of P and T for which *three* phases coexist in dynamic equilibrium

- **Critical point** → High-temperature terminus of the liquid-vapor phase boundary

- **Critical temperature, T_c** → Temperature above which a gas cannot condense into a liquid
  Only one phase is observed above T_c.

Note: Critical point and critical temperature are introduced in Section 9.7 in the text. The definitions given above are repeated later.

Phase Diagram of Water

Note the regions (bounded areas) labeled ice (solid), liquid, and vapor (gas). In each area, only a single phase (ice, water, or water vapor) is stable. Within each region, pressure and/or temperature may be varied independently with no accompanying phase transition. Two independent variables, or *two* degrees of freedom, characterize such a region. The three lines that separate the regions define the phase boundaries of solid-liquid, liquid-vapor, and solid-vapor. At a boundary, only P or T may be varied if the two phases are to remain in equilibrium. Only one independent variable, or *one* degree of freedom, exists for water in states corresponding to the boundary. All three phases coexist at the *triple point* with its specific P and T, or *zero* degrees of freedom.

- **Slope of the solid-liquid phase boundary line**
  - → Positive for most substances:
    Solid sinks because it is more dense than liquid.
    At constant temperature, a pressure increase yields no phase change for the solid.
    At constant temperature, a pressure increase may cause the liquid to solidify.
  - → Negative for H_2O:
    Because the solid is less dense than the liquid, ice floats.
    At constant temperature, a pressure increase may cause ice to melt.
    At constant temperature, a pressure increase yields no phase change for the liquid.

- **Increasing the pressure on graphite to make diamond and liquid carbon**
  - → What occurs when the pressure on graphite at 3000 K is increased from 1 bar to 500 kbar?

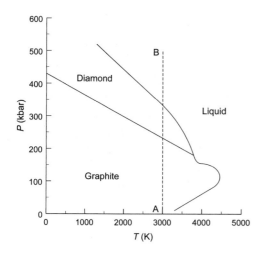

In the initial state of graphite (point A), one phase exists. Temperature is held constant (3000 K), and the pressure is allowed to vary freely. The C(graphite)-C(diamond) phase boundary is reached at approximately 230 kbar. At this point, C(diamond) forms and exists in equilibrium with C(graphite). All the C(graphite) is eventually converted to C(diamond). (This process may be quite slow.) In the diamond region, one phase exists. Additional increases in the pressure on C(diamond) result in the equilibrium of C(diamond) and liquid carbon at about 400 kbar. When all the diamond has melted, the pressure on C(l) is free to increase to attain the final state (point B).

Note: There are three allotropes of carbon: graphite, diamond, and buckminsterfullerene (C_{60}); the latter, discovered in 1985, is composed of soccer-ball-shaped molecules. The thermodynamic stability of buckminsterfullerene has not yet been determined. The validity of its inclusion on the C phase diagram is, therefore, uncertain. (Metastable phases, such as supercooled water, do not appear on phase diagrams.) The crystal structure is face-centered cubic with C_{60} molecules at the corners and faces of a cubic unit cell. The unit cell is shown below.

9.7 Critical Properties

- **Critical point** → Terminus of the liquid-gas phase boundary at high temperature

- **Critical temperature** → The temperature above which a vapor (gas) cannot condense into a liquid. Only one phase is observed above T_c.

- **Supercritical fluid** → Substance above its critical temperature, T_c

Phase Diagram of Water

This discontinuous diagram shows both low and high pressure regions. Point A represents typical "room temperature" conditions. Liquid water has a vapor pressure of 23.76 Torr (0.031 26 atm) at 25°C (298.15 K). At 200°C (473.15 K), the vapor pressure is 15 atm (point B). As the temperature is increased further, the densities of the liquid and vapor in equilibrium approach one another and become nearly equal at the critical point (point C). Above the critical temperature, only one phase with the properties of a very dense vapor remains. The *critical pressure*, P_c, is the vapor pressure measured at the critical temperature. The temperature, pressure, and density values for water at the critical point are $T_c = 374.1$°C (647 K), $P_c = 218.3$ atm, and $d_c = 0.32$ g·cm^{-3}, respectively.

- **Intermolecular forces** → Both T_c and T_b have a tendency to *increase* with the *increasing* strength of intermolecular forces.

Substance	T_c (K)	T_b (K)
He (helium)	5.2	4.3
Ar (argon)	150	88
Xe (xenon)	290	166
NH$_3$ (ammonia)	405	240
H$_2$O (water)	647	373

Note: The strength of the London forces *increases* with increasing molar mass (number of electrons) of a noble gas atom. With *stronger* intermolecular forces, *higher* critical and normal boiling temperatures are expected. Ammonia and water form strong hydrogen bonds. Water forms more hydrogen bonds per molecule than does ammonia. Therefore, water has critical and normal boiling temperatures *higher* than those of ammonia. In making these comparisons, care must be taken.

SG-104

SOLUBILITY (Sections 9.8–9.13)

9.8 The Limits of Solubility

- **Two component solution** → Contains one solvent and one solute species
  Example: NaCl (solute) in water (solvent)

- **Interactions** → Solvent-solvent, solute-solute, and solute-solvent

- **Unsaturated solution**
  - → All solute added to solvent dissolves.
  - → Amount of solute dissolved in the solvent is *less* than the equilibrium amount.

- **Saturated solution**
  - → Solubility limit of solute has been reached, with any additional solute present as a precipitate.
  - → The *equilibrium* amount of solute has been dissolved in solvent.
  - → Dissolved and undissolved solute molecules are in *dynamic* equilibrium; in other words, the rate of dissolution equals the rate of precipitation.

- **Supersaturated solution**
  - → Under certain conditions, an amount of solute greater than the equilibrium amount (solubility limit) can be dissolved in a solvent.
  - → Because *more* than the equilibrium amount is dissolved, the system is thermodynamically unstable.
  - → A slight disturbance (seed crystal) causes precipitation and rapid return to equilibrium.

- **Solubility limit** → Depends on the nature of both the solute and the solvent

- **Molar solubility** → Molar concentration of a saturated solution of a substance
  Units: (moles of solute)/(liter of solution) or $mol \cdot L^{-1} \equiv M$
  Example: A saturated aqueous solution of AgCl has a molar concentration of 1.33×10^{-5} $mol \cdot L^{-1}$ or 1.33×10^{-5} M at 25°C.

- **Gram solubility** → Mass concentration of a saturated solution of a substance
  Units: (grams of solute)/(liter of solution) or $g \cdot L^{-1}$
  Example: A saturated aqueous solution of AgCl has a mass concentration of 1.91×10^{-3} $g \cdot L^{-1}$ at 25°C.

- **Molal solubility** → Molality of a saturated solution of a substance
  Units: (moles of solute)/(kg of solvent) or $mol \cdot kg^{-1}$
  Example: A saturated aqueous solution of AgCl has a molality of 1.33×10^{-5} $mol \cdot kg^{-1}$ at 25°C.

  Note: For dilute solutions, the molar and molal solubilities will be essentially equal as in the examples above. This is because 1 L of water has a mass very nearly equal to 1 kg at 25°C and the mass of the solute is very small compared with that of the solvent. In concentrated solutions, the mass of the solute becomes an appreciable fraction of the mass of the solution, and a liter of solution contains substantially less than 1 kg of water. In such a solution, molality and molarity can differ considerably.

9.9 The Like-Dissolves-Like Rule

- **Rule** $\rightarrow$ If solute-solute and solvent-solvent intermolecular forces (London, dipole, hydrogen-bonding, or ionic) are similar, larger solubilities are predicted. *Lesser* solubilities are expected if these forces are dissimilar. **Or,** *like dissolves like.*

- **Use** $\rightarrow$ A *qualitative* guide to predict and understand the solubility of various solute species in different solvents

    Example: Oil is composed of long chain hydrocarbons. Oil and water do not mix. Oil molecules are held together by London forces, whereas water associates primarily by hydrogen bonds. The like-dissolves-like rule suggests a solvent with only cohesive London forces is needed to dissolve oil. Gasoline, a mixture of shorter chain hydrocarbons such as heptane and octane, is a possible solvent, as is benzene, an unsaturated cyclic hydrocarbon.

    Example: Glucose, $C_6H_{12}O_6$, has five $-OH$ groups capable of forming hydrogen bonds. It is soluble in hydrogen-bonding solvents such as water, and insoluble in nonpolar solvents such as hexane.

    Example: Potassium iodide, KI, is an ionic compound. Because both water and ammonia are highly polar molecules, they are expected to be effective in solvating both K^+ and I^- ions. Ethyl alcohol molecules are less polar than water, so KI is expected to be less soluble in ethanol than in water. A similar argument can be used to explain the *slight* solubility of KI in acetone.

- **Hydrophilic** $\rightarrow$ Water-attracting

- **Hydrophobic** $\rightarrow$ Water-repelling

- **Soaps** $\rightarrow$ Long chain molecules, with hydrophobic and hydrophilic ends, which are soluble in both polar and nonpolar solvents

- **Surfactant** $\rightarrow$ Surface-active molecule with a hydrophilic head and a hydrophobic tail
    Examples: Soaps and detergents are surfactants.

- **Micelle** $\rightarrow$ Spherical aggregation of surfactant molecules with hydrophobic ends in the interior and hydrophilic ends on the surface

    Note: The term hydrophobic is somewhat of a misnomer. In actuality, an *attraction* between solute and solvent molecules *always* exists. In the case of water, the solute-solvent interaction disrupts the local structure of water, which is dominated by hydrogen bonding. The stronger the solute$-H_2O$ interaction, the greater the likelihood that the local solvent structure will be disrupted by hydrated solute molecules, increasing the solubility of the solute.

9.10 Pressure and Gas Solubility: Henry's Law

- **Pressure dependence of gas solubility:**
  Qualitative features
    $\rightarrow$ For a gas and liquid in a container, an increase in gas pressure leads to an increase in its solubility in the liquid.
    $\rightarrow$ Gas molecules strike the liquid surface and some dissolve. An increase in gas pressure leads to an increase in the number of impacts per unit time, thereby increasing solubility.

→ In a gas mixture, the solubility of each component depends on its partial pressure because molecules strike the surface independently of one another.

Quantitative features

→ Henry's law $\boxed{s = k_H P}$

→ The solubility, s, of a gas in a liquid is directly proportional to the partial pressure, P, of the gas above the liquid.

→ $s \equiv$ molar solubility (see Section 9.8)

→ $k_H \equiv$ Henry's law constant, a function of T (see Section 9.11), the gas, and the solvent

→ Units: s (mol·L^{-1}), k_H (mol·L^{-1}·atm^{-1}), and P (atm)

Example: Estimate the molar solubility and gram solubility of O_2 in dry air dissolved in a liter of water open to the atmosphere at 20°C. Assume that air is 20.95% O_2 by volume. The pressure of the atmosphere is 1 atm, thus the partial pressure of O_2 is 0.2095 atm.

$$s(O_2) = (1.3 \times 10^{-3} \text{ mol·L}^{-1}\text{·atm}^{-1})(0.2095 \text{ atm}) = 2.7 \times 10^{-4} \text{ mol·L}^{-1}$$

$$s \times M = (2.7 \times 10^{-4} \text{ mol·L}^{-1})(32.0 \text{ g·mol}^{-1}) = 8.6 \times 10^{-3} \text{ g·L}^{-1}$$

9.11 Temperature and Solubility

• **Dependence of molar solubility on temperature**

→ For solid and liquid solutes, solubility *usually* increases with increasing temperature, but for some salts, for example Li_2CO_3, solubility decreases with increasing temperature (see Fig. 9.22 in the text).

→ For gaseous solutes, solubility *usually* decreases with increasing temperature.

→ Despite these trends, solubility behavior can sometimes appear complex, as with sodium sulfate, Na_2SO_4, whose solubility in water increases then decreases (see Fig. 9.22 in the text). In this instance, complex solubility behavior arises because different hydrated forms of sodium sulfate precipitate at different temperatures.

Examples:

1. The solute that precipitates from a saturated solution may be different from that of the solid initially dissolved. At room temperature, anhydrous potassium hydroxide, KOH(s), readily dissolves in water. The solid that precipitates from a saturated solution of KOH is the dihydrate, KOH·2 H_2O. Therefore, the chemical equilibrium in the saturated solution is

    $$KOH \cdot 2\,H_2O(s) \rightleftharpoons K^+(aq, \text{ saturated}) + OH^-(aq, \text{ saturated}).$$

2. From the diagram on the next page, notice that between −50 and 160°C, the solubility of LiCl displays three discontinuities, corresponding to temperatures at which one solid hydrated form is transformed into another. The four regions between the three discontinuities correspond, from low to high temperature, to precipitation of LiCl·3 H_2O, LiCl·2 H_2O, LiCl·H_2O, and LiCl, respectively. Note also that saturated aqueous solutions freeze at lower temperatures and boil at higher temperatures than pure water (see Section 9.16), accounting for the large temperature range of these solutions.

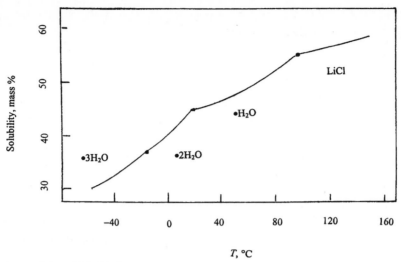

9.12 The Enthalpy of Solution

- **Solutions of ionic substances** $\rightarrow A_mB_n(s) \rightarrow mA^{n+}(aq) + nB^{m-}(aq)$

- **Enthalpy of solution, ΔH_{sol}**
  - $\rightarrow$ Enthalpy change per mole of substance dissolved
  - $\rightarrow$ Depends on *concentration* of solute

- **Limiting enthalpy of solution**
  - $\rightarrow$ Refers to the formation of a very dilute solution
    Values are given in Table 9.6 of the text.
  - $\rightarrow$ Use of limiting enthalpy of solution avoids complications arising from interionic interactions that occur in more concentrated solutions because ions are far apart in very dilute solutions.
  - **Note:** All topics that follow refer to the limiting enthalpy condition.

- **Nature of the formation of solutions of ionic substances**
  - $\rightarrow$ Conceptualized as a two-step process: sublimation of an ionic solid to form gaseous ions followed by the solvation of the gaseous ions to form an ionic solution
  - $\rightarrow$ The enthalpy change for the sublimation step is designated the lattice enthalpy, ΔH_L (see Table 7.6 in the text).
  - $\rightarrow$ The enthalpy change for the second step (formation of hydrated ions from the gas phase ions) is designated as the enthalpy of hydration, ΔH_{hyd} (see Table 9.7 in the text).
  - $\rightarrow$ ΔH_L *always* has a positive value, whereas ΔH_{hyd} *always* has a negative value.
  - $\rightarrow$ $\Delta H_{sol} = \Delta H_L + \Delta H_{hyd}$ is sum of two numbers of opposite sign, both of which are typically quite large.
  - $\rightarrow$ As a consequence, **limiting enthalpies of solution, ΔH_{sol}, may be positive or negative.**

- **Endothermic process:** $\Delta H_L > |\Delta H_{hyd}|$ and $\Delta H_{sol} < 0$

- **Exothermic process:** $\Delta H_L < |\Delta H_{hyd}|$ and $\Delta H_{sol} > 0$

  Note: For small, highly charged ions, both ΔH_L and $|\Delta H_{hyd}|$ have large values.

9.13 The Gibbs Free Energy of Solution

- **Solutions of ionic substances** $\rightarrow A_mB_n(s) \rightarrow mA^{n+}(aq) + nB^{m-}(aq)$

- **Free energy of solution, ΔG_{sol}**
  - → Free energy change per mole of substance dissolved
  - → Depends on *concentration* of solute
  - → $\Delta G_{sol} = \Delta H_{sol} - T\Delta S_{sol}$ at constant T

- **Nature of solubility**
  - → A substance will dissolve if $\Delta G_{sol} < 0$ and will continue to dissolve until a saturated solution, for which $\Delta G_{sol} = 0$, is obtained.
  - → Both ΔH_{sol} and ΔS_{sol} change with increasing concentration to make ΔG_{sol} more positive.
  - → If the solute is consumed before ΔG_{sol} reaches 0, an unsaturated solution results, with the potential to dissolve additional solute.
  - → With excess solute present, ΔG_{sol} reaches 0 and a saturated solution results.
  - → For endothermic enthalpies of solution, the increase in entropy of solution drives the solubility process.
  - → A substance with a *large* endothermic enthalpy of solution is *usually* insoluble.

- **Nature of solubility**
  - → $\Delta G_{sol} = \Delta H_{sol} - T\Delta S_{sol}$. Focus on entropy changes.
  - → So, for most ionic solutes, solubility *increases* with increasing temperature.
  - → When gases dissolve, their motion is more constricted in the liquid phase, $\Delta S_{sol} < 0$, and gas solubility *decreases* with increasing T.

COLLIGATIVE PROPERTIES (Sections 9.14–9.17)

- **Colligative Properties**
  - → Depend on the relative amounts of solvent and solute, and not on their type.
  - → Include boiling point elevation, freezing point depression, vapor pressure lowering, and osmosis
- **Molality, molarity, mole fraction** → Different concentration units used for quantitative treatment of colligative properties

9.14 Molality

- **Molality, b**
  - → Moles of solute per kilogram of solvent
  - → Independent of temperature (solute and solvent given as masses or mass equivalent)
  - → Units of moles per kilogram ($mol \cdot kg^{-1}$)
  - → Used when relative number of molecules of components is to be emphasized

    For a binary solution, $b_{solute} = \dfrac{n_{solute}}{kilograms_{solvent}}$

  - **Note:** The letter m is used to designate mass and molality. Care should be taken to avoid confusing the two.

- **Mole fraction, x**
  - → Ratio of moles of solute to the total number of moles of all species in a mixture
  - → Independent of temperature

→ Dimensionless quantity

→ Used when relative number of molecules of components is to be emphasized

$$\text{For a binary solution,} \quad x_{\text{solute}} = \frac{n_{\text{solute}}}{n_{\text{solute}} + n_{\text{solvent}}}$$

- **Molarity, c**

  → Moles of solute divided by total volume of *solution*

  → Temperature dependent (volume of solution changes with temperature)

  → Units of moles per liter (mol·L^{-1})

$$\text{For a binary solution,} \quad c_{\text{solute}} = \frac{n_{\text{solute}}}{V_{\text{solution}}}$$

- **Converting from one concentration unit to another**

  → The molar mass of one component is required to convert *molality* to *mole fraction* or *mole fraction* to *molality*.

  → The density of the solution is required to convert *molarity* to *molality* or *vice versa*.

  Note: When converting from one concentration unit to another, it is convenient to assume one of the following: (1) the solution has a volume of one liter (*molarity* ⇔ *molality*), (2) the total amount of solvent and solute is one mole (*mole fraction* ⇔ *molality*), or (3) the mass of the solvent in the solution is one kilogram (*molality* ⇔ *mole fraction*).

9.15 Vapor-Pressure Lowering

- **Qualitative features**

  → The vapor pressure of a solvent in equilibrium with a solution containing a *nonvolatile solute* (for example, sucrose or sodium chloride in water) is lower than that of the pure solvent.

  → For an *ideal solution* or a sufficiently dilute *real* solution, the vapor pressure of any volatile component is proportional to its mole fraction in solution (Raoult's law in words).

- **Quantitative features**

  → Raoult's law $\boxed{P = x_{\text{solvent}} P_{\text{pure}}}$

  → The vapor pressure P of a solvent is equal to the product of its mole fraction in solution, x_{solvent}, and its vapor pressure when pure, P_{pure}. The quantities P and P_{pure} are measured at the same temperature.

- **Ideal solution**

  → A hypothetical solution that obeys Raoult's law exactly for all concentrations of solute

  → Solute-solvent, solute-solute, and solvent-solvent interactions are all the same; therefore, the enthalpy of solution, ΔH_{sol}, is zero.

  → Entropy of solution, $\Delta S_{\text{sol}} > 0$

  → Free energy of solution, $\Delta G_{\text{sol}} < 0$, leading to a lowering of vapor pressure of the solvent (see Section 9.3).

  → The process of solution formation is entropy driven.

→ Solutions approximating ideality are typically formed by similar solute and solvent species, such as hexane and heptane.

→ Real solutions do not obey Raoult's law at all concentrations, but they do follow Raoult's law in the limit of low solute concentration (dilute solution).

→ Raoult's law is a limiting law.

- **Nonideal (real) solution**

  → Solution that does not obey Raoult's law at a certain concentration

  → Solute-solvent interactions differ from solvent-solvent interactions.

  → ΔH_{sol} is not equal to zero.

  → Real solutions approximate ideal behavior at concentrations below 10^{-1} mol·kg^{-1} for *nonelectrolyte* solutions and below 10^{-2} mol·kg^{-1} for *electrolyte* solutions.

  → Real solutions tend to behave ideally as the solute concentration approaches zero.

9.16 Boiling-Point Elevation and Freezing-Point Depression

- **Boiling-point elevation**

  → A nonvolatile solute lowers the vapor pressure of the solvent. As a result, a solution containing a nonvolatile solute will not boil at the normal boiling point of the pure solvent. The temperature must be increased above that value to bring the vapor pressure of the solution to atmospheric pressure. Therefore, boiling is achieved at a higher temperature (the boiling point is elevated).

  → Arises from the influence of the solute on the entropy of the solvent

  → Quantitatively, $\boxed{\text{boiling-point elevation} = k_b \times \text{molality} = k_b \times b_{solute}}$

  → The boiling-point constant, k_b, depends on the solvent and has units of K·kg·mol^{-1} (Table 9.8).

  → The boiling-point elevation equation holds for nonvolatile solutes in dilute solutions that are approximately ideal.

- **Freezing-point depression**

  → Vapor-pressure lowering of a solution decreases the triple-point temperature, the intersection of the liquid-vapor (vapor pressure) and solid-vapor phase boundaries. The solid-liquid phase boundary originating at the triple point is moved slightly to the left on the phase diagram (the solid-vapor boundary is unchanged). The freezing temperature of the solution is thereby lowered (the freezing-point is depressed).

  → Arises from the influence of the solute on the entropy of the solvent

  → Proportional to the molality of the solute

  → Quantitatively, $\boxed{\text{freezing-point depression} = k_f \times \text{molality} = k_f \times b_{solute}}$

  → The freezing-point constant, k_f, depends on the solvent and has units of K·kg·mol^{-1} (Table 9.8).

  → The freezing-point depression equation holds for nonvolatile solutes in dilute solutions that are approximately ideal.

- **Freezing-point depression corrected for ionization or aggregation of the solute**

  → $\boxed{\text{freezing-point depression} = i\,k_f \times \text{molality} = i\,k_f \times b_{solute}}$

→ The van't Hoff factor, i, determined experimentally, is an adjustment used to treat *electrolytes* (e.g., *ionic solids,* strong *acids* and *bases*) that *dissociate* and molecular solutes that *aggregate* (e.g., acetic acid dimers).

→ For electrolytes, i is the number of moles of ions formed by each mole of solute dissolved in 1 kg of solvent if all ions behave independently. In aqueous solutions, this only occurs in very dilute solutions; with increasing concentration, interionic effects cause i to be smaller than the value expected for complete ionization.

→ For molecular aggregates, i may be used to estimate the average size of the aggregates.

9.17 Osmosis

- **Qualitative features**

  → *Osmosis:* The tendency of a solvent to flow through a membrane into a more concentrated solution.

  → Used to determine an unknown molar mass, particularly for large molecules such as polymers and proteins.

  → *Osmometry* is a technique used to determine the molar mass of a solute if the mass concentration is measured.

  → A *semipermeable membrane* allows only certain types of molecules to pass through. Typical membranes allow passage of water and small molecules, but not large molecules or ions.

  → In osmosis, the free energy of the solution is lower than that of the solvent; hence, dilution is a *spontaneous process*. The free energy of the solution can be increased by applying pressure on the solution. The increased pressure at equilibrium is called the *osmotic pressure, Π*.

  → In a static apparatus open to the atmosphere, pressure is applied by the increased height of the raised column of solution caused by the flow of solvent through the membrane into the solution. In a dynamic apparatus in a closed system, pressure is applied by the increased force of a piston confining the solution (preventing solvent flow).

  → If the external pressure $P < \Pi$, osmosis (dilution) is *spontaneous*.
  If $P = \Pi$, the system is at equilibrium (no net flow).
  If $P > \Pi$, *reverse osmosis* occurs (flow of solvent in solution to the pure solvent).
  Reverse osmosis is used to purify seawater.

- **Quantitative features**

  → van't Hoff equation $\boxed{\Pi = iRTc_{\text{solute}}}$

  → c_{solute} = molarity of the solution = $\dfrac{\text{moles of solute}}{\text{liters of solution}}$

  i = van't Hoff factor, R = gas constant, T = temperature in kelvin
  Use units of Π in atm and of R in $\text{L·atm·K}^{-1}\text{·mol}^{-1}$; i is dimensionless.

BINARY LIQUID MIXTURES (Sections 9.18–9.20)

9.18 The Vapor Pressure of a Binary Liquid Mixture

- **Ideal solution with two volatile components A and B; liquid and vapor phases in equilibrium**
  - → Let $P_{A,pure}$ = vapor pressure of pure A and $P_{B,pure}$ = vapor pressure of pure B.
  - → In an ideal solution, each component obeys Raoult's law at all concentrations.
  - → $P_A = x_{A,liquid} P_{A,pure}$, where $x_{A,liquid}$ = mole fraction of A in the liquid mixture
  - → $P_B = x_{B,liquid} P_{B,pure}$, where $x_{B,liquid}$ = mole fraction of B in the liquid mixture
  - → For a binary mixture, $x_{A,liquid} + x_{B,liquid} = 1$.
  - → P_A and P_B are the partial pressures of A and B, respectively, in the vapor above the solution. P_{total} is the total pressure above the solution. Dalton's law of partial pressures for ideal gases is $P_{total} = P_A + P_B$ (see Section 4.8).
  - → Combination of Raoult's law and Dalton's law produces the expression

$$P_{total} = P_A + P_B = x_{A,liquid} P_{A,pure} + x_{B,liquid} P_{B,pure}$$

  - → Nearly ideal solutions with two volatile components are formed when the components are very similar (hexane/octane and benzene/toluene), and both components are completely soluble in each other (miscible). The composition may range from $0 < x < 1$; the labels "solute" and "solvent" are not usually used.

- **Composition of the vapor above a binary ideal solution**
  - → Each vapor component obeys Dalton's law.
  - → $P_A = x_{A,vapor} P$, where $x_{A,vapor}$ = mole fraction of A in the vapor mixture and $P = P_{total}$
  - → $P_B = x_{B,vapor} P$, where $x_{B,vapor}$ = mole fraction of B in the vapor mixture
  - → For a binary mixture, $x_{A,vapor} + x_{B,vapor} = 1$.
  - → Combination of Raoult's law and Dalton's law yields an expression for $x_{A,vapor}$

$$x_{A,vapor} = \frac{P_A}{P} = \frac{P_A}{P_A + P_B} = \frac{x_{A,liquid} P_{A,pure}}{x_{A,liquid} P_{A,pure} + x_{B,liquid} P_{B,pure}}$$

  - → If $P_{A,pure} \neq P_{B,pure}$, then $x_{A,vapor} \neq x_{A,liquid}$ and $x_{B,vapor} \neq x_{B,liquid}$.
    The liquid and vapor compositions differ and the vapor is always *richer* in the more volatile component.

9.19 Distillation

- **Temperature-composition phase diagram (ideal solution)**
  - → Plot of the temperature of an equilibrium mixture *vs.* composition (mole fraction of one component)
  - → Two curves are plotted on the same diagram, T *vs.* x_{liquid} and T *vs.* x_{vapor}.
  - → At a given temperature, the composition of each phase at equilibrium is given.
  - → See Fig. 9.37 in the text for a graphical illustration of a temperature-composition diagram.

- **Distillation** → Purification of a liquid by evaporation and condensation
- **Distillate** → Vapor produced during distillation that is condensed and collected in the final stage
- **Fractional distillation**
  - → Continuous separation (purification) of two or more liquids by repeated evaporation and condensation steps on a vertical column
  - → Each step takes place on a fractionating column in small increments.
  - → Liquid and vapor are in equilibrium at each point in the column, but their compositions vary with height, as does the temperature.
  - → At the top of the column, condensed vapor (distillate) is collected in a series of samples or *fractions*; the most volatile fraction is collected first.
  - → Progress may be displayed on a temperature-composition phase diagram.

9.20 Azeotropes

- **Nonideal solutions with volatile components**
  - → Most mixtures of liquids are not ideal.
  - → For *nonideal solutions* with *volatile* components: Raoult's law is not obeyed, the enthalpy of mixing, $\Delta H_{mix} \neq 0$, and solute-solvent interactions are different from solvent-solvent interactions.

- **Azeotrope**
  - → A solution that, like a pure liquid, distills at a constant temperature without a change in composition.
  - → At the azeotrope temperature, $x_{liquid} = x_{vapor}$ for each component

- **Positive deviation from Raoult's law**
  - → Vapor pressure of the mixture is *greater* than the value predicted by Raoult's law.
  - → Enthalpy of mixing is endothermic, $\Delta H_{mix} > 0$.
  - → Solute-solvent interactions are *weaker* than solute-solute interactions.
  - → See Fig. 9.41(a) for an illustration of the vapor-pressure behavior of a mixture of ethanol and benzene. To make an ethanol-benzene solution, strong hydrogen bonds in ethanol are broken and replaced by weaker ethanol-benzene London interactions. Weaker interactions *increase* the vapor pressure and *decrease* the boiling temperature.

- **Minimum-boiling azeotrope**
  - → Forms if the minimum boiling temperature is less than that of each pure liquid
  - → Distillation yields azeotrope as the distillate (see Fig. 9.42 in text).

- **Negative deviation from Raoult's law**
  - → Vapor pressure of the mixture is *smaller* than the value predicted by Raoult's law.
  - → Enthalpy of mixing is exothermic, $\Delta H_{mix} < 0$.
  - → Solute-solvent interactions are *stronger* than solute-solute interactions

→ See Fig. 9.41(b) in the text for an illustration of the vapor-pressure behavior of a mixture of acetone and chloroform. In the acetone-chloroform solution, strong interactions, similar to hydrogen bonding, occur between a lone pair of electrons on the O atom of acetone and the H atom of chloroform. Stronger solute-solvent interactions *decrease* the vapor pressure and *increase* the boiling temperature.

- **Maximum-boiling azeotrope**
  → Forms if the maximum boiling temperature is greater than each of the pure liquids
  → Distillation yields one pure component as the distillate (see Fig. 9.43 in the text).

IMPACT ON BIOLOGY AND MATERIALS (Sections 9.21–9.22)

9.21 Colloids

- **Colloids**
  → Dispersion of large particles in a solvent
  → Particles (*dispersed phase*) with lengths or diameters between 1 nm and 1 μm *dispersed* or *suspended* in a *dispersion medium* (gas, liquid, or solid solvent)
  → *Classification of colloids* is given in Table 9.9 in the text.
  → *Aerosols* are solids dispersed in a gas (smoke) *or* liquids dispersed in a gas (hairspray, mist, fog).
  → *Sols* or *gels* are solids dispersed in a liquid (printing ink, paint).
  → *Emulsions* are liquids dispersed in a liquid (milk, mayonnaise).
  → *Solid emulsions* are liquids dispersed in a solid (ice cream).
  → *Foams* are gases dispersed in a liquid (soapsuds). *Solid foams* are gases dispersed in a solid (Styrofoam).
  → *Solid dispersions* are solids dispersed in a solid (ruby glass, some alloys).

- **Aqueous colloids**
  → *Hydrophilic colloids* contain molecules with *polar groups*.
    Examples: Proteins that form gels and puddings
  → *Hydrophobic colloids* contain molecules with *nonpolar groups* only weakly attracted to water.
    Examples: Fats that form emulsions (milk and mayonnaise).
  → Rapid mixing of silver nitrate and sodium bromide solutions may produce a *hydrophobic colloidal suspension* rather than a precipitate of silver bromide. The tiny silver bromide particles are kept from further aggregation by *Brownian motion,* the motion of small particles resulting from constant collisions with solvent molecules. The *sol* is further stabilized by *adsorption* of ions on the surfaces of the particles. The adsorbed ions are hydrated by surrounding water molecules and help prevent further aggregation.

9.22 Bio-based Materials and Biomimetic Materials

- **Bio-based materials**
  → Materials taken from or made from natural materials in living things
  → Packing pellets made from corn and soybeans, polylactic acid (a polymer used to make plastic packaging), and various kinds of pharmaceuticals

Text Example: *Hyaluronic acid* (Fig. 9.48), a major component of the fluid that lubricates joints; plays a role in the repair of tissues, especially the skin. With its many –OH groups, hyaluronic acid forms hydrogen bonds with water, and as it moves through the body, many water molecules move with it. This material is used in sport medicine to heal injuries by reducing inflammation.

- **Biomimetic materials**
    - → Materials that are modeled after naturally occurring materials
    - → Gels or flexible polymers modeled after natural membranes and tissues
    - → Surfactant compounds called *phospholipids* are found in fats and form the membranes of living cells. These membranes are double layers of phospholipid molecules that line up with their hydrocarbon tails pointing into the membrane and their polar head groups forming the membrane surface.
    - → *Liposomes* are artificial membranes shaped like tiny bags, which can be used to encapsulate drug molecules for delivery to different organs and regions of the body.

Chapter 10   CHEMICAL EQUILIBRIA

REACTIONS AT EQUILIBRIUM (Sections 10.1–10.5)

10.1 The Reversibility of Reactions

- **Chemical reaction**
    - → Reaction mixture approaches a state of dynamic equilibrium.
    - → At equilibrium, the rates of the forward and reverse reactions are equal.

    Example: $H_2(g)$, $I_2(g)$, and $HI(g)$ in a closed container

Forward reaction:	$H_2(g) + I_2(g) \rightarrow 2HI(g)$
Reverse reaction:	$2HI(g) \rightarrow H_2(g) + I_2(g)$
Dynamic equilibrium:	$H_2(g) + I_2(g) \rightleftharpoons 2HI(g)$ *or* $2HI(g) \rightleftharpoons H_2(g) + I_2(g)$

    At chemical equilibrium, the reaction may be represented either way.

- **Note:** The approach to equilibrium may be too slow to measure. Such a system may appear to be at equilibrium, but is not.

    Examples: The sugar glucose, exposed to the air, is not in equilibrium with its combustion products, CO_2 and H_2O.

    Methane burning on a gas range is not at equilibrium with combustion products CO_2 and H_2O. Methane is supplied continuously and the reverse reaction does not occur.

10.2 Equilibrium and the Law of Mass Action

- **Law of mass action**
    - → The composition of a reaction mixture can be expressed in terms of an *equilibrium constant K*, which is unitless.
    - → The equilibrium constant K has the form:

    $$K = \left\{ \frac{\text{activities of products}}{\text{activities of reactants}} \right\}_{\text{equilibrium}}$$

- *Activity* **of substance J, a_J**
    - → Partial pressure or concentration of a substance relative to its standard value
    - → Pure number that is unitless
    - → Idealized systems

    $$a_J = \frac{P_J}{P_J{}^\circ} \quad \text{for an ideal gas}$$

    partial pressure of the gas (bar) divided by the pressure of the gas in its standard state (1 bar)

    $$a_J = \frac{[J]}{[J]^\circ} \quad \text{for a solute in a dilute solution}$$

    molarity of substance J divided its molarity in the standard state (1 mol·L^{-1})

    $a_J = 1$ for a pure solid or liquid

→ Because the standard state values of activity are all defined as unity, they are omitted in the expression for activity and in equilibrium constant expressions.

$$a_J = \frac{P_J}{P_J{}^\circ} = \frac{P_J}{1} = P_J \quad \text{or} \quad \boxed{a_J = P_J \ \text{for an ideal gas}}$$

Similarly, $\boxed{a_J = [J] \ \text{for a solute in a dilute solution}}$ and

$$\boxed{a_J = 1 \ \text{for a pure solid or liquid}}$$

- *Activity coefficient, γ_J*
  → Pure number that is unitless
  → Equal to 1 for ideal gases and solutes in dilute solutions
  → Accounts for intermolecular interactions
  → Used for concentrated solutions and gases, which do not behave ideally

  $a_J = \gamma_J P_J \quad$ for a real gas

  $a_J = \gamma_J [J] \quad$ for any concentration of solute in a solution

  $a_J = 1 \qquad$ for a pure solid or liquid

- **Form of the equilibrium constant**
  → For a general reaction: $a\,\mathrm{A} + b\,\mathrm{B} \rightarrow c\,\mathrm{C} + d\,\mathrm{D}$

  $$\boxed{K = \left(\frac{a_C^c\, a_D^d}{a_A^a\, a_B^b} \right)_{\text{equilibrium}}} \qquad \text{law of mass action, } K, \text{ is unitless}$$

  → K is the ratio of the activities of the products raised to powers equal to their stoichiometric coefficient to the corresponding expression for the reactants.
  → Use activities given in the boxed equations above to write expressions for K.
  → Distinguish between *homogeneous* and *heterogeneous* equilibria.

- **Homogeneous equilibria** → Reactants and products all in the same phase

- **Heterogeneous equilibria** → Reactants and products with different phases
  Activities of pure solids and liquids are set equal to 1.

- **Note:** For reactions in solution, expressions for equilibrium constants should be obtained by using *net ionic equations*.

  Example: Write the equilibrium constant for the reaction $2\,\mathrm{HI\,(g)} \rightleftharpoons \mathrm{H_2\,(g)} + \mathrm{I_2\,(g)}$. Reactants and products are all gases whose activities are equal to their partial pressures, P_j. Therefore,

  $$K = \left(\frac{P_{H_2} P_{I_2}}{(P_{HI})^2} \right)_{\text{at equilibrium}}$$

10.3 The Thermodynamic Origin of Equilibrium Constants

- **ΔG_r as a function of the composition of the reaction mixture**
  → $G_m(J)$ is the *molar free energy* of the reactant or product J.
  → $G_m{}^\circ(J)$ is the *molar free energy* of the reactant or product J *in the standard state*.

$\rightarrow$ $\boxed{G_m(J) = G_m^\circ(J) + RT \ln a_J}$ for any substance in any state

$\rightarrow$ $\boxed{G_m(J) = G_m^\circ(J) + RT \ln P_J}$ for ideal gases (from Section 9.3)

$\rightarrow$ $\boxed{G_m(J) = G_m^\circ(J) + RT \ln[J]}$ for solutes in dilute solution

$\rightarrow$ $\boxed{G_m(J) = G_m^\circ(J)}$ for pure solids and liquids ($a = 1$)

$\rightarrow$ In *dilute* aqueous solutions, the activity of water may be approximated as $a(H_2O) = 1$.

- **ΔG_m for changes in pressure or concentration in idealized systems**
  - $\rightarrow$ Standard state of a substance is its pure form at a pressure of 1 bar.
  - $\rightarrow$ For a solute, the standard state is for a concentration of 1 mol·L^{-1}.

- **ΔG_r for chemical reactions (see Section 8.13)**
  - $\rightarrow$ Generalized chemical reaction: $a A + b B \rightarrow c C + d D$

  $\rightarrow$ $\boxed{\Delta G_r^\circ = \sum n G_m^\circ(\text{products}) - \sum n G_m^\circ(\text{reactants})}$

  - $\rightarrow$ ΔG_r°, the *standard reaction free energy*, is the difference in molar free energies of the products and reactants in their standard states.

  - $\rightarrow$ The stoichiometric coefficients, n, are used as pure numbers (molar convention) to obtain ΔG_r° in kJ·mol^{-1}. The same convention is used for ΔG_r below.

  $\rightarrow$ Evaluated by using $\boxed{\Delta G_r^\circ = \sum n \Delta G_f^\circ(\text{products}) - \sum n \Delta G_f^\circ(\text{reactants})}$

  - $\rightarrow$ Standard free energies of formation, ΔG_f°, for a variety of substances are given in Appendix 2A.

  $\rightarrow$ $\boxed{\Delta G_r = \sum n G_m(\text{products}) - \sum n G_m(\text{reactants})}$

  - $\rightarrow$ ΔG_r is the reaction free energy at any definite, fixed composition of the reaction mixture.

  $\rightarrow$ Evaluated by using $\boxed{\Delta G_r = \Delta G_r^\circ + RT \ln Q}$

  - $\rightarrow$ Q is called the **reaction quotient.**

- **Expressions for Q for a general reaction: $a A + b B \rightarrow c C + d D$**

  $\blacktriangleright$ Real systems: $\boxed{Q = \dfrac{a_C^c a_D^d}{a_A^a a_B^b}}$

  $\rightarrow$ For ideal gases or dilute solutions, Q reduces to:

  $\blacktriangleright$ Ideal gases: $\boxed{Q = \dfrac{P_C^c P_D^d}{P_A^a P_B^b}}$ Q is unitless.

  $\blacktriangleright$ Dilute solutions: $\boxed{Q = \dfrac{[C]^c [D]^d}{[A]^a [B]^b}}$ Q is unitless.

- **Chemical equilibrium**
  - → At equilibrium, $\Delta G_r = 0$ and $Q = K$ (equilibrium constant)
  - → K has the same form as Q, but it is determined uniquely by the equilibrium composition of the reaction system.

  - → Real systems: $$K = \left(\frac{a_C^c a_D^d}{a_A^a a_B^b} \right)_{equilibrium}$$ law of mass action, K is dimensionless (no units).

  - → Activities can be expressed in concentration units multiplied by an activity coefficient γ.

  - → For a gas, $a_J = \frac{\gamma_J P_J}{P^\circ} = \gamma_J P_J$ numerically. For a solute, $a_J = \frac{\gamma_J [J]}{[J]^\circ} = [J]$ numerically.

- **Evaluation of K from ΔG_r° and *vice versa*** → $\boxed{\Delta G_r^\circ = -RT \ln K}$

10.4 The Extent of Reaction

- **Thermodynamically,**

$$\Delta G_r^\circ = \Delta H_r^\circ - T\Delta S_r^\circ = -RT \ln K; \quad \ln K = -\frac{\Delta H_r^\circ}{RT} + \frac{\Delta S_r^\circ}{R} \quad \textbf{Solve for } K$$

$$K = e^{-\Delta H_r^\circ / RT} e^{\Delta S_r^\circ / R}$$

- **In a chemical reaction, products are favored if**
  - → K is large.
  - → The reaction is exothermic in the standard state (negative ΔH_r°, the more negative the better).
  - → The reaction has a positive standard entropy of reaction ($\Delta S_r^\circ > 0$, final state more random).
  - → Similarly, reactants are favored for endothermic reactions (positive ΔH_r°) with negative entropy change ($\Delta S_r^\circ < 0$)

- **Guidelines for attaining equilibrium**
  - → If $K > 10^3$, products are favored.
  - → If $10^{-3} < K < 10^3$, neither reactants nor products are strongly favored.
  - → If $K < 10^{-3}$, reactants are favored.

    Example: At a certain temperature, suppose that the equilibrium concentrations of nitrogen and ammonia in the reaction

    $$N_2(g) + 3 H_2(g) \rightleftharpoons 2 NH_3(g)$$

    were found to be 0.11 M and 1.5 M, respectively. What is the equilibrium concentration of H_2 if $K_c = 4.4 \times 10^4$?

    The value of K_c is in the range that favors products. Solving the equilibrium constant expression requires the evaluation of the cube root of the concentration of H_2. Use the y^x key on your calculator.

    $$K_c = \left(\frac{[NH_3]^2}{[N_2][H_2]^3} \right)_{eq} \quad \text{and} \quad [H_2]^3 = \frac{[NH_3]^2}{K_c[N_2]}$$

$$[H_2] = \sqrt[3]{\frac{[NH_3]^2}{K_c[N_2]}} = \left(\frac{(1.5)^2}{(4.4 \times 10^4)(0.11)}\right)^{1/3} = \left(4.65 \times 10^{-4}\right)^{1/3} = 0.077 \text{ M}$$

Note that the product ammonia is favored as expected.

10.5 The Direction of Reaction

- **For the general reaction** $a \text{A}(g) + b \text{B}(g) \rightleftharpoons c \text{C}(g) + d \text{D}(g)$

$$K = \left(\frac{a_C^c a_D^d}{a_A^a a_B^b}\right)_{\text{equilibrium}}$$ and $$Q = \left(\frac{a_C^c a_D^d}{a_A^a a_B^b}\right)_{\text{not at equilibrium}}$$

- **Guidelines for approach to equilibrium**

  → Compare Q with K.

  → If $Q > K$, the amount of products is too high or the amount of reactants is too low. Reaction proceeds in the reverse direction, toward reactants (←).

  → If $Q < K$, the amount of reactants is too high or the amount of products is too low. Reaction proceeds in the forward direction to form products (→).

  → If $Q = K$, reactants and products are at equilibrium and no observable change occurs (⇌).

EQUILIBRIUM CALCULATIONS (Sections 10.6–10.8)

10.6 The Equilibrium Constant in Terms of Molar Concentrations of Gases

- **Gas-phase equilibria (ideal gas reaction mixture)**

  → $a \text{A}(g) + b \text{B}(g) \rightleftharpoons c \text{C}(g) + d \text{D}(g)$ (general gas phase reaction at equilibrium)

  → $$K = \left(\frac{P_C^c P_D^d}{P_A^a P_B^b}\right)_{\text{equilibrium}}$$ ideal gases (activity = partial pressure)

  → $$K_c = \left(\frac{[C]^c [D]^d}{[A]^a [B]^b}\right)_{\text{equilibrium}}$$ ideal gases (activity = molar concentration)

  → Both K and K_c are unitless..

  → In general, $P_J = \dfrac{n_J RT}{V} = RT \dfrac{n_J}{V} = RT [J]$.

  → Relationship between K and K_c as derived in the text: $\boxed{K = (RT)^{\Delta n} K_c}$

  → $\Delta n = (c + d) - (a + b) = \Sigma(\text{coefficients of products}) - \Sigma(\text{coefficients of reactants})$

- **Mixed-phase equilibria**

  → $a\,A\,(g) + b\,B\,(aq) \rightleftharpoons c\,C\,(g) + d\,D\,(aq)$ (some gases, some solution species at equilibrium)

  → $\boxed{K = \left(\dfrac{P_C^{\,c}\,[D]^d}{P_A^{\,a}\,[B]^b} \right)_{\text{equilibrium}}}$ $\qquad K = (RT)^{\Delta n_g}\,K_c$

  → $\Delta n_g = c - a = \sum(\text{coefficients of } gaseous \text{ products}) - \sum(\text{coefficients of } gaseous \text{ reactants})$

  Note: Δn_g can be positive, negative, or zero.

10.7 Alternative Forms of the Equilibrium Constant

- **Dependence of K on the direction of reaction and the balancing coefficients**

  → $a\,A\,(g) + b\,B\,(g) \rightleftharpoons c\,C\,(g) + d\,D\,(g)$ $\Rightarrow$ $K_1 = \left(\dfrac{a_C^{\,c}\,a_D^{\,d}}{a_A^{\,a}\,a_B^{\,b}} \right)_{\text{equilibrium}}$

  → $c\,C\,(g) + d\,D\,(g) \rightleftharpoons a\,A\,(g) + b\,B\,(g)$ $\Rightarrow$ $K_2 = \left(\dfrac{a_A^{\,a}\,a_B^{\,b}}{a_C^{\,c}\,a_D^{\,d}} \right)_{\text{equilibrium}} = \dfrac{1}{K_1} = K_1^{-1}$

  → $na\,A\,(g) + nb\,B\,(g) \rightleftharpoons nc\,C\,(g) + nd\,D\,(g)$ $\Rightarrow$ $K_3 = \left(\dfrac{a_C^{\,nc}\,a_D^{\,nd}}{a_A^{\,na}\,a_B^{\,nb}} \right)_{\text{equilibrium}} = K_1^{\,n}$

  → Multiply all coefficients of a reaction by n, raise the equilibrium constant to the nth power

- **Combining chemical reactions**
  → Adding reaction 1 with equilibrium constant K_1 to reaction 2 with equilibrium constant K_2 yields a combined reaction with equilibrium constant K that is the *product* of K_1 and K_2.
  $(\Delta G_1° + \Delta G_2°)$ leads to $K = K_1 \times K_2$
  → Subtracting reaction 2 with individual equilibrium constant K_2 from reaction 1 with equilibrium constant K_1 yields a combined reaction with equilibrium constant K that is the *quotient* of K_1 and K_2.

  $(\Delta G_1° - \Delta G_2°)$ leads to $K = \dfrac{K_1}{K_2}$

- **Summary:** Add two reactions $(1 + 2)$, multiply equilibrium constants $(K_1 \times K_2)$.
  Subtract two reactions $(1 - 2)$, divide equilibrium constants (K_1 / K_2).

10.8 Using Equilibrium Constants

- **Equilibrium table** → Very helpful in solving all types of equilibrium problems
- **Format and use of the equilibrium table:**
  Step 1 Reactant and product species taking part in the reaction are identified.
  Step 2 The initial composition of the system is listed.
  Step 3 Changes needed to reach equilibrium are listed in terms of *one unknown quantity*.
  Step 4 The equilibrium composition in terms of one unknown is entered on the last line.

Step 5   The equilibrium composition is entered into the equilibrium constant expression and the unknown quantity (equilibrium constant or reaction species concentration) is determined.

Step 6   Approximations may sometimes be used to simplify the calculation of the unknown quantity in step 5.

- **Calculating equilibrium constants from pressures or concentrations**

    Example: A sample of $HI(g)$ is placed in a flask at a pressure of 0.100 bar. After equilibrium is attained, the partial pressure of $HI(g)$ is 0.050 bar. Evaluate K for the reaction

    $$2\,HI(g) \rightleftharpoons H_2(g) + I_2(g)$$

    (a) Set up an equilibrium table. In the initial system, line 1, there are no products so some must form to reach equilibrium.

    (b) On line 2, let x be the change in pressure of I_2 required to reach equilibrium. From the stoichiometric coefficients, the change in the pressure of H_2 is x while the change in the pressure of HI is $-2x$ (HI *disappears twice as fast* as iodine appears).

    (c) Lines 1 and 2 of the table are added to give the equilibrium pressures of the reactants and products in terms of x.

    (d) The value of x is determined from the change in pressure of HI ($-2x$) from the initial to the equilibrium value.

    (e) Using the calculated value of x, the equilibrium pressures from line 3 of the equilibrium table are calculated and entered into the equilibrium constant expression to obtain K.

	Species		
	HI	H_2	I_2
1. Initial pressure	0.100	0	0
2. Change in pressure	$-2x$	$+x$	$+x$
3. Equilibrium pressure	$0.100 - 2x$	x	x

    $$P_{HI} = 0.100 - 2x = 0.050 \quad \text{and} \quad x = 0.025 = P_{H_2} = P_{I_2}$$

    $$K = \frac{P_{H_2} P_{I_2}}{P_{HI}^2} = \frac{(0.025)^2}{(0.050)^2} = 0.25$$

- **Calculating the equilibrium composition**

    Example: A sample of ethane, $C_2H_6(g)$, is placed in a flask at a pressure of 2.00 bar at 900 K. The equilibrium constant K for the gas phase dehydrogenation of ethane to form ethylene

    $$C_2H_6(g) \rightleftharpoons C_2H_4(g) + H_2(g)$$

    is 0.050 at 900 K. Determine the composition of the equilibrium mixture and the percent decomposition of C_2H_6.

    (a) Set up an equilibrium table and let x be the unknown equilibrium pressure of each product, C_2H_4 and H_2. The equilibrium pressure of C_2H_6 is then $2.00 - x$.

    (b) Solve for x by rearranging the equilibrium constant expression.

    (c) The new form is a quadratic equation, which is solved exactly by using the quadratic formula.

(d) The equilibrium pressures can then be obtained.

(e) The percent decomposition is the loss of reactant divided by the initial pressure times 100%, or $\left(\dfrac{x}{2.00}\right)(100)\%$.

	Species		
	C_2H_6	C_2H_4	H_2
1. Initial pressure	2.00	0	0
2. Change in pressure	$-x$	$+x$	$+x$
3. Equilibrium pressure	$2.00 - x$	x	x

$$K = \frac{P_{C_2H_4}P_{H_2}}{P_{C_2H_6}} = \frac{(x)^2}{(2.00 - x)} = 0.050$$

Rearranging yields a quadratic equation: $x^2 + (0.050)x - 0.10 = 0$

Recall: $ax^2 + bx + c = 0 \Rightarrow x = \dfrac{-b \pm \sqrt{b^2 - 4ac}}{2a}$ (exact solution)

Then, $a = 1$, $b = 0.050$, and $c = -0.10$.

$$x = \frac{-(0.050) \pm \sqrt{(0.050)^2 - 4(1)(-0.10)}}{2(1)} = \frac{-(0.050) \pm \sqrt{0.4025}}{2}$$

$$= \frac{-(0.050) \pm (0.6344)}{2} = \frac{-(0.050) + (0.6344)}{2} = \frac{0.5844}{2} = 0.29$$

The negative root is rejected because it is unphysical.

$P_{C_2H_4} = P_{H_2} = x = 0.29$ bar   and $P_{C_2H_6} = 2.00 - x = 1.71$ bar

Percent decomposition $= \dfrac{\text{change in pressure}}{\text{initial pressure}} \times 100\% = \dfrac{0.29}{2.00} \times 100\% = 14.5\%$

Note: If K is less than 10^{-3} in a problem of this type, the adjustment to equilibrium is small. The pressures of reactants then change by only a small amount. In such a case, an approximate solution that avoids the quadratic equation is possible.

THE RESPONSE OF EQUILIBRIA TO CHANGES IN CONDITIONS
(Sections 10.9–10.13)

10.9 Adding and Removing Reagents

- **Le Chatelier's principle** → When a stress is applied to a system in dynamic equilibrium, the equilibrium tends to adjust to minimize the effect of the stress.

- **Qualitative features**
  - → Use Le Chatelier's principle to predict the qualitative adjustments needed to obtain a new equilibrium composition that relieves the stress.
  - → Assessing the effects of the applied stress on Q relative to K is an alternative approach.

Example: Use Le Chatelier's principle and reaction quotient arguments to describe the effect on the equilibrium

$$2\,SO_2\,(g) + O_2\,(g) \rightleftharpoons 2\,SO_3\,(g) \qquad Q = \left(\frac{P_{SO_3}^2}{P_{SO_2}^2\,P_{O_2}} \right)$$

of (a) addition of SO_2, (b) removal of O_2, (c) addition of SO_3, and (d) removal of SO_3. Both reactant and product pressures appear in the equilibrium constant expression, and in Q. Assuming no change in volume or temperature, adding reactant (or removing product) shifts the equilibrium composition to produce more product molecules, thereby helping to relieve the stress. Adding product (or removing reactant) shifts the equilibrium composition to produce more reactant molecules, thereby helping to relieve the stress.

(a) Adding reactant SO_2 results in $Q < K$. The system adjusts by producing more product SO_3 (consuming SO_2 and O_2) until $Q = K$ ($\rightarrow$).

(b) Removing reactant O_2 results in $Q > K$. The system adjusts by consuming product SO_3 (producing more SO_2 and O_2) until $Q = K$ ($\leftarrow$).

(c) Adding SO_3 results in $Q > K$. The system adjusts by producing more SO_2 and O_2 (consuming SO_3) until $Q = K$ ($\leftarrow$).

(d) Removing SO_3 results in $Q < K$. The system adjusts by consuming SO_2 and O_2 (producing more SO_3) until $Q = K$ ($\rightarrow$).

Note: In this example, the stress changes only the value of Q. The composition adjusts to return the value of Q to the original value of K.

The equilibrium constant does not change. The equilibrium constant has, however, a very weak pressure dependence that is usually safe to ignore.

- **Quantitative features** $\rightarrow$ Use the equilibrium table method to calculate the new composition of an equilibrium system after a stress is applied.

Example: Assume that the reaction $2\,HI\,(g) \rightleftharpoons H_2\,(g) + I_2\,(g)$ has the equilibrium composition $P_{HI} = 0.050$ bar, $P_{H_2} = P_{I_2} = 0.025$ bar, and $K = 0.25$.

If HI, H_2, and I_2 are added to the system in amounts equivalent to 0.008 bar for HI, 0.008 bar for H_2, and 0.008 bar for I_2, predict the direction of change and calculate the new equilibrium composition.

After adding 0.008 bar to each reactant and product, the *nonequlibrium* pressures are $P_{HI} = 0.058$ bar, $P_{H_2} = 0.033$ bar, and $P_{I_2} = 0.033$ bar.

$$Q = \frac{(0.033)(0.033)}{(0.058)^2} = 0.324 > K \quad \text{and the adjustment to the new equilibrium is in the}$$

direction toward the reactants.

Quantitative treatment: Set up an equilibrium table, letting $-x$ be the unknown adjustment to the initial *nonequilibrium* pressures of each product H_2 and I_2. The value of $+2x$ is then the adjustment to the initial *nonequilibrium* pressure of HI.

A negative value of x is expected from our qualitative analysis, which suggests that reactants should form from products to relieve the stress on the system.

	Species		
	HI	H_2	I_2
1. Initial pressure	0.058	0.033	0.033
2. Change in pressure	$+2x$	$-x$	$-x$
3. Equilibrium pressure	$0.058 + 2x$	$0.033 - x$	$0.033 - x$

$$K = \frac{P_{H_2}P_{I_2}}{P_{HI}^2} = \frac{(0.033 - x)^2}{(0.058 + 2x)^2} = 0.25$$

This expression is relatively easy to solve for x by taking the square-root of both sides:

$$\sqrt{K} = \frac{(0.033 - x)}{(0.058 + 2x)} = \sqrt{0.25} = 0.50$$

$0.033 - x = 0.029 + x$; $2x = 0.004$; $x = 0.002$

$P_{H_2} = P_{I_2} = 0.033 - 0.002 = 0.031$ bar *and* $P_{HI} = 0.058 + 0.004 = 0.062$ bar

Equilibrium expression check: $K = \dfrac{(0.031)^2}{(0.062)^2} = 0.25$ as expected

10.10 Compressing a Reaction Mixture

- **Decrease in volume of a reaction mixture at constant temperature**

  → The partial pressure of each component J of the reaction mixture increases;

  recall $P_J = \dfrac{n_J RT}{V}$ (see Section 4.8 in this study guide and in the text).

  → To minimize the increase in pressure, the equilibrium shifts to form fewer gas molecules, if possible.

- **Increase in volume of a reaction mixture at constant temperature**

  → The partial pressure of each member J of the reaction mixture decreases; recall $P_J = \dfrac{n_J RT}{V}$.

  → To compensate for the decrease in pressure, the equilibrium shifts to form more gas molecules, if possible.

- **Increase in pressure of a reaction mixture by introducing an inert gas at constant T and V, as in a steel vessel**

  → The partial pressure of each member J of the reaction mixture does not change;

  recall $P_J = \dfrac{n_J RT}{V}$.

  → The equilibrium composition is unaffected. The equilibrium constant has a weak pressure dependence that is safe to ignore unless the pressure change is large.

- **Increase in volume of a reaction mixture by introducing an inert gas at constant T and *total* pressure P, as in a piston**

  → Increases the *volume* of the system without changing the amount of reactants or products

  → The partial pressure of each member J of the reaction mixture decreases; recall $P_J = \dfrac{n_J RT}{V}$.

→ The return to equilibrium is equivalent to that which occurs by increasing the volume or decreasing the pressure of the system. The equilibrium shifts to form more gas molecules, if possible.

10.11 Temperature and Equilibrium

- **Qualitative features** → Use Le Chatelier's principle to predict the qualitative adjustments needed to obtain a new equilibrium composition that relieves the stress.

- **For *endothermic* reactions, heat is treated as a *reactant*.**
    → *Increasing* the temperature supplies heat to the reaction mixture and the composition of the mixture adjusts to form more product molecules.
    → The value of the equilibrium constant, K, *increases*.
    → *Decreasing* the temperature removes heat from the reaction mixture and the composition of the mixture adjusts to form more reactant molecules.
    → The value of the equilibrium constant, K, *decreases*.

- **For *exothermic* reactions, heat is treated as a *product*.**
    → *Increasing* the temperature supplies heat to the reaction mixture and the composition of the mixture adjusts to form more reactant molecules.
    → The value of the equilibrium constant, K, *decreases*.
    → *Decreasing* the temperature removes heat from the reaction mixture and the composition of the mixture adjusts to form more product molecules.
    → The value of the equilibrium constant, K, *increases*.

- **Quantitative features**

    → van't Hoff equation
    $$\ln \frac{K_2}{K_1} = \frac{\Delta H_r^{\circ}}{R} \left(\frac{1}{T_1} - \frac{1}{T_2} \right)$$
    (see Section 10.11 in the text for the derivation)

    → Gives the qualitative results above (The equilibrium constant is K, not K_c.)
    → A major assumption in the derivation is that both ΔH_r° and ΔS_r° are independent of temperature between T_1 and T_2.

- **Use the van't Hoff equation to calculate**
    → ΔH_r° if the equilibrium constant K is known at two temperatures.
    → The equilibrium constant at any temperature T_2 if ΔH_r° and K at any other temperature are known.

10.12 Catalysts and Haber's Achievement

- **Catalyst increases the rate at which a reaction approaches equilibrium.**
    → Rate of forward reaction is increased.
    → Rate of reverse reaction is increased.
    → Dynamic equilibrium is unaffected
    → The equilibrium constant is not changed.
    → The identity of the catalyst is unchanged.
    → The catalyst is *neither* a reactant *nor* a product. It does *not* appear in the chemical equation.

- **Haber's achievement**
    → Production of ammonia: $3H_2(g) + N_2(g) \rightleftharpoons 2NH_3(g)$
    At 298.15 K, $\Delta H_r^{\circ} = -92.22$ kJ·mol^{-1}, $\Delta G_r^{\circ} = -32.90$ kJ·mol^{-1}, and $K = 5.80 \times 10^5$

→ Although thermodynamically favorable ($\Delta G_r° < 0$), the reaction is very slow at room temperature and pressure.

→ Compression of the gas mixture favors product ammonia (Le Chatelier).

→ Removal of ammonia as it is formed encourages more to be formed (Le Chatelier).

→ Increasing the temperature is required to increase the slow rate of the reaction at 298.15 K, but it adversely affects the position of the equilibrium (van't Hoff equation).

→ To help circumvent the problem, an appropriate catalyst (a mixture of Fe_2O_3 and Fe_3O_4) was discovered at about the time of World War I. It is still used today.

10.13 The Impact on Biology: Homeostasis

- **Homeostasis**
  → Mechanism similar to chemical equilibrium, governed by Le Chatelier's principle
  → Helps living organisms to maintain constant internal conditions
  → Allows living organisms to control biological processes at a constant level
  → An example is the equilibrium involving oxygen and hemoglobin, Hb.

$$Hb(aq) + O_2(aq) \rightleftharpoons HbO_2(aq)$$

[See Section 10.13 in the text for a discussion of the ways in which hemoglobin and myoglobin provide needed oxygen to our bodies under varying physiological conditions.]

Chapter 11   ACIDS AND BASES

THE NATURE OF ACIDS AND BASES (Sections 11.1–11.6)

11.1 Brønsted–Lowry Acids and Bases

- **Brønsted–Lowry definitions**
  - → *Acid:* proton donor; *base:* proton acceptor
  - → General theory for *any* solvent (for example, H_2O or NH_3) or *no* solvent at all
  - → Acid (*proton donor*) is *deprotonated* when reacting with a base.
  - → Base (*proton acceptor*) is *protonated* when reacting with an acid.
  - → Acids and bases react with each other in proton transfer reactions.
  - → A strong acid is *completely deprotonated.*
    - **Example:** $HBr(aq) + H_2O(l) \rightarrow H_3O^+(aq) + Br^-(aq)$ (*strong* acid: reaction goes essentially to completion; proton is transferred from $HBr(aq)$ to $H_2O(l)$).
  - → A strong base is completely *protonated.*
    - **Example:** $NH_2^-(aq) + H_2O(l) \rightarrow NH_3(aq) + OH^-(aq)$ (*strong* base: reaction goes essentially to completion; $NH_2^-(aq)$ accepts a proton from H_2O).
  - → Strong Brønsted acids and bases: a single arrow ($\rightarrow$) is used to represent their reaction.
  - → A weak acid is *partially deprotonated.*
    - **Example:** $HF(aq) + H_2O(l) \rightleftharpoons H_3O^+(aq) + F^-(aq)$ (weak acid reaction, a *chemical equilibrium* is established).
  - → A weak base is *partially protonated.*
    - **Example:** $NH_3(aq) + H_2O(l) \rightleftharpoons NH_4^+(aq) + OH^-(aq)$ (*weak* base reaction, a *chemical equilibrium* is established).
  - → Weak Brønsted acids and bases: a double arrow ($\rightleftharpoons$) is used to represent their reaction.
  - → For strong acids and bases, the equilibrium lies far to right, justifying use of the single arrow for their reaction with the water or other solvents.
- **Conjugate base**
  - → Formed from acid after the proton is donated
  - → Acid $\Rightarrow$ deprotonation $\Rightarrow$ conjugate base
    - **Examples:** The conjugate base of HBr is Br^-; the conjugate base of H_3PO_4 is $H_2PO_4^-$.
- **Conjugate acid**
  - → Formed from base after the proton is accepted
  - → Base $\Rightarrow$ protonation $\Rightarrow$ conjugate acid
    - **Examples:** The conjugate acid of NH_3 is NH_4^+; the conjugate acid of CH_3COO^- is CH_3COOH.
- **Conjugate acid-base pair**   → Differ only by the presence of an H^+ species:
    - HCN, CN^- ; H_3O^+, H_2O ; H_2O, OH^- ; H_2S, HS^-
- **Solvent**
  - → Need not be water. Other solvents include $HF(l)$, $H_2SO_4(l)$, $CH_3COOH(l)$, $C_2H_5OH(l)$, $NH_3(l)$.
  - → Acid-base reactions can occur with *no* solvent as in: $HCl(g) + NH_3(g) \rightarrow NH_4Cl(s)$.

11.2 Lewis Acids and Bases

- **Lewis definitions**
    - → A Lewis acid is an electron pair *acceptor*. **Examples:** $BF_3(g)$, $Fe^{2+}(aq)$, $H_2O(l)$
    - → A Lewis base is an electron pair *donor*. **Examples:** $NH_3(aq)$, $O^{2-}(aq)$
    - → Most general theory (holds for any solvent and includes metal cations)
- **Comparison of Lewis and Brønsted–Lowry Acids and Bases**
    - → Proton (H^+) transfer (Brønsted–Lowry theory) is a special type of Lewis acid-base reaction.
    - → In a proton transfer reaction, H^+ acts as a Lewis acid, accepting an electron pair from a Lewis base.
    - → Brønsted acid: supplier of one particular Lewis acid, a proton
    - → Brønsted base: a particular type of Lewis base that can use a lone pair to bond a proton
        - **Example:** The reaction $O^{2-}(aq) + H_2O(l) \rightarrow 2\,OH^-(aq)$ can be viewed as a Brønsted–Lowry reaction (H_2O is the proton donor, O^{2-} is the proton acceptor) or as a Lewis acid-base reaction (O^{2-} is the electron pair donor (Lewis base)), H^+ on water is the electron pair acceptor (Lewis acid), as shown in the diagram below.

- **Note:** Curly arrows extend from a Lewis base electron pair (tail) to a Lewis acid (head) and indicate a Lewis acid-base reaction

11.3 Acidic, Basic, and Amphoteric Oxides

- **Oxides** → React with water as Lewis acids or bases, or both
- **Oxides of nonmetals**
    - → Called *acidic oxides*
    - → Tend to act as Lewis acids **Examples:** CO_2, SO_3, Cl_2O_7, N_2O_5
    - → React with water to form a Brønsted acid **Examples:**

$$CO_2(g) + H_2O(l) \rightarrow H_2CO_3(aq)$$
$$SO_3(g) + H_2O(l) \rightarrow H_2SO_4(aq)$$

Lewis acid   *Brønsted acid*

- → React with bases to form a salt and water
    - **Example:** $H_2CO_3(aq) + 2\,NaOH(aq) \rightarrow Na_2CO_3(aq) + 2\,H_2O(l)$
- **Oxides of metals**
    - → Called *basic oxides*
    - → Tend to act as Lewis bases **Examples:** Na_2O, MgO, CaO
    - → React with water to form Brønsted bases **Example:** $CaO(s) + H_2O(l) \rightarrow Ca(OH)_2(aq)$

     Lewis base   *Brønsted base*

- → React with acids to form a salt and water **Example:** $CaO(s) + 2\,HBr(aq) \rightarrow CaBr_2(aq) + H_2O(l)$
- **Amphoteric oxides**
    - → Exhibit both acid and base character **Examples:** Al_2O_3, SnO_2
    - → React with both acids and bases
        - **Example:** See the text for reactions of Al_2O_3 as an acid and as a base.

- **Trends in acidity for the oxides of main-group elements**

Increasing acidity ⟶

1	2	13/III	14/IV	15/V	16/VI	17/VII
Li_2O	**BeO**	B_2O_3	CO_2	N_2O_5	(O_2)	OF_2
Na_2O	MgO	**Al_2O_3**	SiO_2	P_4O_{10}	SO_3	Cl_2O_7
K_2O	CaO	**Ga_2O_3**	GeO_2	As_2O_5	SeO_3	Br_2O_7
Rb_2O	SrO	In_2O_3	**SnO_2**	Sb_2O_5	TeO_3	I_2O_7
Cs_2O	BaO	Tl_2O_3	**PbO_2**	Bi_2O_5	PoO_3	At_2O_7

(Increasing acidity ↑)

Note: Acidity of oxides in water tends to increase as shown in the preceding table. Amphoteric oxides are in bold type, basic oxides are to their left and acidic oxides to their right. Exceptions are O_2, which is neutral, and OF_2, which is only weakly acidic.

11.4 Proton Exchange Between Water Molecules

- **Amphiprotic species** → A molecule or ion that can act as either a Brønsted acid (H^+ *donor*) or a Brønsted base (H^+ *acceptor*)

  Examples: H_2O: H^+ *donor* ($\rightarrow OH^-$) or H^+ *acceptor* ($\rightarrow H_3O^+$)

  NH_3: H^+ *donor* ($\rightarrow NH_2^-$) or H^+ *acceptor* ($\rightarrow NH_4^+$)

- **Autoprotolysis** → A reaction in which one molecule transfers a proton to another molecule of the same kind

  Examples: Autoprotolysis of water: $2\,H_2O(l) \rightleftharpoons H_3O^+(aq) + OH^-(aq)$

  Autoprotolysis of ammonia: $2\,NH_3(l) \rightleftharpoons NH_4^+(am) + NH_2^-(am)$

- **Equilibrium constant** → Autoprotolysis of water:
  $$2\,H_2O(l) \rightleftharpoons H_3O^+(aq) + OH^-(aq); \quad K_w = [H_3O^+][OH^-] = 1 \times 10^{-14} \text{ at } 25°C$$
  Like any equilibrium constant $K_w = K_w(T)$

- **Pure water at 25°C** → $[H_3O^+] = [OH^-] = 1.0 \times 10^{-7}$ mol·L^{-1}

- **Neutral solution at 25°C** → $[H_3O^+] = [OH^-] = 1.0 \times 10^{-7}$ mol·L^{-1}

- **Acidic solution 25°C** → $[H_3O^+] > 1.0 \times 10^{-7}$ mol·L^{-1} and $[OH^-] < 1.0 \times 10^{-7}$ mol·L^{-1}

- **Basic solution 25°C** → $[H_3O^+] < 1.0 \times 10^{-7}$ mol·L^{-1} and $[OH^-] > 1.0 \times 10^{-7}$ mol·L^{-1}

11.5 The pH Scale

- **Concentration variation of H_3O^+** → Values higher than 1 mol·L^{-1} and less than 10^{-14} mol·L^{-1} are not commonly encountered.

- **Logarithmic scale**
  → $pH = -\log[H_3O^+]$ *and* $[H_3O^+] = 10^{-pH}$
  pH values range between 1 and 14 in most cases.
  → Pure water at 25°C: pH = 7.00;   ⟹ Neutral solution: pH = 7.00
  → Acidic solution: pH < 7.00;   ⟹ Basic solution: pH > 7.00

  Note: pH has *no* units, but $[H_3O^+]$ has units of molarity.

- **Measurement of pH**

  → *Universal indicator paper* turns different colors at different pH values.

  → *pH meter* measures a difference in electrical potential (proportional to pH) across electrodes that dip into the solution (see Chapter 13).

  Examples: The pH of the solutions for which the $[H_3O^+] = 5.4 \times 10^{-4}$ mol·L^{-1} is:

  $$pH = -\log[H_3O^+] = -(-3.27) = 3.27 \quad (\textit{acidic} \text{ solution})$$

  The $[H_3O^+]$ for a solution that has a pH value of 1.70 is: $[H_3O^+] = 10^{-pH} = 10^{-1.70} = 2.0 \times 10^{-2}$ mol·L^{-1}.

  → pH values for common materials are given in Fig. 11.12 in the text.

11.6 The pOH of Solutions

- **pX as a generalization of pH**

  → $pX = -\log X$ (X = anything)

  → $pOH = -\log[OH^-]$ *and* $[OH^-] = 10^{-pOH}$

  → $pK_w = -\log K_w = -\log(1.0 \times 10^{-14}) = 14.00$ at 25°C

- **Relationship between pH and pOH of a solution**

  → $[H_3O^+][OH^-] = K_w = 1.0 \times 10^{-14}$ at 25°C

  → $\log([H_3O^+][OH^-]) = \log[H_3O^+] + \log[OH^-] = \log K_w$

  → $-\log[H_3O^+] - \log[OH^-] = -\log K_w$

  → **pH + pOH = 14.00 at 25°C**

WEAK ACIDS AND BASES (Sections 11.7–11.10)

11.7 Acidity and Basicity Constants

- **Weak acid** → pH is *larger* than that of a strong acid with the same molarity.
  Incomplete deprotonation

- **General weak acid HA**

  → $HA(aq) + H_2O(l) \rightleftharpoons H_3O^+(aq) + A^-(aq)$; $K_a = \dfrac{a_{H_3O^+} a_{A^-}}{a_{HA} a_{H_2O}}$

  But, $a_{H_2O} = 1$, $a_{HA} \approx [HA]$, $a_{H_3O^+} \approx [H_3O^+]$, and $a_{A^-} \approx [A^-]$.

  → Thus, $K_a = \dfrac{[H_3O^+][A^-]}{[HA]}$ K_a is the *acidity constant;* $pK_a = -\log K_a$ is *commonly used.*

Some Weak Acids and Their Acidity Constants

	Species	Formula	K_a	pK_a
Neutral acids	Hydrofluoric acid	HF	3.5×10^{-4}	3.45
	Acetic acid	CH_3COOH	1.8×10^{-5}	4.75
	Hydrocyanic acid	HCN	4.9×10^{-10}	9.31
Cation acids	Pyridinium	$C_5H_5^+$	5.6×10^{-6}	5.24
	Ammonium	NH_4^+	5.6×10^{-10}	9.25
	Ethylammonium	$C_2H_5NH_3^+$	1.5×10^{-11}	10.82
Anion acids	Hydrogen sulfate	HSO_4^-	1.2×10^{-2}	1.92
	Dihydrogen phosphate	$H_2PO_4^-$	6.2×10^{-8}	7.21

- **In the preceding table**
  - $\rightarrow$ Hydrogen sulfate is the strongest acid (largest K_a, *smallest* pK_a).
  - $\rightarrow$ Ethylammonium is the weakest acid (smallest K_a, *largest* pK_a).

- **Weak base** $\rightarrow$ pH is *smaller* than that of a strong base with the same molarity.
                Incomplete protonation

- **General weak base B**

  - $\rightarrow$ $B(aq) + H_2O(l) \rightleftharpoons BH^+(aq) + OH^-(aq);$ $K_b = \dfrac{a_{BH^+} a_{OH^-}}{a_B a_{H_2O}}$

  But $a_{H_2O} = 1$, $a_B \approx [B]$, $a_{BH^+} \approx [BH^+]$, and $a_{OH^-} \approx [OH^-]$.

  - $\rightarrow$ Thus, $K_b = \dfrac{[BH^+][OH^-]}{[B]}$ K_b is the *basicity constant*; $pK_b = -\log K_b$ is *commonly used.*

Weak Bases and Their Basicity Constants

	Species	Formula	K_b	pK_b
Neutral bases	Ethylamine	$C_2H_5NH_2$	6.5×10^{-4}	3.19
	Methylamine	CH_3NH_2	3.6×10^{-4}	3.44
	Ammonia	NH_3	1.8×10^{-5}	4.75
	Hydroxylamine	NH_2OH	1.1×10^{-8}	7.97
Anion bases	Phosphate	PO_4^{3-}	4.8×10^{-2}	1.32
	Fluoride	F^-	2.9×10^{-11}	10.54

- **In the preceding table**
  - $\rightarrow$ Phosphate, PO_4^{3-}, is the strongest base (largest K_b, *smallest* pK_b).
  - $\rightarrow$ Fluoride, F^-, is the weakest base (smallest K_b, *largest* pK_b).

- **Acid-base strength**
  - → The *larger* the value of K, the *stronger* the acid or base
  - → The *smaller* the value of pK, the *stronger* the acid or base

11.8 The Conjugate Seesaw

- **Reciprocity of conjugate acid-base pairs**

- **Conjugate acid-base pairs of weak acids**
  - → The conjugate base of a *weak* acid is a *weak* base.
  - → The conjugate acid of a *weak* base is a *weak* acid.
  - → For our purposes, *"weak"* is defined by pK values between 1 and 14.

- **Conjugate seesaw (qualitative)**
  - → The *stronger* an acid, the *weaker* its conjugate base
  - → The *stronger* a base, the *weaker* its conjugate acid

- **Conjugate seesaw (quantitative)**
  - → Consider a general *weak* acid HA and its conjugate base A^-.

$$HA(aq) + H_2O(l) \rightleftharpoons H_3O^+(aq) + A^-(aq) \quad K_a = \frac{[H_3O^+][A^-]}{[HA]}$$

$$A^-(aq) + H_2O(l) \rightleftharpoons HA(aq) + OH^-(aq) \quad K_b = \frac{[HA][OH^-]}{[A^-]}$$

$$K_a \times K_b = \frac{[H_3O^+][A^-]}{[HA]} \times \frac{[HA][OH^-]}{[A^-]} = [H_3O^+][OH^-] = K_w$$

  - → Or, for any conjugate pair, $\boxed{K_a \times K_b = K_w}$ and $pK_a + pK_b = pK_w$.
  - → If $K_a > 1 \times 10^{-7}$ ($pK_a < 7$), then the acid is *stronger* than its conjugate base.
  - → If $K_b > 1 \times 10^{-7}$ ($pK_b < 7$), then the base is *stronger* than its conjugate acid.
    See Table 11.3 in the text.

    Example: Sulfurous acid, H_2SO_3, is a stronger acid than nitrous acid, HNO_2. Consequently, nitrite, NO_2^-, is a stronger base than hydrogen sulfite, HSO_3^-.

- **Solvent leveling** → An acid is strong if it is a stronger proton donor than the conjugate acid of the solvent.

- **Strong acid in water**
  - → *Stronger* proton donor than H_3O^+ ($K_a > 1$; $pK_a < 0$)
  - → $H_3O^+(aq) + H_2O(l) \rightleftharpoons H_2O(l) + H_3O^+(aq) \qquad K_a = 1$
  - → All acids with $K_a > 1$ appear equally strong in water.
  - → Strong acids are *leveled* to the strength of the acid H_3O^+.

    Example: HCl, HBr, HI, HNO_3, and $HClO_4$ are strong acids in water. When dissolved in water, they behave as though they were solutions of the acid H_3O^+.

  - → Similar arguments hold for bases.

11.9 Molecular Structure and Acid Strength

- **Binary acids, H_nX**

  → For elements in the *same* period, the more *polar* the H–A bond, the *stronger* the acid.
  Example: HF is a stronger acid than H_2O, which is stronger than NH_3.

  → For elements in the *same* group, the *weaker* the H–A bond, the *stronger* the acid.
  Example: HBr is a stronger acid than HCl, which is stronger than HF.

- **Summary**

  → For binary acids, acid strength increases from left to right across a period (polarity of H–A bond increases), from top to bottom down a group (H–A bond enthalpy decreases).

  → The more polar or the weaker the H–A bond, the stronger the acid

  → See the text for a deeper thermodynamic discussion of these trends.

11.10 The Strengths of Oxoacids and Carboxylic Acids

- **Oxoacids (also called oxyacids)**

  → The *greater* the number of oxygen atoms attached to the *same* central atom, the *stronger* the acid.

  → This trend also corresponds to the *increasing* oxidation number of the *same* central atom.
  Example: $HClO_3$ (Cl ox # 5) is a stronger acid than $HClO_2$ (Cl ox # 3), which is stronger than HClO (Cl ox # 1)

  → For the *same* number of O atoms attached to the central atom, the *greater* the electronegativity of the central atom, the *stronger* the acid.
  Example: HClO is a stronger acid than HBrO, which is stronger than HIO. The halogen is considered the central atom for comparison to other oxoacids.

- **Carboxylic acids** → The *greater* the electronegativities of the groups attached to the carboxyl (COOH) group, the *stronger* the acid.
  Example: $CHBr_2COOH$ is a stronger acid than $CH_2BrCOOH$, which is stronger than CH_3COOH.

Note: The preceding rules are summarized in Tables 11.5, 11.6, and 11.7 in the text.

THE pH OF SOLUTIONS OF WEAK ACIDS AND BASES
(Sections 11.11–11.13)

11.11 Solutions of Weak Acids

- **Initial concentration**

  → Initial concentration of an acid ($[HA]_{initial}$) or base ($[B]_{initial}$) as prepared, assuming no proton transfer has occurred.

  → Primary equilibrium for weak acid: $HA(aq) + H_2O(l) \rightleftharpoons H_3O^+(aq) + A^-(aq)$

  → Secondary equilibrium: $2H_2O(l) \rightleftharpoons H_3O^+(aq) + OH^-(aq)$

- **General result for a dilute solution of the *weak* acid HA**

Main equilibrium	$HA(aq)$	$+$	$H_2O(l)$	$\rightleftharpoons$	$H_3O^+(aq)$	$+$	$A^-(aq)$
Initial molarity	$[HA]_{initial}$		—		0		0
Change in molarity	$-x$				$+x$		$+x$
Equilibrium molarity	$[HA]_{initial} - x$				x		x

Acidity constant expression: $K_a = \dfrac{[H_3O^+][A^-]}{[HA]} = \dfrac{x^2}{[HA]_{initial} - x}$

Quadratic equation: $x^2 + (K_a)x - (K_a[HA]_{initial}) = 0$ *and* $x > 0$ (positive root)

Positive root: $x = \dfrac{-K_a + \sqrt{K_a^2 + 4K_a[HA]_{initial}}}{2} = [H_3O^+] = [A^-] \geq 10^{-6}$ M   (1)

$[HA] = [HA]_{initial} - x$ $[OH^-] = K_w/[H_3O^+]$ from the secondary equilibrium

Percentage *deprotonated* $= \dfrac{x}{[HA]_{initial}} \times 100\%$ (2)

- **Approximation to the general result**

Assume $[HA]_{initial} - x \approx [HA]_{initial}$, then $K_a = \dfrac{x^2}{[HA]_{initial}}$ and $x = \sqrt{K_a[HA]_{initial}}$. (3)

Assumption is justified if percentage deprotonated < 5%.

> **Example:** For acetic acid, CH_3COOH, $K_a = 1.8 \times 10^{-5}$. The value of $[H_3O^+]$ in a 0.100 M solution of CH_3COOH is given by the preceding equation labeled (1).
> With $[HA]_{initial} = 0.1$, the result is $x = [H_3O^+] = 1.33 \times 10^{-3}$ M.
> Using the 5% approximation method in the equation labeled (3) yields $x = [H_3O^+] =$
> $x = \sqrt{K_a[HA]_{initial}} = [(1.8 \times 10^{-5})(0.100)]^{1/2} = 1.34 \times 10^{-3}$ M.
> *The approximation is excellent in this case.*

11.12 Solutions of Weak Bases

- **General result for a *weak* base B at *dilute* concentrations**

Main equilibrium	$B(aq)$	$+$	$H_2O(l)$	$\rightleftharpoons$	$BH^+(aq)$	$+$	$OH^-(aq)$
Initial molarity	$[B]_{initial}$		—		0		0
Change in molarity	$-x$				$+x$		$+x$
Equilibrium molarity	$[B]_{initial} - x$				x		x

Basicity constant expression: $K_b = \dfrac{[BH^+][OH^-]}{[B]} = \dfrac{x^2}{[B]_{initial} - x}$

Quadratic equation: $x^2 + (K_b)x - (K_b[B]_{initial}) = 0$ *and* $x > 0$ (positive root)

Positive root: $x = \dfrac{-K_b + \sqrt{K_b^2 + 4K_b[B]_{initial}}}{2} = [BH^+] = [OH^-] \geq 10^{-6}$ M   (4)

$[B] = [B]_{initial} - x$ $[H_3O^+] = K_w/[OH^-]$ secondary equilibrium

Percentage *protonated* $= \dfrac{x}{[\text{B}]_{\text{initial}}} \times 100\%$ (5)

- **Approximation to the general result**

  Assume $[\text{B}]_{\text{initial}} - x \approx [\text{B}]_{\text{initial}}$, then $K_b = \dfrac{x^2}{[\text{B}]_{\text{initial}}}$ and $x = \sqrt{K_b[\text{B}]_{\text{initial}}}$ (6)

  Assumption is justified if percentage protonated < 5%.

  > **Example:** K_b for the weak base methylamine, CH_3NH_2, is 3.6×10^{-4}. In a 0.200 M solution of methylamine, the $[OH^-]$ is given by the equation labeled (4), with $[\text{B}]_{\text{initial}} = 0.200$ M. Substituting the values of K_b and $[\text{B}]_{\text{initial}}$ yields $x = [OH^-] = [CH_3NH_2^+]$ $= 8.31 \times 10^{-3}$ M.
  >
  > Using the approximate solution in the equation labeled (6), $x = [OH^-] = [CH_3NH_2^+] = 8.49 \times 10^{-3}$ M.
  >
  > The percent difference between the exact and approximate answers is less than 5%, so the approximation is justified.

11.13 The pH of Salt Solutions

- **Acidic cations**
  - → The conjugate acids of *weak* bases   **Examples:** Ammonium, NH_4^+; anilinium, $C_6H_5NH_3^+$
  - → Certain small, highly charged metal cations   **Examples:** Iron(III), Fe^{3+}; copper(II), Cu^{2+}
- **Neutral cations** → Group 1 and 2 cations; cations with +1 charge   **Examples:** Li^+, Mg^{2+}, Ag^+
- **Basic anions** → Conjugate bases of weak acids
  Examples: Acetate, CH_3COO^-; fluoride, F^-; sulfide, S^{2-}
- **Neutral anions** → Conjugate bases of strong acids   **Examples:** Cl^-, ClO_4^-; NO_3^-.
- **Acidic anions** → Only a few, such as HSO_4^-, $H_2PO_4^-$
- **Acid salt**
  - → Typically, a salt with an acidic cation
    Examples: $FeCl_3$, NH_4NO_3 (in these examples, the anions are neutral — see above)
  - → Dissolves in water to yield an acidic solution
    Example: $NH_4NO_3(aq) \rightarrow NH_4^+(aq) + NO_3^-(aq)$ (dissociation of salt in water)
    $NH_4^+(aq) + H_2O(l) \rightleftharpoons H_3O^+(aq) + NH_3(aq)$ (acidic cation reacts with water to give an acidic solution). $NO_3^-(aq)$ is the anion of a strong acid, HNO_3, and has no basic character in water (conjugate seesaw).

- **Basic salt**
  - → A salt with a basic anion
    Examples: Sodium acetate, NaO_2CCH_3; potassium cyanide, KCN
    (In these examples, the cations are neutral – see above.)
  - → Dissolves in water to yield a basic solution
    Example: $NaO_2CCH_3(aq) \rightarrow Na^+(aq) + CH_3COO^-(aq)$ (dissociation of salt in water)
    $CH_3COO^-(aq) + H_2O(l) \rightleftharpoons OH^-(aq) + CH_3COOH(aq)$
    (basic cation reacts with water to yield a *basic solution*)

- **Neutral salt**
    - → A salt with neutral cations and anions

        Examples: Sodium chloride, NaCl; potassium nitrate, KNO_3; magnesium iodide, MgI_2

    - → Dissolves in water to yield a neutral solution

- **Method of equilibrium solution (main equilibrium)**
    - → For acid salts, use the same method as for *weak* acids, but replace HA with BH^+.
    - → For basic salts, use the same method used for *weak* bases, but replace B with A^-. (All are *anions*.)

        Example: Iron(III) chloride, $FeCl_3$, is the salt of an acid cation and a neutral anion. It dissolves in water to give an *acidic* solution:

        $FeCl_3(s) + 6 H_2O(l) \rightarrow Fe(H_2O)_6^{3+}(aq) + 3 Cl^-(aq)$ (dissolution of iron(III) chloride)
        Fe^{3+} is surrounded by six water molecules. The +3 charge enhances the positive charges on the H atoms, making them more acidic than those in H_2O.
        $Fe(H_2O)_6^{3+}(aq) + H_2O(l) \rightleftharpoons Fe(H_2O)_5(OH)^{2+}(aq) + H_3O^+(aq)$;
            (reaction of a weak acid with water)
        For the above equation, $K_a = 3.5 \times 10^{-3}$ (see Table 11.8 in the text).

        The cation is a weak acid and the equation above is the main equilibrium; use the methodology in Section 11.11 above to calculate concentration. For example, the $[H_3O^+]$ of a 0.010 M solution of $FeCl_3(s)$ is given by equation (1):

        $$x = \frac{-K_a + \sqrt{K_a^2 + 4 K_a [HA]_{initial}}}{2} = [H_3O^+]$$
        $$= \frac{-(3.5 \times 10^{-3}) + \sqrt{(3.5 \times 10^{-3})^2 + 4(3.5 \times 10^{-3})(0.010)}}{2}$$

        $x = [H_3O^+] = 4.42 \times 10^{-3}$ $pH = -\log(4.42 \times 10^{-3}) = 2.35 \approx 2.4$

        The solution is quite acidic and would be considered corrosive.

POLYPROTIC ACIDS AND BASES (Sections 11.14–11.17)

11.14 The pH of a Polyprotic Acid Solution
- **Successive deprotonations of a polyprotic acid**
    - → A polyprotic acid is a species that can donate more than one proton.
    - → Stepwise equilibria must be considered if successive K_a values differ by a factor of $\geq 10^3$, which is normally the case. Values of K_a are given in Table 11.10 in the text.
    - → Polyprotic bases (species that can accept more than one proton) also exist.

- **Polyprotic acids and pH**
    - → Use the main equilibrium for a dilute solution of a polyprotic acid to calculate the pH. Neglect any secondary equilibria except for H_2SO_4, whose first deprotonation is complete.
    - → The calculation is identical to that for any weak monoprotic acid.

Example: The pH of a 0.025 M solution of diprotic carbonic acid, H_2CO_3, is calculated using the first dissociation step and equation (1) in Section 10.11:

$$H_2CO_3(aq) + H_2O(l) \rightleftharpoons H_3O^+(aq) + HCO_3^-(aq) \quad K_{a1} = 4.3 \times 10^{-7} \text{ (main equilibrium)}$$

$$x = \frac{-K_{a1} + \sqrt{K_{a1}^2 + 4K_a[HA]_{initial}}}{2} = [H_3O^+]$$

$$= \frac{-(4.3 \times 10^{-7}) + \sqrt{(4.3 \times 10^{-7})^2 + 4(4.3 \times 10^{-7})(0.025)}}{2}$$

$$x = [H_3O^+] = 1.03 \times 10^{-4} \quad pH = -\log(1.03 \times 10^{-4}) = 3.9935 \approx 4.0$$

11.15 Solutions of Salts of Polyprotic Acids

- **pH of a solution of an amphiprotic anion HA^-**

  → Given by $pH = \frac{1}{2}(pK_{a1} + pK_{a2})$

  → Independent of the concentration of the anion

  → Above equation is valid if $s \gg K_w/K_{a2}$ and $s \gg K_{a1}$. Here s = the initial concentration of the salt.

  → When these approximations are not valid, a more complex expression for the pH is needed (see text website for the equation and its derivation).

- **Solution of basic anion of polyprotic acid : A^{2-} or A^{3-} Examples:** Solutions of Na_2S or K_3PO_4

  → Main equilibrium is hydrolysis of the basic anion: $A^{2-}(aq) + H_2O(aq) \rightleftharpoons OH^-(aq) + HA^-(aq)$

  → Treat using methodology for solutions for basic anions: [Section 11.12, Eqs. (4) – (6), this guide]

11.16 The Concentrations of Solute Species

- **Assumptions:**

  → The polyprotic acid is the species present in largest amount

  → Only the first deprotonation contributes significantly to $[H_3O^+]$

  → The autoprotolysis of water doesn't contribute significantly to $[H_3O^+]$ or $[OH^-]$

- **Determination of the concentration of all species for a solution of a weak *diprotic* acid, H_2A**

  → Solve the *main* equilibrium (K_{a1}) as for any weak acid.

  → This yields $[H_2A]$, $[H_3O^+]$, and $[HA^-]$.

  → Solve the *secondary* equilibrium (K_{a2}) for $[A^{2-}]$ using the $[H_3O^+]$ and $[HA^-]$ values determined above.

  → Determine $[OH^-]$ using K_w and $[H_3O^+]$.

- **Determination of the concentration of all species for a solution of a weak *triprotic* acid, H_3A**

  → Solve the *main* equilibrium (K_{a1}) as for any weak acid.

  → This yields $[H_3A]$, $[H_3O^+]$, and $[H_2A^-]$.

  → Solve the *secondary* equilibrium (K_{a2}) for $[HA^{2-}]$ using the $[H_3O^+]$ and $[H_2A^-]$ values from above.

  → Solve the *tertiary* equilibrium (K_{a3}) for $[A^{3-}]$ using the $[H_3O^+]$ and $[HA^{2-}]$ values from above.

  → Determine $[OH^-]$ using K_w and $[H_3O^+]$.

11.17 Composition and pH

- **Polyprotic acid** $\rightarrow$ Concentrations of the several solution species vary with the pH of the solution. As the pH is increased, the fraction of deprotonated species increases.

- **Composition and pH** $\rightarrow$ Solution to the stepwise equilibrium expressions for a polyprotic acid as a function of pH

- **Diprotic acid (H_2A) — qualitative considerations** (See Fig. 11.21 in the text, reproduced below)

$$H_2A(aq) + H_2O(l) \rightleftharpoons H_3O^+(aq) + HA^-(aq) \qquad K_{a1} = \frac{[H_3O^+][HA^-]}{[H_2A]} \quad (7)$$

$$HA^-(aq) + H_2O(l) \rightleftharpoons H_3O^+(aq) + A^{2-}(aq) \qquad K_{a2} = \frac{[H_3O^+][A^{2-}]}{[HA^-]} \quad (8)$$

- $\rightarrow$ At low pH, equilibria represented by Equations (7) and (8) shift to the left (Le Chatelier) and the solution contains a high concentration of H_2A and very little A^{2-}.
- $\rightarrow$ At high pH, equilibria represented by Equations (7) and (8) shift to the right and the solution contains a high concentration of A^{2-} and very little H_2A.
- $\rightarrow$ At intermediate pH values, the solution contains a relatively high concentration of the amphiprotic anion HA^-. (See figure below)

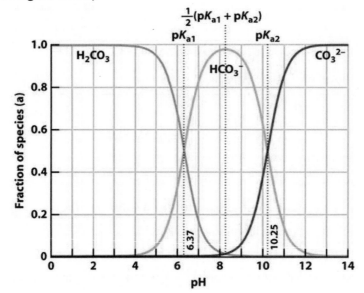

- **Diprotic acid (H_2A) $\Rightarrow$ quantitative considerations**

- $\rightarrow$ *Fraction* (α) of H_2A, HA^-, and A^{2-} in solution as a function of $[H_3O^+]$

- $\rightarrow$ $\alpha(H_2A) = \dfrac{[H_3O^+]^2}{f} \qquad \alpha(HA^-) = \dfrac{[H_3O^+]K_{a1}}{f} \qquad \alpha(A^{2-}) = \dfrac{K_{a1}K_{a2}}{f}$

- $\rightarrow$ where $f = [H_3O^+]^2 + [H_3O^+]K_{a1} + K_{a1}K_{a2}$

AUTOPROTOLYSIS AND pH (Sections 11.18–11.19)

11.18 Very Dilute Solutions of Strong Acids and Bases

- **Very dilute solution** $\rightarrow$ Concentration of a strong acid or base is less than 10^{-6} M.

- **General result for a *strong* acid HA at *very dilute* concentrations**

$$HA(aq) + H_2O(l) \;\rightarrow\; H_3O^+(aq) + A^-(aq)$$

Three equations in *three* unknowns ($[H_3O^+]$, $[OH^-]$, and $[A^-]$)

Autoprotolysis equilibrium: $\quad K_w = [H_3O^+][OH^-] = 1.0 \times 10^{-14}$

Charge balance: $\qquad\qquad [H_3O^+] = [OH^-] + [A^-]$

Material balance: $\qquad\qquad [A^-] = [HA]_{initial}\quad$ (complete deprotonation)

Solution: $\qquad\qquad\qquad [OH^-] = [H_3O^+] - [HA]_{initial}$ (from charge and material balance)

$\qquad\qquad\qquad\qquad K_w = [H_3O^+]([H_3O^+] - [HA]_{initial})$

$\qquad\qquad\qquad\qquad K_w = [H_3O^+]^2 - [HA]_{initial}[H_3O^+], \quad$ let $x = [H_3O^+]$

Quadratic equation: $\qquad x^2 - [HA]_{initial}\, x - K_w = 0 \; and \; x > 0$ (positive root)

Positive root: $\qquad x = \dfrac{[HA]_{initial} + \sqrt{[HA]_{initial}^2 + 4K_w}}{2} = [H_3O^+] \quad$ (9)

$\qquad\qquad\qquad and\; [OH^-] = K_w/[H_3O^+] \quad$ (10)

- **General result for a *strong* base MOH at *very dilute* concentrations**

$$MOH(aq) + H_2O(l) \;\rightarrow\; M^+(aq) + OH^-(aq)$$

Three equations in *three* unknowns

Autoprotolysis equilibrium: $\quad K_w = [H_3O^+][OH^-] = 1.0 \times 10^{-14}$

Charge balance: $\qquad\qquad [M^+] + [H_3O^+] = [OH^-]$

Material balance: $\qquad\qquad [M^+] = [MOH]_{initial}\quad$ (complete dissociation)

Solution: $\qquad\qquad\qquad [OH^-] = [H_3O^+] + [MOH]_{initial}$

$\qquad\qquad\qquad\qquad K_w = [H_3O^+]\{[H_3O^+] + [MOH]_{initial}\}$

$\qquad\qquad\qquad\qquad K_w = [H_3O^+]^2 + [MOH]_{initial}[H_3O^+], \;$ let $x = [H_3O^+]$

Quadratic equation: $\qquad x^2 + [MOH]_{initial}\, x - K_w = 0 \; and \; x > 0$ (positive root)

Positive root: $$x = \frac{-[MOH]_{initial} + \sqrt{[MOH]^2_{initial} + 4K_w}}{2} = [H_3O^+] \quad (10)$$

$$and \quad [OH^-] = K_w/[H_3O^+] \quad (11)$$

Note: Strong bases such as O^{2-} and CH_3^- produce aqueous solutions that are equivalent to MOH.

Example: The pH of a 1.5×10^{-8} M solution of the strong base KOH can be calculated from equations (10) and (11).

From (10), $$[H_3O^+] = \frac{-[MOH]_{initial} + \sqrt{[MOH]^2_{initial} + 4K_w}}{2} = 9.28 \times 10^{-8} \text{ M}.$$

From (11), $pH = -\log[H_3O^+] = -\log(9.28 \times 10^{-8}) = 7.03$.
(slightly *basic* solution as expected)

11.19 Very Dilute Solutions of Weak Acids

- **Very dilute solution** $\rightarrow$ Concentration of a weak acid or base is less than about 10^{-3} M.

- **General result for a *weak* acid HA at *very dilute* concentrations**
  Four equations in *four* unknowns

  1) Weak acid equilibrium: $HA(aq) + H_2O(l) \rightleftharpoons H_3O^+(aq) + A^-(aq)$; $\quad K_a = \dfrac{[H_3O^+][A^-]}{[HA]}$

  2) Autoprotolysis equilibrium: $\quad K_w = [H_3O^+][OH^-] = 1.0 \times 10^{-14}$

  3) Charge balance: $\quad [H_3O^+] = [OH^-] + [A^-]$

  4) Material balance: $\quad [HA]_{initial} = [HA] + [A^-]$ (incomplete deprotonation)

  Solution: $$[A^-] = [H_3O^+] - [OH^-] = [H_3O^+] - \frac{K_w}{[H_3O^+]} \text{ (from charge balance)}$$

  $$[HA] = [HA]_{initial} - [A^-] = [HA]_{initial} - [H_3O^+] + \frac{K_w}{[H_3O^+]}$$

  Substitute expressions for [HA] and [A⁻] into K_a.

  $$K_a = \frac{[H_3O^+]\left([H_3O^+] - \dfrac{K_w}{[H_3O^+]}\right)}{[HA]_{initial} - [H_3O^+] + \dfrac{K_w}{[H_3O^+]}} \quad (12) \qquad \text{Let } x = [H_3O^+] \text{ and then}$$

  $$K_a = \frac{x\left(x - \dfrac{K_w}{x}\right)}{[HA]_{initial} - x + \dfrac{K_w}{x}} \qquad \text{Rearrange to a cubic form.}$$

Cubic equation: $x^3 + K_a x^2 - (K_w + K_a[HA]_{initial})x - K_a K_w = 0$

Solve by using a graphing calculator, trial and error, or the mathematical software present on text website: http://bcs.whfreeman.com/chemicalprinciples5e/.
Only one of the three roots will be physically meaningful.

Alternatively, consider the following approximations:

If $[H_3O^+] > 10^{-6}$ M, then $\dfrac{K_w}{[H_3O^+]} < 10^{-8}$, and this term can be neglected in (12) to yield

$$K_a = \frac{[H_3O^+]^2}{[HA]_{initial} - [H_3O^+]} = \frac{x^2}{[HA]_{initial} - x} \qquad (13)$$

Further, if $[H_3O^+] > 10^{-6}$ M *and* $[H_3O^+] << [HA]_{initial}$, then (13) reduces to

$$K_a = \frac{[H_3O^+]^2}{[HA]_{initial}} = \frac{x^2}{[HA]_{initial}} \qquad (14)$$

Use Equations (12), (13), *or* (14) as appropriate to solve problems.

Note: Some software programs do not work when numbers in the polynomial equation differ by many orders of magnitude.

Chapter 12 AQUEOUS EQUILIBRIA

MIXED SOLUTIONS AND BUFFERS (Sections 12.1–12.3)

- **Common theme in Chapter 12** → Many equilibria can be treated in the following way:
    1. Identify the solute species in solution.
    2. Identify the equilibrium relations among the solute species.
    3. Use the relations in number 2 above to determine the concentrations of the solute species.

12.1 Buffer Action

- **Mixed solution** → A solution containing a weak acid or base and one of its salts

    Example: A solution of hydrofluoric acid and sodium fluoride

- **Buffer**

    → A mixed solution containing weak conjugate acid–base pairs that stabilizes the pH of a solution

    → Provides a source and a sink for protons

- **Acid buffer**

    → Consists of a weak acid and its conjugate base provided as a salt

    Examples: Acetic acid (CH_3COOH) and sodium acetate ($NaCH_3COO$);
    phosphorous acid (H_3PO_3) and potassium dihydrogenphosphite (KH_2PO_3)

    → Buffers a solution on the acid side of neutrality (pH < 7), if $pK_a < 7$

- **Base buffer**

    → Consists of a weak base and its conjugate acid provided as a salt

    Examples: Ammonia (NH_3) and ammonium chloride (NH_4Cl);
    sodium dihydrogenphosphate (NaH_2PO_4) and
    sodium hydrogenphosphate (Na_2HPO_4)

    → Buffers a solution on the base side of neutrality (pH > 7) if $pK_b < 7$

12.2 Designing a Buffer

- **pH of a buffer**

    → The pH of a buffer solution is given approximately by the Henderson–Hasselbalch equation

$$pH \approx pK_a + \log\left(\frac{[\text{base}]_{\text{initial}}}{[\text{acid}]_{\text{initial}}}\right)$$

    → Adjustment from initial concentrations to the equilibrium ones of the conjugate acid–base pair is *usually* negligible. If it is not, the pH can be adjusted to the desired value by adding additional acid or base.

    → Use of this equation is equivalent to setting up a table of concentrations and solving an equilibrium problem if adjustment from initial concentrations is negligible.

    → To minimize approximations and adjustments, an equilibrium table can be prepared and more exact concentrations for the conjugate pair can be obtained directly.

- **Preparation of an acid buffer**
  - → Use an acid (HA) with a pK_a value close to the desired pH.
  - → Add the acid and its conjugate base (A^-) in appropriate amounts to satisfy the Henderson–Hasselbalch equation.
  - → Measure the pH and make any necessary adjustments to the acid or base concentrations to achieve the desired pH.

- **Preparation of a base buffer**
  - → Use a base (B) whose conjugate acid (BH^+) has a pK_a value close to the desired pH.
  - → Add the base and its conjugate acid in appropriate amounts to satisfy the Henderson–Hasselbalch equation.
  - → Measure the pH and make any necessary adjustments to the acid or base concentrations to achieve the desired pH.

12.3 Buffer Capacity

- **Buffer capacity**
  - → The amount of acid or base that can be added before the buffer loses its ability to resist a change in pH
  - → Determined by its concentration and pH
  - → The *greater* the concentrations of the conjugate acid-base pair in the buffer system, the *greater* the buffer capacity

- **Exceeding the capacity of a buffer**
  - → Buffers function by providing a weak acid to react with added base and a weak base to react with added acid.
  - → Adding sufficient base to react with *all* of the weak acid or adding sufficient acid to react with *all* of the weak base exhausts (completely destroys) the buffer.

- **Limits of buffer action**
  - → A buffer has a practical limit to its buffering ability that is reached before the buffer is completely exhausted.
  - → As a *rule of thumb,* a buffer acts most effectively in the range

$$10 > \frac{[\text{base}]_{\text{initial}}}{[\text{acid}]_{\text{initial}}} > 0.1$$

  - → The molarity ratios above yield the following expression for the *effective pH range of a buffer*

$$pH = pK_a \pm 1$$

TITRATIONS (Sections 12.4–12.7)

- **Working with titrations** → Full chemical equation must be kept in mind to ensure the correct stoichiometry.

  Example: The neutralization of calcium hydroxide with hydrochloric acid requiring two moles of acid for each mole of base neutralized:

  $$2\,HCl(aq) + Ca(OH)_2 \rightarrow CaCl_2(aq) + 2\,H_2O(l)$$

- **Analyte** → Unknown acid or base sample to be analyzed
- **Titrant** → Base or acid solution of known concentration added to neutralize the analyte
- **Stoichiometric point** → Stage in a titration when exactly the correct volume of titrant has been added to the analyte to complete the reaction.

12.4 Strong Acid–Strong Base Titrations

- **pH curve** → Plot of the pH of the *analyte* solution as a function of the volume of the *titrant* added during a titration

- **For all strong base–strong acid or strong acid–strong base titrations**
  - → The stoichiometric point has a pH value of 7.
  - → The net ionic equation for the reaction is $H_3O^+(aq) + OH^0(aq) \rightarrow 2\,H_2O(l)$ for all strong base–strong acid or strong acid–strong base titrations.

- **Titration of a strong base (analyte) with a strong acid (titrant) (see Fig. 12.4 in the text)**
  - → The pH *changes (drops) slowly* with added titrant until the stoichiometric point is approached.
  - → The pH then *changes (drops) rapidly* near the stoichiometric point.
  - → At the stoichiometric point, the base is neutralized, at which point the titration is normally ended.
  - → In the region well beyond the stoichiometric point, the pH approaches the value of the titrant solution.

- **Titration of a strong acid (analyte) with a strong base (titrant) (see Fig. 12.5 in the text)**
  - → The pH *changes (rises) slowly* with added titrant until the stoichiometric point is approached.
  - → The pH then *changes (rises) rapidly* near the stoichiometric point.
  - → At the stoichiometric point, all the acid is neutralized.
  - → The pH approaches the value of the titrant solution in the region well beyond the stoichiometric point.

- **pH curve calculations**
  - → As a titration proceeds, both the volume of the analyte solution and the amount of acid or base present in it change.
  - → For calculations, it is often best to calculate amounts (moles) and volume separately, and to divide them to obtain concentrations.
  - → To calculate $[H_3O^+]$ in the analyte solution, the amount of acid or base remaining should be determined and divided by the total volume of the solution.
  - → Before the stoichiometric point is reached, follow the procedure outlined in Example 12.4 in the text.
  - → For a strong acid–strong base titration, the pH at the stoichiometric point is 7.
  - → After the stoichiometric point, follow the procedure outlined in Toolbox 12.1 in the text.

→ Always assume that the volumes of the titrant and analyte solutions are additive:

$$(V_{total} = V_{analyte} + V_{titrant\ added})$$

→ If a solid is added to a solution, its volume is usually neglected.

12.5 Strong Acid–Weak Base and Weak Acid–Strong Base Titrations

- **Characteristics of the titrations**
  - → The titrant is the *strong* acid or base, the analyte is the *weak* base or acid.
  - → Stoichiometric point has a pH value different from 7.

- **Strong base–weak acid titration**
  - → A strong base *dominates* the weak acid and the pH at the stoichiometric point is *greater* than 7.
  - → A strong base converts a weak acid, HA, into its conjugate base form, A⁻.
  - → At the stoichiometric point, the pH is that of the resulting basic salt solution.

- **Strong acid–weak base titration**
  - → A strong acid *dominates* the weak base and the pH at the stoichiometric point is *less* than 7.
  - → A strong acid converts a weak base, B, into its conjugate acid form, BH^+.
  - → At the stoichiometric point, the pH is that of the resulting acidic salt solution.

- **Shape of the pH curve**
  - → *Similar* to that of a strong acid–strong base titration
  - → *Differences* include:
    - ➤ A buffer region appears between the initial analyte solution and the stoichiometric point.
    - ➤ The pH of the stoichiometric point is different from 7.
    - ➤ There is a less abrupt pH change in the region of the stoichiometric point.

- **pH curve calculations**
  - → The pH is governed by the main species in solution.
  - → Calculate amounts and solution volume separately; divide to obtain concentrations.
  - → Before the titrant is added, the analyte is a solution of a weak acid or base in water. Determine the pH of the solution by following the procedures in Chapter 11 in the text.
  - → Before the stoichiometric point is reached, the solution is a buffer with differing amounts of acid/conjugate base. To calculate the pH, follow the procedure outlined in Example 12.6 in the text.
  - → At the stoichiometric point, the analyte solution consists of the salt of a weak acid or of a weak base. To calculate the pH, follow the procedure outlined in Example 12.5 in the text.
  - → Well after the stoichiometric point is passed, the pH of the solution approaches the value of the titrant, appropriately diluted.
  - → Assume that the volumes of the titrant solution added and the analyte solution are additive:

$$(V_{total} = V_{analyte} + V_{titrant\ added})$$

  If the sample is a solid, its volume is usually neglected.

12.6 Acid–Base Indicators

- **Acid–base indicator**
  - → A water-soluble dye with different *characteristic colors* associated with its acid and base forms

→ Exhibits a color change over a *narrow* pH range

→ Added in low concentration to the analyte so that a titration is primarily that of the analyte, not the indicator

→ Selected to monitor the stoichiometric or *end point* of a titration

→ An indicator is a weak acid. It takes part in the following proton-transfer equilibrium:

$$\mathrm{HIn(aq)} + \mathrm{H_2O(l)} \rightleftharpoons \mathrm{H_3O^+(aq)} + \mathrm{In^-(aq)} \qquad K_{\mathrm{In}} = \frac{[\mathrm{H_3O^+}][\mathrm{In^-}]}{[\mathrm{HIn}]}$$

Values of pK_{In} are given in Table 12.3 in the text.

→ The color change is usually most apparent when $[\mathrm{HIn}] = [\mathrm{In^-}]$ and $\mathrm{pH} = pK_{\mathrm{In}}$. The color change begins typically within one pH unit before pK_{In} and is essentially complete about 1 pH unit after pK_{In}. The pK_{In} value of an indicator is usually chosen to be within 1 pH unit of the stoichiometric point.

$$\boxed{pK_{\mathrm{In}} \approx \mathrm{pH(stoichiometric\ point)} \pm 1}$$

- **End point of a titration** → Stage of a titration where the color of the indicator in the analyte is midway between its acid and base colors

- **An indicator should be chosen such that its end point is close to the stoichiometric point.**

12.7 Stoichiometry of Polyprotic Acid Titrations

- **Polyprotic acids**
  - → Review the list of the common polyprotic acids in Table 11.10 of the text.
  - → All the acids in Table 11.10 are *diprotic,* except for phosphoric acid, which is *triprotic.*
  - → Phosphorous acid, H_3PO_3, is *diprotic.* The major Lewis structures are

The proton bonded to phosphorus is *not* acidic.

- **Titration curve of a polyprotic acid**
  - → Has stoichiometric points corresponding to the removal of each acidic hydrogen atom
  - → Has buffer regions between successive stoichiometric points
  - → At each stoichiometric point, the analyte is the salt of a conjugate base of the acid or one of its hydrogen-bearing anions.
  - → See the examples of pH curves for H_3PO_4, phosphoric acid, in Fig. 12.13 and $H_2C_2O_4$, oxalic acid, in Fig. 12.14 of the text.

- **pH calculations** → The pH curve can be estimated at any point by considering the primary species in solution and the *main* proton-transfer equilibrium that determines the pH.

SOLUBILITY EQUILIBRIA (Sections 12.8–12.14)

12.8 The Solubility Product

- **Solubility product, K_{sp}**
  - → Equilibrium constant for the equilibrium between a solid and its dissolved form (saturated solution)

    Example: $Sb_2S_3(s) \rightleftharpoons 2\,Sb^{3+}(aq) + 3\,S^{2-}(aq)$ $K_{sp} = \dfrac{a_{Sb^{3+}}^{2}\, a_{S^{2-}}^{3}}{a_{Sb_2S_3(s)}}$

  - → The activity of a pure solid is 1, and, for *dilute* solutions (sparingly soluble salts), the activity of a solute species can be replaced by its molarity.
  - → To a fair approximation, $K_{sp} = [Sb^{3+}]^2[S^{2-}]^3$.   (saturated solution)

- **Molar solubility, s** → Maximum amount of solute that dissolves in enough water to make one liter of solution

- **Relationship between s and K_{sp}**
  - → From the example above, the stoichiometric relationships are

    1 mol $Sb_2S_3 \triangleq 2$ mol Sb^{3+} and 1 mol $Sb_2S_3 \triangleq 3$ mol S^{2-}
  - → The molar solubility, s, of the salt $Sb_2S_3(s)$ is related to the ion concentrations at equilibrium as follows:

    $[Sb^{3+}] = 2s$ and $[S^{2-}] = 3s$.
  - → The relationship between s and K_{sp} is then

    $$K_{sp} = [Sb^{3+}]^2\,[S^{2-}]^3 = (2s)^2(3s)^3 = (4s^2)(27s^3) = 108s^5 \text{ and } s = \left(\frac{K_{sp}}{108}\right)^{1/5}$$

  - → The relationship $s = \left(\dfrac{K_{sp}}{108}\right)^{1/5}$ holds for any salt whose general formula is X_2Y_3.

  - → Similar relationships can be generated for other type of salts; for example, for an AB salt such as AgCl, $s = K_{sp}^{1/2}$.
  - → Because of interionic effects and incomplete dissociation of some salts (PbI_2, for example), calculations using K_{sp} are only approximate.

12.9 The Common-Ion Effect

- **Common-ion effect**
  - → A decrease in the solubility of a salt caused by the presence of one of the ions of the salt in solution
  - → Follows from Le Chatelier's principle
  - → Used to help remove unwanted ions from solution

    Example: For the equilibrium $Sb_2S_3(s) \rightleftharpoons 2\,Sb^{3+}(aq) + 3\,S^{2-}(aq)$, increasing the concentration of Sb^{3+} ions by addition of $Sb(NO_3)_3$ shifts the equilibrium to the left, causing some precipitation of $Sb_2S_3(s)$. Thus, less $Sb_2S_3(s)$ is dissolved in the presence of an external source of Sb^{3+} (or S^{2-}) than in pure water.

- **Quantitative features**
  - → Use the solubility product expression to estimate the reduction in solubility resulting from the addition of a common ion.
  - → For the equilibrium $Sb_2S_3(s) \rightleftharpoons 2\,Sb^{3+}(aq) + 3\,S^{2-}(aq)$, $K_{sp} = [Sb^{3+}]^2\,[S^{2-}]^3$.

    If $Sb_2S_3(s)$ is added to a solution already containing Sb^{3+} ions with a concentration $[Sb^{3+}]$, then dissolution of this *sparingly soluble salt* will add a negligible amount to the cation concentration. The solubility, s, and its relationship to K_{sp} is then

    $$K_{sp} = [Sb^{3+}]^2\,(3s)^3 \quad \text{and} \quad s = \left(\frac{[Sb^{3+}]^2}{27}\right)^{1/3}$$

    Note: Owing to interionic effects, the activities of ions differ markedly from their concentrations. As a result, simple equilibrium calculations for these systems using concentrations are at best "semi-quantitative." They become more accurate as the solubility of the salt decreases.

12.10 Predicting Precipitation

- **If Q_{sp} is greater than K_{sp}, then a salt precipitates.**

- **To predict whether precipitation occurs**
  - → Calculate Q_{sp} using actual or predicted ion concentrations.
  - → Compare Q_{sp} to K_{sp}.
  - → If $Q_{sp} > K_{sp}$, the salt will precipitate from solution.
  - → If $Q_{sp} = K_{sp}$, the solution is saturated with respect to the ions in the expression for K_{sp}. Any additional amount of these ions will cause precipitation.
  - → If $Q_{sp} < K_{sp}$, the solution is unsaturated and no precipitate will form.

12.11 Selective Precipitation

- **Separation of ions in a mixture** → Separated by adding a soluble salt containing an oppositely charged ion that forms salts having very different solubilities with the original ions

- **Separation of different cations in solution**
  - → Add a soluble salt containing an *anion* with which the cations form insoluble salts.
  - → If the insoluble salts have sufficiently different solubilities, they will precipitate at different anion concentrations and can be collected separately.

    Example: Careful addition of sodium fluoride, $NaF(s)$ to a solution containing Ca^{2+} and Pb^{2+} would lead to some separation of the cations. CaF_2 ($K_{sp} = 4.0 \times 10^{-11}$) would precipitate before PbF_2 ($K_{sp} = 3.7 \times 10^{-8}$).

12.12 Dissolving Precipitates

- **Ion removal**
  - → Removal from solution of one of the ions in the solubility equilibrium *increases* the solubility of the solid.
  - → A precipitate will dissolve completely if a sufficient number of such ions are removed.
  - → In this section, we focus on the removal of *anions*.

- **Removal of basic anions**
  - → Addition of a strong acid, such as HCl, neutralizes a basic anion.

    Example: Hydroxides react to form water:
    $$H_3O^+(aq) + OH^-(aq) \rightarrow 2H_2O(l)$$

  - → In some cases, gases are evolved.

    Example: Carbonates react with acid to form carbon dioxide gas:
    $$2H_3O^+(aq) + CO_3^{2-}(aq) \rightarrow CO_2(g) + 3H_2O(l)$$

- **Removal by oxidation of anions**

    Example: $S^{2-}(aq)$ from a metal sulfide is oxidized to $S(s)$ with nitric acid:
    $$3S^{2-}(aq) + 8H_3O^+(aq) + 2NO_3^-(aq) \rightarrow 3S(s) + 2NO(g) + 12H_2O(l)$$

- **Summary** → The solubility of a salt may be increased by removing one of its ions from solution. An acid can be used to dissolve a hydroxide, sulfide, sulfite, or carbonate salt. Nitric acid can be used to oxidize metal sulfides to sulfur and a soluble salt.

12.13 Complex Ion Formation

- **Complex ions**
  - → If a salt can form a complex ion with other species in solution, its solubility increases.
  - → Formed by the reaction of a metal cation, acting as a Lewis acid, with a Lewis base
  - → May disturb a solubility equilibrium by reducing the concentration of metal ions
  - → Net result is an increase in the solubility of salts.

    Example: $Fe^{2+}(aq) + 6CN^-(aq) \rightleftharpoons Fe(CN)_6^{4-}(aq)$

    In the above reaction, Fe^{2+} acts as a Lewis acid and CN^- acts as a Lewis base. Complexation reduces the concentration of Fe^{2+} in solution, thereby causing a shift in any solubility equilibrium in which Fe^{2+} is involved.

  - → A number of formation reactions are listed in Table 12.5 in the text.
  - → Electron pair acceptors (Lewis acids) that form complex ions include metal cations such as $Fe^{3+}(aq)$, $Cu^{2+}(aq)$, and $Ag^+(aq)$.
  - → Electron pair donors (Lewis bases) that form complex ions include basic anions or neutral bases such as $Cl^-(aq)$, $NH_3(aq)$, $CN^-(aq)$, and the thiosulfate ion $S_2O_3^{2-}(aq)$.

Quantitative features
  - → The equilibrium for a *formation reaction* is characterized by the *formation constant, K_f* .

    Example: $Fe^{2+}(aq) + 6CN^-(aq) \rightleftharpoons Fe(CN)_6^{4-}(aq)$ $K_f = \dfrac{[Fe(CN)_6^{4-}]}{[Fe^{2+}][CN^-]^6}$

  - → Values of K_f at 25°C for several complex ion equilibria are given in Table 12.5 in the text.
  - → Formation and solubility reactions may be combined to predict the solubility of a salt in the presence of a species that forms a complex with the metal ion.

Example: $FeS(s) \rightleftharpoons Fe^{2+}(aq) + S^{2-}(aq)$ $\quad K_{sp} = [Fe^{2+}][S^{2-}]$ $\quad$ *solubility reaction*

$$Fe^{2+}(aq) + 6\,CN^-(aq) \rightleftharpoons Fe(CN)_6^{4-}(aq) \quad K_f = \frac{[Fe(CN)_6^{4-}]}{[Fe^{2+}][CN^-]^6} \quad \text{formation reaction}$$

Combining the above two reactions produces the following:

$$FeS(s) + 6\,CN^-(aq) \rightleftharpoons Fe(CN)_6^{4-}(aq) + S^{2-}(aq)$$

$$K = K_{sp} \times K_f = \frac{[Fe(CN)_6^{4-}][S^{2-}]}{[CN^-]^6} = (6.3 \times 10^{-18})(7.7 \times 10^{36}) = 4.9 \times 10^{19} \gg 1$$

The equilibrium lies far to the right indicating a large solubility for FeS(s) in a cyanide solution.

Because 1 mol FeS $\hat{=}$ 1 mol S^{2-}, the molar solubility, s, is equal numerically to the sulfide ion concentration.

12.14 Qualitative Analysis

- **Qualitative analysis**
  - → Method used to identify and separate ions present in an unknown solution (for example, seawater)
  - → Utilizes complex formation, selective precipitation, and pH control
  - → Uses solubility equilibria to remove and identify ions selectively
- **Outline of the method**
  - → The method is outlined in Fig. 12.20 in the text and is summarized below.

Partial Qualitative Analysis Scheme

Step	Possible precipitate	K_{sp}
1) Add HCl(aq)	AgCl	1.6×10^{-10}
	Hg_2Cl_2	1.3×10^{-18}
	$PbCl_2$	1.6×10^{-5}
2) Add H_2S(aq) (in acid solution, there is a low S^{2-} concentration)	Bi_2S_3	1.0×10^{-97}
	CdS	4.0×10^{-29}
	CuS	7.9×10^{-45}
	HgS	1.6×10^{-52}
	Sb_2S_3	1.6×10^{-93}
3) Add base to H_2S(aq) (in basic solution, there is a higher S^{2-} concentration)	FeS	6.3×10^{-18}
	MnS	1.3×10^{-15}
	NiS	1.3×10^{-24}
	ZnS	1.6×10^{-24}

Note: In solubility calculations, the fairly strong basic character of S^{2-} must be considered.
$$S^{2-}(aq) + H_2O(l) \rightleftharpoons HS^-(aq) + OH^-(aq) \quad K_b = 1.4$$

→ A solution may contain any or all of the cations given in the preceding table. We wish to determine which ones are present and which are absent.

- **Procedure 1: Add HCl to the solution.**
  → Most chlorides are soluble and will not precipitate as the $[Cl^-]$ increases.
  → For AgCl and Hg_2Cl_2, we expect their Q_{sp} values to be exceeded and for them to precipitate.
  → $PbCl_2$ is slightly more soluble, but it should also precipitate.
  → The precipitate obtained, if any, using Procedure 1 is collected and the remaining solution tested for additional ions using Procedure 2.

- **Procedure 2: Add H_2S to the solution.**
  → Because the solution is highly acidic after Procedure 1, the concentration of sulfide will be very low:
  $$H_2S(aq) + 2H_2O(l) \rightleftharpoons 2H_3O^+(aq) + S^{2-}(aq)$$
  → Thus, only sulfide salts with very low solubility products, such as CuS or HgS, will precipitate.
  → The resulting precipitate, if any, is collected and the remaining solution tested for additional ions using Procedure 3.

- **Procedure 3: Add base to the solution.**
  → Base reduces $[H_3O^+]$ and, as seen from the equilibrium above, increases $[S^{2-}]$.
  → At higher sulfide concentration, more soluble sulfide salts such as FeS and ZnS precipitate.
  → The resulting precipitate, if any, is collected.

- **Precipitates obtained in Procedures 1, 2, and 3 are analyzed separately for the presence of each cation in the group.**

- **Determining the presence or absence of Ag^+, Hg_2^{2+}, and Pb^{2+} in Procedure 1**
  → Of the three possible chlorides, $PbCl_2$ is the most soluble and its solubility increases with increasing temperature.
  → Rinse the precipitate with hot water, collect the liquid, and test for Pb^{2+} by adding chromate.
  → A precipitate of insoluble yellow lead(II) chromate indicates that Pb^{2+} is present in the initial solution:
  $$Pb^{2+}(aq) + CrO_4^{2-}(aq) \rightarrow PbCrO_4(s)$$
  → Add aqueous ammonia to the remaining precipitate.
  → Silver(I) will dissolve, forming the diammine complex:
  $$Ag^+(aq) + 2NH_3(aq) \rightarrow Ag(NH_3)_2^+(aq)$$
  → In aqueous ammonia, mercury(I) will disproportionate (a single element is simultaneously oxidized and reduced) to form a gray mixture of liquid mercury and solid $HgNH_2Cl$:
  $$Hg_2Cl_2(s) + 2NH_3(aq) \rightarrow Hg(l) + HgNH_2Cl(s) + NH_4^+(aq) + Cl^-(aq)$$
  → Formation of the gray mixture confirms the presence of mercury(I) in the original solution.
  → Silver(I) is confirmed by adding HCl to the ammine solution to precipitate white AgCl(s):
  $$Ag(NH_3)_2^+(aq) + Cl^-(aq) + 2H_3O^+(aq) \rightarrow AgCl(s) + 2NH_4^+(aq) + 2H_2O(l)$$
  → Similar procedures can be used to identify the presence of the remaining cations.

Chapter 13 ELECTROCHEMISTRY

REPRESENTING REDOX REACTIONS (Sections 13.1–13.2)

13.1 Half-Reactions

- **Half-reaction**
  - → *Conceptual* way of reporting an oxidation or reduction process
  - → Important for understanding oxidation-reduction (*redox*) reactions and electrochemical cells

- **Oxidation**
  - → *Removal* of electrons from a species
  - → Increase in oxidation number of a species
  - → Represented by an oxidation half-reaction
  - **Example:** $Zn(s) \rightarrow Zn^{2+}(aq) + 2\,e^-$. Zinc metal loses two electrons to form zinc(II) ions.

- **Reduction**
  - → *Gain* of electrons by a species
  - → Decrease in oxidation number of a species
  - → Represented by a reduction half-reaction
  - **Example:** $Cu^{2+}(aq) + 2\,e^- \rightarrow Cu(s)$. Copper(II) ions gain two electrons each to form copper metal.

- **Redox couple** → Ox/Red. Oxidized (Ox) and reduced (Red) species in a half-reaction with the oxidized form listed first by convention
  - **Example:** The Zn^{2+}/Zn redox couple

- **Half-reactions**
  - → Express separately the oxidation and reduction contributions to a redox reaction
  - → May be added to obtain a redox reaction if the electrons cancel
  - **Example:** From the half reactions above, the following redox reaction is obtained:

Oxidation half-reaction:	$Zn(s) \rightarrow Zn^{2+}(aq) + 2\,e^-$
Reduction half-reaction:	$Cu^{2+}(aq) + 2\,e^- \rightarrow Cu(s)$
Redox reaction:	$Zn(s) + Cu^{2+}(aq) \rightarrow Zn^{2+}(aq) + Cu(s)$

13.2 Balancing Redox Equations

- **Balancing methods**
  - → Two methods are commonly used to balance redox equations: the *half-reaction method* and the oxidation number method.
  - → We will use the half-reaction method.
  - → In either method, mass and charge are balanced separately.
  - → In redox reactions, it is conventional to represent hydronium ions, $H_3O^+(aq)$, as $H^+(aq)$.

- **Half-reaction method**
  - → A six-step procedure (see also Toolbox 13.1 in the text)
    1. Identify all species being oxidized and reduced from changes in their oxidation numbers.
    2. Write unbalanced skeletal equations for each half-reaction.

3. Balance all elements in each half-reaction except O and H.

4. a) For an *acidic* solution, balance O by using H_2O, then balance H by adding H^+.

   b) For a *basic* solution, balance O by using H_2O, then balance H by adding H_2O to the side of each half-reaction that needs H and adding OH^- to the other side.

   Note: Adding H_2O to one side and OH^- to the other side has the *net* effect of adding H atoms to balance hydrogen if charge is ignored. Charge is balanced in the next step.

5. Balance charge by adding electrons to the appropriate side of each half-reaction.

6. a) Multiply the half-reactions by factors that give an equal number of electrons in each reaction.

   b) Add the two half-reactions to obtain the balanced equation.

→ An alternative procedure for *basic* solutions is to balance the redox equation for *acid* solutions first. Then, add

$$nH^+(aq) + nOH^-(aq) \rightarrow nH_2O(l) \quad or \quad nH_2O(l) \rightarrow nH^+(aq) + nOH^-(aq)$$

to the redox equation to eliminate nH^+ from the product *or* reactant side, whichever is necessary.

GALVANIC CELLS (Sections 13.3–13.10)

13.3 The Structure of Galvanic Cells

- **Galvanic cell** → Electrochemical cell in which a spontaneous reaction produces an electric current

- **Galvanic cell components**

  → Two *electrodes* (metallic conductors and/or conducting solids) make electrical contact with the cell contents, the conducting medium.

  → An *electrolyte* (an ionic solution, paste, or crystal) is the conducting medium.

  → *Anode*, labeled (−) or ⊖: the electrode at which *oxidation* occurs.

  → *Cathode*, labeled (+) or ⊕: the electrode at which *reduction* occurs.

  → If the anode and cathode do not share a common electrolyte, the oxidation reaction and reduction reaction compartments are separated to prevent direct mixing of the electrolyte solutions. Electrical contact between these solutions is maintained by a *salt bridge*, a gel containing an electrolyte such as KCl(aq), or an equivalent device.

- **Hydrogen electrode or half cell**

  → Inert Pt wire, used to conduct electricity, immersed in an acid solution, through which hydrogen gas is bubbled

  → Half-reaction: $2H^+(aq) + 2e^- \rightarrow H_2(g)$ (reduction) or $H_2(g) \rightarrow 2H^+(aq) + 2e^-$ (oxidation)

- **The Daniell cell, an example of a galvanic cell**

  → Zinc metal reacts spontaneously with Cu(II) ions in aqueous solution.

  → Copper metal precipitates and Zn(II) ions enter the solution.

→ The reaction is exothermic, as suggested by the standard enthalpy of the cell reaction:

$$Zn(s) + Cu^{2+}(aq) \rightarrow Zn^{2+}(aq) + Cu(s) \qquad \Delta H_r^\circ = -225.56 \text{ kJ} \cdot \text{mol}^{-1}$$

Oxidation
$Zn(s) \rightarrow Zn^{2+}(aq) + 2e^-$

Reduction
$Cu^{2+}(aq) + 2e^- \rightarrow Cu(s)$

The galvanic cell to the left contains *anode* and *cathode compartments* linked by a salt bridge to prevent the mixing of ions.

The anode compartment contains a zinc electrode in an electrolyte solution such as KCl or $ZnCl_2$.

The cathode compartment contains an electrode (not necessarily Cu) in an electrolyte solution containing $Cu^{2+}(aq)$ ions.

The salt bridge contains an electrolyte such as KCl in an aqueous gelatinous medium to control the flow of $K^+(aq)$ and $Cl^-(aq)$ ions. When current is drawn from the cell, the electrodes are linked by an external wire to a load (examples include a light bulb or motor). Current (electrons) flows through the wire from anode to cathode as indicated and the spontaneous chemical *cell reaction* $Zn(s) + Cu^{2+}(aq) \rightarrow Cu(s) + Zn^{2+}(aq)$ occurs.

As the reaction proceeds, $Zn(s)$ is oxidized to $Zn^{2+}(aq)$ in the anode compartment (oxidation) and $Cu^{2+}(aq)$ is reduced to $Cu(s)$ in the cathode compartment (reduction). The $Cu(s)$ deposits on the cathode, and the $Zn(s)$ anode slowly disappears. The closed line labeled *circuit* traces the direction that current flows in the cell. Current is carried by *electrons* in the electrodes and external connections, and by *ions* in the conducting medium.

13.4 Cell Potential and Reaction Free Energy

• **Cell potential and emf**

→ *Cell potential, E:* a measure of a cell reaction's ability to move electrons through the external circuit

→ *Electromotive force* (emf): cell potential when a cell is operated reversibly

→ For our purposes, cell potential E will be taken to mean emf unless otherwise noted.

→ The amount of useful work a cell can perform depends on the cell potential and the amount of the limiting reactant present.

• **Units and definitions**

→ Current: *ampere*, A, the SI base unit of electric current

→ Charge: *coulomb*, $C \equiv A \cdot s$, the SI derived unit of charge

→ Energy: *joule*, $J \equiv kg \cdot m^2 \cdot s^{-2}$, the SI derived unit of energy

→ Potential: *volt*, $V \equiv J \cdot C^{-1}$, SI derived unit of the cell potential

→ A charge of one *coulomb* falling through a potential difference of one *volt* produces one *joule* of energy.

→ Cell potential: Measured using a voltmeter connected to the two electrodes of a cell (see Fig. 13.4 in the text)

→ Convention: Write galvanic cells with anode on the left and cathode on the right. This arrangement gives a positive cell potential.

- **Examples of galvanic cell reactions with measured cell potentials**
  - $\rightarrow \quad Mg(s) + Zn^{2+}(aq, 1\ M) \rightarrow Zn(s) + Mg^{2+}(aq, 1\ M)$ $\qquad E° = 1.60\ V$
  - $\rightarrow \quad 2H^+(aq, 1\ M) + Zn(s) \rightarrow Zn^{2+}(aq, 1\ M) + H_2(g, 1\ bar)$ $\qquad E° = 0.76\ V$

  Note: For the preceding cells, all reactants and products are in their standard states and the potentials measured are called standard potentials, $E°$.

- **Relation between reaction free energy, ΔG_r, and cell potential, E**
  - $\rightarrow$ Reaction free energy, ΔG_r: *maximum nonexpansion* work, w_e, obtainable from a reaction at constant T and P: $\Delta G_r = w_e$
  - $\rightarrow$ When n electrons move through a potential difference, E, the work done is the total charge times the potential difference: $w_e = QV$.
  - $\rightarrow$ The charge on one mole of electrons is N_A times the charge of a single electron, $-e = -1.602\ 177 \times 10^{-19}\ C$: $Q = -eN_A$.
  - $\rightarrow$ The Faraday constant is the *magnitude* of the charge of *one mole of electrons*:
    $F = -eN_A = 9.648\ 531 \times 10^4\ C \cdot mol^{-1}$.
  - $\rightarrow$ The work done and reaction free energy are given by the product of the total charge, $-nF$, and the cell potential, E: $\Delta G_r = w_e = -nFE$.
  - $\rightarrow$ In the expression above, n is the stoichiometric coefficient of the *electrons* in the oxidation and reduction half-reactions that are combined to determine the cell reaction.
  - $\rightarrow$ *Maximum* nonexpansion work can only be obtained if the cell operates *reversibly*.
  - $\rightarrow$ *Reversibility* is only realized when the applied potential equals the potential generated by the cell.
  - $\rightarrow$ All working cells operate *irreversibly* and produce lower potentials than the emf.
  - $\rightarrow$ In sum, the relationship between the free energy and the cell potential for *reversible* electrochemical cells is one of the most important concepts in this chapter and is given by

  $$\boxed{\Delta G_r = -nFE}$$

- **Standard cell potential, $E°$**
  - $\rightarrow$ The standard state of a substance (s, l, or g) is the pure substance at a pressure of 1 bar.
  - $\rightarrow$ For solutes in solution, we take a molar concentration of $1\ mol \cdot L^{-1}$ as the standard-state activity.
  - $\rightarrow$ The standard cell potential, $E°$, is the cell potential measured when all reactants and products are in their standard states.
  - $\rightarrow$ The standard reaction free energy can be obtained from the standard cell potential (emf) and *vice versa:*

  $$\boxed{\Delta G_r° = -nFE°}$$

13.5 The Notation for Cells

- **Cell diagram**
  - $\rightarrow$ A symbolic representation of the cell components.
  - $\rightarrow$ A single vertical line | represents a phase boundary.
  - $\rightarrow$ A double vertical line || represents a salt bridge.
  - $\rightarrow$ Reactants and products are represented by chemical symbols.
  - $\rightarrow$ Components in the same phase are separated by commas.

→ The anode components are written first, followed by the salt bridge, if present, and then the cathode components.

→ Phase symbols (s, l, g) are normally used.

→ Concentrations of solutes and pressures of gases may also be given.

- **Cell diagrams with concentrations of reactants and products**

  → Often, the concentrations of reactants and products are included in the cell diagram to give a more complete description of the cell contents.

    Example: $Mg(s) + Zn^{2+}(aq, 1\ M) \rightarrow Zn(s) + Mg^{2+}(aq, 0.5\ M)$ cell reaction

    $Mg(s)\,|\,Mg^{2+}(aq, 0.5\ M)\,\|\,Zn^{2+}(aq, 1\ M)\,|\,Zn(s)$ cell diagram

    Standard cell diagram showing anode, cathode, and cell potential (see Section 13.6 below)

    $Mg(s)\,|\,Mg^{2+}(aq, 1\ M)\,\|\,Zn^{2+}(aq, 1\ M)\,|\,Zn(s)$
    anode (-)   cathode (+) $E\degree = 1.60$ V

    Electrons flow spontaneously from the anode (Mg) to the cathode (Zn) and the cell potential is positive: $\Delta G_r\degree = -nFE\degree < 0$ for spontaneity.

13.6 Standard Potentials

- **Standard cell potential**

  → *Difference* between the two standard electrodes of an electrochemical cell

  → The standard cell potential, $E\degree$, is the difference between the standard potentials of the electrode on the right side of the diagram and the electrode on the left side of the diagram.

  > $E\degree = E\degree(\text{electrode on right of cell diagram}) - E\degree(\text{electode on left of cell diagram})$
  >
  > or
  >
  > $E\degree = E_R\degree - E_L\degree$

  → If $E\degree > 0$, the cell reaction is spontaneous ($K_{eq} > 1$) under standard conditions as written.

  → If $E\degree < 0$, the reverse cell reaction is spontaneous ($K_{eq} < 1$) under standard conditions as written.

  → The standard potential of an electrode is sometimes called the *standard electrode potential* or the *standard reduction potential*.

- **Cell potential**

  → Measured with an electronic voltmeter

  → Voltmeter draws negligible current

  → Measured voltage is positive (+) when the + terminal of the meter is connected to the cathode

- **Standard electrode potentials, $E\degree$**

  → The potential of a single electrode cannot be measured.

  → A relative scale is required.

  → By convention, the standard potential of the hydrogen electrode is assigned a value of 0 *at all temperatures: $E\degree(H^+/H_2) \equiv 0$.*

→ For the standard hydrogen electrode (SHE), the half-reaction for reduction is
$2H^+(aq, 1 M) + 2e^- \rightarrow H_2(g, 1 bar)$, and the half-cell diagram is

$$H^+(aq, 1 M) \,|\, H_2(g, 1 bar) \,|\, Pt(s) \qquad E°(H^+/H_2) \equiv 0$$

→ If an electrode is found to be the anode in combination with the SHE, it is assigned a *negative potential*. If it is the cathode, its standard potential is *positive*.

→ Values of standard electrode potentials measured at 25°C are given in Table 13.1 and Appendix 2B in the text.

- **Standard electrode potentials and free energy**
  - → The relation between free energy and potential also applies to standard electrode potentials:

$$\boxed{\Delta G_r° = -nFE°}$$

  - → $\Delta G_f°$ values in Appendix 2A can be used to determine $\Delta G_r°$, and then an unknown standard potential using the equation directly above.

- **Calculating the standard potential of a redox couple from those of two related couples**
  - → If two redox couples are added or subtracted to give a third redox couple:
    - ➤ The standard potentials in general cannot be added to give the unknown potential.
    - ➤ This is because potential is an intensive property.
    - ➤ Free energies are extensive and can be added.
  - → To calculate the unknown potential, we convert potential values to free energy values.

- **Summary of the process of combining half-reactions**
  - → Combining two half-reactions (A and B) to obtain a third half-reaction (C) is accomplished by addition of the associated free energy changes.
  - → The equation, $\Delta G_r° = -nFE°$, applies to each half-reaction and leads to the following equation:

$$\boxed{E_C° = \dfrac{n_A E_A° + n_B E_B°}{n_C}}$$

- **Standard electrode potentials of half-reactions**
  - → Consider the half-reaction $M^{2+}(aq, 1 M) + 2e^- \rightarrow M(s) \qquad E = E°(M^{2+}/M)$.
  - → Metals with *negative* standard potentials $(E = E°(M^{2+}/M) < 0)$ have a thermodynamic tendency to reduce $H_3O^+(aq)$ to $H_2(g)$ under standard conditions (1 M acid).
  - → Metals with *positive* standard potentials $((E = E°(M^{2+}/M) >)$ cannot reduce hydrogen ions under standard conditions; in this instance, hydrogen gas is the stronger reducing agent, and will itself be oxidized.
  - → In general:
    - ➤ The more *negative* the standard electrode potential, the greater the tendency of a metal to *reduce* H^+.
    - ➤ The more *positive* the standard electrode potential, the greater the tendency of a metal ion to *oxidize* H_2.

13.7 The Electrochemical Series

- **Electrochemical series**

  → Standard half-reactions arranged in order of *decreasing* standard electrode potential and alphabetically (see both Table 13.1 and Appendix 2B in the text)

  → Table of relative strengths of oxidizing and reducing agents (see the table below)

  → In going *up the table,* the oxidizing strength of *reactants* increases.

  Example: $F_2(g)$ is an exceptionally *strong oxidizing agent* with a strong tendency to gain electrons and be reduced. $F_2(g)$ will oxidize any of the *product species* on the right side of the table in the reactions listed below it.

  → In going *down the table,* the reducing strength of *products* increases.

  Example: Li(s) is an exceptionally *strong reducing agent* with a large tendency to lose electrons and be oxidized. Li(s) will reduce any of the *reactant species* on the left side of the table in the reactions listed above it.

  → *Standard* electrode potentials have values ranging from about +3 V to −3 V, a difference of about 6 V.

  → No single *standard* galvanic cell may have a potential larger than about 6 V because there are no half-reactions that can give a larger potential.

	Reduction half-reaction		$E°$(V)
	$F_2(g) + 2e^- \rightarrow 2F^-(aq)$	Reducing	+2.87
	$Mn^{3+}(aq) + e^- \rightarrow Mn^{2+}(aq)$	strength	+1.51
	$I_2(s) + 2e^- \rightarrow 2I^-(aq)$		+0.54
	$2H^+(aq) + 2e^- \rightarrow H_2(g)$		0
	$Fe^{2+}(aq) + 2e^- \rightarrow Fe(s)$		−0.44
	$Al^{3+}(aq) + 3e^- \rightarrow Al(s)$		−1.66
Oxidizing	$Na^+(aq) + e^- \rightarrow Na(s)$		−2.71
strength	$Li^+(aq) + e^- \rightarrow Li(s)$		−3.05

- **Viewing half-reactions as conjugate pairs**

  → For a Brønsted-Lowry acid-base conjugate pair, the stronger the conjugate acid, the weaker the conjugate base.

  → For a given electrochemical half-reaction, the stronger the *conjugate* oxidizing agent, the weaker the *conjugate* reducing agent. $F_2(g)$ is an extremely powerful oxidizing agent, whereas $F^-(aq)$ is an extremely weak reducing agent.

- **Uses of the electrochemical series**

  → Predict whether a reaction has $K > 1$ or $K < 1$ for species in their standard states.

  → Predict *relative* reducing and oxidizing strengths.

  → Predict which reactants in their standard states may react spontaneously in a redox reaction.

  → Calculate standard cell potentials and equilibrium constants.

13.8 Standard Potentials and Equilibrium Constants

- **Equilibrium constants**

  → Obtained from standard cell potentials

  → For redox, acid-base, dissolution/precipitation, and dilution (change in concentration) reactions

- **Quantitative aspects**

  → Combining $\Delta G_r^{\circ} = -RT \ln K$ (from Chapter 10) with $\Delta G_r^{\circ} = -nFE^{\circ}$ yields

$$\ln K = \frac{nFE^{\circ}}{RT} = \frac{nE^{\circ}}{0.025\,693\ \text{V}} \quad \text{at } T = 298.15\ \text{K}$$

  → In the equation above, we have used the "molar convention" so that n is dimensionless (a pure number). See Section 10.3 in the text for a discussion of this convention.

  → Standard electrode potentials may be used to calculate E° values.

  → Because ΔG_r° applies to *all* reactions, the equation above may also be used for acid-base reactions, precipitation reactions, and dilution reactions.

  → For non-redox reactions, the overall "cell" reaction will not show the electron transfer that occurs in the half-reactions explicitly.

13.9 The Nernst Equation

- **Properties of a galvanic cell**

  → As a galvanic cell discharges, the cell potential, E, decreases and reactants form products.

  → When $E = 0$, the cell is completely discharged and the cell reaction is at equilibrium.

- **Quantitative aspects**

  → From Section 10.3 in the text, $\Delta G_r = \Delta G_r^{\circ} + RT \ln Q$, where Q is the reaction quotient. (Note that the *identical* symbol Q is also used for the quantity of electricity.)

  → From this chapter, $\Delta G_r = -nFE$ and $\Delta G_r^{\circ} = -nFE^{\circ}$.

  → Combining the above equations leads to the *Nernst equation:*

$$E = E^{\circ} - \frac{RT}{nF} \ln Q = E^{\circ} - \frac{(0.025\,693\ \text{V})}{n} \ln Q \quad \text{at } 298.15\ \text{K}$$

  → In the Nernst equation, "the molar convention" is used and n is a pure number (dimensionless).

- **The Nernst equation**

  → Describes the quantitative relationship between cell potential and the chemical composition of the cell

  → Is used to estimate the potential for the following:
  - ➤ Cell from its chemical composition
  - ➤ Half-cell *not* in its standard state

13.10 Ion-Selective Electrodes

- **Ion-selective electrode** → An electrode sensitive to the concentration of a particular ion

    Example: A metal wire in a solution containing the metal ion. The electrode potential for $E(Ag^+, Ag)$ or $E(Cu^{2+}, Cu)$ is sensitive to the concentration of Ag^+ or Cu^{2+}, respectively.

- **Measurement of pH**

    → An important application of the Nernst equation

    → Utilizes a galvanic cell containing an electrode sensitive to the concentration of the H^+ ion

    → A calomel electrode connected by a salt bridge to a hydrogen electrode:

Calomel electrode: $Hg_2Cl_2(s) + 2e^- \rightarrow 2Hg(l) + 2Cl^-(aq)$	$E° = +0.27$ V
Hydrogen electrode: $H_2(g) \rightarrow 2H^+(aq) + 2e$	$E° = 0.00$ V

Cell reaction: $Hg_2Cl_2(s) + H_2(g) \rightarrow 2H^+(aq) + 2Hg(l) + 2Cl^-(aq)$	$E° = +0.27$ V

    → Q in the Nernst equation for the cell reaction above will depend only on the unknown concentration of H^+ if the Cl^- concentration in the calomel electrode is fixed at a value determined for a *saturated* solution of KCl and the pressure of H_2 gas is kept constant at 1 bar.

- **Glass electrode**

    → A thin-walled glass bulb containing an electrolyte with a potential proportional to pH

    → Used to replace the (messy) hydrogen electrode in modern pH meters

    → Modern pH meters use glass and calomel electrodes in a single unit (probe) for convenience. The probe contacts the test solution (unknown pH) through a small salt bridge.

- **pX meters**

    → Devices sensitive to other ions such as Na^+ and CN^-

    → Used in industrial applications and in pollution control

---

ELECTROLYTIC CELLS (Sections 13.11–13.12)

13.11 Electrolysis

- **Electrolysis**

    → Process of driving a reaction in a nonspontaneous direction by using an electric current

    → Conducted in an electrolytic cell

- **Electrolytic cell**

    → An electrochemical cell in which electrolysis occurs

    → Different from a galvanic cell in design

    → Both electrodes normally share the same compartment.

    → As in a galvanic cell, oxidation occurs at the anode and reduction occurs at the cathode.

    → Unlike a galvanic cell, the anode has a *positive* charge and the cathode is *negative*.

    → In an electrolytic cell, electrons are forced to flow from the anode where oxidation occurs to the cathode where reduction occurs.

- **Example of an electrolytic cell**

  → Sodium metal is produced by the electrolysis of molten rock salt (impure sodium chloride) and calcium chloride. Calcium chloride lowers the melting point of sodium chloride (recall freezing point depression), increasing the energy efficiency of the process.

  → A schematic diagram of an electrolytic cell for the production of Na(l) and $Cl_2(g)$ is shown below (also see text Fig. 13.15).

  → The cell has one compartment that contains both the anode and cathode. The electrode material is often an inert metal such as platinum. Electrical current is supplied by an external source, such as a battery or power supply to drive this nonspontaneous reaction. This current *forces* reduction to occur at the cathode and oxidation to occur at the anode. The cathode is placed on the right side as in a galvanic cell. Sodium ions are reduced at the cathode and chloride ions are oxidized at the anode:

  cathode: $Na^+(l) + e^- \rightarrow Na(l)$

  anode: $2\,Cl^-(l) \rightarrow Cl_2(g) + 2\,e^-$

  Note that liquid sodium metal is formed in the electrolysis because the cell temperature of 600°C is above the melting point of sodium (98°C).

  Notes: In the *molten* salt mixture, Na^+ is reduced in preference to Ca^{2+}. In *aqueous* salt solutions, care must be taken in predicting the products of electrolysis because of the possibility of oxidizing or reducing water instead of either the anion or cation, respectively.

  $2\,Cl^-(l) \rightarrow Cl_2(g) + 2\,e^-$

  $Na^+(l) + e^- \rightarrow Na(l)$

- **Overpotential and the products of electrolysis**

  → For electrolysis to occur, an external potential *at least as great as* that of the spontaneous cell reaction must be applied to an electrolytic cell.

  → The *actual* potential required is often *greater* than this minimum value. The additional voltage required is called *overpotential*.

  → Overpotential depends on the structure of a solution near an electrode, which differs from that in the bulk solution. Near an electrode, electrons are transferred across an electrode-solution interface that depends on the solution and the condition of the electrode.

  → Because of overpotential, the observed electrolysis products may differ from those predicted using standard electrode potentials and the Nernst equation.

  → In a solution with more than one *reducible* species, the *reactant* with the *most positive* reduction potential will be reduced.

  → In a solution with more than one *oxidizable* species, the *product* with the *most negative* reduction potential will be oxidized.

13.12 The Products of Electrolysis

- **Faraday's law of electrolysis**

  → The number of moles of product formed is *stoichiometrically equivalent* to the number of moles of electrons supplied.

→ The amount of product depends on the current, the time it is applied, and the number of moles of electrons in the half-reaction.

- **Quantitative aspects of Faraday's law**
  → Determine the stoichiometry of a half-reaction; for example, $M^{n+} + ne^- \rightarrow M(s)$.
  → The quantity of electricity, Q, passing through the electrolysis cell is determined by the electric current, I, and the time, t, of current flow. Recall that 1 ampere (A) = 1 coulomb (C) s^{-1}.

$$\text{Charge supplied (C)} = \text{current (A)} \times \text{time (s)} \quad \text{or} \quad Q = It$$

  → Because $Q = nF$, where n is the number of moles of electrons and F is the Faraday constant, then

$$\text{Moles of electrons} = \frac{\text{charge supplied (C)}}{F} = \frac{\text{current } (I) \times \text{time } (t)}{F} \quad \text{or} \quad n(e^-) = \frac{It}{F}$$

  → The moles of product and the mass of product produced are determined from the number of moles of electrons passing through the cell, $n(e^-)$, the half-reaction, and the molar mass, M.

$$\text{Moles of product} = n(\text{p}) = \frac{It}{n(e^-)F}$$

$$\text{Mass of product} = m(\text{p}) = \left(\frac{It}{n(e^-)F} \right) M(\text{p})$$

  where $M(\text{p})$ is the molar mass of the product p.

THE IMPACT ON MATERIALS (Sections 13.13–13.15)

13.13 Applications of Electrolysis

- **Uses of electrolysis**
  → Extracting metals from their salts. Examples include Al(s), Na (see Fig. 13.15 in the text), and Mg.
  → Preparation of fluorine, chlorine, and sodium hydroxide
  → Refining (purifying) metals such as copper
  → Electroplating metals
- **Electroplating**
  → Electrolytic deposition of a thin metal film on an object
  → Object to be plated is made the cathode of an electrolysis cell.
  → Cell electrolyte is a salt of the metal to be plated.
  → Cations are supplied either by the added salt, or from oxidation of the anode, which is then made of the plating metal. For example, silver, gold, and chromium plating (see Fig. 13.16 in the text)

13.14 Corrosion

- **Corrosion**
  → Unwanted oxidation of a metal
  → An electrochemical process that is destructive and costly

- **Retarding corrosion**

  → Coating the metal with paint (**painting**) or plastic. Non-uniform coverage or uneven bonding leads to deterioration of the coating and rust.

  → **Passivation,** the formation of a nonreactive surface layer (Example: Al_2O_3 on aluminum metal)

  → **Galvanization**, coating the metal with an unbroken layer of zinc. Zinc is preferentially oxidized and the zinc oxide that forms provides a protective coating (passivation).

  → Use of a **sacrificial anode,** a metal more easily oxidized than the one to be protected, attached to it. Used for large structures. A disadvantage is that sacrificial anodes must be replaced when completely oxidized (think of an underground pipeline).

  → **Cationic electrodeposition** coatings. A sacrificial metal (yttrium) is used as the primer for automobile and truck bodies to reduce corrosion.

- **Mechanism of corrosion**

  → Exposed metal surfaces can act as anodes or cathodes, points where oxidation or reduction can occur.

  → If the metal is wet (from dew, rain, or groundwater), a film of water containing dissolved ions may link the anode and cathode, acting as a salt bridge.

  → The circuit is completed by electrons flowing through the metal, which acts as a wire in the external circuit of a galvanic cell.

- **Mechanism of rust formation on iron metal**

  → At the *anode,* iron is oxidized to iron(II) ions.

  (1) $Fe(s) \rightarrow Fe^{2+}(aq) + 2e^-$ $\qquad\qquad -E° = -(-0.44 \text{ V}) = +0.44 \text{ V}$

  → At the *cathode,* oxygen is reduced to water.

  (2) $O_2(g) + 4H^+(aq) + 4e^- \rightarrow 2H_2O(l)$ $\qquad E° = +1.23 \text{ V}$

  → Iron(II) ions migrate to the cathode where they may be oxidized by oxygen to iron(III).

  (3) $Fe^{2+}(aq) \rightarrow Fe^{3+}(aq) + e^-$ $\qquad\qquad -E° = -(+0.77 \text{ V}) = -0.77 \text{ V}$

  → The overall redox equation is obtained by adding reaction (1) four times, reaction (3) four times, and reaction (2) three times. The resulting cell reaction has $n = 12$.

  (4) $4Fe(s) + 3O_2(g) + 12H^+(aq) \rightarrow 4Fe^{3+}(aq) + 6H_2O(l)$ $\qquad E° = +1.27 \text{ V}$

  → Next, the iron(III) ions in reaction (4) precipitate as a hydrated iron(III) oxide or rust. The coefficient, x, of the waters of hydration in the oxide is not well defined.

  (5) $4Fe^{3+}(aq) + 2(3 + x)H_2O(l) \rightarrow 2Fe_2O_3 \cdot xH_2O(s) + 12H^+(aq)$

  → Finally, the overall reaction for the formation of rust is obtained by adding reaction (4) and (5):

  (6) $4Fe(s) + 3O_2(g) + 2xH_2O(l) \rightarrow 2Fe_2O_3 \cdot xH_2O(s)$

  → From this mechanism, summarized in reaction (6), we expect corrosion to be accelerated by moisture, oxygen, and salts, which enhance the rate of ion transport.

13.15 Practical Cells

- **Practical galvanic cells**

  → Commonly called batteries

  → Should be inexpensive, portable, safe, and environmentally benign

  → Should have a high *specific energy* (reaction energy per kilogram) and stable current

- **Primary cells**
  - $\rightarrow$ Cannot be recharged
  - $\rightarrow$ Some types of primary cells:
    - **Dry cells** — (A, AA, D, etc., batteries), used in flashlights, toys, remote controls, etc.

      Cell diagram: $Zn(s) \mid ZnCl_2(aq), NH_4Cl(aq) \mid MnO(OH)(s) \mid MnO_2(s) \mid C(gr)$, 1.5 V

      Cell reaction: $Zn(s) + 2\,NH_4^+(aq) + 2\,MnO_2(s) \rightarrow Zn^{2+}(aq) + 2\,MnO(OH)(s) + 2\,NH_3(g)$
    - **Alkaline cells** — similar to dry cells but with an alkaline electrolyte that increases battery life. Used in backup (emergency) power supplies and smoke detectors

      Cell diagram: $Zn(s) \mid ZnO(s) \mid OH^-(aq) \mid Mn(OH)_2(s) \mid MnO_2(s) \mid C(gr)$, 1.5 V

      Cell reaction: $Zn(s) + MnO_2(s) + H_2O(l) \rightarrow ZnO(s) + Mn(OH)_2(s)$
    - **Silver cells** — have only solid reactants and products, high emf, long life; used in pacemakers, hearing aids, and cameras

      Cell diagram: $Zn(s) \mid ZnO(s) \mid KOH(aq) \mid Ag_2O(s) \mid Ag(s) \mid steel$, 1.6 V

      Cell reaction: $Zn(s) + Ag_2O(s) \rightarrow ZnO(s) + 2\,Ag(s)$

- **Secondary cells**
  - $\rightarrow$ Can be recharged
  - $\rightarrow$ Must be charged before initial use
  - $\rightarrow$ Some types of secondary cells:
    - **Lead-acid cell** (automobile battery) — has low specific energy but generates high current over small time period to start vehicles

      Cell diagram: $Pb(s) \mid PbSO_4(s) \mid H^+(aq), HSO_4^-(aq) \mid PbO_2(s) \mid PbSO_4(s) \mid Pb(s)$, 2 V

      Cell reaction: $Pb(s) + PbO_2(s) + 2\,H_2SO_4(aq) \rightarrow 2\,PbSO_4(s) + 2\,H_2O(l)$
    - **Lithium-ion cell** — high energy density, high emf, can be recharged many times; used in laptop computers.
    - **Sodium-sulfur cell** — used to power electric vehicles, has all *liquid* reactants and products, a solid electrolyte, and relatively high voltage.

      Cell diagram: $Na(l) \mid Na^+(solution) \parallel S^{2-}(solution) \mid S_8(l)$, 2.2 V

      Cell reaction: $16\,Na(l) + S_8(l) \rightarrow 16\,Na^+(solution) + 8\,S^{2-}(solution)$
    - **Nickel-metal hydride (NiMH) cell** — used in hybrid vehicles as a supplemental energy source; utilizes hydrogen storage in metal alloys, has a low mass, high energy density, high current load capability, and is nontoxic (environmentally friendly)

- **Fuel cells**
  - $\rightarrow$ Galvanic cells designed for continuous reaction
  - $\rightarrow$ Require a continuous supply of reactants
  - $\rightarrow$ **Alkali-fuel cell** — used in the space shuttle

    Cell diagram: $Ni(s) \mid H_2(g) \mid KOH(aq) \mid O_2(g) \mid Ni(s)$, 1.23 V

    Cell reaction: $2\,H_2(g) + O_2(g) \rightarrow 2\,H_2O(l)$

Chapter 14   CHEMICAL KINETICS

REACTION RATES   (Sections 14.1–14.3)

14.1 Concentration and Reaction Rate

- **Definition of rate**

  General $\rightarrow$ Rate is the change in a property divided by the time required for the change to occur.

  Chemical kinetics $\rightarrow$ Reaction rate is the change in molar concentration of a reactant or product divided by the time required for the change to occur.

- **Average *vs.* instantaneous rates: an analogy to travel**

  $\rightarrow$ The *average* speed of an automobile trip is the length of the journey divided by the total time for the journey.

  $\rightarrow$ The *instantaneous* speed is obtained if the car is timed over a very short distance at some point in the journey.

- **Average reaction rate**

  $\rightarrow$ For a general chemical reaction, $a\text{A} + b\text{B} \rightarrow c\text{C} + d\text{D}$, the average reaction rate is the *change* in molar concentration of a reactant or product divided by the time interval required for the change to occur.

  $\rightarrow$ The average reaction rate may be determined for *any* reactant or product.

  $\rightarrow$ Rates for different reactants and products may have different numerical values.

- **Unique average reaction rate**

  $\rightarrow$ For a general chemical reaction, $a\text{A} + b\text{B} \rightarrow c\text{C} + d\text{D}$, the unique average reaction rate is the average reaction rate of reactant or product divided by its stoichiometric coefficient used as a *pure number*.

  $\rightarrow$ The unique average reaction rate may be determined from *any* reactant or product.

  $\rightarrow$ The unique average reaction rate is the same (has the same numerical value) for any reactant or product.

  $\rightarrow$ For the reaction above, the unique average reaction rate is defined as:

  $$\text{Unique average reaction rate} = -\frac{1}{a}\frac{\Delta[\text{A}]}{\Delta t} = -\frac{1}{b}\frac{\Delta[\text{B}]}{\Delta t} = \frac{1}{c}\frac{\Delta[\text{C}]}{\Delta t} = \frac{1}{d}\frac{\Delta[\text{D}]}{\Delta t}$$

  Note: Two critical underlying assumptions for generating a unique reaction rate are that the overall reaction time is slow with respect to the buildup and decay of any intermediate and that the stoichiometry of the reaction is maintained throughout.

- **Reaction rates**

  $\rightarrow$ The time required to measure the concentration changes of reactants and products varies considerably according to the reaction.

  $\rightarrow$ Spectroscopic techniques are often used to monitor concentration, particularly for fast reactions. An important example is the *stopped-flow technique* shown in Fig. 14.3 in the text. The fastest reactions occur on a time scale of femtoseconds (10^{-15} s).

14.2 The Instantaneous Rate of Reaction

- **Reaction rates**

  → Most reactions slow down as they proceed and reactants are depleted.

  → If equilibrium is reached, the forward and reverse reaction rates are equal.

  → The reaction rate at a specific time is called the *instantaneous reaction rate.*

- **Instantaneous reaction rate**

  → For a product, the *slope* of a tangent line of a graph of concentration *vs.* time. The slope is always positive because product concentrations increase with time.

  → For a reactant, the *negative of the slope* of a tangent line on a graph of concentration *vs.* time. The slope is always negative because reactant concentrations decrease with time.

- **Mathematical form of the instantaneous rate**

  → The slope of the concentration *vs.* time curve is the derivative of the concentration with respect to time.

  → For the instantaneous disappearance of a reactant, R,

  $$\text{Reaction rate} = -\frac{d[R]}{dt}$$

  → For the instantaneous appearance of a product, P,

  $$\text{Reaction rate} = \frac{d[P]}{dt}$$

  → For the general reaction, $a\text{A} + b\text{B} \rightarrow c\text{C} + d\text{D}$,

  $$\boxed{\text{Unique instantaneous reaction rate} = -\frac{1}{a}\frac{d[A]}{dt} = -\frac{1}{b}\frac{d[B]}{dt} = \frac{1}{c}\frac{d[C]}{dt} = \frac{1}{d}\frac{d[D]}{dt}}$$

  → In the remainder of the chapter, the term reaction rate is used specifically to mean the *unique* instantaneous reaction rate written here.

  → The two conditions relating to *unique average* rates described in Section 14.1 of the *Study Guide* also apply here.

14.3 Rate Laws and Reaction Order

- **Initial reaction rate**

  → Instantaneous rate at $t = 0$

  → In what is called the method of initial rates, analysis is simpler at $t = 0$ because products that may affect reaction rates are not present.

  → The method of initial rates is often used to determine the *rate law* of a reaction.

- **Rate law of a reaction**

  → Expression for the instantaneous reaction rate in terms of the concentrations of species that affect the rate such as reactants or products

  → Always determined by experiment and usually difficult to predict

  → May have relatively simple or complex mathematical form

  → The exponents in a rate expression are *not* necessarily the stoichiometric coefficients of a reaction; therefore, we use m and n in place of a and b in rate equations.

- **Simple rate laws**

  → Form of a simple rate law: $\boxed{\text{Rate} = k[A]^m}$

  → k = rate constant, which is the reaction rate when all the concentrations appearing in the rate law are 1 M (recall that the symbol M stands for the units of $\text{mol}\cdot\text{L}^{-1}$).

  → k is *independent* of concentration(s), but is *dependent* on temperature, $k(T)$.

  → A refers to the reactant, and m, the power of $[A]$, is called the order in reactant A and is determined experimentally.

  → Order is *not related to the reaction stoichiometry*. Exceptions include a special class of reactions, which are called *elementary* (see Section 14.7).

  → Typical orders for reactions with one reactant are 0, 1, and 2. Fractional values are also common.

Zero-order reaction	$(m = 0)$:	$\text{Rate} = k$
First-order reaction	$(m = 1)$:	$\text{Rate} = k[A]$
Second-order reaction	$(m = 2)$:	$\text{Rate} = k[A]^2$

- **More complex rate laws**

  → Many reactions have experimental rate laws containing concentrations of more than one species.

  → Form of a complex rate law: $\boxed{\text{Rate} = k[A]^m[B]^n\ldots}$

  The overall order is the sum of the powers $m + n + \ldots$.

  → The powers may also be negative values, and forms of even greater complexity with no overall order are found (see table below for an example and also Section 14.16).

  → More complex rate laws are treated in Section 14.8 in the text and this study guide.

  → The following table lists experimental rate laws for a variety of reactions:

Reaction	Rate law	Order	Units of k
$2\,NH_3(g) \rightarrow N_2(g) + 3\,H_2(g)$	$\text{Rate} = k$	Zero	$\text{mol}\cdot\text{L}^{-1}\cdot\text{s}^{-1}$
$2\,N_2O_5(g) \rightarrow 4\,NO_2(g) + O_2(g)$	$\text{Rate} = k[N_2O_5]$	First	s^{-1}
$2\,NO_2(g) \rightarrow NO(g) + NO_3(g)$	$\text{Rate} = k[NO_2]^2$	Second	$\text{L}\cdot\text{mol}^{-1}\cdot\text{s}^{-1}$
$2\,NO(g) + O_2(g) \rightarrow 2\,NO_2(g)$	$\text{Rate} = k[NO]^2[O_2]$	First in O_2 Second in NO Third overall	$\text{L}^2\cdot\text{mol}^{-2}\cdot\text{s}^{-1}$
$I_3^-(aq) + 2\,N_3^-(aq)$ $\rightarrow 3\,I^-(aq) + 3\,N_2(g)$ in the presence of $CS_2(aq)$	$\text{Rate} = k[CS_2][N_3^-]$	First in CS_2 First in N_3^- Second overall	$\text{L}\cdot\text{mol}^{-1}\cdot\text{s}^{-1}$
$2\,O_3(g) \rightarrow 3\,O_2(g)$	$\text{Rate} = k\dfrac{[O_3]^2}{[O_2]}$ $= k\,[O_3]^2[O_2]^{-1}$	Second in O_3 Minus one in O_2 First overall	s^{-1}

→ In general, the reaction order does not follow from the stoichiometry of the chemical equation.

→ In the case of the *catalytic* decomposition of ammonia on a hot platinum wire (first entry in the table on the previous page), the reaction order is zero initially because the reaction occurs on the surface of the wire, and the surface coverage is independent of concentration. The rate of a zero-order reaction is independent of concentration until the reactant is nearly exhausted or until equilibrium is reached.

→ Reaction rates may depend on species that do not appear in the overall chemical equation.

→ The reaction of triodide ion with nitride ion, for example, is accelerated by the presence of carbon disulfide, neither a reactant nor product. The concentration of carbon disulfide, a *catalyst,* appears in the rate law. Catalysis is discussed in Section 14.14 in the text and this study guide.

→ In the last example in the table, the order with respect to the product O_2 is -1.

→ Negative orders often arise from reverse reactions in which products reform reactants, thereby slowing the overall reaction rate. Oxygen is present at the beginning of this reaction as commonly studied because ozone is made by an electrical discharge in oxygen gas. It is possible to study the kinetics of the decomposition of pure ozone. In the absence of oxygen, the initial rate law is quite different from the one listed in the above table.

- **Procedure for determining the reaction orders and rate law**

  → The reaction order for a given species is determined by varying its initial concentration and measuring the initial reaction rate, while keeping the concentrations of all other species constant.

  → The procedure is repeated for the other species until all the reaction orders are obtained.

  → Once the form of the rate law is known, a value of the rate constant is calculated for each set of initial concentrations. The reported rate constant is typically the average value for all the measurements. See Table 14.1 in the text for examples of rate laws and rate constants.

- **Pseudo-order reactions**

  → If a species in the rate-law equation is present in great excess, its concentration is effectively constant as the reaction proceeds.

  → The constant concentration may be incorporated into the rate constant and the apparent order of the reaction changes.

  → Reactions in which the solvent appears in the rate law often show pseudo-order behavior because the solvent is usually present in large excess.

  → A technique called the *isolation method* is often used to determine the order in a particular reactant by keeping the concentration of other reactants in excess. The method is repeated for all reactants and the complete rate law is obtained.

  → A hidden danger is that occasionally the rate law is *not* the same in the extreme ranges of the concentrations of reactants. Different reaction mechanisms that yield different rate laws may occur under the extreme conditions.

CONCENTRATION AND TIME (Sections 14.4–14.6)

14.4 First-Order Integrated Rate Laws

- **Integration of a first-order rate law**

  → First-order rate law:
  $$\boxed{-\frac{d[A]}{dt} = k[A]}$$ (1)

  → Integration with limits:
  $$-\int_{[A]_0}^{[A]_t} \frac{d[A]}{[A]} = \int_0^t k\,dt$$ (2)

  → Result in logarithmic form:
  $$\boxed{\ln\left(\frac{[A]_t}{[A]_0}\right) = -kt}$$ (3)

  → Result in exponential form:
  $$\boxed{[A]_t = [A]_0 e^{-kt}}$$ (4)

  → If a plot of $\ln[A]_t$ vs. t gives a straight line, the rate law is first-order in [A].

  → The rearranged logarithmic expression, $\ln[A]_t = \ln[A]_0 - kt$, shows the straight-line behavior, $y = \text{intercept} + (\text{slope} \times x)$, of a plot of $\ln[A]_t$ vs. t.

  → Plot of $\ln[A]_t$ vs. t: slope $= -k$ *and* intercept $= \ln[A]_0$

- **Integrated first-order rate law (first equation above)**

  → [A] decays exponentially with time [see eq. (4) above].

  → Used to confirm that a reaction is first order and to measure its rate constant

  → Known k and $[A]_0$ values can be used to predict the value of [A] at any time t.

14.5 Half-Lives for First-Order Reactions

- **Half-life expression**

  → The half-life, $t_{1/2}$, of a substance is the time required for its concentration to fall to one-half its initial value.

  → At $t = t_{1/2}$, $[A]_{t_{1/2}} = \frac{1}{2}[A]_0$. The ratio of concentrations is $\dfrac{[A]_0}{[A]_{t_{1/2}}} = \dfrac{[A]_0}{\frac{1}{2}[A]_0} = 2$.

  → Solve for $t_{1/2}$ by rearranging the logarithmic form of the integrated rate law, eq. (3) above, to give

  $$t_{1/2} = \frac{1}{k}\ln\left(\frac{[A]_0}{[A]_{t_{1/2}}}\right).$$ Then substitute $\dfrac{[A]_0}{[A]_{t_{1/2}}} = 2$.

  → Half-life expression:
  $$\boxed{t_{1/2} = \frac{\ln 2}{k}}$$ (5)

- **Half-life of a first-order reaction**

  → Independent of the initial concentration

  → Characteristic of the reaction

→ Inversely proportional to the rate constant, k

→ The logarithmic form of the rate law may be modified as follows:

$$\ln\left(\frac{[A]_0}{[A]_t}\right) = kt = (\ln 2)\left(\frac{t}{t_{1/2}}\right) \qquad (6)$$

14.6 Second-Order Integrated Rate Laws

- **Integration of a second-order rate law**

  → Second-order rate law:
  $$-\frac{d[A]}{dt} = k[A]^2 \qquad (7)$$

  → Integration with limits:
  $$-\int_{[A]_0}^{[A]_t}\frac{d[A]}{[A]^2} = \int_0^t k\, dt \qquad (8)$$

  → Integrated form:
  $$\frac{1}{[A]_t} - \frac{1}{[A]_0} = kt \qquad (9)$$

  → Alternative form:
  $$[A]_t = \frac{[A]_0}{1 + [A]_0 kt} \qquad (10)$$

  → If a plot of $1/[A]_t$ *vs. t* gives a straight line, the rate law is second-order in [A].

  → The rearranged expression, $1/[A]_t = 1/[A]_0 + kt$, reveals the linear behavior of a plot of $1/[A]_t$ *vs. t*.

  → Plot of $1/[A]_t$ *vs. t:*   slope $= +k$ *and*   intercept $= 1/[A]_0$

- **Integrated second-order rate law**

  → For the same initial rates, [A] decays more slowly with time for a second-order process than for a first-order one (see Fig. 14.14 in the text).

  → Used to confirm that a reaction is second order and to measure its rate constant

  → If k and $[A]_0$ are known, the value of [A] at any time t can be predicted.

- **Summary of rate laws** → See Table 14.2 in the text.

- **Pseudo reaction order** → Treated in Section 14.3 in this study guide.

REACTION MECHANISMS (Sections 14.7–14.10)

14.7 Elementary Reactions

- **Reactions**

  → A net chemical reaction is assumed to occur at the molecular level in a series of separable steps, called *elementary reactions*.

→ A *reaction mechanism* is a proposed group of elementary reactions that accounts for the reaction's overall stoichiometry and experimental rate law.

- **Elementary reaction**

  → A single-step reaction, written *without* state symbols, that describes the behavior of *individual* atoms and molecules taking part in a chemical reaction

  → The *molecularity* of an elementary reaction specifies the number of *reactant* molecules that take part in it.

  → *Unimolecular:* Only one reactant species is written (molecularity = 1).

  C_3H_6 (cyclopropane) $\rightarrow$ CH_2=CH$-CH_3$ (propene)

  Assumption: *Cyclopropane molecule decomposes spontaneously to form a propene molecule.*

  → *Bimolecular:* Two reactant species combine to form products (molecularity = 2).

  H_2 + Br $\rightarrow$ HBr + H

  Assumption: *A hydrogen molecule and a bromine atom collide to form a hydrogen bromide molecule and a hydrogen atom.*

  → *Termolecular:* Three reactant species combine to form products (molecularity = 3).

  I + I + H_2 $\rightarrow$ HI + HI

  Assumption: *Two iodine atoms and a hydrogen molecule collide simultaneously to form two hydrogen iodide molecules.*

- **Reaction mechanism**

  → *Sequence* of elementary reactions which, when added together, gives the net chemical reaction and reproduces its rate law

  → *Reaction intermediate, not* a reactant or a product, is a species that appears in one or more of the elementary reactions in a proposed mechanism. It usually has a small concentration and does not appear in the overall rate law.

  → The plausibility of a reaction mechanism may be tested, but it cannot be proved.

  → The presence and behavior of a reaction intermediate are sometimes testable features of a proposed reaction mechanism.

  → Reverse elementary reactions can also be proposed in a reaction mechanism. They provide a mechanism for a reaction to reach equilibrium.

  Example: C_3H_6 (cyclopropane) $\rightarrow$ CH_2=CH$-CH_3$ (forward elementary reaction)

  CH_2=CH$-CH_3$ $\rightarrow$ C_3H_6 (cyclopropane) (reverse elementary reaction)

  At equilibrium, the forward and reverse reaction rates are equal.

14.8 The Rate Laws of Elementary Reactions

- **Forms of rate laws for elementary reactions**

  → Order follows from the stoichiometry of the *reactants* in an *elementary* reaction. Note that products do *not* appear in the rate law of an *elementary* reaction. See Table 14.3 in the text.

  → *Unimolecular* elementary reactions are first-order.

  Reaction: A $\rightarrow$ products   Rate = k [A]

→ *Bimolecular* elementary reactions are second-order overall. If there are two different species present, the reaction is first-order in each one.

Reaction: $A + A \rightarrow$ products   Rate $= k\,[A]^2$

Reaction: $A + B \rightarrow$ products   Rate $= k\,[A][B]$

→ *Termolecular* elementary reactions are third-order overall. If there are two different species, the reaction is second-order in one of them and first-order in the other. If there are three different species, the reaction is first-order in each one.

Reaction: $A + A + A \rightarrow$ products   Rate $= k\,[A]^3$

Reaction: $A + A + B \rightarrow$ products   Rate $= k\,[A]^2[B]$

Reaction: $A + B + C \rightarrow$ products   Rate $= k\,[A][B][C]$

→ The molecularity of an elementary reaction that is written in the reverse direction also follows from its stoichiometry. The *products* have become *reactants*.

Forward reaction: $A + B \rightarrow P + Q + R$ Forward rate $= k\,[A][B]$

Reverse reaction: $P + Q + R \rightarrow A + B$ Reverse rate $= k'\,[P][Q][R]$

Note: The prime symbol, ′, is used here and in the following sections to denote a rate constant written for a *reverse* elementary reaction. The same symbol is used in Section 14.11 to denote a rate constant for the *same* reaction (elementary or otherwise) measured at a *different* temperature. *Do not confuse these two very different meanings.*

- **Rate laws from reaction mechanisms**

  → The time evolution of each reactant, intermediate, and product may be determined by integrating the rate laws for the elementary reactions in a proposed reaction mechanism.

  → Recall that a successful mechanism must account for the stoichiometry of the overall reaction as well as the observed rate law.

  → Mathematical solution of kinetic equations often requires numerical methods for solving simultaneous differential equations. In a few cases, exact analytical solutions exist.

  → For multistep mechanisms, approximate methods are often used to determine a rate law consistent with a proposed reaction mechanism. If the derived rate law is in agreement with the experimental one, the plausibility of the mechanism is tested by further experimentation, if possible.

- **Approximate methods for determining rate laws**

  → Characteristics of reactions commonly encountered allow for the determination of rate laws with the use of three types of approximations.

  → The *rate-determining step* approximation is made to determine a rate law for a mechanism in which one step occurs at a rate *substantially* slower than any others.

  → The slow step is a bottleneck, and the overall reaction rate cannot be larger than the slow step.

  → The rate law for the rate-determining step is written first. If a reaction intermediate appears as a reactant in this step, its concentration term must be eliminated from the rate law.

  → The final rate law only has concentration terms for reactants and products.

  → The *pre-equilibrium condition* describes situations in which reaction intermediates are formed and removed in steps prior to the rate-determining one.

$\rightarrow$ If the formation and removal of the intermediate in prior steps is rapid, an equilibrium concentration is established. If the intermediate appears in the rate-determining step, its relatively slow reaction in that step does not change its equilibrium concentration. The pre-equilibrium condition is sometimes called a *fast equilibrium*.

$\rightarrow$ The **steady-state approximation** is a more general method for solving reaction mechanisms. The net rate of formation of any intermediate in the reaction mechanism is set equal to 0.

$\rightarrow$ An intermediate is assumed to attain its steady-state concentration instantaneously, decaying slowly as reactants are consumed. An expression is obtained for the steady-state concentration of each intermediate in terms of the rate constants of elementary reactions and the concentrations of reactants and products.

$\rightarrow$ The rate law for an elementary step that leads directly to product formation is usually chosen. The concentrations of all intermediates are removed from the chosen rate law, and a final rate law for the formation of product that reflects the concentrations of reactants and products is obtained.

$\rightarrow$ Often, a predicted rate law does not quite agree with the experimental one. Reaction conditions that modify the form of the rate law predicted by the mechanism may lead to final agreement (see Section 14.16 in the text and this study guide for an example). If complete exploration of conditions fails to reproduce the experimental result, the proposed reaction mechanism is rejected.

14.9 Chain Reactions

- **Chain reaction**

    $\rightarrow$ Series of linked elementary reactions that *propagate* in chain cycles

    $\rightarrow$ A reaction intermediate is produced in an *initiation* step.

    $\rightarrow$ Reaction intermediates are the *chain carriers,* which are often radicals.

    $\rightarrow$ In one propagation cycle, a reaction intermediate typically reacts to form a different intermediate, which in turn reacts to regenerate the original one.

    $\rightarrow$ In each cycle, some product is usually formed.

    $\rightarrow$ The cycle is eventually broken by a *termination* step that consumes the intermediate.

    $\rightarrow$ In *chain branching,* a chain carrier reacts to form two or more carriers in a single step.

    $\rightarrow$ Chain branching often leads to chemical explosions.

    $\rightarrow$ Branched chain reaction: $H^\bullet + O_2 \rightarrow HO^\bullet + {}^\bullet O^\bullet$ (two carriers are produced from one)

- **Example in text of a chain reaction**

    $\rightarrow$ Reaction: $\quad H_2(g) + Br_2(g) \rightarrow 2HBr(g)$

    $\rightarrow$ Initiation: $\quad Br_2(g) \rightarrow Br^\bullet + Br^\bullet$ (Br$^\bullet$ is a chain carrier)

    $\rightarrow$ Propagation: $\quad Br^\bullet + H_2 \rightarrow HBr + H^\bullet$ (H$^\bullet$ is also a chain carrier)

    $\qquad\qquad\qquad H^\bullet + Br_2 \rightarrow HBr + Br^\bullet$

    $\rightarrow$ Termination: $\quad Br^\bullet + Br^\bullet \rightarrow Br_2; \quad Br^\bullet + H^\bullet \rightarrow HBr; \quad H^\bullet + H^\bullet \rightarrow H_2$

14.10 Rates and Equilibrium

- **Elementary reaction**

  → At *equilibrium,* the forward and reverse reaction rates for an *elementary* reaction are equal. For a given elementary reaction at equilibrium, $A + B \rightleftharpoons P + Q$.

  Forward rate $= k[A][B] =$ reverse rate $= k'[P][Q]$

  → The *equilibrium constant* for an *elementary* reaction is equal to the ratio of the rate constants for the forward and reverse reactions.

  In the above example, $K = \dfrac{[P][Q]}{[A][B]} = \dfrac{(\text{reverse rate})/k'}{(\text{forward rate})/k} = \dfrac{k}{k'}$

  $$\boxed{K = \dfrac{k}{k'}} \qquad \text{for any elementary reaction}$$

- **Equilibrium constants from reaction mechanisms**

  → Consider an overall reaction with a multistep mechanism in which the rate constants for the elementary reactions are $k_1, k_2, k_3, \ldots$ in the forward direction and $k_1', k_2', k_3', \ldots$ in the reverse direction.

  → If each elementary step is in equilibrium, the overall reaction is obtained by summing the elementary reactions, and the equilibrium constant for the overall reaction is obtained by multiplying the equilibrium constants for the individual steps.

  → For the reaction at equilibrium, the above procedure gives

  $$\boxed{K = \dfrac{k_1}{k_1'} \times \dfrac{k_2}{k_2'} \times \dfrac{k_3}{k_3'} \times \ldots}$$

  → Large equilibrium constants ($K \gg 1$) are expected if forward rate constants are much larger than reverse rate constants.

  → Note that that we have gained an understanding of when to expect large or small equilibrium constants from kinetic data alone, with no knowledge of the Gibbs free energy.

MODELS OF REACTIONS (Sections 14.11–14.13)

14.11 The Effect of Temperature

- **Qualitative aspects**

  → A change in temperature results in a change of the rate constant of a reaction, and, therefore, its rate.

  → For *most* reactions, *increasing* the temperature results in a *larger value* of the rate constant.

  → For *many* reactions in organic chemistry, reaction rates *approximately* double for a 10°C increase in temperature.

- **Quantitative aspects**

  → The temperature dependence of the rate constant of many reactions is given by the Arrhenius equation, shown here in two forms:

  Exponential form:
  $$k = A e^{-\frac{E_a}{RT}}$$

  Logarithmic form:
  $$\ln k = \ln A - \frac{E_a}{RT}$$

  R is the universal gas constant ($8.314\,47$ J$\cdot$K$^{-1}\cdot$mol^{-1}).
  T is the absolute temperature (K).

  → The two constants, A and E_a, called the *Arrhenius parameters,* are *nearly* independent of temperature.

  → The parameter A is the *pre-exponential factor.* The parameter E_a is the *activation energy.* The two parameters depend on the reaction being studied.

  → The Arrhenius equation applies to all types of reactions, gas phase or in solution.

- **Obtaining Arrhenius parameters**

  → Plot $\ln k$ *vs.* $1/T$.

  → If the plot displays straight-line behavior, the *slope* of the line is $-E_a/R$ and the *intercept* is $\ln A$. Note that the *intercept* at ($1/T = 0$) implies $T = \infty$.

  → The parameter A *always* has a positive value. The *larger* the value of A, the *larger* is the value of k.

  → The parameter E_a *almost always* has a positive value. The *larger* the value of E_a, the *smaller* the value of k. If $E_a > 0$, the value of k *increases* as temperature *increases*.

  → The rate constant increases more quickly with increasing temperature if E_a is large.

  → If E_a and k are known at a given temperature, T, the logarithmic form of the Arrhenius equation can be used to calculate the rate constant, k', at another temperature, T'. Or alternatively, a value of E_a can be estimated from values of k' at T' and k at T.

  $$\ln \frac{k'}{k} = \frac{E_a}{R}\left(\frac{1}{T} - \frac{1}{T'}\right)$$

- **Units of the Arrhenius parameters**

  → A has the same units as the rate constant of the reaction.

  → E_a has units of energy, and values are usually reported in kJ$\cdot$mol^{-1}.

  Note: An equivalent SI unit for the liter is dm^3. Rate constants and Arrhenius pre-exponential factors incorporating volume units are sometimes reported using dm^3 or cm^3 instead of liters. Those incorporating amount units are sometimes reported in units of molecules instead of moles. For example, a second-order rate constant of 1.2×10^{11} L$\cdot$mol$^{-1}\cdot$s^{-1} is equivalent to 1.2×10^{11} dm$^3\cdot$mol$^{-1}\cdot$s^{-1} *or* 2.0×10^{-10} cm$^3\cdot$molecule$^{-1}\cdot$s^{-1}.

- **Selected Arrhenius Parameters** $\rightarrow$ See Table 14.4 in the text.

- **Temperature dependence of the equilibrium constant of an elementary reaction**

  - $\rightarrow$ Follows from the temperature dependence of the ratio of the rate constant for the forward reaction to that of the reverse reaction, with both expressed in Arrhenius form

  - $\rightarrow$ If the reaction is *exothermic,* the activation energy in the forward direction is smaller than in the reverse one (see the two figures in this study guide in Section 14.13 or Fig. 14.27 in the text).

    - ➤ The rate constant with the larger activation energy increases more rapidly as temperature increases than the one with smaller energy.

    - ➤ As temperature is *increased,* k' is predicted to increase faster than k and the equilibrium constant *decreases.*

    - ➤ *Reactants* are favored.

  - $\rightarrow$ If the reaction is *endothermic,* the activation energy is larger in the forward direction than in the reverse one.

    - ➤ The rate constant with the larger activation energy increases more rapidly as temperature increases than the one with smaller energy.

    - ➤ As temperature is *increased,* k is predicted to increase faster than k' and the equilibrium constant *increases.*

    - ➤ *Products* are favored.

14.12 Collision Theory

- **Collision theory**
  - $\rightarrow$ Model for how reactions occur at the molecular level
  - $\rightarrow$ Applies to gases
  - $\rightarrow$ Accounts for rate constants and the exponential form of the Arrhenius equation
  - $\rightarrow$ Reveals the significance of the Arrhenius parameters A and E_a

- **Assumptions of collision theory**
  - $\rightarrow$ Molecules must collide in order to react.
  - $\rightarrow$ Only collisions with kinetic energy in excess of some minimum value, E_{min}, lead to reaction.
  - $\rightarrow$ Only molecules with the correct orientation with respect to each other may react.

- **Total rate of collisions in a gas mixture**
  - $\rightarrow$ Determined quantitatively given from the kinetic model of a gas
  - $\rightarrow$ Rate of collision = total number of collisions per unit volume per second

$$\boxed{\text{Rate of collision} = \sigma \bar{v}_{rel} N_A^2 [A][B]}$$

  $\sigma =$ collision cross-section (area a molecule presents as a target during collision)
  Larger molecules are more likely to collide with other molecules than are smaller ones.

  $\bar{v}_{rel} =$ mean speed at which molecules approach each other in a gas

→ For a gas mixture of A and B with molar masses M_A and M_B, respectively,

$$\overline{v}_{rel} = \sqrt{\frac{8RT}{\pi\mu}} \qquad \mu = \frac{M_A M_B}{M_A + M_B}$$

→ Molecules at high temperature collide more often than at low temperature.

→ N_A = Avogadro constant

→ [A] and [B] are the molar concentrations of A and B, respectively.

- **Total rate of reaction in a gas mixture**
  - → Rate of reaction = number of collisions that lead to reaction per unit volume per second

  $$\text{Rate of reaction} = P\sigma \overline{v}_{rel} N_A^2 [A][B] \times e^{-E_{min}/RT}$$

  - → E_{min} = minimum kinetic energy required for a collision to lead to reaction. Collisions with energy less than the minimum do not lead to reaction.

  - → $e^{-E_{min}/RT}$ = fraction of collisions with at least the energy E_{min}. Derived from the *Boltzmann distribution* (see Fig. 14.26 in the text)

  - → P = steric factor < 1. Reactive collisions often require preferred directions of approach.

- **Reaction rate constant**
  - → Rate constant = $k = \dfrac{\text{rate of reaction}}{[A][B]}$

  $$k = P\sigma \overline{v}_{rel} N_A^2 e^{-E_{min}/RT}$$

  - → Comparison with the Arrhenius equation in exponential form, $k = Ae^{-\frac{E_a}{RT}}$:

    $E_a \approx E_{min}$ and $A \approx P\sigma \overline{v}_{rel} N_A^2$

  - → Because the speed is proportional to $T^{1/2}$, the model predicts that the pre-exponential factor *and* activation energy both have a *slight* temperature dependence.

- **Summary**
  - → According to collision theory, reaction occurs only if reactant molecules collide with a kinetic energy equal to or greater than the Arrhenius activation energy.
  - → The reaction rate constant increases with the size and speed of the colliding species (molecules).
  - → The steric factor accounts for collisions that lead to no reaction because the orientation of the molecules during such a collision is unfavorable. See Table 14.5 in the text for some P values.

14.13 Transition State Theory

- **Transition state theory (also often called Activated Complex Theory)**
  - → Applies to gas phase and solution reactions
  - → Reacting molecules collide and distort.
  - → During an encounter, the kinetic energy, E_K, of reactants decreases, the potential energy increases, the original chemical bonds lengthen, and new bonds begin to form.

→ Products may form if E_K for the collision $\geq E_a$ and if the reactants have the proper orientation.

→ Reactants reform if $E_K < E_a$, regardless of orientation.

→ If $E_K \geq E_a$, the molecules form either reactants or products. The lowest kinetic energy that may produce products is equal to E_a.

→ *Activated complex:* Combination of two reacting molecules at the transition point between reactants and products

- **Reaction profile**

  → Shows how energy changes as colliding reactants form a *transition state* or an *activated complex* and then form products along the *minimum energy pathway.*

  → An analogous process is hiking through a mountain range and choosing the pathway of lowest elevation through it.

  → Progress of reaction refers to the relevant spatial coordinates of the molecules that show bond breaking as reactants approach the energy maximum and bond formation as products form.

  → Reaction profiles for endothermic and exothermic reactions are shown in schematic drawings.

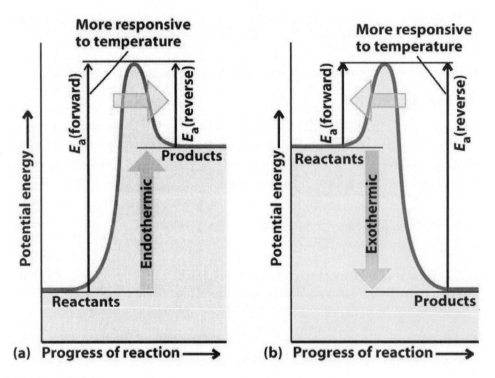

- **Potential energy surface**

  → Three-dimensional plot in which potential energy is plotted on the z-axis, and the other two axes correspond to interatomic distances. See Fig. 14.31 in the text for a potential energy surface.

  → Similar to a mountain pass, the *saddle point* corresponds to an activated complex with the minimum activation energy, E_a, required for reactants to form products.

ACCELERATING REACTIONS (Sections 14.14–14.16)

14.14 Catalysis

- **Nature of a catalyst**
  - → Accelerates a reaction rate without being consumed
  - → Does not appear in the overall reaction
  - → Lowers the activation energy of a reaction by changing the pathway (mechanism)
  - → Accelerates *both* the forward and reverse reaction rates
  - → Does *not* change the final equilibrium composition of the reaction mixture
  - → May appear in the rate equation, but usually does not
- **Homogeneous catalyst** → Present in the same phase as the *reactants*
- **Heterogeneous catalyst** → Present in a different phase from the *reactants*

14.15 Industrial Catalysts

- **Automobile Exhaust**
  - → Catalytic converters of automobiles use catalysts to bring about the complete and rapid combustion of unburned fuel.
  - → Mixture of gases leaving an engine includes carbon monoxide, unburned hydrocarbons, and the nitrogen oxides collectively referred to as NO_x.
  - → Air pollution is decreased if the carbon compounds are oxidized to carbon dioxide and the NO_x reduced, by another catalyst, to nitrogen.
  - → An existing challenge is to find a catalyst or a combination of catalysts that will accelerate both the oxidation and the reduction reactions and be active when the car is first started and the engine is cool.

- **Microporous Catalysts**
  - → Heterogeneous catalysts used in catalytic converters and for many other specialized applications, because of their very large surface areas and reaction specificity.
  - → **Zeolites** are microporous aluminosilicates with three-dimensional structures characterized by hexagonal channels connected by tunnels.
  - → The enclosed active sites in **zeolites** gives them a distinct advantage over other heterogeneous catalysts, because an intermediate can be held in place inside the channels until the products are formed.
  - → Channels allow products to grow only to a particular size.
- **Catalyst ZSM-5**
  - → Used to convert methanol to gasoline
  - → Pores of the **zeolite** are just large enough to allow product hydrocarbons consisting of about eight carbon atoms to escape, so the chains do not grow too long and equilibrium is never reached.

- **Poisoning a catalyst**
  - → Catalysts can be poisoned or inactivated.
  - → For a heterogeneous catalyst, irreversible adsorption of a substance on the catalyst's surface may prevent reactants from reaching the surface and reacting on it.

→ For a homogeneous catalyst, a reactive site on a molecule may be similarly "poisoned" by the binding of a substance directly to the site or nearby either in a permanent or reversible way. The properties of the site are significantly altered to prevent catalytic action.

→ Some heavy metals, especially lead, are very potent poisons for heterogeneous catalysts, which is why lead-free gasoline must be used in engines fitted with catalytic converters.

→ The elimination of lead has the further benefit of decreasing the amount of poisonous lead in the environment.

14.16 Living Catalysts: Enzymes

- **Enzyme, E**

  → A biological catalyst

  → Typically, a large molecule (protein) with crevices, pockets, and ridges on its surface (see Fig. 14.39 in the text)

  → Catalyzes the reaction of a reactant known as the *substrate*, S, that produces a *product*, P:

  $$E + S \rightarrow E + P$$

  → Capable of increasing the rate of specific reactions enormously

- **Induced-fit mechanism**

  → Enzymes are highly specific catalysts that function by an *induced-fit mechanism* in which the substrate approaches a correctly-shaped pocket of the enzyme, known as the *active site*.

  → On binding, the enzyme distorts to accommodate the substrate, allowing the reaction to occur.

  → The shape of the pocket determines which substrate can react.

  → The induced-fit mechanism is a more realistic version of the lock and key mechanism, which assumes that the substrate and enzyme fit together like a lock and key.

- **Michaelis–Menten mechanism of enzyme reaction**

  → A two-step mechanism that accounts for the observed dependence of the reaction rate of many enzyme-substrate reactions

  → For an enzyme, E, and a substrate, S, the overall reaction and the Michaelis–Menten mechanism are

  Overall reaction: $\qquad E + S \rightarrow E + P$

  Mechanism: $\quad$ Step 1 $\quad E + S \rightleftharpoons ES$
  $\qquad\qquad\qquad$ Step 2 $\quad ES \rightarrow E + P$

  → In step 1, the enzyme and substrate form an enzyme-substrate complex, ES, that can dissociate to reform substrate (reverse of step 1) or decay to form product P (step 2). The rate of formation of product is given by step 2.

  Rate of formation of $P = k_2[ES]$

  → The steady-state approximation may be applied to the intermediate, ES, and its concentration is

  $$[ES] = \frac{k_1[E][S]}{k_1' + k_2}$$

→ Because the free enzyme concentration may be substantially diminished during the reaction, it is customary to express the rate in terms of the total enzyme concentration, $[E]_0 = [E] + [ES]$. Then, $[E]$ is replaced by $[E]_0 - [ES]$ in the steady-state result, leading to

$$[ES] = \frac{k_1[E]_0[S]}{k_1' + k_2 + k_1[S]}$$

$$\text{Rate of formation of } P = \frac{k_1 k_2[E]_0[S]}{k_1' + k_2 + k_1[S]} = \frac{k_2[E]_0[S]}{K_M + [S]}$$

→ $K_M = \dfrac{k_1' + k_2}{k_1}$ is known as the Michaelis constant.

→ At low concentration of substrate, $[S] \ll K_M$, and the rate law is given by

$$\text{Rate of formation of } P = \frac{k_2[E]_0[S]}{K_M}$$

→ The rate varies *linearly* with the concentration of substrate in this region.

Physically, in this region, most enzyme sites lack substrate molecules, so doubling the substrate concentration doubles the number of occupied sites and the rate as well.

→ At high concentration of substrate, $[S] \gg K_M$ and the rate law becomes

$$\text{Rate of formation of } P = \frac{k_2[E]_0[S]}{[S]} = k_2[E]_0$$

→ In this region, the rate is *independent* of the substrate concentration (see Fig. 14.41 in the text).

→ *Physically, all the enzyme sites are occupied with substrate molecules, so adding substrate has no effect on the rate.*

- **Summary**
  → *Enzymes* are large proteins that function as *biological catalysts*.
  → *Substrate* molecules (reactants) bind to *active sites* on enzyme molecules.
  → The model of enzyme *binding* and *action* is called the *induced-fit mechanism*.
  → Substrate molecules undergo reaction at the active site to form *product* molecules.
  → Product molecules are released and the active site is restored.

Chapter 15 THE ELEMENTS:
THE MAIN GROUP ELEMENTS

PERIODIC TRENDS (Sections 15.1–15.2)

15.1 Atomic Properties

- **Atomic properties mainly responsible for the properties of an element:**
  - $\rightarrow$ Atomic radius; first ionization energy; electron affinity; electronegativity; polarizability

- **Atomic and ionic radii**
  - $\rightarrow$ Typically *decrease* from left to right across a period and *increase* down a group

- **First ionization energy**
  - $\rightarrow$ Typically *increases* from left to right across a period and *decreases* down a group
  - $\rightarrow$ Increase across a period: Caused by increasing attraction of nuclear charge to electrons in the same valence shell
  - $\rightarrow$ Trend down a group: Concentric shells of electrons are progressively more distant from the nucleus.

- **Electron affinity (E_{ea})**
  - $\rightarrow$ Measure of the energy *released* when an anion is formed. The greater the E_{ea}, the more stable the anion formed.
  - $\rightarrow$ *Highest* values of E_{ea} are found at the top right of the periodic table. Note that Cl has a greater *electron affinity* value than F, but F has a greater *electronegativity* value.
  - $\rightarrow$ Elements with high E_{ea} values are present as *anions* in compounds with *metallic* elements and commonly have *negative* oxidation states in covalent compounds with other *nonmetallic* elements. **Examples**: NaCl, an ionic salt; sulfur hexafluoride SF_6, in which F has an oxidation state of -1.

- **Electronegativity (χ)**
  - $\rightarrow$ Measure of the tendency of an atom to attract bonding electrons when part of a compound
  - $\rightarrow$ Typically *increases* from left to right across a period and *decreases* down a group
  - $\rightarrow$ A useful guide to the type of bond an element may form
  - $\rightarrow$ A predictor of the *polarity* of chemical bonds between nonmetallic elements

- **Polarizability**
  - $\rightarrow$ A measure of the ease of distorting an atom's electron cloud
  - $\rightarrow$ Typically *decreases* across a period (left to right) and *increases* within a group (top to bottom)
  - $\rightarrow$ Greatest for the more massive atoms in a group
  - $\rightarrow$ Anions are more polarizable than parent atoms.
  - $\rightarrow$ Diagonal neighbors in the periodic table have similar polarizability values and a similar amount of covalent character in the bonds they form.

- **Polarizing power**
  - $\rightarrow$ Associated with ions of small size and high charge, such as Al^{3+} or Be^{2+}
  - $\rightarrow$ Cations with high polarizing power tend to have covalent character in bonds formed with highly polarizable anions. **Examples**: $AlCl_3$ and BeI_2

15.2 Bonding Trends

- **Elements bonded to other elements**
  - → Period 2 elements: Valence is determined by the number of valence-shell electrons and the octet rule.
  - → Period 3 or higher ($Z \geq 14$): The valence of an element may be increased through octet expansion allowed by access to empty d-orbitals.

    Examples: PCl_5 and SF_6 (see Section 2.10 of the text)
  - → *Inert-pair effect:* Elements at the bottom of the p block may display an oxidation number two less than the group number suggests.

    Example: Lead has an oxidation number of +2 in PbO and +4 in PbO_2. (see Section 1.19 of the text)

- **Pure elements**
  - → With *low* ionization energies tend to have *metallic* bonds, for example, Li(s) and Mg(s)
  - → With *high* ionization energies are typically *molecular* and form *covalent* bonds

    Examples: $N_2(g)$, $O_2(g)$, and $F_2(g)$
  - → In or near the diagonal band of metalloids (intermediate ionization energies and three, four, or five valence electrons), elements tend to form network structures in which each atom is bonded to three or more other atoms.

    Example: The *network* structure of *black phosphorus,* in which each P atom is bonded to three others

Chemical Properties: Hydrides

- **Binary hydrides**
  - → Binary compounds with hydrogen (hydrides)
  - → Formed by all main-group elements except the noble gases, possibly In and Tl, and several d-block elements
  - → Properties show periodic behavior.
  - → Formulas directly related to the group number

    Examples: SiH_4, NH_3, H_2S, and HCl in Groups 14, 15, 16, and 17, respectively

- **Saline (or salt-like) hydrides**
  - → Formed by all members of the s block except Be
  - → Ionic salts of strongly electropositive metals with hydrogen, present as the hydride ion, H^-
  - → White, high-melting solids with crystal structures resembling those of the corresponding halides
  - → Made by heating the metal in hydrogen

    Example: $2\,Na(s) + H_2(g) \rightarrow 2\,NaH(s)$
  - → React readily with water (H^- is a strong base) forming a basic solution and releasing $H_2(g)$

    Example: $H^-(aq) + H_2O(l) \rightarrow H_2(g) + OH^-(aq)$

- **Metallic hydrides**
  - → Black, powdery, electrically conducting solids formed by certain d-block elements
  - → Release hydrogen when heated or when treated with acids. A typical reaction in acid is:

$$CuH(s) + H_3O^+(aq) \rightarrow Cu^+(aq) + H_2(g) + H_2O(l)$$

- **Molecular hydrides**
  - → Formed by nonmetals and consist of discrete molecules

→ Volatile, many are Brønsted acids or bases.
→ Gaseous compounds include the base $NH_3(g)$, the hydrogen halides $HX(g)$, X= F, Cl, Br, and I, and the lighter hydrocarbons, such as CH_4 and C_2H_4 (ethene).
→ Liquids include water and heavier hydrocarbons, such as benzene, C_6H_6, and octane, C_8H_{18}.
→ With some exceptions hydrides follow the pattern in the table below.

Block	s	d	p
Type	saline	metalllic	molecular

Chemical Properties: Oxides

- **Binary oxides**
  → Formed by all the main-group elements except the noble gases
  → Main-group binary oxides of *metals* are basic.
  Examples: MgO, Na_2O
  → Main-group binary oxides of *nonmetals* are acidic.
  Examples: NO_2, SO_3
  Exception: Al_2O_3 is *amphoteric*. Other *amphoteric oxides* include BeO, Ga_2O_3, SnO_2, and PbO_2.
  → Soluble ionic oxides: Formed from elements on the left side of the periodic table
  → Insoluble oxides with high melting points: Formed from elements in the left of the p block
  → Oxides with low melting points, often gaseous: Formed from elements on the right of the p block

- **Metallic elements and metalloids**
  → Ionic oxides: Formed from metallic elements with low ionization energies
  Example: Na_2O
  → Amphoteric oxides: Formed from metallic elements with intermediate ionization energies (Be, B, and Al) and from metalloids
  → Amphoteric oxides: Do not react with water but dissolve in both acidic and basic solutions

- **Nonmetallic elements**
  → Many oxides of nonmetals are gaseous molecular compounds. Most can act as Lewis acids. They form acidic solutions in water and are called *acid anhydrides*.
  Examples: The acids HNO_3 and H_2SO_4 are derived from the oxides N_2O_5 and SO_3, respectively.
  → Oxides that do *not* react with water are called *formal anhydrides* of acids. The formal anhydride is obtained by removing the elements of water (H, H, and O) from the molecular formula of the acid, for example, the formal anhydride of acetic acid, CH_3COOH, is CH_2CO (ketene), whose Lewis structure is given on the right.
  → With some exceptions hydrides follow the pattern in the table below.
  → Some general properties of oxides are also summarized in the table below.

Block	s	d		p
Type:	ionic	ionic		covalent
acid/base character	basic	basic to amphoteric		amphoteric to acidic

HYDROGEN (Sections 15.3–15.4)

15.3 The Element

- **Hydrogen**
  - → One valence electron but few similarities to the alkali metals
  - → A nonmetal that resembles the halogens but has some different, unique properties
  - → Does not fit clearly into any group and is therefore not assigned to one
- **Commercial production of hydrogen**
  - → Formed 1) as a by-product of the petroleum *refining* and 2) by the *electrolysis* of water
  - → In the refining process, a Ni-catalyzed *re-forming reaction* produces products called *synthesis gas:*

$$CH_4(g) + H_2O(g) \rightarrow CO(g) + 3H_2(g)$$

  In a second step, a *shift reaction* employing an Fe/Cu catalyst yields hydrogen gas:

$$CO(g) + H_2O(g) \rightarrow CO_2(g) + H_2(g)$$

  - → *Electrolysis of water* utilizes a salt solution to improve conductivity: $2H_2O(l) \rightarrow 2H_2(g) + O_2(g)$
- **Laboratory production and uses of hydrogen**
  - → Produced by reaction of an active metal such as Zn with a strong acid such as HCl:

$$Zn(s) + 2H_3O^+(aq) \rightarrow Zn^{2+}(aq) + H_2(g) + 2H_2O(l)$$

  - → Produced by reaction of a hydride with water:

$$CaH_2(s) + 2H_2O(l) \rightarrow Ca^{2+}(aq) + 2OH^-(aq) + 2H_2(g)$$

- **Liquid hydrogen** → Has a very low density and the highest specific density of any known fuel

15.4 Compounds of Hydrogen

- **Unusual properties of hydrogen**
  - → It can form a cation, H^+, or an anion, hydride, H^-.
  - → It can form covalent bonds with many other elements because of its intermediate electronegativity value of $\chi = 2$.
- **The hydride ion, H^-**
  - → Large, with a radius (154 pm) between those of F^- (133 pm) and Cl^- (181 pm) ions
  - → Highly polarizable, which adds covalent character in its bonds to cations
  - → The two electrons in H^- are weakly attached to the single proton, and an electron is easily lost, making H^- a good reducing agent.
  - → Saline hydrides are very strong reducing agents. The reduction half-reaction and standard reduction potential for the reduction of hydrogen are $H_2(g) + 2e^- \rightarrow 2H^-(aq)$ and $E° = -2.25$ V, respectively.
  - → The alkali metals are also strong reducing agents with reduction potentials similar to the hydride value.
  - → Hydride ions in saline hydrides reduce water upon contact: $H^-(aq) + H_2O(l) \rightarrow H_2(g) + OH^-(aq)$.
- **Hydrogen bonds**
  - → Relatively strong intermolecular bonds
  - → Formed by hydrides of N, O, and F

→ Can be understood in terms of the coulombic attraction between the partial positive charge on a hydrogen atom and the partial negative charge of another atom

→ The H atom is covalently bonded to a very electronegative N, O, or F atom, giving it a partial positive charge.

→ The partial negative charge is often the lone-pair electrons on a different N, O, or F atom.

→ A hydrogen bond is represented as three dots between atoms *not* covalently bonded to each other: [O–H$\cdots$:O].

→ A hydrogen bond is typically about 5% as strong as a covalent bond between the same types of atoms.

Example: Hydrogen bonding in HF   H—$\ddot{\underset{\cdot\cdot}{F}}$: $\cdots$ H—$\ddot{\underset{\cdot\cdot}{F}}$: $\cdots$ H—$\ddot{\underset{\cdot\cdot}{F}}$:

GROUP 1: THE ALKALI METALS (Sections 15.5–15.6)

15.5 The Group 1 Elements

- **Alkali metals**
  - → Have the valence electron configuration ns^1, where n is the period number
- **Preparation**
  - → Pure alkali metals are obtained by electrolysis of the molten salts.
  - → **Exception:** K, prepared by reaction of molten KCl and Na vapor at 750°C:

$$KCl(l) + Na(g) \rightarrow NaCl(s) + K(g)$$

  This reaction is driven to form products by the condensation (removal) of K(g) because the equilibrium constant is unfavorable. Recall Le Chatelier's principle (see Section 10.9 of the text).

- **Properties of alkali metals**
  - → Dominated by the ease with which their single valence electron is lost
  - → Soft, silver-gray metals with low melting points, boiling points, and densities
  - → Melting and boiling points decrease down the group. Cs (melting point 28°C) is very reactive and is transported in sealed ampoules. Fr is quite radioactive and little is known about its properties.
  - → All alkali metals are highly reactive.

- **Applications**
  - → Li metal is used in the rechargeable lithium-ion battery, which holds a charge for a long time.
  - → Li is also used in thermonuclear weapons.

- **Alkali metals**
  - → Strong reducing agents (low first ionization energies lead to their ease of oxidation)

    Example: Reduction of Ti(IV): $TiCl_4 + 4\,Na(l) \rightarrow 4\,NaCl(s) + Ti(s)$
  - → Reduce water to form basic solutions, releasing hydrogen gas

    Example: $2\,Na(s) + 2\,H_2O(l) \rightarrow 2\,NaOH(aq) + H_2(g)$
  - → Dissolve in NH_3(l) to yield solvated electrons, which are used to reduce organic compounds
  - → React directly with almost all nonmetals, noble gases excluded
  - → However, only one alkali metal, Li, reacts with nitrogen to form the nitride (Li_3N).

- **Formation of compounds containing oxygen**
  - → Li forms mainly the oxide Li_2O.
  - → Na forms a pale yellow peroxide Na_2O_2.
  - → K, Rb, and Cs form mainly a superoxide, for example, KO_2.

15.6 Compounds of Lithium, Sodium, and Potassium

- **Lithium**
  - → Differs significantly from the other Group 1 elements, a common occurrence for elements at the head of a group.
  - → For Group I, the differences originate in part from the small size of Li^+.
  - → Cations have strong polarizing power and a tendency toward covalency in their bonding.
  - → Has a diagonal relationship with Mg and is found in the minerals of Mg
  - → Is found in compounds with oxidation number +1; its compounds are used in ceramics, lubricants, and medicine.

      Examples: Lithium carbonate is an effective drug for manic-depressive (bipolar) behavior. Lithium soaps, which have higher melting points than sodium and potassium soaps, are used as thickeners in lubricating greases for high-temperature applications.

- **Sodium compounds**
  - → Important in part because they are *plentiful, inexpensive,* and water *soluble*
  - → NaCl is mined as *rock salt* and is used in the electrolytic production of $Cl_2(g)$ and NaOH.
  - → NaOH is an inexpensive starting material for the production of other sodium salts.
  - → $Na_2CO_3 \cdot 10\,H_2O$ was once used as *washing soda* to precipitate Ca^{2+} and Mg^{2+} ions.
  - → $NaHCO_3$ is also known as *sodium bicarbonate, bicarbonate of soda,* or *baking soda.*
  - → Weak acids [lactic acid (milk), citric acid (lemons), acetic acid (vinegar)] react with the hydrogen carbonate anion to release carbon dioxide gas:

      $$HCO_3^-(aq) + HA(aq) \rightarrow A^-(aq) + H_2O(l) + CO_2(g).$$
  - → Double-acting *baking powder* contains a solid weak acid as well as the hydrogen carbonate ion.

- **Potassium compounds**
  - → Usually more expensive than the corresponding sodium compounds
  - → Use may be justified because potassium compounds are generally less hygroscopic than sodium compounds.
  - → The K^+ cation is larger and is less strongly hydrated by H_2O molecules.
  - → KNO_3 releases $O_2(g)$ when heated and is used to help matches ignite.
  - → KNO_3 is also the oxidizing agent in black gunpowder (75% KNO_3, 15% charcoal, and 10% sulfur).
  - → KCl is used directly in some fertilizers as a source of potassium, but KNO_3 is required for some crops that cannot tolerate high chloride ion concentrations.
  - → Mineral sources of potassium are *sylvite,* KCl, and *carnallite,* $KCl \cdot MgCl_2 \cdot 6\,H_2O$.

GROUP 2: THE ALKALINE EARTH METALS (Sections 15.7–15.8)

15.7 The Group 2 Elements

- **Alkaline earth metals**
  - → Have the valence electron configuration ns^2, where n is the period number
  - → Ca, Sr, and Ba are called alkaline earths because their oxides are basic (alkaline)
  - → The name *alkaline earth* has been extended to all the Group 2 elements.

- **Preparation**
  - → Pure metals: obtained by electrolysis of the molten salts (all) or by chemical reduction (except Be)
  - → Ca(s), Sr(s), and Ba(s) may be obtained by reduction with Al(s) in a variation of the thermite reaction.

    Example: $3\,CaO(s) + 2\,Al(s) \rightarrow Al_2O_3(s) + 3\,Ca(s)$.
  - → Be occurs in nature as *beryl,* $3\,BeO \cdot Al_2O_3 \cdot 6\,SiO_2$.
  - → Mg is found in seawater as $Mg^{2+}(aq)$ and in the mineral dolomite, $CaCO_3 \cdot MgCO_3$.

- **Properties**
  - → Silver-gray metals with much higher melting points, boiling points, and densities than those of the preceding alkali metals in the same period
  - → Melting points decrease down the group with the exception of Mg, which has the lowest value. The trend in boiling point is irregular (see Table 15.4 in the text).
  - → All Group 2 elements except Be reduce water.

    Example: $Sr(s) + 2\,H_2O(l) \rightarrow Sr^{2+}(aq) + 2\,OH^-(aq) + H_2(g)$
  - → Be does not react with water. Mg reacts with hot water, and Ca reacts with cold water.
  - → Be and Mg do not dissolve in HNO_3 because they develop a protective oxide film.
  - → Mg burns vigorously in air because it reacts with $O_2(g)$, $N_2(g)$, and $CO_2(g)$.
  - → Reactivity of the Group 2 metals with water and oxygen increases down the group, just like the Group 1 metals.

- **Applications of beryllium**
  - → Has a low density, making it useful for missile and satellite fabrication
  - → Used as a "window" for x-ray tubes
  - → Added in small amounts to Cu to increase its rigidity
  - → Be/Cu alloys are used in nonsparking tools (oil refineries and grain elevators) and for nonmagnetic parts that resist deformation and corrosion (electronics industry).

- **Applications of magnesium**
  - → A silver-white metal protected from air oxidation by a white oxide film, making it appear dull gray
  - → Light and very soft, but its alloys have great strength and are used in applications where lightness and toughness are desired (airplanes)
  - → Limitation in widespread usage is related to its expense, low melting temperature, and difficulty in machining.

- **Applications of calcium, strontium, and barium**
  - → Group 2 metals can be detected in burning compounds by the colors they give to flames. Ca burns orange-red, Sr crimson, and Ba yellow-green. Fireworks are made from their salts.

15.8 Compounds of Beryllium, Magnesium, and Calcium

- **Beryllium**
  - → Differs significantly from the other Group 2 elements, as is typical of an element at the head of its group
  - → The difference originates in part from the small size of the Be^{2+} cation, which has strong polarizing power and a tendency toward covalency in its bonding.
  - → Has a diagonal relationship with Al. It is the only member of the group that, like Al, reacts in aqueous NaOH. In strongly basic solution, it forms the *beryllate ion,* $[Be(OH)_4]^{2-}$.
  - → Compounds are extremely toxic.
  - → The high polarizing power of Be^{2+} produces moderately covalent compounds, while its small size allows no more than four groups to be attached, accounting for the prominence of the tetrahedral BeX_4 unit.
  - → $BeH_2(g)$ and $BeCl_2(g)$ condense to form solids with chains of tetrahedral BeH_4 and $BeCl_4$ units, respectively.
  - → In $BeCl_2$, the Be atoms act as Lewis acids and accept electron pairs from the Cl atoms, forming a chain of tetrahedral $BeCl_4$ units in the solid.

- **Magnesium**
  - → More metallic properties than Be
  - → Compounds are primarily ionic with some covalent character.
  - → MgO and Mg_3N_2 are formed when magnesium burns in air.
  - → MgO dissolves very slowly and only slightly in water. It can withstand high temperatures (*refractory*) because it melts at 2800°C.
  - → MgO conducts heat well, but electricity only poorly, and is used as an insulator in electric heaters.
  - → $Mg(OH)_2$ is a base. It is not very soluble in water but forms instead a white colloidal suspension, which is used as a stomach antacid called *milk of magnesia*. A side effect of stomach acid neutralization is the formation of $MgCl_2$, which acts as a purgative.
  - → $MgSO_4$ (*Epsom Salts*) is also a common purgative.
  - → Chlorophyll is the most important compound of Mg. Magnesium ions in chlorophyll enable photosynthesis to occur.
  - → Magnesium also plays an important role in energy generation in living cells. It is involved in the contraction of muscles.

- **Calcium**
  - → Ca is more metallic in character than Mg, but its compounds share some similar properties.
  - → $CaCO_3$ occurs naturally as chalk and limestone. Marble is a very dense form with colored impurities, most commonly Fe cations.
  - → The most common forms of pure calcium carbonate are *calcite* and *aragonite*. All carbonates are the fossilized remains of marine life.
  - → CaO is formed by heating $CaCO_3$: $CaCO_3(s) \rightarrow CaO(s) + CO_2(g)$.
    CaO is called *quicklime* because it reacts rapidly and exothermically with water:

    $$CaO(s) + H_2O(l) \rightarrow Ca^{2+}(aq) + 2\,OH^-(aq)$$

  - → $Ca(OH)_2$, the product of the equation above, is known as *slaked lime* because in this form the thirst for water has been quenched, or slaked.

→ An aqueous solution of $Ca(OH)_2$, which is only slightly soluble in water, is called *lime water,* which is used as a test for carbon dioxide:

$$Ca(OH)_2(aq) + CO_2(g) \rightarrow CaCO_3(s) + H_2O(l).$$

- **Carbide, sulfate, and phosphate**

  → CaO may be converted into calcium carbide: $CaO(s) + 3\,C(s) \rightarrow CaC_2(s) + CO(g).$

  → Calcium sulfate dihydrate (*gypsum*), $CaSO_4 \cdot 2\,H_2O$, is used in construction materials.

  → Ca is found in the rigid structural components of living organisms, either as $CaCO_3$ (shellfish shells) or as $Ca_3(PO_4)_2$ (bone).

  → Tooth enamel is a *hydroxyapatite,* $Ca_5(PO_4)_3OH$, which is subject to attack by acids produced when bacteria act on food residues. A more resistant coating is formed when the OH^- ions are replaced by F^- ions. The resulting mineral is called *fluoroapatite:*

$$Ca_5(PO_4)_3OH(s) + F^-(aq) \rightarrow Ca_5(PO_4)_3F(s) + OH^-(aq)$$

  Tooth enamel is strengthened by the addition of fluorides (fluoridation) to drinking water, and by the use of fluoridated toothpaste containing tin(II) fluoride, SnF_2, or sodium monofluorophosphate (MFP), Na_2FPO_3.

GROUP 13 / III: THE BORON FAMILY (Sections 15.9–15.11)

15.9 The Group 13/III Elements

  → Have the valence electron configuration $ns^2\,np^1$

  → Common (Maximum) oxidation number is +3, but heavier elements in the group, such as indium and thallium, also exhibit oxidation number +1 (inert pair effect – see Section 1.19 of the text).

- **Preparation of boron**

  → Mined as the hydrates *borax* and *kernite,* $Na_2B_4O_7 \cdot xH_2O$, with $x = 10$ and 4, respectively

  → The ore is converted to the oxide, B_2O_3, and extracted in its amorphous elemental form by reaction of the oxide with $Mg(s)$:

$$B_2O_3(s) + 3\,Mg(s) \rightarrow 2\,B(s) + 3\,MgO(s).$$

  → A purer form of B is obtained by reduction of volatile $BBr_3(g)$ or $BCl_3(g)$ with $H_2(g)$.

- **Preparation of aluminum**

  → Mined as *bauxite,* $Al_2O_3 \cdot xH_2O$, where x ranges to a maximum of 3

  → The ore, which contains considerable iron oxide impurity, is processed to obtain alumina, Al_2O_3.

  → Al is then extracted by an electrolytic process (*Hall process*). Al_2O_3 is mixed with *cryolite,* Na_3AlF_6, to lower the melting point from 2050°C (alumina) to 950°C (mixture). The half-cell and cell reactions are:

Cathode:	$Al^{3+}(melt) + 3\,e^- \rightarrow Al(l)$
Anode:	$2\,O^{2-}(melt) + C(s,gr) \rightarrow CO_2(g) + 4\,e^-$
Cell:	$4\,Al^{3+}(melt) + 6\,O^{2-}(melt) + 3\,C(s,gr) \rightarrow 4\,Al(l) + 3\,CO_2(g)$

- **Properties of boron**
  - → First element of the *p* block
  - → Element is attacked by only the strongest oxidizing agents.
  - → A nonmetal in most of its chemical properties
  - → Has acidic oxides and forms a fascinating array of binary molecular hydrides
  - → It exists in several allotropic forms in which B atoms attempt to share eight electrons, despite the small size of B and its ability to contribute only three electrons. In this sense, B is regarded as electron deficient.
  - → One common form of B is a gray-black, nonmetallic, high-melting solid.
  - → Another common form is a dark brown powder with a structure (icosahedral with 20 faces) that has clusters of 12 atoms as shown on the right. The bonds create a very hard, three-dimensional structure.

- **Properties of aluminum**
  - → Low density, but is a strong metal with excellent electrical conductivity
  - → Easily oxidized, but its surface is passivated by a protective oxide film
  - → If the thickness of the oxide layer is increased electrolytically, *anodized aluminum* results.
  - → Al is amphoteric, reacting with both *nonoxidizing* acids and hot aqueous bases. The reactions are:

$$2\,Al(s) + 6\,H^+(aq) \rightarrow 2\,Al^{3+}(aq) + 3\,H_2(g)$$

$$2\,Al(s) + 2\,OH^-(aq) + 6\,H_2O(l) \rightarrow 2\,Al(OH)_4^-(aq) + 3\,H_2(g)$$

- **Applications**
  - → Boron fibers are incorporated into plastics, forming resilient material stiffer than steel and lighter than Al(s).
  - → Boron carbide is similar in hardness to diamond, and boron nitride is similar in structure and mechanical properties to graphite, but, unlike graphite, boron nitride does not conduct electricity.
  - → Al has widespread use in construction and aerospace industries. Because it is a soft metal, its strength is improved by alloy formation with Cu and Si.
  - → Because of its high electrical conductivity, Al is used in overhead power lines. Its high negative electrode potential has led to its use in fuel cells.

- **Gallium**
  - → Produced as a by-product in the production of Al
  - → Common doping agent for semiconductors
  - → GaAs, gallium arsenide, is used in light emitting diodes (LEDs).

- **Thallium**
  - → Poisonous heavy metal sometimes used as a rat poison

15.10 Group 13/III Oxides and Halides

- **Boron oxide**
  - → Boron, like most nonmetals, has *acidic* oxides. Hydration of B_2O_3 yields boric acid, H_3BO_3 or $B(OH)_3$, in which B has an incomplete octet and can therefore act as a Lewis acid:

$$(OH)_3B + :OH_2 \rightarrow (OH)_3B{-}OH_2$$

The complex formed is a weak *monoprotic* acid ($pK_a = 9.14$).

→ H_3BO_3 is a mild antiseptic and pesticide. It is also used as a fire retardant in home insulation and clothing.

→ The main use of H_3BO_3, is as a source of boron oxide, the anhydride of boric acid:

$$2\,H_3BO_3(s) \xrightarrow{\text{heat}} B_2O_3(s) + 3\,H_2O(l)$$

B_2O_3, which melts at 450°C and dissolves many metal oxides, is used as a flux for soldering or welding. It is also used in the manufacture of fiberglass and borosilicate glass, such as Pyrex.

- **Aluminum oxide**

  → Al_2O_3, known as *alumina,* exists in several crystal forms, each of which is used in science and in commerce.

  → α-Alumina is a very hard substance, *corundum,* used as an abrasive known as *emery.*

  → γ-Alumina is absorbent and is used as the stationary phase in chromatography (see Major Technique 4 in Chapter 9 of the text). It is produced by heating $Al(OH)_3$ and is moderately reactive and amphoteric:

  $$Al_2O_3(s) + 6\,H_3O^+(aq) + 3\,H_2O(l) \rightarrow 2\,[Al(H_2O)_6^{3+}](aq)$$

  $$Al_2O_3(s) + 2\,OH^-(aq) + 3\,H_2O(l) \rightarrow 2\,Al(OH)_4^-(aq)$$

  The strong polarizing effect of the small, highly charged Al^{3+} ion on the water molecules surrounding it gives the $[Al(H_2O)_6^{3+}]$ ion acidic properties.

  → Aluminum sulfate is prepared by reaction of Al_2O_3 with sulfuric acid:

  $$Al_2O_3(s) + 3\,H_2SO_4(aq) \rightarrow Al_2(SO_4)_3(aq) + 3\,H_2O(l)$$

  $Al_2(SO_4)_3$ (*papermaker's alum*) is used in the preparation of paper.

  → Sodium aluminate, $NaAl(OH)_4$, is used along with $Al_2(SO_4)_3$ in water purification.

- **Halides**

  → Boron halides are made either by direct reaction of the elements or from the oxide.

  Examples: $B_2O_3(s) + 3\,CaF_2(s) + 3\,H_2SO_4(l) \xrightarrow{\text{heat}} 2\,BF_3(g) + 3\,CaSO_4(s) + 3\,H_2O(l)$

  $\qquad\qquad\quad B_2O_3(s) + 3\,C(s) + 3\,Cl_2(g) \xrightarrow{\text{heat}} 2\,BCl_3(g) + 3\,CO(g)$

  → All boron halides are trigonal-planar, covalently bonded, and electron-deficient, with an incomplete octet on B, as a consequence of which they are strong Lewis acids.

  → $AlCl_3$ is formed as an ionic compound by direct reaction of the elements, or of alumina with chlorine in the presence of carbon:

  $$2\,Al(s) + 3\,Cl_2(g) \rightarrow 2\,AlCl_3(s)$$

  $$Al_2O_3(s) + 3\,Cl_2(g) + 3\,C(s,\ gr) \rightarrow 2\,AlCl_3(s) + 3\,CO(g)$$

  Ionic $AlCl_3$ melts at 192°C to form a molecular liquid, Al_2Cl_6. This dimer of $AlCl_3$, formed by donation of an unshared electron pair on Cl to a vacant orbital on an adjacent Al atom, contains $AlCl_4$ tetrahedral units and two bridging Cl atoms, as shown on the right.

  → Aluminum halides react with water in a highly exothermic reaction.

  → The white solid, lithium aluminum hydride, prepared from $AlCl_3$ is an important reducing agent in organic chemistry:

  $$4\,LiH + AlCl_3 \rightarrow LiAlH_4 + 3\,LiCl.$$

15.11 Boranes and Borohydrides

- **Boranes**
    - → Compounds of boron and hydrogen, such as B_2H_6 and $B_{10}H_{14}$
    - → Electron-deficient compounds with three-center, two-electron B–H–B bonds
    - → Valid Lewis structures cannot be written.
    - → Molecular Orbital theory provides a framework for understanding the delocalized behavior of an electron pair associated with three atoms, as shown to the right.

    - → Diborane, B_2H_6, is highly reactive. When heated, it decomposes to hydrogen and boron:

$$B_2H_6(g) \rightarrow 2B(s) + 3H_2(g)$$

    It reacts with water, reducing hydrogen and forming boric acid:

$$B_2H_6(g) + 6H_2O(l) \rightarrow 2B(OH)_3(aq) + 6H_2(g)$$

- **Borohydrides**
    - → Anionic versions of boranes. The most important borohydride is BH_4^-, as in sodium borohydride, $NaBH_4$, which is produced by the reaction of NaH with BCl_3:

$$4NaH + BCl_3 \rightarrow NaBH_4 + 3NaCl$$

    The borohydride anion BH_4^- is an effective and important reducing agent.

    - → Diborane is produced by the reaction of sodium borohydride with boron trifluoride:

$$3BH_4^- + 4BF_3 \rightarrow 3BF_4^- + 4B_2H_6$$

- **Borides (see Section 6.9 in the text)**

GROUP 14 / IV: THE CARBON FAMILY (Sections 15.12–15.14)

15.12 The Group 14/IV Elements
- → Have the valence electron configuration $ns^2 np^2$
- → The lighter elements have all four valence electrons available for bonding.
- → Heavier elements (Sn, Pb) display the inert-pair effect and exhibit oxidation numbers 2 and 4.
- → For Pb, the most common oxidation number is 2.
- → Metallic character increases from C to Pb.

- **General properties**
    - → Carbon forms covalent compounds with nonmetals and ionic compounds with metals.
    - → The oxides of C and Si are acidic.
    - → Ge, a metalloid, exhibits metallic or nonmetallic properties, depending on the compound.
    - → Sn and Pb are classified as metals, but tin has some intermediate properties.
    - → Sn is amphoteric and reacts with hot, concentrated hydrochloric acid and hot alkali:

$$Sn(s) + 2H_3O^+(aq) \rightarrow Sn^{2+}(aq) + H_2(g) + 2H_2O(l)$$

$$Sn(s) + 2OH^-(aq) + 2H_2O(l) \rightarrow Sn(OH)_4^{2-}(aq) + H_2(g)$$

- → Carbon, at the top of the group, differs greatly in its properties from the other members.
- → Carbon tends to form multiple bonds, whereas silicon does not.
- → CO_2 is a molecular gas, while $SiO_2(s)$ (the mineral silica) contains networks of $-O-Si-O-$ groups.
- → Si compounds can act as Lewis acids, whereas C compounds usually cannot. Si may expand its octet, whereas C does not.

- **Graphite**
  - → The thermodynamically stable allotrope of carbon (Soot contains small crystals of graphite.)
  - → Produced pure commercially by heating C rods in an electric furnace for several days
  - → Contains sp^2-hybridized C atoms
  - → Consists structurally of large sheets of fused benzene-like hexagonal units. A π-bonding network of delocalized electrons accounts for the high electrical conductivity of graphite.
  - → When certain impurities are present, graphitic sheets can slip past one another, and graphite becomes an excellent dry lubricant.
  - → Only soluble in a few liquid metals
  - → Soot and carbon black have commercial applications in rubber and inks. Activated charcoal is an important, versatile purifier. Unwanted compounds are adsorbed onto its microcrystalline surface.

- **Diamond**
  - → The hardest substance known and an excellent conductor of heat
  - → An excellent abrasive, and the heat generated by friction is rapidly conducted away
  - → Natural diamonds are uncommon. Synthetic diamonds are produced from graphite at high pressures (> 80 kbar) and temperatures (> 1500°C) and by thermal decomposition of methane.
  - → Soluble in liquid metals, such as Cr and Fe, but less soluble than graphite. This solubility difference is utilized in synthesizing diamond at high pressure and high temperature.
  - → Carbon in diamond is sp^3-hybridized, and each C atom in a diamond crystal is bonded directly to four other C atoms and indirectly interconnected to all of the other C atoms in the crystal through C–C single σ-bonds.
  - → The σ-bonding network of localized electrons accounts for the electrical insulating properties of diamond.

- **Fullerenes**
  - → Buckminsterfullerene, C_{60}, is a soccer-ball-like molecule first identified in 1985. Numerous fullerenes, containing 44 to 84 carbon atoms, were discovered quickly thereafter.
  - → Fullerenes are molecular electrical insulators and are soluble in organic solvents.
  - → Have relatively low ionization energies *and* large electron affinities, so they readily lose or gain electrons to form ions.
  - → Contain an even number of carbon atoms, differ by a C_2 unit, and have pentagonal and hexagonal ring structures. Solid samples of fullerenes are called *fullerites*.
  - → C_{60} has the atomic arrangement of a *truncated icosahedron* (soccer ball), with 32 faces, 12 pentagons, and 20 hexagons. The interior of the molecule is large, and other atoms may be inserted to produce compounds with unusual properties. The crystal structure of C_{60} is face-centered cubic.
  - → The anion C_{60}^{3-} is very stable, and K_3C_{60} is a superconductor below 18 K.

- **Carbon nanotubes** (see Section 6.19 in the text)
  - → Concentric tubes with graphene-like walls rolled into cylinders.

→ Strong, conducting fibers with large surface areas

- **Silicon**
  → Occurs naturally as silicon dioxide $SiO_2(s)$ and as silicates, which contain SiO_4^{4-} units
  → Forms of SiO_2 include quartz, quartzite, and sand.
  → Pure Si, which is widely used in semiconductors, is obtained from quartzite in a three-step process. Silicon produced in the first step is crude and requires further purification:

$$SiO_2(s) + 2\,C(s) \rightarrow Si(s) + 2\,CO_2(g)$$

$$Si(s) + 2\,Cl_2(g) \rightarrow SiCl_4(l)$$

$$SiCl_4(l) + 2\,H_2(g) \rightarrow Si(s) + 4\,HCl(g)$$

  → To be used in semiconductors, silicon must be purified further by zone refining (see Fig. 15.29 in the text) to obtain ultrapure Si.
  → Amorphous Si, used in photovoltaic devices, is prepared from the decomposition of silane, SiH_4, in an electric discharge.

- **Germanium**
  → Mendeleev predicted the existence of Ge, "eka-silicon," before its discovery.
  → Occurs as an impurity in Zn ores
  → Used mainly in the semiconductor industry

- **Tin**
  → Reduced from its ore, *cassiterite* (SnO_2), through reaction with C at 1200°C:

$$SnO_2(s) + C(s) \rightarrow Sn(l) + CO_2(g)$$

  → Used to plate other materials and for the production of alloys
  → Expensive, not very strong, but resists corrosion

- **Lead**
  → The principal ore is *galena,* PbS, which is roasted in air to yield the oxide, which in turn is reduced with coke to produce the metal:

$$2\,PbS(s) + 3\,O_2(g) \rightarrow 2\,PbO(s) + 2\,SO_2(g)$$

$$PbO(s) + C(s) \rightarrow Pb(s) + CO(g)$$

  → Dense, malleable, and chemically inert (passivated)
  → Used as an electrode material in car batteries, to transport concentrated sulfuric acid, and as a shield for radiation.

15.13 Oxides of Carbon and Silicon

- **Carbon dioxide, CO_2**
  → A nonmetal oxide and the acid anhydride of carbonic acid, H_2CO_3 (The equilibrium between CO_2 and H_2CO_3 favors CO_2.)
  → Used to make carbonated beverages with high partial pressures of CO_2
  → $CO_2(aq)$ is an equilibrium mixture of CO_2, H_2CO_3, HCO_3^-, and a very small amount of CO_3^{2-}.
  → Solid CO_2 (*dry ice*) sublimes to the gas phase, making it useful as a refrigerant and cold pack.

- **Carbon monoxide, CO**
  → A highly toxic colorless, odorless, flammable gas that is nearly insoluble in water.
  → Has an extremely large bond enthalpy (1074 kJ·mol^{-1}).

→ Produced when an organic compound or C itself is burned in a limited amount of oxygen.

→ Commercially produced as *synthesis gas:*

$$CH_4(g) + H_2O(g) \rightarrow CO(g) + 3H_2(g)$$

→ CO is the *formal* anhydride of formic acid and is produced in the laboratory by dehydration of formic acid with hot, concentrated sulfuric acid:

$$HCOOH(l) \rightarrow CO(g) + H_2O(l)$$

→ CO is a moderately good Lewis base. One example is the reaction of CO with a *d*-block metal or ion:

$$Ni(s) + 4CO(g) \rightarrow Ni(CO)_4(l)$$

→ CO can also act as a Lewis acid by accepting electrons into its vacant antibonding π-orbitals.

→ Complex formation is responsible for the toxicity of CO, which attaches more strongly than O_2 to iron in hemoglobin, thereby blocking the acceptance of O_2 by red blood cells.

→ CO can act as a reducing agent, yielding CO_2.

- **Silica or silicon dioxide, SiO_2**

    → Hard rigid, water insoluble network solid

    → Occurs naturally as quartz and sand.

    → Derives its strength from its covalently bonded network structure

    → Occurs naturally in pure form only as quartz and sand

    → Flint is silica with carbon impurities that make it black.

- **Silicates, compounds containing SiO_4^{4-} units (see Section 6.7 in the text)**

    → Orthosilicic acid (H_4SiO_4) and metasilicic acid (H_2SiO_3) are weak acids.

    → When sodium orthosilicate is acidified in solution, *silica gel*, used as a drying agent, a support for catalysts, and packing for chromatography columns, *etc.*, is produced:

$$4H_3O^+(aq) + SiO_4^-(aq) + xH_2O(l) \rightarrow SiO_2(s) \cdot xH_2O(gel) + 6H_2O(l)$$

- **Aluminosilicates and cements (see Sections 6.7 and 6,8, respectively, in the text)**

15.14 Other Important Group 14/IV Compounds

- **Carbides**

    → Can be saline (*saltlike*), covalent, or interstitial

    → See Section 6.9 in the text for covalent and interstitial carbides.

    → *Saline carbides* are most commonly formed from Group 1 and 2 metals, Al, and a few other metals. *s*-Block metals form saline carbides when their oxides are heated with C.

    → Anions present in carbides are normally *methanide*, C^{4-}, or *acetylide,* C_2^{2-}, both very strong Brønsted bases. Methanides react with water to produce a basic solution and methane gas:

$$Al_4C_3(s) + 12H_2O(l) \rightarrow 4Al(OH)_3(s) + 3CH_4(g)$$

Acetylides release acetylene, HC≡CH, upon reaction with water:

$$CaC_2(s) + 2H_2O(l) \rightarrow Ca(OH)_2(s) + C_2H_2(g)$$

- **Tetrachlorides**

    → All Group 14/IV elements form *liquid* molecular tetrachlorides, of which $PbCl_4$ is the least stable. Carbon tetrachloride, a carcinogenic liquid, is formed by the reaction of chlorine and methane:

$$CH_4(g) + 4Cl_2(g) \rightarrow CCl_4(l) + 4HCl(g)$$

→ Si reacts directly with chlorine to form silicon tetrachloride. Unlike CCl_4, $SiCl_4$ reacts with water as a Lewis acid, accepting a lone pair of electrons from H_2O:

$$SiCl_4(l) + 2 H_2O(l) \rightarrow SiO_2(s) + 4 HCl(aq)$$

- **Cyanides**
  → Cyanides are strong Lewis bases that form a range of complexes with *d*-block metal ions. They are extremely poisonous, combining with cytochromes involved with the transfer of electrons and the supply of energy in cells.
  → The cyanide ion, CN^-, is the conjugate base of the acid HCN, which is made by heating methane, ammonia, and air in the presence of a Pt catalyst at 1100°C:

$$2 CH_4(g) + 2 NH_3(g) + 3 O_2(g) \rightarrow 2 HCN(g) + 6 H_2O(g)$$

- **Methane and silane**
  → C bonds readily with itself to form chains and rings and forms a multitude of compounds with hydrogen (hydrocarbons). (see Chapter 18 in the text)
  → Si forms a much smaller number of compounds with hydrogen; the simplest is the analogue of methane, *silane,* SiH_4, formed by the reaction of the reducing agent $LiAlH_4$ with silicon halides in ether:

$$SiCl_4 + LiAlH_4 \rightarrow SiH_4 + LiCl + AlCl_3$$

  → SiH_4 is more reactive than CH_4 and ignites spontaneously in air. Higher silanes are unstable and decompose on standing.

GROUP 15 / V: THE NITROGEN FAMILY
(Sections 15.15–15.18)

15.15 The Group 15 / V Elements

- **Nitrogen**
  → *Nitrogen* differs greatly from other group members owing to its high electronegativity ($\chi = 3.0$), small size (radius of 74 pm), ability to form multiple bonds, and lack of available *d*-orbitals.
  → Found with oxidation numbers from −3 to +5.
  → *Elemental nitrogen,* N_2, is found in air (78.1% by volume) and is prepared by fractional distillation of liquid air.
  → The strong triple bond in N_2 ($\Delta H_B = 944$ kJ·mol^{-1}) makes it very stable.
  → *Nitrogen fixation* is the process by which N_2 is converted into compounds usable by plants.
  → The *Haber synthesis* of ammonia is the major industrial method for fixing nitrogen, but it requires high temperatures and pressures and is, therefore, expensive.
  → Lightning produces some oxides of nitrogen, which are washed into soil by rain.
  → Bacteria found in the root nodules of legumes also fix N_2. An important area of current research is the search for catalysts that can mimic bacteria and fix nitrogen under ordinary temperatures.
  → The chemical and physical properties of nitrogen are different from those of its congeners because of the small size and unavailability of *d*-orbitals in nitrogen.

→ Nitrogen is highly electronegative ($\chi = 3.0$), forms strong hydrogen bonds, and forms multiple bonds with other Period 2 elements (*e.g.*, cyanide, $C\equiv N^-$).

- **Phosphorus**
  → *Phosphorus* differs from nitrogen in its lower electronegativity ($\chi = 2.2$), larger size (radius of 110 pm), and the presence of available *d*-orbitals. Phosphorus may form as many as six bonds, whereas nitrogen can form a maximum of four.
  → *Phosphorus* is prepared from the *apatites,* mineral forms of *calcium phosphate,* $Ca_3(PO_4)_2$. The rocks are heated in an electric furnace with carbon and sand. The phosphorus vapor formed condenses as *white phosphorus,* a soft, white, poisonous molecular solid consisting of tetrahedral P_4 molecules. It bursts into flame upon exposure to air and is normally stored under water.
  → *Red phosphorus* is less reactive than white phosphorus and is used in match heads because it can be ignited by friction. It is a network solid consisting, most likely, of linked P_4 tetrahedral units.
  → *Red phosphorus* is the most thermodynamically stable form of phosphorus, yet *white phosphorus* crystallizes from the vapor or from the liquid phase. It is the form chosen to have an enthalpy and free energy of formation equal to 0. For *red phosphorus,* $\Delta H_f^\circ = -17.6$ kJ·mol^{-1} and $\Delta G_f^\circ = -12.1$ kJ·mol^{-1} at 25°C.
  → *Black phosphorus* is a third form of the element; it has a network structure in which each P atom is bonded to three others at approximately right angles.

- **Arsenic and antimony**
  → *Arsenic* and *antimony* are metalloids. They are elements known since ancient times because they are easily reduced from their ores.
  → As elements, *arsenic* and *antimony* are used in lead alloys in the electrodes of storage batteries and in the semiconductor industry. *Gallium arsenide* is used as a light-detecting material with near infrared response and in lasers, including ones used for CD players.

- **Bismuth**
  → *Bismuth* is a metal with large, weakly bonded atoms. Bismuth has a low melting point, and it is used in alloys that serve as fire detectors in sprinkler systems.
  → *Bismuth* is also used to make low-temperature castings. Like ice, solid bismuth is less dense than the liquid. Bismuth ($Z = 83$) is the last element in the periodic table with a stable isotope, ^{209}Bi.

15.16 Compounds with Hydrogen and the Halogens

- **Ammonia, NH_3, and phosphine, PH_3**
  → *Ammonia* is prepared in large amounts by the Haber process.
  → Ammonia is a pungent, toxic gas that condenses to a colorless liquid at −33°C. It acts as a solvent for a wide range of substances.
  → Autoprotolysis occurs to a much smaller extent in liquid ammonia (am) than in pure water.

$$2\,NH_3(am) \rightleftharpoons NH_4^+(am) + NH_2^-(am) \qquad K_{am} = 1 \times 10^{-33} \text{ at } -35°C$$

  Very strong bases that are protonated by water survive in liquid ammonia.
  → *Ammonia* is a weak Brønsted base in water; it is also a fairly strong Lewis base, particularly toward *d*-block elements. One reaction is:

$$Cu^{2+}(aq) + 4\,NH_3(aq) \rightarrow Cu(NH_3)_4^{2+}(aq).$$

  In this reaction, unshared electron pairs on nitrogen interact with the Lewis acid Cu^{2+} to form four $Cu-NH_3$ bonds.

→ *Ammonium salts* decompose when heated; if the salt contains a nonoxidizing anion, NH_3 is produced. The reaction is:
$$(NH_4)_2SO_4(s) \rightarrow 2\,NH_3(g) + H_2SO_4(l).$$
Decomposing ammonium carbonate has a pungent odor and is used as a "smelling salt" to revive people who have fainted.

→ *Phosphine* is a poisonous gas that smells faintly of garlic and bursts into flame in air if it is impure. It is much less stable than ammonia. Because it does not form hydrogen bonds, it is not very soluble in water and is a very weak base ($pK_b = 27.4$). Its aqueous solution is nearly neutral.

- **Hydrazine, N_2H_4**
  → *Hydrazine* is an oily, colorless liquid. It is prepared by gentle oxidation of ammonia with alkaline hypochlorite solution:
  $$2\,NH_3(aq) + ClO^-(aq) \rightarrow N_2H_4(aq) + Cl^-(aq) + H_2O(l)$$
  Its physical properties are similar to water; its melting point is 1.5°C and its boiling point is 113°C. It is dangerously explosive and is stored in aqueous solution.

  → A mixture of *methylhydrazine* and liquid N_2O_4 is used as a rocket fuel because these two liquids ignite on contact and produce a large volume of gas.
  $$4\,CH_3NHNH_2(l) + 5\,N_2O_4(l) \rightarrow 9\,N_2(g) + 12\,H_2O(g) + 4\,CO_2(g)$$

- **Nitrides, N^{3-}, and phosphides, P^{3-}**
  → *Nitrides* are solids that contain the nitride ion. Nitrides are stable only with small cations such as lithium, magnesium, or aluminum. Nitrides dissolve in water to form ammonia and the corresponding hydroxide:
  $$Zn_3N_2(s) + 6\,H_2O(l) \rightarrow 2\,NH_3(g) + 3\,Zn(OH)_2(aq)$$

  → *Phosphides* are solids that contain the phosphide ion. Because PH_3 is a very weak parent acid of the strong Brønsted base P^{3-}, water is a sufficiently strong proton donor to react with phosphides to form phosphine:
  $$2\,P^{3-}(s) + 6\,H_2O(l) \rightarrow 2\,PH_3(g) + 6\,OH^-(aq)$$

- **Azides, N_3^-**
  → The *azide ion* is a highly reactive anion. Sodium azide is prepared by the reaction of dinitrogen oxide, N_2O, with molten sodium amide, $NaNH_2$.

  → Some *azides*, such as AgN_3, $Cu(N_3)_2$, and $Pb(N_3)_2$, are shock sensitive. The azide ion is a weak base, and its conjugate acid, hydrazoic acid, HN_3, is a weak acid similar in strength to acetic acid.

- **Halides**
  → *Nitrogen* has an oxidation number of +3 in the *nitrogen halides*.

  → Nitrogen trifluoride, NF_3, is the most stable and does not react with water. Nitrogen trichloride, however, reacts with water to form ammonia and hypochlorous acid, $HClO$. Nitrogen triiodide is so unstable that it decomposes explosively upon contact with light.

  → *Phosphorus* forms chlorides with an oxidation number of +3 in PCl_3 and +5 in PCl_5.

  → *Phosphorus trichloride,* a liquid, is formed by direct chlorination of phosphorus. It is a major intermediate in the production of pesticides, oil additives, and flame retardants.

  → *Phosphorus pentachloride,* a solid, is made by allowing the trichloride to react with additional chlorine. It exists in the solid as tetrahedral PCl_4^+ cations and octahedral PCl_6^- anions, but it vaporizes to a gas of trigonal bipyramidal PCl_5 molecules.

→ *Phosphorus pentabromide* is also molecular in the vapor and ionic as a solid; but in the solid, the anions are Br^- anions, presumably because of the difficulty in fitting six large Br atoms around a central P atom.

→ Both *phosphorus trichloride* and *phosphorus pentachloride* react with water in a *hydrolysis reaction,* a reaction with water in which new element-oxygen bonds are formed and there is no change in oxidation state. An *oxoacid* and *hydrogen chloride* gas are formed in the reaction.

$$PCl_3(l) + 3\,H_2O(l) \rightarrow H_3PO_3(s) + 3\,HCl(g)$$

$$PCl_5(s) + 4\,H_2O(l) \rightarrow H_3PO_4(l) + 5\,HCl(g)$$

15.17 Nitrogen Oxides and Oxoacids

- **Dinitrogen oxide, N_2O**

  → *Dinitrogen oxide* is commonly called *nitrous oxide* (laughing gas), and it is the oxide of nitrogen with the lowest oxidation number for nitrogen (+1). It is formed by gentle heating of ammonium nitrate:

  $$NH_4NO_3(s) \rightarrow N_2O(g) + 2\,H_2O(g)$$

  → N_2O is fairly unreactive, but it is toxic if inhaled in large amounts.

- **Nitrogen oxide, NO**

  → *Nitrogen oxide* is commonly called *nitric oxide;* the nitrogen atom has an oxidation number of +2. Nitrogen oxide, a colorless gas, is produced industrially by the catalytic oxidation of ammonia at 1000°C:

  $$4\,NH_3(g) + 5\,O_2(g) \rightarrow 4\,NO(g) + 6\,H_2O(g)$$

  → The endothermic formation of NO from the oxidation of N_2 occurs readily at the high temperatures that exist in automobile engines and turbine engine exhausts:

  $$N_2(g) + O_2(g) \rightarrow 2\,NO(g)$$

  → NO is further oxidized to NO_2 upon exposure to air. In this form, it contributes to smog, acid rain, and the destruction of the stratospheric ozone layer.

- **Nitrogen dioxide, NO_2**

  → *Nitrogen dioxide* is a choking, poisonous, brown gas that contributes to the color and odor of smog. The oxidation number of nitrogen in NO_2 is +4. NO_2, like NO, is paramagnetic. It disproportionates in water to form nitric acid and nitrogen oxide:

  $$3\,NO_2(g) + H_2O(l) \rightarrow 2\,HNO_3(aq) + NO(g)$$

  → *Nitrogen dioxide* is prepared in the laboratory by heating lead(II) nitrate:

  $$2\,Pb(NO_3)_2(s) \rightarrow 4\,NO_2(g) + 2\,PbO(s) + O_2(g)$$

  And it exists in equilibrium with its colorless dimer, dinitrogen tetroxide:

  $$2\,NO_2(g) \rightleftharpoons N_2O_4(g)$$

  Only the dimer exists in the solid, so the brown gas condenses to a colorless solid.

- **Nitrous acid, HNO_2**

  → *Nitrous acid* can be produced in aqueous solution by mixing its anhydride, dinitrogen trioxide, with water. The oxidation number of nitrogen in HNO_2 is +3:

  $$N_2O_3(g) + H_2O(l) \rightarrow 2\,HNO_2(aq)$$

  → *Nitrous acid* has not been isolated as a pure compound, but it has many uses in aqueous solution. It is a weak acid with $pK_a = 3.4$. The conjugate base, nitrite ion, NO_2^-, forms ionic solids that are soluble in water and mildly toxic.

- **Nitric acid, HNO_3**
    - → *Nitric acid,* a widely used industrial and laboratory acid, is produced by the three-step *Ostwald process*:

$$4NH_3(g) + 2O_2(g) \rightarrow 4NO(g) + 6H_2(g)$$
$$2NO(g) + O_2(g) \rightarrow 2NO_2(g)$$
$$3NO_2(g) + H_2O(l) \rightarrow 3HNO_3(aq) + NO(g)$$

    - → The oxidation number of nitrogen in HNO_3 is +5.
    - → Nitric acid has been isolated as a pure compound; it is a colorless liquid with a boiling point of 83°C. It is usually used in aqueous solution. It is an excellent oxidizing agent as well as a strong acid.

15.18 Phosphorus Oxides and Oxoacids

- **Phosphorus(III) oxide, P_4O_6, and phosphorous acid, H_3PO_3**
    - → *Phosphorus(III) oxide* is made by heating white phosphorus in a limited supply of air. P_4O_6 molecules consist of P_4 tetrahedral units, with each oxygen atom lying between two phosphorus corner atoms. Six oxygen atoms lie along the six edges of the tetrahedron, one per edge. The edge bonds are represented as $-P-O-P-$.
    - → P_4O_6 is the anhydride of *phosphorous acid:* $\quad 4H_3PO_3(aq) \rightarrow P_4O_6(s) + 6H_2O(l)$ Phosphorous acid is a *diprotic acid.*

- **Phosphorus(V) oxide, P_4O_{10}, and phosphoric acid, H_3PO_4**
    - → *Phosphorus(V) oxide* is made by heating white phosphorus in an excess supply of air. P_4O_{10} molecules are similar to P_4O_6 but have an additional oxygen attached to each phosphorus at the apices of the tetrahedron.
    - → P_4O_{10} is the *anhydride* of *phosphoric acid,* H_3PO_4. It traps and reacts with water very efficiently and is widely used as a drying agent. Phosphoric acid is a *triprotic acid.* It is only mildly oxidizing, despite the fact that phosphorus is in a high oxidation state.
    - → Phosphoric acid is the parent acid of phosphate salts that contain the tetrahedral phosphate ion, PO_4^{3-}. Phosphates are generally not very soluble, which makes them suitable structural material for bones and teeth.

- **Polyphosphates**
    - → *Polyphosphates* are compounds made up of linked PO_4^{3-} tetrahedral units. The simplest structure is the *pyrophosphate ion,* $P_2O_7^{4-}$, in which two phosphate ions are linked by an oxygen atom through $-O-P-O-$ bonds.
    - → More complicated structures that form with longer chains or rings also exist. The biochemically most important polyphosphate is *adenosine triphosphate,* ATP, which contains three phosphorus tetrahedral units linked by $-O-P-O-$ bonds. Hydrolysis of ATP to *adenosine diphosphate,* ADP, by the rupture of an O–P bond releases energy that is used by cells to drive biochemical reactions within the cell:

$$ATP + H_2O \rightarrow ADP + HPO_4^{2-} \qquad \Delta H = -41 \text{ kJ}$$

    ATP is often called the "fuel of life" because it provides the energy for many of the biochemical reactions occurring in cells.

GROUP 16 / VI: THE OXYGEN FAMILY
(Sections 15.19–15.21)

15.19 The Group 16 / VI Elements

- **Group 16**
    - → Valence shell configuration ns^2np^4
    - → Anions favored; only 2 electrons need to complete valence shell.

- **Oxygen**
    - → *Elemental oxygen*, O_2, is found in air (21.0% by volume) and is prepared by fractional distillation of liquid air.
    - → Oxygen is a colorless, odorless, *paramagnetic* gas.
    - → A satisfactory Lewis structure for oxygen cannot be drawn. The Molecular Orbital theory accounts for the paramagnetic properties (see Section 3.10 of the text).
    - → *Elemental oxygen* exists in two allotropic forms, O_2 and O_3 (*ozone*). Inorganic chemists often use the term *dioxygen* to refer to molecular oxygen.
    - → Ozone is produced by an electric discharge in oxygen gas; its pungent odor is apparent near sparking electrical equipment and lightning strikes.

- **Sulfur**
    - → *Sulfur* is found in ores and as elemental sulfur.
    - → *Sulfur* is widely distributed as sulfide ores, which include *galena*, PbS; *cinnabar*, HgS; *iron pyrite*, FeS_2 (fool's gold); and *spahlerite*, ZnS.
    - → *Sulfur* is also found in elemental form as deposits, called *brimstone*, which are created by bacterial action on H_2S.
    - → *Elemental sulfur* is a yellow, tasteless, almost odorless, insoluble, nonmetallic molecular solid of crownlike S_8 rings.
    - → The more stable allotrope, *rhombic sulfur*, forms beautiful yellow crystals, whereas the less stable allotrope, *monoclinic sulfur*, forms needlelike crystals. The two allotropes differ in the manner in which the S_8 rings are stacked together.
    - → For simplicity, the elemental form of sulfur is often represented as S(s) rather than S_8(s).
    - → One method of recovering sulfur is the *Claus process*, in which some of the H_2S that occurs in oil and natural gas is first oxidized to sulfur dioxide:
    $$2\,H_2S(g) + 3\,O_2(g) \rightarrow 2\,SO_2(g) + 2\,H_2O(l)$$
    - → Sulfur dioxide oxidizes the remaining hydrogen sulfide and both are converted to elemental sulfur at 300°C in the presence of an alumina catalyst:
    $$2\,H_2S(g) + SO_2(g) \rightarrow 3\,S(s) + 2\,H_2O(l)$$
    - → *Sulfur* is used primarily to produce *sulfuric acid* and to vulcanize rubber.
    - → Vulcanization increases the toughness of rubber by introducing cross-links between the polymer chains of natural rubber.
    - → An important property of sulfur is its ability to form chains of atoms, *catenation*. The –S–S– links that connect different parts of the chains of amino acids in proteins are an important example. These *disulfide links* contribute to the shapes of proteins (see Chapter 19 in the text).

- **Selenium and tellurium**
  - $\rightarrow$ *Selenium* and *tellurium* are found in sulfide ores; they are also recovered from the anode sludge formed during the electrolytic refining of copper.
  - $\rightarrow$ Both elements have several allotropic forms; the most stable one consists of long *zigzag chains* of atoms.
  - $\rightarrow$ The allotropes look like a silver-white metal, but electrical conductivity is poor. Exposing selenium to light increases its electrical conductivity, so it is used in solar cells.
- **Polonium**
  - $\rightarrow$ *Polonium* is a radioactive, low-melting metalloid. It was identified in uranium ores in 1898 by the Curies. Its most stable isotope, ^{209}Po, has a half-life of 103 y.
  - $\rightarrow$ *Polonium* decays by emission of an alpha particle, ^{4}He^{2+}. It is used in antistatic devices in textile mills to counteract the buildup of negative charges on fast moving fabric.

15.20 Compounds with Hydrogen

- **Water, H_2O, and hydrogen sulfide, H_2S**
  - $\rightarrow$ *Water* is a remarkable compound possessing a unique set of physical and chemical properties.
  - $\rightarrow$ *Water* is purified in domestic water supplies in a multi-step procedure that includes *aeration,* addition of *slaked lime,* removal of particles by *coagulation* and *flocculation,* addition of CO_2, *filtration,* reduction of pH, and addition of chlorine *disinfectant* to kill bacteria. High-purity water is obtained by *distillation.*
  - $\rightarrow$ *Water* has a higher boiling point than expected on the basis of its molar mass because of the extensive hydrogen bonding between H_2O molecules.
  - $\rightarrow$ Hydrogen bonding also causes a more open structure and, therefore, a lower density for the solid than for the liquid. Because of its high polarity, water is an excellent solvent for ionic compounds.
  - $\rightarrow$ Chemically, water is *amphiprotic,* and it can both donate and accept protons; so it is both a *Brønsted acid* and a *Brønsted base.*

  $$H_2O(l, acid) + NH_3(aq) \rightarrow NH_4^+(aq) + OH^-(aq)$$
  $$CH_3COOH(aq) + H_2O(l, base) \rightarrow H_3O^+(aq) + CH_3COO^-(aq)$$

  - $\rightarrow$ *Water* can act as a *Lewis base* by donating its unshared pair of electrons on the oxygen atom:

  $$Fe^{3+}(aq) + 6 H_2O(l) \rightarrow Fe(H_2O)_6^{3+}(aq)$$

  - $\rightarrow$ *Water* can act as an *oxidizing agent* and a *reducing agent.*

  $$2 Na(s) + 2 H_2O(l, ox.\ agent) \rightarrow 2 NaOH(aq) + H_2(g)$$
  $$2 H_2O(l, red.\ agent) + 2 F_2(g) \rightarrow 4 HF(aq) + O_2(g)$$

  - $\rightarrow$ *Hydrolysis reaction:* Water as a reactant forms a bond between oxygen and another element. Hydrolysis can occur *with* or *without* a change in oxidation number:

  $$PCl_5(s) + 4 H_2O(l) \rightarrow H_3PO_4(aq) + 5 HCl(aq) \qquad (1)$$
  $$Cl_2(g) + H_2O(l) \rightarrow ClOH(aq) + HCl(aq) \qquad (2)$$

  Reaction (1) has *no* change in oxidation number. Reaction (2) is a *disproportionation reaction,* in which the oxidation number of chlorine changes from 0 in Cl_2 to +1 in HClO and −1 in HCl.

  - $\rightarrow$ *Hydrogen sulfide* is an example of a Group 16/ VI binary compound with hydrogen. It is a toxic gas with an offensive odor (rotten eggs). Egg proteins contain sulfur and release H_2S when they decompose. Iron(II) sulfide sometimes appears as a pale green discoloration at the boundary between the white and the yolk.

→ *Hydrogen sulfide* is prepared by the direct reaction of hydrogen and sulfur at 600°C or by protonation of the sulfide ion, which is a Brønsted base:

$$FeS(s) + 2\,HCl(aq) \rightarrow FeCl_2(aq) + H_2S(g)$$

→ *Hydrogen sulfide* dissolves in water and is slowly oxidized by dissolved air to form a colloidal dispersion of particles of sulfur.

→ *Hydrogen sulfide* is a *weak diprotic aci,* and the parent acid of the hydrogen sulfides, HS^-, and the sulfides, S^{2-}.

→ Sulfides of *s*-block elements are moderately soluble in water, whereas the sulfides of the heavy *p*- and *d*-block metals are generally very insoluble.

- **Hydrogen peroxide, H_2O_2, and polysulfanes, $HS-S_n-SH$**

→ *Hydrogen peroxide* is a highly reactive, pale blue liquid that is appreciably more dense $(1.44\ g\cdot mL^{-1})$ than water. Its melting point is $-0.4°C$ and its boiling point is $152°C$.

→ Chemically, *hydrogen peroxide* is more acidic than water ($pK_a = 11.75$). It is a strong oxidizing agent, but it can also function as a reducing agent.

$$2\,Fe^{2+}(aq) + H_2O_2(aq,\ ox.\ agent) + 2\,H^+(aq) \rightarrow 2\,Fe^{3+}(aq) + 2\,H_2O(l)$$
$$Cl_2(g) + H_2O_2(aq,\ red.\ agent) + 2\,OH^-(aq) \rightarrow 2\,Cl^-(aq) + 2\,H_2O(l) + O_2(g)$$

→ The O–O bond in *hydrogen peroxide* is very weak. The oxidation state of oxygen in hydrogen peroxide is -1.

→ A *polysulfane* is a *catenated* molecular compound of composition $HS-S_n-SH$, where n can take values from 0 through 6.

→ The sulfur analogue of hydrogen peroxide occurs with $n = 0$.

→ Two polysulfide ions obtained from polysulfanes are found to occur in the mineral *lapis lazuli* (see Fig. 15.46 in the text); its color derives from impurities of S_2^- (blue) and S_3^- (hint of green).

→ The S_2^- ion is an analogue of the *superoxide* ion, O_2^-; and S_3^- is an analogue of the *ozonide* ion, O_3^-.

15.21 Sulfur Oxides and Oxoacids

- **Sulfur dioxide, SO_2, sulfurous acid, H_2SO_3, and sulfite, SO_3^{2-}**

→ *Sulfur dioxide,* a poisonous gas, is made by burning sulfur in air. Volcanic activity, combustion of fuels contaminated with sulfur, and oxidation of hydrogen sulfide are the major sources of atmospheric SO_2. The equation for oxidation of hydrogen sulfide in the atmosphere is:

$$2\,H_2S(g) + 3\,O_2(g) \rightarrow 2\,SO_2(g) + 2\,H_2O(g)$$

→ *Sulfurous acid* is formed by the reaction of its acid anhydride, SO_2, and water:

$$SO_2(g) + H_2O(l) \rightarrow H_2SO_3(aq)$$

→ *Sulfurous acid* is an equilibrium mixture of $(HO)SHO_2$ and $(HO)SO(OH)$. These molecules are also in equilibrium with SO_2 molecules, each of which is surrounded by a cage of water molecules.

→ When the solution is cooled, crystals with a composition of roughly $SO_2 \cdot 7\,H_2O$ form. This substance is an example of a *clathrate,* in which a molecule sits in a cage of other molecules. Methane, carbon dioxide, and the noble gases also form clathrates with water.

→ In *sulfite ions* and *sulfur dioxide,* the oxidation number of the sulfur atom is $+4$. Because this value is intermediate in sulfur's range of -2 to $+6$, these compounds can act as either oxidizing agents or reducing agents.

→ The most important reaction of sulfur dioxide is its slow oxidation to sulfur trioxide, in which the oxidation number of the sulfur atom is +6.

$$2 SO_2(g) + O_2(g) \rightarrow 2 SO_3(g)$$

This reaction is catalyzed by metal cations in droplets of water or by certain surfaces. Indirect pathways also exist in the atmosphere for the conversion of SO_2 to SO_3.

- **Sulfur trioxide, SO_3, sulfuric acid, H_2SO_4, and sulfate, SO_4^{2-}**

  → At normal temperatures, *sulfur trioxide* is a volatile liquid with a boiling point of 45°C. Its shape is trigonal planar with bond angles of 120°.

  → In the solid, and to some extent in the liquid, the molecules aggregate to form *trimers,* S_3O_9, as well as larger groupings.

  → *Sulfuric acid* is produced by the *contact process,* in which sulfur is first burned in oxygen at 1000°C to form sulfur dioxide, which then reacts with oxygen to form sulfur trioxide over a V_2O_5 catalyst at 500°C.

  → Because sulfur trioxide forms a corrosive acid with water, it is absorbed in 98% concentrated sulfuric acid to give a dense, oily liquid called *oleum,* $H_2S_2O_7$:

  $$SO_3(g) + H_2SO_4(l) \rightarrow H_2S_2O_7(l)$$

  → *Sulfuric acid* is a colorless, corrosive, oily liquid that boils (decomposes) at 300°C. It is a strong *Brønsted acid,* a powerful *dehydrating agent,* and a mild *oxidizing agent.*

  → In water, the first ionization is complete, but HSO_4^- is a *weak acid* with $pK_a = 1.92$. As a *dehydrating agent,* sulfuric acid is able to extract water from a variety of compounds, including sucrose, $C_{12}H_{22}O_{11}$, and formic acid, HCOOH.

  $$C_{12}H_{22}O_{11}(s) \rightarrow 12 C(s) + 11 H_2O(l)$$

  $$HCOOH(l) \rightarrow CO(g) + H_2O(l)$$

  → Because of the large exothermicity associated with dilution, mixing sulfuric acid with water can cause violent splashing, so for reasons of safety, sulfuric acid (which is more dense than water) is always carefully added to water, not the water to acid.

  → The *sulfate ion* is a mild *oxidizing agent.*

  $$4 H^+(aq) + SO_4^{2-}(aq) + 2 e^- \rightarrow SO_2(g) + 2 H_2O(l) \quad E° = 0.17 \text{ V}$$

GROUP 17 / VII: THE HALOGENS
(Sections 15.22–15.23)

15.22 The Group 17 / VII Elements

- **Fluorine**

  → *Fluorine* is a reactive, almost colorless gas of F_2 molecules.

  → The properties follow from its high electronegativity, small size, and lack of available *d*-orbitals for bonding.

  → *Fluorine* is the most electronegative element and has an oxidation number of −1 in all its compounds.

→ Elements combined with fluorine are often found with their highest oxidation numbers, such as +7 for I in IF_7.

→ The small size of the fluoride ion results in high lattice enthalpies for fluorides, a property that makes them less soluble than other halides.

→ *Fluorine* occurs widely in many minerals including *fluorspar,* CaF_2; *cryolite,* Na_3AlF_6; and the *fluoroapatites,* $Ca_5F(PO_4)_3$.

→ *Elemental fluorine* is obtained by the electrolysis of an anhydrous, molten KF–HF mixture at 75°C, using a carbon electrode.

→ Fluorine is highly reactive and highly oxidizing; it is used in the production of SF_6 for electrical equipment, Teflon (polytetrafluoroethylene), and UF_6 for isotope enrichment.

→ The strong bonds formed by *fluorine* make many of its compounds relatively inert.

→ Fluorine's ability to form *hydrogen bonds* results in relatively high melting points, boiling points, and enthalpies of vaporization for many of its compounds.

- **Chlorine**
  → *Chlorine* is a reactive, pale yellow-green gas of Cl_2 molecules that condenses at −34°C.

  → *Chlorine* reacts directly with all the elements except carbon, nitrogen, oxygen, and the noble gases.

  → Used in a large number of industrial processes, including the manufacture of plastics, solvents, and pesticides

  → Also used as a disinfectant to treat water supplies.

  → *Elemental chlorine* is obtained by the electrolysis of molten or aqueous NaCl. It is a strong oxidizing agent and oxidizes metals to high oxidation states:
  $$2\,Fe(s) + 3\,Cl_2(g) \rightarrow 2\,FeCl_3(s)$$

- **Bromine**
  → *Bromine* is a corrosive, reddish brown liquid of Br_2 molecules. It has a penetrating odor.

  → Organic bromides are used in textiles as fire retardants and in pesticides. Inorganic bromides, particularly silver bromide, are used in photographic emulsions.

  → *Elemental bromine* is produced from brine wells by the oxidation of Br^- by Cl_2.
  $$2\,Br^-(aq) + Cl_2(g) \rightarrow Br_2(l) + 2\,Cl^-(aq)$$

- **Iodine**
  → A blue-black, lustrous solid of I_2 molecules that readily sublimes to form a purple vapor

  → Occurs as I^- in seawater and as an impurity in Chile saltpeter, KNO_3

  → The best source is the brine from oil wells; *elemental iodine* is produced from brine wells by the oxidation of I^- by Cl_2.

  → *Elemental iodine* is slightly soluble in water, but it dissolves well in iodide solutions because it reacts with aqueous I^- to form the triodide ion, I_3^-.

  → *Iodine* is an essential trace element in human nutrition, and *iodides* are often added to table salt. This "iodized salt" prevents iodine deficiency, which leads to the enlargement of the thyroid gland in the neck (a condition called *goiter*).

- **Astatine**
  → *Astatine* is a radioactive element that occurs in uranium ores, but only to a tiny extent. Its most stable isotope, ^{210}At, has a half-life of 8.3 h. The isotopes formed in uranium ores have much shorter lifetimes. The properties of astatine are surmised from spectroscopic measurements.

  → *Astatine* is created by bombarding bismuth with alpha particles in a cyclotron, which accelerates particles to high speed.

15.23 Compounds of the Halogens

- **Interhalogens**

  → *Interhalogen* compounds consist of a heavier halogen atom at the center of the compound and an odd number of lighter ones bonded to it on the periphery. An exception is I_2Cl_6, which contains two (–I–Cl–I–) bridges between the two central iodine atoms. The two bridging chlorine atoms act as Lewis bases (electron pair donors), in a manner similar to Al_2Cl_6.

  → *Interhalogens* are typically formed by reaction between stoichiometric amounts of the elements:

  $$I_2(g) + 7F_2(g) \rightarrow 2IF_7(g)$$

  → As the size of the central atom in an *interhalogen* compound increases, the bond enthalpies also increase, *usually* resulting in lower reactivity of the compound.

  (X–F): ClF_5 (154 kJ·mol^{-1}) < BrF_5 (187 kJ·mol^{-1}) < IF_5 (269 kJ·mol^{-1})

  → Table 15.11 in the text gives a compilation of the known *interhalogens*.

- **Hydrogen halides and metal halides**

  → *Hydrogen halides* are prepared by direct reaction of the elements or by the action of nonvolatile acids on *metal halide:.*

  $$Cl_2(g) + H_2(g) \rightarrow 2HCl(g)$$
  $$CaF_2(s) + H_2SO_4(aq, \text{conc.}) \rightarrow Ca(HSO_4)_2(aq) + 2HF(g)$$

  → Because Br$^-$ and I$^-$ are oxidized by sulfuric acid, phosphoric acid is used in the preparation of HBr and HI.

  → *Hydrogen halides* are pungent, colorless gases, but *hydrogen fluoride* is a liquid below 20°C. Its low volatility is a sign of extensive hydrogen bonding, and short zigzag chains up to about (HF)$_5$ persist in the vapor.

  → *Hydrogen halides* dissolve readily in water to produce acidic solutions. *Hydrogen fluoride* has the unusual property of attacking and dissolving silica glasses, and it is used for glass etching (see Fig. 6.22 in the text).

  → Anhydrous *metal halides* may be formed by direct reaction of the elements:

  $$2Fe(s) + 3Cl_2(g) \rightarrow 2FeCl_3(s)$$

  Halides of metals tend to be ionic unless the metal has an oxidation number greater than +2.

  Examples: Sodium chloride and copper(II) chloride are ionic compounds with high melting points, whereas titanium(IV) chloride and iron(III) chloride sublime as molecules.

- **Halogen oxides and oxoacids**

  → *Hypohalous acids,* HXO *or* HOX, consist of the halogen atom as a *terminal* atom with an oxidation number of +1. They are prepared by the direct reaction of a halogen with water; the corresponding *hypohalite* (XO$^-$) salts are prepared by the reaction of a halogen with aqueous alkali solution:

  $$Cl_2(g) + H_2O(l) \rightarrow HOCl(aq) + HCl(aq)$$
  $$Cl_2(g) + NaOH(aq) \rightarrow NaOCl(aq) + HCl(aq)$$

  → *Hypochlorites* cause the oxidation of organic material in water by producing oxygen in aqueous solution, which readily oxidizes organic materials. The production of O_2 occurs in two steps; the net result is:

  $$2OCl^-(aq) \rightarrow 2Cl^-(aq) + O_2(g).$$

→ *Chlorates*, ClO_3^-, contain a *central* chlorine atom with an oxidation number of +5. They are prepared by the reaction of chlorine with hot aqueous alkali:

$$3\,Cl_2(g) + 6\,OH^-(aq) \rightarrow ClO_3^-(aq) + 5\,Cl^-(aq) + 3\,H_2O(l)$$

→ *Chlorates* decompose upon heating; the final product is determined by the presence of a catalyst:

$$4\,KClO_3(l) \rightarrow 3\,KClO_4(s) + KCl(s) \quad \textit{no catalyst}$$

$$2\,KClO_3(l) \rightarrow 2\,KCl(s) + 3\,O_2(g) \quad \textit{MnO}_2 \textit{ catalyst}$$

→ *Chlorates* are good oxidizing agents and are also used to produce the important compound, ClO_2. Sulfur dioxide is a convenient reducing agent for this reaction:

$$2\,NaClO_3(l) + SO_2(g) + H_2SO_4(aq, \text{dilute}) \rightarrow 2\,NaHSO_4(aq) + 2\,ClO_2(g)$$

→ *Chlorine dioxide*, ClO_2, contains a *central* chlorine atom with an oxidation number of +4. It has an odd number of electrons and is a paramagnetic yellow gas that is used to bleach paper pulp. It is highly reactive and may explode violently under the right conditions.

→ *Perchlorates*, ClO_4^-, contain a *central* chlorine atom with an oxidation number of +7. They are prepared by the electrolysis of aqueous chlorates. The half-reaction is:

$$ClO_3^-(aq) + H_2O(l) \rightarrow ClO_4^-(aq) + 2\,H^+(aq) + 2\,e^-$$

→ *Perchlorates* and *perchloric acid*, $HClO_4$, are powerful oxidizing agents; *perchloric acid* in contact with small amounts of organic materials may explode.

→ The *oxidizing strength* and *acidity* of oxoacids both increase as the oxidation number of the halogen increases.

→ For oxoacids with the same number of oxygen atoms, the oxidizing and acid strength both increase as the electronegativity value of the halogen increases. The general rules for acid strength of oxoacids are summarized in Section 11.10 of the text.

GROUP 18 / VIII: THE NOBLE GASES
(Sections 15.24–15.25)

15.24 The Group 18 / VIII Elements

- **Helium**

  → *Helium* (Greek word *helios,* sun) is the second most abundant element in the universe after hydrogen. Helium atoms are light and travel at velocities sufficient to escape from the Earth's atmosphere. The first evidence for helium was discovered in the solar spectrum taken during a solar eclipse in 1868.

  → Alpha particles, $^4He^{2+}$, are released by nuclear decay of naturally occurring uranium and thorium ores. They are high-energy helium-4 nuclei, which capture two electrons to become helium atoms. Low density and lack of flammability of helium make it suitable for lighter-than-air airships such as blimps. Helium is found in the atmosphere with a concentration of 5.24 ppm (parts per million by volume).

  → *Helium* has the lowest melting point of any substance, only freezes under pressure, and is the only substance with two liquid phases. Below 2 K, liquid helium-II exhibits superfluidity, the ability to flow without viscosity. It is used as a cryogen for studies at low temperature.

- **Neon**
  - → *Neon* (Greek word *neos*, new) is widely used in advertising display signs because it emits an orange-red glow when an electric current passes through it.
  - → Of all the noble gases, the discharge of neon is the most intense at ordinary voltages and currents. The neon atom is the source of laser light in the helium-neon laser.
  - → *Neon* has over 40 times more refrigerating capacity per unit volume than liquid helium and more than three times that of liquid hydrogen. It is compact, inert, and less expensive than helium when used as a *cryogenic* coolant.
  - → Neon is found in the atmosphere with a concentration of 18.2 ppm.
- **Argon**
  - → *Argon* (Greek word *argos,* inactive) is mainly used to fill electric light bulbs at a pressure of about 400 Pa, where it conducts heat away from the filament. It is also used to fill fluorescent tubes.
  - → *Argon* is also used as an inert gas shield (to prevent oxidation) for arc welding and cutting, and as a protective atmosphere for growing silicon and germanium crystals.
  - → Argon is the most abundant noble gas with an atmospheric concentration of about 0.93% (or about 1 part per hundred by volume).
- **Krypton**
  - → *Krypton* (Greek word *kryptos,* hidden) gives an intense white light when an electric discharge is passed through it, and it is used in airport runway lighting.
  - → *Krypton* is produced in nuclear fission, and its atmospheric abundance is a measure of worldwide nuclear activity.
  - → Krypton is found in the atmosphere with a concentration of about 1 ppm.
- **Xenon**
  - → *Xenon* (Greek word *xenos,* stranger) is used in halogen lamps for automobile headlights and in high-speed photographic flash tubes.
  - → *Xenon* is used in the nuclear energy area in bubble chambers, probes, and other applications where a high atomic mass is desirable. Currently, the anesthetic properties of xenon are being explored.
  - → Xenon is found in the atmosphere with a concentration of about 0.087 ppm.
- **Radon**
  - → *Radon* (name derived from the element *radium*) is a radioactive gas. It is formed by radioactive decay processes deep in the Earth.
  - → Uranium-238 decays very slowly to radium-226, which decays further by alpha particle emission to radon-222, and ultimately to lead-206 (see Fig. 17.16 in the text). The half-life of radon-222 is 3.825 days. Other shorter-lived isotopes are formed from the decay of thorium-232 and uranium-235.
  - → Every square mile of soil to a depth of 6 inches is estimated to contain about 1 g of radium, which releases radon in tiny amounts into the atmosphere. Accumulation of radon in buildings can be dangerous to human health.
  - → *Radon* is used in implant seeds for the therapeutic treatment of localized tumors. Radon is found in the atmosphere with a concentration estimated to be 1 in 10^{21} parts of air.

15.25 Compounds of the Noble Gases

- **Noble gases** → Ionization energies of the noble gases are very high and decrease down the group.

- **Helium, neon, and argon**
  - → The ionization energies of helium, neon, and argon are so high that they form no stable *neutral* compounds. However, He and Ne have been trapped inside buckminsterfullerene cages.
  - → If electrons are removed from noble gas atoms, the resulting ions are radicals, and they may form stable molecular ions such as $NeAr^+$, ArH^+, and $HeNe^+$ in the gas phase. If an electron is added to them, these ions rapidly dissociate into neutral atoms.
- **Krypton**
  - → *Krypton* forms a thermodynamically unstable neutral compound, KrF_2 (recall that diamond is also thermodynamically unstable). It is a volatile white solid, which decomposes slowly at room temperature.
  - → In 1988, a compound with a Kr–N bond was discovered, but it is stable only at temperatures below $-50°C$.
- **Xenon**
  - → *Xenon* is the noble gas element with the richest chemistry. It forms several compounds with fluorine and oxygen, and even compounds with Xe–N and Xe–C bonds. Xenon is found in compounds with oxidation numbers of +2, +4, +6, and +8, and XeF_2 is thermodynamically stable.
  - → Direct reaction of *fluorine* with *xenon* at high temperature yields compounds with oxidation numbers of +2 (XeF_2, thermodynamically stable), +4 (XeF_4), and +6 (XeF_6). In the case of XeF_6, high pressure is required. All three fluorides are crystalline solids. In the gas phase, all are molecular compounds. Solid *xenon hexafluoride* is an ionic compound with a complex structure of XeF_5^+ cations bridged by F^- anions.
  - → *Xenon fluorides* are powerful *fluorinating agents,* reagents that attach fluorine to other substances. *Xenon trioxide* is synthesized by the hydrolysis of *xenon tetrafluoride.*
    $$6\,XeF_4(s) + 12\,H_2O(l) \rightarrow 2\,XeO_3(aq) + 4\,Xe(g) + 3\,O_2(g) + 24\,HF(aq)$$
  - → The trioxide is the anhydride of *xenic acid,* H_2XeO_4. In aqueous basic solution, the acid forms the *hydrogen xenate ion,* $HXeO_4^-$, which further disproportionates to the *perxenate ion,* XeO_6^{4-}, in which xenon attains its highest oxidation number of +8.
    $$2\,HXeO_4^-(aq) + 2\,OH^-(aq) \rightarrow XeO_6^{4-}(aq) + Xe(g) + O_2(g) + 2\,H_2O(l)$$
- **Radon**
  - → *Radon* chemistry is difficult to study because all of its isotopes are radioactive.
  - → A *radon fluoride* of unknown composition is found to form readily, but it decomposes rapidly during attempts to vaporize it for further study.

Chapter 16 THE ELEMENTS: THE *d* BLOCK

THE *d*-BLOCK ELEMENTS AND THEIR COMPOUNDS
(Sections 16.1–16.2)

16.1 Trends in Physical Properties

- **d-Block elements**
  - → Electron configurations: $ns^2(n-1)d^m$, m = 1 − 10 (with ten exceptions).
  - → For most *d*-block elements, the $(n-1)d$ orbitals fill after ns^2, but the *ns* electrons are removed before $(n-1)d$ electrons when cations are formed (there are also some exceptions to this rule for cations with a +1 charge).
  - → All are metals, most are good electrical conductors, particularly Ag, Cu, and Au.
  - → Most are malleable, ductile, lustrous, and silver-white in color.
    Exceptions: Cu (red-brown) and Au (yellow)
  - → Most have higher melting and boiling points than main group elements.
    A major exception is Hg, which is a liquid at room temperature.

- **Shapes of the *d*-orbitals**
  - → Some properties of the *d*-block elements follow from the shapes of the *d*-orbitals.
  - → *d*-orbital *lobes* are relatively far apart from each other, so electrons in different *d*-orbitals on the same atom repel each other weakly.
  - → Electron density in *d*-orbitals is low near the nucleus. Because *d*-electrons are relatively far from the nucleus, they are not very effective in shielding other electrons from the positive charge of the nucleus.

- **Trends in atomic radii**

  Across a Period
  - → *Nuclear charge* and the *number* of *d*-electrons both increase across a row from left to right.
  - → The first five electrons added to a *d*-subshell are placed in different orbitals (Hund's rule).
  - → Since repulsion between *d*-electrons in different orbitals is small, increasing *effective nuclear charge* is the dominant factor, so atoms tend to become smaller as Z increases in the first half of the *d* block. **Exception:** Mn
  - → Further across the block, radii begin to *increase* slightly with Z because *electron-electron repulsion* in doubly-filled orbitals outweighs the effect of increasing *effective nuclear charge*.
  - → Attractions and repulsions are finely balanced, and the range of atomic *d*-metal radii is small.
  - → *d*-Metals form many alloys because atoms of one metal can easily replace atoms of another metal.

  Down a group – Periods 4 and 5
  - → *d*-metals in *Period 5* are typically larger than those in *Period 4*.
  - → The effective nuclear charge on the outer electrons is roughly the same, whereas the *n* quantum number increases down the group. The usual pattern is an increase in size for group members from top to bottom. This pattern is followed in *Periods 4* and *5* for the *d*-metals. **Exception:** Mn

  Across Period 6 – the lanthanide contraction.
  - → *d*-metal radii in *Period 6* are *approximately the same* as those in *Period 4*. They are smaller than expected because of the *lanthanide contraction,* the decrease in radius along the first row of the *f* block.

→ There are 14 *f*-block elements before the first *6d*-block element in *Period 6,* Lu. At Lu, the atomic radius has fallen from 224 pm for Ba to 173 pm for Lu.

→ *Period 6 d*-block elements are substantially more dense than those in *Period 5* owing to the *lanthanide* contraction. The radii are about equal, but the masses are almost twice as large. Ir and Os are the two most dense elements (approximately 22.6 $g \cdot cm^{-3}$).

→ A second effect of the *contraction* is the *low reactivity* of Pt and Au, whose valence electrons are so tightly bound that they are not readily available for chemical reaction.

16.2 Trends in Chemical Properties

- **Oxidation numbers (states)**

   → We will use *oxidation state* and *oxidation number* interchangeably.

   → Most *d*-block elements have more than one common oxidation number and one or more less common oxidation numbers (see Figure 16.6 in the text).

   → Oxidation numbers range from negative values to +8.

   Example: In $[Fe(CO)_4]^{2-}$, Fe has an oxidation number of −2, whereas in OsO_4, Os has an oxidation number of +8.

   → Oxidation numbers show greatest variability for elements in the middle of the *d* block.

   Example: Mn, at the center of its row, has seven oxidation numbers, whereas Sc, at the beginning of the same row, has only one (in addition to zero).

   → *Period 5* and *Period 6 d*-block elements tend to have higher oxidation numbers than those in *Period 4*.

   Example: The highest oxidation number of Ni (Group 10, *Period 4*) is +4, whereas that of Pt (Group 10, *Period 6*) is +6.

- **Chemical properties and oxidation states**

   → Species containing *d*-block elements with *high oxidation numbers* tend to be good *oxidizing agents*.

   Examples: MnO_4^- with Mn +7 and CrO_4^{2-} with Cr +6

   → Species containing *d*-block elements with *high oxidation numbers* tend to exhibit *covalent bonding*.

   Example: Mn_2O_7 with Mn +7 is a covalent *liquid* at room temperature.

   → Species containing *d*-block elements with *high oxidation numbers* tend to have *oxides* that are *acid anhydrides*.

   Example: CrO_3 (with Cr +6) + H_2O → 2 H_2CrO_4 (acidic)

   → Species containing *d*-block elements with *low oxidation numbers* tend to be good *reducing agents*.
   Examples: CrO with Cr +2 and $FeCl_2$ with Fe +2

   → Species containing *d*-block elements with *low oxidation numbers* tend to exhibit *ionic bonding*.
   Example: Mn_3O_4, with Mn+2 and Mn +3 is an ionic solid.

   → Species containing *d*-block elements with *low oxidation numbers* tend to have *oxides* that are *basic anhydrides*.

   Example: CrO with Cr +2 is a *basic* oxide

SELECTED ELEMENTS: A SURVEY (Sections 16.3–16.4)

- **Properties** $\rightarrow$ Chemical properties of the *d*-block elements are more diverse than their physical properties.

16.3 Scandium Through Nickel

- **Properties of the metals (see Table 16.1 in the text)**
  - $\rightarrow$ Densities increase from Sc (2.99 g·cm^{-3}) to Ni (8.91 g·cm^{-3}).
  - $\rightarrow$ Melting points increase from Sc (1540 K) to V (1920 K), then decrease uniformly to Ni (1455 K). **Exception:** Mn has an anomalously low mp, 1250 K.
  - $\rightarrow$ Boiling points increase from Sc (2800 K) to V (3400K), then show irregular behavior to Ni (2150 K).
  - $\rightarrow$ Valence electron configurations follow the filling rule: $4s$ fills before $3d$ from Sc, $[Ar]3d^1 4s^2$, to Ni, $[Ar]3d^8 4s^2$, with the exception of Cr, $[Ar]3d^5 4s^1$.

- **Scandium, Sc, $[Ar]3d^1 4s^2$**
  - $\rightarrow$ A highly reactive metal that reacts with water as vigorously as Ca
  - $\rightarrow$ Has only a few commercial uses and is thought *not* to be essential to life
  - $\rightarrow$ Has one oxidation state, +3, in compounds
  - $\rightarrow$ Small, highly charged Sc^{3+} is strongly hydrated in water, forming the $Sc(H_2O)_6^{3+}$ ion, which is a weak acid (about as strong as acetic acid).

- **Titanium, Ti, $[Ar]3d^2 4s^2$**
  - $\rightarrow$ A light, strong metal passivated by an oxide coating, which masks its inherent reactivity
  - $\rightarrow$ Commonly found with the +4 oxidation number in its ores: *rutile,* TiO_2, and *ilmenite,* $FeTiO_2$
  - $\rightarrow$ Requires strong reducing agents to extract from its ores
  - $\rightarrow$ The metal is used in jet engines and dental appliances.
  - $\rightarrow$ Titanium oxide, TiO_2, is used as a paint pigment and as a semiconductor in solar cells.
  - $\rightarrow$ Forms a series of oxides called *titanates*
  - $\rightarrow$ *Barium titanate,* $BaTiO_3$, is a *piezoelectric* material (develops an electrical signal when stressed).

- **Vanadium, V, $[Ar]3d^3 4s^2$**
  - $\rightarrow$ A soft, silver-gray metal
  - $\rightarrow$ V metal is used in *ferroalloys* (Fe, V, C) to make tough steels for automotive springs
  - $\rightarrow$ Exhibit positive oxidation numbers ranging from +1 to +5
  - $\rightarrow$ Common oxidation numbers are +4 and +5.
  - $\rightarrow$ Many colored compounds containing vanadium species are used as ceramic glazes.
  - $\rightarrow$ Orange-yellow *Vanadium(V) oxide,* V_2O_5, is used as an oxidizing agent/catalyst in the contact process for the manufacture of sulfuric acid.

- **Chromium, Cr, $[Ar]3d^5 4s^1$**
  - $\rightarrow$ A bright, corrosion-resistant metal
  - $\rightarrow$ Used to make stainless steel and for chromium plating
  - $\rightarrow$ Obtained from *chromite,* $FeCr_2O_4$, by reduction with C in an electric furnace:
    $$FeCr_2O_4(s) + 4\ C(s) \rightarrow Fe(l) + 2\ Cr(l) + 4\ CO(g)$$
  - $\rightarrow$ An exception to the normal filling order of electrons ($3d^5 4s^1$, not $3d^4 4s^2$)

- **Compounds of Cr**
  - → Positive oxidation numbers range from +1 to +6.
  - → Common oxidation numbers are +3 and +6.
  - → *Chromium(IV) oxide,* CrO_2, is used as a *ferromagnetic* coating for "chrome" magnetic tapes.
  - → *Sodium chromate,* $Na_2CrO_4(s)$, is used to prepare many other Cr compounds. In acid solution, CrO_4^{2-} forms the *dichromate ion,* $Cr_2O_7^{2-}$; in both ions, the oxidation number of Cr is +6.
    $$2\,CrO_4^{2-}(aq) + 2\,H_3O^+(aq) \rightarrow Cr_2O_7^{2-}(aq) + 3\,H_2O(l)$$
  - → Cr compounds are used as pigments, corrosion inhibitors, fungicides, and ceramic glazes.
- **Manganese, Mn, [Ar] $3d^5 4s^2$**
  - → A gray metal that corrodes easily and is rarely used alone
  - → Important in alloys such as steel and bronze
  - → Obtained from the ore, *pyrolusite,* mostly MnO_2, at high temperature:
    $$3MnO_2(s) + 4Al(s) \rightarrow 3Mn(l) + 2Al_2O_3(s)$$
- **Compounds of Mn**
  - → Positive oxidation numbers range from +1 to +7.
  - → The most stable oxidation number is +2, but +4, +7, and, to some extent, +3 are also common.
  - → The most important compound of Mn, MnO_2, or *manganese dioxide,* is a brown-black solid used as a decolorizer in glass, to prepare other Mn compounds, and in dry cells.
  - → The *permanganate ion,* MnO_4^-, with Mn in the +7 oxidation state, is an important oxidizing agent and a mild disinfectant.
- **Iron, Fe, [Ar] $3d^6 4s^2$**
  - → The most abundant element on Earth and the most widely used *d*-metal
  - → Obtained from the principal ores *hematite,* Fe_2O_3, and *magnetite,* Fe_3O_4
  - → Reactive metal that readily corrodes in moist air, forming rust
  - → Passivated by oxidizing acids such as HNO_3, but reacts with nonoxidizing acids
  - → Forms corrosion-resistant alloys and steels (see Table 16.2 in the text)
  - → A *ferromagnetic* metal, as are its oxides and many of its alloys
- **Pig iron and steel (see also Section 6.3 in the text and this study guide)**
  - → Steels are very important industrial products and a measure of a country's economic power.
  - → *Pig iron,* a precursor to steel, contains 3-5% C, ~2% Si, and other impurities and is produced by reduction of iron ores, such as Fe_2O_3, in a blast furnace. The ore is reduced through a series of redox and Lewis acid-base reactions.
  - → To make steel, *pig iron* is purified to *lower* the carbon content and to remove other impurities. Then, appropriate metals are *added* to form the desired steel.
  - → Steels typically contain 2% or less carbon; their hardness, tensile strength, and ductility depend on the carbon content. Higher carbon content produces harder steel.
- **Compounds of Fe**
  - → Positive oxidation numbers range from +2 to +6.
  - → Common oxidation numbers are +2 and +3.
  - → Salts of iron vary in color from pale yellow to dark green.
  - → Color in aqueous solution is dominated by $[FeOH(H_2O)_5]^{2+}$, the conjugate base of $Fe(H_2O)_6]^{3+}$.
  - → *Fe(II)* is readily oxidized to *Fe(III);* the reaction is slow in acid solution and rapid in base.

- **Cobalt, Co, [Ar] $3d^7 4s^2$**
  - → A silver-gray metal used mainly in Fe alloys
  - → Permanent magnets such as those used in loudspeakers are made of *alnico steel,* an alloy of Fe, Ni, Co, and Al.
  - → Co steels are hard and are used for drill bits and surgical tools.
- **Compounds of Co**
  - → Positive oxidation numbers range from +1 to +4.
  - → Common oxidation numbers are +2 and +3.
  - → *Co(II) oxide,* CoO(s), is a deep blue salt used to color glass and ceramic glazes.
- **Nickel, Ni, [Ar] $3d^8 4s^2$**
  - → A hard, silver-white metal used mainly in stainless steel
  - → Used as a catalyst for the *hydrogenation* of organic molecules
  - → Obtained as a pure metal by the *Mond process* (the first and third steps require heat):

$$NiO(s,ore) + H_2(g) \rightarrow Ni(s, \text{impure}) + H_2O(g)$$

$$Ni(s, \text{impure}) + 4CO(g) \rightarrow Ni(CO)_4(l, \text{pure})$$

$$Ni(CO)_4(l, \text{pure}) \rightarrow Ni(s, \text{pure}) + 4CO(g)$$

- **Compounds of Ni**
  - → Positive oxidation numbers range from +1 to +4.
  - → Common oxidation numbers are +2 and +3; +2 is the most stable.
  - → The green color of aqueous solutions of nickel salts arises from $[Ni(H_2O)_6]^{2+}$ ions.
  - → In nickel-cadmium (nicad) batteries, *Ni(III)* is reduced to *Ni(II).*
- **Carbonyl compounds**
  - → Many transition metals form compounds with CO.
  - → In *carbonyl compounds,* both the metal and carbonyl (CO) groups are regarded as *nearly* neutral species, and the metal is assigned an oxidation state of 0.
  - → When Fe is heated in CO, it reacts to form trigonal-bipyramidal *iron pentacarbonyl,* $Fe(CO)_5$, a yellow molecular liquid that melts at $-20°C$, boils at $103°C$, and decomposes in visible light.
  - → Other examples of transition metal carbonyls are *nickel tetracarbonyl,* $Ni(CO)_4$, a colorless, toxic, flammable liquid that boils at $43°C$ and *chromium hexacarbonyl,* $Cr(CO)_6$, a colorless crystal that sublimes readily. $Ni(CO)_4$ is *tetrahedral;* and $Cr(CO)_6$ is *octahedral.*
  - → Some metal carbonyl compounds can be reduced to produce anions in which the metal has a *negative* oxidation number.
    Example: In the $[Fe(CO)_4]^{2-}$ anion, produced by reduction of $Fe(CO)_5$, Fe has an oxidation number of -2:

$$Fe(CO)_5 \xrightarrow[\text{tetrahydrofuran}]{Na} [Fe(CO)_4]^{2-} + CO$$

16.4 Groups 11 and 12

- **Properties of the metals (see Table 16.3 in the text)**
  - → All the elements of Groups 11 and 12 are metals with completely filled *d*-subshells.
  - → The Group 11 metals, *copper, silver, and gold,* are called the *coinage metals* and have valence electron configurations of $(n-1)d^{10}ns^1$.

→ The low reactivity of the coinage metals derives partly from the poor shielding abilities of the *d*-electrons, and hence the strong attraction of the nucleus on the outermost electrons.

→ The effect is enhanced in Period 6 by lanthanide contraction, accounting for the inertness of gold.

→ The Group 12 metals, *zinc, cadmium, and mercury,* have valence electron configurations of $(n-1)d^{10}ns^2$. Zinc and, to a lesser extent, cadmium resemble beryllium or magnesium in their chemistry.

- **Copper, Cu, [Ar]** $3d^{10}4s^1$
  → Found as the metal and in sulfide ores such as *chalcopyrite,* $CuFeS_2$

  → Can be extracted from its ores by *pyrometallurgical* (high temperature) or *hydrometallurgical* (aqueous solution) methods:

  → *Pyrometallurgical example:* $2\,CuFeS_2(s) + 3\,O_2(g) \rightarrow 2\,CuS(s) + 2\,FeO(s) + 2\,SO_2(g)$
  The CuS is smelted to obtain blister copper: $CuS(s) + O_2(g) \rightarrow Cu(l) + SO_2(g)$.
  Both steps require high temperature.

  → *Hydrometallurgical example:* Sulfuric acid oxidizes Cu in ores to give soluble Cu^{2+} salts, which are then reduced: $Cu^{2+}(aq) + H_2(g) \rightarrow Cu(s) + 2\,H^+(aq)$ $\varDelta G^o = -65$ kJ

  → Used as an electrical conductor in wires

  → Forms alloys such as brass and bronze, which are used in plumbing and casting

  → Corrodes in moist air forming green *basic copper carbonate:*

  $$2\,Cu(s) + H_2O(l) + O_2(g) + CO_2(g) \rightarrow Cu_2(OH)_2CO_3(s)$$

  → Is oxidized by *oxidizing* acids such as HNO_3

- **Compounds of Cu**
  → Common oxidation numbers are +1 and +2; +2 is more stable.

  → *Cu(I)* disproportionates in water to form metallic copper and copper(II).

  → *Cu(II)* forms a pale-blue hydrated ion in water, $[Cu(H_2O)_6]^{2+}$.

  → Cu is essential in animal metabolism. In mammals, Cu-containing enzymes are required for healthy nerve and connective tissue. In species such as octopi and lobsters, Cu (not Fe) is used to transport oxygen in the blood.

- **Silver, Ag, [Kr]** $4d^{10}5s^1$
  → Rarely found as the free metal and obtained as a by-product of the refining of Cu and Pb

  → Reacts with sulfur to produce a black tarnish on silver dishes and cutlery

  → Like Cu, Ag is oxidized by *oxidizing* acids:

  $$3\,Ag(s) + 4\,H_3O^+(aq) + NO_3^-(aq) \rightarrow 3\,Ag^+(aq) + NO(g) + 6\,H_2O(l)$$

- **Compounds of Ag**
  → Oxidation numbers are +1, +2, and +3, but +1 is most stable.

  → Ag(I) does *not* disproportionate in water.

  → Except for $AgNO_3$ and AgF, Ag salts are insoluble in water.

  → Silver halides are used in photographic film.

- **Gold, Au, [Xe]** $4f^{14}5d^{10}6s^1$
  → So inert that it is usually found as the free metal

  → Highly malleable, easily made into thin foil (gold leaf)

→ *Cannot* be oxidized by nitric acid, but it dissolves in *aqua regia,* a mixture of sulfuric and hydrochloric acids:

$$Au(s) + 6H^+(aq) + 3NO_3^-(aq) + 4Cl^-(aq) \rightarrow AuCl_4^-(aq) + 3NO_2(g) + 3H_2O(l)$$

→ A carat is 1/24; 10- and 14-carat gold contains 10/24 and 14/24 parts of gold(s) by mass, respectively.

→ Positive oxidation numbers are +1, +2, and +3, but +1 and +3 are most common.

- **Zinc, Zn, [Ar] $3d^{10}4s^2$**

  → A silvery, reactive metal, found mainly in the sulfide ore, *sphalerite,* ZnS

  → Obtained from the ore by *froth flotation,* followed by *smelting* with coke, both at high *T:*
  $$2ZnS(s) + 3O_2(g) \rightarrow 2ZnO(s) + 2SO_2(g)$$
  $$ZnO(s) + C(s) \rightarrow Zn(l) + CO(g)$$

  → Used primarily for *galvanizing* Fe

  → Like Cu, Zn is protected by a hard film of *basic carbonate,* $Zn_2(OH)_2CO_3$.

- **Compounds of Zn**

  → Common positive oxidation number is +2.

  → An *amphoteric* metal, Zn dissolves in both acidic and basic solutions:

  Acid: $Zn(s) + 2H_3O^+(aq) \rightarrow Zn^{2+}(aq) + H_2(g) + 2H_2O(l)$

  Base: $Zn(s) + 2OH^-(aq) + 2H_2O(l) \rightarrow [Zn(OH)_4]^{2-} + H_2(g)$

  $Zn(OH)_4^{2-}$ is called the *zincate ion.* The reactions above suggest that galvanized containers should not be used to transport either acids or alkalis.

  → Zn is an essential element for human health; it occurs in many enzymes.

  → Zn is toxic only in very large amounts.

- **Cadmium, Cd, [Kr] $4d^{10}5s^2$**

  → A silvery, reactive metal

- **Compounds of Cd**

  → Common positive oxidation number is +2.

  → *Cadmiate ions* (analogous to *zincate ions*) are known, but Cd does not react with strong bases. Like Zn, Cd reacts with nonoxidizing acids.

  → Unlike Zn, Cd salts are *deadly poisons.*

  → Cd disrupts human metabolism by replacing other essential metals in the body, such as Zn and Ca, leading to soft bones, and to kidney and lung disorders.

  → Cd and Zn are alike chemically in many ways, but both are quite different from Hg.

- **Mercury, Hg, [Xe] $4f^{14}5d^{10}6s^2$**

  → A volatile, silvery metal, and a liquid at room temperature (Ga and Cs are liquid on *warm* days)

  → Found mainly in the sulfide ore, *cinnebar,* HgS

  → Obtained from the ore by *froth flotation,* followed by *roasting* in air; *vapor is poisonous.*

  → Combines with many metals to form mercury alloys called *amalgams*

  → The liquid temperature range (from −39°C to 357°C) is unusually wide and makes Hg well suited for use in thermometers, silent electrical switches, and high-vacuum pumps.

- **Compounds of Hg**

  → Positive oxidation numbers are +1, +2.

→ The Hg(I) cation, $[\text{Hg–Hg}]^{2+}$, is diatomic, with a covalent bond between the nuclei. It is usually written as Hg_2^{2+}.

→ Hg does not react with nonoxidizing acids or with bases, but does react with oxidizing acids in a reaction similar to those for Cu and Ag:

$$3\,\text{Hg(l)} + 8\,\text{H}^+\text{(aq)} + 2\,\text{NO}_3^-\text{(aq)} \rightarrow 3\,\text{Hg}^{2+}\text{(aq)} + 2\,\text{NO(g)} + 4\,\text{H}_2\text{O(l)}$$

→ Hg compounds, particularly organic ones, are acutely poisonous. Frequent exposure to low levels of Hg(g) results in the accumulation of Hg in the body. Damaging effects after such chronic exposure include impaired neurological function and hearing loss.

COORDINATION COMPOUNDS (Sections 16.5–16.7)

16.5 Coordination Complexes

• **Understanding complexes (terminology)**

→ *Coordination number* is the number of links formed by a central metal atom or ion, where a link corresponds to a single or multiple bond.

→ *Ligand* is a molecule or ion attached to a metal atom or ion by one or more coordinate-covalent bonds.

→ A ligand *coordinates* to the metal when it attaches.

→ For ionic complexes, ligands attached *directly* to a central atom are enclosed in *brackets*.

   Example: $[\text{ZnCl}_4]^2$

→ The *charge* of the ion is displayed in the upper right corner as usual.

→ Ligands attached to a metal define the *coordination "sphere"* of the central atom or ion.

→ Common coordination spheres include octahedral (almost all 6-coordinate complexes), tetrahedral (4-coordinate complexes), or square planar (also 4-coordinate).

→ The *isocyano* ligand is attached to the metal by the electron pair on the nitrogen end.

• **Naming *d*-metal complexes and coordination compounds**

→ Use the formulas and names for the common ligands given in text Table 16.4 of the text.

→ Follow the procedure outlined in text Toolbox 16.1.

• **Ligand substitution reaction**

→ A reaction in which one Lewis base takes the place of another

→ Sometimes all the ligands are replaced, but often the substitution is incomplete.

   Example: $[\text{Ni(H}_2\text{O)}_6]^{2+}\text{(aq)} + 6\,\text{NH}_3\text{(g)} \rightarrow [\text{Ni(NH}_3)_6]^{2+}\text{(aq)} + 6\,\text{H}_2\text{O(l)}$

16.6 The Shapes of Complexes

• **Complexes** → Exhibit a variety of shapes (coordination geometries) depending on the *coordination number*

• **Coordination number 6**

→ Complexes are usually *octahedral,* with ligands at the vertices of the octahedron and the metal at the center.

→ Some examples are hexamminecobalt(III), $[Co(NH_3)_6]^{3+}$, and hexacyanoferrate(II), $[Fe(CN)_6]^{4-}$.

- **Coordination number 4**

    → Two shapes are common, *tetrahedral* and *square planar*.
    Examples: $[Zn(OH)_4]^{2-}$ and $TiCl_4$ are *tetrahedral* complexes.

    → *Square planar* complexes are most common for metal species with a d^8 electronic configuration, such as Ni^{2+}, Pt^{2+}, and Au^{3+}.
    Examples: $[PtCl_4]^{2-}$ and $[Ni(CN)_4]^{2-}$ are *square planar* complexes.

- **Coordination number 2**

    → Complexes are *linear*.
    Examples: Dimethylmercury(0), $Hg(CH_3)_2$, and diamminesilver(I), $[Ag(NH_3)_2]^+$

- **Metallocenes**

    → "Sandwich" compounds with aromatic rings bonded by $2p\pi$-electrons to a metal atom or ion. The aromatic rings act as ligands.
    Examples: Dibenzenechromium, $Cr(C_6H_6)_2$, and ferrocene, $[Fe(C_5H_5)_2]$, both shown on the right.

    → The oxidation numbers of Cr and Fe in the metallocenes above are 0 and +2, respectively.

- **Other coordination geometries**

    → Complexes with coordination number 5 (trigonal-bipyramidal or square-pyramidal)

    → Coordination number 8 (square-antiprism or dodecahedral) is also known (see Section 16.6 in the text).

- **Monodentate and polydentate ligands**

    → A *monodentate* ligand can bind to a metal at only one site.
    Examples: Chloro, Cl^-; cyano, CN^-; carbonyl, CO; and methyl, CH_3.

    → A *polydentate* ligand can bind to a metal at more than one site simultaneously.

    → A *bidentate* ligand can bind to a metal at two sites.
    Examples: Oxalato (ox), $C_2O_4^{2-}$, and ethylenediamine (en), $NH_2CH_2CH_2NH_2$

Oxalato   Ethylenediamine

    → The binding sites in the ligands above, ox and en, are indicated by lobes corresponding to lone pairs of electrons

    → *Tridentate* ligands may bind at three sites.

    → *Chelates* — polydentate ligands that can form rings with a metal ion.
    Examples: The ox and en ligands above can form five-membered rings with metal ions.

    → Chelating agents are used to remove (sequester) *d*-block metals by binding to them and removing them from solution.

16.7 Isomers

- **Isomers** $\rightarrow$ Species that contain the same numbers of the same atoms but in different arrangements.
- **Two major classes of isomers:**
  - $\rightarrow$ 1) *Structural isomers,* in which *some* atoms have different partners (different connectivity)
  - $\rightarrow$ 2) *Stereoisomers,* in which *all* atoms have the same partners (same connectivity) but *some* atoms are arranged differently in space.
  - $\rightarrow$ Types of *structural isomers*: *ionization, hydrates, linkage,* and *coordination.*
  - $\rightarrow$ Types of *stereoisomers*: *geometrical* and *optical.*
- **Ionization isomers**
  - $\rightarrow$ Differ by the exchange of a ligand with an anion or molecule *outside* the coordination sphere
  - $\rightarrow$ Form *different* ions in solution
    Example: $[CoBr(NH_3)_5]SO_4$ and $[CoSO_4(NH_3)_5]Br$ are ionization isomers.
- **Hydrate isomers**
  - $\rightarrow$ Special type of ionization isomer
  - $\rightarrow$ Differ by the exchange of a H_2O molecule and another ligand in the coordination sphere
    Example: The solid hexahydrate of chromium(III) chloride, $CrCl_3 \cdot 6\,H_2O$, is known in three forms: $[Cr(H_2O)_6]Cl_3$, $[CrCl(H_2O)_5]Cl_2 \cdot H_2O$, and $[CrCl_2(H_2O)_4]Cl \cdot 2\,H_2O$.
- **Linkage isomers**
  - $\rightarrow$ Differ in the identity of the atom in a given ligand attached to the metal
  - $\rightarrow$ A ligand may have more than one atom with a lone pair that can bond to the metal, but because of its size or shape, only one atom from the ligand may bond to the metal at one time.
    Examples: NO_2^- (M–NO_2 *nitro* and M–ONO *nitrito*); CN^- (M–CN *cyano* and M–NC *isocyano*)
    The Lewis structure for CN^- clearly shows the lone pairs on C and N that can bond to a metal ion. $[:C\equiv N:]^-$
- **Coordination isomers**
  - $\rightarrow$ Occur when one or more ligands is exchanged between a complex that is a cation and one that is an anion in a coordination compound
    Example: $[Cu(NH_3)_4][PtCl_4]$ and $[Pt(NH_3)_4][CuCl_4]$
- **Geometrical isomers**
  - $\rightarrow$ Atoms bonded to the same neighbors, but with different orientations
  - $\rightarrow$ *Geometrical isomers* are also *stereoisomers.*
    Example: Cis (adjacent) and trans (across) isomers of the square-planar complex $[Pt(NH_3)_2Cl_2]$

- **Optical isomers**
  - $\rightarrow$ Special type of *stereoisomers:* They are *nonsuperimposable* mirror images of each other.
    Example: CHBrClF has a *tetrahedral shape* and *two* optical isomers:

  - $\rightarrow$ Molecules (a) and (b) are not superimposable; therefore, they are different compounds. They are related to each other as the left hand is related to the right hand (*mirror images*).

- **Chirality and optical activity**
  - → A *chiral species* is *not* identical to its mirror image. All *optical isomers* are *chiral*.
  - → An *achiral species* is identical to its mirror image.
  - → A *chiral species* and its mirror image form a pair of *enantiomers*. Molecules (a) and (b) shown on the previous page are *enantiomers*.
  - → *Chiral molecules* display *optical activity,* the ability to rotate the plane of polarization of light. For a given pair, one enantiomer rotates the plane clockwise, whereas its mirror image rotates it by the same amount counterclockwise.
  - → A *racemic mixture* is a mixture of enantiomers in equal molar proportions.
  - → Because enantiomers rotate light equally in opposite directions, a racemic sample is *not* optically active.

THE ELECTRONIC STRUCTURES OF COMPLEXES
(Sections 16.8–16.12)

16.8 Crystal Field Theory

- → Accounts for the optical and magnetic properties of complexes
- → Ligands are considered to be point *negative* charges.
- ***d*-orbital splitting in an octahedral complex, ML_6**
  - → Ligands approach the *positively charged* metal ion along the x-, y-, and z-axes, and the complex is stabilized by coulombic attraction. But the ligands, represented by negative charges, also repel electrons in the d-orbitals.
  - → This repulsion causes the d-orbitals to be split into two sets: t_{2g} (the three orbitals whose lobes lie between the ligands $(d_{xy}, d_{xz},$ and $d_{yz})$, and e_g (the two orbitals whose lobes point essentially towards the ligands $(d_{x^2-y^2}$ and $d_{z^2})$.
  - → Both the t_{2g} and e_g orbitals are raised in energy, but the e_g orbitals are raised more than the t_{2g} orbitals because they are closer to the point negative charges.
  - → The separation in energy between the t_{2g}- and e_g-orbitals is the *octahedral ligand field splitting,* Δ_O, where the O denotes octahedral.
  - → With respect to the *average energy of the d-orbitals*, the three t_{2g}-orbitals are *lowered* in energy by $(2/5)\Delta_O$ and the two e_g-orbitals and are *raised* in energy by $(3/5)\Delta_O$.

- ***d*-orbital splitting in a tetrahedral complex, ML_4**
  - → Ligands approach the *positively charged* metal ion along the corner directions of a tetrahedron.
  - → The d-orbitals are split into two sets: t_2 (three orbitals, $d_{xy}, d_{xz},$ and d_{yz}) and e (two orbitals, $d_{x^2-y^2}$ and d_{z^2}).
  - → In this instance, with respect to the average energy of the d-orbitals, the three t_2-orbitals $(d_{xy}, d_{xz},$ and $d_{yz})$ are *raised* in energy by $(2/5)\Delta_T$ and the two e-orbitals $(d_{x^2-y^2}$ and $d_{z^2})$ are *lowered* in energy by $(3/5)\Delta_T$.

→ The separation in energy between the t_2- and e-orbitals is the *tetrahedral ligand field splitting*, Δ_T, where the T denotes tetrahedral.

- **Comparison of octahedral and tetrahedral complexes**

  → Tetrahedral ligand-field splitting values tends to be *smaller* than corresponding octahedral splitting values, in part because there are fewer ligands to repel electrons in the d-orbitals in the tetrahedral case.

  → The relative energies of the t_2 and e sets of orbitals are reversed in octahedral and tetrahedral complexes.

  → In tetrahedral complexes, the subscript g is *not* used to describe the orbitals.

16.9 The Spectrochemical Series

- **Relative strength of the ligand field splitting, Δ**

  → For a given d-metal ion in a given oxidation state, Δ depends on the ligand.

  → Some ligands produce larger values of Δ than others.

  → The *relative* values of Δ are *approximately* in the same order for all d-metal ions.

- **Spectrochemical series**

  → Ligands are arranged according to the magnitude of the ligand-field splitting they produce.

  → *Weak field ligands* produce small values of Δ: $I^- < Br^- < Cl^- \approx SCN^- < F^- < OH^- < (ox) < H_2O$; (ox) = the bidentate oxylate ligand.

  → *Strong field ligands* produce large values of Δ: $NH_3 < (en) < NO_2^- < CN^- \approx CO$; (en) = the bidentate ethylenediamine ligand.

- **Electron configurations in isolated atoms**

  → In an isolated atom, the five d-orbitals in a subshell have the same energy.

  → According to Hund's rule, electrons occupy each degenerate orbital separately with spins *parallel* until each has one electron.

- **Octahedral complexes**

  → Similar rules apply to d-metal complexes; however, the five d-orbitals in a subshell are divided into two groups.

  → In octahedral complexes, the lower energy t_{2g}-orbitals fill first; degenerate orbitals (d_{xy}, d_{yz}, d_{zx}) fill with spins parallel until the t_{2g}-orbitals are half full (d^3 complex).

  → The electron configurations of d^4 through d^7 complexes depend on the magnitude of Δ_O compared with the energy required to pair an electron in a t_{2g}-orbital.

  → If Δ_O is *large* compared with the spin-pairing energy (*strong-field ligands*), electrons will enter the lower-energy t_{2g}-orbitals with spins paired before filling the e_g-orbitals.

  → If Δ_O is *small* compared with the spin-pairing energy (*weak-field ligands*), electrons will enter the higher-energy e_g-orbitals with spins parallel before pairing begins in the t_{2g}-orbitals.

  → For d^1 through d^3 and d^8 through d^{10} complexes, the orbitals filled are the same for strong-field and weak-field ligands.

  → A similar procedure is followed for *tetrahedral complexes,* but the two groups of d-orbitals are reversed in energy, and the e-orbitals fill first.

  → Electron configurations of d^n complexes are given in Table 16.5 in the text.

- **High- and low-spin complexes**
  - → A d^n complex with the maximum number of unpaired electrons is called a *high-spin complex*. High-spin complexes are expected for *weak-field ligands*.
  - → A d^n complex with the minimum number of unpaired electrons is called a *low-spin complex*. Low-spin complexes are expected for *strong-field ligands*.
  - → The d^4 through d^7 octahedral complexes may be high- or low-spin.
  - → Because Δ_T is small with respect to the energy required to pair an electron in the *e*-orbitals, t_2-orbitals are always accessible and *tetrahedral complexes* are *almost always* high-spin.

16.10 The Colors of Complexes

- **Absorption of light and transmission or reflection**
  - → Visible light contains all wavelengths from about 400 nm (blue) to 800 nm (red).
  - → Transition metal complexes often *absorb* some visible light and *transmit* or *reflect* the rest.
  - → *Transmitted* or *reflected* light (the *complementary color* of the light absorbed) is the light we see and determines the color of an object.
  - → The color wheel (see Fig. 16.31 in the text) provides a simple way of relating color *absorbed* to color *seen* in some systems.

- **Electronic transitions in complexes**
  - → Light absorption in complexes is often associated with *d–d transitions*. An electron is excited from a t_{2g}- to an e_g-orbital in an *octahedral complex* or from an *e*- to a t_2-orbital in a *tetrahedral complex*.
  - → *Charge-transfer transitions* may also occur in which an electron is excited from a ligand-centered orbital to a metal-centered orbital or *vice versa*. These transitions are often quite intense.
    Example: The deep purple color of the permanganate ion, MnO_4^-
  - → Color arises when a substance absorbs light from a portion of the visible spectrum and transmits or reflects the rest. Both *d–d* and charge-transfer transitions can occur in the visible region and either or both processes can contribute to the color of a transition metal complex.

- **Color, *d–d* transitions, and the spectrochemical series**
  - → *Weak-field* complexes have *small* Δ values and absorb *low-energy* (*red* or *infrared*) radiation.
  - → Absent strong charge-transfer transitions, solutions of these complexes appear *green* (the complementary color of *red*) or *colorless* (if the absorption is in the *infrared*).
  - → *Strong-field* complexes have *large* Δ values and absorb *high-energy* (*blue, violet,* or *ultraviolet*) radiation.
  - → In the absence of complicating charge-transfer transitions, solutions of these complexes appear *orange* or *yellow* (the complementary colors of *blue* and *violet,* respectively) or *colorless* (if absorption is in the *ultraviolet*).
  - → The ligand field splitting Δ can be determined from the electronic spectrum of a given complex, but, in most cases, it is not related simply to the color of its solution.

16.11 Magnetic Properties of Complexes

- **Complexes may be diamagnetic or paramagnetic**
  - → *Diamagnetic complexes* have no unpaired electrons and are repelled by a magnetic field.
  - → *Paramagnetic complexes* have one or more unpaired electrons and are attracted into a magnetic field. The greater the number of unpaired electrons, the greater the attraction.
- **Magnetic properties of complexes**
  - → Determined from the *d*-electron configuration of the complex
  - → Strong-field ligands produce weakly paramagnetic low-spin complexes.
  - → Weak-field ligands produce strongly paramagnetic high-spin complexes.

16.12 Ligand Field Theory

- **Ligand field theory**
  - → Molecular Orbitals (MOs) are generated from the available Atomic Orbitals (AOs) in a complex.
  - → Only one Atomic Orbital (AO) is used for each monodentate ligand (two for bidentate, etc.).
    Examples: For a chloride ligand, a Cl 3*p*-orbital, directed toward the metal, is used. For an ammonia ligand, the sp^3 lone-pair orbital is chosen. For bidentate ethylenediamine(en), two nitrogen sp^3 lone-pair orbitals are chosen.
  - → The treatment of π-*bonding* in complexes provides further insight into the *spectrochemical series*.
  - → Ligand field theory explains why uncharged species such as CO are strong-field ligands, whereas negatively charged species such as Cl⁻ are weak-field ligands.
  - → Based on electrostatic considerations, we would expect the *opposite* to be true.
- **Octahedral complexes**
  - → For first-row transition elements, the 4*s*-, 4*p*-, and 3*d*-orbitals (nine total orbitals) of the metal ion are chosen because they have similar energies.
  - → The nine metal and six ligand orbitals (one from each ligand) yield a total of 15 AOs, which overlap to form 15 MOs (see Fig. 16.37 in the text).
  - → The final result is that six MOs are bonding, six are antibonding, and three are nonbonding.
  - → The 12 ligand electrons and the available *d*-valence electrons are placed in the MOs in accordance with the building-up principle to determine the ground-state electron configuration of the complex.
  - → In transition metal ions, the valence *s*- and *p*-orbitals are typically vacant, whereas the *d*-orbitals are partially occupied.
  - → The first 12 electrons form six metal-ligand sigma bonds. The *d*-electrons of the metal enter the t_{2g}- and e_g-orbitals in the same way they did in the crystal field theory. Ligand field theory, however, identifies the t_{2g}-orbitals as *nonbonding* and the e_g-orbitals as *antibonding*.
  - → π-*bonding* in octahedral complexes arises from overlap of the metal t_{2g}-orbitals with *p*- or π-ligand orbitals to form *bonding* and *antibonding* MOs, π and π*, respectively.
  - → The details of this interaction are shown in Fig. 16.38 in the text.
- **Weak-field ligand (see Fig. 16.38a in the text)**
  - → The *antibonding* ligand π*-orbital is too high in energy to take part in bonding.
  - → The *bonding* ligand π-orbital combines with a metal t_{2g}-orbital and the combination *raises* the energy of the MO that the metal *d*-electron occupies.
  - → The electron from the metal enters an *antibonding* MO. The energy of the e_g-orbitals is unchanged, but the splitting between the t_{2g}- and e_g-orbitals is reduced.
  - → The ligand field splitting, Δ_O, is *decreased* in this case.

→ Some examples of weak-field ligands are Cl^- and Br^-, which have a valence shell p-orbital with two electrons in it, oriented perpendicular to the metal-ligand sigma bond.

- **Strong-field ligand (see Fig. 16.38b in the text)**

  → The bonding π-orbital of the ligand is too low in energy to take part in the bonding. The *empty* antibonding π^*-orbital of the ligand combines with a t_{2g}-orbital of the metal, and the combination *lowers* the energy of the MO that the metal d-electron occupies. The electron from the metal enters a *bonding* MO.

  → The energy of the e_g-orbitals is unchanged, and the splitting between the t_{2g}- and e_g-orbitals is increased.

  → The ligand field splitting, Δ_O, is *increased* in this case.

  → Some examples of strong-field ligands are CO and CN^-, which each have an unoccupied valence shell π^*-orbital oriented perpendicular to the metal-ligand sigma bond.

Chapter 17 NUCLEAR CHEMISTRY

NUCLEAR DECAY (Sections 17.1–17.5)

17.1 The Evidence for Spontaneous Nuclear Decay

- **Radioactivity**
  - → Spontaneous emission of radiation or particles by a nucleus undergoing decay
  - → Includes the emission of α particles, β particles, and γ radiation
  - → These three types, the most common forms of radioactivity, were the first ones to be discovered.
- **Nuclear radiation (see Table 17.1 in the text):**

  α particles
  - → Helium-4 nuclei, $^4_2\text{He}^{2+}$, traveling at speeds equal to about 10% of the speed of light, c
  - → Deflected by electric and magnetic fields
  - → When α particles interact with matter, their speed is reduced and they are neutralized.
  - → Denoted by $^4_2\alpha$ or α

  β particles
  - → Rapidly moving *electrons,* e^-, emitted by nuclei at speeds less than 90% of the speed of light
  - → Deflected by electric and magnetic fields, sometimes called *negatrons*
  - → When traveling between positively and negatively charged plates, α and β particles are deflected in opposite directions.
  - → Denoted by $^0_{-1}\text{e}$, β^-, or β (charge of -1, negligible mass compared with protons and neutrons)

  γ radiation
  - → High-energy *photons* (electromagnetic radiation) traveling at the speed of light
  - → Uncharged and not deflected by electric and magnetic fields
  - → Denoted by γ or $^0_0\gamma$ (no mass and no charge)

  β^+ particles
  - → Called *positrons*
  - → Have the mass of an electron but carry a positive charge
  - → When an electron and a positron, its *antiparticle,* meet, they are annihilated and are completely transformed into energy, mostly γ radiation.
  - → Denoted by $^0_{+1}\text{e}$ or β^+

  p particles
  - → *Protons* are positively charged subatomic particles found in nuclei. When emitted in nuclear decay, protons typically travel at speeds equal to about 10% of the speed of light.
  - → Denoted by $^1_1\text{H}^+$, ^1_1p, or p

  n particles
  - → *Neutrons* are uncharged subatomic particles found in nuclei. The mass of a neutron is slightly larger than the mass of a proton (see Table B.1 in the text). When emitted in nuclear decay, neutrons typically travel at speeds less than about 10% of the speed of light.
  - → Denoted by ^1_0n or n

- **Penetrating power of nuclear radiation**
  - → Penetrating power into matter decreases with increased charge and mass of the particle.
  - → Uncharged γ radiation and neutrons are more penetrating than charged positrons or alpha particles.
  - → Singly charged β particles (electrons) are more penetrating than the more massive, doubly charged α particles, which are virtually nonpenetrating.
- **Antiparticles**
  - → Subatomic particles with equal mass and opposite charge
  - → Annihilate each other upon encounter, producing energy

17.2 Nuclear Reactions

- **Nucleons and nuclides**
  - → A *nucleon* is a proton or a neutron in the nucleus of an atom.
  - → A *nuclide* is an atom characterized by its atomic number, mass number, and nuclear energy state.
  - → Nuclides are often denoted by ${}_{Z}^{A}E$, where Z is the atomic number, A is the mass number (protons plus neutrons), and E is the symbol for the element. Note that $A - Z$ equals the number of neutrons in the nuclide.

    Examples: ${}_{6}^{12}C$ and ${}_{92}^{235}U$
  - → Nuclear isomers are atomic nuclei of a particular nuclide that have equal proton number and equal mass number, differ in energy content, and are long-lived.

    Examples: The two states of ${}_{43}^{99}Tc$ observed in decay schemes. The most long-lived theoretically unstable nuclear isomer is tantalum-180m, which has a half-life in excess of 1000 trillion years, and has never been observed to decay to tantalum-180.
  - → *A nuclide is a neutral atom and therefore is uncharged.* There are about 256 nuclides in nature, which are so stable that they have never been observed to decay. They occur among the 82 different elements which have one or more stable nuclides.
  - → *Radioactive nuclei* can change their structure spontaneously by *nuclear decay,* the partial breakup of a nucleus.
- **Nuclear reactions**
  - → Any transformations that a nucleus undergoes
  - → Differ from chemical reactions in three important ways:
    1) Isotopes of a given element often have different nuclear properties but always have similar chemical properties.

       Example: ${}^{12}C$ and ${}^{14}C$ are difficult to separate chemically because they have similar chemical properties, yet the two isotopes have different nuclear stability. Carbon-12 is a stable nuclide, whereas carbon-14 decays with a half-life of 5730 y (see Section 17.7 in the text or in this study guide).
    2) Nuclear reactions often produce a different element.

       Example: Titanium-44 captures one of its electrons to produce the nuclide scandium-44.
    3) Nuclear reactions involve enormous energies relative to chemical reactions.

- **Balancing nuclear reactions**
  - → Like chemical reactions, *nuclear reactions* must be balanced with respect to both *charge* and *mass*.
  - → **Charge balance:** For the several nuclides $^A_Z E$ in a nuclear reaction, the sum of the atomic numbers Z of the reactants must equal the sum of the Z values of the products.
  - → **Mass balance:** The sum of the mass numbers A of the reactants must equal the sum of the mass numbers A of the products.
  - → The result of *most* nuclear decay reactions is *nuclear transmutation,* the conversion of one element into another.
  - → For nuclear decay reactions
    - ➤ A reactant nuclide is called a *parent nuclide.*
    - ➤ A product nuclide is called a *daughter nuclide.*

17.3 The Pattern of Nuclear Stability

- **Characterization of nuclei (see Fig. 17.12 in the text):**

  Even-even
  - → Nuclei with an even number of protons and neutrons, consequently A and Z are even numbers
  - → 157 even-even nuclides are stable according to the text.

    Examples: helium-4, $^4_2 He$; carbon-12, $^{12}_6 C$; oxygen-16, $^{16}_8 O$; and neon-22, $^{12}_{10} Ne$

  Even-odd
  - → Nuclei with an even number of protons and an odd number of neutrons, consequently Z is even and A is odd
  - → 53 even-odd nuclides are stable according to the text.

    Examples: helium-3, $^3_2 He$; berylium-9, $^9_4 Be$; carbon-13, $^{13}_6 C$; and neon-21, $^{21}_{10} Ne$

  Odd-even
  - → Nuclei with an odd number of protons and an even number of neutrons, consequently both Z and A are odd numbers
  - → 50 odd-even nuclides are stable according to the text.

    Examples: hydrogen-1, $^1_1 H$; lithium-7, $^7_3 Li$; boron-11, $^{11}_5 B$; and nitrogen-15, $^{15}_7 N$

  Odd-odd
  - → Nuclei with an odd number of protons and neutrons, consequently Z is odd and A is even
  - → *Only 4 odd-odd nuclides are stable.*
  - → The stable odd-odd nuclides are hydrogen-2 (deuterium), $^2_1 H$; lithium-6, $^6_3 Li$; boron-10, $^{10}_5 B$; and nitrogen-14, $^{14}_7 N$.

  Summary
  - → Even-even nuclides are the most stable and most abundant nuclides.
  - → Odd-odd nuclides are the least stable and least abundant nuclides.

- **Strong force**
  - → An attractive force that holds nucleons together in the nucleus
  - → Overcomes the coulomb repulsion of protons
  - → Acts only over a very short distance, approximately the diameter of the nucleus

- **Magic numbers**
  - → Are 2, 8, 20, 50, 82, 114, 126, and 184 for *either* protons or neutrons
  - → Nuclei with *magic numbers* of protons and/or neutrons are more likely to be stable.
  - → Analogous to the pattern of electronic stability in atoms associated with electrons in filled subshells
  - → *Doubly magic* nuclides are very stable.

    Examples: helium-4, $_2^4\text{He}$; oxygen-16, $_8^{16}\text{O}$; calcium-40, $_{20}^{40}\text{Ca}$; and lead-208, $_{82}^{208}\text{Pb}$

- **Band of stability and sea of instability (see Fig. 17.13 in the text)**
  - → On a plot of nuclide mass A (y-axis) vs. atomic number Z (x-axis)
  - → Stable nuclides are found in a narrow band, the *band of stability,* that ends at $Z = 83$ (bismuth).
  - → All nuclides with $Z > 83$ are unstable.
  - → Unstable nuclides are found in the *sea of instability,* a region above and below the *band of stability.*
  - → Nuclides *above* the band of stability are *neutron rich.*
  - → Neutron rich nuclei are likely to emit β particles (neutron → proton + β).
  - → Nuclides *below* the band of stability are *proton rich.*
  - → Proton rich nuclei are likely to emit positrons β$^+$ particles (proton → neutron + β$^+$) or to capture electrons (proton + β → neutron).
  - → Heavier nuclides below the band of stability and those with $Z > 83$ may also decay by emitting α particles.

- **Models of nuclear structure (not in the text)**
  - → Three common models of nuclear structure
  - → Description and order of sophistication:
    1) The *liquid drop model,* in which nucleons are considered to be packed together in the nucleus like molecules in a liquid
    2) The *independent particle model,* in which nucleons are described by quantum numbers and assigned to shells. Nucleons in the outermost shells are most easily lost as a result of radioactive decay.
    3) The *collective model,* in which nucleons are considered to occupy quantized energy levels and to interact with each other by the strong force and the electrostatic (coulomb) force

17.4 Predicting the Type of Nuclear Decay

- **Massive nuclei with $Z > 83$**
  - → Lose protons to reduce their atomic number and generally lose neutrons as well
  - → Stepwise decay gives rise to a *radioactive series,* a characteristic sequence of nuclides.

  Examples of radioactive series:
  1) The uranium-238 series (see Fig. 17.16 in the text) is one in which α (alpha) and β (beta) particles are successively ejected. The final step is the formation of a stable isotope of lead (magic number 82), lead-206.
  2) The uranium-235 series proceeds similarly and ends at lead-207. This series is sometimes called the actinium series.
  3) The thorium-232 series ends at lead-208.

- **Neutron rich nuclides**
  - → Nuclides that have high neutron to proton (n/p) ratios with respect to the band of stability
  - → Tend to decay by reducing the number of neutrons
  - → Commonly undergo β decay, which decreases the n/p ratio

- **Proton rich nuclides**
  - → Nuclides that have high proton to neutron (p/n) ratios with respect to the band of stability
  - → Tend to decay by reducing the number of protons
  - → Commonly undergo electron capture, positron emission, or proton emission (least likely), which decreases the p/n ratio

17.5 Nucleosynthesis

- **Formation of the elements**
  - → Elements are formed by *nucleosynthesis*.
  - → Nucleosynthesis occurs when particles collide vigorously, as in stars or particle accelerators.
  - → *Transmutation* is the conversion of one element into another.
  - → The first artificial transmutation was observed in 1919 by Rutherford, who converted $^{14}_{7}N$ to $^{17}_{8}O$ by bombarding the parent nuclide with high-speed $\mapsto$ particles:

$$^{14}_{7}N + ^{4}_{2}\alpha \rightarrow ^{17}_{8}O + ^{1}_{1}p$$

   The reactants must collide with enough energy to overcome the repulsive coulomb force between positively charged nuclei.
  - → Nucleosynthesis reactions are commonly written in shorthand as:

      target (incoming species, ejected species) product

   In the Rutherford example above, the shorthand notation is $^{14}_{7}N\,(\alpha,p)\,^{17}_{8}O$.
  - → Nucleosynthesis can be effected using a variety of projectile species including γ radiation, protons, neutrons, deuterium nuclides, α particles, and heavier nuclides.
  - → Because they are uncharged, neutrons can travel relatively slowly and still react with nuclides.

- **Transuranium elements**
  - → Elements following uranium ($Z = 92$) in the periodic table
  - → Elements from rutherfordium (Rf, $Z = 104$) to meitnerium (Mt, $Z = 109$) were given official names in 1997.
  - → Darmstadtium (Ds, $Z = 110$) was given an official name in 2003 and roentgenium (Rg, $Z = 111$ in 2004.

- **Transmeitnerium elements**
  - → Elements following meitnerium; two now have official names and several others have been reported as of this writing.
  - → Temporary names are based on shorthand numbers derived from Latin (see Table 17.2 in the text).

NUCLEAR RADIATION (Sections 17.6–17.8)

17.6 The Biological Effects of Radiation

- **Nuclear radiation** $\rightarrow$ Sometimes called ionizing radiation because it can eject electrons from atoms

- **Penetrating power**

    - $\rightarrow$ α, β, and γ particles are forms of *ionizing radiation* with different *penetration power*.

    - $\rightarrow$ Different amount of shielding is required to prevent harmful effects (see Table 17.3 in the text).

    - $\rightarrow$ α particles have the least penetrating power and are absorbed by paper and the outer layers of skin. They are extremely dangerous if inhaled or ingested.

    - $\rightarrow$ β particles are 100 times more penetrating than α particles and are absorbed by 1 cm of flesh or 3 mm of aluminum, for example.

    - $\rightarrow$ γ radiation is 100 times more penetrating than β particles. It can pass through buildings and bodies, leaving a trail of ionized or damaged molecules, and is absorbed by lead bricks or thick concrete.

    - $\rightarrow$ The order of *increasing* penetrating power is $\alpha < \beta < \gamma$.

- **Radiation damage to tissue**

    - $\rightarrow$ Depends on the *type* and *strength* of the radiation, the *length* of exposure, and the *extent* to which the radiation can reach sensitive tissue

    - $\rightarrow$ For example, Pu^{4+} is an α emitter and, if ingested, replaces Fe^{3+} in the body, resulting in inhibition of the production of red blood cells.

- **Absorbed dose of radiation**

    - $\rightarrow$ Energy deposited in a sample, such as a human body, when exposed to radiation

    - $\rightarrow$ Units of the absorbed dose of radiation are:

        1 rad = amount of radiation that deposits 0.01 J of energy per kg of tissue
        1 gray (Gy) = an energy deposit of $1 \; J \cdot kg^{-1}$ (SI unit)
        1 Gy = 100 rad

    Note: The name *rad* stands for "radiation absorbed dose."

- **Dose equivalent**

    - $\rightarrow$ Dose modified to account for different destructive powers of radiation

    - $\rightarrow$ Relative destructive power is called the *relative biological effectiveness, Q.* For β^- and γ radiation, $Q \equiv 1$. For α radiation, $Q \approx 20$.

    - $\rightarrow$ The natural unit of dose equivalent is the rem (*roentgen equivalent man*).

    - $\rightarrow$ Definition: Dose equivalent in rem = $Q \times$ absorbed dose in rad

    - $\rightarrow$ SI Units: Dose equivalent in sievert (Sv) = $Q \times$ absorbed dose in gray (Gy)
        1 Sv = 100 rem

- **Common radiation dose and harmful effects**

    - $\rightarrow$ Typical annual human dose equivalent from natural sources is $0.2 \; rem \cdot y^{-1}$, a number with a wide range depending upon habitat and lifestyle.

    - $\rightarrow$ A typical chest x-ray gives a dose equivalent of about 7 millirem.

    - $\rightarrow$ 30 rad (30 rem) of γ radiation may cause a reduction in white blood cell count.

    - $\rightarrow$ 30 rad (600 rem) of α radiation causes death.

17.7   Measuring the Rate of Nuclear Decay

- **Activity of a sample**
  - → Number of nuclear disintegrations per second
  - → One nuclear disintegration per second is called 1 becquerel, Bq (SI unit).
  - → The curie, Ci, an older unit for activity, is equal to 3.7×10^{10} Bq, which is equal to the radioactive output of 1 g of radium-226.
  - → Because the curie represents a large value of decay, the activity of typical samples is given in millicuries (mCi) or microcuries (μCi).

    Note: See Table 17.4 in the text for a list of the radiation units.

- **Nuclear decay**
  - → The decay of a nucleus can be written as:

    Parent nucleus → daughter nucleus + radiation
  - → This form is the same as a *unimolecular elementary reaction,* with an unstable nucleus taking the place of an excited molecule.
  - → The rate of nuclear decay depends only on the identity of the nucleus (isotope), not on its chemical form or temperature.

- **Law of radioactive decay**
  - → The rate of nuclear decay is proportional to the number of radioactive nuclei N:

    $$\boxed{\text{Activity} = \text{rate of decay} = k \times N}$$

  - → k is the *decay constant* (or rate constant for the reaction).
  - → The units of the *decay constant* are $(\text{time})^{-1}$, the same as for any *first-order* rate constant.

- **Integrated rate law for nuclear decay and half-life**

  - → Exponential form: $\boxed{N = N_0 e^{-kt}}$ The number of radioactive nuclei N decays exponentially with time. Large values of k correspond to more rapid decay.

  - → Logarithmic form: $\boxed{\ln\left(\dfrac{N}{N_0}\right) = -kt}$

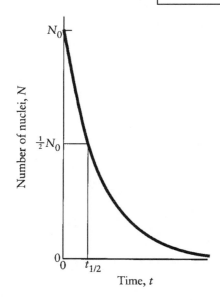

On the left is a plot of N vs. t for nuclear decay. N_0 is the number of radioactive nuclei present at $t = 0$. N is the number of radioactive nuclei present at a later time t. The time required for the nuclei to decay to half their initial number is equal to the half-life, $t_{1/2}$. At this time,

$$N = \frac{1}{2} N_0$$

Substituting this value into the integrated rate law yields:

$$\boxed{t_{1/2} = \frac{\ln 2}{k}}$$

Half-lives of radioactive nuclides span an enormous range, from picoseconds (^{215}Fr, 120 ps) to billions of years (^{238}U, 4.5×10^9 y). See Table 17.5 in the text for a list of half-lives of common radioactive isotopes.

- **Isotopic dating**
  - → Used to determine the ages of rocks and of archeological artifacts
  - → Carried out by measuring the activity of a radioactive isotope in the sample
  - → Useful isotopes for dating are uranium-238, potassium-40, tritium, $_1^3 H$, and carbon-14.

- **Radiocarbon dating**
  - → Most important example of isotopic dating
  - → Uses the beta decay of carbon-14, for which the half-life is 5730 y

- **Radioactive ^{14}C**
  - → Formed in the atmosphere by neutron bombardment of ^{14}N
  $$^{14}_{7}N + ^{1}_{0}n \rightarrow ^{14}_{6}C + ^{1}_{1}p$$
  - → Enters living organisms as $^{14}CO_2$ through photosynthesis and digestion
  - → Leaves living organisms by excretion and respiration
  - → Achieves a steady-state concentration in living organism with an activity of 15 disintegrations per minute per gram of *total* carbon
  - → Shows decreasing activity in dead organisms because they no longer ingest ^{14}C from the atmosphere
  - → The time of death of an organism can be estimated by measuring the activity of the sample and using the integrated rate law.

17.8 Uses of Radioisotopes

- **Radioisotopes**
  - → Used to help cure disease, preserve food, and trace the mechanisms of chemical reactions
  - → Used to power spacecraft, locate sources of water, and determine the age of nonliving materials including wood (^{14}C dating), rocks ($^{238}U/^{206}Pb$ ratio), and ground water (tritium dating using the $^1H/^3H$ ratio)

- **Radioactive tracers**
  - → Radioactive isotopes that are used to track changes and locations

    Example: Phosphorus-32, a β emitter with $t_{1/2} = 14.28$ d, can be incorporated into phosphate-containing fertilizer to follow the mechanism of plant growth.
  - → In chemical reactions, tracers can help determine the mechanism of the reaction. Nonradioactive isotopes are also used for this purpose.

NUCLEAR ENERGY (Sections 17.9–17.12)

17.9 Mass-Energy Conversion

- **Nuclear binding energy**
  - → Energy *released* when protons and neutrons join together to form a *nucleus:*
  $$(Z)_1^1 p + (A-Z) \, _0^1 n \rightarrow \, _Z^A E^{Z+}$$

→ Since *nuclides* are *neutral* atoms, the energy released is also given *approximately* by:

$$(Z)_1^1\text{H} + (A - Z)\,_0^1\text{n} \rightarrow\ _Z^A\text{E}$$

→ The binding energy is then given by applying the Einstein mass-energy equation to each reactant and product species:

$$E_{\text{bind}} = |\Delta E| = |\sum E(\text{products}) - \sum E(\text{reactants})|$$
$$= |E(\text{product}) - \sum E(\text{reactants})| = |mc^2(\text{product}) - \sum mc^2(\text{reactants})|$$
$$= |m(\text{product}) - \sum m(\text{reactants})| \times c^2$$

$$\boxed{E_{\text{bind}} = |\Delta m| \times c^2}$$

→ *Binding energy* is a measure of the stability of the nucleus.

→ The *binding energy per nucleon* is defined as E_{bind}/A and is a better measure of the stability of the nucleus.

A plot of the binding energy per nucleon *vs.* atomic mass number is shown on the left (see also Fig. 17.20 in the text). Nucleons are bound together most strongly in the elements near iron and nickel, which accounts for the high abundance of iron and nickel in meteorites and on planets similar to the Earth. From the figure, it appears that light nuclei become more stable when they "fuse" together to form heavier ones. Heavy nuclei become more stable when they undergo "fission" and split into lighter ones. Also, note the increased stability of the nuclei with magic numbers, particularly oxygen-16, which is *doubly magic*.

• **Units used in nuclear calculations**

→ *Mass* is usually reported in *atomic mass units,* u.

→ One atomic mass unit is defined as *exactly* $\frac{1}{12}$ the mass of one atom (nuclide) of ^{12}C:

$$\boxed{1\ \text{u} = 1.660\,54 \times 10^{-27}\ \text{kg}}$$

→ The *atomic mass unit,* u, is numerically equivalent to the atomic mass constant m_u and this older notation is sometimes used.

→ *Binding energy* is usually reported in electronvolts, eV, or millions (mega) of electronvolts, MeV.

→ An electronvolt is the change in potential energy of an electron (charge of $1.602\,18 \times 10^{-19}$ C) when it is moved through a potential difference of 1 V:

$$\boxed{1\ \text{eV} = 1.602\,18 \times 10^{-19}\ \text{J}}$$

17.10 Nuclear Fission

- **Nuclear fission**
  - $\rightarrow$ Occurs when a nucleus breaks into two or more smaller nuclei
  - $\rightarrow$ Releases a large amount of energy
  - $\rightarrow$ May be *spontaneous* or *induced*

- **Spontaneous nuclear fission**
  - $\rightarrow$ Occurs when oscillations in heavy nuclei lead them to break into two smaller nuclei of similar mass
  - $\rightarrow$ Yields a variety of products for a given nuclide (see Fig. 17.22 in the text)

    Example: Consider the spontaneous fission of americium-244. Two of the daughter nuclides formed are iodine-134 and molybdenum-107. The reaction is:

    $$^{244}_{95}\text{Am} \rightarrow {}^{134}_{53}\text{I} + {}^{107}_{42}\text{Mo} + 3\,{}^{1}_{0}\text{n}$$

- **Induced nuclear fission**
  - $\rightarrow$ Caused by bombarding a heavy nucleus with neutrons
  - $\rightarrow$ Yields a variety of daughter nuclides for a given parent

    Example: Consider the induced fission of plutonium-239 by neutron bombardment. A number of products are formed. Two important reactions are:

    $$^{239}_{94}\text{Pu} + {}^{1}_{0}\text{n} \rightarrow {}^{98}_{42}\text{Mo} + {}^{138}_{52}\text{Te} + 4\,{}^{1}_{0}\text{n}$$

    $$^{239}_{94}\text{Pu} + {}^{1}_{0}\text{n} \rightarrow {}^{100}_{43}\text{Tc} + {}^{135}_{51}\text{Sb} + 5\,{}^{1}_{0}\text{n}$$

- **Energy changes in fission reactions**
  - $\rightarrow$ Energy released during fission is calculated by using Einstein's equation.
  - $\rightarrow$ For a balanced nuclear reaction,

    $$\boxed{\Delta m = \sum m(\text{products}) - \sum m(\text{reactants})} \quad \text{and} \quad \boxed{\Delta E = \Delta m \times c^2}$$

- **Fissionable and fissile nuclei**
  - $\rightarrow$ *Fissionable nuclei* are nuclei that can undergo induced fission.
  - $\rightarrow$ *Fissile nuclei* are fissionable nuclei that can undergo induced fission *with slow-moving neutrons.*

    Examples of fissile nuclei: ${}^{235}_{92}\text{U}$, ${}^{233}_{92}\text{U}$, and ${}^{239}_{94}\text{Pu}$, which are nuclear power plant fuels.

    The nuclide, ${}^{238}_{92}\text{U}$, is fissionable by fast-moving neutrons but is *not* fissile.

- **Nuclear branched chain reactions**
  - $\rightarrow$ Neutrons are *chain carriers* in a branched chain reaction (see Section 14.9 in the text).
  - $\rightarrow$ Chain branching occurs when an induced fission reaction produces two or more neutrons.

    Example: ${}^{235}_{92}\text{U} + {}^{1}_{0}\text{n} \rightarrow {}^{97}_{40}\text{Zr} + {}^{137}_{52}\text{Te} + 2\,{}^{1}_{0}\text{n}$, in which two neutrons are produced for each uranium-235 nuclide that reacts. The two neutrons can either escape into the surroundings or be captured by (react with) other uranium-235 nuclides.
  - $\rightarrow$ A *critical mass* is the mass of fissionable material above which so few neutrons escape from the sample that the fission chain reaction is sustained.

→ The critical mass for pure plutonium of normal density is about 15 kg.

→ A sample of plutonium with mass greater than 15 kg is *supercritical;* the reaction is self-sustaining and may result in an explosion.

→ A *subcritical* sample is one that has less than the critical mass for its density.

- **Controlled and uncontrolled (explosive) nuclear reactions**

  → Explosive nuclear reactions occur when a *subcritical* amount of fissile material is made *supercritical* so rapidly that the chain reaction occurs uniformly throughout the material.

  → Controlled nuclear reactions occur in nuclear reactors.

  → Controlled reactions are not explosive because there is a subcritical amount of fuel.

  → The rate of reaction is controlled by *moderators,* such as graphite rods.

  → Moderators slow the emitted neutrons so that a greater fraction can induce fission.

  → The difference between a fission explosion and a controlled fission is shown in the figure below. Each symbol ⊗ represents a fission event. The neutrons from the fission are shown but products are not.

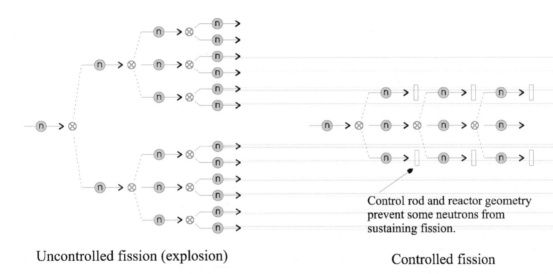

Uncontrolled fission (explosion)　　　　　　　　Controlled fission

Note:　The number of neutrons released in an event is variable as is the identity of the products (see Fig. 17.24 in the text).

- **Breeder reactors**

  → Nuclear reactors used to synthesize fissile nuclides for fuel and weapon use.

  → Breeder reactors run very hot and fast because no moderator is present. As a result, they are more dangerous than nuclear reactors used for power generation.

17.11 Nuclear Fusion

- **Nuclear fusion**

  → Occurs when lighter nuclei fuse together to form a heavier nucleus

  → Is accompanied by a large release of energy because heavier nuclei have larger binding energies per nucleon

→ Occurs in stars and in the hydrogen bomb, in which the fusion reaction is uncontrolled

→ Is difficult to achieve because positively charged nuclei must collide with tremendous kinetic energies to fuse

→ Fusion is essentially free of radioactive waste.

→ Fusion has not yet been sustained for an appreciable time in a controlled reaction on Earth.

→ A safe, sustained fusion reaction could provide an almost unlimited source of energy.

- **Energy changes in fusion reactions**

  → Energy released during a fusion reaction is calculated by using Einstein's equation.

  → The procedure is the same as for fission reactions.

17.12 The Chemistry of Nuclear Power

- **Chemistry**

  → The key to the safe use of nuclear power

  → Used to prepare nuclear fuel

  → Used to recover important fission products

  → Used to dispose of nuclear waste safely

- **Uranium**

  → The fuel of nuclear reactors

  → Obtained primarily from the ore pitchblende, UO_2, by reducing the oxide to the metal and then *enriching* or increasing the fraction of the fissile nuclide, uranium-235

  → To be useful as a fuel, the percentage of ^{235}U must be increased from its natural abundance of 0.7% to about 3%.

- **Enrichment**

  → Exploits the mass difference between ^{235}U and ^{238}U

  → Separation is accomplished by repeated effusion of $^{235}UF_6$ and $^{238}UF_6$ vapor (see Graham's law of effusion in Section 4.9 of the text).

$$\frac{\text{Rate of effusion of } ^{235}UF_6}{\text{Rate of effusion of } ^{238}UF_6} = \sqrt{\frac{M_{238}}{M_{235}}} = \sqrt{\frac{352.1}{349.0}} = 1.004$$

Note: Because the ratio is close to one, repeated effusion steps (hundreds) are required to get the desired separation.

- **Nuclear waste**

  → Spent nuclear fuel called *nuclear waste* is still radioactive.

  → Waste is a mixture of uranium and fission products.

  → Nuclear waste must be stored safely for about 10 half-lives, and is generally buried underground.

  → For underground burial, incorporation of the highly radioactive fission (HRF) products into a glass or ceramic material is better than placing it in metal storage drums.

  → Storage drums can corrode, allowing the waste to seep into aquifers and/or contaminate large areas of soil.

Chapter 18 ORGANIC CHEMISTRY I: THE HYDROCARBONS

ALIPHATIC HYDROCARBONS (Sections 18.1–18.6)

18.1 Types of Aliphatic Hydrocarbons

- **Types of hydrocarbons (compounds containing only C and H)**

  → *Aromatic* hydrocarbon: hydrocarbon with one or more benzene rings.

    Examples: Benzene (C_6H_6), toluene ($C_6H_5CH_3$), and naphthalene ($C_{10}H_8$), all shown on the right

  → *Aliphatic* hydrocarbon: hydrocarbon without a benzene ring

  → *Saturated hydrocarbon:* an aliphatic hydrocarbon with no carbon-carbon multiple bonds

    Example: Butane ($CH_3CH_2CH_2CH_3$)

  → *Unsaturated hydrocarbon:* an aliphatic hydrocarbon with at least one carbon-carbon multiple bond.

    Examples: Butene (CH_2=$CHCH_2CH_3$) and butyne (CH≡CCH_2CH_3)

- **Aliphatic hydrocarbons: structural formula** → Shows chemical connectivity, how many atoms are attached to one another

    Example: Butane and methylpropane have the same *empirical* (C_2H_5) and *molecular* (C_4H_{10}) formulas, respectively, but different *structural formulas*.

Butane   Methylpropane

    In the structure of methylpropane, an abbreviated notation for the methyl group (CH_3 −) attached to the propane chain is used (the three C–H bonds are *not* shown). A structure with all three terminal methyl groups in abbreviated form is shown on the right.

- **Condensed structural formula**

  → Simple way to represent structures of complicated organic molecules

  → Shows in an abbreviated form how the atoms are grouped together

  → The *condensed structural formula* for butane is $CH_3CH_2CH_2CH_3$ or $CH_3(CH_2)_2CH_3$: the latter is often written when the chain is lengthy.

  → The *condensed structural formula* for methylpropane is $CH_3CH(CH_3)CH_3$ or $(CH_3)_3CH$ because the methyl groups are equivalent. The first formula is the one chosen to name the compound (methylpropane *not* trimethylmethane).

  → Many organic compounds have more than one name, although one, the IUPAC (International Union of Pure and Applied Chemistry) name, is often preferred. Methylpropane above is also known as tertiary butane.

  → We have seen similar complexities with units.

- **Line structure**

  → A *line structure* represents a chain of carbon atoms as a zigzag line.

  → The end of each short line represents a carbon atom.

  → C–H bonds and H atoms are not shown but are added "mentally" to the structure.

  → Other atoms, such as O and N atoms, are written explicitly, along with any H atoms to which they are bonded.

Example: The zigzag *structural formula* and the *line structure* of methylpropane are shown on the right.

→ A benzene ring is represented by a circle inside a hexagon or by a Kekulé structure. The presence of a H atom attached to each C atom is implied. These forms of the benzene ring are also used in structural formulas. The respective *structural formulas* and *line structures* of ethylbenzene, $C_6H_5CH_2CH_3$, are shown below.

Nomenclature of Hydrocarbons

- **Alkanes**
  - → Alkanes are saturated hydrocarbons.
  - → In naming single-chain *alkanes,* add a prefix denoting the number of C atoms in the chain to the suffix *-ane* (see Table 18.1 in the text).
  - → The first three members of the alkane family are methane (CH_4), ethane (CH_3CH_3), and propane ($CH_3CH_2CH_3$)
  - → If a *hydrogen atom* is removed from an *alkane,* the name of the *alkyl* group that results is derived by replacing the suffix *-ane* with *-yl.*
  - → The first three members of the alkyl family are the methyl (CH_3-), ethyl (CH_3CH_2-), and propyl ($CH_3CH_2CH_2-$) groups.
  - → Methylpropane ($CH_3CH(CH_3)CH_3$) is a propane molecule in which a H atom on the middle C atom is replaced by a methyl (CH_3) group.
  - → *Cycloalkanes* are alkanes that contain *rings* of $-CH_2-$ units. The first three members are cyclopropane (C_3H_6), cyclobutane (C_4H_8), and cyclopentane (C_5H_{10}), all shown on the right.

- **Properties of alkanes**
  - → Structures may be written as linear, but they are actually linked tetrahedral units. Rotation about C–C bonds is fairly unrestricted, so the chains often roll up into a ball in gases and liquids to maximize the attraction between different parts of the chain.
  - → Electronegativity values of C and H are similar, so, to a first approximation, hydrocarbon molecules can be regarded as nonpolar.
  - → The dominant interaction between the molecules is therefore the London interaction, whose strength increases with the number of electrons in the molecule.
  - → Thus, alkanes should become less volatile with increasing molar mass (see Fig. 18.4 in the text).
  - → Many alkane molecules appear in forms with the same atoms bonded together in different arrangements, such as the isomers butane and methylpropane shown on the previous page.
  - → In general, *isomers* are different compounds with the same molecular formula.

- **Alkenes**
  - → *Alkenes* are a series of compounds with formulas derived from $H_2C=CH_2$ by inserting CH_2 groups. The next member of the family is propene, $H-CH_2-CH=CH_2$ ($CH_3CH=CH_2$).
  - → Ethene or ethylene, C_2H_4 or $H_2C=CH_2$, is the simplest *unsaturated* hydrocarbon.
  - → Alkenes are named in a manner analogous to that of alkanes except the ending is *-ene.*

→ The double bond is located by numbering the C atoms in the chain and writing the lower of the two numbers of the two C atoms joined by the double bond (see Toolbox 18.1 in the text).

- **Alkynes**
  - → *Alkynes* are aliphatic hydrocarbons with at least one carbon-carbon triple bond.
  - → Ethyne or acetylene, HC≡CH, is the simplest alkyne.
  - → Alkynes are named like the alkenes but with the suffix *-yne*.

- **Nomenclature rules for aliphatic hydrocarbons are given in Toolbox 18.1 in the text.**

- **Alkane nomenclature**
  - → If the C atoms in an alkane are bonded in a row with a formula of the form $CH_3(CH_2)_mCH_3$, the compound, called an *unbranched alkane,* is named according to the number of C atoms present. All are named with a stem name and the suffix *-ane*. A cyclic (ring) alkane has the general formula $(CH_2)_m$ and is designated by the prefix *cyclo-*.

    Examples: Pentane ($CH_3(CH_2)_3CH_3$) and cyclopentane (C_5H_{10})

  - → A *branched alkane* has one or more C atoms not in a single row, and these carbons constitute *side chains*.
  - → Butane is an *unbranched alkane,* and methylpropane is a *branched alkane* with a CH_3 *side chain*. Side chains are treated as *substituents,* species that have been substituted for a H atom.
  - → A systematic procedure for naming alkanes is:

  Step 1. Identify the *longest* chain of C atoms in the molecule.

  Step 2. Number the C atoms on the longest chain so that the carbon atoms with substituents have the *lowest* possible numbers.

  Step 3. Identify each substituent and the number of the C atom in the chain on which it is located.

  Step 4. Name the compound by listing the substituents in *alphabetical order,* with the numbered location of each substituent preceding its name. The name(s) of the substituent(s) is/are followed by the parent name of the alkane, which is determined by the number of C atoms in the longest chain.

  > **Example:** Heptane with a methyl group at the number 3 carbon is named 3-methylheptane.

  Step 5. If two or more of the same substituents are present (such as two methyl groups), a Greek prefix such as *di-, tri-,* or *tetra-* is attached to the name of the group. Numbers of the C atoms to which groups are attached, separated by commas, are included in the name.

  > **Example:** Heptane with methyl groups at C atoms 3 and 4 is named 3,4-dimethylheptane.

  Step 6. As in the example above, numbers in a name are always separated from letters by a hyphen, while numbers are separated from each other by commas.

- **Alkene nomenclature**
  - → Alkenes are named by using the stem name of the corresponding alkane with a number that specifies the location of the double bond. The number is obtained by numbering the longest carbon chain so that the double bond is associated with the lowest numbered carbon possible and assigning to the double bond the lower number of the two double-bonded carbons.

$$\overset{6\quad\;5\quad\;4\quad\;3\quad\;2\quad\;1}{CH_3CH_2CH_2CH=CHCH_3}$$

2-hexene (not 3-hexene)

→ Other substituents are named as for alkanes, with a number specifying the location of the group and an appropriate prefix denoting how many groups of each kind are present.

→ Numbering the longest chain so that the double bond has the lowest number takes precedence over keeping the substituent numbers low.

→ If more than one double bond is present, the number of these bonds is indicated by a Greek prefix. The suffix is *-ene*.

Example: 1,3-hexadiene ($H_2C=CHCH=CHCH_2CH_3$) or

- **Alkyne nomenclature**

→ Alkynes are named by using the stem name of the corresponding alkane with a number specifying the location of the triple bond. The numbering convention is the same as that used for double bonds. If more than one triple bond is present, the number of these bonds is indicated by a Greek prefix. The suffix is *-yne*.

→ When numbering atoms in the chain, the lowest numbers are given preferentially to (a) functional groups named by (c) suffixes, (b) double bonds, (c) triple bonds, and (d) groups named by prefixes.

Example: 4-chloro-2-hexyne ($CH_3CH_2CH(Cl)C\equiv CCH_3$)

18.2 Isomers

- **Structural isomers**

→ Molecules constructed from the same atoms connected differently

→ Have the *same* molecular formula, but *different* structural formulas

Example: Butane ($CH_3(CH_2)_2CH_3$) and methylpropane ($CH_3CH(CH_3)CH_3$) are structural isomers with the same molecular formula (C_4H_{10}). The molecules have a different *connectivity*.

- **Stereoisomers**

→ Molecules with the same connectivity but with some of their atoms arranged differently in space

→ *Geometrical isomers* are *stereoisomers* with different arrangements in space on either side of a double bond, or above and below the ring of a cycloalkane.

→ *Optical isomers* are *stereoisomers* in which each isomer is the mirror image of the other, *and* the images are nonsuperimposable.

- **Geometrical isomers**

→ Compounds with the same molecular formula and with atoms bonded to the same neighbors but with a different arrangement of atoms in space

→ For molecules with a double bond, effective lack of rotation about the double bond makes this type of isomer possible.

→ *Geometrical isomers* of organic compounds are distinguished by the italicized prefixes: *cis* (from the Latin word for *on this side*) and *trans* (from the Latin word for *across*).

cis-2-butene   *trans*-2-butene

Example: Consider 2-butene, which exists as *cis* and *trans* geometrical isomers. In the trans isomer, the two methyl groups lie across the double bond from each other. In the cis isomer, they are on the same side of the double bond.

Rotation about the double bond does not normally occur because the π-bond is rigid, and the two isomers are distinct compounds with different chemical and physical properties. If sufficient energy is added to either compound, a *cis–trans isomerization* reaction may occur in which the π-bond is partially broken and part of the sample is converted to the other isomer.

- **Optical isomers**
  - → *Nonsuperimposable* mirror images
  - → A *chiral* molecule has a mirror image that is nonsuperimposable. Your hand is a chiral object; its mirror image is not superimposable.
  - → An organic molecule is chiral if at least one of its C atoms has four *different* groups attached to it. Such a C atom is called a *stereogenic center*.
  - → A pair of *enantiomers* consists of a chiral molecule and its mirror image.
  - → Enantiomers have identical chemical properties (melting point, solubility, *etc.*), except when they react with other chiral species.
  - → Enantiomers differ in only one physical property; chiral molecules display *optical activity,* the ability to rotate the plane of polarization of light.
  - → Mixtures of enantiomers in equal proportions are *racemic mixtures,* which are not optically active.
  - → A molecule that has a superimposable mirror image is said to be *achiral.*
- **Summary** → The three different types of isomerism are displayed in Fig. 18.1 in the text.

18.3 Properties of Alkanes

- **Physical properties**
  - → *Alkanes* are best regarded as nonpolar molecules held together primarily by London forces, which increase with the number of electrons in a molecule.
  - → Consequently, for unbranched *alkanes,* melting and boiling points increase with chain length.
  - → Unbranched *alkanes* tend to have higher melting points, boiling points, and heats of vaporization than their branched structural isomers.
  - → Molecules with unbranched chains can get closer together than molecules with branched chains. As a result, molecules with branched chains have *weaker* intermolecular forces than their isomers with unbranched chains.

    Example: Unbranched butane has a higher boiling point (−0.5°C) than its branched-chain structural isomer, methylpropane (−11.6°C).

- **Chemical properties**
  - → *Alkanes* are not very reactive chemically. They were once called *paraffins,* derived from the Latin "little affinity."
  - → *Alkanes* are unaffected by concentrated sulfuric acid, boiling nitric acid, strong oxidizing agents such as $KMnO_4$, and by boiling aqueous NaOH.
  - → The C–C bond enthalpy (348 kJ·mol^{-1}) and the C–H bond enthalpy (412 kJ·mol^{-1}) are large, so there is little energy advantage in replacing them with most other bonds.
  - → Notable exceptions are C=O (743 kJ·mol^{-1}), C–OH (360 kJ·mol^{-1}), and C–F (484 kJ·mol^{-1}).
- **Reactivity** → Alkanes undergo catalyzed reactions in the refining of petroleum (see Sections 18.4 and 18.9) as well as combustion and substitution reactions described below.

- **Combustion (oxidation) reactions**
  - → *Alkanes* are used as fuels because their enthalpies of combustion are high.
  - → The products of combustion are carbon dioxide and water. Strong C–H bonds are replaced by even stronger O–H bonds in H_2O, and the O=O bonds are replaced by two strong C=O bonds in CO_2.

    Example: Combustion of one mole of octane releases 5471 kJ of heat:

    $$C_8H_{18}(l) + 12.5\ O_2(g) \rightarrow 8\ CO_2(g) + 9\ H_2O(l) \qquad \Delta H_r^\circ = -5471\ \text{kJ·mol}^{-1}$$

18.4 Alkane Substitution Reactions

- **Substitution reactions**
  - → *Alkanes* are used as raw materials for the synthesis of many reactive organic compounds.
  - → Organic chemists introduce reactive groups into alkane molecules in a process called *functionalization*.
  - → Functionalization of alkanes is achieved by a *substitution reaction,* in which an atom or group of atoms replaces an atom in the original molecule (hydrogen, in the case of alkanes).
  - → Reaction of methane (CH_4) and chlorine (Cl_2) is an example of a substitution reaction. In the presence of ultraviolet light or temperatures above 300°C, the gases react explosively:

  $$CH_4(g) + Cl_2(g) \xrightarrow{\text{light or heat}} CH_3Cl(g) + HCl(g)$$

  Chloromethane (CH_3Cl) is only one of four products; the others are dichloromethane (CH_2Cl_2), trichloromethane ($CHCl_3$), and tetrachloromethane (CCl_4), which is carcinogenic.

- **Radical chain mechanism**
  - → Kinetic studies suggest that alkane substitution reactions such as the one above proceed by a *radical chain* mechanism.
  - → The *initiation step* is the dissociation of chlorine:

  $$Cl_2 \xrightarrow{\text{light or heat}} 2\,Cl\cdot$$

  - → Chlorine atoms proceed to attack methane molecules and abstract a hydrogen atom:

  $$Cl\cdot + CH_4 \rightarrow HCl + \cdot CH_3$$

  Because one of the products is a radical, this reaction is a *propagation step*.
  - → In a second propagation step, the methyl radical may react with a chlorine molecule:

  $$Cl_2 + \cdot CH_3 \rightarrow CH_3Cl + Cl\cdot$$

  The chlorine atom formed may take part in the other *propagation step* or attack a CH_3Cl molecule to eventually form CH_2Cl_2. $CHCl_3$ and CCl_4 can also be formed by a continuation of this process.
  - → A termination step occurs when two radicals combine to form a nonradical product, as in the reaction: $Cl\cdot + \cdot CH_3 \rightarrow CH_3Cl$

  The substitution reaction is not very clean and the product is usually a mixture of compounds. One may limit the production of the more highly substituted alkanes by using a large excess of the alkane.

18.5 Properties of Alkenes

- **Double bond**
  - → The carbon-carbon double bond, C=C, consists of a σ-bond and a π-bond.
  - → Each carbon atom is sp^2-hybridized and one of the hybrid orbitals is used to form the σ-bond.
  - → The unhybridized *p*-orbitals on each atom overlap with each other to form a π-bond.
  - → All four atoms attached to the C=C group lie in the same plane and are fixed in that arrangement by resistance to twisting of the π-bond (see Fig. 18.7 in the text).
  - → Alkenes cannot roll up into a compact arrangement as alkanes can, so alkenes have lower melting points.
  - → In C=C, the π-bond is weaker than the σ-bond. A consequence of this weakness is the reaction most common in alkenes: replacement of the π-bond by two new σ-bonds, as in an elimination reaction.

- **Formation of alkenes by elimination reactions**
  - → In the petrochemical industry, alkanes are converted to more reactive alkenes by a catalytic process called *dehydrogenation:*

  $$CH_3CH_3(g) \xrightarrow{\ Cr_2O_3\ } CH_2{=}CH_2(g) + H_2(g)$$

  This is an example of an *elimination reaction,* one in which two groups on neighboring C atoms are removed from a molecule, leaving a multiple bond (see Fig. 18.8 in the text).
  - → In the laboratory, alkenes are produced by *dehydrohalogenation* of haloalkanes, the removal of a hydrogen atom and a halogen atom from neighboring carbon atoms:

  $$CH_3CH_2Br \xrightarrow{\ CH_3CH_2O^-\ \text{in ethanol at}\ 70°C\ } CH_2{=}CH_2 + HBr$$

  This elimination reaction is carried out in hot ethanol containing sodium ethoxide, CH_3CH_2ONa.
  - → *Dehydrohalogenation* occurs by attack of the ethoxide ion on a hydrogen atom of the methyl group. A H atom is removed as a proton and $CH_3CH_2O{-}H$ is formed. When the methyl C atom forms a second bond to its neighbor, the Br^- ion departs.
  - → States of reactants and products in organic reactions are often not given because the reaction may take place on a catalyst surface or in a non-aqueous solvent, as in the reaction above.

18.6 Electrophilic Addition

- **Addition reaction**
  - → Characteristic reaction of alkenes, in which atoms supplied by the reactant form σ-bonds to the two C atoms joined by the π-bond, which is broken (see Fig. 18.9 in the text).
  - → Almost all addition reactions are exothermic.
  - → A *hydrogenation* reaction is the addition of two hydrogen atoms at a double bond:

  $$CH_3CH{=}CHCH_3 + H_2 \rightarrow CH_3CH_2{-}CH_2CH_3$$

  - → A *halogenation* reaction is the addition of two halogen atoms at a double bond:

  $$CH_3CH{=}CHCH_3 + Cl_2 \rightarrow CH_3CHCl{-}CHClCH_3$$

  - → A *hydrohalogenation* reaction is the addition of a hydrogen atom and a halogen atom at a double bond: $CH_3CH{=}CHCH_3 + HCl \rightarrow CH_3CH_2{-}CHClCH_3$

- **Estimating the reaction enthalpy of an addition reaction**
  - → Use bond enthalpies for the bonds that are broken and formed. Note that bond enthalpies strictly apply only to gas-phase reactions.
  - → Recall:

  $$\Delta H_r° \approx \sum_{\text{reactants}} n\Delta H_B(\text{bonds broken}) - \sum_{\text{products}} n\Delta H_B(\text{bonds formed})$$

- **Mechanism of addition reactions**
  - → Double bonds contain a high density of high energy electrons associated with the π-bond. This region of high electron density is attractive to positively charged reactants.
  - → An *electrophile* is a reactant that is attracted to a region of high electron density. It may be a positively charged species or one that has or can acquire a *partial* positive charge during the reaction.
  - → An example treated in the text is the bromination of ethene to form dibromoethane:

  $$H_2C{=}CH_2 + Br_2 \rightarrow CH_2BrCH_2Br$$

- → Bromine molecules are polarizable; a partial positive charge builds up on the bromine atom closest to the double bond. Bromine acts as an *electrophile*. The partial charge becomes a full charge and a bromine cation attaches to the double bond, leaving a Br⁻ ion behind. The cyclic intermediate is called a *bromonium ion*.

- → A Br⁻ ion is attracted by the positive charge of the bromonium ion. It forms a bond to one C atom, and the bromine atom in the cyclic ion forms another bond, giving 1,2-dibromoethane.

- **Hydrogenation reaction**

  - → Hydrogen can be added across a double bond with the help of a solid-state catalyst:

$$H_2(g) + \cdots C = C \cdots \xrightarrow{\text{catalyst}} \cdots CH - CH \cdots$$

  - → This reaction is used in the food industry, for example, to convert liquid oils to solids. Recall that alkenes have lower melting points than alkanes, other things equal.

AROMATIC COMPOUNDS (Sections 18.7–18.8)

18.7 Nomenclature of Arenes

- **Definition**

  - → Aromatic hydrocarbons are called *arenes*.

  - → *Benzene* (C_6H_6) is the parent compound. The benzene ring is also called the *phenyl* group, as in 2-phenyl-*trans*-2-butene ($CH_3C(C_6H_5)=CHCH_3$), shown on the right.

  - → Aromatic compounds include those with fused benzene rings, such as naphthalene ($C_{10}H_8$), anthracene ($C_{14}H_{10}$), and phenanthrene ($C_{14}H_{10}$).

Naphthalene   Anthracene   Phenanthrene

- **Benzene ring numbering**

  - → Benzene ring substituents are designated by numbering the C atoms from 1 to 6 around the ring. Compounds are named by counting around the ring in the direction that gives the smallest numbers to the substituents. The prefixes *ortho-*, *meta-*, and *para-* are commonly used to denote substituents at C atoms 2, 3, and 4, respectively, relative to another substituent at C atom 1.

    Example: The three possible dichlorobenzene isomers are shown in the diagram on the right.

  - → Numbering carbon atoms in fused ring systems is somewhat more complicated and is not covered in the text

1,2-Dichlorobenzene
ortho-Dichlorobenzene

1,3-Dichlorobenzene
meta-Dichlorobenzene

1,4-Dichlorobenzene
para-Dichlorobenzene

18.8 Electrophilic Substitution

- **Substitution reactions** → *Arenes* have delocalized π-electrons; but, unlike alkenes, arenes undergo predominantly *substitution* reactions, with the π-bonds of the ring unaffected.

  Example: The reaction of benzene (C_6H_6) with chlorine produces chlorobenzene when one chlorine atom substitutes for a hydrogen atom: $C_6H_6 + Cl_2 \xrightarrow{\text{Fe}} C_6H_5Cl + HCl$

- **Electrophilic substitution** → If the mechanism of substitution of a benzene ring involves electrophilic attack, the reaction is called *electrophilic substitution*.

- **Halogenation**
  - → In the halogenation of benzene, an iron catalyst is used. The iron is converted to iron(III) halide (FeX_3). Iron(III) halide reacts further to polarize the X_2 molecule.
  - → The commonly accepted mechanism for aromatic halogenation, X_2, is then

    $$FeX_3 + X\text{–}X \rightleftharpoons X_3Fe^{\delta-}\cdots X\text{–}X^{\delta+} \qquad \textit{fast equilibrium}$$

    $$X_3Fe^{\delta-}\cdots X\text{–}X^{\delta+} + C_6H_6 \rightarrow FeX_4^- + C_6H_6X^+ \qquad \textit{slow}$$

    $$C_6H_6X^+ + FeX_4^- \rightarrow C_6H_5X + HX + FeX_3 \qquad \textit{fast}$$

  - → In the last step, the hydrogen atom is easily removed from the ring, for in that way the π-electron delocalization and stabilization, lost in the slow step, is regained.

- **Nitration**
  - → A mixture of HNO_3 and concentrated H_2SO_4 slowly converts benzene into nitrobenzene.
  - → The nitronium ion, NO_2^+, is the electrophile and the nitrating agent.
  - → The commonly accepted mechanism for aromatic nitration is:

    $$HNO_3 + 2\,H_2SO_4 \rightleftharpoons NO_2^+ + H_3O^+ + 2\,HSO_4^-$$

    $$NO_2^+ + C_6H_6 \rightarrow C_6H_6NO_2^+ \qquad \textit{slow}$$

    $$C_6H_6NO_2^+ + HSO_4^- \rightarrow C_6H_5NO_2 + H_2SO_4 \qquad \textit{fast}$$

  - → In the last step, the hydrogen atom is removed from the ring, restoring aromaticity to it.

- **Ortho- and para-directing activators**
  - → Electrons are donated into the delocalized Molecular Orbitals of the ring (*resonance effect*).
  - → The reaction rate is *much* faster than in unsubstituted benzene, and the substituent is called an *activator*.
  - → Products of substitution reactions favor the *ortho* and *para* positions of the ring because these locations have more electron density than the *meta* positions.
  - → Examples of *ortho-* and *para*-directing activators include –OH, –NH$_2$, and substituted amines.
  - → The atom bonded to the benzene ring has a *nonbonding* pair of electrons.

  Note: Alkyl groups have no nonbonding electron pair, yet they are also *ortho-* and *para*-directing activators. The mechanism in this case is a different one (see any organic chemistry text).

- **Ortho- and para-directing deactivators**
  - → Electrons are donated into the delocalized Molecular Orbitals of the ring (*resonance effect*), but the substituent is very electronegative.
  - → The reaction rate is *slightly* slower than in unsubstituted benzene, and the substituent is called a *deactivator*. Deactivation occurs when the substituent is *highly* electronegative and withdraws some electron density from the ring.

→ Products of a substitution reaction still favor the *ortho* and *para* positions of the ring because these positions have *relatively* more electron density than the *meta* positions.

→ The *only* examples are –F, –Cl, –Br, and –I.

→ The atom bonded to the benzene ring has a *nonbonding* pair of electrons.

- **Meta-directing deactivators**

  → Highly electronegative substances that can withdraw electrons partially and/or a substituent that *removes* electrons by resonance

  → The reaction rate is *much* slower than in unsubstituted benzene, and the substituent is called a *deactivator*.

  → Products of a substitution reaction favor the *meta* position because the *ortho* and *para* positions have greatly decreased electron density.

  → Examples include –COOH, $-NO_2$, $-CF_3$, and –C≡N.

  → *All* electron pairs on the atom bonded to the benzene ring are *bonding* pairs.

IMPACT ON MATERIALS: FUELS (Sections 18.9–18.10)

18.9 Gasoline

- **Gasoline**
  → Derived from petroleum by fractional distillation (see Section 9.19) and further processing
  → Contains primarily C_5 to C_{11} hydrocarbons
  → Petroleum is refined to increase the quantity and quality of gasoline.

- **Quantity** → The quantity of gasoline is increased by *cracking* (breaking down long hydrocarbon chains) and by *alkylation* (combining small molecules to make larger ones).

  Examples: 1) Cracking fuel oil to obtain an octane/octane isomeric mixture

  2) Alkylation converts a butane/butene mixture to octane.

  $$C_4H_{10} + C_4H_8 \xrightarrow{\text{catalyst}} C_8H_{18}$$

- **Quality**
  → Octane rating: used to measure the quality of gasoline
  → Other things equal, branched hydrocarbons have higher octane ratings than unbranched ones.
  → *Isomerization* and *aromatization* are used to improve octane rating of gasoline.
  → *Isomerization* is used to convert straight chain hydrocarbons to branched ones:

  $$CH_3(CH_2)_6CH_3 \xrightarrow{\text{AlCl}_3} CH_3C(CH_3)_2CH_2CH(CH_3)CH_3$$

  → *Aromatization* converts alkanes to arenes, as in the conversion of heptane to methylbenzene:

  $$CH_3(CH_2)_5CH_3 \xrightarrow{\text{AlCl}_3,\ \text{Cr}_2\text{O}_3} CH_3C_6H_5 + 4\ H_2$$

  → Ethanol, a renewable fuel, also increases the octane rating of gasoline.

18.10 Coal

- **Coal and its uses**

  → Coal contains a much lower ratio of hydrogen to carbon than petroleum and is harder to purify, to work, and to transport.

  → End product of anaerobic decay of vegetable matter

  → Less environmentally friendly than gasoline

  → When it burns, coal also releases a large amount of pollution in the form of particulate matter (primarily ash) and sulfur and nitrogen oxides.

  → Contains many aromatic rings and is primarily aromatic in nature

  → Coal fragments, when heated in the absence of oxygen, yield *coal tar*, which contains many aromatic hydrocarbons and their derivatives.

  → Consequently, coal is used as a raw material for other chemicals.

  → Many pharmaceuticals, fertilizers, and dyes are derived from *coal tar*

  Examples: Naphthalene for making indigo dyes (for blue jeans) and benzene, used to make nylon, detergents, and pesticides

- **Coal gas**

  → When coal is destructively distilled—heated in the absence of oxygen so that it decomposes and vaporizes—its sheetlike molecules break up, and the fragments include the aromatic hydrocarbons and their derivatives.

  → *Coal gas,* which is given off first, contains carbon monoxide, hydrogen, methane, and small amounts of other gases. The complex liquid mixture that remains is called *coal tar.*

  → *Coal gas* is the starting point for a number of alternative fuels. The mixture of carbon monoxide and hydrogen can be used as is, but that mixture is highly toxic.

  → Therefore, coal gas is used to make gaseous fuels such as gaseous methane and hydrogen and liquid fuels such as methanol.

  → These fuels are highly desired because they burn cleanly, producing little air pollution.

Chapter 19 ORGANIC CHEMISTRY II: POLYMERS AND BIOLOGICAL COMPOUNDS

COMMON FUNCTIONAL GROUPS (Sections 19.1–19.8)

19.1 Haloalkanes

- **Haloalkanes**
  - → Alkanes in which a *halogen atom,* X, replaces one or more H atoms
  - → Insoluble in water
  - → Some are highly toxic and environmentally unfriendly.
    Example: The chlorofluorocarbon (CFC) 1,2-dichloro-1-fluoroethane, $HFClC–CClH_2$, is partly responsible for depletion of the ozone layer.
  - → The C–X bonds in haloalkanes are *polar;* C carries a partial positive charge and the halogen, X, a partial negative one.
  - → C–X bond polarity governs much of the chemistry of haloalkanes.
  - → Haloalkanes are susceptible to *nucleophilic substitution,* in which a reactant that seeks out centers of positive charge in a molecule (a *nucleophile*) replaces a halogen atom.
  - → *Nucleophiles* include anions such as OH^- and Lewis bases with lone pairs such as NH_3.
    Example: Hydroxide acts as a nucleophile in the following *hydrolysis* reaction:
    $$CH_3Br + OH^- \rightarrow CH_3OH + Br^-$$

19.2 Alcohols

- **Definitions**
  - → *Hydroxyl group:* An –OH group covalently bonded to a C atom
  - → *Alcohol:* An organic compound that contains a *hydroxyl* group not directly bonded to an aromatic ring or to a carbonyl group, $\overset{\diagdown}{\underset{\diagup}{C}}=O$
  - → Alcohols are named by adding the suffix *–ol* to the stem of the parent hydrocarbon.
    Examples: Methanol for CH_3OH; ethanol for CH_3CH_2OH
  - → To identify the location of the –OH group, the number of the C atom attached to it is given.
    Examples: 1-propanol for $CH_3CH_2CH_2OH$; 2-propanol for $CH_3CH(OH)CH_3$
  - → A *diol* is an organic compound with two hydroxyl groups.
    Example: 1,2-ethanediol for $HOCH_2CH_2OH$ (ethylene glycol)

- **Classes of alcohols**
  - → *Primary alcohol:* RCH_2–OH, where R can be any group
    Examples: Methanol and 1-propanol
  - → *Secondary alcohol:* R_2CH–OH, where the R groups can be the same or different
    Examples: 2-propanol, $(CH_3)_2CH$–OH (same R groups); 2-butanol, $CH_3CH_2CH(–OH)CH_3$ (different R groups)

→ *Tertiary alcohol*: R_3C-OH, where the R groups can be the same or different

    Examples: 2-methyl-2-propanol, $(CH_3)_3C-OH$ (*same* R groups, CH_3-); also called tertiary butyl alcohol or *t*-butyl alcohol for short.

    3-methyl-3-pentanol, $CH_3CH_2C(CH_3)(-OH)CH_2CH_3$ (*different* R groups, CH_3CH_2- and CH_3-). A structure for the latter is shown above:

- **Properties of alcohols**

  → Alcohols are polar molecules that can lose the $-OH$ proton in certain solvents, but typically not in water.

  → Alcohols have relatively high boiling points and low volatility because of hydrogen bond formation via the $-OH$ group. In this way, they are similar to water.

  → Compare the normal boiling point of ethanol (78.2°C) with a molar mass of 46.07 g·mol^{-1} to that of pentane (36.0°C) with a molar mass of 72.14 g·mol^{-1}.

  → Methanol is prepared from synthesis gas, a mixture of carbon monoxide and hydrogen:

  $$CO(g) + 2H_2(g) \xrightarrow{\text{catalyst, 250°C, 50-100 atm}} CH_3OH(g)$$

  → Ethanol is prepared by carbohydrate fermentation and by hydration of ethene (addition reaction):

  $$CH_2=CH_2(g) + H_2O(g) \xrightarrow{\text{catalyst, 300°C}} CH_3CH_2OH(g)$$

19.3 Ethers

- **Ethers**

  → Organic compounds of the form $R-O-R$, where R is any alkyl group and the two R groups may be the same or different

  → More volatile than alcohols with the same molar mass because ethers do not form hydrogen bonds with each other

  → Act, however, as hydrogen-bond acceptors by using the lone pairs of electrons on the O atom

  → Useful solvents for other organic compounds because they have low polarity and low reactivity

  → Quite flammable and must be handled with care

      Examples: Diethyl ether $(CH_3CH_2-O-CH_2CH_3)$;
                    1-butyl methyl ether $(CH_3CH_2CH_2CH_2-O-CH_3)$

- **Crown ethers**

  → Cyclic polyethers of formula $(CH_2CH_2-O)_n$

  → Name reflects the crownlike shape of the molecules.

  → Bind strongly to alkali metal ions such as Na^+ and K^+, allowing inorganic salts to be dissolved in organic solvents

      Example: The common oxidizing agent potassium permanganate $(KMnO_4)$ is insoluble in nonpolar solvents such as benzene (C_6H_6). In the presence of [18]-crown-6, it dissolves in benzene according to the reaction on the right. The name indicates an 18-membered ring with 6 ether oxygen atoms.

19.4 Phenols

- **Phenols**

  → Organic compounds with a *hydroxyl group* attached *directly* to an aromatic ring

  → The parent compound, *phenol,* is a white, crystalline, molecular solid.

  Phenol, C_6H_5OH: or   Melting point: 40.9°C

  → Substituted phenols occur naturally, and some are responsible for the fragrances of plants.

  Examples: *Thymol* is the active ingredient in oil of thyme, and *eugenol* provides the scent and flavor in oil of cloves.

- **Acid-base properties of phenols**

  → Phenols are generally *weak* acids in contrast to alcohols, which typically are *not* acidic.

  → The acidity of phenols can be understood on the basis of resonance stabilization (delocalization) of the negative charge of the conjugate base of phenol ($C_6H_5O^-$).

  → Because of resonance stabilization, the phenoxide ($C_6H_5O^-$) anion is a weaker conjugate base than the corresponding conjugate bases of typical alcohols.

  Example: The ethoxide ($CH_3CH_2O^-$) anion is a stronger base than the phenoxide anion shown above.

19.5 Aldehydes and Ketones

- **Aldehydes**

  → Organic compounds of the form shown on the right, where R is a H atom, an aliphatic group, or an aromatic group

  → The group characteristic of aldehydes is written as –CHO, as in formaldehyde (HCHO), the first member of the family in which R = H.

  → Named systematically by replacing the ending *–e* by *–al,* but many common names exist, such as formaldehyde for methanal and acetaldehyde for ethanal

  → The C atom on the carbonyl group is *included* in the count of C atoms when determining the alkane from which the aldehyde is derived.

  → A few common aldehydes are shown below:

  Ethanal   Benzaldehyde   Methanal

  → Aldehydes generally contribute to the flavor of fruits and nuts, and to the odors of plants.

  Example: Benzaldehyde provides part of the aroma of almonds and cherries.

→ Aldehydes can be prepared by the *mild* oxidation of *primary* alcohols.

Example: Formaldehyde is prepared industrially by oxidizing methanol with a Ag catalyst. The overall reaction is:

$$2\,CH_3OH(g) + O_2(g) \xrightarrow{\text{600°C, Ag}} 2\,HCHO(g) + 2\,H_2O(g)$$

- **Ketones**

  → Organic compounds of the form shown on the right, where the R groups (alkyl or aryl) may be the same or different

  → The *carbonyl group,* −C=O, characteristic of ketones, is written −CO, as in propanone, CH_3COCH_3 (acetone), the first member of the family.

  → Named systematically by replacing the ending *-e* by *-one,* but many common names exist, such as acetone or dimethyl ketone for propanone, and methyl ethyl ketone for butanone

  → The C atom on the carbonyl group is *included* in the count of C atoms when determining the alkane from which the ketone is derived.

  → A number is used to locate the C atom in the carbonyl group to avoid ambiguity.

  → A few common ketones are shown on the right:

  Propanone Benzophenone 2-Pentanone

  → *Ketones* contribute to flavors and fragrances.

  Example: *Carvone,* shown on the right, is the essential oil in spearmint.

  → Ketones can be prepared by the oxidation of *secondary* alcohols.

  → There is less risk of further oxidation than with aldehydes, so stronger oxidizing agents are used. Dichromate oxidation of secondary alcohols produces ketones with little excess oxidation and in good yields.

  Example: The oxidation of 2-propanol to propanone is shown schematically on the right:

19.6 Carboxylic Acids

- **Carboxylic acids**

  → Organic compounds of the form shown on the right, where R is a H atom, an alkyl, or an aryl group

  → The *carboxyl group* characteristic of carboxylic acids is written −COOH as in formic acid (HCOOH), the first member of the family.

  → The *carboxyl group* contains a *hydroxyl group,* −OH, attached to a *carbonyl group,* −C=O.

  → Named systematically by replacing the ending *-e* by *-oic acid,* but many common names are used, such as formic acid for methanoic acid and acetic acid for ethanoic acid

  → The carbonyl C atom is included in the C atom count to determine the parent hydrocarbon molecule.

→ Some common carboxylic acids are shown on the right:

Methanoic acid   Ethanoic acid   Benzoic acid   Malonic acid

Note: Malonic acid contains two –COOH groups; it is called a *diacid*.

→ Carboxylic acids contain hydroxyl groups, –OH, and the carbonyl group, –C=O, which can take part in hydrogen bonding.

→ Carboxylic acids can be prepared by oxidation of aldehydes or of *primary* alcohols in an acidified solution containing a strong oxidizing agent such as $KMnO_4$, or $Na_2Cr_2O_7$:

Aldehyde   Primary alcohol

→ In some cases, alkyl groups can be oxidized *directly* to carboxyl groups.

→ An industrial example is the oxidation of the methyl groups on *p*-xylene by a cobalt(III) catalyst to form *terephthalic acid,* which is used in the production of artificial fibers.

19.7 Esters

• **Esters**

→ An *ester* is the product of a *condensation* reaction between a carboxylic acid and an alcohol. A water molecule is also produced.

Carboxylic acid   Alcohol   Ester

→ *Esterification* reactions can be acid catalyzed as in the example above.

→ *Condensation* reaction: one in which two molecules combine to form a large one, and a small molecule is eliminated.

→ Many *esters* have fragrant aromas and contribute to flavors of fruits.

Examples: The esters *n*-amyl acetate and *n*-octyl acetate are responsible for the aromas of bananas and oranges, respectively.

n-Amyl acetate   n-Octyl acetate

→ Naturally occurring esters also include fats and oils such as tristearin, a component of beef fat.

Tristearin

19.8 Amines, Amino Acids, and Amides

- **Amines**
  - → Derivatives of NH_3, formed by replacing one or more H atoms with organic groups, R
  - → The *amino group* ($-NH_2$) is the functional group of *amines.*
  - → Named by specifying the groups attached to the nitrogen atom (N) alphabetically, followed by the suffix *amine*
  - → Designated as *primary, secondary,* or *tertiary,* depending on the number of R groups attached to the N atom
  - → Ammonia and several representative amines are shown on the right:

- **Properties of amines**
  - → Characterized by four sp^3 hybrid orbitals on the N atom; three participate in three single bonds and the fourth has a lone pair of electrons
  - → *Amines* are widespread in nature.
  - → They often have disagreeable odors
    Example: *Putrescine,* $NH_2(CH_2)_4NH_2$
  - → Similar to ammonia itself, amines are weak bases.

- **Quaternary ammonium ions**
  - → Tetrahedral ions of formula R_4N^+
  - → Four R groups, which may be the same or different, are bonded to the central nitrogen atom.
  - → Negligible acid or base properties and little effect on pH
  - → Isolated as quaternary ammonium salts

    Example: Dimethylethylpropylammonium chloride, with four R groups, is a quaternary ammonium salt.

- **Amino acids**
  - → Carboxylic acids with an *amino* group and the *carboxyl* group separated by *one* C atom

→ Technically, these are α-amino acids. Amino acids with the *amino* and the *carboxyl* groups separated by *two* C atoms are β-amino acids, *etc.*

→ A more precise name for an amino acid is an *aminocarboxylic acid*.

→ Have the general formula $R(NH_2)CH(COOH)$

Examples: *Glycine, methionine,* and *phenylalanine*

Note: In amino acids, the central C atom is *stereogenic* and the molecule is *chiral* except for *glycine*.

Glycine   Methionine   Phenylalanine

- **Properties of amino acids**

  → Building blocks of proteins (see Section 19.13 in the text)

  → Essential for human health

  → Form *hydrogen bonds* with *both* the *amino* and *carboxyl* groups

  → Form *double ions* (*zwitterions*) by transferring a proton from the *carboxyl* to the *amino* group as shown on the right:

  Amino acid   Double ion form

  → In aqueous solution, amino acids exist in four forms whose concentrations are pH dependent.

  1   2   3   4

  → The concentrations of these species can be determined as with polyprotic acids (see Section 11.17 in the text).

  → Form 1 predominates at low pH while form 4 predominates at high pH. At neutral pH, form 2 predominates.

  → The neutral form 3 exists only in trace amounts in aqueous solutions.

- **Amides**

  → Molecules with the general formula:

  → Formed by the condensation reaction of carboxylic acids with amines, eliminating water

  → Near room temperature, the reaction mixture of an amine and a carboxylic acid forms an ammonium salt that reacts further to form the amide at higher temperature.

  → The general reaction is:

  Acid   Amine   Amide

Note: The amide shown above has an N–H group that may participate in *intermolecular* hydrogen bonding. The atoms that form the product water molecule are shown in a box.

SG-263

THE IMPACT ON TECHNOLOGY (Sections 19.9–19.12)

19.9 Addition Polymerization

- **Addition polymers**
  - → Form when *alkene* monomers react with themselves with no net loss of atoms to form *polymers*
  - → Prepared by *radical polymerization,* a radical chain reaction
  - → Radical polymerization reactions are started by using an *initiator* such as an organic peroxide, R–O–O–R, which decomposes when heated to form two free radicals.
  - → A radical polymerization mechanism is shown for ROOR and $CH_2=CHX$:

  Initiation: $\quad$ R—O—O—R $\xrightarrow{\text{heat}}$ R—O• + •O—R

  Propagation:

  Termination:

  → Radical polymerization reactions include *initiation* and *propagation* steps as shown above.

  → The reaction terminates when all the monomer is consumed or when two radical chains of any length react to form a single diamagnetic (nonradical) species.

  → One possible *termination* step is shown in the scheme above.

- **Stereoregular polymer**
  - → Each unit or pair of repeating units in the polymer has the same relative orientation.
  - → These polymers pack well together, are relatively strong, dense, and impact-resistant.

- **Isotactic polymer**
  - → A *stereoregular* polymer with all substituents on the same side of the extended carbon chain

    Example: A portion of an isotactic polypropylene chain:

Isotactic configuration - same side

- **Syndiotactic polymer**
  - → A *stereoregular* polymer with substituents alternating regularly on either side of the extended carbon chain

    Example: A portion of a syndiotactic polypropylene chain with alternating methyl groups:

    Syndiotactic configuration - alternating sides

- **Atactic polymer**
  - → A polymer with substituents randomly oriented with respect to the extended carbon chain
  - → *Atactic* polymers are *not* stereoregular. They tend to be amorphous and less well suited for many applications.

    Example: An example of an *atactic* polypropylene chain with randomly oriented methyl side groups:

    Atactic configuration - random sides

- **Ziegler–Natta catalysts**
  - → Consist of aluminum- and titanium-containing compounds, such as titanium tetrachloride, $TiCl_4$, and triethyl aluminum, $(CH_3CH_2)_3Al$
  - → Used in the production of *stereoregular* polymers, including synthetic rubber
  - → Chemists were unable to synthesize rubber with useful mechanical properties until the discovery of Ziegler–Natta catalysts in 1953.

19.10 Condensation Polymerization

- **Condensation polymers**
  - → Are formed
    - ➤ By a series of *condensation* reactions
    - ➤ From reactants, each of which has *two* functional groups
    - ➤ From stoichiometric amounts of the reactants
  - → Encompass the classes of polymers known as *polyesters* and *polyamides*
  - → Are characterized by reactions proceeding without the need for an initiator
  - → Typically have shorter chain lengths than addition polymers because each monomer can initiate the reaction

- **Polyesters**

  → Form from the condensation of a *diacid* and a *diol*

  → For polyester polymerization, it is necessary to have two functional groups on each monomer and to mix stoichiometric amounts of the reactants.

  Example: Kodel polyester is formed from the esterification of terephthalic acid and 1,4-bis(hydroxymethyl)-cyclohexane as shown by the reactions below.

  Terephthalic acid 1,4-Bis(hydroxymethyl)-cyclohexane

  Kodel polyester

- **Polyamides**

  → Are formed from the condensation of a *diacid* with a *diamine*

  → For polyamide polymerization, it is necessary to have two functional groups on each monomer and to mix stoichiometric amounts of the reactants.

  Example: Nylon-66 forms by condensation of 1,6-hexanedioic acid and 1,6-hexanediamine. The reactants first form a salt, which then condenses to the polyamide.

  1,6-Hexanedioic acid 1,6-Hexanediamine

  Nylon-66

→ N-H···O=C hydrogen bonding between chains strengthens nylons and helps them absorb moisture.

19.11 Copolymers

- **Copolymers**
  - → Polymers with more than one type of repeating unit
  - → Produced from more than one type of monomer
  - → Four different forms: *alternating, block, random,* and *graft* copolymers (see Fig. 19.14 in the text)
- **Alternating copolymers**
  - → Follow the pattern –A–B–A–B–A–B–A–B–, where A and B are monomer units
  - → Nylon-66 (see above) is an *alternating* copolymer. The index (66) indicates the number of C atoms (6) in each type of monomer.
- **Block copolymers**
  - → Follow the pattern –A–A–A–A–B–B–B–B–A–A–A–
  - → Long segments of one monomer, A, are followed by long segments of monomer B.

    Example: *High-impact polystyrene* is a block copolymer of *styrene* and *butadiene:*

  Styrene   1,3-Butadiene

- **Random copolymers**
  - → Follow no particular pattern

    –A–A–B–A–B–A–B–B–A–B–A–B–B–B–A–B–A–A–B–B–B–

    Example: Radical polymerization of the monomers styrene and 3-methylstyrene is expected to form a random copolymer:

  Styrene   3-Methylstyrene

- **Graft copolymers**
  - → Consist of long chains of one monomer, A, with pendant chains of the second monomer, B:

    Example: Soft contact lenses are composed of a graft copolymer that has a backbone of nonpolar monomers but side groups of a different water-absorbing monomer.

  Graft copolymer

- **Properties of copolymers**
  - → The different types of copolymers extend the range of physical properties obtainable for materials.

    Example: Soft polyurethane foams formed from *diisocyanates* and *glycols* are used for insulation and for furniture stuffing. The formation of a *polyurethane* is shown below:

Toluene-2,6-diisocyanate   Ethylene glycol   A polyurethane

- **Composite materials (see Section 6.17 in the text)**
  - → Consist of two or more materials solidified together
  - → Combine the advantages of the component materials and exhibit superior properties
  - → Differ from copolymers because their components retain their identities in what can be large, separate regions.
    - **Example:** Fiberglass, a material of great strength and flexibility, consists of inorganic materials in a polymer matrix.

19.12 Physical Properties of Polymers

- **Polymers can be designed to have specific properties for a specific application**
- **Synthetic polymers**
  - → Have no definite molar mass, only an *average* value
  - → Tend to soften gradually upon heating and have no definite melting point
  - → Contain chains of various lengths mixed together
  - → Longer *average* chain length leads to higher viscosity and a higher softening point.

- **Properties of polymers**
  - → Depend on the *average* chain length
  - → Depend on the *polarity* of the functional groups
    - **Example:** Nylons with polar side groups capable of hydrogen bonding are strong and absorb water; polypropylene, with H and CH_3 side groups, is virtually impervious to water.
  - → Polar groups are associated with stronger intermolecular forces and tend to increase both softening points and mechanical strength. Longer chain length also increases mechanical strength.
  - → Depend on the manner in which chains pack
  - → Long unbranched chains form crystalline regions that lead to strong, dense materials.
  - → Branched-chain polymers exhibit more tangled arrangements; they are less likely to form crystalline regions and tend to be weaker, less dense materials.

- **Elasticity**
  - → Ability of a polymer to return to its original shape after being stretched
  - → *Elastomers* are materials that return easily to their original shape after stretching.
  - → *Elasticity* of natural rubber is improved by vulcanization (heating with S), which forms disulfide (−S−S−) links between chains, increasing the resilience of the polymer (see Fig. 19.17 in the text).
  - → Extensive cross-linking provides a rigid network of interlinked polymer chains and leads to very hard materials.
  - → Most polymers are electrical insulators, but conducting polymers are known (see Box 19.1 in the text).

- **Thermoplastic polymer**
  - → Can be softened after it is molded.
  - → Can be recycled by melting and reprocessing
  - → Are often made of addition polymers
    - **Example:** Polyethylene

- **Thermosetting polymer**
  - → Takes a permanent shape when molded; does not soften when heated.

    Examples: Vulcanized rubber in car tires and urea formaldehyde foam
- **Silicones**
  - → Synthetic polymers based on silicone
  - → Contain $-O-Si-O-Si-O-$ chains with pendant organic groups, such as CH_3 (Fig 19.18 in the text).
  - → Used as waterproofing materials, implants, drug delivery, and cosmetics.

THE IMPACT ON BIOLOGY (Sections 19.13–19.15)

19.13 Proteins

- **Proteins**
  - → Are condensation copolymers of *up to* 20 naturally occurring amino acids (see Table 19.4 in the text)
  - → Nine of the 20 amino acids, known as *essential amino acids,* cannot be produced by the human body.
  - → Perform highly specific functions in the human body
  - → Enzymes (globular proteins) act as specific and efficient catalysts.

    Example: The enzyme *alcohol dehydrogenase,* a globular protein, oxidizes ethanol to ethanal.

- **Peptides**
  - → Are molecules formed from two or more amino acids
  - → Named starting with the amino acid on the left (N-terminus)
  - → The $-CO-NH-$ link is called a *peptide bond,* and each amino acid in a peptide is called a *residue.*
  - → Typical proteins contain *polypeptide chains* of more than a hundred residues joined through peptide bonds and arranged in a particular order.
  - → *Oligopeptides* are molecules with only a few amino acid residues.

    Example: Reaction of the naturally occurring amino acids, aspartic acid (Asp) and phenylalanine (Phe), produces the artificial sweetener aspartame (Asp-Phe), a *dipeptide* which contains two *residues.* Notice the peptide link, $-CO-NH-$, shown in the box.

Aspartame, a dipeptide

- **Structure of proteins**
  - → *Primary structure*
    - ➤ Sequence of residues in the peptide chain

    Example: The *primary* structure of *aspartame* is Asp-Phe. The *dipeptide* with the N- and C-termini interchanged is Phe-Asp, a *different* oligopeptide.

  - → *Secondary structure*
    - ➤ Describes the shape of the polypeptide chain
    - ➤ Controlled by *intramolecular* interactions between different parts of the peptide chain
    - ➤ Common *secondary* structures include the α *helix* and the β *sheet* (see Figs. 19.20 and 19.21 in the text).
    - ➤ The α *helix* is a helical portion of a polypeptide held together by hydrogen bonding.
    - ➤ The β-*pleated sheet* is a portion of a polypeptide in which the segments of the chain lie side by side to form nearly flat sheets.

  - → *Tertiary structure*
    - ➤ Describes the overall three-dimensional shape of the polypeptide, including the way in which α helix and β sheet regions fold together to shape the macromolecule.
    - ➤ Folding is a consequence of the hydrophobic and hydrophilic interactions between residues lying in different parts of the primary structure. One important link responsible for *tertiary* structure is the *disulfide link* (−S−S−) between amino acids containing sulfur.
    - ➤ Proteins are classified as *globular* (ball-shaped) or *fibrous* (hair-shaped with long chains of polypeptides that occur in bundles).
    - ➤ Globular proteins are soluble in water; fibrous proteins are not. Essentially, all enzymes are globular proteins.

    Example: *Hemoglobin* (responsible for oxygen transport in blood) and *cytochrome c* (a component of electron transport chains in mitochondria and bacteria) are globular proteins; *fibroin* (the protein of silk) is a fibrous protein.

    - ➤ Spider silk is one of the strongest known fibers. Artificial spider silk (see Fig 19.24 in the text) is being used to make lightweight, thin, and flexible body armor.

  - → *Quaternary structure*
    - ➤ Describes the arrangement of *subunits* in proteins with more than one polypeptide chain
    - ➤ Not all proteins contain more than one subunit, so not all have a *quaternary* structure.

    Example: *Hemoglobin* (see Fig. 19.21 in the text) contains four polypeptide units and has a *quaternary* structure.

- **Denaturation**
  - → Loss of structure of proteins
  - → Occurs when a protein loses quaternary, tertiary, or secondary structure (disruption of *noncovalent* interactions)
  - → *Denaturation* is caused by heating and other means, is often irreversible, and is accompanied by loss of function of the protein.

19.14 Carbohydrates

- **Carbohydrates**
  - → Most abundant class of naturally occurring organic compounds
  - → Constitute more than 50% of the Earth's biomass
  - → Members include starches, cellulose, and sugars.
  - → Often have the empirical formula CH_2O (hence the name *carbohydrates*)

- **D-Glucose**
  - → Most abundant carbohydrate
  - → Forms starch and cellulose by polymerization reactions
  - → Has the molecular formula $C_6H_{12}O_6$
  - → Is classified as both an alcohol *and* an aldehyde (*pentahydroxyl aldehyde*)
  - → Exists in an acyclic (open-chain) and in two cyclic (closed-chain) forms called *anomers*
  - → In water, the cyclic structures, also called *glucopyranoses*, are favored.

- **Structure of D-glucose**
  - → The straight-chain structure of D-glucose and the interconversion to the two cyclic *anomers* are shown below:

α-D-Glucose
36%

D-Glucose
0.02%

β-D-Glucose
64%

  - → The percentages indicate proportions of the several species present in water solution.
  - → The anomers, α-D-glucose and β-D-glucose, differ in the stereochemistry at the carbon atom derived from the aldehyde C atom.
  - → More realistic representations of the cyclic forms of D-glucose are shown on the right:

α-D-Glucose

β-D-Glucose

Notes: 1) L-Glucose, not found in nature, is the mirror image of D-glucose, and together they form a pair of enantiomers, which differ in the direction of rotation of polarized light.

2) The symbols D and L refer to a configuration relationship to the compound D-glyceraldehyde and not necessarily to the sign of rotation of light.

- **Polysaccharides**
  - → Polymers of glucose
  - → Include starch (digestible by humans) and cellulose (not digestible by humans)
    - **Example:** Cellulose, the structural material of plants and wood, is a condensation polymer of β-D-glucose. A three subunit chain of cellulose is shown below. The polymer is formed formally by elimination of H_2O from glucose monomers.

19.15 Nucleic Acids

- **DNA**
  - → Abbreviation for *deoxyribonucleic acid*
  - → DNA molecules
    - ➤ Carry genetic information from generation to generation
    - ➤ Control the production of proteins
    - ➤ Serve as the template for the synthesis of RNA (see RNA subsection)

- **Structure of DNA**
  - → DNA molecules are condensation copolymers of enormous size.
  - → DNA is composed of a sugar phosphate backbone and pendent bases as shown below:

  - → A base-sugar unit is called a *nucleoside*.
  - → A base-sugar-phosphate grouping is called a *nucleotide*.

- **Sugar and bases**
  - → In DNA, the sugar is *deoxyribose*.
  - → There are four possible bases: *cytosine* (C), *guanine* (G), *adenine* (A), and *thymine* (T).
  - → The sugar and four bases are shown below:

Deoxyribose   Cytosine (C)   Guanine (G)   Adenine (A)   Thymine (T)

- **Nucleosides**
  - → Formed in DNA by condensation involving the OH group at carbon 1 in deoxyribose and an appropriate amine hydrogen atom of a base
  - → Cytosine, guanine, adenine, and thymine nucleosides are shown below:

Cytosine nucleoside   Guanine nucleoside   Adenine nucleoside   Thymine nucleoside

- **Nucleotides**
  - → Condense at carbon 3 and carbon 5, eliminating water, to form the DNA molecule (nucleic acid), a *polynucleotide*
  - → A trinucleotide segment of a polynucleotide is shown on the right:

- **DNA double helix**
  - → Polynucleotide (nucleic acid) strands link to each other in pairs to form the well-known double helix structure (see Fig. 19.30 in the text).
  - → Association between strands of two polynucleotides occurs when hydrogen bonds between bases on *different* strands are formed.
  - → In DNA, only two types of base pairs (G with C and A with T) occur.
  - → These are shown below with the hydrogen bonds indicated by dashed lines:

Cytosine   Guanine   Thymine   Adenine

Note: CG and AT base pairs have approximately the same size and shape, which reduces distortions in the double helix structure.

- **RNA**
  - → The abbreviation for *ribonucleic acid*
  - → Is a polynucleotide, like DNA
  - → Is shorter than DNA, and generally single-stranded

- **Three types of RNA**
  - → *Messenger RNA* (mRNA) carries genetic information from a cell nucleus to the cytoplasm where translation to protein takes place.
  - → *Ribosomal RNA* (rRNA) appears to act as a catalyst in the biosynthesis of proteins.
  - → *Transfer RNA* (tRNA) is used to transport amino acids during protein synthesis.

- **Structure of RNA**
  - → RNA contains four base pairs as in DNA, but *uracil* (U) substitutes for *thymine* (T).
  - → *Uracil* lacks the methyl group of *thymine* at the carbon-5 position. The structures of thymine and uracil are shown below:

Thymine (T)   Uracil (U)

Note: The position of the methyl group on thymine does not interfere with its hydrogen bonding with adenine (A) (see figure on the previous page of an AT base pair).

- → The sugar in RNA is ribose (deoxyribose is shown for comparison):

Ribose   Deoxyribose

- → Although RNA molecules are generally single-stranded, they contain regions of a double helix produced by the formation of *hairpin loops*.
- → In these regions, the usual base pairing is A with U and G with C.
- → However, in RNA, imperfections in base pairing are common.

SOLUTIONS MANUAL

FUNDAMENTALS

A.1   (a) chemical;   (b) physical;   (c) physical

A.3   The temperature of the injured camper, the evaporation and condensation of water are physical properties. The ignition of propane is a chemical change.

A.5.   (a) Physical;   (b) Chemical;   (c) Chemical

A.7   (a) intensive;   (b) intensive;   (c) extensive;   (d) extensive

A.9   (a) $1000. \text{ grain} \times \left(\dfrac{1 \quad kilograin}{1,000 \; grain} \right) = 1 \text{ kilograin}$

(b) $0.01 \text{ batman} \times \left(\dfrac{100 \; centibatman}{1 \; batman} \right) = 1 \text{ centibatman}$

(c) $1 \times 10^{6} \text{ mutchkin} \times \left(\dfrac{1 \; megamutchkin}{10^{6} \; mutchkin} \right) = 1 \text{ megamutchkin}$

A.11 $1.00 \text{ cup} \times \left(\dfrac{1 \text{ pint}}{2 \text{ cups}} \right) \times \left(\dfrac{1 \text{ quart}}{2 \text{ pints}} \right) \times \left(\dfrac{0.946 \text{ L}}{1 \text{ quart}} \right) \times \left(\dfrac{1000 \text{ mL}}{1 \text{ L}} \right) = 236 \text{ mL}$

A.13 $d = \dfrac{m}{V}$

$= \left(\dfrac{112.32 \text{ g}}{29.27 \text{ mL} - 23.45 \text{ mL}} \right) \left(\dfrac{1 \text{ mL}}{1 \text{ cm}^{3}} \right)$

$= 19.3 \text{ g} \cdot \text{cm}^{-3}$

A.15 $d = \dfrac{m}{V}$, rearranging gives $V = \dfrac{m}{d}$

$$= \left(\frac{0.750 \text{ carat}}{3.51 \text{ g} \cdot \text{cm}^{-3}} \right) \left(\frac{200 \text{ mg}}{1 \text{ carat}} \right) \left(\frac{1 \text{ g}}{1000 \text{ mg}} \right)$$

$$= 0.0427 \text{ cm}^3$$

A.17 $105.50 \text{ g} - 43.50 \text{ g} = 62.00 \text{ g} = m_{H_2O}$

$$d = \frac{m}{V} \qquad V = \frac{m}{d} = \frac{m_{H_2O}}{d_{H_2O}} = \frac{62.00 \text{ g}}{0.9999 \text{ g} \cdot \text{cm}^{-3}} = 62.00 \text{ cm}^3$$

$$d_{liquid} = \left(\frac{96.75 \text{ g} - 43.50 \text{ g}}{62.00 \text{ cm}^3} \right) = 0.8589 \text{ g} \cdot \text{cm}^{-3}$$

A.19 $d = \dfrac{m}{V}$, rearrange gives $V = \dfrac{m}{d}$

$$V = \frac{20.0 \text{ g}}{2.70 \text{ g .cm}^3} = 7.41 \text{ cm}^3$$

Since the area is one centimeter, the thickness must be 7.41 cm

A.21   The result should have 3 significant figures because the number 3.25 has the least significant figures. The results should be 0.989

A.23   (a) $4.82 \text{ nm} \times \left(\dfrac{1000 \text{pm}}{1 \text{nm}} \right) = 4.82 \times 10^3 \text{ pm}$

(b) $\left(\dfrac{1.83 \text{mL}}{\text{min}} \right) \times \left(\dfrac{1 \text{cm}^3}{1 \text{mL}} \right) \times \left(\dfrac{1000 \text{mm}^3}{1 \text{cm}^3} \right) \times \left(\dfrac{1 \text{min}}{60 s} \right) = 30.5 \text{ mm}^3/\text{s}$

(c) $1.88 \text{ ng} \times \left(\dfrac{1 \text{g}}{10^9 \text{ng}} \right) \times \left(\dfrac{1 \text{kg}}{10^3 \text{g}} \right) = 1.88 \times 10^{-12} \text{ kg}$

(d) $\left(\dfrac{2.66 \text{g}}{\text{cm}^3} \right) \times \left(\dfrac{1 \text{kg}}{1000 \text{g}} \right) \times \left(\dfrac{10^6 \text{cm}^3}{1 \text{m}^3} \right) = 2.66 \times 10^3 \text{ kg/m}^3$

(e) $\left(\dfrac{0.044 \text{g}}{\text{L}} \right) \times \left(\dfrac{1000 \text{mg}}{1 \text{g}} \right) \times \left(\dfrac{1 \text{L}}{1000 \text{cm}^3} \right) = 0.044 \text{ mg/cm}^3$

A.25 (a) $d = \dfrac{m}{V}$

$$= \left(\frac{0.213g}{1.100cm \times 0.531cm \times 0.212cm} \right) = \left(\frac{0.213g}{0.1238cm^3} \right)$$

$$= 1.72 \text{g} \cdot \text{cm}^{-3}$$

This determination is more precise because the volume is not limited to two significant figures as it is in part (b)

(b) $d = \dfrac{m}{V}$

$$= \left(\frac{41.003g - 39.753g}{20.37\text{mL} - 19.65\text{mL}} \right)$$

$$= \left(\frac{1.250g}{0.72\text{mL}} \right) \left(\frac{1\text{mL}}{1\text{cm}^3} \right)$$

$$= 1.7 \text{g} \cdot \text{cm}^{-3}$$

A.27 (a) Formula: $°X = 50 + 2°C$

(b) $°X = 50 + 2 \times 22°C = 94°X$

A.29 $E_K = \dfrac{1}{2}mv^2$

$$= \frac{1}{2}(4.2 \text{ kg})(14 \text{ km} \cdot \text{h}^{-1})^2 \left(\frac{1 \text{ h}}{3600 \text{ s}} \right)^2 \left(\frac{1000 \text{ m}}{1 \text{ km}} \right)^2$$

$$= 32 \text{ kg} \cdot \text{m}^2 \cdot \text{s}^{-2}$$

$$= 32 \text{ J}$$

A.31 $m = 2.8$ metric tons, $v_i = 100 \text{ km} \cdot \text{hr}^{-1}$, $v_f = 50 \text{ km} \cdot \text{hr}^{-1}$

$$E_K = \frac{1}{2}mv^2$$

$$E_{K(init)} = \frac{1}{2}(2.8 \text{ metric tons})\left(\frac{10^3 \text{ kg}}{1 \text{ metric ton}}\right)$$

$$\left[\left(\frac{100 \text{ km}}{1 \text{ hr}}\right)\left(\frac{1 \text{ hr}}{60 \text{ min.}}\right)\left(\frac{1 \text{ min.}}{60 \text{ sec.}}\right)\right]^2$$

$$= 1.08 \text{ kg} \cdot \text{km}^2 \cdot \text{s}^{-2} = 1.08 \times 10^6 \text{ kg} \cdot \text{m}^2 \cdot \text{s}^{-2} = 1,080 \text{ kJ}$$

$$E_{K(final)} = \frac{1}{2}(2.8 \text{ metric tons})\left(\frac{10^3 \text{ kg}}{1 \text{ metric ton}}\right)$$

$$\left[\left(\frac{50 \text{ km}}{1 \text{ hr}}\right)\left(\frac{1 \text{ hr}}{60 \text{ min.}}\right)\left(\frac{1 \text{ min.}}{60 \text{ sec.}}\right)\right]^2$$

$$= 0.27 \text{ kg} \cdot \text{km}^2 \cdot \text{s}^{-2} = 0.27 \times 10^6 \text{ kg} \cdot \text{m}^2 \cdot \text{s}^{-2} = 270 \text{ kJ}$$

$$E_{K(init)} - E_{K(final)} = (1,080 - 270) \text{ kJ} = 810 \text{ kJ} = 8.1 \times 10^2 \text{ kJ}$$

This amount of energy could have been recovered, neglecting friction and other losses, or used to drive the vehicle up a hill.

$$E_P = mgh \quad g = 9.81 \text{ ms}^{-2}$$

Setting potential energy equal to $8.1 \times 10^2 \text{ kJ} = 8.1 \times 10^5 \text{ kg m}^2 \text{ s}^{-2}$ and solving for height gives

$$h = \frac{E_p}{mg} = \left(\frac{8.1 \times 10^5 \text{ kg} \cdot \text{m}^2 \cdot \text{s}^{-2}}{(2800 \text{ kg})(9.81 \text{ ms}^{-2})}\right) = 29. \text{ m}$$

A.33 $E_P = mgh$

$$= (40.0 \text{ g})(9.81 \text{ m} \cdot \text{s}^{-2})(0.50 \text{ m})\left(\frac{1 \text{ kg}}{1000 \text{ g}}\right)$$

$$= 0.20 \text{ kg} \cdot \text{m}^2 \cdot \text{s}^{-2} \text{ for one raise of a fork.}$$

For 30 raises, $(30)(0.20 \text{ kg} \cdot \text{m}^2 \cdot \text{s}^{-2}) = 6.0 \text{ J}$

A.35 Since $R = R_E + h,$

$$E_p = \Delta E_P = \frac{-Gm_E m}{R_E + h} - \left(\frac{-Gm_E m}{R_E} \right)$$

$$= \frac{-Gm_E m}{R_E} \left(\frac{1}{1 + \dfrac{h}{R_E}} - 1 \right) = \frac{Gm_E m}{R_E} \left(1 - \frac{1}{1 + \dfrac{h}{R_E}} \right)$$

If $x = \dfrac{h}{R_E}$, then $\dfrac{1}{1 + \dfrac{h}{R_E}} = 1 - \dfrac{h}{R_E}$ from the expansion in x.

Then

$$E_p = \frac{Gm_E m}{R_E} \left(1 - \left(1 - \frac{h}{R_E} \right) \right) = \frac{Gm_E mh}{R_E^2} = mgh$$

$$\text{where } g = \frac{Gm_E}{R_E^2}$$

B.1 number of beryllium atoms $= \dfrac{\text{mass of sample}}{\text{mass of one atom}}$

$$= \left(\frac{0.210 \text{ g}}{1.50 \times 10^{-26} \text{ kg} \cdot \text{atom}^{-1}} \right) \left(\frac{1 \text{ kg}}{1000 \text{ g}} \right)$$

$$= 1.40 \times 10^{22} \text{ atoms}$$

B.3 (a) 5p, 6n, 5e; (b) 5p, 5n, 5e; (c) 15p, 16n, 15e; (d) 92p, 146n, 92e

B.5 (a) ^{194}Ir ; (b) ^{22}Ne ; (c) ^{51}V

B.7

Element	Symbol	Protons	Neutrons	Electrons	Mass number
Chlorine	^{36}Cl	17	19	17	36
Zinc	^{65}Zn	30	35	30	65
Calcium	^{40}Ca	20	20	20	40
Lanthanum	^{137}La	57	80	57	137

B.9   (a) They all have the same mass; (b) they have differing numbers of protons, neutrons, and electrons

B.11   (a) 0.5519; (b) 0.4479; (c) 2.459×10^{-4}; (d) 551.9 kg

B.13   (a) Scandium is a Group 3 metal.   (b) Strontium is a Group 2 metal.   (c) Sulfur is a Group 16 nonmetal. (d) Antimony is a Group 15 metalloid.

B.15   (a) Sr, metal;   (b) Xe, nonmetal;   (c) Si, metalloid

B.17   Fluorine, F, Z=9, gas; Chlorine, Cl, Z=17, gas; Bromine, Br, Z=35, liquid; Iodine, I, Z=53, solid

B.19   (a) d block; (b) p block; (c) d block; (d) s block; (e) p block; (f) d block

B.21   The element is Pb. It is located in group IVa (or 14) and period 6. It is a metal.

C.1   Container (a) holds a mixture (one is single compound and another is single element); Container (b) holds a single element.

C.3   The chemical formula of xanthophyll is $C_{40}H_{56}O_2$.

C.5   (a) $C_3H_7O_2N$;   (b) C_2H_7N

C.7   (a) Cesium is a metal in Group 1; it will form Cs^+ ions.   (b) Iodine is a nonmetal in Group 17/VII and will form I^- ions.   (c) Selenium is a Group 16/VI nonmetal and will form Se^{2-} ions.   (d) Calcium is a Group 2 metal and will form Ca^{2+} ions.

C.9 (a) $^{10}Be^{2+}$ has 4 protons, 6 neutrons, and 2 electrons. (b) $^{17}O^{2-}$ has 8 protons, 9 neutrons, and 10 electrons. (c) $^{80}Br^-$ has 35 protons, 45 neutrons, and 36 electrons. (d) $^{75}As^{3-}$ has 33 protons, 42 neutrons, and 36 electrons.

C.11 (a) $^{19}F^-$; (b) $^{24}Mg^{2+}$; (c) $^{128}Te^{2-}$; (d) $^{86}Rb^+$

C.13 (a) Aluminum forms Al^{3+} ions; tellurium forms Te^{2-} ions. Two aluminum atoms produce a charge of $2 \times +3 = +6$. Three tellurium atoms produce a charge of $3 \times -2 = -6$. The formula for aluminum telluride is Al_2Te_3. (b) Magnesium forms Mg^{2+} ions and oxygen forms O^{2-} ions. A magnesium ion produces a charge of $+2$, which is required to balance the charge on one O^{2-} ion. The formula for magnesium oxide is MgO. (c) Sodium forms $+1$ ions; sulfur forms -2 ions. The formula for sodium sulfide is Na_2S. (d) Rubidium forms $+1$ ions and iodine forms -1 ions. One iodide ion is required to balance the charge of one rubidium ion, so the formula is RbI.

C.15 (a) Group III; (b) aluminum, Al

C.17 (a) Mass fraction due to the neutrons: $(48-22)/48 = 0.542$

(b) Mass fraction due to the protons: $22/48 = 0.458$

(c) Mass fraction due to the electrons: $(22 \times \dfrac{1}{1840})/48 = 2.49 \times 10^{-4}$

(can be ignored)

(d) $30. \text{ kg} \times \left(\dfrac{22}{48}\right) = 14. \text{ kg}$

C.19 (a) Na_2HPO_3

(b) $(NH_4)_2SO_4$

(c) Since the copper ion has the same charge as one magnesium ion, the copper cation carries a charge of +2.

(d) Since three tin ions have the same charge as six potassium ions, each tin ion carries a charge of +2.

C.21 (a) HCl, molecular compound (in the gas phase); (b) S_8, element (molecular substance); (c) CoS, ionic compound; (d) Ar, element; (e) CS_2, molecular compound; (f) $SrBr_2$, ionic compound

D.1 (a) $MnCl_2$. Mn forms 2+ ions and chlorine forms -1 ions.

(b) $Ca_3(PO_4)_2$. Calcium forms +2 ions and the phosphate ion is PO_4^{3-}.

(c) $Al_2(SO_3)_3$. Aluminum forms +3 ions and the sulfite ion is SO_3^{2-}.

(d) Mg_3N_2. Magnesium forms 2+ ions and the nitride ion is N^{3-}.

D.3 (a) calcium phosphate; (b) tin(IV) sulfide, stannic sulfide; (c) vanadium(V) oxide; (d) copper(I) oxide, cuprous oxide

D.5 (a) TiO_2; (b) $SiCl_4$; (c) CS_2; (d) SF_4; (e) Li_2S; (f) SbF_5;

(g) N_2O_5; (h) IF_7

D.7 (a) sulfur hexafluoride; (b) dinitrogen pentoxide; (c) nitrogen triiodide; (d) xenon tetrafluoride; (e) arsenic tribromide; (f) chlorine dioxide

D.9 (a) hydrochloric acid; (b) sulfuric acid; (c) nitric acid; (d) acetic acid; (e) sulfurous acid; (f) phosphoric acid

D.11 (a) ZnF_2 (b) $Ba(NO_3)_2$ (c) AgI (d) Li_3N (e) Cr_2S_3

D. 13 (a) $BaCl_2$; (b) Ionic

D.15 (a) sodium sulfite;   (b) iron(III) oxide or ferric oxide;   (c) iron(II) oxide or ferrous oxide;   (d) magnesium hydroxide;   (e) nickel(II) sulfate hexahydrate;   (f) phosphorus pentachloride; (g) chromium(III) dihydrogen phosphate;   (h) diarsenic trioxide;   (i) ruthenium(II) chloride

D.17 (a) $CuCO_3$ Copper (II) carbonate; (b) K_2SO_3 potassium sulfite; (c) LiCl, lithium chloride

D.19 (a) heptane;   (b) propane;   (c) pentane;   (d) butane

D.21 (a) cobalt(III) oxide monohydrate; $Co_2O_3 \cdot H_2O$; (b) cobalt(II) hydroxide; $Co(OH)_2$

D.23 E=Si; SiH_4. Silicon tetrahydride; Na_4Si, sodium silicide

D.25 (a) lithium aluminum hydride, ionic (with a molecular anion); (b) sodium hydride, ionic

D.27 (a) selenic acid;   (b) sodium arsenate;   (c) calcium tellurite;   (d) barium arsenate;   (e) antimonic acid;   (f) nickel(III) selenate

D.29 (a) alcohol; (b) carboxylic acid; (c) haloalkane

E.1 $1.0 \text{ mole} \times \dfrac{6.022 \times 10^{23} \text{ Ag atom}}{1.0 \text{ mole Ag}} \times \dfrac{144 \text{pm}}{1 \text{ Ag atom}} \times \dfrac{1\text{m}}{10^{12} \text{ pm}}$

$= 8.67 \times 10^{13} \text{ m} = 8.67 \times 10^{10} \text{ km}$

E.3 You will to add 18 Ca atoms to balance the 9 Br atoms on the left pan because a Br atom is twice as massive as a Ca atom.

E.5 (a) moles of people $= \dfrac{6.0 \times 10^9 \text{ people}}{6.022 \times 10^{23} \text{ people} \cdot \text{mol}^{-1}}$

$$= 1.0 \times 10^{-14} \text{ mol}$$

(b) time $= \dfrac{1 \text{ mol peas}}{1.0 \times 10^{-14} \text{ mol} \cdot \text{s}^{-1}} = 1.0 \times 10^{14} \text{ s}$

$$(1.0 \times 10^{14} \text{ s}) \left(\dfrac{1 \text{ h}}{3600 \text{ s}} \right) \left(\dfrac{1 \text{ day}}{24 \text{ h}} \right) \left(\dfrac{1 \text{ yr}}{365 \text{ days}} \right) = 3.2 \times 10^6 \text{ years}$$

E.7 (a) $MgSO_4 \cdot 7 H_2O$ formula mass $= 246.48 \text{ g} \cdot \text{mol}^{-1}$

atoms of O $= \left(\dfrac{5.15 \text{ g}}{246.48 \text{ g} \cdot \text{mol}^{-1}} \right) \left(\dfrac{11 \text{ mol O atoms}}{\text{mol } MgSO_4 \cdot 7 H_2O} \right)$

$$(6.022 \times 10^{23} \text{ atoms} \cdot \text{mol}^{-1}) = 1.38 \times 10^{23}$$

(b) formula units $= \left(\dfrac{5.15 \text{ g}}{246.48 \text{ g} \cdot \text{mol}^{-1}} \right) (6.022 \times 10^{23} \text{ atoms} \cdot \text{mol}^{-1})$

$$= 1.26 \times 10^{22}$$

(c) moles of $H_2O = 7 \left(\dfrac{5.15 \text{ g}}{246.48 \text{ g} \cdot \text{mol}^{-1}} \right) = 0.146 \text{ mol}$

E.9 (a) mass of average Li atom

$$= \left(\dfrac{7.42}{100} \right) (9.988 \times 10^{-24} \text{ g}) + \left(\dfrac{92.58}{100} \right) (1.165 \times 10^{-23} \text{ g})$$

$$= 1.153 \times 10^{-23} \text{ g} \cdot \text{atom}^{-1}$$

molar mass $= (1.153 \times 10^{-23} \text{ g} \cdot \text{atom}^{-1}) (6.022 \times 10^{23} \text{ atoms} \cdot \text{mol}^{-1})$

$$= 6.94 \text{ g} \cdot \text{mol}^{-1}$$

(b) mass of average Li atom

$$= \left(\dfrac{5.67}{100} \right) (9.988 \times 10^{-24} \text{ g}) + \left(\dfrac{100 - 5.67}{100} \right) (1.165 \times 10^{-23} \text{ g})$$

$$= 1.1556 \times 10^{-23} \text{ g} \cdot \text{atom}^{-1}$$

molar mass $= (1.1556 \times 10^{-23} \text{ g} \cdot \text{atom}^{-1}) (6.022 \times 10^{23} \text{ atoms} \cdot \text{mol}^{-1})$

$$= 6.96 \text{ g} \cdot \text{mol}^{-1}$$

E.11   The percentage $^{10}B = 100 -$ percentage ^{11}B

$$\text{molar mass} = \left(\frac{\% \ ^{10}B}{100\%}\right)(\text{mass } ^{10}B) + \left(\frac{\% \ ^{11}B}{100\%}\right)(\text{mass } ^{11}B)$$

$$= \left(\frac{100\% - \% \ ^{11}B}{100\%}\right)(\text{mass } ^{10}B) + \left(\frac{\% \ ^{11}B}{100\%}\right)(\text{mass } ^{11}B)$$

Rearranging gives

$$\% \ ^{11}B = \frac{100 \cdot \text{molar mass} - 100 \cdot \text{mass } ^{10}B}{\text{mass } ^{11}B - \text{mass } ^{10}B}$$

$$= \frac{100(10.81 \text{ g} \cdot \text{mol}^{-1}) - 100(10.013 \text{ g} \cdot \text{mol}^{-1})}{11.093 \text{ g} \cdot \text{mol}^{-1} - 10.013 \text{ g} \cdot \text{mol}^{-1}}$$

$$= 73.8\%$$

$$\% \ ^{10}B = 26.2\%$$

E.13   (a) $\dfrac{75 \text{ g}}{114.82 \text{ g} \cdot \text{mol}^{-1}} = 0.65 \text{ mol In}$

$\dfrac{80 \text{ g}}{127.60 \text{ g} \cdot \text{mol}^{-1}} = 0.63 \text{ mol Te}$

75 g of indium contains more moles of atoms than 80. g of tellurium

(b) $\dfrac{15.0 \text{ g}}{30.97 \text{ g} \cdot \text{mol}^{-1}} = 0.484 \text{ mol P}$

$\dfrac{15.0 \text{ g}}{32.07 \text{ g} \cdot \text{mol}^{-1}} = 0.468 \text{ mol S}$

15.0 g of P has slightly more atoms than 15.0 g of S.

(c) Because the two samples have the same number of atoms, they will have the same number of moles, which is given by

$$\left(\frac{7.36 \times 10^{27} \text{ atoms}}{6.022 \times 10^{23} \text{ atoms} \cdot \text{mol}^{-1}}\right) = 1.22 \times 10^{4} \text{ moles}$$

E.15   (a) molar mass of D_2O: $(2.014 \text{ g/mol}) \times 2 + 15.999 \text{ g/mol} = 20.027 \text{ g/mol}$

(b) Assuming the volume occupied by a D_2O molecule is the same as that of H_2O, the density will be proportional to the molar mass, and the density of D_2O is: $d = \dfrac{molar\ mass\ of\ D_2O}{molar\ mass\ of\ H_2O} \times$ density of water

$$= \left(\dfrac{20.027\ g \cdot mol^{-1}}{18.016\ g \cdot mol^{-1}}\right) \times 1.00\ g \cdot cm^{-3} = 1.11\ g \cdot cm^{-3}$$

(c) The volume of the spherical tank $V = \dfrac{4}{3}\pi r^3 = \dfrac{4}{3}\pi \times (6\ m)^3 = 905.\ m^3$

$= 9.05 \times 10^8\ cm^3$; the volume calculated from the density data:

$$v = \dfrac{m}{d} = \dfrac{1.00 \times 10^9\ g}{1.11\ g \cdot cm^{-3}} = 9.01 \times 10^8\ cm^3 = 901\ m^3$$

The volume calculated from the density data is slightly smaller than real tank volume.

(d) The reported D_2O mass was not accurate.

(e) The assumption in part (b) is reasonable because the nuclei are so small and the volume difference between a D_2O molecule and a H_2O can be ignored.

E.17 (a) molar mass of $Al_2O_3 = 101.96\ g \cdot mol^{-1}$

$$n_{Al_2O_3} = \dfrac{10.0\ g}{101.96\ g \cdot mol^{-1}} = 0.0981\ mol$$

$N_{Al_2O_3} = (0.0981\ mol)(6.022 \times 10^{23}\ atoms \cdot mol^{-1}) = 5.91 \times 10^{22}$ molecules

(b) molar mass of $HF = 20.01\ g \cdot mol^{-1}$

$$n_{HF} = \dfrac{25.92 \times 10^{-3}\ g}{20.01\ g \cdot mol^{-1}} = 1.30 \times 10^{-3}\ mol$$

$N_{HF} = (1.30 \times 10^{-3}\ mol)(6.022 \times 10^{23}\ atoms \cdot mol^{-1})$

$= 7.83 \times 10^{20}$ molecules

(c) molar mass of hydrogen peroxide $= 34.02\ g \cdot mol^{-1}$

$$n_{H_2O_2} = \dfrac{1.55 \times 10^{-3}\ g}{34.02\ g \cdot mol^{-1}} = 4.56 \times 10^{-5}\ mol$$

$N_{H_2O_2} = (4.56 \times 10^{-5}\ mol)(6.022 \times 10^{23}\ atoms \cdot mol^{-1})$

$= 2.75 \times 10^{19}$ molecules

(d) molar mass of glucose = 180.15 $g \cdot mol^{-1}$

$$n_{glucose} = \frac{1250 \text{ g}}{180.15 \text{ g} \cdot mol^{-1}} = 6.94 \text{ mol}$$

$N_{glucose} = (6.94 \text{ mol})(6.022 \times 10^{23} \text{ atoms} \cdot mol^{-1}) = 4.18 \times 10^{24}$ molecules

(e) molar mass of N atoms = 14.01 $g \cdot mol^{-1}$

$$n_N = \frac{4.37 \text{ g}}{14.01 \text{ g} \cdot mol^{-1}} = 0.312 \text{ mol}$$

$N_N = (0.312 \text{ mol})(6.022 \times 10^{23} \text{ atoms} \cdot mol^{-1}) = 1.88 \times 10^{23}$ atoms

molar mass of N_2 molecules = 28.02 $g \cdot mol^{-1}$

$$n_{N_2} = \frac{4.37 \text{ g}}{28.02 \text{ g} \cdot mol^{-1}} = 0.156 \text{ mol}$$

$N_{N_2} = (0.156 \text{ mol})(6.022 \times 10^{23} \text{ atoms} \cdot mol^{-1}) = 9.39 \times 10^{22}$ molecules

E.19 (a) molar mass of $CuBr_2$ = 223.35 $g \cdot mol^{-1}$

$$n_{Cu^{2+}} = \left(\frac{3.00 \text{ g CuBr}_2}{223.35 \text{ g} \cdot mol^{-1} \text{ CuBr}_2} \right) \times \left(\frac{1 \text{ mol Cu}^{2+}}{1 \text{ mol CuBr}_2} \right) = 0.0134 \text{ mol Cu}^{2+}$$

(b) molar mass of SO_3 = 80.06 $g \cdot mol^{-1}$

$$n_{SO_3} = \left(\frac{0.700 \text{ g SO}_3}{80.06 \text{ g} \cdot mol^{-1} \text{ SO}_3} \right) = 8.74 \times 10^{-3} \text{ mol SO}_3$$

(c) molar mass of UF_6 = 352.03 $g \cdot mol^{-1}$

$$n_F = 25.2 \text{ kg} \times \left(\frac{1000 \text{ g}}{1 \text{ kg}} \right) \times \left(\frac{1 \text{ mol UF}_6}{352.03 \text{ g UF}_6} \right) \times \left(\frac{6 \text{ mol F}^-}{1 \text{ mol UF}_6} \right) = 430. \text{ mol F}^-$$

(d) molar mass of $Na_2CO_3 \cdot 10H_2O$ = 286.21 $g \, mol^{-1}$

$$n_{H_2O} = \left(\frac{2.00 \text{ g Na}_2CO_3 \bullet 10H_2O}{286.21 \text{ g} \cdot mol^{-1} \text{ Na}_2CO_3 \bullet 10H_2O} \right) \times \left(\frac{10 \text{ mol H}_2O}{1 \text{ mol Na}_2CO_3 \bullet 10H_2O} \right)$$
$$= 0.0699 \text{ mol H}_2O$$

E.21 (a) number of formula units = $(0.750 \text{ mol})(6.022 \times 10^{23}$ formula units

$\cdot mol^{-1})$

$= 4.52 \times 10^{23}$ formula units

(b) molar mass of $Ag_2SO_4 = 311.80 \ g \cdot mol^{-1}$

$$\left(\frac{2.39 \times 10^{20} \ \text{formula units}}{6.022 \times 10^{23} \ \text{formula units} \cdot mol^{-1}} \right) (311.80 \ g \cdot mol^{-1}) \left(\frac{1000 \ mg}{1 \ g} \right)$$
$= 124 \ mg$

(c) molar mass of $NaHCO_2 = 68.01 \ g \cdot mol^{-1}$

$$\left(\frac{3.429 \ g}{68.01 \ g \cdot mol^{-1}} \right) (6.022 \times 10^{23} \ \text{formula units} \cdot mol^{-1})$$
$= 3.036 \times 10^{22}$ formula units

E.23 (a) molar mass of $H_2O = 18.02 \ g \cdot mol^{-1}$

$$\left(\frac{18.02 \ g \cdot mol^{-1}}{6.022 \times 10^{23} \ \text{molecules} \cdot mol^{-1}} \right) = 2.992 \times 10^{-23} \ g \cdot \text{molecule}^{-1}$$

(b) $N_{H_2O} = \left(\frac{1000 \ g}{18.02 \ g \cdot mol^{-1}} \right) (6.022 \times 10^{23} \ \text{molecules} \cdot mol^{-1})$

$= 3.34 \times 10^{25}$ molecules

E.25 (a) molar mass of $CuCl_2 \cdot 4H_2O = 206.53 \ g \ mol^{-1}$

$$n = \left(\frac{8.61 \ g \ CuCl_2 \bullet 4H_2O}{206.53 \ g \cdot mol^{-1} \ CuCl_2 \bullet 4H_2O} \right) = 0.0417 \ mol \ CuCl_2 \bullet 4H_2O$$

(b) Because there are 2 mol Cl^- per mole of compound, the number of moles will be twice the amount in part (a), 0.0834 mol Cl^-

(c) There are 4 mole of H_2O for each mole compound. Therefore,
Number of water molecules $= (0.0417 \times 4)$ mole $H_2O \times (6.022 \times 10^{23}$ molecules $\cdot \ mol^{-1}) = 1.00 \times 10^{23} \ H_2O$ molecules

(d) fraction of mass due to O $= \dfrac{4(16.00 \ g \cdot mol^{-1})}{206.53 \ g \cdot mol^{-1}} = 0.3099$

E.27 (a) $\left(\dfrac{10\ mol\ H_2O}{1\ mol\ hydrated\ cpd}\right)\left(\dfrac{18.01\ g\ H_2O}{1\ mol\ H_2O}\right)\left(\dfrac{1\ mol\ hydrated\ cpd}{286.148\ g\ hydrated\ cpd}\right)\times 100$

$= 62.94\%\ H_2O$

Therefore 1.6 kg out of 2.5 kg was water.

$\dfrac{?\$}{L\ H_2O} = \left(\dfrac{\$175.00}{2.5\ kg\ hydratedcpd}\right)\left(\dfrac{2.5\ kg\ hydratedcpd}{1.6\ kg\ H_2O}\right)\left(\dfrac{1\ kg\ H_2O}{1\ L\ H_2O}\right)$

$= \$109.28\ per\ L\ H_2O$

(b) Since the anhydrous compound costs $195.00 for 2.5kg, or

$78.00/kg. In the hydrous compound, only 2.5-1.6=0.9 kg is Na_2CO_3,

a fair price would have been $\left(\dfrac{\$78.00}{kg}\right)\times 0.9 = \$70.20$.

E.29 $Zr\ atoms/mole = (90.0000\ g\ ^{90}Zr)\left(\dfrac{1\ atom}{1.4929\ x10^{-22}\ g}\right)$

$= 6.028535\ x\ 10^{23}$ atoms $= 1$ mole

Since there are more atoms per mole, each mole will weigh more and all of the molar masses will be proportionately higher by

$\left(\dfrac{6.028535\,x10^{23} - 6.02214\,x10^{23}}{6.02214\,x10^{23}}\,x100\right) = 0.10619\%$

(a) $^{12}C \rightarrow 12.0000....\ g\,mol^{-1}\ x\ 1.0010619 = 12.013\ g\,mol^{-1}$

(b) $^{196.97}Au \rightarrow 196.97....\ g\,mol^{-1}\ x\ 1.0010619 = 197.18\ g\,mol^{-1}$

E.31 Solve by factor label (dimensional analysis).

$?\ mole\ H\ atoms = 28.0\ cm^3\ NaBH_4\left(\dfrac{1.074\ g}{1\ cm^3}\right)\left(\dfrac{2.50\times10^{23}\ H\ atoms}{3.93\ g}\right)$

$\times\left(\dfrac{1\ mol}{6.022\times10^{23}\ H\ atoms}\right)$

$= 3.18$ moles H atoms

E.33 The molar mass of $Ca_5(PO_4)_3OH$ is 502.31 g/mol;

The molar mass of $Ca_5(PO_4)_3F$ is 504.30 g/mol;

$$\% \text{ increase in mass} = \left(\frac{504.30 - 502.31}{504.30}\right) \times 100 = 0.39\%$$

F.1 $C_{10}H_{16}O$

$10 \times 12.01 \text{ g} = 120.1 \text{ g}$ $\qquad \times \ (\dfrac{100}{152.2264}) = 78.90\%$

$16 \times 1.0079 \text{ g} = 16.1264 \text{ g}$ $\qquad \times \ (\dfrac{100}{152.2264}) = 10.59\%$

$1 \times 16.00 \text{ g} = 16.00 \text{ g}$ $\qquad \times \ (\dfrac{100}{152.2264}) = 10.51\%$

F.3 $C_7H_{15}NO_3$

$7 \times 12.01 \text{ g} = 84.07 \text{ g}$ $\qquad \times \left(\dfrac{100}{161.20 \text{ g}}\right) = 52.15\%$

$15 \times 1.0079 \text{ g} = 15.1185 \text{ g}$ $\quad \times \left(\dfrac{100}{161.20 \text{ g}}\right) = 9.3787\%$

$1 \times 14.01 \text{ g} = 14.01 \text{ g}$ $\qquad \times \left(\dfrac{100}{161.20 \text{ g}}\right) = 8.691\%$

$3 \times 16.00 \text{ g} = \underline{48.00 \text{ g}}$ $\qquad \times \left(\dfrac{100}{161.20 \text{ g}}\right) = \underline{29.78\%}$

$\qquad\qquad\qquad 161.20 \text{ g}$ $\qquad\qquad\qquad\qquad 100.00\%$

F.5   (a) M_2O 88.8% M   For 100 g of compound, 88.8 g is M, 11.2 g is O.

$$\frac{? \text{ g } M_2O}{\text{mole } M_2O} = \left(\frac{100 \text{ g } M_2O}{11.2 \text{ g O}}\right)\left(\frac{16.00 \text{ g O}}{1 \text{ mol O}}\right)\left(\frac{1 \text{ mol O}}{1 \text{ mol } M_2O}\right)$$

$$= 143 \text{ g} \cdot \text{mol}^{-1} \ M_2O$$

Therefore, 143-16=127 g/mol are due to M. Since there are 2 moles of M per mole of M_2O, the molar mass of M=63.4 g/mol. That molar mass matches Cu. (b) copper(I) oxide

F.7   (a) For 100 g of compound,

$$\text{moles of Na} = \frac{32.79 \text{ g}}{22.99 \text{ g} \cdot \text{mol}^{-1}} = 1.426 \text{ mol}$$

$$\text{moles of Al} = \frac{13.02 \text{ g}}{26.98 \text{ g} \cdot \text{mol}^{-1}} = 0.4826 \text{ mol}$$

$$\text{moles of F} = \frac{54.19 \text{ g}}{19.00 \text{ g} \cdot \text{mol}^{-1}} = 2.852 \text{ mol}$$

Dividing each number by 0.4826 gives a ratio of 1 Al : 2.95 Na : 5.91 F.
The formula is Na_3AlF_6.

(b) For 100 g of compound,

$$\text{moles of K} = \frac{31.91 \text{ g}}{39.10 \text{ g} \cdot \text{mol}^{-1}} = 0.8161 \text{ mol}$$

$$\text{moles of Cl} = \frac{28.93 \text{ g}}{35.45 \text{ g} \cdot \text{mol}^{-1}} = 0.8161 \text{ mol}$$

mass of O is obtained by difference:

$$\text{moles of O} = \frac{100 \text{ g} - 31.91 \text{ g} - 28.93 \text{ g}}{16.00 \text{ g} \cdot \text{mol}^{-1}} = 2.448 \text{ mol}$$

Dividing each number by 0.8161 gives a ratio of 1.00 K : 1 Cl : 3.00 O.
The formula is $KClO_3$.

(c) For 100 g of compound,

$$\text{moles of N} = \frac{12.2 \text{ g}}{14.01 \text{ g} \cdot \text{mol}^{-1}} = 0.871 \text{ mol}$$

$$\text{moles of H} = \frac{5.26 \text{ g}}{1.0079 \text{ g} \cdot \text{mol}^{-1}} = 5.22 \text{ mol}$$

$$\text{moles of P} = \frac{26.9 \text{ g}}{30.97 \text{ g} \cdot \text{mol}^{-1}} = 0.869 \text{ mol}$$

$$\text{moles of O} = \frac{55.6 \text{ g}}{16.00 \text{ g} \cdot \text{mol}^{-1}} = 3.475 \text{ mol}$$

Dividing each number by 0.869 gives a ratio of 1.00 N : 6.01 H : 1.00 P :
4.00 O. The formula is NH_6PO_4 or $[NH_4][H_2PO_4]$, ammonium
dihydrogen phosphate.

F.9   (a) moles of P $= \dfrac{4.14\ \text{g}}{30.97\ \text{g} \cdot \text{mol}^{-1}} = 0.134\ \text{mol}$

moles of Cl $= \dfrac{27.8\ \text{g} - 4.14\ \text{g}}{35.45\ \text{g} \cdot \text{mol}^{-1}} = 0.667\ \text{mol}$

Dividing each number by 0.134 mol gives a ratio of 4.98 Cl : 1 P. The formula is PCl_5.

(b) Name is Phosphorus pentachloride.

F.11   For 100 g of compound,

Moles of C $= \left(\dfrac{67.49\ g}{12.01\ g \cdot mol^{-1}} \right) = 5.619\ \text{mol}$

Moles of H $= \left(\dfrac{4.60\ g}{1.008\ g \cdot mol^{-1}} \right) = 4.563\ \text{mol}$

Moles of Cl $= \left(\dfrac{12.45\ g}{35.45\ g \cdot mol^{-1}} \right) = 0.3512\ \text{mol}$

Moles of N $= \left(\dfrac{9.84\ g}{14.01\ g \cdot mol^{-1}} \right) = 0.7024\ \text{mol}$

Moles of O $= \left(\dfrac{5.62\ g}{16.00\ g \cdot mol^{-1}} \right) = 0.3512\ \text{mol}$

Dividing each number by 0.3512 mol gives a ratio of 16.00 C : 13.00 H : 1.00 Cl : 2.00N : 1.00 O. The formula is $C_{16}H_{13}ClN_2O$

F.13   For 100 g of the osmium carbonyl compound,

moles of C $= \dfrac{15.89\ \text{g}}{12.01\ \text{g} \cdot \text{mol}^{-1}} = 1.323\ \text{mol}$

moles of O $= \dfrac{21.18\ \text{g}}{16.00\ \text{g} \cdot \text{mol}^{-1}} = 1.324\ \text{mol}$

moles of Os $= \dfrac{62.93\ \text{g}}{190.2\ \text{g} \cdot \text{mol}^{-1}} = 0.3309\ \text{mol}$

Dividing each number by 0.3309 mol gives a ratio of 4.00 C : 4.00 O : 1.00 Os. (a) The empirical formula is OsC_4O_4. (b) The formula mass of OsC_4O_4 is 302.24 g·mol^{-1}. The molar mass is 907 g·mol^{-1} which is 3 times the formula mass, so the molecular formula is $Os_3C_{12}O_{12}$.

F.15 For 100 g of caffeine,

$$\text{moles of C} = \frac{49.48\text{ g}}{12.01\text{ g·mol}^{-1}} = 4.12\text{ mol}$$

$$\text{moles of H} = \frac{5.19\text{ g}}{1.0079\text{ g·mol}^{-1}} = 5.15\text{ mol}$$

$$\text{moles of N} = \frac{28.85\text{ g}}{14.01\text{ g·mol}^{-1}} = 2.059\text{ mol}$$

$$\text{moles of O} = \frac{16.48\text{ g}}{16.00\text{ g·mol}^{-1}} = 1.03\text{ mol}$$

Dividing each number by 1.03 mol gives a ratio of 4.00 C : 5.00 H : 2.00 N : 1.00 O. The formula is $C_4H_5N_2O$ with a molar formula mass of 97.10 g·mol^{-1}. Because the molecular molar mass is twice this value, the actual formula will be $C_8H_{10}N_4O_2$.

F.17 For the 1.78 mg sample of didemnin-A,

$$\text{mmoles of C} = \frac{1.11\text{ mg}}{12.01\text{ g·mol}^{-1}} = 0.0924\text{ mmol}$$

$$\text{mmoles of H} = \frac{0.148\text{ mg}}{1.0079\text{ g·mol}^{-1}} = 0.147\text{ mmol}$$

$$\text{mmoles of N} = \frac{0.159\text{ mg}}{14.01\text{ g·mol}^{-1}} = 0.0113\text{ mmol}$$

$$\text{mmoles of O} = \frac{0.363\text{ mg}}{16.00\text{ g·mol}^{-1}} = 0.0227\text{ mmol}$$

Dividing by 0.0113 mmol gives 8.18 C : 13.0 H : 1.00 N : 2.00 O.

Multiplying by 7 gives the formula $C_{49}H_{78}N_6O_{12}$ with a molar mass of 942 g·mol^{-1}.

F.19 Calculate the mass percent carbon for each fuel from its formula.

ethene, C_2H_4 $\dfrac{2(12.01)}{2(12.01)+4(1.0079)} \times 100 = 85.63\%$ C

propanol, C_3H_7OH $\dfrac{3(12.01)}{3(12.01)+8(1.0079)+16.00} \times 100 = 59.96\%$ C

heptane, C_7H_{16} $\dfrac{7(12.01)}{7(12.01)+16(1.0079)} \times 100 = 83.91\%$ C

ethene (85.63%) > heptane (83.91%) > propanol (59.96%)

F.21 (a) empirical formula: C_2H_3Cl, molecular formula: $C_4H_6Cl_2$

(b) empirical formula: CH_4N, molecular formula: $C_2H_8N_2$

F.23 This problem requires that we relate unknowns to each other appropriately by writing a balanced chemical equation and using other information in the problem.

$$x\ NaNO_3 + y\ Na_2SO_4 \rightarrow (x+2y)\ Na^+ + x\ NO_3^- + y\ SO_4^{2-}$$

$$\frac{1.61\ g\ Na^+}{22.99\ g \cdot mol^{-1}Na^+} = 0.07003\ mol\ Na^+ = x + 2y$$

5.37 g total-1.61 g Na$^+$ =3.76 g =(62.01 g·mol^{-1})x +(96.07 g·mol^{-1})y

Rearrange and substitute:

3.76=62.01(0.07003-2y)+96.07y

0.06065 = 0.07003 − 2y + 1.548y

0.009385 = 0.4516y

y = 0.02078 moles sulfate, x=0.02847 moles nitrate

Therefore, the mass of sodium nitrate in the mixture was

$$0.02847 \text{ mol} \left(\frac{85.00 \text{ g NaNO}_3}{1 \text{ mol}} \right) = 2.42 \text{ g NaNO}_3$$

$$\frac{2.42 \text{ g NaNO}_3}{5.37 \text{ g total}} \times 100 = 45.1\% \text{ NaNO}_3$$

G.1 (a) False. They can only be separated by chemical methods.

(b) False. The composition of a solution is fixed.

(c) False. The properties of a compound are unlike those of the elements that compose it.

G.3 (a) heterogeneous, decanting; (b) heterogeneous, dissolving followed by filtration and distillation; (c) homogeneous, distillation

G.5 mass of K_2NO_3 = 510. g × 5.45% = 27.8 g K_2NO_3.

Mass of H_2O = 510. g – 27.8 g = 482.2 g H_2O

Measuring 482.2 g H_2O on a balance and pure it into a beaker. Then you weigh 27.8 g of K_2NO_3 and mix it with the water until it totally dissolve.

G.7 mass of $AgNO_3$ = $(0.179 \text{ mol} \cdot \text{L}^{-1})(0.5000 \text{ L})(169.88 \text{ g} \cdot \text{mol}^{-1}) = 15.2$ g

G.9 (a) molarity of $Na_2CO_3 = \dfrac{2.111 \text{ g}}{(105.99 \text{ g} \cdot \text{mol}^{-1})(0.2500 \text{ L})} = 0.079\ 67$ M

Na_2CO_3

$$V = \frac{(2.15 \times 10^{-3} \text{ mol Na}^+ (1 \text{ mol Na}_2\text{CO}_2)}{(0.079\ 67 \text{ mol} \cdot \text{L}^{-1} \text{ Na}_2\text{CO}_3)(2 \text{ mol Na}^+)} = 1.35 \times 10^{-2} \text{ L or 13.5 mL}$$

(b) $V = \dfrac{(4.98 \times 10^{-3} \text{ mol CO}_3^{2-})(1 \text{ mol Na}_2\text{CO}_2)}{(0.079\ 67 \text{ mol} \cdot \text{L}^{-1} \text{ Na}_2\text{CO}_3)(1 \text{ mol CO}_3^{2-})}$

$= 6.25 \times 10^{-2}$ L or 62.5 mL

(c) $V = \dfrac{(50.0 \times 10^{-3} \text{ g Na}_2\text{CO}_3)}{(105.99 \text{ g} \cdot \text{mol}^{-1})(0.079\ 67 \text{ mol} \cdot \text{L}^{-1} \text{ Na}_2\text{CO}_3)}$

$= 5.92 \times 10^{-3}$ L or 5.92 mL

G.11 (a) Weigh 1.6 g (0.010 mol, molar mass of $KMnO_4 = 158.04 \ g \cdot mol^{-1}$) into a 1.0-L volumetric flask and add water to give a total volume of 1.0 L. Smaller (or larger) volumes could also be prepared by using a proportionally smaller (or larger) mass of $KMnO_4$.

(b) Starting with $0.050 \ mol \cdot L^{-1} \ KMnO_4$, add four volumes of water to one volume of starting solution, because the concentration desired is one-fifth of the starting solution. This relation can be derived from the expression

$$V_i \times molarity_i = V_f \times molarity_f$$

where i represents the initial solution and f the final solution. But $V_f = V_i + V_d$ where V_d represents the volume of solvent that must be added to dilute the initial solution. Rearranging the first equation gives

$$\frac{V_i}{V_f} = \frac{molarity_f}{molarity_i}$$

$$\frac{V_i}{V_i + V_d} = \frac{molarity_f}{molarity_i}$$

So if the ratio of final molarity to initial molarity is 1 : 5, we can write

$$\frac{V_i}{V_i + V_d} = \frac{1}{5}$$
$$5V_i = V_i + V_d$$
$$4V_i = V_d$$

For example, to prepare 50 mL of solution, you would add 40 mL of water to 10 mL of $0.050 \ mol \cdot L^{-1} \ KMnO_4$.

G.13 (a) $V(0.778 \ mol \cdot L^{-1}) = (0.1500 \ L)(0.0234 \ mol \cdot L^{-1})$

$V = 4.51 \times 10^{-3} \ L$ or 4.51 mL

(b) The concentration desired is one-fifth of the starting NaOH solution, so the stockroom attendant will need to add four volumes of water to one volume of the $2.5 \text{ mol} \cdot \text{L}^{-1}$ solution. To prepare 60.0 mL of solution, divide 60.0 by 5; so 12.0 mL of $2.5 \text{ mol} \cdot \text{L}^{-1}$ NaOH solution are added to 48.0 mL of water. (See the solution to G.11)

G.15 (a) mass of $CuSO_4 = (0.20 \text{ mol} \cdot \text{L}^{-1})(0.250 \text{ L})(159.60 \text{ g} \cdot \text{mol}^{-1})$
$$= 8.0 \text{ g}$$

(b) mass of $CuSO_4 \cdot 5 H_2O = (0.20 \text{ mol} \cdot \text{L}^{-1})(0.250 \text{ L})(249.68 \text{ g} \cdot \text{mol}^{-1})$
$$= 12 \text{ g}$$

G.17 (a) Total moles of K^+:

$$\left(\frac{0.500 \, g \, KCl}{74.55 \, g \cdot mol^{-1}} \right) + \left(\frac{0.500 \, g \, K_2S}{110.26 \, g \cdot mol^{-1}} \right) \left(\frac{2 \, mol \, K^+}{1 \, mol \, K_2S} \right) +$$

$$\left(\frac{0.500 \, g \, K_3PO_4}{212.27 \, g \cdot mol^{-1}} \right) \left(\frac{3 \, mol \, K^+}{1 \, mol \, K_3PO_4} \right)$$

$$= 6.71 \times 10^{-3} \text{ mol} + 9.07 \times 10^{-3} \text{ mol} + 7.07 \times 10^{-3} \text{ mol} = 2.29 \times 10^{-2} \text{ mol}$$

Molarity of $K^+ = \left(\frac{2.29 \times 10^{-2} \, mol \, K^+}{0.500 L} \right) = 4.58 \times 10^{-2} \text{ M}$

(b) molarity of S^{2-}:

$$\left(\frac{0.500 \, g \, K_2S}{110.26 \, g \cdot mol^{-1}} \right) \left(\frac{1 \, mol \, S^{2-}}{1 \, mol \, K_2S} \right) \left(\frac{1}{0.500 \, L} \right) = 9.07 \times 10^{-3} \text{ M}$$

G.19 (a) mass of $K_2SO_4 = (0.125 \text{ mol} \cdot \text{L}^{-1})(1.00 \text{ L})(174.26 \text{ g} \cdot \text{mol}^{-1}) = 21.8 \text{ g}$

(b) mass of $NaF = (0.015 \text{ mol} \cdot \text{L}^{-1})(0.375 \text{ L})(41.99 \text{ g} \cdot \text{mol}^{-1}) = 0.24 \text{ g}$

(c) mass of $C_{12}H_{22}O_{11} = (0.35 \text{ mol} \cdot \text{L}^{-1})(0.500 \text{ L})(342.29 \text{ g} \cdot \text{mol}^{-1})$
$$= 60. \text{ g}$$

G.21 The total molar concentration of Cl^-:

$$\left[\left(\frac{0.50 \, g \, NaCl}{58.44 \, g \cdot mol^{-1}} \right) + \left(\frac{0.30 \, g \, KCl}{74.55 \, g \cdot mol^{-1}} \right) \right] \times \left(\frac{1}{0.100 \, L} \right) = 0.13 \text{ M } Cl^-$$

G.23 We can show that fewer than 1 molecule of X would be left after only 70 doublings.

$$10 \text{ mL} \times \frac{0.10 \text{ mol}}{1000 \text{ mL}} \times \frac{6.02214 \times 10^{23} \text{ molecules}}{1 \text{ mol}} = 6.0 \times 10^{20} \text{ molecules are}$$

in the first 10 mL aliquot of the solution. In order to find the number of times the volume must be doubled to get to one molecule, we can solve for the number of times this amount of molecules must be cut in half until it equals 1.

$$6.0 \times 10^{20} \text{ molecules} \times \left(\frac{1}{2}\right)^n = 1 \text{ molecule}$$

$$\log(6.0 \times 10^{20}) + n \log\left(\frac{1}{2}\right) = \log(1)$$

$$20.8 + n(-0.301) = 0$$

$$20.8 = 0.301n$$

$$n = 69$$

The other 21 additional doublings involve solutions with no X remaining. There can be no health benefits if there are no molecules of the active substance, X, left in the solution.

G.25 Find the volume of concentrated HCl solution that is equivalent to 10.0 L of the dilute HCl solution with respect to the number of moles of solute present in each volume.

? mL con. HCl(aq)

$$= 10.0 \text{ L dilute HCl(aq)} \left(\frac{0.7436 \text{ mol HCl}}{1 \text{ L dilute HCl(aq)}}\right) \left(\frac{36.46 \text{ g HCl}}{1 \text{ mol HCl}}\right)$$

$$\times \left(\frac{100 \text{ g con. HCl(aq)}}{37.50 \text{ g HCl}}\right) \left(\frac{1 \text{ cm}^3}{1.205 \text{ g}}\right)$$

$$= 600. \text{ mL con. HCl}$$

Therefore 600. mL of concentrated HCl(aq) must be diluted up to a final volume of 10.0 L by adding water in order to form 0.7436 M HCl(aq).

G.27 The level of instrument can detect:

$$\left(\frac{1.00 \times 10^{-3} \; mol \; Pb}{1.00 \; L \; solution}\right)\left(\frac{207.2 \; g \; Pb}{mol \; Pb}\right)\left(\frac{1 \; L \; solution}{1000 \; mL \; solution}\right)$$

$$\left(\frac{1.00 \; mL \; solution}{1.00 \; g \; solution}\right) \times 1.00 \times 10^9 \; solution = 2.07 \; ppb < 10 \; ppb$$

The instrument can detect lower than 10 ppb. It is satisfactory.

H.1 (a) You cannot add a different compound or element to the chemical equation which is not produced (or involved) in the chemical reaction. In this case, O atom is not produced by the defined reaction.

(b) $2 \; Cu \; + \; SO_2 \;\; \rightarrow \;\; 2 \; CuO \; + \; S$

H.3 $2 \; SiH_4 \; + \; 4 \; H_2O \;\; \rightarrow \;\; 2 \; SiO_2 \; + \; 8 \; H_2$

H.5 (a) $NaBH_4(s) \; + \; 2 \; H_2O(l) \; \rightarrow \; NaBO_2(aq) \; + \; 4 \; H_2(g)$

(b) $Mg(N_3)_2(s) \; + \; 2 \; H_2O(l) \; \rightarrow \; Mg(OH)_2(aq) \; + \; 2 \; HN_3(aq)$

(c) $2 \; NaCl(aq) + SO_3(g) \; + H_2O(l) \rightarrow Na_2SO_4(aq) + 2 \; HCl(aq)$

(d) $4 \; Fe_2P(s) + 18 \; S(s) \rightarrow P_4S_{10}(s) + 8 \; FeS(s)$

H.7 (a) $Ca(s) \; + 2 \; H_2O(l) \; \rightarrow \; H_2(g) \; + \; Ca(OH)_2(aq)$

(b) $Na_2O(s) + H_2O(l) \rightarrow 2 \; NaOH(aq)$

(c) $3 \; Mg(s) \; + \; N_2(g) \; \rightarrow \; Mg_3N_2(s)$

(d) $4NH_3(g) \; + \; 7 \; O_2(g) \; \rightarrow \; 6 \; H_2O(g) \; + 4 \; NO_2(g)$

H.9 (I) $3 \; Fe_2O_3(s) + CO(g) \rightarrow 2 \; Fe_3O_4(s) + CO_2(g)$

(II) $Fe_3O_4(s) + 4 \; CO(g) \rightarrow 3 \; Fe(s) + 4 \; CO_2(g)$

H.11 (I) $N_2(g) + O_2(g) \rightarrow 2 \; NO(g)$

(II) $2 \; NO(g) + O_2(g) \rightarrow 2 \; NO_2(g)$

H.13 $4\,HF(aq) + SiO_2(s) \rightarrow SiF_4(aq) + 2\,H_2O(l)$

H.15 $C_7H_{16}(l) + 11\,O_2(g) \rightarrow 7\,CO_2(g) + 8\,H_2O(g)$

H.17 $C_{14}H_{18}N_2O_5(s) + 16\,O_2(g) \rightarrow 14\,CO_2(g) + 9\,H_2O(l) + N_2(g)$

H.19 $2\,C_{10}H_{15}N(s) + 26\,O_2(g) \rightarrow 19\,CO_2(g) + 13\,H_2O(l) + CH_4N_2O(aq)$

H.21 (I) $H_2S(g) + 2\,NaOH(s) \rightarrow Na_2S(aq) + 2\,H_2O(l)$

(II) $4\,H_2S(g) + Na_2S(alc) \rightarrow Na_2S_5(alc) + 4\,H_2(g)$

(III)

$2\,Na_2S_5(alc) + 9\,O_2(g) + 10\,H_2O(l) \rightarrow 2\,Na_2S_2O_3 \cdot 5\,H_2O(s) + 6\,SO_2(g)$

H.23 (a) We can find the empirical formulas from the percent compositions.

First oxide:

P 43.64 g ÷ 30.97 g/mol = 1.409 mol

O 56.36 g ÷ 16.00 g/mol = 3.523 mol

$\dfrac{3.523\ \text{mol}}{1.409\ \text{mol}} = 2.500$ 1:2.5 or 2:5 P_2O_5

Second oxide:

P 56.34 g ÷ 30.97 g/mol = 1.819 mol

O 43.66 g ÷ 16.00 g/mol = 2.729 mol

$\dfrac{2.729\ \text{mol}}{1.819\ \text{mol}} = 1.500$ 1:1.5 or 2:3 P_2O_3

These empirical formulas could be named diphosphorus pentoxide and diphosphorus trioxide. The names according to the Stock system are given below with the formulas of the actual compounds.

(b) P_4O_{10} (phosphorus (V) oxide), P_4O_6 (phosphorus (III) oxide);

Since the molar masses of the empirical formulas are 142 g/mol and 110 g/mol respectively, and these masses are both half as big as the molar masses of the compounds, both molecular formulas are twice the empirical formulas.

(c) $P_4(s) + 3\ O_2(g) \rightarrow P_4O_6(s);\ \ P_4(s) + 5\ O_2(g) \rightarrow P_4O_{10}(s)$

I.1 The picture would show a precipitate, $CaSO_4(s)$, at the bottom of the flask composed of red and pink particles in a 1:1 ratio. Sodium (black) and chloride (blue) ions, $NaCl(aq)$, would remain throughout the solution.

I.3 You should use the reagent to only react with one of the ions and form precipitate, so that two ions can be separated.

(a) Diluted H_2SO_4 solution will be used as reagent.

$$Pb^{2+}(aq)\ +\ SO_4^{2-}(aq)\ \rightarrow\ PbSO_4(s)$$

(b) H_2S solution will also be used as reagent.

$$Mg^{2+}(aq)\ +\ S^{2-}(aq)\ \rightarrow\ MgS(s)$$

I.5 (a) CH_3OH, nonelectrolyte;   (b) $BaCl_2$, strong electrolyte;   (c) KF, strong electrolyte

I.7 (a) soluble;   (b) slightly soluble;   (c) insoluble;   (d) insoluble

I.9 (a) $Na^+(aq)$ and $I^-(aq)$;   (b)

$Ag^+(aq)$ and $CO_3^{2-}(aq)$, Ag_2CO_3 is insoluble. The very small amount that does go into solution will be present as Ag^+ and CO_3^{2-} ions.

(c) $NH_4^+(aq)$ and $PO_4^{3-}(aq)$ (d) $Fe^{2+}(aq)$ and $SO_4^{2-}(aq)$

I.11 (a) $Fe(OH)_3$, precipitate; (b) Ag_2CO_3, precipitate forms; (c) No precipitate will form because all possible products are soluble in water.

I.13 (a) net ionic equation: $Fe^{2+}(aq) + S^{2-}(aq) \rightarrow FeS(s)$;

spectator ions: Na^+, Cl^-

(b) net ionic equation: $Pb^{2+}(aq) + 2\,I^-(aq) \rightarrow PbI_2(s)$;

spectator ions: K^+, NO_3^-

(c) net ionic equation: $Ca^{2+}(aq) + SO_4^{2-}(aq) \rightarrow CaSO_4(s)$;

spectator ions: NO_3^-, K^+

(d) net ionic equation: $Pb^{2+}(aq) + CrO_4^{2-}(aq) \rightarrow PbCrO_4(s)$;

spectator ions: Na^+, NO_3^-

(e) net ionic equation: $Hg_2^{2+}(aq) + SO_4^{2-}(aq) \rightarrow Hg_2SO_4(s)$;

spectator ions: K^+, NO_3^-

I.15 (a) overall equation: $(NH_4)_2CrO_4(aq) + BaCl_2(aq)$
$\rightarrow BaCrO_4(s) + 2\,NH_4Cl(aq)$

complete ionic equation:

$2\,NH_4^+(aq) + CrO_4^{2-}(aq) + Ba^{2+}(aq) + 2\,Cl^-(aq)$
$\rightarrow BaCrO_4(s) + 2\,NH_4^+(aq) + 2\,Cl^-(aq)$

net ionic equation: $Ba^{2+}(aq) + CrO_4^{2-}(aq) \rightarrow BaCrO_4(s)$;

spectator ions: NH_4^+, Cl^-

(b) $CuSO_4(aq) + Na_2S(aq) \rightarrow CuS(s) + Na_2SO_4(aq)$

complete ionic equation:

$Cu^{2+}(aq) + SO_4^{2-}(aq) + 2\,Na^+(aq) + S^{2-}(aq)$
$\rightarrow CuS(s) + 2\,Na^+(aq) + SO_4^{2-}(aq)$

net ionic equation: $Cu^{2+}(aq) + S^{2-}(aq) \rightarrow CuS(s)$;

spectator ions: Na^+, SO_4^{2-}

(c) $3 \, FeCl_2(aq) + 2 \, (NH_4)_3PO_4(aq)$

$\rightarrow Fe_3(PO_4)_2(s) + 6 \, NH_4Cl(aq)$

complete ionic equation:

$3 \, Fe^{2+}(aq) + 6 \, Cl^-(aq) + 6 \, NH_4^+(aq) + 2 \, PO_4^{3-}(aq)$

$\rightarrow Fe_3(PO_4)_2(s) + 6 \, NH_4^+(aq) + 6 \, Cl^-(aq)$

net ionic equation: $3 \, Fe^{2+}(aq) + 2 \, PO_4^{3-}(aq) \rightarrow Fe_3(PO_4)_2(s)$;

spectator ions: Cl^-, NH_4^+

(d) $K_2C_2O_4(aq) + Ca(NO_3)_2(aq) \rightarrow CaC_2O_4(s) + 2 \, KNO_3(aq)$

complete ionic equation:

$2 \, K^+(aq) + C_2O_4^{2-}(aq) + Ca^{2+}(aq) + 2 \, NO_3^-(aq)$

$\rightarrow CaC_2O_4(s) + 2 \, K^+(aq) + 2 \, NO_3^-(aq)$

net ionic equation: $Ca^{2+}(aq) + C_2O_4^{2-}(aq) \rightarrow CaC_2O_4(s)$;

spectator ions: K^+, NO_3^-

(e) $NiSO_4(aq) + Ba(NO_3)_2(aq) \rightarrow Ni(NO_3)_2(aq) + BaSO_4(s)$

complete ionic equation:

$Ni^{2+}(aq) + SO_4^{2-}(aq) + Ba^{2+}(aq) + 2 \, NO_3^-(aq)$

$\rightarrow Ni^{2+}(aq) + 2 \, NO_3^-(aq) + BaSO_4(s)$

net ionic equation: $Ba^{2+}(aq) + SO_4^{2-}(aq) \rightarrow BaSO_4(s)$;

spectator ions: Ni^{2+}, NO_3^-

I.17 (a) $Ba(CH_3CO_2)_2(aq) + Li_2CO_3(aq) \rightarrow BaCO_3(s) + 2 \, LiCH_3CO_2(aq)$

$Ba^{2+}(aq) + 2 \, CH_3CO_2^-(aq) + 2 \, Li^+(aq) + CO_3^{2-}(aq) \rightarrow$

$BaCO_3(s) + 2 \, Li^+(aq) + 2CH_3CO_2^-(aq)$

net ionic equation: $Ba^{2+}(aq) + CO_3^{2-}(aq) \rightarrow BaCO_3(s)$

(b) $2 \, NH_4Cl(aq) + Hg_2(NO_3)_2(aq) \rightarrow 2 \, NH_4NO_3(aq) + Hg_2Cl_2(s)$

$2 \, NH_4^+(aq) + 2 \, Cl^-(aq) + Hg_2^{2+}(aq) + 2 \, NO_3^-(aq) \rightarrow$

$Hg_2Cl_2(s) + 2\ NH_4^+(aq) + 2\ NO_3^-(aq)$

net ionic equation: $Hg_2^{2+}(aq) + 2\ Cl^-(aq) \rightarrow Hg_2Cl_2(s)$

(c) $Cu(NO_3)_2(aq) + Ba(OH)_2(aq) \rightarrow Cu(OH)_2(s) + Ba(NO_3)_2(aq)$

$Cu^{2+}(aq) + 2\ NO_3^-(aq) + Ba^{2+}(aq) + 2\ OH^-(aq) \rightarrow$

$Cu(OH)_2(s) + Ba^{2+}(aq) + 2\ NO_3^-(aq)$

net ionic equation: $Cu^{2+}(aq) + 2\ OH^-(aq) \rightarrow Cu(OH)_2(s)$

I.19   (a) $AgNO_3$ and Na_2CrO_4

     (b) $CaCl_2$ and Na_2CO_3

     (c) $Cd(ClO_4)_2$ and $(NH_4)_2S$

I.21   (a) $2\ Ag^+(aq) + SO_4^{2-}(aq) \rightarrow Ag_2SO_4(s)$

     (b) $Hg^{2+}(aq) + S^{2-}(aq) \rightarrow HgS(s)$

     (c) $3\ Ca^{2+}(aq) + 2\ PO_4^{3-}(aq) \rightarrow Ca_3(PO_4)_2(s)$

     (d) $AgNO_3$ and Na_2SO_4; Na^+, NO_3^-;

     $Hg(CH_3CO_2)_2$ and Li_2S; Li^+, $CH_3CO_2^-$

     $CaCl_2$ and K_3PO_4; K^+, Cl^-

I.23   white ppt. = $AgCl(s)$, Ag^+; no ppt. with H_2SO_4, no Ca^{2+};
     black ppt.=ZnS, Zn^{2+}

I.25   (a) $2\ NaOH(aq) + Cu(NO_3)_2(aq) \rightarrow Cu(OH)_2(s) + 2\ NaNO_3(aq)$

     complete ionic equation:

       $2\ Na^+(aq) + 2\ OH^-(aq) + Cu^{2+}(aq) + 2\ NO_3^-(aq)$

         $\rightarrow Cu(OH)_2(s) + 2\ Na^+(aq) + 2\ NO_3^-(aq)$

     net ionic equation: $Cu^{2+}(aq) + 2\ OH^-(aq) \rightarrow Cu(OH)_2(s)$

(b) $M_{Na^+} = \left(\dfrac{40.0 \text{ mL} \times 0.100 \text{ M}}{50.0 \text{ mL}} \right) = 0.0800$ M

I.27 (a) Find the number of moles of potassium chromate per liter of solution.

$$\left(\frac{? \text{ mol } K_2CrO_4}{L \text{ solution}} \right)\left(\frac{5.20 \text{ g } K_2CrO_4}{75.0 \text{ mL solution}} \right)\left(\frac{1 \text{ mol } K_2CrO_4}{194.20 \text{ g } K_2CrO_4} \right)\left(\frac{1000 \text{ mL}}{1 \text{ L}} \right)$$

$$= 0.357 \text{ M } K_2CrO_4$$

(b) Find the number of grams of potassium in the amount of potassium chromate that was dissolved.

$? \text{ g } K^+ = 5.20 \text{ g } K_2CrO_4$

$$\times \left(\frac{1 \text{ mol } K_2CrO_4}{194.20 \text{ g } K_2CrO_4} \right)\left(\frac{2 \text{ mol } K^+}{1 \text{ mol } K_2CrO_4} \right)\left(\frac{39.098 \text{ g } K^+}{1 \text{ mol } K^+} \right) = 2.09 \text{ g } K^+$$

(c) $MgCrO_4$

J.1 (a) base; (b) acid; (c) base; (d) acid; (e) base

J.3 (a) overall equation: $HF(aq) + NaOH(aq) \rightarrow NaF(aq) + H_2O(l)$

complete ionic equation:

$HF(aq) + Na^+(aq) + OH^-(aq) \rightarrow Na^+(aq) + F^-(aq) + H_2O(l)$

net ionic equation:　　$HF(aq) + OH^-(aq) \rightarrow F^-(aq) + H_2O(l)$

(b) overall equation: $(CH_3)_3N(aq) + HNO_3(aq) \rightarrow (CH_3)_3NHNO_3(aq)$

complete ionic equation:
$(CH_3)_3N(aq) + H_3O^+(aq) + NO_3^-(aq)$
$\rightarrow (CH_3)_3NH^+(aq) + NO_3^-(aq) + H_2O(l)$

net ionic equation:

$(CH_3)_3N(aq) + H_3O^+(aq) \rightarrow (CH_3)_3NH^+(aq) + H_2O(l)$

(c) overall equation: $LiOH(aq) + HI(aq) \rightarrow LiI(aq) + H_2O(l)$

complete ionic equation:

$$Li^+(aq) + OH^-(aq) + H_3O^+(aq) + I^-(aq) \rightarrow Li^+(aq) + I^-(aq) + 2\,H_2O(l)$$

net ionic equation: $OH^-(aq) + H_3O^+(aq) \rightarrow 2\,H_2O(l)$

J.5 (a) $HBr(aq) + KOH(aq) \rightarrow KBr(aq) + H_2O(l)$

(b) $Zn(OH)_2(aq) + 2\,HNO_2(aq) \rightarrow Zn(NO_2)_2(aq) + 2\,H_2O(l)$

(c) $Ca(OH)_2(aq) + 2\,HCN(aq) \rightarrow Ca(CN)_2(aq) + 2\,H_2O(l)$

(d) $3\,KOH(aq) + H_3PO_4(aq) \rightarrow K_3PO_4(aq) + 3\,H_2O(l)$

J.7 (a) acid: $H_3O^+(aq)$; base: $CH_3NH_2(aq)$; (b) acid: $HCl(aq)$;

base: $C_2H_5NH_2(aq)$;

(c) acid: $HI(aq)$; base: $CaO(s)$

J.9 Since X turns litmus red and conducts electricity poorly, it is a weak acid.
We can find the empirical formula from the percent composition.

C 26.68 g $\div$ 12.01 g/mol = 2.221 mol

H 2.239 g $\div$ 1.0079 g/mol = 2.221 mol

O 71.081 g $\div$ 16.00 g/mol = 4.443 mol

So the subscripts are 1:1:2 on the empirical formula.

(a) CHO_2;

(b) Since the molar mass of the empirical formula is 45.0 g·mol^{-1} while the molar mass of X is 90.0 g·mol^{-1}, the molecular formula is twice the empirical formula or $C_2H_2O_4$.

(c) The weak acid whose formula matches the one given in part (b) is oxalic acid.

$$(COOH)_2(aq) + 2\,NaOH(aq) \rightarrow Na_2C_2O_4(aq) + 2\,H_2O(l)$$

net ionic equation: $(COOH)_2(aq) + 2\,OH^- \rightarrow C_2O_4^{2-}(aq) + 2\,H_2O(l)$

J.11 (a) $C_6H_5O^-(aq) + H_2O(l) \rightarrow C_6H_5OH(aq) + OH^-(aq)$

(b) $ClO^-(aq) + H_2O(l) \rightarrow HClO(aq) + OH^-(aq)$

(c) $C_5H_5NH^+(aq) + H_2O(l) \rightarrow C_5H_5N(aq) + H_3O^+(aq)$

(d) $NH_4^+(aq) + H_2O(l) \rightarrow NH_3(aq) + H_3O^+(aq)$

J.13 (a) $AsO_4^{3-}(aq) + H_2O(l) \rightarrow HAsO_4^{2-}(aq) + OH^-(aq)$

$HAsO_4^{2-}(aq) + H_2O(l) \rightarrow H_2AsO_4^-(aq) + OH^-(aq)$

$H_2AsO_4^-(aq) + H_2O(l) \rightarrow H_3AsO_4(aq) + OH^-(aq)$

In each equation, H_2O is the acid.

(b)

$$? \text{ mol Na}^+ = 35.0 \text{ g Na}_3\text{AsO}_4\left(\frac{1 \text{ mol Na}_3\text{AsO}_4}{207.89 \text{ g Na}_3\text{AsO}_4}\right)$$
$$\times\left(\frac{3 \text{ mol Na}^+}{1 \text{ mol Na}_3\text{AsO}_4}\right)$$
$$= 0.505 \text{ mol Na}^+$$

J.15 (a) $CO_2(g) + H_2O(l) \rightarrow H_2CO_3(aq)$ (carbonic acid)

(b) $SO_3(g) + H_2O(l) \rightarrow H_2SO_4(aq)$ (sulfuric acid)

K.1 (a) $2\,NO_2(g) + O_3(g) \rightarrow N_2O_5(g) + O_2(g)$

(b) $S_8(s) + 16\,Na(s) \rightarrow 8\,Na_2S(s)$

(c) $2\,Cr^{2+}(aq) + Sn^{4+}(aq) \rightarrow 2\,Cr^{3+}(aq) + Sn^{2+}(aq)$

(d) $2\,As(s) + 3\,Cl_2(g) \rightarrow 2\,AsCl_3(l)$

K.3 (a) $Mg(s) + Cu^{2+}(aq) \rightarrow Mg^{2+}(aq) + Cu(s)$

(b) $Fe^{2+}(aq) + Ce^{4+}(aq) \rightarrow Fe^{3+}(aq) + Ce^{3+}(aq)$

(c) $H_2(g) + Cl_2(g) \rightarrow 2\,HCl(g)$

(d) $4\,Fe(s) + 3\,O_2(g) \rightarrow 2\,Fe_2O_3(s)$

K.5 (a) +4;　(b) +4;　(c) −2;　(d) +5;　(e) +1;　(f) 0

K.7   (a) +2;   (b) +2;   (c) +6;   (d) +4;   (e) +1

K.9   (a) Methanol CH_3OH (aq) is oxidized to formic acid (the carbon atom goes from an oxidation number of +2 to +4). The O_2 (g) is reduced to O^{2-} present in water.   (b) Mo is reduced from +5 to +4, while *some* sulfur (that which ends up as S(s)) is oxidized from –2 to 0. The sulfur present in MoS_2 (s) remains in the –2 oxidation state.

(c) Tl^+ is both oxidized and reduced. The product Tl(s) is a reduction of Tl^+ (from +1 to 0) while the Tl^{3+} is produced via an oxidation of Tl^+. A reaction in which a single substance is both oxidized and reduced is known as a *disproportionation reaction.*

K.11   (a) Cl_2 will be reduced more easily and is therefore a stronger oxidizing agent than Cl^-.
(b) N_2O_5 will be a stronger oxidizing agent because it will be readily reduced. N^{5+} will accept e^- more readily than will N^+.

K.13   (a) oxidizing agent: H^+ in HCl(aq); reducing agent: Zn(s)
(b) oxidizing agent: SO_2 (g); reducing agent: H_2S(g)
(c) oxidizing agent: B_2O_3 (s); reducing agent: Mg(s)

K.15   (a) $ClO_4^- \rightarrow ClO_2$, Cl goes from +7 to +4; need a reducing agent
(b) $SO_4^{2-} \rightarrow SO_2$, S goes from +6 to +4; need a reducing agent

K.17 CO_2(g) + 4 H_2(g) $\rightarrow$ CH_4(g)   + 2 H_2O(l)
Oxidation-reduction reaction;
CO_2 is oxidizing reagent; H_2 is reducing reagent

K.19 (a) oxidizing agent: $WO_3(s)$; reducing agent: $H_2(g)$

(b) oxidizing agent: HCl reducing agent: Mg(s)

(c) oxidizing agent: $SnO_2(s)$; reducing agent: C(s)

(d) oxidizing agent: $N_2O_4(g)$; reducing agent: $N_2H_4(g)$

K.21 (a) $3\ N_2H_4(l) \rightarrow 4\ NH_3(g) + N_2(g)$; (b) −2 in N_2H_4; −3 in NH_3, 0 in N_2;

(c) N_2H_4 is both oxidizing and reducing agent;

(d) Factor label (dimensional analysis) can be used to find the volume of nitrogen.

$$? \text{ L } N_2(g) = 1.0 \text{ L } N_2H_4(l)\left(\frac{1000 \text{ cm}^3}{1 \text{ L}}\right)\left(\frac{1.004 \text{ g}}{1 \text{ cm}^3}\right)\left(\frac{1 \text{ mol}}{32.0 \text{ g}}\right)$$

$$\times\left(\frac{1 \text{ mol } N_2}{3 \text{ mol } N_2H_4}\right)\left(\frac{28.0 \text{ g}}{1 \text{ mol}}\right)\left(\frac{24 \text{ L}}{28 \text{ g}}\right)$$

$$= 2.5 \times 10^2 \text{ L } N_2(g)$$

K.23 (a) $2\ Cr^{2+} + Cu^{2+} \rightarrow 2\ Cr^{3+} + Cu(s)$; (b) $2\ e^-$ transferred;

(c) Use factor label (dimensional analysis):

$$\frac{? \text{ mol } NO_3^-}{\text{L solution}} = \left(\frac{50.5 \text{ g } Cr(NO_3)_2}{250.0 \text{ mL solution}}\right)\left(\frac{1 \text{ mol } Cr(NO_3)_2}{176.0 \text{ g } Cr(NO_3)_2}\right)$$

$$\times\left(\frac{2 \text{ mol } NO_3^-}{1 \text{ mol } Cr(NO_3)_2}\right)\left(\frac{1000 \text{ mL}}{1 \text{ L}}\right)$$

$$= 2.27 \text{ M } NO_3^-$$

$$\frac{? \text{ mol } SO_4^{2-}}{\text{L solution}} = \left(\frac{60.0 \text{ g } CuSO_4}{250.0 \text{ mL solution}}\right)\left(\frac{1 \text{ mol } CuSO_4}{159.62 \text{ g } CuSO_4}\right)$$

$$\times\left(\frac{1 \text{ mol } SO_4^{2-}}{1 \text{ mol } CuSO_4}\right)\left(\frac{1000 \text{ mL}}{1 \text{ L}}\right)$$

$$= 1.50 \text{ M } SO_4^{2-}$$

K.25 (a) $SiCl_4(l) + 2\ H_2(g) \rightarrow Si(s) + 4\ HCl(g)$; silicon tetrachloride; Si^{4+}

(b) $SnO_2(s) + C(s) \rightarrow Sn(l) + CO_2$; tin(IV) oxide; Sn^{4+}

(c) $V_2O_5(s) + 5\,Ca(l) \rightarrow 2\,V(s) + 5\,CaO(s)$; vanadium(V) oxide; V^{5+}

(d) $B_2O_3(s) + 3\,Mg(s) \rightarrow 2\,B(s) + 3\,MgO(s)$; boron oxide; B^{3+}

K.27 (a) $Cl_2(g)\ +\ H_2O(l)\ \rightarrow\ HClO(aq)\ +\ HCl(aq)$

Oxidizing agent: $Cl_2(g)$; Reducing agent: $Cl_2(g)$

(b) $4\,NaClO_3(aq) +\ 2\,SO_2(g) + 2\,H_2SO_4(aq,\ dilute)$

$\rightarrow 4\,NaHSO_4(aq)\ + 4\,ClO_2(g)$

Oxidizing agent: $NaClO_3(aq)$; Reducing agent: $SO_2(g)$

(c) $2\,CuI(aq)\ \rightarrow\ 2\,Cu(s)\ +\ I_2(s)$

Oxidizing agent: $CuI(aq)$; Reducing agent: $CuI(aq)$

L.1 $Li_3N(s)\ + 2\,H_2(g)\ \rightarrow\ LiNH_2(s)\ +\ 2\,LiH(s)$

(a) moles of H_2 needed to react with 1.5 mg Li_3N:

$$= 1.5 \times 10^{-3}\ g\ Li_3N \left(\frac{1\ mol\ Li_3N}{34.83\ g\ Li_3N} \right)\left(\frac{2\ mol\ H_2}{1\ mol\ Li_3N} \right) = 8.6 \times 10^{-5}\ mol\ H_2$$

(b) mass of Li_3N to produce 0.650 mol LiH:

$$= 0.650\ mol\ LiH \left(\frac{1\ mol\ Li_3N}{2\ mol\ LiH} \right)\left(\frac{34.83\ g\ Li_3N}{1\ mol\ Li_3N} \right) = 11.3\ g\ Li_3N$$

L.3 $6\,NH_4ClO_4(s) + 10\,Al(s) \rightarrow 5\,Al_2O_3(s) + 3\,N_2(g) + 6\,HCl(g) + 9\,H_2O(g)$

(a) $(1.325\ kg\ NH_4ClO_4)\left(\dfrac{1\ mol\ NH_4ClO_4}{117.49\ g\ NH_4ClO_4} \right)\left(\dfrac{1000\ g}{1\ kg} \right)$

$\left(\dfrac{10\ mol\ Al}{6\ mol\ NH_4ClO_4} \right)\left(\dfrac{26.98\ g\ Al}{1\ mol\ Al} \right) = 507.1\ g\ Al$

(b) $3500\ kg\ Al \left(\dfrac{1000\ g\ Al}{1\ kg\ Al} \right)\left(\dfrac{1\ mol\ Al}{26.98\ g\ Al} \right)$

$\times \left(\dfrac{5\ mol\ Al_2O_3}{10\ mol\ Al} \right)\left(\dfrac{101.96\ g\ Al_2O_3}{1\ mol\ Al_2\ O_3} \right)$

$= 6.613 \times 10^6\ g\ Al_2O_3$ or $6.613 \times 10^3\ kg\ Al_2O_3$

L.5 $2\,C_{57}H_{110}O_6(s) + 163\,O_2(g) \rightarrow 114\,CO_2(g) + 110\,H_2O(l)$

(a) $(454\text{ g fat})\left(\dfrac{1\text{ mol fat}}{891.44\text{ g fat}}\right)\left(\dfrac{110\text{ mol }H_2O}{2\text{ mol fat}}\right)\left(\dfrac{18.02\text{ g }H_2O}{1\text{ mol }H_2O}\right)$

$= 505\text{ g }H_2O$

(b) $(454\text{ g fat})\left(\dfrac{1\text{ mol fat}}{891.44\text{ g}}\right)\left(\dfrac{163\text{ mol }O_2}{2\text{ mol fat}}\right)\left(\dfrac{32.00\text{ g }O_2}{1\text{ mol }O_2}\right)$

$= 1.33 \times 10^3\text{ g }O_2$

L.7 $2\,C_8H_{18}(l) + 25\,O_2(g) \rightarrow 16\,CO_2(g) + 18\,H_2O(l)$

$d = 0.79\text{ g}\cdot mL^{-1}$, density of gasoline

$(3.785\text{ L gas})\left(\dfrac{1000\text{ mL}}{1\text{ L}}\right)\left(\dfrac{0.79\text{ g gas}}{1\text{ mL}}\right)\left(\dfrac{1\text{ mol gas}}{114.22\text{ g gas}}\right)\left(\dfrac{18\text{ mol }H_2O}{2\text{ mol }C_8H_{18}}\right)$

$\left(\dfrac{18.02\text{ g }H_2O}{1\text{ mol }H_2O}\right) = 4.246 \times 10^3\text{ g }H_2O \text{ or } 4.2\text{ kg }H_2O$

L.9 $CaCO_3(s) + 2\,HCl(aq) \rightarrow CaCl_2(aq) + H_2O(l) + CO_2(g)$

$Mg(OH)_2(s) + 2\,HCl(aq) \rightarrow MgCl_2(aq) + 2\,H_2O(l)$

The mass of HCl can be neutralized by 400. mg $CaCO_3$:

400. mg $CaCO_3$

$\left(\dfrac{1\,g}{1000\,mg}\right)\left(\dfrac{1\,mol\,CaCO_3}{99.087\,g\,CaCO_3}\right)\left(\dfrac{2\,mol\,HCl}{1\,mol\,CaCO_3}\right)\left(\dfrac{36.461\,g\,HCl}{1\,mol\,HCl}\right)$

$= 0.294\text{ g HCl}$

The mass of HCl can be neutralized by 150. mg $Mg(OH)_2$:

150. mg $Mg(OH)_2 \times$

$\left(\dfrac{1\,g}{1000\,mg}\right)\left(\dfrac{1\,mol\,Mg(OH)_2}{58.319\,g\,Mg(OH)_2}\right)\left(\dfrac{2\,mol\,HCl}{1\,mol\,Mg(OH)_2}\right)\left(\dfrac{36.461\,g\,HCl}{1\,mol\,HCl}\right)$

$= 0.188\text{ g HCl}$

Total mass of HCl can be neutralized: $0.294\text{ g} + 0.188\text{ g} = 0.482\text{ g HCl}$

L.11 (a) $HCl + NaOH \rightarrow NaCl + H_2O$

$$17.40 \text{ mL} \left(\frac{0.234 \text{ mol HCl}}{1000 \text{ mL}} \right) \left(\frac{1 \text{ mol NaOH}}{1 \text{ mol HCl}} \right) = 0.00407 \text{ mol}$$

$$\text{concentration of NaOH} = \frac{0.00407 \text{ mol}}{15.00 \times 10^{-3} \text{ L}} = 0.271 \text{ mol} \cdot \text{L}^{-1}$$

(b) $(0.271 \text{ mol} \cdot \text{L}^{-1})(0.01500 \text{ L})(40.00 \text{ g} \cdot \text{mol}^{-1}) = 0.163 \text{ g NaOH}$

L.13 (a) $Ba(OH)_2(aq) + 2\,HNO_3(aq) \rightarrow Ba(NO_3)_2(aq) + 2\,H_2O(l)$

$$\frac{?\text{ mol HNO}_3}{\text{L HNO}_3(aq)} = \left(11.56 \text{ mL Ba(OH)}_2(aq) \right) \left(\frac{9.670 \text{ g Ba(OH)}_2}{250.\text{ mL Ba(OH)}_2(aq)} \right)$$

$$\times \frac{\left(\dfrac{1 \text{ mol Ba(OH)}_2}{171.36 \text{ g Ba(OH)}_2} \right) \left(\dfrac{2 \text{ mol HNO}_3}{1 \text{ mol Ba(OH)}_2} \right)}{(25.0 \text{ mL HNO}_3(aq)) \left(\dfrac{1 \text{ L}}{1000 \text{ mL}} \right)}$$

$$= 0.209 \text{ mol} \cdot \text{L}^{-1}$$

(b) mass of HNO_3 in solution:

$$\left(\frac{0.209 \text{ mol}}{1000 \text{ mL}} \right)(25.0 \text{ mL}) \left(\frac{63.02 \text{ g HNO}_3}{1 \text{ mol HNO}_3} \right) = 0.329 \text{ g}$$

L.15 $HX(aq) + NaOH(aq) \rightarrow NaX(aq) + H_2O(l)$

$$(68.8 \text{ mL}) \left(\frac{0.750 \text{ mol NaOH}}{1000 \text{ mL NaOH}} \right) = 0.0516 \text{ mol NaOH}$$

3.25 g HX corresponds to 0.0516 mol NaOH used

$$\frac{3.25 \text{ g}}{0.0516 \text{ mol}} = 63.0 \text{ g} \cdot \text{mol}^{-1} = \text{molar mass of acid}$$

L.17 $NaI(aq) + AgNO_3(aq) \rightarrow AgI(s) + NaNO_3(aq)$

$$\text{M of AgNO}_3 = \left(\frac{1.76 \text{ g AgI}}{234.77 \text{ g} \cdot \text{mol}^{-1}} \right) \left(\frac{1 \text{ mol AgNO}_3}{1 \text{ mol AgI}} \right) \left(\frac{1}{0.0500 \text{ L}} \right) = 0.150 \text{ M}$$

L.19 (a) $Na_2CO_3(aq) + 2\,HCl(aq) \rightarrow 2\,NaCl(aq) + H_2CO_3(aq)$

(b) First find the concentration of the diluted acid.

$$? \text{ M HCl(aq) dilute} = \left(\frac{0.832 \text{ g Na}_2\text{CO}_3}{0.100 \text{ L base solution}}\right)\left(\frac{1 \text{ mol Na}_2\text{CO}_3}{105.99 \text{ g Na}_2\text{CO}_3}\right)$$

$$\times \left(\frac{0.025 \text{ L base solution}}{0.031\,25 \text{ L acid solution}}\right)$$

$$\times \left(\frac{2 \text{ mol HCl}}{1 \text{ mol Na}_2\text{CO}_3}\right) = 0.126 \text{ M HCl(aq) dilute}$$

The original HCl solution is 100 times more concentrated than the solution used for titration (diluted 10.00 mL to 1000 mL), so the original concentration of the HCl solution is $12.6 \text{ mol} \cdot \text{L}^{-1}$.

L.21 $I_3^- + SnCl_x + (y-x)Cl^- \rightarrow 3\,I^- + SnCl_y$

The information given can be used to find the molar mass of the reactant in order to identify it.

$$25.00 \text{ mL}\left(\frac{0.120 \text{ mol I}_3^-}{1000 \text{ mL}}\right) = 3.00 \times 10^{-3} \text{ mol I}_3^-$$

$$30.00 \text{ mL}\left(\frac{19.0 \text{ g tin chloride}}{1000 \text{ mL}}\right) = 0.570 \text{ g tin chloride}$$

If the reaction is 1:1 then the # moles of I_3^- is the same as the number of moles of $SnCl_x$. In that case, the molar mass of the tin chloride reactant is

$$\frac{0.570 \text{ g}}{3.00 \times 10^{-3} \text{ mol}} = 190. \text{ g} \cdot \text{mol}^{-1}. \text{ This molar mass matches that of SnCl}_2,$$

$189.61 \text{ g} \cdot \text{mol}^{-1}$. Tin (II) chloride also has the correct mass percent tin.

$$\frac{118.71 \text{ g} \cdot \text{mol}^{-1} \text{ Sn}}{189.61 \text{ g} \cdot \text{mol}^{-1} \text{ SnCl}_2} \times 100 = 62.6\%$$

Since the product compound is oxidized relative to the reactant, we can expect it to be Sn(IV). The net ionic equation for the reaction is

$I_3^- + Sn^{2+} \rightarrow 3\, I^- + Sn^{4+}$. Another way to write a balanced reaction would be $I_3^- + SnCl_2(aq) + 2\, Cl^- \rightarrow 3\, I^- + SnCl_4(aq)$.

L.23 (a) $S_2O_3^{2-}$ is both oxidized and reduced.

(b) Find the number of grams of thiosulfate ion in 10.1 mL of solution.

$$? \text{ g } S_2O_3^{2-} = 10.1 \text{ mL } HSO_3^-(aq)\left(\frac{1.45 \text{ g } HSO_3^-(aq)}{1 \text{ mL } HSO_3^-(aq)}\right)\left(\frac{55.0 \text{ g } HSO_3^-}{100 \text{ g } HSO_3^-(aq)}\right)$$

$$\times \left(\frac{1 \text{ mol } HSO_3^-}{81.0 \text{ g } HSO_3^-}\right)\left(\frac{1 \text{ mol } S_2O_3^{2-}}{1 \text{ mol } HSO_3^-}\right)\left(\frac{112.0 \text{ g } S_2O_3^{2-}}{1 \text{ mol } S_2O_3^{2-}}\right)$$

$$= 11.1 \text{ g } S_2O_3^{2-} \text{ present initially}$$

L.25 $XCl_4 + 2\, NH_3 \rightarrow XCl_2(NH_3)_2 + Cl_2$

The reactant and product that contain X are in a 1:1 ratio, so 3.571 g of the reactant is equivalent to 3.180 g of the product. The molar mass of the reactant is $x + 4(35.453 \text{ g/mol})$ while that of the product is $x + 2(35.453$ g/mol)+2(14.01 g/mol) + 6(1.0079 g/mol). Therefore we can set up the following proportion in order to solve for x in g/mol:

$$\frac{x + 141.8}{x + 104.97} = \frac{3.571}{3.180}$$

$$1.1230x - x = 23.923$$

$$x = 194.6 \text{ g} \cdot \text{mol}^{-1}, \text{ or Pt}$$

L.27 The number of moles of product is

$2.27 \text{ g} \div 208.23 \text{ g} \cdot \text{mol}^{-1} = 0.0109 \text{ moles } BaCl_2$. An equivalent number of moles is represented by 3.25 g of $BaBr_x$, so its molar mass is

$3.25 \text{ g} \div 0.0109 \text{ moles} = 298 \text{ g} \cdot \text{mol}^{-1} \, BaBr_x$. Since 137.33 g is

attributable to Ba, 161 g must be Br. Each Br has a mass of 80.4 g/mol so there must be 2 moles of Br for each mole of Ba in the reactant.

$x = 2, \, BaBr_2 + Cl_2 \rightarrow BaCl_2 + Br_2$

L.29   First equation must be balanced as following:

$$Fe + Br_2 \rightarrow FeBr_2$$

$$3\,FeBr_2 + Br_2 \rightarrow Fe_3Br_8$$

$$Fe_3Br_8 + 4\,Na_2CO_3 \rightarrow 8\,NaBr + 4\,CO_2 + Fe_3O_4$$

The masses of Fe are needed to produce 2.5 t of NaBr:

$$2.50\ t \times \left(\frac{1000\ kg}{1\ t}\right)\left(\frac{1000\ g}{1\ kg}\right)\left(\frac{1\ mol\ NaBr}{102.9\ g\ NaBr}\right)\left(\frac{1\ mol\ Fe_3Br_8}{8\ mol\ NaBr}\right)$$

$$\times \left(\frac{3\ mol\ FeBr_2}{1\ mol\ Fe_3Br_8}\right)\left(\frac{1\ mol\ Fe}{1\ mol\ FeBr_2}\right)\left(\frac{55.84\ g\ Fe}{1\ mol\ Fe}\right) = 5.09 \times 10^5\ g\ Fe$$

$$= 509\ kg\ Fe$$

L.31   (a) $M_1V_1 = M_2V_2$ is used for calculation of dilution problems:

$$V_1 = \left(\frac{1.00\ L \times 0.50\ M}{16.0\ M}\right) = 0.031\ L = 31\ mL$$

Pipette 31 mL of 16 M HNO_3 into a 1.00 L volumetric flask which contains about 800 mL of H_2O. Dilute to the mark with H_2O. Shake the flask to mix the solution thoroughly.

(b) $V_{NaOH} = \left(\dfrac{100\ mL \times 0.50\ M}{0.20\ M}\right) = 2.5 \times 10^2\ mL$

L.33   Total mass of tin oxide is: 28.35g − 26.45 g = 1.90 g

(a) $1.50\ g\ Sn \times \left(\dfrac{1\ mol\ Sn}{118.71\ g\ Sn}\right) = 1.264 \times 10^{-2}\ mol\ Sn$

Moles of O: $(1.90-1.50)g \times \left(\dfrac{1\ mol\ O}{16.00\ g\ O}\right) = 0.025\ mol\ O$

Mole ratio of Sn:O is 1:2

Empirical formula: SnO_2

(b) tin(IV) oxide

L.35   (a) It will not affect the reported KOH concentration. The reaction is mole-to-mole interaction (HCl/KOH) (as long as the volume of KOH is accurately measured).

(b) It will make the reported KOH concentration too high.

(c) It will make the reported KOH concentration too high.

(d) It will make the reported KOH concentration too high.

M.1 $CaCO_3(s) \rightarrow CaO(s) + CO_2(g)$

theoretical yield:

$$(42.73 \text{ g } CaCO_3) \left(\frac{1 \text{ mol } CaCO_3}{100.09 \text{ g } CaCO_3} \right) \left(\frac{1 \text{ mol } CO_2}{1 \text{ mol } CaCO_3} \right) \left(\frac{44.01 \text{ g } CO_2}{1 \text{ mol } CO_2} \right)$$

$$= 18.79 \text{ g } CO_2$$

actual yield:

$$\frac{17.5 \text{ g}}{18.79 \text{ g}} \times 100\% = 93.1\% \text{ yield}$$

M.3 (a) $B_2O_3(s) + 3 \text{ Mg(s)} \rightarrow 3 \text{ MgO(s)} + 2 \text{ B(s)}$

(b) mass of B produced from 125 kg of B_2O_3:

$$125 \times 10^3 \text{g } B_2O_3 \left(\frac{1 \text{ mol } B_2O_3}{69.619 \text{ g } B_2O_3} \right) \left(\frac{2 \text{ mol } B}{1 \text{ mol } B_2O_3} \right) \left(\frac{10.811 \text{ g } B}{1 \text{ mol } B} \right)$$

$$= 3.88 \times 10^4 \text{ g B}$$

mass of B produced from 125 kg of Mg:

$$125 \times 10^3 \text{g Mg} \left(\frac{1 \text{ mol } Mg}{24.305 \text{ g } Mg} \right) \left(\frac{2 \text{ mol } B}{3 \text{ mol } Mg} \right) \left(\frac{10.811 \text{ g } B}{1 \text{ mol } B} \right)$$

$$= 3.71 \times 10^4 \text{ g B}$$

Mg is limiting. 3.71×10^4 g B can be produced.

M.5 $C_xH_yCl_z + (x + \frac{y}{4}) O_2 \rightarrow x \, CO_2 + \frac{y}{2} H_2O + \frac{z}{2} Cl_2$

$$\frac{1.52 \text{ g}}{360.88 \text{ g} \cdot \text{mol}^{-1}} = 4.21 \times 10^{-3} \text{ mol Arochlor yields } \frac{2.224 \text{ g}}{44.0 \text{ g} \cdot \text{mol}^{-1}}$$

$$= 5.055 \times 10^{-2} \text{ mol } CO_2$$

$$\frac{2.53 \text{ g}}{360.88 \text{ g} \cdot \text{mol}^{-1}} = 7.01 \times 10^{-3} \text{ mol Arochlor yields } \frac{0.2530 \text{ g}}{18.01 \text{ g} \cdot \text{mol}^{-1}}$$

$$= 1.405 \times 10^{-2} \text{ mol H}_2\text{O}$$

Therefore, $x = \dfrac{5.055 \times 10^{-2}}{4.21 \times 10^{-3}} = 12.0$ and $y = \dfrac{1.405 \times 10^{-2}}{7.01 \times 10^{-3}} = 2.00$

$$12.011x + 1.0079y + 35.453z = 360.88$$

$$12.011(12.0) + 1.0079(2.00) + 35.453z = 360.88$$

$$35.453z = 214.7$$

$$z = 6.06$$

Since the number or Cl atoms per Arochlor 1254 molecule must be a whole number, the number of chlorine atoms is 6.

M.7   (a) $P_4(s) + 3 O_2(g) \rightarrow P_4O_6(s)$

$P_4O_6(s) + 2 O_2(g) \rightarrow P_4O_{10}(s)$

In the first reaction, 5.77 g P_4 uses

$$(5.77 \text{ g P}_4)\left(\frac{1 \text{ mol P}_4}{123.88 \text{ g P}_4}\right)\left(\frac{3 \text{ mol O}_2}{1 \text{ mol P}_4}\right)\left(\frac{32.00 \text{ g O}_2}{1 \text{ mol O}_2}\right) = 4.47 \text{ g O}_2 \text{ (g)}$$

excess $O_2 = 5.77 \text{ g} - 4.47 \text{ g O}_2 = 1.30 \text{ g O}_2$

In the second reaction, 5.77 g P_4 uses

$$\left(\frac{5.77 \text{ g P}_4}{123.88 \text{ g} \cdot \text{mol}^{-1} \text{ P}_4}\right)\left(\frac{1 \text{ mol P}_4O_6}{1 \text{ mol P}_4}\right)\left(\frac{2 \text{ mol O}_2}{1 \text{ mol P}_4O_6}\right)\left(\frac{32.00 \text{ g O}_2}{1 \text{ mol O}_2}\right)$$
$$= 2.98 \text{ g O}_2$$

limiting reagent: O_2

(b) $\left(\dfrac{1.30 \text{ g O}_2}{32.00 \text{ g} \cdot \text{mol}^{-1} \text{ O}_2}\right)\left(\dfrac{1 \text{ mol P}_4O_{10}}{2 \text{ mol O}_2}\right)\left(\dfrac{283.88 \text{ g P}_4O_{10}}{1 \text{ mol P}_4O_{10}}\right) = 5.77 \text{ g P}_4O_{10}$

(c) $\left(\dfrac{1.30 \text{ g O}_2}{32.00 \text{ g} \cdot \text{mol}^{-1} \text{ O}_2}\right)\left(\dfrac{1 \text{ mol P}_4O_6}{2 \text{ mol O}_2}\right)\left(\dfrac{219.88 \text{ g P}_4O_6}{1 \text{ mol P}_4O_6}\right)$
$= 4.47 \text{ g P}_4O_6 \text{ used}$

In the first reaction, 5.77 g P_4 produces

$$\left(\frac{5.77 \text{ g P}_4}{123.88 \text{ g} \cdot \text{mol}^{-1}}\right)\left(\frac{219.88 \text{ g P}_4\text{O}_6}{1 \text{ mol P}_4\text{O}_6}\right)\left(\frac{1 \text{ mol P}_4\text{O}_6}{1 \text{ mol P}_4}\right) = 10.2 \text{ g P}_4\text{O}_6$$

excess reagent: $10.2 \text{ g} - 4.47 \text{ g} = 5.7 \text{ g P}_4\text{O}_6$

M.9   (a) $Cu^{2+}(aq) + 2 OH^-(aq) \rightarrow Cu(OH)_2(s)$

(b) $2.00 \text{ g NaOH} \times \left(\dfrac{1 \text{ mol NaOH}}{40.0 \text{ g NaOH}}\right) = 0.0500 \text{ mol NaOH}$

$(0.0800 \text{ L})(0.500 \text{ M}) = 0.0400 \text{ mol Cu(NO}_3)_2$

Moles of NaOH required to react with 0.04 mol $Cu(NO_3)_2$:

$$0.0400 \text{ mol Cu(NO}_3)_2 \times \left(\frac{2 \text{ mol NaOH}}{1 \text{ mol Cu(NO}_3)_2}\right) = 0.0800 \text{ mol NaOH}$$

0.0800 mol > 0.0500 mol; therefore, NaOH is limiting reagent.

The mass of $Cu(OH)_2(s)$:

$$0.0500 \text{ mol NaOH} \times \left(\frac{1 \text{ mol Cu(OH)}_2}{2 \text{ mol NaOH}}\right) \times \left(\frac{97.57 \text{ g Cu(OH)}_2}{1 \text{ mol Cu(OH)}_2}\right) = 2.44 \text{ g}$$

M.11 $(0.682 \text{ g CO}_2)\left(\dfrac{1 \text{ mol CO}_2}{44.01 \text{ g CO}_2}\right)\left(\dfrac{1 \text{ mol C}}{1 \text{ mol CO}_2}\right) = 0.0155 \text{ mol C}$

$(0.0155 \text{ mol C})(12.01 \text{ g} \cdot \text{mol}^{-1} \text{ C}) = 0.186 \text{ g C}$

$(0.174 \text{ g H}_2\text{O})\left(\dfrac{1 \text{ mol H}_2\text{O}}{18.02 \text{ g H}_2\text{O}}\right)\left(\dfrac{2 \text{ mol H}}{1 \text{ mol H}_2\text{O}}\right) = 0.0193 \text{ mol H}$

$(0.0193 \text{ mol H})(1.0079 \text{ g} \cdot \text{mol}^{-1} \text{ H}) = 0.0195 \text{ g H}$

$(0.110 \text{ g N}_2)\left(\dfrac{1 \text{ mol N}_2}{28.02 \text{ g N}_2}\right)\left(\dfrac{2 \text{ mol N}}{1 \text{ mol N}_2}\right) = 0.007\,85 \text{ mol N}$

$(0.007\,85 \text{ mol N})(14.01 \text{ g} \cdot \text{mol}^{-1} \text{ N}) = 0.110 \text{ g N}$

mass of O $= 0.376 \text{ g} - (0.186 \text{ g} + 0.0193 \text{ g} + 0.110 \text{ g}) = 0.061 \text{ g O}$

$\dfrac{0.061 \text{ g O}}{16.00 \text{ g O}} = 0.0038 \text{ mol O}$

Dividing each amount by 0.0038 gives C : H : N : O ratios = 4.1 : 5.1 : 2.1 : 1. The empirical formula is $C_4H_5N_2O$.

The molecular mass of caffeine is $194\ \mathrm{g \cdot mol^{-1}}$. Its empirical mass is $97.10\ \mathrm{g \cdot mol^{-1}}$.

$$\text{molecular formula} = 2 \times \text{empirical formula} = C_8H_{10}N_4O_2$$

$$2\ C_8H_{10}N_4O_2(s) + 19\ O_2(g) \rightarrow 16\ CO_2(g) + 10\ H_2O(l) + 4\ N_2(g)$$

M.13 Calculate the mass percentage composition; doing so eases the comparison of data from multiple analyses.

$$2.20\ \mathrm{g\ CO_2} \times \left(\frac{1\ \mathrm{mol\ CO_2}}{44.01\ \mathrm{g\ CO_2}}\right) \times \left(\frac{1\ \mathrm{mol\ C}}{1\ \mathrm{mol\ CO_2}}\right) \times \left(\frac{12.01\ \mathrm{g\ C}}{1\ \mathrm{mol\ C}}\right) = 0.600\ \mathrm{g\ C}$$

$$\%C = \frac{0.600\ \mathrm{g\ C}}{1.35\ \mathrm{g\ unknown}} \times 100\% = 44.5\%\ C$$

$$0.901\ \mathrm{g\ H_2O} \times \left(\frac{1\ \mathrm{mol\ H_2O}}{18.02\ \mathrm{g\ H_2O}}\right) \times \left(\frac{2\ \mathrm{mol\ H}}{1\ \mathrm{mol\ H_2O}}\right) \times \left(\frac{1.0079\ \mathrm{g\ H}}{1\ \mathrm{mol\ H}}\right) = 0.101\ \mathrm{g\ H}$$

$$\%H = \frac{0.101\ \mathrm{g\ H}}{1.35\ \mathrm{g\ unknown}} \times 100\% = 7.48\%\ H$$

$$0.130\ \mathrm{g\ N_2} \times \left(\frac{1\ \mathrm{mol\ N_2}}{28.02\ \mathrm{g\ N_2}}\right) \times \left(\frac{2\ \mathrm{mol\ N}}{1\ \mathrm{mol\ N_2}}\right) \times \left(\frac{14.01\ \mathrm{g\ N}}{1\ \mathrm{mol\ N}}\right) = 0.130\ \mathrm{g\ N}$$

$$\%N = \frac{0.130\ \mathrm{g\ N}}{0.500\ \mathrm{g\ unknown}} \times 100\% = 26.0\%\ N$$

Oxygen must be present in the compound because the percentages of C, H, and N only account for 78.0% of its composition. Combustion analysis does not generate data directly for oxygen; we calculate it by difference.

$$\%O = 100 - 78.0 = 22\%\ O$$

To find the empirical formula, assume a sample size of 100 g and find the mole ratios.

$$44.5 \text{ g C} \times \left(\frac{1 \text{ mol C}}{12.01 \text{ g C}} \right) = 3.71 \text{ mol} \times \left(\frac{1}{1.38 \text{ mol}} \right) = 2.69 \times 3 = 8.07$$

$$7.48 \text{ g H} \times \left(\frac{1 \text{ mol H}}{1.0079 \text{ g H}} \right) = 7.42 \text{ mol} \times \left(\frac{1}{1.38 \text{ mol}} \right) = 5.38 \times 3 = 16.1$$

$$26.0 \text{ g N} \times \left(\frac{1 \text{ mol N}}{14.01 \text{ g N}} \right) = 1.86 \text{ mol} \times \left(\frac{1}{1.38 \text{ mol}} \right) = 1.35 \times 3 = 4.05$$

$$22.0 \text{ g O} \times \left(\frac{1 \text{ mol O}}{16.00 \text{ g O}} \right) = 1.38 \text{ mol} \times \left(\frac{1}{1.38 \text{ mol}} \right) = 1.00 \times 3 = 3.00$$

The empirical formula is $C_8H_{16}N_4O_3$.

M.15 $3 \text{ Ca(NO}_3)_2 \text{(aq)} + 2 \text{ H}_3\text{PO}_4 \text{(aq)} \rightarrow \text{Ca}_3(\text{PO}_4)_2 \text{(s)} + 6 \text{ HNO}_3 \text{(aq)}$

(a) The solid is calcium phosphate, $\text{Ca}_3(\text{PO}_4)_2$.

(b) $(206 \text{ g Ca(NO}_3)_2) \left(\frac{1 \text{ mol Ca(NO}_3)_2}{164.10 \text{ g Ca(NO}_3)_2} \right) \left(\frac{2 \text{ mol H}_3\text{PO}_4}{3 \text{ mol Ca(NO}_3)_2} \right)$

$\left(\frac{97.99 \text{ g H}_3\text{PO}_4}{1 \text{ mol H}_3\text{PO}_4} \right) = 82.01 \text{ g H}_3\text{PO}_4$

Therefore $\text{Ca(NO}_3)_2$ is the limiting reagent.

$(206 \text{ g Ca(NO}_3)_2) \left(\frac{1 \text{ mol Ca(NO}_3)_2}{164.10 \text{ g Ca(NO}_3)_2} \right) \left(\frac{1 \text{ mol Ca}_3(\text{PO}_4)_2}{3 \text{ mol Ca(NO}_3)_2} \right)$

$\left(\frac{310.18 \text{ g Ca}_3(\text{PO}_4)_2}{1 \text{ mol Ca}_3(\text{PO}_4)_2} \right) = 130. \text{ g Ca}_3(\text{PO}_4)_2$

M.17 If the 2-naphthol $(144.16 \text{ g} \cdot \text{mol}^{-1})$ were pure, it would give the following combustion analysis:

$$\%C = \frac{10(12.01 \text{ g} \cdot \text{mol}^{-1})}{(144.16 \text{ g} \cdot \text{mol}^{-1} \text{ naphthol})} \times 100\% = 83.31\% \text{ C}$$

$$\%H = \frac{8(1.0079 \text{ g} \cdot \text{mol}^{-1})}{(144.16 \text{ g} \cdot \text{mol}^{-1} \text{ naphthol})} \times 100\% = 5.59\% \text{ H}$$

The observed percentages are low as is expected for a sample contaminated with a substance that contains no C or H. Because the sample does not contain C or H, the percent purity can be easily obtained by

$$\%\text{purity (based on C)} = \frac{\% \text{ found}}{\% \text{ theoretical}} = \frac{77.48\% \text{ mixture}}{83.31\% \text{ pure naphthol}} \times 100\%$$

$$= 93.00\%$$

$$\%\text{purity (based on H)} = \frac{\% \text{ found}}{\% \text{ theoretical}} = \frac{5.20\% \text{ mixture}}{5.59\% \text{ pure naphthol}} \times 100\%$$

$$= 93.0\%$$

M.19 (a) $C_xH_yCl_z + (x + \frac{y}{4} - \frac{z}{2}) O_2 \rightarrow x\,CO_2 + \frac{y}{2} H_2O$

$$\frac{2.492 \text{ g CO}_2}{44.0 \text{ g} \cdot \text{mol}^{-1}} = 5.664 \times 10^{-2} \text{ mol CO}_2 = \text{mol C}$$

$$\frac{0.6495 \text{ g H}_2\text{O}}{18.01 \text{ g} \cdot \text{mol}^{-1}} = 3.608 \times 10^{-2} \text{ mol H}_2\text{O} = 7.216 \times 10^{-2} \text{ mol H}$$

$$(5.664 \times 10^{-2} \text{ mol C}) \left(\frac{12.011 \text{ g}}{\text{mol}} \right) = 0.6803 \text{ g C}$$

$$(7.216 \times 10^{-2} \text{ mol H}) \left(\frac{1.0079 \text{ g}}{\text{mol}} \right) = 0.07273 \text{ g H}$$

Mass of oxygen: 1.000 g compound $-(0.6803 \text{ g} + 0.07273 \text{ g}) = 0.2470$ g

0.2470 g O $\div 16.00$ g $\cdot$ mol$^{-1} = 1.544 \times 10^{-2}$ mol O

$$\frac{5.664 \times 10^{-2}}{1.544 \times 10^{-2}} = 3.67 \qquad \frac{7.216 \times 10^{-2}}{1.544 \times 10^{-2}} = 4.67$$

The mole ratio of C:H:O is 3.67:4.67:1 or 11:14:3, so the empirical formula is $C_{11}H_{14}O_3$.

(b) The molar mass of the empirical formula is 194 g/mol, which is half of 388.46 g/mol. Therefore, the molecular formula of the compound is $C_{22}H_{28}O_6$.

M.21 Determine the number of moles of each element present in the compound then find their ratios to get the subscripts for the empirical formula.

$$\frac{0.055 \text{ g Cl}}{35.453 \text{ g} \cdot \text{mol}^{-1}} = 1.55 \times 10^{-3} \text{ mol Cl}$$

$$\frac{0.0682 \text{ g CO}_2}{44.0 \text{ g} \cdot \text{mol}^{-1}} = 1.55 \times 10^{-3} \text{ mol CO}_2 = \text{mol C}$$

$$\frac{0.0140 \text{ g H}_2\text{O}}{18.01 \text{ g} \cdot \text{mol}^{-1}} = 7.78 \times 10^{-4} \text{ mol H}_2\text{O} = 1.56 \times 10^{-3} \text{ mol H}$$

0.100 g compound

$$-\left(0.055 \text{ g Cl} + 1.55 \times 10^{-3} \text{ mol} \cdot \frac{12.0 \text{ g C}}{\text{mol}} + 1.55 \times 10^{-3} \text{ mol} \cdot \frac{1.0079 \text{ g H}}{\text{mol}} \right)$$

$$= 0.0247 \text{ g O}$$

$$\times \left(\frac{1 \text{ mol}}{16.00 \text{ g}} \right) = 1.55 \times 10^{-3} \text{ mol O}$$

The mole ratio is 1:1:1:1, so the empirical formula is CHOCl.

As 0.100 g of the compound contains 1.55×10^{-3} moles of each element, its

molar mass is $\dfrac{0.100 \text{ g}}{1.55 \times 10^{-3} \text{ mol}} = 64.52 \text{ g} \cdot \text{mol}^{-1}$. The molar mass of the

empirical formula is 64.47 g·mol^{-1}, so the molecular formula is CHOCl.

M.23   HA + XOH → H$_2$O   +   XA

Moles of HA: $\left(\dfrac{2.45 \text{ g HA}}{231 \text{ g} \cdot \text{mol}^{-1}} \right) = 0.0106 \text{ mol}$

Moles of XOH: $\left(\dfrac{1.50 \text{ g XOH}}{125 \text{ g} \cdot \text{mol}^{-1}} \right) = 0.0120 \text{ mol}$

Molar mass of XA = (231−1.0) + (125−17) = 338 g·mol^{-1}

Mole of XA produced: $\left(\dfrac{2.91 \text{ g XA}}{338 \text{ g} \cdot \text{mol}^{-1}} \right) = 8.61 \times 10^{-3} \text{ mol}$

Theoretical yield: (based on the limiting reagent, HA)

$$0.0106 \text{ mol HA} \times \left(\frac{1 \text{ mol XA}}{1 \text{ mole HA}} \right) = 0.0106 \text{ mol XA}$$

Percentage yield: $\left(\dfrac{8.61 \times 10^{-3} \text{ mol XA}}{0.0106 \text{ mole XA}} \right) \times 100\% = 81.2\%$

M.25 (a) $2 \, Al(s) \; + \; 3 \, Cl_2(g) \; \rightarrow \; 2 \, AlCl_3(s)$

(b) Mass of $AlCl_3$ produced based on 255. g Al:

$$255. \text{ g Al} \left(\frac{1 \; mol \; Al}{26.982 \; g \; Al} \right) \left(\frac{2 \; mol \; AlCl_3}{2 \; mol \; Al} \right) \left(\frac{133.341 \; g \; AlCl_3}{1 \; mol \; AlCl_3} \right)$$

$= 1.26 \times 10^3$ g $AlCl_3$

Mass of $AlCl_3$ produced based on 535. g Cl_2:

$$535. \text{ g Cl}_2 \left(\frac{1 \; mol \; Cl_2}{70.906 \; g \; Cl_2} \right) \left(\frac{2 \; mol \; AlCl_3}{3 \; mol \; Cl_2} \right) \left(\frac{133.341 \; g \; AlCl_3}{1 \; mol \; AlCl_3} \right)$$

$= 671.$ g $AlCl_3$

Cl_2 is limiting. Maximum amount of $AlCl_3$ can be produced $= 671.$ g

(c) %yield $= \dfrac{300}{671} \times 100 = 44.7\%$

CHAPTER 1

ATOMS: THE QUANTUM WORLD

1.1 (a) Radiation may pass through a metal foil. (b) All light (electromagnetic radiation) travels at the same speed; the slower speed supports the particle model. (c) This observation supports the radiation model. (d) This observation supports the particle model; electromagnetic radiation has no mass and no charge.

1.3 microwaves < visible light < ultraviolet light < x-rays < γ-rays

1.5 All of these can be determined using $E=h\nu$ and $\nu\lambda=c$. For example, in the first entry frequency is given, so:

$$\lambda = \frac{c}{\nu} = \frac{2.998\times10^8 \text{ m}\cdot\text{s}^{-1}}{8.7\times10^{14}\text{ s}^{-1}} = 3.4\times10^{-7}\text{ m} = 340 \text{ nm} ;$$

and $E = h\nu = (6.626\times10^{-34}\text{ J}\cdot\text{s})(8.7\times10^{-14}\text{ s}^{-1}) = 5.8\times10^{-19}\text{ J}$

Frequency (2 s.f.)	Wavelength (2 s.f.)	Energy of photon (2 s.f.)	Event
8.7×10^{14} Hz	340 nm	5.8×10^{-19} J	Suntan
5.0×10^{14} Hz	600 nm	3.3×10^{-19} J	Reading
300 MHz	1 m	2×10^{-25} J	Microwave popcorn
1.2×10^{17} Hz	2.5 nm	7.9×10^{-17} J	Dental X-ray

1.7 (a) The Rydberg equation gives v when $\mathfrak{R} = 3.29 \times 10^{15}$ s^{-1}, from which one can calculate λ from the relationship $c = v\lambda$.

$$v = \mathfrak{R}\left(\frac{1}{n_1^2} - \frac{1}{n_2^2}\right)$$

and $c = v\lambda = 2.99792 \times 10^8$ $m \cdot s^{-1}$

$$c = \mathfrak{R}\left(\frac{1}{n_1^2} - \frac{1}{n_2^2}\right)\lambda$$

$$2.99792 \times 10^8 \ m \cdot s^{-1} = (3.29 \times 10^{15} \ s^{-1})\left(\frac{1}{4} - \frac{1}{16}\right)\lambda$$

$$\lambda = 4.86 \times 10^{-7} \ m = 486 \ nm$$

(b) Balmer series

(c) visible, blue

1.9 In each of these series, the principal quantum number for the lower energy level involved is the same for each absorption line. Thus, for the Lyman series, the lower energy level is $n = 1$; for the Balmer series, $n = 2$; for Paschen series, $n = 3$; and for the Brackett series, $n = 4$.

1.11 For hydrogen-like one-electron ions, we use the Z-dependent Rydberg relation with the relationship $c = \lambda v$ to determine the transition wavelength. For He$^+$, $Z = 2$.

$$v = Z^2\mathfrak{R}\left(\frac{1}{n_1^2} - \frac{1}{n_2^2}\right) = (2^2)(3.29 \times 10^{15} \ s^{-1})\left(\frac{1}{1^2} - \frac{1}{2^2}\right) = 9.87 \times 10^{15} \ \text{Hz}$$

$$\lambda = \frac{c}{v} = \frac{2.99792 \times 10^8 \ m \cdot s^{-1}}{9.87 \times 10^{15} \ s^{-1}} = 3.04 \times 10^{-8} \ m = 30.4 \ nm$$

1.13 (a) false. The total intensity is proportional to T^4. (Stefan-Boltzmann Law) (b) true; (c) false. Photons of radio-frequency radiation are lower in energy than photons of ultraviolet radiation.

1.15 Wien's law states that $T\lambda_{max} = \text{constant} = 2.88 \times 10^{-3}$ K·m.

If $T/\text{K} = 1540°\text{C} + 273°\text{C} = 1813$ K, then $\lambda_{max} = \dfrac{2.88 \times 10^{-3} \text{ K} \cdot \text{m}}{1813 \text{ K}}$

$\lambda_{max} = 1.59 \times 10^{-6}$ m, or 1590 nm

1.17 The energy is first converted from eV to joules:

$E = (140.511 \times 10^{3} \text{ eV})(1.6022 \times 10^{-19} \text{ J} \cdot \text{eV}^{-1}) = 2.2513 \times 10^{-14}$ J

From $E = hv$ and $c = v\lambda$ we can write

$$\lambda = \frac{hc}{E}$$
$$= \frac{(6.626\,08 \times 10^{-34} \text{ J} \cdot \text{s})(2.997\,92 \times 10^{8} \text{ m} \cdot \text{s}^{-1})}{2.2513 \times 10^{-14} \text{ J}}$$
$$= 8.8237 \times 10^{-12} \text{ m or } 8.8237 \text{ pm}$$

1.19 (a) From $c = v\lambda$ and $E = hv$, we can write

$$E = hc\lambda^{-1}$$
$$= (6.626\,08 \times 10^{-34} \text{ J} \cdot \text{s})(2.997\,92 \times 10^{8})(589 \times 10^{-9} \text{ m})^{-1}$$
$$= 3.37 \times 10^{-19} \text{ J}$$

(b) $E = \left(\dfrac{5.00 \times 10^{-3} \text{ g Na}}{22.99 \text{ g} \cdot \text{mol}^{-1} \text{ Na}} \right)(6.022 \times 10^{23} \text{ atoms} \cdot \text{mol}^{-1})$

$\qquad \times (3.37 \times 10^{-19} \text{ J} \cdot \text{atom}^{-1})$

$\qquad = 44.1 \text{ J}$

(c) $E = (6.022 \times 10^{23} \text{ atoms} \cdot \text{mol}^{-1})(3.37 \times 10^{-19} \text{ J} \cdot \text{atom}^{-1})$

$\qquad = 2.03 \times 10^{5} \text{ J or } 203 \text{ kJ}$

1.21 32 W = 32 J·sec^{-1}, so in 2 seconds 64 J will be emitted.

For violet light ($\lambda = 420$ nm $= 420 \times 10^{-9}$ m) the energy per photon is:

$$E = hc\lambda^{-1}$$
$$= (6.626\ 08 \times 10^{-34}\ \text{J} \cdot \text{s})\ (2.997\ 92 \times 10^{8})\ (420 \times 10^{-9}\ \text{m})^{-1}$$
$$= 4.7 \times 10^{-19}\ \text{J} \cdot \text{photon}^{-1}$$

$$\text{number of photons} = \left(\frac{64\text{J}}{1}\right)\left(\frac{\text{photon}}{4.7 \times 10^{-19}\,\text{J}}\right)$$
$$= 1.4 \times 10^{20}\ \text{photons}$$

$$\text{number of photons} = (64\ \text{J})\ (4.7 \times 10^{-19}\ \text{J} \cdot \text{photon}^{-1})^{-1}$$
$$= 1.4 \times 10^{20}\ \text{photons}$$

1.23 (a) Use the de Broglie relationship, $\lambda = hp^{-1} = h(mv)^{-1}$.

$$m_{\text{e}} = (9.109\ 39 \times 10^{-28}\ \text{g})\ (1\ \text{kg}/1000\ \text{g}) = 9.109\ 39 \times 10^{-31}\ \text{kg}$$
$$(3.6 \times 10^{3}\ \text{km} \cdot \text{s}^{-1})\ (1000\ \text{m} \cdot \text{km}^{-1}) = 3.6 \times 10^{6}\ \text{m} \cdot \text{s}^{-1}$$

$$\lambda = h(mv)^{-1}$$
$$= \frac{6.626\ 08 \times 10^{-34}\ \text{J} \cdot \text{s}}{(9.109\ 39 \times 10^{-31}\ \text{kg})\ (3.6 \times 10^{6}\ \text{m} \cdot \text{s}^{-1})}$$
$$= 2.0 \times 10^{-10}\ \text{m}$$

(b) $E = h\nu$
$$= (6.626\ 08 \times 10^{-34}\ \text{J} \cdot \text{s})\ (2.50 \times 10^{16}\ \text{s}^{-1})$$
$$= 1.66 \times 10^{-17}\ \text{J}$$

(c) The photon needs to contain enough energy to eject the electron from the surface as well as to cause it to move at $3.6 \times 10^{3}\ \text{km} \cdot \text{s}^{-1}$. The energy involved is the kinetic energy of the electron, which equals $\frac{1}{2}mv^{2}$.

$$E_{photon} = 1.66 \times 10^{-17} \text{ J} + \frac{1}{2} mv^2$$

$$= 1.66 \times 10^{-17} \text{ J} + \frac{1}{2} (9.109\,39 \times 10^{-31} \text{ kg}) (3.6 \times 10^6 \text{ m} \cdot \text{s}^{-1})^2$$

$$= 1.66 \times 10^{-17} \text{ J} + 5.9 \times 10^{-18} \text{ J}$$

$$= 2.25 \times 10^{-17} \text{ J}$$

But we are asked for the wavelength of the photon, which we can get from $E = h\nu$ and $c = \nu\lambda$ or $E = hc\lambda^{-1}$.

$$2.25 \times 10^{-17} \text{ kg} \cdot m^2 \cdot s^{-2} = (6.62608 \times 10^{-34} \text{ kg} \cdot m^2 \cdot s^{-1})$$
$$\times (2.99792 \times 10^8 \text{ m} \cdot s^{-1}) \lambda^{-1}$$
$$\lambda = 8.8 \times 10^{-9} \text{ m}$$
$$= 8.8 \text{ nm}$$

(d) 8.6 nm is in the x-ray/gamma ray region.

1.25 To answer this question, we need to convert the quantities to a consistent set of units, in this case, SI units.

$$(5.15 \text{ ounce}) (28.3 \text{ g} \cdot \text{ounce}^{-1}) (1 \text{ kg}/1000 \text{ g}) = 0.146 \text{ kg}$$

$$\left(\frac{92 \text{ mi}}{\text{h}} \right) \left(\frac{1 \text{ h}}{3600 \text{ s}} \right) \left(\frac{1 \text{ km}}{0.6214 \text{ mi}} \right) \left(\frac{1000 \text{ m}}{1 \text{ km}} \right) = 41 \text{ m} \cdot \text{s}^{-1}$$

Use the de Broglie relationship.

$$\lambda = hp^{-1} = h(mv)^{-1}$$
$$= h(mv)^{-1}$$
$$= \frac{6.626\,08 \times 10^{-34} \text{ J} \cdot \text{s}}{(0.146 \text{ kg}) (0.041 \text{ km} \cdot \text{s}^{-1})}$$
$$= \frac{6.626\,08 \times 10^{-34} \text{ kg} \cdot m^2 \cdot s^{-1}}{(0.146 \text{ kg}) (41 \text{ m} \cdot \text{s}^{-1})}$$
$$= 1.1 \times 10^{-34} \text{ m}$$

1.27 From the de Broglie relationship, $p = h\lambda^{-1}$ or $h = mv\lambda$, we can calculate the velocity of the neutron:

$$v = \frac{h}{m\lambda}$$

$$= \frac{(6.626\ 08 \times 10^{-34}\ \text{kg} \cdot \text{m}^2 \cdot \text{s}^{-1})}{(1.674\ 93 \times 10^{-27}\ \text{kg})(100 \times 10^{-12}\ \text{m})} \quad \text{(remember that } 1\ \text{J} = 1\ \text{kg} \cdot \text{m}^2 \cdot \text{s}^{-2})$$

$$= 3.96 \times 10^3\ \text{m} \cdot \text{s}^{-1}$$

1.29 The uncertainty principle states that $\Delta p \Delta x = \frac{1}{2}\hbar$; so, for an electron

$\Delta p = m_e \Delta v$, then $m_e \Delta v \Delta x = \frac{1}{2}\hbar$ and $\Delta v = \frac{1}{2}\frac{\hbar}{m_e \Delta x}$; if we assume that the

uncertainty in the position of the electron is equal to the diameter of the

lead atom $\Delta x = 350\ \text{pm} = 350 \times 10^{-12}\ \text{m}$.

Remembering that $1\ \text{J} = 1\ \text{kg} \cdot \text{m}^2 \cdot \text{s}^{-2}$, gives

$$\hbar = \left(1.054\quad 457 \times 10^{-34}\ \text{J} \cdot \text{s}\right)\left(\frac{1\text{kg} \cdot \text{m}^2 \cdot \text{s}^{-2}}{\text{J}}\right)$$

$$= = 1.054\quad 457 \times 10^{-34}\ \text{kg} \cdot \text{m}^2 \cdot \text{s}^{-1}$$

Then

$$\Delta v = \frac{1}{2}\left(\frac{1.054\ 457 \times 10^{-34}\ \text{kg} \cdot \text{m}^2 \cdot \text{s}^{-1}}{(9.109\ 38 \times 10^{-31}\text{kg})(350 \times 10^{-12}\ \text{m})}\right)$$

$$\Delta v = 1.65 \times 10^5\ \text{m} \cdot \text{s}^{-1}$$

1.31 Yes there are degenerate levels. The first three cases of degenerate levels
are:

$n_1 = 1,\ n_2 = 2$ is degenerate with $n_1 = 2,\ n_2 = 1$

$n_1 = 1,\ n_2 = 3$ is degenerate with $n_1 = 3,\ n_2 = 1$

$n_1 = 2,\ n_2 = 3$ is degenerate with $n_1 = 3,\ n_2 = 2$

1.33 (a) This problem is the same as that solved in Example 1.5, but the
electron is moving between different energy levels. For movement

between energy levels separated by a difference of 1 in principal quantum number, the expression is

$$\Delta E = E_{n+1} - E_n = \frac{(n+1)^2 h^2}{8mL^2} - \frac{n^2 h^2}{8mL^2} = \frac{(2n+1)h^2}{8mL^2}$$

For $n = 2$ and $n + 1 = 3$, $\Delta E = \dfrac{5h^2}{8mL^2}$

Then $\lambda_{3,2} = \dfrac{hc}{E} = \dfrac{8mhcL^2}{5h^2} = \dfrac{8mcL^2}{5h}$

For an electron in a 150-pm box, the expression becomes

$$\lambda_{3,2} = \frac{8(9.109\,39 \times 10^{-31}\ \text{kg})\,(2.997\,92 \times 10^8\ \text{m} \cdot \text{s}^{-1})\,(150 \times 10^{-12}\ \text{m})^2}{5(6.626\,08 \times 10^{-34}\ \text{J} \cdot \text{s})}$$

$$= 1.48 \times 10^{-8}\ \text{m}$$

(b) We need to remember that the equation for ΔE was originally determined for energy separations between successive energy levels, so the expression needs to be altered to make it general for energy levels two units apart:

$$\Delta E = E_{n+2} - E_n = \frac{(n+2)^2 h^2}{8mL^2} - \frac{n^2 h^2}{8mL^2} = \frac{(n^2 + 4n + 4 - n^2)h^2}{8mL^2} = \frac{(4n+4)h^2}{8mL^2}$$

$$\lambda = \frac{hc}{\Delta E} = \frac{hc(8mL^2)}{h^2[4n+4]} = \frac{8mcL^2}{h^2[4n+4]}$$

For $n = 2$, the expression becomes

$$\lambda = \frac{8mcL^2}{h[(4 \times 2) + 4]} = \frac{8mcL^2}{12h}$$

$$= \frac{8(9.10939 \times 10^{-31}\ \text{kg})\,(2.99792 \times 10^8\ \text{m} \cdot \text{s}^{-1})\,(150 \times 10^{-12}\ \text{m})^2}{12(6.62608 \times 10^{-34}\ \text{kg} \cdot \text{m}^2 \cdot \text{s}^{-1})}$$

$$= 6.18 \times 10^{-9}\ \text{m}$$

1.35   (a) Refer to the plot below for parts (a) thru (d); nodes are where the wavefunction is zero:

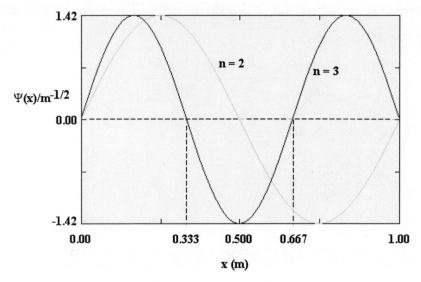

(b) for $n = 2$ there is one node at $x = 0.500$ m.

(c) for $n = 3$ there are two nodes, one at $x = 0.333$ and 0.667 m .

(d) the number of nodes is equal to $n - 1$

Refer to the plot below for parts (e) and (f):

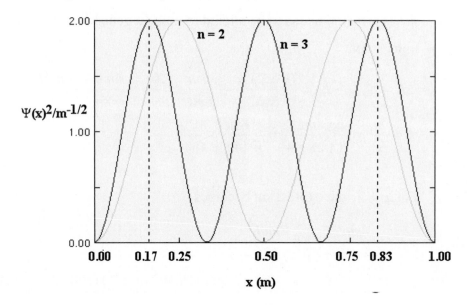

(e) for $n = 2$ a particle is most likely to be found at $x = 0.25$ m and $x = 0.75$ m.

(f) for $n = 3$ a particle is most likely to be found at $x = 0.17$, 0.50 and 0.83 m.

1.37 (a) Integrate over the "left half of the box" or from 0 to ½ L:

$$\int_0^{\frac{L}{2}} \Psi^2 = \frac{2}{L} \int_0^{\frac{L}{2}} \left(\sin \frac{n\pi x}{L} \right)^2 dx$$

$$= \frac{2}{L} \left[\left(\frac{-1}{2n\pi} \cdot \cos \frac{n\pi x}{L} \cdot \sin \frac{n\pi x}{L} + \frac{x}{2} \right)_0^{\frac{L}{2}} \right]$$

given n is an integer:

$$= \frac{2}{L} \left[\left(\frac{L/2}{2} \right) - 0 \right] = \frac{1}{2}$$

(b) Integrate over the "left third of the box" or from 0 to 1/3 L:

$$\int_0^{\frac{L}{2}} \Psi^2 = \frac{2}{L} \int_0^{\frac{L}{2}} \left(\sin \frac{n\pi x}{L} \right)^2 dx$$

$$= \frac{2}{L} \left[\left(\frac{-1}{2n\pi} \cdot \cos \frac{n\pi x}{L} \cdot \sin \frac{n\pi x}{L} + \frac{x}{2} \right)_0^{\frac{L}{3}} \right]$$

$$= \frac{-1}{L \cdot n \cdot \pi} \cos \frac{n\pi}{3} \cdot \sin \frac{n\pi}{3} + \frac{1}{3}$$

if n is a multiple of 3, the first term in this sum is zero and the probability of finding an electron in the left third of the box is 1/3. Also, as n becomes large the probability of finding the electron in the left third of the box approaches 1/3.

1.39 The equation derived in Illustration 1.4 can be used:

$$\frac{\psi^2(r = 0.55a_0, \theta, \phi)}{\psi^2(0, \theta, \phi)} = \frac{\dfrac{e^{-2(0.55a_0)/a_0}}{\pi a_0^{\,3}}}{\left(\dfrac{1}{\pi a_0^{\,3}} \right)} = 0.33$$

1.41 To show that three p orbitals taken together are spherically symmetric, sum the three probability distributions (the wavefunctions squared) and show that the magnitude of the sum is not a function of θ or ϕ.

$$p_x = R(r)C\sin\theta\cos\phi$$
$$p_y = R(r)C\sin\theta\sin\phi$$
$$p_z = R(r)C\cos\theta$$

where $C = \left(\dfrac{3}{4\pi}\right)^{\frac{1}{2}}$

Squaring the three wavefunctions and summing them

$$R(r)^2 C^2 \sin^2\theta\cos^2\phi + R(r)^2 C^2 \sin^2\theta\sin^2\phi + R(r)^2 C^2 \cos^2\theta$$
$$= R(r)^2 C^2 \left(\sin^2\theta\cos^2\phi + \sin^2\theta\sin^2\phi + \cos^2\theta\right)$$
$$= R(r)^2 C^2 \left(\sin^2\theta\left(\cos^2\phi + \sin^2\phi\right) + \cos^2\theta\right)$$

Using the identity $\cos^2 x + \sin^2 x = 1$ this becomes

$$R(r)^2 C^2 \left(\sin^2\theta + \cos^2\theta\right) = R(r)^2 C^2$$

With one electron in each p orbital, the electron distribution is not a fuction of θ or ϕ and is, therefore, spherically symmetric.

1.43 (a) The probability (P) of finding an electron within a sphere of radius a_o may be determined by integrating the appropriate wavefunction squared from 0 to a_o :

$$P = \frac{4}{a_o^{\,2}} \int_0^{a_o} r^2 \exp\left(-\frac{2r}{a_o}\right) dr$$

This integral is easier to evaluate if we allow the following change of variables:

$$z = \frac{2r}{a_o} \quad \therefore \ z = 2 \text{ when } r = a_o, z = 0 \text{ when } r = 0, \text{ and } dr = \left(\frac{a_o}{2}\right) dz$$

$$P = \frac{1}{2}\int_0^2 z^2 \exp(-z)dz = -\frac{1}{2}(z^2 + 2z + 2)\exp(-z)\Big|_0^2$$

$$= -\frac{1}{2}\Big[((4+4+2)\exp(-2))-2\Big]$$
$$= 0.323 \text{ or } 32.3\%$$

(b) Following the answer developed in (a) changing the integration limits to 0 to 2 a_o:

$z = 4$ when $r = 2a_o$, $z = 0$ when $r = 0$, and $dr = \left(\dfrac{a_o}{2}\right)dz$

$$P = \frac{1}{2}\int_0^4 z^2 \exp(-z)dz = -\frac{1}{2}(z^2 + 2z + 2)\exp(-z)\Big|_0^4$$

$$= -\frac{1}{2}\big[(26\exp(-4)) - 4\big]$$

$$= 0.761 \text{ or } 76.1\%$$

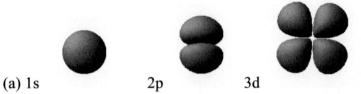

1.45 (a) 1s   2p   3d

(b) A node is a region in space where the wavefunction ψ passes through 0. (c) The simplest s-orbital has 0 nodes, the simplest p-orbital has 1 nodal plane, and the simplest d-orbital has 2 nodal planes. (d) Given the increase in number of nodes, an f-orbital would be expected to have 3 nodal planes.

1.47 The p_x orbital will have its lobes oriented along the x axis, the p_y orbital will have its lobes oriented along the y axis, and the p_z orbital will have its lobes oriented along the z axis.

1.49 (a) 1 orbital; (b) 5 orbitals; (c) 3 orbitals; (d) 7 orbitals

1.51 (a) 7 values: 0, 1, 2, 3, 4, 5, 6; (b) 5 values; $-2, -1, 0, 1, 2$; (c) 3 values: $-1, 0, 1$; (d) 4 subshells: $4s, 4p, 4d,$ and $4f$

1.53 (a) $n = 6; l = 1$; (b) $n = 3; l = 2$; (c) $n = 2; l = 1$; (d) $n = 5; l = 3$

1.55 (a) $-1, 0, +1$; (b) $-2, -1, 0, +1, +2$; (c) $-1, 0, +1$;

(d) $-3, -2, -1, 0, +1, +2, +3$.

1.57 (a) 6 electrons; (b) 10 electrons; (c) 2 electrons; (d) 14 electrons

1.59 (a) $5d$, five; (b) $1s$, one; (c) $6f$, seven; (d) $2p$, three

1.61 (a) six; (b) two; (c) eight; (d) two

1.63 (a) cannot exist; (b) exists; (c) cannot exist; (d) exists

1.65 (a) The total Coulomb potential energy $V(r)$ is the sum of the individual coulombic attractions and repulsions. There will be one attraction between the nucleus and each electron plus a repulsive term to represent the interaction between each pair of electrons. For lithium, there are three protons in the nucleus and three electrons. Each attractive Coulomb potential will be equal to

$$\frac{(-e)(+3e)}{4\pi\varepsilon_0 r} = \frac{-3e^2}{4\pi\varepsilon_0 r}$$

where $-e$ is the charge on the electron and $+3e$ is the charge on the nucleus, ε_0 is the vacuum permittivity, and r is the distance from the electron to the nucleus. The total attractive potential will thus be

$$\left(\frac{-3e^2}{4\pi\varepsilon_0 r_1}\right) + \left(\frac{-3e^2}{4\pi\varepsilon_0 r_2}\right) + \left(\frac{-3e^2}{4\pi\varepsilon_0 r_3}\right) = \left(\frac{-3e^2}{4\pi\varepsilon_0}\right)\left(\frac{1}{r_1} + \frac{1}{r_2} + \frac{1}{r_3}\right)$$

The repulsive terms will have the form

$$\frac{(-e)(-e)}{4\pi\varepsilon_0 r_{ab}} = \frac{e^2}{4\pi\varepsilon_0 r_{ab}}$$

where r_{ab} represents the distance between two electrons a and b. The total repulsive term will thus be

$$\frac{e^2}{4\pi\varepsilon_0 r_{12}} + \frac{e^2}{4\pi\varepsilon_0 r_{13}} + \frac{e^2}{4\pi\varepsilon_0 r_{23}} = \frac{e^2}{4\pi\varepsilon_0}\left(\frac{1}{r_{12}} + \frac{1}{r_{13}} + \frac{1}{r_{23}}\right)$$

This gives

$$V(r) = \left(\frac{-3e^2}{4\pi\varepsilon_0}\right)\left(\frac{1}{r_2} + \frac{1}{r_2} + \frac{1}{r_3}\right) + \frac{e^2}{4\pi\varepsilon_0}\left(\frac{1}{r_{12}} + \frac{1}{r_{13}} + \frac{1}{r_{23}}\right)$$

(b) The first term represents the coulombic attractions between the nucleus and each electron, and the second term represents the coulombic repulsions between each pair of electrons.

1.67 (a) false. Z_{eff} is considerably affected by the total number of electrons present in the atom because the electrons in the lower energy orbitals will "shield" the electrons in the higher energy orbitals from the nucleus. This effect arises because the *e-e* repulsions tend to offset the attraction of the electron to the nucleus. (b) true; (c) false. The electrons are increasingly less able to penetrate to the nucleus as l increases. (d) true.

1.69 Only (d) is the configuration expected for a ground-state atom; the others all represent excited-state configurations.

1.71 (a) This configuration is possible. (b) This configuration is not possible because $l = 0$ here, so m_l must also equal 0. (c) This configuration is not possible because the maximum value l can have is $n - 1$; $n = 4$, so $l_{max} = 3$.

1.73 (a) silver $\quad\quad\quad\quad$ $[Kr]4d^{10}5s^1$

(b) beryllium $\quad\quad\;$ $[He]2s^2$

(c) antimony $\quad\quad$ $[Kr]4d^{10}5s^2 5p^3$

(d) gallium $\quad\quad\;\;$ $[Ar]3d^{10}4s^2 4p^1$

(e) tungsten $\quad\quad$ $[Xe]4f^{14}5d^4 6s^2$

(f) iodine $[Kr]4d^{10}5s^25p^5$

1.75 (a) tellurium; (b) vanadium; (c) carbon; (d) thorium

1.77 (a) $4p$; (b) $4s$; (c) $6s$; (d) $6s$

1.79 (a) 5; (b) 11; (c) 5; (d) 20

1.81 (a) 3; (b) 2; (c) 3; (d) 2

1.83

Element	Electron Configuration	Unpaired electrons
Ga	$[Ar]\,3d^{10}4s^24p^1$	1
Ge	$[Ar]\,3d^{10}4s^24p^2$	2
As	$[Ar]\,3d^{10}4s^24p^3$	3
Se	$[Ar]\,3d^{10}4s^24p^4$	2
Br	$[Ar]\,3d^{10}4s^24p^5$	1

1.85 (a) ns^1; (b) ns^2np^3; (c) $(n-1)d^5ns^2$; (d) $(n-1)d^{10}ns^1$

1.87 (a) silicon (118 pm) > sulfur (104 pm) > chlorine (99 pm);

(b) titanium (147 pm) > chromium (129 pm) > cobalt (125 pm);

(c) mercury (155 pm) > cadmium (152 pm) > zinc (137 pm);

(d) bismuth (182 pm) > antimony (141 pm) > phosphorus (110 pm)

1.89 $P^{3-} > S^{2-} > Cl^-$

1.91 (a) Ca; (b) Na ; (c) Na

1.93 (a) oxygen (1310 kJ · mol^{-1}) > selenium (941 kJ · mol^{-1}) > tellurium (870 kJ · mol^{-1}); ionization energies generally decrease as one goes down a group. (b) gold (890 kJ · mol^{-1}) > osmium (840 kJ · mol^{-1}) > tantalum (761 kJ · mol^{-1}); ionization energies generally decrease as one goes from right to left in the periodic table. (c) lead (716 kJ · mol^{-1}) > barium (502 kJ · mol^{-1}) > cesium (376 kJ · mol^{-1}); ionization energies generally decrease as one goes from right to left in the periodic table.

1.95 (a) fluorine; (b) carbon; (c) chlorine; (d) lithium.

1.97 (a) The inert-pair effect is the term used to describe the fact that heavy (period 5 and greater) *p*-block elements have a tendency to form ions that are two units lower in charge than that expected based on their group number. (b) The inert-pair effect is presumed to only be observed for heavy elements because of the poor shielding ability of the *d* electrons in these elements, enhancing the ability of the *s* electrons to penetrate to the nucleus and therefore be bound tighter than expected.

1.99 (a) A diagonal relationship is a similarity in chemical properties between an element in the periodic table and one lying one period lower and one group to the right. (b) It is caused by the similarity in size of the ions. The lower-right element in the pair would generally be larger because it lies in a higher period, but it also will have a higher oxidation state, which will cause the ion to be smaller. (c) For example, Al^{3+} and Ge^{4+} compounds show the diagonal relationship, as do Li^+ and Mg^{2+}.

1.101 Only (b) Li and Mg exhibit a diagonal relationship.

1.103 The ionization energies of the s-block metals are considerably lower, thus making it easier for them to lose electrons in chemical reactions.

1.105 (a) metal;   (b) nonmetal;   (c) metal;   (d) metalloid;   (e) metalloid; (f) metal

1.107 The peaks observed in the PES spectra correspond to orbital energies; for each energy value seen, a corresponding orbital is present. Thus, if two values are seen in the PES spectrum, that atom has two orbitals (a $1s$ and a $2s$); one can then use ionization energies to determine the identity of the element. See Figure 1.52 and Appendix 2 for the successive ionization energies of the elements.

(a) The observed values ($7.30 \text{ MJ} \cdot \text{mol}^{-1}$ and $0.52 \text{ MJ} \cdot \text{mol}^{-1}$) correspond respectively to the second ($7300 \text{ kJ} \cdot \text{mol}^{-1}$) and first ($519 \text{ kJ} \cdot \text{mol}^{-1}$) ionization energies of Li ($1s^2 2s^1$).

(b) The PES values observed ($1.6 \text{ MJ} \cdot \text{mol}^{-1}$ and $0.90 \text{ MJ} \cdot \text{mol}^{-1}$) correspond respectively to the second ($1760 \text{ kJ} \cdot \text{mol}^{-1}$) and first ($900 \text{ kJ} \cdot \text{mol}^{-1}$) ionization energies of Be ($1s^2 2s^2$).

1.109 (a) $\dfrac{v}{c} = 3600 \text{ cm}^{-1}$

$v = c(3600 \text{ cm}^{-1})$
$v = (2.997\,92 \times 10^8 \text{ m} \cdot \text{s}^{-1})(3600 \text{ cm}^{-1})$
$v = (2.997\,92 \times 10^{10} \text{ cm} \cdot \text{s}^{-1})(3600 \text{ cm}^{-1})$
$v = 1.1 \times 10^{14} \text{ s}^{-1}$

(b) From $E = hv$: $E = (6.626\,08 \times 10^{-34} \text{ J} \cdot \text{s})(1.079 \times 10^{14} \text{ s}^{-1})$

$= 7.2 \times 10^{-20} \text{ J.}$

(c) 1.00 mol of molecules $= 6.022 \times 10^{23}$ molecules, so the energy absorbed by 1.00 mol will be

$$(7.151 \times 10^{-20} \text{ J} \cdot \text{molecule}^{-1})(6.022 \times 10^{23} \text{ molecules} \cdot \text{mol}^{-1})$$

$$= 4.3 \times 10^4 \text{ J} \cdot \text{mol}^{-1} \text{ or } 43 \text{ kJ} \cdot \text{mol}^{-1}.$$

1.111 A ground-state oxygen atom has four electrons in the *p*-orbitals. This configuration means that as one goes across the periodic table in Period 2, oxygen is the first element encountered in which the *p*-electrons must be paired. This added electron-electron repulsion energy causes the ionization potential to be lower.

1.113 molar volume $(\text{cm}^3 \cdot \text{mol}^{-1})$ = molar mass $(\text{g} \cdot \text{mol}^{-1})$/density $(\text{g} \cdot \text{cm}^{-3})$

Element	Molar vol.	Element	Molar vol.
Li	13	Na	24
Be	4.87	Mg	14.0
B	4.38	Al	9.99
C	5.29	Si	12.1
N	16	P	17.0
O	14.0	S	15.3
F	17.1	Cl	21.4
Ne	16.7	Ar	24.1

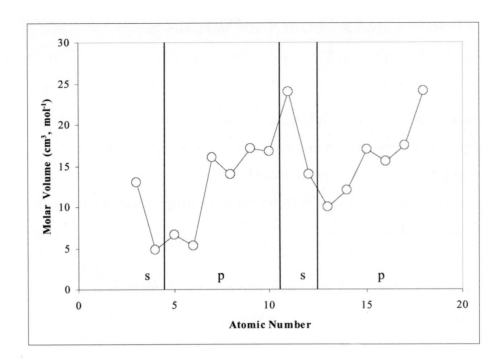

The molar volume roughly parallels atomic size (volume), which decreases as the *s*-sublevel begins to fill and subsequently increases as the *p*-sublevel fills (refer to the text discussion of periodic variation of atomic radii). In the above plot, this effect is most clearly seen in passing from Ne(10) to Na(11) and Mg(12), then to Al(13) and Si(14). Ne has a filled 2*p*-sublevel; the 3*s*-sublevel fills with Na and Mg; and the 3*p*-sublevel begins to fill with Al.

1.115 In general, as the principal quantum number increases, the energy spacing between orbitals becomes smaller. This trend indicates that it does not take very much change in electronic structure to cause the normal orbital energy pattern to rearrange.

1.117 (a)

$$\int_0^L \left(\sin \frac{\pi \cdot x}{L} \right) \cdot \left(\sin \frac{2\pi \cdot x}{L} \right) dx$$

$$= \frac{L}{2\pi} \left(\sin \frac{\pi \cdot x}{L} \right) - \frac{L}{6\pi} \left(\sin \frac{3\pi \cdot x}{L} \right) \Bigg|_0^L = 0$$

(b) Below is a plot of the first two wavefunctions describing the one-dimensional particle-in-a-box and the product of these two wavefunctions. Notice that the area above zero in the product exactly cancels the area below zero, making the integral of the product zero. This happens whenever a wavefunction that is unaltered by a reflection through the center of the box (wavefunctions with odd n) is multiplied by a wavefunction that changes sign everywhere when reflected through the center of the box (wavefunctions with even n).

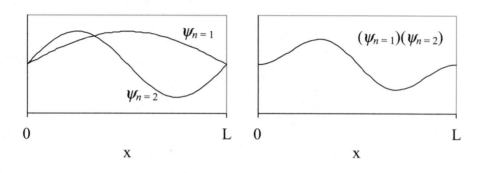

1.119 (a) Evaluating the integral for an $n = 1$ to $n = 3$ transition:

$$\int_0^L \sin\left(\frac{\pi x}{L}\right) \cdot x \cdot \sin\left(\frac{3\pi x}{L}\right)$$

$$= \frac{L^2}{8\pi^2}\left(\cos\left(\frac{2\pi x}{L}\right) + \frac{2\pi x}{L}\sin\frac{2\pi x}{L}\right) -$$

$$\frac{L^2}{32\pi^2}\left(\cos\left(\frac{4\pi x}{L}\right) + \frac{4\pi x}{L}\sin\frac{4\pi x}{L}\right)\Bigg|_0^L$$

$$= \frac{L^2}{8\pi^2}(1+0) - \frac{L^2}{32\pi^2}(1+0) - \left[\frac{L^2}{8\pi^2}(1+0) - \frac{L^2}{32\pi^2}(1+0)\right]$$

$$= 0$$

Because the integral is zero, one would not expect to observe a transition between the $n = 1$ and $n = 3$ states.

(b) Again, evaluating the integral:

$$\int_0^L \sin\left(\frac{\pi x}{L}\right) \cdot x \cdot \sin\left(\frac{2\pi x}{L}\right)$$

$$= \frac{L^2}{\pi^2}\left(\frac{1}{2}\left(\cos\left(\frac{\pi x}{L}\right) + \frac{\pi x}{L}\sin\left(\frac{\pi x}{L}\right)\right) - \frac{1}{18}\left(\cos\left(\frac{3\pi x}{L}\right) + \frac{3\pi x}{L}\sin\left(\frac{3\pi x}{L}\right)\right)\right)\Bigg|_0^L$$

Given the L^2 term, we see that the integral, and therefore I, will increase as the length of the box increases.

1.121 The allowed energies of a particle of mass m in a one-dimensional box of length L are determined using $E_n = \dfrac{n^2 h^2}{8mL^2}$.

(a) if $L = 139$ pm $= 1.39 \times 10^{-10}$ m for a $C - C$ bond, and $m_e = 9.109 \times 10^{-31}$ kg, then

$$\Delta E = \frac{\left(6.626 \times 10^{-34}\,\text{J} \cdot \text{sec}\right)^2}{8\left(9.109 \times 10^{-31}\,\text{kg}\right)\left(1.39 \times 10^{-10}\,\text{m}\right)^2}\left(2^2 - 1^2\right) = 9.35 \times 10^{-18}\,\text{J}$$

(b) This energy corresponds to a wavelength of 2.13×10^{-8} m or 21.3 nm. This falls within the X-ray region.

(c) if the chain is extended to be 1000 carbons long, then $L = 1.39 \times 10^{-7}$ m (the length of 999 $C - C$ bonds) and $m_e = 9.109 \times 10^{-31}$ kg. Inserting these numbers in place of the m and L values used in part (a) gives $\Delta E = 9.35 \times 10^{-24}$ J.

(d) Comparing the result of part (a) with that of part (c) indicates that the individual energy levels get closer as the chain gets longer.

1.123 (a) In copper it is energetically favorable for an electron to be promoted from the 4s orbital to a 3d orbital, giving a completely filled 3d subshell. In the case of Cr, it is energetically favorable for an electron to be promoted from the 4s orbital to a 3d orbital to exactly ½ fill the 3d subshell.

(b) From Appendix 2C, the other elements for which anomalous electron configurations exist are Nb, Mo, Ru, Rh, Pd, Ag, Pt and Au. Of these, the explanation used for chromium and copper is valid for Mo, Pd, Ag and Au.

(c) Because the np orbitals are so much lower in energy than the $(n+1)s$ orbitals no elements are known where the $(n+1)s$ orbitals fill in preference to the np orbitals.

1.125 Based on (a), (b) and (c), the element must be a member of either Group 4/IVB (the titanium family) or Group 14/IVA (the carbon family). From (d) we can calculate n:

$$\Delta E = E_{n-1} - E_n = -2.18 \times 10^{-18}\ J \left(\frac{1}{(n-1)^2} - \frac{1}{n^2} \right) = -4.9 \times 10^{-20}\ J;$$

$$\left(\frac{1}{(n-1)^2} - \frac{1}{n^2} \right) = 0.0225 \text{ and } n = 5. \text{ The element must be either tin or}$$

hafnium.

1.127 By examining at the trends followed by the other Group 1 elements, one can arrive at estimates of various properties of francium:

(a) radius of neutral atom = 285 pm (approx. 20 pm larger than Cs);

(b) radius of +1 ion = 194 pm (approx. 20 pm larger than Cs^+);

(c) $IE_1 = 356$ kJ·mol^{-1} (approx. 20 kJ less than Cs)

1.129 $\underline{A}$ = Na; $\underline{B}$ = Cl; $\underline{C}$ = Na^+; $\underline{D}$ = Cl^-

The assignments can be made by looking at neutral atom and ionic radii (Figures 1.46 and 1.48)

1.131 (a) The electron configuration of atomic chlorine is $[Ne]3s^2 3p^5$; it has one unpaired electron. The electron configuration of a chloride ion is $[Ne]3s^2 3p^6$; this configuration is identical to neutral argon.

(b) Assuming a one quantum level jump, an excited chlorine should have an electron configuration of $[Ne]3s^2 3p^4 4s^1$.

(c) The energy of a given level n in a non-hydrogen atom can be estimated by $E_n = -\dfrac{Z_{eff}^2 h\Re}{n^2} = -\dfrac{Z_{eff}^2 \left(2.18 \times 10^{-18}\, J\right)}{n^2}$. For chlorine, Z_{eff} is approximately equal to 6 (Fig. 1.45). For an electron to jump from the $n = 3$ to $n = 4$ quantum level the energy needed is: $\Delta E = E_4 - E_3$, so

$$\Delta E = -(6^2)2.18 \times 10^{-18}\, J\left(\frac{1}{4^2} - \frac{1}{3^2}\right) = 3.82 \times 10^{-18}\, J. \text{ This energy}$$

corresponds to a wavelength of 52.0 nm (the X-ray region).

(d) This amount of energy corresponds to 2.30×10^3 kJ·mol^{-1} or 23.8 eV per chlorine atom

(e) If the proportion of ^{37}Cl in a sample of chlorine atoms is reduced to 37.89% (half of its typical value), the proportion of ^{35}Cl will be increased to 62.11%. Based on this, the average mass of a chlorine atom will be:

$$Cl_{ave\ mass} = \left(\frac{37.89\%}{100}\right)6.139 \times 10^{-23}g + \left(\frac{62.11\%}{100}\right)5.807 \times 10^{-23}g$$

$$Cl_{ave\ mass} = 5.933 \times 10^{-23} g/atom = 35.73 \text{ g/mol}$$

(f) thru (h): refer to the table below:

Compound	Chlorine oxidation number	Name
ClO_2	+4	Chlorine dioxide
$NaClO$	+1	Sodium hypochlorite
$KClO_3$	+5	Potassium chlorate
$NaClO_4$	+7	Sodium perchlorate

CHAPTER 2

CHEMICAL BONDS

2.1 (a) 5; (b) 4; (c) 7; (d) 3

2.3 (a) [Ar]; (b) $[Ar]3d^{10}4s^2$; (c) $[Kr]4d^5$; (d) $[Ar]3d^{10}4s^2$

2.5 (a) $[Ar]3d^{10}$; (b) $[Xe]4f^{14}5d^{10}6s^2$; (c) $[Ar]3d^{10}$; (d) $[Xe]4f^{14}5d^{10}$

2.7 (a) $[Kr]4d^{10}5s^2$; same; In^+ and Sn^{2+} lose $5p$ valence electrons.
(b) none
(c) $[Kr]4d^{10}$; Pd

2.9 (a) Co^{2+}; (b) Fe^{2+}; (c) Mo^{2+}; (d) Nb^{2+}

2.11 (a) Co^{3+}; (b) Fe^{3+}; (c) Ru^{3+}; (d) Mo^{3+}

2.13 (a) $4s$; (b) $3p$; (c) $3p$; (d) $4s$

2.15 (a) −2; (b) −2; (c) +1; (d) +3; (e) +2

2.17 (a) 3; (b) 6; (c) 6; (d) 2

2.19 (a) $[Kr]4d^{10}5s^2$; no unpaired electrons; (b) $[Kr]4d^{10}$; no unpaired electrons; (c) $[Xe]4f^{14}5d^4$; four unpaired electrons; (d) [Kr]; no unpaired electrons; (e) $[Ar]3d^8$; two unpaired electrons

2.21 (a) [Ar]; no unpaired electrons; (b) $[Kr]4d^{10}5s^2$; no unpaired electrons; (c) [Xe]; no unpaired electrons; (d) $[Kr]4d^{10}$; no unpaired electrons.

2.23 (a) +7; (b) −1; (c) [Ne] for +7, [Ar] for −1; (d) electrons are lost or added to give noble-gas configuration.

2.25 (a) Mg_3As_2; (b) In_2S_3; (c) AlH_3; (d) H_2Te; (e) BiF_3

2.27 (a) $TlCl_3$; (b) Al_2S_3; (c) MnO_2

2.29 The coulombic attraction is directly proportional to the charge on each ion (Equation 1) so the ions with the higher charges will give the greater coulombic attraction. The answer is therefore (b) Ga^{3+}, O^{2-}.

2.31 The Li^+ ion is smaller than the Rb^+ ion (58 vs 149 pm). Because the lattice energy is related to the coulombic attraction between the ions, it will be inversely proportional to the distance between the ions (see Equation 2). Hence the larger rubidium ion will have the lower lattice energy for a given anion.

2.33

2.35

(a)

(b)

(c)

2.37

(a)

(b)

(c)

2.39

(a)

(b)

(c)

2.41 The structure has a total of 32 electrons; of these, 21 are accounted for by the chlorines (3 Cl's × 7 valence electrons each); of the 11 electrons remaining, 6 come from the oxygen. This leaves 5 electrons unaccounted for; these must come from E. Therefore, E must be a member of the nitrogen family and since it is a third period element E must be phosphorous (P)

2.43

(a) $\left[\begin{array}{c} H \\ | \\ H-N-H \\ | \\ H \end{array}\right]^{+}$:Cl̈:⁻   (b) $\begin{array}{c} K^{+} \\ \left[:\ddot{P}:\right]^{3-} K^{+} \\ K^{+} \end{array}$ (c) Na⁺ $\left[:\ddot{C}l-\ddot{O}:\right]^{-}$

2.45 There are four possible resonance structures one can draw for anthracene (the atoms have been numbered to aid you in seeing the differences between them):

2.47

2.49

(a) $\left[:N \equiv O: \right]^+$ with formal charges 0, +1

(b) $:N \equiv N:$ with formal charges 0, 0

(c) $:C \equiv O:$ with formal charges -1, +1

(d) $\left[:C \equiv C: \right]^{-2}$ with formal charges -1, -1

(e) $\left[:C \equiv N: \right]^-$ with formal charges -1, 0

2.51

(a)

$\begin{array}{c} ^0O = Cl - O:^0 \\ \| \\ :O:_0 \quad H \end{array}$ (with Cl = 0)

lower energy

$\begin{array}{c} ^{-1}:O - Cl - O:^0 \\ | \\ :O:_{-1} \quad H \end{array}$ (with Cl = +2)

(b)

$^0O = C = S:^0$ (with C = 0)

lower energy

$^{-1}:O - C \equiv S:^{+1}$ (with C = 0)

(c)

$^0H - C \equiv N:^0$ (with C = 0)

lower energy

$^0H - \overset{-1}{C} = \overset{+1}{N}:$

2.53 Radicals are species with an unpaired electron, therefore only (b) and (c) are radicals since they have an odd number of electrons while (a) and (d) have an even number of electrons allowing Lewis structures to be drawn with all electrons paired.

2.55   (a) The sulfite ion has one Lewis structure that obeys the octet rule:

and three with an expanded octet:

The structures with expanded octets have lower formal charges.

(b) There is one Lewis structure that obeys the octet rule:

The formal charge at sulfur can be reduced to 0 by including one double bond contribution. This change gives rise to two expanded octet structures.

Notice that, unlike the sulfite ion, which has three resonance forms, the presence of the hydrogen ion restricts the electrons to the oxygen atom to which it is attached. Because H is electropositive, its placement near an oxygen atom makes it less likely for that oxygen atom to donate a lone pair to an adjacent atom.

(c) The perchlorate ion has one Lewis structure that obeys the octet rule:

$$\left[\begin{array}{c} :\ddot{O}: \\ | \\ :\ddot{O}-Cl-\ddot{O}: \\ | \\ :\ddot{O}: \end{array} \right]^{-}$$

The formal charge at Cl can be reduced to 0 by including three double-bond contributions, thereby giving rise to four resonance forms.

$$\left[\begin{array}{c} :O: \\ \| \\ \ddot{O}=Cl=\ddot{O} \\ | \\ :\ddot{O}: \end{array} \right]^{-} \qquad \left[\begin{array}{c} :O: \\ \| \\ :\ddot{O}-Cl=\ddot{O} \\ \| \\ :\ddot{O}: \end{array} \right]^{-}$$

$$\left[\begin{array}{c} :\ddot{O}: \\ | \\ \ddot{O}=Cl=\ddot{O} \\ \| \\ :\ddot{O}: \end{array} \right]^{-} \qquad \left[\begin{array}{c} :O: \\ \| \\ \ddot{O}=Cl-\ddot{O}: \\ \| \\ :O: \end{array} \right]^{-}$$

For the nitrite ion, there are two resonance forms, both of which obey the octet rule:

$$\left. \ddot{O}=\ddot{N}-\ddot{O}: \right]^{-} \qquad \left. :\ddot{O}-\ddot{N}=\ddot{O} \right]^{-}$$

2.57   The Lewis Structures are

(a) $:\overset{\cdot}{\underset{\cdot\cdot}{C}l}-\ddot{O}:$ (b) $:\ddot{\underset{\cdot\cdot}{C}l}-\ddot{O}-\ddot{O}-\ddot{\underset{\cdot\cdot}{C}l}:$

(c) $\ddot{O}=N-\ddot{O}-\ddot{\underset{\cdot\cdot}{C}l}:$ with $:\ddot{O}:$ above N

Radicals are species with an unpaired electron, therefore only (a) is a radical.

2.59

(a) $\left[:\ddot{C}l - \ddot{I} - \ddot{C}l: \right]^{+}$

I has 2 bonding pairs and 2 lone pairs

(b) $\left[\begin{array}{c} :\ddot{C}l: \\ :\ddot{C}l - \overset{.}{\underset{.}{I}} \\ :\ddot{C}l: \end{array} \ddot{C}l: \right]^{-}$

I has 4 bonding pairs and 2 lone pairs

(c)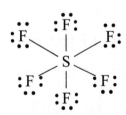

I has 3 bonding pairs and 2 lone pairs

(d)

I has 5 bonding pairs and 1 lone pair

2.61

(a)

12 electrons

(b)

10 electrons

(c)

12 electrons

(d)

10 electrons

2.63

(a)   (b)   (c)

2 lone pairs   2 lone pairs   1 lone pair

2.65 (a) In $BeCl_2$, there are 4 electrons around the central beryllium:

$$:\ddot{C}l \!-\! Be \!-\! \ddot{C}l:$$

(b) In ClO_2 there are an odd number of electrons around the central chlorine:

$$:\ddot{O} \!-\! \dot{\ddot{C}l} \!-\! \ddot{O}:$$

2.67

(a)

$$\left[\;:\overset{-1}{\ddot{O}} \!-\! \overset{0}{\overset{H}{\underset{\overset{\|}{:\underset{0}{\ddot{O}}:}}{S}} \!-\! \overset{0}{\ddot{O}}: \; \right]^{1-}$$
lower energy

$$\left[\;:\overset{-1}{\ddot{O}} \!-\! \overset{+1}{\overset{H}{\underset{\overset{|}{:\underset{-1}{\ddot{O}}:}}{S}} \!-\! \overset{0}{\ddot{O}}: \; \right]^{1-}$$

(b)

$$\left[\;:\overset{-1}{\ddot{O}} \!-\! \overset{\overset{0}{:\ddot{O}:}}{\overset{\|}{\underset{\overset{|}{\underset{-1}{:\ddot{O}:}}}{S}}}{}^{+1} \!-\! \overset{0}{\ddot{O}}: \; \right]^{1-}$$
lower energy

$$\left[\;:\overset{-1}{\ddot{O}} \!-\! \overset{\overset{-1}{:\ddot{O}:}}{\underset{\overset{|}{\underset{-1}{:\ddot{O}:}}}{S}}{}^{+2} \!-\! \overset{0}{\ddot{O}}: \; \right]^{1-}$$

2.69 (a) In the first structure, the formal charges at Xe and F are 0, whereas, in the second structure, Xe is −1, one F is 0, and the other F is +1. The first structure is favored on the basis of formal charges.　(b) In the first structure, all of the atoms have formal charges of 0, whereas, in the second structure, one O atom has a formal charge of +1 and the other O has a formal charge of −1. The first structure is thus preferred.

2.71　In (1.8) < Sn (2.0) < Sb (2.1) < Se (2.6)

2.73　Electronegativity decreases down a group, so the heavier alkali metals and alkaline earth metals are the most electropositive. In order of increasing *electropositive* character: Li (1.0) < Na (0.93) < K (0.82) ~ Rb (0.82) < Cs (0.79); Be (1.6) < Mg (1.3) ~ Ca (1.3) < Sr (0.95) < Ba (0.89) ~ Ra (0.9)

2.75　(a)　The bond in HCl would be more ionic. The electronegativity difference is greater between H and Cl then between H and I, making the H—Cl bond more ionic.

(b)　The bonds in CF_4 would be more ionic. The electronegativity difference is greater between C and F than between C and H, making the C—F bonds more ionic.

(c)　C and S have nearly identical electronegativities, so the C—S bonds would be expected to be almost completely covalent, whereas the C—O bonds would be more ionic.

2.77　(a) The electronegativity difference between K and Cl (2.4) is greater than that between Al and Cl (1.6), so KCl should be more soluble in water than $AlCl_3$. (b) The electronegativity difference between Mg and O is 2.1 while it is 2.5 when comparing Ba and O, therefore BaO should be more water soluble than MgO.

2.79 $Rb^+ < Sr^{2+} < Be^{2+}$; smaller, more highly charged cations have greater polarizing power. The ionic radii are 149 pm, 116 pm, 27 pm, respectively.

2.81 $O^{2-} < N^{3-} < Cl^- < Br^-$; the polarizability increases as the ion gets larger and less electronegative. The ionic radii for these species are 140 pm, 171 pm, 181 pm, 196 pm, respectively.

2.83   (a) $CO_3^{2-} > CO_2 > CO$

CO_3^{2-} will have the longest C—O bond length. In CO there is a triple bond and in CO_2 the C—O bonds are double bonds. In carbonate, the bond is an average of three Lewis structures in which the bond is double in one form and single in two of the forms. We would thus expect the bond order to be approximately 1.3. Because the bond length is inversely related to the number of bonds between the atoms, we expect the bond length to be longest in carbonate.

(b) $SO_3^{2-} > SO_2 \sim SO_3$

Similar arguments can be used for these molecules as in part (a). In SO_2 and SO_3, the Lewis structures with the lowest formal charge at S have double bonds between S and each O. In the sulfite ion, however, there are three Lewis structures that have a zero formal charge at S. Each has one S—O double bond and two S—O single bonds. Because these S—O bonds would have a substantial amount of single bond character, they would be expected to be longer than those in SO_2 or SO_3. This is consistent with the experimental data that show the S—O bond lengths in SO_2 and SO_3 to be 143 pm, whereas those in SO_3^{2-} range from about 145 pm to 152 pm depending on the compound.

(c) $CH_3NH_2 > CH_2NH > HCN$

The C—N bond in HCN is a triple bond, in CH_2NH it is a double bond, and in CH_3NH_2 it is a single bond. The C—N bond in the last molecule would, therefore, be expected to be the longest.

2.85 (a) The covalent radius of N is 75 pm, so the N—N single bond in hydrazine would be expected to be ca. 150 pm. The experimental value is 145 pm.

(b) The C—O bonds in carbon dioxide are double bonds. The covalent radius for doubly bonded carbon is 67 pm and that of O is 60 pm. Thus we predict the C=O in CO_2 to be ca. 127 pm. The experimental bond length is 116.3 pm.

(c) The C—O bond is a double bond so it would be expected to be the same as in (b), 127 ppm. This is the experimentally found value. The C—N bonds are single bonds and so one might expect the bond distance to be the sum of the single bond C radius and the single bond N radius (77 plus 75 pm) which is 152 pm. However, because the C atom is involved in a multiple bond, its radius is actually smaller. The sum of that radius (67 pm) and the N single bond radius gives 142 pm, which is close to the experimental value of 133 pm.

(d) The N—N bond is a double bond so we expect the bond distance to be two times the double bond covalent radius of N, which is 2 × (60 pm) or 120 pm. The experimental value is 123.0 pm.

2.87 (a) 77 pm + 72 pm = 149 pm (b) 111 pm + 72 pm = 183 pm

(c) 141 pm + 72 pm = 213 pm. Bond distance increases with size going down Group 14/IV.

2.89

(a)

$$\left[\begin{array}{c} \overset{0}{:}\!\overset{..}{O}: \\ \underset{-1}{:}\!\overset{..}{O}\!-\!\overset{C}{\underset{0}{}}\!-\!\overset{..}{O}:^{-1} \\ \underset{0}{C} \\ :\overset{..}{O}:_{0} \end{array} \right]^{-2}$$

$$\left[\begin{array}{c} :\overset{..}{O}: \\ :\overset{..}{O}\!-\!C\!-\!\overset{..}{O}: \\ C \\ :\overset{..}{O}: \end{array} \right]^{-2}$$

$$\left[\begin{array}{c} :\overset{..}{O}: \\ :\overset{..}{O}\!=\!C\!=\!\overset{..}{O}: \\ C \\ :\overset{..}{O}: \end{array} \right]^{-2}$$

$$\left[\begin{array}{c} :\overset{..}{O}: \\ :\overset{..}{O}\!=\!C\!-\!\overset{..}{O}: \\ C \\ :\overset{..}{O}: \end{array} \right]^{-2}$$

(b) $\left[:\overset{..}{\underset{..}{Br}}\!\overset{-1}{=}\!\overset{0}{\underset{..}{O}} \right]^{-}$

(c) $\left[:\overset{-1}{C}\!\equiv\!\overset{-1}{C}: \right]^{-2}$

2.91

$$\left[\begin{array}{c} \overset{..}{C}H_2 \\ C \\ H_2C\!=\!\quad \overset{..}{C}H_2 \end{array} \right]^{2-}$$

$$\left[\begin{array}{c} \overset{..}{C}H_2 \\ C \\ H_2\overset{..}{C}\!=\!\quad CH_2 \end{array} \right]^{2-}$$

$$\left[\begin{array}{c} CH_2 \\ \| \\ C \\ H_2\overset{..}{C}\!=\!\quad \overset{..}{C}H_2 \end{array} \right]^{2-}$$

2.93

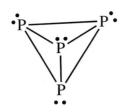

Yes, the molecule obeys the octet rule.

2.95   The Lewis structure for N_5^+ is possible without having multiple formal charges being present. The first of the three resonance structures shown below is the most important Lewis structure, since in it no two like charges are near one another.

$$\overset{+1}{:N}\equiv N-\overset{-1}{\underset{\cdot\cdot}{\overset{\cdot\cdot}{N}}}-\overset{+1}{N}\equiv N: \qquad \overset{-1}{\underset{\cdot\cdot}{\overset{\cdot\cdot}{N}}}=N=\overset{+1}{\underset{\cdot\cdot}{N}}-\overset{+1}{N}\equiv N:$$

$$:N\equiv N-\overset{+1}{\underset{\cdot\cdot}{N}}=N=\overset{-1}{\underset{\cdot\cdot}{\overset{\cdot\cdot}{N}}}$$

2.97   (a) $H-C\equiv C-H \qquad H-C\equiv Si-H$

$H-Si\equiv Si-H \qquad H-C\equiv N: \qquad :N\equiv N:$

(b)

SM-88

+ suitable resonance form

2.99 (a)

(b) All the atoms have formal charge 0 except the two oxygen atoms, which are -1. The negative charge is most likely to be concentrated at the oxygen atoms.

(c) The protons will bond to the oxygen atoms. Oxygen atoms are the most negative sites in the molecule and act as Lewis bases due to their lone pairs of electrons. The resulting compound is named hydroquinone.

2.101

Z	Configuration	Number of unpaired e^-	Element	Charge	Energy state
26	[Ar]$3d^6$	4	Fe	2+	ground
52	[Kr] $4d^{10}5s^25p^56s^1$	2	Te	2-	excited
16	[Ne]$3s^23p^6$	0	S	2-	ground
39	[Kr]$4d^1$	1	Y	2+	ground
30	[Ar] $3d^84s^2$	2	Zn	2+	excited

2.103

(a)

$$H-\ddot{O}-\ddot{N}=C=\ddot{O} \qquad\qquad H-\ddot{O}-\overset{+1}{N}\equiv C-\overset{-1}{\ddot{O}}:$$

$$H-\overset{+1}{\ddot{O}}=\ddot{N}-\overset{-1}{C}=\ddot{O}$$

(b)

$$\underset{H}{\overset{H}{>}}C=\ddot{S}=\ddot{O} \qquad\qquad \underset{H}{\overset{H}{>}}C=\overset{+1}{\ddot{S}}-\overset{-1}{\ddot{O}}:$$

(c)

$$\underset{H}{\overset{H}{>}}C=\overset{+1}{N}=\overset{-1}{\ddot{N}}$$

(d)

$$\ddot{O}=\overset{+1}{N}=C=\overset{-1}{\ddot{N}} \qquad\qquad \ddot{O}=N-C\equiv\ddot{N}$$

2.105

	d(K – X)	Lattice Energy, kJ/mol
Fluoride	271 pm	826
Chloride	319 pm	717
Iodide	358 pm	645

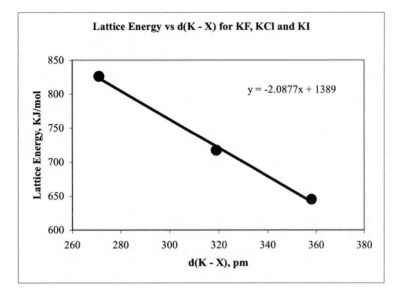

The data fit a straight line with a correlation coefficient of greater than 99%.   (b) From the equation derived for the straight line relationship

Lattice Energy $= -1.984\, d_{M-X} + 1356$

and the value of $d_{K-Br} = 338$ pm , we can estimate the lattice energy of KBr to be 693 $kJ \cdot mol^{-1}$.

(c) The experimental value for the lattice energy for KBr is 689 $kJ \cdot mol^{-1}$, so the agreement is very good.

2.107

2.109 (a) I: Tl_2O_3; II: Tl_2O; (b) $+3$; $+1$; (c) $[Xe]4f^{14}5d^{10}$; $[Xe]\,4f^{14}5d^{10}6s^2$;

(d) Because compound II has a lower melting point, it is probably more covalent, which is consistent with the fact that the $+3$ ion is more polarizing.

2.111 (a)

The four structures with three double bonds (third row) and the one with four double bonds are the most plausible Lewis structures according to formal charge arguments because these five structures minimize the formal charges. (b) The structure with four double bonds fits these observations best since its bond lengths would all be 140 pm, or only 4 pm shorter than the observed length. However, the four structures with three double bonds also fit because, if the double bonds are delocalized by resonance, we can estimate the average bond length to be

$\frac{1}{4}(170 \text{ pm}) + \frac{3}{4}(140 \text{ pm}) = 147.5$ pm, or just 3.5 pm longer than observed.

(c) +7; The structure with all single bonds fits this criterion best.

(d) Approaches (a) and (b) are consistent but approach (c) is not. This result is reasonable because oxidation numbers are assigned by assuming ionic bonding.

2.113 The alkyne group has the stiffer C—H bond because a large force constant, k, results in a higher-frequency absorption.

2.115 Look at the Lewis structures for the molecules:

(a) :C≡O: (b) Ö=Ö   (c) :Ö—Ö=Ö

(d) :Ö—S̈=Ö   (e) :N̈=N=Ö   (f) :Är:

Of these, a, c, d, and e can all function as greenhouse gases

2.117 (a) Due to resonance, ozone (O_3) has only one observed O—O bond length, that being 128 pm. Oxygen (O_2) has a bond length of 121 pm (Table 2.5).

(b) The bond length in ozone is longer than that of oxygen; this can be explained by looking at the Lewis structures for these two species:

Ö=Ö—Ö: ⟷ :Ö—Ö=Ö   Ö=Ö

Ozone (two resonance forms)   Oxygen

Because of resonance, the double bond in ozone can alternate between the two distal oxygens; as a result one would expect to see that the actual bond

SM-93

length is somewhere between that of an O—O single bond and an O—O double bond, which is observed.

2.119 (a) Compare the length of a S-S bond to that of a Cl-Cl bond. From Table 2.5, the Cl-Cl bond length is 199 pm; here the S-S bond length in S_2F_2 is reported to be 190 pm. From Table 2.5 it can be determined that, on average, an X-Y bond is approximately 20 pm longer that a X=Y bond; this suggests that the S-S bond (which is 9 pm shorter than a Cl-Cl single bond) has some double bond character.

(b) and (c) The Lewis structures for the two possible S_2F_2 are:

If resonance is occurring then one would expect that the S-S bond length is indeed between a single and a double bond in length.

2.121 (a) The Lewis structures of NO and NO_2 are:

From Table 2.4, the average bond dissociation energy of a N=O bond is 630 kJ mol^{-1}, which is right in line with the Lewis structure of NO. The bond dissociation energy of each NO bond in NO_2 is 469 kJ mol^{-1}, which is about half-way between a N-O double and an N-O single bond, suggesting that the resonance picture of NO_2 is a reasonable one.

(b) An N-O single bond should have a bond length of 149 pm while an NO double bond should have a bond length of 120 pm (values are obtained by summing the respective covalent radii values from Figure 2.21). From Table 2.5 the length of a N-O triple bond can be estimated to be between 105 and 110 pm. Since NO itself has a bond length of 115 pm, this suggests that its actual bond order is somewhere between that of a double and a triple bond.

(c)

$$0 \quad \overset{\cdot\cdot}{N} = \overset{\cdot\cdot}{\underset{\cdot\cdot}{O}} \quad 0$$

$$\overset{\cdot\cdot}{\underset{\cdot\cdot}{O}} \underset{0}{=} \underset{+1}{N} - \overset{\cdot\cdot}{\underset{\cdot\cdot}{\underset{-1}{O}}}:$$

(d)

$$:\overset{0}{\underset{\cdot\cdot}{O}}: \qquad :\overset{0}{\underset{\cdot\cdot}{O}}:$$

$$\underset{+1}{N} \overset{0}{-} \overset{\cdot\cdot}{\underset{\cdot\cdot}{O}} \overset{}{-} \underset{+1}{N}$$

$$:\overset{\cdot\cdot}{\underset{\cdot\cdot}{O}}: \qquad :\overset{\cdot\cdot}{\underset{\cdot\cdot}{O}}:$$

$$\quad -1 \qquad\qquad -1$$

(e) $N_2O_5(g) + H_2O(l) \rightarrow 2\ HNO_3(aq)$; Nitric acid is produced

(f)

$$4.05\text{ g }N_2O_5 \times \frac{1\text{ mol }N_2O_5}{108.02\text{ g }N_2O_5} \times \frac{2\text{ mol }HNO_3}{1\text{ mol }N_2O_5} = 7.50 \times 10^{-2}\text{ mol }HNO_3$$

$$\text{Molarity} = \frac{7.50 \times 10^{-2}\text{ mol }HNO_3}{1.00\text{ L}} = 7.50 \times 10^{-2}\text{ mol·L}^{-1}\ HNO_3$$

(g) The oxidation number of nitrogen for the various nitrogen oxides are as follows: NO: +2; NO_2: +4; N_2O_3: +3; and N_2O_5: +5. An oxidizing agent is a species that wants to gain electrons; based on oxidation number, N_2O_5 should be the most potent oxidizing agent of the nitrogen oxides, as it possesses the most positive oxidation number for N of the group.

CHAPTER 3

MOLECULAR SHAPE AND STRUCTURE

3.1 (a) Must have lone pairs;

(b) May have lone pairs.

3.3 (a) The shape is angular; the electron pair on the central Cl atom results in a trigonal planar arrangement. (b) The bond angle will be slightly less than $120°$.

3.5 (a) The shape of the thionyl chloride molecule is trigonal pyramidal. (b) The O—S—Cl angles are identical. The lone electron pair repels the bonded electron pairs equally and thus, all O—S—Cl bond angles are compressed equally. (c) The expected bond angle is slightly less than $109.5°$.

3.7 (a) linear; (b) slightly less than $180°$

3.9

(a) The sulfur atom will have five pairs of electrons about it: one nonbonding pair and four bonding pairs to chlorine atoms. The arrangement of electron pairs will be trigonal bipyramidal; the nonbonding pair of electrons will prefer to lie in an equatorial position, because in that location the e-e repulsions will be lowest. The actual structure is described as a seesaw. AX_4E

(b) Like the sulfur atom in (a), the iodine in iodine trichloride has five pairs of electrons about it, but here there are two lone pairs and three bonding pairs. The arrangement of electron pairs will be the same as in (a), and again the lone pairs will occupy the equatorial positions. Because the name of the molecule ignores the lone pairs, it will be classified as T-shaped. AX_3E_2

(c) There are six pairs of electrons about the central iodine atom in IF_4^-. Of these, two are lone pairs and four are bonding pairs. The pairs will be placed about the central atom in an octahedral arrangement with the lone pairs opposite each other. This will minimize repulsions between them. The name given to the structure is square planar. AX_4E_2

(d) In determining the shape of a molecule, double bonds count the same as single bonds. The XeO_3 structure has four "objects" about the central Xe atom: three bonds and one lone pair. These will be placed in a tetrahedral arrangement. Because the lone pair is ignored in naming the molecule, it will be classified as trigonal pyramidal. AX_3E

3.11 The Lewis structures are:

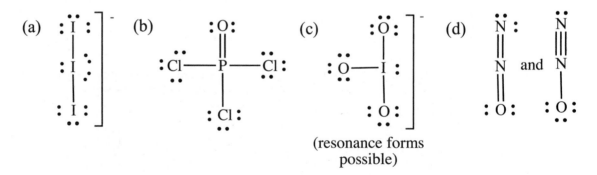

(resonance forms possible)

(a) The I_3^- molecule is predicted to be linear, so the I—I—I angle should equal 180°. AX_2E_3

(b) The $POCl_3$ molecule is tetrahedral. All bond angles should be 109.5°. AX_4

(c) The shape of IO_3^- will be a trigonal pyramid, so the O—I—O bond angles should be less than 109.5°. AX_3E

(d) The structure of N_2O is linear with a bond angle of 180°. AX_2

3.13 The Lewis structures are

(a) (b) (c) d)

(a) The shape of CF_3Cl is tetrahedral; all halogen—C—halogen angles should be approximately $109.5°$. AX_4;

(b) $TeCl_4$ molecules will be see saw shaped with Cl—Te—Cl bond angles of approximately $90°$ and $120°$. AX_4E;

(c) COF_2 molecules will be trigonal planar with F—C—F and O—C—F angles of $120°$. AX_3;

(d) CH_3^- ions will be trigonal pyramidal with H—C—H angles of slightly less than $109.5°$. AX_3E

3.15 (a) The angles represented by a and b are expected to be about $120°$, while c is expected to be about $109.5°$ in 2, 4-pentanedione. All of the angles are expected to be about $120°$ in the acetylacetonate ion.

(b) The major difference arises at the C of the original sp^3- hybridized CH_2 group, which upon deprotonation and resonance goes to sp^2 hybridization with only three groups attached (the double-headed arrow is read as "in resonance with"):

3.17 (a) slightly less than $120°$; (b) $180°$; (c) $180°$; (d) slightly less then $109.5°$

3.19 The Lewis structures are

(a) H–C–Cl structure with H, C, Cl and two additional Cl
(b) CCl structure with four Cl around central C
(c) S C S structure
(d) F S F structure with F atoms

Molecules (a) and (d) are polar; (b) and (c) are nonpolar.

3.21 (a) pyridine: polar

(b) ethane: nonpolar

(c) trichloromethane: polar

3.23 (a) Of the three forms, **1** and **2** are polar; only **3** is nonpolar. This is because the C—Cl bond dipoles are pointing in exactly opposite directions in **3**. (b) The dipole moment for **1** would be the largest because the C—Cl bond vectors are pointing most nearly in the same direction in **1** ($60°$ apart) whereas in **2** the C—Cl vectors point more away from each other ($120°$), giving a larger cancellation of dipole.

3.25   H   H
         C C C N:
       H

The first two carbons (CH_2 and CH) are sp^2 hybridized with H—C—H and C—C—H angles of 120°. The third carbon (bonded to N) is sp hybridized with a C—C—N angle of 180°.

3.27 (a) tetrahedral, bond angle of 109.5°

(b) Tetrahedral about the carbon atoms ($109.5°$); C—Be—C angle of 180°.

(c) angular, H—B—H angle slightly less than 120°

(d) angular, Cl—Sn—Cl angle slightly less than 120°

3.29 (a) H—C—H and H—C—C angles of 120°.

   H   H
    C C
   H   H

(b) linear, 180°. $Cl\ C\ N\!:$

(c) Tetrahedral, 109.5°.

$$:\!\ddot{C}l\!:$$
$$\qquad\qquad |$$
$$:\!\ddot{O}\!-\!P\!-\!\ddot{C}l\!:\qquad\quad \ddot{O}\!=\!P\!-\!\ddot{C}l\!:$$
$$\qquad\quad |\qquad\qquad\qquad |$$
$$:\!\ddot{C}l\!:\qquad\qquad :\!\ddot{C}l\!:$$

(Note: Both are acceptable Lewis structures; the presence of the double bond makes the second molecule the more stable Lewis structure.)

(d) The arrangement of atoms about each N is trigonal pyramidal giving H—N—H and H—N—N bond angles of approximately 107°.

$$H\ \ddot{N}\ \ddot{N}\ H$$
$$\qquad H\ \ H$$

3.31 (a) tetrahedral:

$$:\!\ddot{C}l\!:$$
$$:\!\ddot{O}\quad Sb\quad \ddot{C}l\!:$$
$$:\!\ddot{C}l\!:$$

(b) tetrahedral:

$$:\!\ddot{C}l\!:$$
$$:\!\ddot{O}\quad S\quad \ddot{C}l\!:$$
$$:\!\ddot{O}\!:$$

(c) seesaw:

$$\left[\ :\!\ddot{O}\ \ \overset{:\!\ddot{F}\!:}{\underset{:\!\ddot{O}\!:}{I^{\bullet}\ddot{F}\!:}}\ \right]^{-}$$

Note: While all are acceptable Lewis structures, those with double bonds are the more stable structures.

3.33 (a) sp^3, orbitals oriented toward corners of a tetrahedron (109.5° apart);

(b) sp, orbitals oriented directly opposite to each other (180° apart);

(c) sp^3d^2, orbitals oriented toward the corners of an octahedron (interorbital angles of 90° and 180°);

(d) sp^2, orbitals oriented toward the corners of an equilateral triangle trigonal planar array (angles $= 120°$); trigonal planar.

3.35 (a) sp;   (b) sp^2;   (c) sp^3;   (d) sp^3

3.37 (a) sp^2;   (b) sp^3;   (c) sp^3d;   (d) sp^3

3.39   (a) sp^3;   (b) sp^3d^2;   (c) sp^3d;   (d) sp^3

3.41   The Lewis structure of tetrahedral P_4 is:

(a) Each phosphorous is attached to three other phosphorous atoms and a single lone pair, therefore its hybridization is sp^3.

(b) While each P within this structure is polar (due to the presence of a single lone pair on each phosphorous), P_4 is nonpolar due to the 3D orientations of those lone pairs.

3.43   As the s-character of a hybrid orbital increases, the bond angle increases.

3.45   In formaldehyde, both the C and the O are sp^2 hybridized. The H—C—O bond angle is 120°; the molecule has three sigma bonds (one each connecting the two H's and O to the C) and one pi bond (between the C and the O).

3.47   Atomic orbitals a and b are mutually orthogonal if $\int a \cdot b \, d\tau = 0$ (assuming $a \neq b$) where the integration is over all space. Furthermore, an orbital, a, is normalized if $\int a^2 \, d\tau = 1$.

In this problem, the two hybrid orbitals

are: $h_1 = s + p_x + p_y + p_z$ and $h_2 = s - p_x + p_y - p_z$. Therefore, to show these two orbitals are orthogonal we must show $\int h_1 h_2 \, d\tau = 0$.

$$\int h_1 h_2 \, d\tau = \int (s + p_x + p_y + p_z)(s - p_x + p_y - p_z) \, d\tau =$$

$$\int (s^2 - sp_x + sp_y - sp_z + sp_x - p_x^2 + p_x p_y - p_x p_z + sp_y -$$

$$p_x p_y + p_y^2 - p_y p_z + sp_z - p_z p_x + p_z p_y - p_z^2) \, d\tau$$

Of course, this integral of a sum may be written as a sum of integrals:

$$\int s^2 \, d\tau - \int sp_x \, d\tau + \int sp_y \, d\tau - \int sp_z \, d\tau + \ldots$$

Because the hydrogen wavefunctions are mutually orthogonal, the members of this sum which are integrals of a product of two different wavefunctions are zero. Therefore, this sum of integrals simplifies to:

$$\int s^2 \, d\tau - \int p_x^2 \, d\tau + \int p_y^2 \, d\tau - \int p_z^2 \, d\tau = 1 - 1 + 1 - 1 = 0$$

(recall that the integral of the square of a normalized wavefunction is one.)

3.49 We are given: $\lambda = -\dfrac{\cos\theta}{\cos^2(\frac{1}{2}\theta)}$. In the H_2O molecule, the bond angle is $104.5°$.

Therefore, $\lambda = 0.67$ and the hybridization is $sp^{0.67}$.

3.51 The molecular orbital diagrams are as follows (only the valence electrons are shown):

(a) σ_{2p}^* (b) σ_{2p}^* (c) σ_{2p}^*

π_{2s}^* π_{2s}^* π_{2s}^*

σ_{2p} σ_{2p} σ_{2p}

π_{2s} π_{2s} π_{2s}

σ_{2s}^* σ_{2s}^* σ_{2s}^*

σ_{2s} σ_{2s} σ_{2s}

(a) Li_2 $BO = \frac{1}{2}(2) = 1$

diamagnetic, no unpaired electrons

(b) Li_2^+ $BO = \frac{1}{2}(1) = \frac{1}{2}$

paramagnetic, one unpaired electron

(c) Li_2^- $BO = \frac{1}{2}(2-1) = \frac{1}{2}$

paramagnetic, one unpaired electron

3.53   (a) (1) $\left(\sigma_{2s}\right)^2\left(\sigma_{2s}{}^*\right)^2\left(\sigma_{2p}\right)^2\left(\pi_{2p_x}\right)^2\left(\pi_{2p_y}\right)^2\left(\pi_{2p_x}{}^*\right)^2\left(\pi_{2p_y}{}^*\right)^2\left(\sigma_{2p}{}^*\right)^1$

(2) $\left(\sigma_{2s}\right)^2\left(\sigma_{2s}{}^*\right)^2\left(\sigma_{2p}\right)^2\left(\pi_{2p_x}\right)^2\left(\pi_{2p_y}\right)^2\left(\pi_{2p_x}{}^*\right)^2\left(\pi_{2p_y}{}^*\right)^1$

(3) $\left(\sigma_{2s}\right)^2\left(\sigma_{2s}{}^*\right)^2\left(\sigma_{2p}\right)^2\left(\pi_{2p_x}\right)^2\left(\pi_{2p_y}\right)^2\left(\pi_{2p_x}{}^*\right)^2\left(\pi_{2p_y}{}^*\right)^2\left(\sigma_{2p}{}^*\right)^2$

(b) (1) 0.5;   (2) 1.5;   (3) 0

(c) (1) and   (2) are paramagnetic, with one unpaired electron each

(d) σ for (1) and (3), π for (2).

3.55   The charge on C_2^{n-} is -2 and the bond order is 3 $\left(BO = \frac{1}{2}(2+4+2-2)=3 \right)$.

3.57   (a) The energy level diagram for N_2 is as follows:

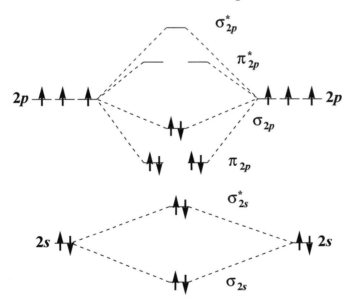

(b) The oxygen atom is more electronegative, which will make its orbitals lower in energy than those of N. The revised energy-level diagram is shown below. This will make all of the bonding orbitals closer to O than to N in energy and will make all the antibonding orbitals closer to N than to O in energy.

Energy level diagram for NO^+

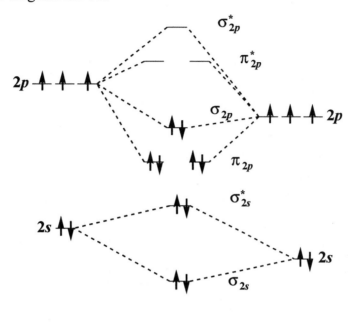

Orbitals on N   Orbitals on O

(c) The electrons in the bonding orbitals will have a higher probability of being at O because O is more electronegative and its orbitals are lower in energy.

3.59   (a) B_2 (6 valence electrons): $(\sigma_{2s})^2 (\sigma_{2s}^*)^2 (\pi_{2p_x})^1 (\pi_{2p_y})^1$,

bond order = 1.

(b) Be_2 (4 valence electrons): $(\sigma_{2s})^2 (\sigma_{2s}^*)^2$, bond order = 0.

(c) F_2 (14 valence electrons):

$(\sigma_{2s})^2 (\sigma_{2s}^*)^2 (\sigma_{2p})^2 (\pi_{2p_x})^2 (\pi_{2p_y})^2 (\pi_{2p_x}^*)^2 (\pi_{2p_y}^*)^2$, bond order = 1.

3.61 CO $(\sigma_{2s})^2 (\sigma_{2s}^*)^2 (\pi_{2p_x})^2 (\pi_{2p_y})^2 (\sigma_{2p})^2$; bond order = 3

$$\text{CO}^+ \quad \left(\sigma_{2s}\right)^2 \left(\sigma_{2s}*\right)^2 \left(\pi_{2p_x}\right)^2 \left(\pi_{2p_y}\right)^2 \left(\sigma_{2p}\right)^1 ; \text{ bond order} = 2.5$$

Due to the higher bond order for CO, it should form a stronger bond and therefore have the higher bond enthalpy

3.63 (a) – (c) All of these molecules possess unpaired electrons and therefore are paramagnetic. B_2^- and B_2^+ have an odd number of electrons and must, therefore, have at least one unpaired electron; indeed, both have only one unpaired electron. B_2 has an even number of electrons, but in its molecular orbital energy level diagram, the HOMO is a degenerate set of π_{2p} orbitals that are each singly occupied, giving this molecule two unpaired electrons. For B_2^-, one more electron will be placed in this degenerate set of orbitals, causing one of the original unpaired electrons to now be paired. B_2^- will therefore have one unpaired electron. Likewise, B_2^+ will have one less electron than B_2; thus one of the originally unpaired electrons will be removed, leaving one unpaired electron in this molecule as well.

3.65 (a) F_2 with 14 valence electrons has a valence electron configuration of $(\sigma_{2s})^2$ $(\sigma_{2s}*)^2 (\sigma_{2p})^2 (\pi_{2px})^2 (\pi_{2py})^2 (\pi_{2px}*)^2 (\pi_{2py}*)^2$ with a bond order of 1. After forming F_2^- from F_2, an electron is added into a σ_{2p}^* orbital. The addition of an electron to this antibonding orbital will result in a reduction of the bond order to 1/2 (See 53). F_2 will have the stronger bond. (b) B_2 will have an electron configuration of $(\sigma_{2s})^2 (\sigma_{2s}*)^2 (\pi_{2px})^1 (\pi_{2py})^1$ with a bond order of 1. Removing one electron to form B_2^+ will eliminate one electron in the bonding orbitals, creating a bond order of 1/2. B_2 will have the stronger bond.

3.67 C_2^+ $\left(\sigma_{2s}\right)^2\left(\sigma_{2s}*\right)^2\left(\pi_{2p_x}\right)^2\left(\pi_{2p_y}\right)^1$;

C_2 $\left(\sigma_{2s}\right)^2\left(\sigma_{2s}*\right)^2\left(\pi_{2p_x}\right)^2\left(\pi_{2p_y}\right)^2$;

C_2^- $\left(\sigma_{2s}\right)^2\left(\sigma_{2s}*\right)^2\left(\pi_{2p_x}\right)^2\left(\pi_{2p_y}\right)^2\left(\sigma_{2p}\right)^1$;

C_2^- is expected to have the lowest ionization energy because its electron is lost from a higher energy MO (σ_{2p}) than either C_2^+ or C_2 (π_{2p_y}).

3.69 Given the overlap integral $S = \int \Psi_{A1s} \Psi_{B1s} d\tau$, the bonding orbital

$\Psi = \Psi_{A1s} + \Psi_{B1s}$, and the fact that the individual atomic orbitals are normalized, we are asked to find the normalization constant N which will normalize the bonding orbital Ψ such that:

$\int N^2 \Psi^2 \, d\tau = N^2 \int (\Psi_{A1s} + \Psi_{B1s})^2 d\tau = 1$

$N^2 \int (\Psi_{A1s} + \Psi_{B1s})^2 d\tau = N^2 \int (\Psi_{A1s}^2 + 2\Psi_{A1s}\Psi_{B1s} + \Psi_{B1s}^2) d\tau$

$= N^2 \left(\int \Psi_{A1s}^2 \, d\tau + 2\int \Psi_{A1s}\Psi_{B1s} d\tau + \int \Psi_{B1s}^2 \, d\tau \right)$

Given the definition of the overlap integral above and the fact that the individual orbitals are normalized, this expression simplifies to:

$N^2(1 + 2S + 1) = 1$

Therefore, $\quad N = \sqrt{\dfrac{1}{2 + 2S}}$

3.71

(a)
$$\left[\begin{array}{c} :\overset{\textstyle ..}{\underset{\textstyle ..}{Cl}}: \\ | \\ :\overset{..}{\underset{..}{Cl}} - Ga - \overset{..}{\underset{..}{Cl}}: \\ | \\ :\overset{..}{\underset{..}{Cl}}: \end{array} \right]^{-}$$

Shape: tetrahedral

Hybridization: sp^3

Bond angles: 109.5°

Non-polar

(b)

Shape: see-saw

Hybridization: dsp^3

Bond angles: $> 90°$ and $>120°$

Polar

(c)

Shape: see-saw

Hybridization: dsp^3

Bond angles: $> 90°$ and $>120°$

Polar

(d)

Shape: tetrahedral

Hybridization: sp^3

Bond angles: $109.5°$

Non-polar

3.73 (a) The Lewis structures are:

(b) All of these species are expected to be diamagnetic. None are radicals.

(c) The predicted bond angles in each species based upon the Lewis structure and VSEPR theory will be

CH_3^+	AX_3	trigonal planar	120°
CH_4	AX_4	tetrahedral	109.5°
CH_3^-	AX_3E	pyramidal	slightly less than 109.5°
CH_2	AX_2E	angular	slightly less than 120°
CH_2^{2+}	AX_2	linear	180°
CH_2^{2-}	AX_2E_2	angular	less than 109.5°, more so than CH_3^- due to the presence of two lone pairs

SM-108

The order of increasing H—C—H bond angle will be

$$CH_2^{2-} < CH_3^- < CH_4 < CH_2 < CH_3^+ < CH_2^{2+}$$

3.75 (a) The elemental composition gives an empirical formula of CH_4O, which agrees with the molar mass. There is only one reasonable Lewis structure; this corresponds to the compound methanol. All of the bond angles about carbon should be $109.5°$. The bond angles about oxygen should be close to $109.5°$ but will be somewhat less, due to the repulsions by the lone pairs. (b) Both carbon and oxygen are sp^3 hybridized. (c) The molecule is polar.

3.77

Bonding	Anti bonding

(a)

(b)

(c) The bonding and antibonding orbitals for HF appear different due to the fact that a p-orbital from the F atom is used to construct bonding and antibonding orbitals whereas in the H_2 molecule s orbitals on each atom are used to construct bonding and antibonding orbitals.

3.79 (a) The expected molecular orbital diagram for CF is

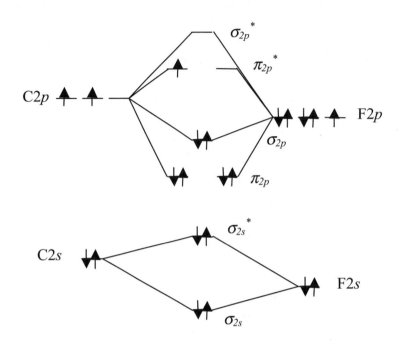

The bond order for the neutral species is 2.5 because one electron occupies a π_{2p}^{*} orbital. Adding an electron to form CF^{-} will reduce the bond order by 1/2 to 2, while removing an electron from form CF^{+} will increase the bond order to 3. Because bond lengths increase as bond order decreases, the C-F bond length varies in these molecules in the following manner: $CF^{+} < CF < CF^{-}$. (b) The CF^{+} ion will be diamagnetic but both CF and CF^{-} will have unpaired electrons (one in the case of CF and two in the case of CF^{-}).

3.81 The Lewis structure of borazine is nearly identical to that of benzene. It is obtained by replacing alternating C atoms in the benzene structure with B and N, as shown. The orbitals at each B and N atom will be sp^{2} hybridized.

3.83 The antibonding molecular orbital is obtained by taking the difference between two atomic orbitals that are proportional to e^{-r/a_o}. Halfway between the two nuclei, the distance from the first nucleus, r_1, is equal to the distance to the second nucleus, r_2, and the antibonding orbital is proportional to:

$$\Psi \propto e^{-r/a_o} - e^{-r/a_o} = 0$$

3.85

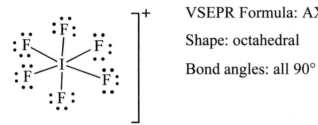

VSEPR Formula: AX_6

Shape: octahedral

Bond angles: all 90°

3.87 The energy of an electron in the nth quantum state of a one-dimensional box is given by: $E = n^2 h^2 / (8mL^2)$, where h is Planck's constant, m is the mass of the electron and L is the length of the box (recall n must be an integer). The lowest energy transition for this system will be from the highest occupied quantum state, which we will identify with the quantum number n_{HO}, to the next highest state or the lowest unoccupied quantum state, n_{LU}. Since each quantum state can hold 2 electrons and each carbon atom contributes one electron, $n_{HO} = N/2$ and $n_{LU} = (N/2) + 1$. (In the case N is odd, $N/2$ must be rounded up to the nearest integer.) Also, the length of the box is given by $L = NR$ where R is the average C—C bond length. Therefore, the lowest energy transition is given by:

$$\Delta E = E_{LU} - E_{HO} = \frac{n_{LU}h^2}{8mN^2R^2} - \frac{n_{HO}h^2}{8mN^2R^2} = \frac{h^2}{8mN^2R^2}\left[\left(\frac{N}{2}+1\right)^2 - \left(\frac{N}{2}\right)^2\right]$$

$$= \frac{h^2(N+1)}{8mN^2R^2}$$

To shift the wavelength of the absorption to longer wavelengths (lower energies) the length of the carbon chain, N, must increase.

3.89 (a)

σ $\qquad\qquad$ π $\qquad\qquad$ δ

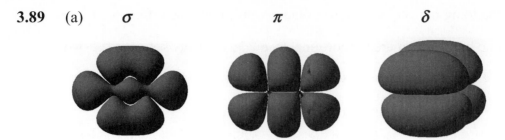

(b) The overlap between the orbitals will decrease as one goes from σ to π to δ, so we expect the bond strengths to decline in the same order.

3.91 The effect of changes (a) and (b) will be similar. The overall bond order will change. In the first case, electrons will be removed from π orbitals so that the net π-bond order will drop from three to two. The same thing will happen in (b), but because two electrons are added to antibonding orbitals, a net total of one π-bond will be broken. Based on this simple model, the ions formed should be paramagnetic because the electrons are added to or taken from doubly degenerate orbitals.

3.93 (a) The Lewis structure of benzyne is

(b) Benzyne would be highly reactive because the two carbon atoms that are *sp* hybridized are constrained to have a very strained structure compared to what their hybridization would like to adopt—namely a linear arrangement. Instead of 180° angles at these carbon atoms, the angles by necessity of being in a six-membered ring are constrained to be close to 120°. A possibility that allows the carbon atoms to adopt more reasonable angles is the formation of a diradical:

3.95 (a) The carbon atoms are all sp^3 hybridized.   (b) The C—C—C,   H—C—H and H—C—C bond angles should be 109.5° based upon the answer to (a).   (c) Because of the ring structure, however, the C—C—C bond angles must be 60°. (d) The σ-bond will have the electron density of the bond located on a line between the two atoms that it joins.   (e) If the C atoms are truly sp^3 hybridized, then the bonding orbitals will not necessarily point directly between the C atoms. (f) The sp^3 hybridized orbitals can still overlap even if they do not point directly between the atoms as shown. Such bonds are sometimes called "bent" bonds, or "banana" bonds. As a result of the situation in the C—C—C bond angles, the H—C—H bond angles are also distorted from 109.5°.

3.97 (a) and (d) have the possibility of n-to-π * transitions because these molecules possess both an atom with a lone pair of electrons (on O's in HCOOH and on N in HCN) and a π-bond to that atom. The other molecules have either a lone pair or a π-bond, but not both.

3.99 (a-b)

All atoms in this molecule have a formal charge of zero.

3.101 (a) A possible Lewis structure for $\left[\text{Sb}_2\text{F}_7\right]^-$ based on the information provided is as follows:

Each Sb has a formal charge of -1 while the bridging F has a formal charge of +1.

(b) Each Sb atom in this structure has a hybridization of sp^3d.

(NOTE: the crystal structure of this ion has been done for both the potassium and the cesium salt; to read more about this controversial ion, see "Crystal Structure of KSb_2F_7. On the Existence of the $\left[\text{Sb}_2\text{F}_7\right]^-$ ion", S.H. Mastin and R.R. Ryan, *Inorganic Chemistry*, **10**, 1757 (1971).

3.103 A possible Lewis structure for $\left[\text{Bi}_2\text{Cl}_4\right]^{2-}$ is as follows:

Formal charges for the bridging Cl atoms and the two Bi are shown.

3.105   A possible Lewis structure for I_5^- that fits the conditions given in the problem is:

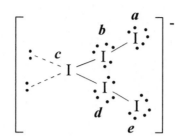

In this structure, two atoms (I_b and I_d) have a formal charge of -1, one atom (I_c) has a formal charge of +1, and two atoms (I_a and I_e) have no formal charge; the ion has an overall charge of -1. To explain the shape, I_b and I_d are sp^3d hybridized (trigonal bipyramid) with the attached atoms occupying axial positions, which results in a bond angle of 180° around each axial atom. I_c has a VSEPR formula of AX_2E_2; its four electron pairs adopt a tetrahedral electron geometry, which is consistent with sp^3 hybridization.

3.107 Al_2Cl_6 is nonpolar; all dipoles in the molecule (due to the Al–Cl bonds) will cancel:

these cancel   these cancel

3.109   (a) sp^3d^3; (b) sp^3d^3f; (c) sp^2d.

3.111   (a) The Lewis structure of hydrogen peroxide is

H —Ö— Ö— H

No atoms have formal charges and the oxidation number of each O is −1.
Oxidation number is more useful than formal charge in predicting the oxidizing

ability of a compound as this will give us a clearer idea of whether a given atom wants to gain or lose an electron.

(b) The bond angles should be about the same as that for water, ca. 104°. All atoms in this molecule are in the same plane but because of unrestricted rotation around the O–O bond it is expected that the hydrogens (and therefore the lone pairs of electrons on the oxygens) can rotate in and out of the plane; one of these rotations will lead to a polar structure (the different forms, arising from rotations around a single σ bond, are called *rotomers*). Because of this, the molecule is expected to be polar, but slightly less so than water.

planar, nonpolar rotomer   planar, polar rotomer   nonplanar, polar rotomer (one of many)

(c) (1) O_2: $\left(\sigma_{2s}\right)^2 \left(\sigma_{2s}{}^*\right)^2 \left(\sigma_{2p}\right)^2 \left(\pi_{2p_x}\right)^2 \left(\pi_{2p_y}\right)^2 \left(\pi_{2p_x}{}^*\right)^1 \left(\pi_{2p_y}{}^*\right)^1$;

Bond order = 2; paramagnetic.

(2) O_2^-: $\left(\sigma_{2s}\right)^2 \left(\sigma_{2s}{}^*\right)^2 \left(\sigma_{2p}\right)^2 \left(\pi_{2p_x}\right)^2 \left(\pi_{2p_y}\right)^2 \left(\pi_{2p_x}{}^*\right)^2 \left(\pi_{2p_y}{}^*\right)^1$;

Bond order = 1.5; paramagnetic.

(3) O_2^+: $\left(\sigma_{2s}\right)^2 \left(\sigma_{2s}{}^*\right)^2 \left(\sigma_{2p}\right)^2 \left(\pi_{2p_x}\right)^2 \left(\pi_{2p_y}\right)^2 \left(\pi_{2p_x}{}^*\right)^1$;

Bond order = 2.5; paramagnetic.

(4) O_2^{2-}: $\left(\sigma_{2s}\right)^2 \left(\sigma_{2s}{}^*\right)^2 \left(\sigma_{2p}\right)^2 \left(\pi_{2p_x}\right)^2 \left(\pi_{2p_y}\right)^2 \left(\pi_{2p_x}{}^*\right)^2 \left(\pi_{2p_y}{}^*\right)^2$;

Bond order = 1; diamagnetic.

(d) Bond length should inversely follow bond order; therefore, the longest bond should belong to the molecule with the lowest bond order $\left(O_2^{2-}\right)$ and the shortest bond should belong to the species with the largest bond order $\left(O_2^+\right)$, which is borne out in the data given.

(e) $2\,Fe^{2+} + H_2O_2 + 2\,H^+ \longrightarrow 2\,Fe^{3+} + 2\,H_2O$

$$54.1\text{ mL }H_2O_2 \times \frac{1\text{ L}}{1000\text{ mL}} \times \frac{0.200\text{ mol }H_2O_2}{1\text{ L }H_2O_2} = 1.08\times10^{-2}\text{ mol }H_2O_2$$

$$1.08\times10^{-2}\text{ mol }H_2O_2 \times \frac{2\text{ mol }Fe^{2+}}{1\text{ mol }H_2O_2} \times \frac{2\text{ mol }Fe^{3+}}{2\text{ mol }Fe^{2+}} \times \frac{55.85\text{ g }Fe^{3+}}{1\text{ mol }Fe^{3+}} = 1.21\text{ g }Fe^{3+}$$

(f) $2\,Fe^{3+} + H_2O_2 + 2\,OH^- \longrightarrow 2\,Fe^{2+} + 2\,H_2O + 2O_2$

$$35.6\text{ mL }H_2O_2 \times \frac{1\text{ L}}{1000\text{ mL}} \times \frac{0.200\text{ mol }H_2O_2}{1\text{ L }H_2O_2} = 7.12\times10^{-3}\text{ mol }H_2O_2$$

$$7.12\times10^{-3}\text{ mol }H_2O_2 \times \frac{2\text{ mol }Fe^{3+}}{1\text{ mol }H_2O_2} \times \frac{2\text{ mol }Fe^{2+}}{2\text{ mol }Fe^{3+}} \times \frac{55.85\text{ g }Fe^{2+}}{1\text{ mol }Fe^{2+}} = 0.795\text{ g }Fe^{2+}$$

(g) Two electrons are transferred.

(h) The homolytic cleavage of hydrogen peroxide to form hydroxyl radicals is:

$$H-\overset{..}{\underset{..}{O}}-\overset{..}{\underset{..}{O}}-H \longrightarrow H-\overset{..}{\underset{..}{O}}\cdot \quad + \quad \cdot\overset{..}{\underset{..}{O}}-H$$

The energy required to break this bond is 157 kJ/mol. To break a single O–O will take:

$$\frac{157\text{ kJ}}{1\text{ mol O}-\text{O bonds}} \times \frac{1\text{ mol O}-\text{O bonds}}{6.022\times10^{23}\text{ O}-\text{O bonds}} \times \frac{1000\text{ J}}{1\text{ kJ}}$$

$$= 2.61\times10^{-19}\text{ J per O}-\text{O bond}$$

$$\upsilon = \frac{E}{h} = \frac{2.61\times10^{-19}\text{ J}}{6.626\times10^{-34}\text{ J}\cdot\text{sec}} = 3.94\times10^{14}\text{ Hz}$$

$$\lambda = \frac{c}{\upsilon} = \frac{2.998\times10^8\text{ m}\cdot\text{s}^{-1}}{3.94\times10^{14}\text{ s}^{-1}} = 7.61\times10^{-7}\text{ m} = 761\text{ nm}$$

CHAPTER 4

THE PROPERTIES OF GASES

4.1 (a) 8×10^9 Pa; (b) 80 kbar; (c) 6×10^7 Torr;

(d) 1×10^6 lb $\cdot$ in^{-2}

4.3 (a) The difference in column height will be equal to the difference in pressure between atmospheric pressure and pressure in the gas bulb. If the pressures were equal, the height of the mercury column on the air side and on the apparatus side would be the same. The pressure in the gas bulb is 0.890 atm or 0.890×760 Torr $\cdot$ atm^{-1} = 676 Torr. The difference would be 762 Torr – 676 Torr = 86 Torr = 86 mm Hg. (b) The side attached to the bulb will be higher because the neon pressure is less than the pressure of the atmosphere. (c) If the student had recorded the level in the atmosphere arm to be higher than the level in the bulb arm by 86 mm Hg then the pressure in the bulb would have been reported as 762 Torr + 86 Torr = 848 Torr.

4.5 $d_1 h_1 = d_2 h_2$

$$73.5 \text{ cm} \times \frac{13.6 \text{ g} \cdot \text{cm}^{-3}}{1.10 \text{ g} \cdot \text{cm}^{-3}} = 909 \text{ cm or } 9.09 \text{ m}$$

4.7 $(20. \text{ in})(10. \text{ in})(14.7 \text{ lb} \cdot \text{in}^{-2}) = 2.9 \times 10^3 \text{ lb}$

4.9 (a) In Boyle's law, $V \propto \dfrac{1}{P}$; $V = \pi r^2 h$; $h \propto V$

Therefore, $\dfrac{h_1}{h_2} = \dfrac{P_2}{P_1}$, since the atmospheric pressure is 29.85 inHg

$P_1 = (29.85 + 12.0)\ \text{inHg} = 41.85\ \text{inHg}$

$P_2 = (29.85 + 30.0)\ \text{inHg} = 59.85\ \text{inHg}$

$h_2 = \dfrac{h_1 P_1}{P_2} = \dfrac{32.0\ \text{in} \times 41.85\ \text{inHg}}{59.85\ \text{inHg}} = 22.4\ \text{in}$

(b) If the atmospheric pressure is 29.85 inHg, the pressure of gas in tube

(1) is: 29.85 inHg + 12.0 inHg = 41.85 inHg

The pressure of gas in tube (2) is: 29.85 inHg + 30.0 inHg = 59.85 inHg

4.11 (a) By using the data provided to make a plot (P vs. T), the slope $= \dfrac{nR}{V}$;

Slope = 0.0041; V = 20 L; or simply use any conditions in the table to

calculate the volume. For example, using the second set of data:

V = nRT/P = (1.0 × 0.082 × 293)/1.20 = 20 L

(b) when additional mole of gases is added into the same volume, the

pressure will be doubled at each T in (a). The plot is:

(c) When pressures (Y axis) are the same for both lines, two lines intersect, at which: $0.0082\,T - 0.0026 = 0.0041T - 0.0013$; $T = 0.32$ K

4.13 (a) Using the ideal gas law with the gas constant R expressed in kPa:

$$P(0.3500 \text{ L}) = (0.1500 \text{ mol}) (8.314\,51 \text{ L} \cdot \text{kPa} \cdot \text{K}^{-1} \cdot \text{mol}^{-1}) (297 \text{ K});$$
$$P = 1.06 \times 10^3 \text{ kPa}$$

(b) BrF_3 has a molar mass of 136.91 g·mol^{-1}. We then substitute into the ideal gas equation:

$$\left(\frac{10.0 \text{ Torr}}{760 \text{ Torr} \cdot \text{atm}^{-1}} \right) V$$

$$= \left(\frac{23.9 \times 10^{-3} \text{ g}}{136.91 \text{ g} \cdot \text{mol}^{-1}} \right) (0.082\,06 \text{ L} \cdot \text{atm} \cdot \text{K}^{-1} \cdot \text{mol}^{-1}) (373 \text{ K})$$

$$V = 4.06 \times 10^2 \text{ mL}$$

(c)
$$(0.77 \text{ atm})(0.1000 \text{ L})$$
$$= \left(\frac{m}{64.06 \text{ g} \cdot \text{mol}^{-1}} \right) (0.082\,06 \text{ L} \cdot \text{atm} \cdot \text{K}^{-1} \cdot \text{mol}^{-1}) (303 \text{ K})$$

$$m = 0.20 \text{ g}$$

(d)
$$(129 \text{ kPa})(6.00 \times 10^3 \text{ m}^3) \left(\frac{1 \times 10^6 \text{ cm}^3}{\text{m}^3} \right) \left(\frac{1 \text{ L}}{1000 \text{ cm}^3} \right)$$
$$= n(8.314\,51 \text{ L} \cdot \text{kPa} \cdot \text{K}^{-1} \cdot \text{mol}^{-1})(287 \text{ K})$$

$$n = 3.24 \times 10^5 \text{ mol CH}_4$$

(e) The number of He atoms is the Avogadro constant N_A multiplied by the number of moles. The number of moles is obtained from the ideal gas equation:

$$PV = nRT$$

$$n = \frac{PV}{RT}; \, n = \frac{N}{N_A}$$

so the number of atoms N will be given by

$$N = N_A \left(\frac{PV}{RT} \right)$$

$$= (6.022 \times 10^{23} \text{ atoms} \cdot \text{mol}^{-1}) \left(\frac{(2.00 \text{ kPa}) (1.0 \times 10^{-6} \text{ L})}{(8.314\,51 \text{ L} \cdot \text{kPa} \cdot \text{K}^{-1} \cdot \text{mol}^{-1}) (158 \text{ K})} \right)$$

$$= 9.2 \times 10^{14} \text{ atoms}$$

4.15 (a)

Volume, L	$\dfrac{nR}{V}$, atm $\cdot$ K^{-1}
0.01	8.21
0.02	4.10
0.03	2.74
0.04	2.05
0.05	1.64

(b) The slope is equal to $\dfrac{nR}{V}$

(c) The intercept is equal to 0.00 for all the plots.

4.17 (a) from $P_1V_1 = P_2V_2$, we have

$(2.0 \times 10^5 \text{ kPa}) (7.50 \text{ mL}) = (P_2) (1000 \text{ mL})$; solving for P_2 we get

1.5×10^3 kPa; (b) similar to (a),

$P_1V_1 = P_2V_2$ or $(643 \text{ Torr}) (54.2 \text{ cm}^3) = (P_2)(7.8 \text{ cm}^3)$, $P_2 = 4.5 \times 10^3$ Torr

4.19 Using $\dfrac{P_1}{T_1} = \dfrac{P_2}{T_2}$, and expressing temperature in Kelvins,

$$\frac{1.10\ atm}{298\ K} = \frac{P_2}{985\ K}; \quad P_2 = 3.64 \text{ atm}$$

4.21 Because P is a constant, we can use $\dfrac{V_1}{T_1} = \dfrac{V_2}{T_2}$, which gives

$$\frac{12.4\ L}{295\ K} = \frac{V_2}{255\ K}; \quad V_2 = 10.7 \text{ L}$$

4.23 If P and T are constant, then $\dfrac{V_1}{n_1} = \dfrac{V_2}{n_2}$, or $\dfrac{V_1}{0.100 \text{ mol}} = \dfrac{V_2}{0.110 \text{ mol}}$.

Solving for V_2 in terms of V_1, we obtain $V_2 = \dfrac{n_2 V_1}{n_1} = \dfrac{0.110 V_1}{0.10} = 1.10 V_1$.

So the volume must be increased by 10.% to keep P and T constant.

4.25 Slope $= \left(\dfrac{V}{T}\right) = 2.88 \times 10^{-4}$ L·K^{-1}. For a given T, we can find a V value,

vice versa. Assuming at T = 25°C = 298 K, V = 0.0858 L

$$P = 0.90 \text{ bar} \times \left(\frac{1\,atm}{1.01325\,bar}\right) = 0.89 \text{ atm};$$

Based on the equation: PV=nRT, $\quad$ n$= \left(\dfrac{mass}{Mol.\ mass}\right)$

$$\text{Mass} = \left(\frac{PV(Mol.mass)}{RT}\right) = \left(\frac{0.89\,atm \times 0.0858\,L \times 16.04\,g \cdot mol^{-1}}{0.08206\,L \cdot atm \cdot mol^{-1} \cdot K^{-1} \times 298K}\right)$$

$= 0.050$ g

4.27 (a) Because P, V, and T all change, we use the relation $\dfrac{P_1 V_1}{T_1} = \dfrac{P_2 V_2}{T_2}$.

Substituting for the appropriate values we get

$$\frac{(0.255 \text{ atm})(35.5 \text{ mL})}{228 \text{ K}} = \frac{(1.00 \text{ atm})(V_2)}{298 \text{ K}}; V_2 = 11.8 \text{ mL}$$

(b) The same relation holds as in (a) but here the final temperature and volume are known:

$$\frac{(0.255 \text{ atm}) (35.5 \text{ mL})}{228 \text{ K}} = \frac{(P_2)(12.0 \text{ mL})}{293 \text{ K}} \cdot P_2 = 0.969 \text{ atm} \quad \text{(c) Similarly,}$$

we can use the same expression, with P and V known and T wanted.

$$\frac{(0.255 \text{ atm})(35.5 \text{ mL})}{228 \text{ K}} = \frac{\left(\dfrac{(500 \text{ Torr})}{760 \text{ Torr} \cdot \text{atm}^{-1}}\right)(12.0 \text{ mL})}{T_2}; T_2 = 199 \text{ K}$$

4.29 (a) $V = \dfrac{nRT}{P} = \dfrac{(1 \text{ mol})(0.082\,06 \text{ L} \cdot \text{atm} \cdot \text{K}^{-1} \cdot \text{mol}^{-1})(773 \text{ K})}{1 \text{ atm}} = 63.4 \text{ L}$

(b) $V = \dfrac{nRT}{P} = \dfrac{(1 \text{ mol})(0.082\,06 \text{ L} \cdot \text{atm} \cdot \text{K}^{-1} \cdot \text{mol}^{-1})(77 \text{ K})}{1 \text{ atm}} = 6.3 \text{ L}$

4.31 Because P, V, and T are state functions, the intermediate conditions are irrelevant to the final states. We can simply use the ideal gas law in the form

$$\frac{P_1 V_1}{T_1} = \frac{P_2 V_2}{T_2}$$

$$\frac{\left(\dfrac{759 \text{ Torr}}{760 \text{ Torr} \cdot \text{atm}^{-1}}\right)(1.00 \text{ L})}{253 \text{ K}} = \frac{\left(\dfrac{252 \text{ Torr}}{760 \text{ Torr} \cdot \text{atm}^{-1}}\right)(V_2)}{1523 \text{ K}}$$

$V_2 = 18.1 \text{ L}$

4.33 Because T is constant, we can use

$P_1 V_1 = P_2 V_2$

$(1.00 \text{ atm}) (1.00 \text{ L}) = P_2 (0.239 \text{ L})$

$P_2 = 4.18 \text{ atm}$

4.35 The volume of each tank is: 3980. cm^3 = 3.980 L, $V_{\text{two\_tanks}}$= 7.960 L

The pressure of each tank, P= 5860. kPa x $\left(\dfrac{1\,atm}{101.325 kPa}\right)$ = 57.83 atm

Based on the equation: PV=nRT, $n= \left(\dfrac{mass}{Mol.\ mass} \right)$

(Mass of O_2 in two tanks)

$= \left(\dfrac{PV(mol.mass)}{RT} \right) = \left(\dfrac{57.83\,atm \times 7.960\,L \times 32.00\,g \cdot mol^{-1}}{0.08206\,L \cdot atm \cdot mol^{-1} \cdot K^{-1} \times 289K} \right) = 621$ g

4.37 $PV = nRT$

$\left(\dfrac{24.5\ \text{kPa}}{101.325\ \text{kPa} \cdot \text{atm}^{-1}} \right)(0.2500\ \text{L})$

$= n(0.082\ 06\ \text{L} \cdot \text{atm} \cdot \text{K}^{-1} \cdot \text{mol}^{-1})(292.7\ \text{K})$

$n = 2.52 \times 10^{-3}$ mol

4.39 (a) $\dfrac{P_1 V_1}{T_1} = \dfrac{P_2 V_2}{T_2}$

$\dfrac{(104\ \text{kPa})(2.0\ \text{m}^3)}{294.3\ \text{K}} = \dfrac{(52\ \text{kPa})\,V_2}{268.2\ \text{K}}$

$V_2 = 3.6\ \text{m}^3$

(b) $\dfrac{P_1 V_1}{T_1} = \dfrac{P_2 V_2}{T_2}$

$\dfrac{(104\ \text{kPa})(2.0\ \text{m}^3)}{294.3\ \text{K}} = \dfrac{(0.880\ \text{kPa})\,V_2}{221.2\ \text{K}}$

$V_2 = 1.8 \times 10^2\ \text{m}^3$

4.41 Density is proportional to the molar mass of the gas as seen from the ideal gas law:

$PV = nRT$

$PV = \dfrac{m}{M} RT$

density = mass per unit volume $= \dfrac{m}{V} = \dfrac{MP}{RT}$

The molar masses of the gases in question are 28.01 g·mol^{-1} for CO(g), 44.01 g·mol^{-1} for CO_2(g), and 34.01 g·mol^{-1} for H_2S(g). The most dense will be the one with the highest molar mass, which in this case is CO_2. The order of increasing density will be $CO < H_2S < CO_2$.

4.43 The pressure of the Ar sample will be given by

$$P_{Ar} = \frac{nRT}{V} = \frac{\left(\dfrac{2.00 \times 10^{-3}\,g}{39.95\,g \cdot mol^{-1}}\right)(0.082\,06\,L \cdot atm \cdot K^{-1} \cdot mol^{-1})(293\,K)}{0.050\,0\,L}$$

$$P_{Kr} = \frac{\left(\dfrac{2.00 \times 10^{-3}\,g}{83.80\,g \cdot mol^{-1}}\right)(0.082\,06\,L \cdot atm \cdot K^{-1} \cdot mol^{-1})(T_2)}{0.050\,0\,L}$$

Because we want the pressure to be the same, we can set these two equal to each other. Because volume, mass of the gases, and the gas constant R are the same on both sides of the equation, they will cancel.

$$\left(\frac{1}{83.80\,g \cdot mol^{-1}}\right)(T_2) = \left(\frac{1}{39.95\,g \cdot mol^{-1}}\right)(293\,K)$$

Solving for T_2, we obtain temperature = 615 K or 342°C.

4.45 (a) Density is proportional to the molar mass of the gas as seen from the ideal gas law. See Section 4.9.

$$d = \frac{(119.37\,g \cdot mol^{-1})\left(\dfrac{200.\,Torr}{760\,Torr \cdot atm^{-1}}\right)}{(0.082\,06\,L \cdot atm \cdot K^{-1} \cdot mol^{-1})(298\,K)} = 1.28\,g \cdot L^{-1}$$

(b) $d = \dfrac{(119.37\,g \cdot mol^{-1})(1.00\,atm)}{(0.082\,06\,L \cdot atm \cdot K^{-1} \cdot mol^{-1})(373\,K)} = 3.90\,g \cdot L^{-1}$

4.47 (a) $$M = \frac{dRT}{P} = \frac{(8.0 \text{ g} \cdot \text{L}^{-1})(0.082\,06 \text{ L} \cdot \text{atm} \cdot \text{K}^{-1} \cdot \text{mol}^{-1})(300. \text{ K})}{2.81 \text{ atm}}$$

$$= 70. \text{ g} \cdot \text{mol}^{-1}$$

(b) The compound is most likely CHF_3 for which $M = 70 \text{ g} \cdot \text{mol}^{-1}$.

(c) You can use the relationship in (a) to calculate the new density, or you can apply the proportionality changes expected from the change in pressure and temperature to the original density:

$$d_2 = (8.0 \text{ g} \cdot \text{L}^{-1})\left(\frac{1.00 \text{ atm}}{2.81 \text{ atm}}\right)\left(\frac{300. \text{ K}}{298 \text{ K}}\right) = 2.9 \text{ g} \cdot \text{L}^{-1}$$

4.49 From the analytical data, an empirical formula of CHCl is calculated. The empirical formula mass is $48.47 \text{ g} \cdot \text{mol}^{-1}$. The problem may be solved using the ideal gas law:

$$PV = nRT$$

$$PV = \frac{m}{M} RT$$

$$M = \frac{mRT}{PV}$$

$$M = \frac{(3.557 \text{ g})(0.082\,06 \text{ L} \cdot \text{atm} \cdot \text{K}^{-1} \cdot \text{mol}^{-1})(273 \text{ K})}{(1.10 \text{ atm})(0.755 \text{ L})} = 95.9 \text{ g} \cdot \text{mol}^{-1}$$

n in the formula $(CHCl)_n$ is equal to

$95.9 \text{g} \cdot \text{mol}^{-1} \div 48.47 \text{g} \cdot \text{mol}^{-1} = 1.98$. The formula is $C_2H_2Cl_2$.

4.51 Density is proportional to the molar mass of the gas as seen from the ideal gas law:

$$PV = nRT$$

$$PV = \frac{m}{M} RT$$

$$\text{density} = \text{mass per unit volume} = \frac{m}{V} = \frac{MP}{RT}$$

$$0.943 \text{ g} \cdot \text{L}^{-1} = \frac{M\left(\dfrac{53.1 \text{ kPa}}{101.325 \text{ kPa} \cdot \text{atm}^{-1}}\right)}{(0.082\,06 \text{ L} \cdot \text{atm} \cdot \text{K}^{-1} \cdot \text{mol}^{-1})(298 \text{ K})}$$

$$M = 44.0 \text{ g} \cdot \text{mol}^{-1}$$

4.53 (a) The number of moles of H_2 needed will be 1.5 times the amount of NH_3 produced, as seen from the balanced equation:

$$\tfrac{1}{2}N_2(g) + \tfrac{3}{2}H_2(g) \rightarrow NH_3(g)$$

or

$$N_2(g) + 3\,H_2(g) \rightarrow 2\,NH_3(g)$$

Once the number of moles is known, the volume can be obtained from the ideal gas law.

$$V = \frac{n_{H_2}RT}{P} = \frac{\left(\tfrac{3}{2}n_{NH_3}\right)RT}{P} = \frac{\left(\left(\dfrac{3 \text{ mol H}_2}{2 \text{ mol NH}_3}\right)\dfrac{(1.0 \times 10^3 \text{ kg})(10^3 \text{ g} \cdot \text{kg}^{-1})}{17.03 \text{ g} \cdot \text{mol}^{-1}}\right)RT}{P}$$

$$= \frac{\left(\dfrac{3}{2}\right)\left(\dfrac{10^6 \text{ g}}{17.03 \text{ g} \cdot \text{mol}^{-1}}\right)(0.082\,06 \text{ L} \cdot \text{atm} \cdot \text{K}^{-1} \cdot \text{mol}^{-1})(623 \text{ K})}{15.00 \text{ atm}}$$

$$= 3.0 \times 10^5 \text{ L}$$

(b) The ideal gas equation

$$\frac{P_1 V_1}{n_1 R T_1} = \frac{P_2 V_2}{n_2 R T_2} \quad \text{simplifies to} \quad \frac{P_1 V_1}{T_1} = \frac{P_2 V_2}{T_2}$$

because R and n are constant for this problem.

$$\frac{(15.00 \text{ atm})(3.0 \times 10^5 \text{ L})}{623 \text{ K}} = \frac{(376 \text{ atm})V_2}{(523 \text{ K})}$$

$$V_2 = 1.0 \times 10^4 \text{ L}$$

4.55 We need to find the number of moles of $CH_4(g)$ present in each case. Because the combustion reaction is the same in both cases, as are the temperature and pressure, the larger number of moles of $CH_4(g)$ should produce the larger volume of $CO_2(g)$. We will use the ideal gas equation to solve for n in the first case:

$$n = \frac{PV}{RT} = \frac{(1.00 \text{ atm})(2.00 \text{ L})}{(0.082\ 06 \text{ L} \cdot \text{atm} \cdot \text{K}^{-1} \cdot \text{mol}^{-1})(348 \text{ K})} = 0.0700 \text{ mol } CH_4$$

2.00 g of CH_4 will be $\dfrac{2.00 \text{ g}}{16.04 \text{ g} \cdot \text{mol}^{-1}} = 0.124 \text{ mol}$

The latter case will have the greater number of moles of CH_4 and should produce the larger amount of $CO_2(g)$.

4.57 The molar mass of glucose is $180.15 \text{ g} \cdot \text{mol}^{-1}$. From this, we can calculate the number of moles of glucose formed and, using the reaction stoichiometry, determine the number of moles of CO_2 needed. With that information and the other information provided in the problem, we can use the ideal gas law to calculate the volume of air that is needed:

$$PV = nRT$$

$$V = \frac{\left[\left(\dfrac{10.0 \text{ g glucose}}{180.15 \text{ g glucose} \cdot \text{mol}^{-1} \text{ glucose}}\right)\left(\dfrac{6 \text{ mol } CO_2}{1 \text{ mol glucose}}\right)\right]}{\left(\dfrac{0.26 \text{ Torr}}{760 \text{ Torr} \cdot \text{atm}^{-1}}\right)}$$

$$\times\ (0.082\ 06 \text{ L} \cdot \text{atm} \cdot \text{K}^{-1} \cdot \text{mol}^{-1})(298 \text{ K})$$

$$= 2.4 \times 10^4 \text{ L}$$

4.59 (a) This is a limiting reactant problem. Our first task is to determine the number of moles of NH_3 and HCl that are present to start with. This can be done from the ideal gas equation:

$$PV = nRT$$

$$n_{NH_3} = \frac{PV}{RT} = \frac{\left(\dfrac{100\ \text{Torr}}{760\ \text{Torr} \cdot \text{atm}^{-1}}\right)(0.0150\ \text{L})}{(0.082\ 06\ \text{L} \cdot \text{atm} \cdot \text{K}^{-1} \cdot \text{mol}^{-1})(303\ \text{K})} = 7.94 \times 10^{-5}\ \text{mol}$$

$$n_{HCl} = \frac{PV}{RT} = \frac{\left(\dfrac{150\ \text{Torr}}{760\ \text{Torr} \cdot \text{atm}^{-1}}\right)(0.0250\ \text{L})}{(0.082\ 06\ \text{L} \cdot \text{atm} \cdot \text{K}^{-1} \cdot \text{mol}^{-1})(298\ \text{K})} = 2.02 \times 10^{-4}\ \text{mol}$$

The ammonia is the limiting reactant. The number of moles of $NH_4Cl(s)$ that form will be equal to the number of moles of NH_3 that react. From the molar mass of NH_4Cl ($53.49\ \text{g} \cdot \text{mol}^{-1}$) and the number of moles, we can calculate the mass of NH_4Cl that forms:

$$(7.94 \times 10^{-5}\ \text{mol}\ NH_4Cl(s))(53.49\ \text{g} \cdot \text{mol}^{-1}) = 4.25 \times 10^{-3}\ \text{g}$$

(b) There will be $(2.02 \times 10^{-4}\ \text{mol} - 7.94 \times 10^{-5}\ \text{mol}) = 1.23 \times 10^{-4}\ \text{mol}\ HCl$ left after the reaction. This quantity will exist in a total volume after mixing of 40.0 mL or 0.0400 L. Again, we use the ideal gas law to determine the final pressure:

$$PV = nRT$$

$$P = \frac{(1.23 \times 10^{-4}\ \text{mol})(0.082\ 06\ \text{L} \cdot \text{atm} \cdot \text{K}^{-1} \cdot \text{mol}^{-1})(300\ \text{K})}{0.0400\ \text{L}} = 0.0757\ \text{atm}$$

4.61 (a) $x_{HCl} = \left(\dfrac{n_{HCl}}{n_{Hcl} + n_{benzene}}\right) = \left(\dfrac{\#\ of\ HCl\ molecule}{Total\ \#\ molecules}\right) = \left(\dfrac{9}{10}\right) = 0.9;$

$X_{benzene} = 1 - 0.9 = 0.1$

(b) $P_{HCl} = 0.9 \times 0.8\ \text{atm} = 0.72\ \text{atm};\ P_{benzene} = 0.1 \times 0.8\ \text{atm} = 0.08\ \text{atm};$

4.63 (a) The molar volume of an ideal gas is 22.4 L at 273.15 K. 1.0 mol of ideal gas will exert a pressure of 1.0 atm under those conditions. The partial pressure of $N_2(g)$ will be 1.0 atm. Because there are 2.0 mol of $H_2(g)$, the partial pressure of $H_2(g)$ will be 2.0 atm. (b) The total pressure will be 1.0 atm + 2.0 atm = 3.0 atm.

4.65 $n_{total} = n_{N_2} + n_{O_2}$; $n_{N_2} = \dfrac{PV}{RT} = \dfrac{(0.50 \text{ bar}) \times \left(\dfrac{1 \text{ atm}}{1.01325 \text{ bar}}\right) \times (1.00 \text{ L})}{(0.08206 \text{ L} \cdot \text{atm} \cdot \text{K}^{-1} \cdot \text{mol}^{-1}) \times (288 \text{ K})}$

$$= 0.021 \text{ mol}$$

$n_{total} = 0.021 + 0.10 = 0.12 \text{ mol}$; $x_{nitrogen} = \left(\dfrac{0.021}{0.12}\right) = 0.18$; $x_{oxygen} = 0.82$

If 0.020 mol were released, then

$n_{left} = 0.10 \text{ mol}$ and $n_{oxygen} = (0.82) \times (0.10 \text{ mol}) = 0.082 \text{ mol}$

$P_{O_2} = \dfrac{nRT}{V} = \dfrac{(0.082 \text{ mol}) \times (0.08206 \text{ L} \cdot \text{atm} \cdot \text{K}^{-1} \cdot \text{mol}^{-1}) \times (288 \text{ K})}{1.00 \text{ L}}$

$= 1.9 \text{ atom}$

4.67 (a) Of the 756.7 Torr measured, 17.54 Torr will be due to water vapor. The pressure due to $H_2(g)$ will, therefore, be

756.7 Torr − 17.54 Torr = 739.2 Torr. (b) $H_2O(l) \rightarrow H_2(g) + \frac{1}{2}O_2(g)$;

(c) To answer this question, we must determine the number of moles of H_2 produced in the reaction. Using the partial pressure of H_2 calculated in part (a) and the ideal gas equation, we can set up the following:

$$\left(\frac{739.2 \text{ Torr}}{760 \text{ Torr} \cdot \text{atm}^{-1}}\right)(0.220 \text{ L}) = n(0.082\ 06 \text{ L} \cdot \text{atm} \cdot \text{K}^{-1} \cdot \text{mol}^{-1})(293 \text{ K})$$

Solving for n, we obtain $n = 0.008\ 90$ mol. According to the stoichiometry of the reaction, half as much oxygen as hydrogen should be produced, so the number of moles of $O_2 = 0.004\ 45$ mol. The mass of O_2 will be given by $(0.004\ 45 \text{ mol})(32.00 \text{ g} \cdot \text{mol}^{-1}) = 0.142$ g.

4.69 No. The reason is that not all the molecules are traveling at the same velocity. Force $= \left(\dfrac{NmAv_x^{\ 2}}{V}\right)$, if v_x varies, force will be different.

4.71 Graham's law of effusion states that the rate of effusion of a gas is inversely proportional to the square root of its molar mass:

$$\text{rate of effusion} = \frac{1}{\sqrt{M}}$$

Diffusion also follows this relationship. If we have two different gases whose rates of diffusion are measured under identical conditions, we can take the ratio

$$\frac{\text{rate}_1}{\text{rate}_2} = \frac{\frac{1}{\sqrt{M_1}}}{\frac{1}{\sqrt{M_2}}} = \sqrt{\frac{M_2}{M_1}}$$

If a compound takes 1.24 times as long to diffuse as Kr gas, the rate of diffusion of Kr is 1.24 times that of the unknown. We can now use the expression to calculate the molar mass of the unknown, given the mass of Kr:

$$\frac{1.24}{1} = \sqrt{\frac{M_2}{83.80 \text{ g} \cdot \text{mol}^{-1}}}$$

$$M_2 = 129 \text{ g} \cdot \text{mol}^{-1}$$

A mass of 129 g · mol^{-1} corresponds to a molecular formula of $C_{10}H_{10}$.

4.73 The rate of effusion is inversely proportional to the square root of the molar mass. Using a ratio as follows allows us to calculate the time of effusion without knowing the exact conditions of pressure and temperature:

$$\frac{\text{rate}_1}{\text{rate}_2} = \frac{\frac{1}{\sqrt{M_1}}}{\frac{1}{\sqrt{M_2}}} = \sqrt{\frac{M_2}{M_1}}$$

The rate will be equal to the number of molecules N that effuse in a given time interval. For the conditions given, N will be the same for argon and for the second gas chosen.

$$\frac{\dfrac{N}{\text{time}}}{\dfrac{N}{147\ \text{s}}} = \frac{\dfrac{1}{\text{time}}}{\dfrac{1}{147\ \text{s}}} = \sqrt{\frac{39.95\ \text{g} \cdot \text{mol}^{-1}}{M_1}}$$

In order to calculate the time of effusion, we need to know only the molar mass of the gases.

(a) For CO_2 with a molar mass of

$$44.01\ \text{g} \cdot \text{mol}^{-1} : \frac{\dfrac{1}{\text{time}_{CO_2}}}{\dfrac{1}{147\ \text{s}}} = \sqrt{\frac{39.95\ \text{g} \cdot \text{mol}^{-1}}{44.01\ \text{g} \cdot \text{mol}^{-1}}}$$

time = 154 s

(b) For C_2H_4 with a molar mass of

$$28.05\ \text{g} \cdot \text{mol}^{-1} : \frac{\dfrac{1}{\text{time}_{C_2H_4}}}{\dfrac{1}{147\ \text{s}}} = \sqrt{\frac{39.95\ \text{g} \cdot \text{mol}^{-1}}{28.05\ \text{g} \cdot \text{mol}^{-1}}}$$

time = 123 s

(c) For H_2 with a molar mass of

$$2.01\ \text{g} \cdot \text{mol}^{-1} : \frac{\dfrac{1}{\text{time}_{CO_2}}}{\dfrac{1}{147\ \text{s}}} = \sqrt{\frac{39.95\ \text{g} \cdot \text{mol}^{-1}}{2.01\ \text{g} \cdot \text{mol}^{-1}}}$$

time = 33.0 s

(d) For SO_2 with a molar mass of

$$64.06\ \text{g} \cdot \text{mol}^{-1} : \frac{\dfrac{1}{\text{time}_{CO_2}}}{\dfrac{1}{147\ \text{s}}} = \sqrt{\frac{39.95\ \text{g} \cdot \text{mol}^{-1}}{64.06\ \text{g} \cdot \text{mol}^{-1}}}$$

time = 186 s

4.75 The formula mass of C_2H_3 is $27.04 \text{ g} \cdot \text{mol}^{-1}$. From the effusion data, we can calculate the molar mass of the sample.

$$\frac{\text{rate}_1}{\text{rate}_2} = \frac{\dfrac{1}{\sqrt{M_1}}}{\dfrac{1}{\sqrt{M_2}}} = \sqrt{\frac{M_2}{M_1}}$$

Because time is inversely proportional to rate, we can write alternatively

$$\frac{\dfrac{1}{349 \text{ s}}}{\dfrac{1}{210 \text{ s}}} = \sqrt{\frac{39.95 \text{ g} \cdot \text{mol}^{-1}}{M_1}}$$

$$\frac{210}{349} = \sqrt{\frac{39.95 \text{ g} \cdot \text{mol}^{-1}}{M_1}}$$

$$M_1 = 110 \text{ g} \cdot \text{mol}^{-1}$$

The molar mass is 4.1 times that of the empirical formula mass, so the molecular formula is C_8H_{12}.

4.77 (a) The average kinetic energy is obtained from the expression: average kinetic energy $= \frac{3}{2}RT$. The value is independent of the nature of the monatomic ideal gas. The numerical values are:

(a) $4103.2 \text{ J} \cdot \text{mol}^{-1}$; (b) $4090.7 \text{ J} \cdot \text{mol}^{-1}$;

(c) $4103.2 \text{ J} \cdot \text{mol}^{-1} - 4090.7 \text{ J} \cdot \text{mol}^{-1} = 12.5 \text{ J} \cdot \text{mol}^{-1}$

4.79 The root mean square speed is calculated from the following equation:

$$v_{\text{rms}} = \sqrt{\frac{3RT}{M}}$$

(a) methane, CH_4, $M = 16.04 \text{ g} \cdot \text{mol}^{-1}$

$$v_{\text{rms}} = \sqrt{\frac{3 \times (8.314 \text{ kg} \cdot \text{m}^2 \cdot \text{s}^{-2} \cdot \text{K}^{-1} \cdot \text{mol}^{-1}) \times (253 \text{ K})}{1.604 \times 10^{-2} \text{ kg} \cdot \text{mol}^{-1}}}$$

$$= 627 \text{ m} \cdot \text{s}^{-1}$$

(b) ethane, C_2H_6, $M = 30.07 \text{ g} \cdot \text{mol}^{-1}$

$$v_{rms} = \sqrt{\frac{3 \times (8.314 \text{ kg} \cdot \text{m}^2 \cdot \text{s}^{-2} \cdot \text{K}^{-1} \cdot \text{mol}^{-1}) \times (253 \text{ K})}{3.007 \times 10^{-2} \text{ kg} \cdot \text{mol}^{-1}}}$$

$$= 458 \text{ m} \cdot \text{s}^{-1}$$

(c) propane, C_3H_8, $M = 44.09 \text{ g} \cdot \text{mol}^{-1}$

$$v_{rms} = \sqrt{\frac{3 \times (8.314 \text{ kg} \cdot \text{m}^2 \cdot \text{s}^{-2} \cdot \text{K}^{-1} \cdot \text{mol}^{-1}) \times (253 \text{ K})}{4.409 \times 10^{-2} \text{ kg} \cdot \text{mol}^{-1}}}$$

$$= 378 \text{ m} \cdot \text{s}^{-1}$$

4.81 Based on equation: $v_{rms} = \left(\dfrac{3RT}{M}\right)^{1/2}$, If T is constant,

$(v_{rms})_1 (M_1)^{1/2} = (v_{rms})_2 (M_2)^{1/2}$, $(550. \text{ m} \cdot \text{s}^{-1})(16.06)^{1/2} = (v_{rms})_{Kr} \text{ x} (83.80)^{1/2}$

$(v_{rms})_{Kr} = 241 \text{ m} \cdot \text{s}^{-1}$

Note: you do not need to change M into kg/mol in this calculation

4.83 Use the expression for the root mean square speed to determine the temperature.

$$v_{rms} = \sqrt{\frac{3RT}{M}}$$

$$T = \frac{v_{rms}^2 \cdot M}{3R} = \frac{(1477 \text{ m} \cdot \text{s}^{-1})^2 \cdot (4.00 \times 10^{-3} \text{ kg} \cdot \text{mol}^{-1})}{3(8.314 \text{ J} \cdot \text{K}^{-1} \cdot \text{mol}^{-1})}$$

$$= 349.9 \text{ K}$$

Use the Maxwell Distribution of Speeds (Equation 4.22) appropriately for both gases:

$$\frac{f(v_{He})}{f(v_{Ar})} = \frac{4\pi N \left(\dfrac{M_{He}}{2RT}\right)^{\frac{3}{2}} (v_{He})^2 e^{-M_{He}(v_{He})^2 / 2RT}}{4\pi N \left(\dfrac{M_{Ar}}{2RT}\right)^{\frac{3}{2}} (v_{Ar})^2 e^{-M_{Ar}(v_{Ar})^2 / 2RT}}$$

$$= \frac{\left(M_{He}\right)^{\frac{3}{2}} (v_{He})^2 e^{-M_{He}(v_{He})^2 / 2RT}}{\left(M_{Ar}\right)^{\frac{3}{2}} (v_{Ar})^2 e^{-M_{Ar}(v_{Ar})^2 / 2RT}}$$

$$= \frac{\left(4.00 \text{ g} \cdot \text{mol}^{-1}\right)^{\frac{3}{2}} (1477 \text{ m} \cdot \text{s}^{-1})^2}{\left(39.95 \text{ g} \cdot \text{mol}^{-1}\right)^{\frac{3}{2}} (467 \text{ m} \cdot \text{s}^{-1})^2} \times \frac{e^{\frac{-(4.00 \times 10^{-3} \text{ kg} \cdot \text{mol}^{-1}) \times (1477 \text{ m} \cdot \text{s}^{-1})^2}{2 \times (8.314 \text{ J} \cdot \text{K}^{-1} \cdot \text{mol}^{-1}) \times (349.9 \text{ K})}}}{e^{\frac{-(39.95 \times 10^{-3} \text{ kg} \cdot \text{mol}^{-1}) \times (467 \text{ m} \cdot \text{s}^{-1})^2}{2 \times (8.314 \text{ J} \cdot \text{K}^{-1} \cdot \text{mol}^{-1}) \times (349.9 \text{ K})}}}$$

$$= \left(\frac{4.00}{39.95}\right)^{\frac{3}{2}} \left(\frac{1477}{467}\right)^{\frac{3}{2}} (0.9977)$$

$$= 0.316$$

4.85 (a) The most probable speed is the one that corresponds to the maximum on the distribution curve. (b) The percentage of molecules having the most probable speed decreases as the temperature is raised (the distribution spreads out).

4.87 Hydrogen bonding is important in HF. At low temperatures, this hydrogen bonding causes the molecules of HF to be attracted to each other more strongly, thus lowering the pressure. As the temperature is increased, the hydrogen bonds are broken and the pressure rises more quickly than for an ideal gas. Dimers (2 HF molecules bonded to each other) and chains of HF molecules are known to form.

4.89 (a) root mean square speed $(v_{rms}) = \sqrt{\dfrac{3RT}{M}}$; the smaller the molecular mass, the larger the v_{rms}. Therefore, H_2 molecules have greater root mean square speed.

(b) NH_3 will have stronger attraction between molecules because the H atom on one NH_3 molecule intends to interact with the lone pair electrons on the N of another NH_3 molecule (H-bonding).

4.91 The pressures are calculated very simply from the ideal gas law:

$$P = \frac{nRT}{V} = \frac{(1.00 \text{ mol})(0.082\,06 \text{ L} \cdot \text{atm} \cdot \text{K}^{-1} \cdot \text{mol}^{-1})(298 \text{ K})}{V}$$

Calculating for the volumes requested, we obtain $P =$ (a) 1.63 atm; (b) 48.9 atm; (c) 489 atm. The calculations can now be repeated using the van der Waals equation:

$$\left(P + \frac{an^2}{V^2}\right)(V - nb) = nRT$$

We can rearrange this to solve for P :

$$P = \left(\frac{nRT}{V - nb}\right) - \left(\frac{an^2}{V^2}\right)$$

$$= \left(\frac{(1.00 \text{ mol})(0.082\ 06 \text{ L} \cdot \text{atm} \cdot \text{K}^{-1} \cdot \text{mol}^{-1})(298 \text{ K})}{V - (1.00 \text{ mol})(0.04267 \text{ L} \cdot \text{mol}^{-1})}\right)$$

$$- \left(\frac{(3.640 \text{ L}^2 \cdot \text{atm} \cdot \text{mol}^{-2})(1.00^2)}{V^2}\right)$$

Using the three values for V, we calculate for $P =$ (a) 1.62; (b) 38.9; (c) 1.88×10^3 atm. Note that at low pressures, the ideal gas law gives essentially the same values as the van der Waals equation, but at high pressures there is a very significant differences.

4.93

α (atm·L^2·mol^{-2})	Substance
17.58	CH_3CN
3.392	CO_2
2.253	CH_4
0.2107	Ne

4.95 Ammonia: $a = 4.225$ L$^2 \cdot$ atm $\cdot$ mol^{-2}; $b = 0.037\ 07$ L $\cdot$ atm^{-1}

Oxygen: $a = 1.378$ L$^2 \cdot$ atm $\cdot$ mol^{-2}; $b = 0.0318\ 83$ L $\cdot$ atm^{-1}

Volume	P, ammonia	P, oxygen	P, ideal
0.05	3581	1897	489
0.1	811	497	245
0.2	256	180	122

0.3	140	106	82
0.4	94	75	61
0.5	70	58	49
0.6	55	47	41
0.7	46	39	35
0.8	39	34	31
0.9	34	30	27
1	30	27	24

Clearly, the greater deviation from the ideal gas law values occurs at low volumes or higher pressures. Ammonia deviates more strongly and its van der Waals constants are larger than those for oxygen. This may likely arise because ammonia is polar and will have stronger intermolecular interactions.

4.97 (a) $x_{Ne} = \left(\dfrac{6}{10}\right) = 0.6$; $x_{Ar} = \left(\dfrac{4}{10}\right) = 0.4$

At constant T and V, $\left(\dfrac{n_1}{P_1}\right) = \left(\dfrac{n_2}{P_2}\right) \rightarrow \left(\dfrac{x_{Ne}}{P_{Ne}}\right) = \left(\dfrac{x_{Ar}}{P_{Ar}}\right)$

$P_{Ar} = \left(\dfrac{(0.4)(420.\ \text{Torr})}{0.6}\right) = 280.\ \text{Torr}$

(b) $P_{total} = (280.\ \text{Torr}) + (420.\ \text{Torr}) = 700.\ \text{Torr}$

4.99 (a) $Cl_2(g) + F_2(g) \rightarrow 2\ ClF(g)$

$2.00\ \text{mol}\ F_2 \left(\dfrac{2\ mol\ ClF}{1\ mol\ F_2}\right) = 4.00\ \text{mol ClF}$

$PV = nRT;\ V = nRT/P = \dfrac{4.00\ mol \times 0.08206 \dfrac{L \cdot atm}{mol \cdot K} \times 298\ K}{1.00\ atm} = 97.8\ L$

(b) $Cl_2(g) + 3\ F_2(g) \rightarrow 2\ ClF_3(g)$

$2.00\ \text{mol}\ F_2 \left(\dfrac{2\ mol\ ClF}{3\ mol\ F_2}\right) = 1.33\ \text{mol ClF}_3$

$$\text{PV=nRT;} \quad V = nRT/P = \frac{1.33 \ mol \times 0.08206 \dfrac{L \cdot atm}{mol \cdot K} \times 298 \ K}{1.00 \ atm} = 32.5 \ L$$

(c) $Cl_2(g) \ + \ 5 \ F_2(g) \ \rightarrow \ 2 \ ClF_5(g)$

$$2.00 \ mol \ F_2 \left(\frac{2 \ mol \ ClF}{5 \ mol \ F_2} \right) = 0.80 \ mol \ ClF_3$$

$$\text{PV=nRT;} \quad V = nRT/P = \frac{0.80 \ mol \times 0.08206 \dfrac{L \cdot atm}{mol \cdot K} \times 298 \ K}{1.00 \ atm} = 19.6 \ L$$

4.101 (a)

$$\ddot{\text{O}} = \dot{\text{N}} - \ddot{\ddot{\text{O}}}: \quad \longleftrightarrow \quad :\ddot{\ddot{\text{O}}} - \dot{\text{N}} = \ddot{\text{O}}$$

(b) Since the wavelength of the absorbed photons is 197 nm, we can find the energy per photon.

$$E_{photon} = \frac{hc}{\lambda} = \frac{(6.626 \times 10^{-34} \, J \cdot s)(2.998 \times 10^8 m \cdot s^{-1})}{197 \ nm} \left(\frac{10^9 \ nm}{m} \right)$$
$$= 1.008 \times 10^{-18} \ J$$

The number of photons in 1.07 mJ must be equal to the number of NO_2 molecules.

$$? \ photons = 1.07 \ mJ \left(\frac{1 \ J}{1000 \ mJ} \right) \left(\frac{1 \ photon}{1.008 \times 10^{-18} \ J} \right)$$
$$= 1.062 \times 10^{15} \ photons = 1.062 \times 10^{15} \ NO_2 \ molecules$$

The pressure is created by all the molecules in the sample, so the ideal gas law can be used to find the total molecules.

$$? \ molecules \ total = N_{AV} \times n_{tot} = N_{AV} \times \frac{P_{tot}V}{RT}$$
$$= (6.022 \times 10^{23} \, molecules \cdot mol^{-1})$$
$$\times \left(\frac{(0.85 \ atm)(2.5 \ L)}{(0.08206 \ L \cdot atm \cdot K^{-1} \cdot mol^{-1})(293 \ K)} \right)$$
$$= 5.32 \times 10^{22} \, molecules \ total$$

Therefore, the proportion of NO_2 molecules in the sample is

$$\frac{1.062 \times 10^{15}\ NO_2\ \text{molecules}}{5.32 \times 10^{22}\ \text{molecules total}} = 0.020\ \text{ppm}$$

4.103 Use the ideal gas law to calculate the number of moles of HCl.

$$n = \frac{PV}{RT} = \frac{(690.\ \text{Torr})(200.\ \text{mL})}{(0.082\ 06\ \text{L} \cdot \text{atm} \cdot \text{K}^{-1} \cdot \text{mL}^{-1})(293\ \text{K})} \cdot \frac{(1\ \text{atm})}{(760\ \text{Torr})} \cdot \frac{(1\ \text{L})}{(1000\ \text{mL})}$$

$$= 7.55 \times 10^{-3}\ \text{mol HCl}$$

Since the reaction between HCl and NaOH occurs in a 1:1 mole ratio, this number of moles of NaOH is also present in the volume of NaOH(aq) required to reach the stoichiometric point of the titration. Therefore, the molarity of the NaOH solution is

$$\frac{\text{moles NaOH}}{1\ \text{L of solution}} = \frac{(7.55 \times 10^{-3}\ \text{mol NaOH})}{(15.7\ \text{mL})} \cdot \frac{1000\ \text{mL}}{1\ \text{L}}$$

$$= 0.481\ \text{M}$$

4.105 (a) $N_2O_4(g) \rightarrow 2\ NO_2(g)$; (b) If all the gas were $N_2O_4(g)$, then the moles can be calculated from the ideal gas equation:

$$P = \frac{nRT}{V}$$

$$= \frac{\left(\dfrac{43.78\ \text{g}}{92.02\ \text{g} \cdot \text{mol}^{-1}}\right)(0.082\ 06\ \text{L} \cdot \text{atm} \cdot \text{K}^{-1} \cdot \text{mol}^{-1})(298\ \text{K})}{5.00\ \text{L}}$$

$$= 2.33\ \text{atm}$$

(c) The only difference in the calculation between part (b) and part (c) is that the molar mass of NO_2 is half that of N_2O_4.

$$P = \frac{nRT}{V}$$

$$= \frac{\left(\dfrac{43.78\ \text{g}}{46.01\ \text{g} \cdot \text{mol}^{-1}}\right)(0.082\ 06\ \text{L} \cdot \text{atm} \cdot \text{K}^{-1} \cdot \text{mol}^{-1})(298\ \text{K})}{5.00\ \text{L}}$$

$$= 4.65\ \text{atm}$$

(d) Because both N_2O_4 and NO_2 are present, we need to determine some way of calculating the relative amounts of each present. This can be done by taking advantage of the gas law relationships. The total pressure at the end of the reaction will give us the total number of moles present:

$$P_{total} = 2.96 \text{ atm}$$

$$(2.96 \text{ atm})(5.00 \text{ L}) = n_{total}(0.082\ 06 \text{ L} \cdot \text{atm} \cdot \text{K}^{-1} \cdot \text{mol}^{-1})(298 \text{ K})$$

$$n_{total} = 0.605 \text{ mol}$$

$$\therefore n_{N_2O4} + n_{NO_2} = 0.605 \text{ mol}$$

This gives us one equation, but we have two unknowns, so another relationship is needed. We can take advantage of knowing the stoichiometry of the reaction. If we assume that all of the gas begins at N_2O_4 and we allow some to react, we can write the following:

Initial amount of N_2O_4 0.476 mol

Amount of N_2O_4 that reacts x (mol)

Amount of NO_2 formed $2x$ (mol)

When the reaction is completed, there will be $0.476 - x$ mole of N_2O_4 and $2x$ mole NO_2. The total number of moles will be given by:

$$(0.476 - x) + 2x = n_{total}$$
$$0.605 \text{ mol} = 0.476 \text{ mol} + x$$

$$x = 0.129 \text{ mol}$$
$$n_{NO_2} = 2x$$
$$= 2(0.129 \text{ mol})$$
$$= 0.258 \text{ mol}$$

$$n_{N_2O_4} = 0.476 \text{ mol} - x$$
$$= 0.347 \text{ mol}$$

$$x_{NO_2} = \frac{0.258 \text{ mol}}{0.605 \text{ mol}} = 0.426$$

$$x_{N_2O_4} = \frac{0.347 \text{ mol}}{0.605 \text{ mol}} = 0.574$$

4.107 (a) The elemental analyses yield an empirical formula of NH_2. The formula unit has a mass of 16.02 $g \cdot mol^{-1}$. The mass, volume, pressure, and temperature data will allow us to calculate the molar mass, using the ideal gas equation:

$$PV = nRT$$

$$PV = \frac{m}{M} RT$$

$$M = \frac{mRT}{PV} = \frac{(0.473 \text{ g})(0.082 \text{ 06 L} \cdot \text{atm} \cdot \text{K}^{-1} \cdot \text{mol}^{-1})(298 \text{ K})}{(1.81 \text{ atm})(0.200 \text{ L})} = 31.9 \text{ g} \cdot \text{mol}^{-1}$$

The molar mass divided by the mass of the empirical formula mass will give the value of n in the formula $(NH_2)_n$. 31.9 $g \cdot mol^{-1} \div 16.02$ $g \cdot mol^{-1} = 1.99$, so the molecular formula is N_2H_4, which corresponds to the molecule known as hydrazine.

(b)

$$H - \overset{\cdot\cdot}{N} - \overset{\cdot\cdot}{N} - H$$
$$\underset{H}{|} \quad \underset{H}{|}$$

(c)

$$\frac{rate_A}{rate_B} = \sqrt{\frac{M_B}{M_A}}$$

$$\frac{\dfrac{3.5 \times 10^{-4} \text{ mol}}{15.0 \text{ min}}}{\dfrac{X}{25.0 \text{ min}}} = \sqrt{\frac{32.05 \text{ g} \cdot \text{mol}^{-1}}{17.03 \text{ g} \cdot \text{mol}^{-1}}}$$

$$X = \left(\frac{3.5 \times 10^{-4} \text{ mol}}{15.0 \text{ min}} \right)(25.0 \text{ min}) \sqrt{\frac{17.03 \text{ g} \cdot \text{mol}^{-1}}{32.05 \text{ g} \cdot \text{mol}^{-1}}}$$

$$= 4.2 \times 10^{-4} \text{ mol}$$

4.109 Using v_{rms} = root mean square speed and Equation 4.22 (the Maxwell Distribution of Speeds):

$$\frac{f(10v_{rms})}{f(v_{rms})} = \frac{4\pi N\left(\dfrac{M}{2RT}\right)^{\frac{3}{2}}(10v_{rms})^2 e^{-M(10v_{rms})^2/2RT}}{4\pi N\left(\dfrac{M}{2RT}\right)^{\frac{3}{2}}(v_{rms})^2 e^{-M(v_{rms})^2/2RT}}$$

$$= \frac{100v_{rms}^2 e^{-M100v_{rms}^2/2RT}}{v_{rms}^2 e^{-Mv_{rms}^2/2RT}} = 100e^{(-M100v_{rms}^2/2RT \,+\, Mv_{rms}^2/2RT)}$$

$$= 100e^{-99Mv_{rms}^2/2RT}$$

The ratio is not independent of temperature since the variable T appears in the denominator of a negative exponent on e. It makes sense that the ratio should become bigger at higher temperatures as the distribution spreads out such that the number of molecules with higher speeds increases while the number with lower speeds decreases.

4.111 The two scents will diffuse according to Equation 4.17a.

$$\frac{\text{rate}_{\text{fruity}}}{\text{rate}_{\text{minty}}} = \sqrt{\frac{M_{C_8H_8O_2}}{M_{C_{10}H_{20}O_2}}} = \sqrt{\frac{136}{172}} = 0.889$$

Let x = the distance traveled by ethyl octanoate (fruity) and y = the distance traveled by p-anisaldehyde (minty) in the same amount of time. Then

$$x + y = 5 \text{ m} \qquad \text{and} \qquad \frac{x}{y} = \frac{0.889 \text{ m}}{1 \text{ m}}, \text{ or } x = 0.889y$$

Substituting for x gives

$$0.889y + y = 5$$
$$1.889y = 5$$
$$y = 2.65$$

So the fruity smell will travel 5-2.65=2.35 m in the same time that the minty smell will travel 2.65 m. A person must stand more than 2.35 m away from the north end of the room where the fruity smell originates in order to smell the minty scent first.

4.113 The molar mass calculation follows from the ideal gas law:

$$PV = nRT$$

$$PV = \frac{m}{M} RT$$

$$M = \frac{mRT}{PV} = \frac{(1.509 \text{ g})(0.082\ 06 \text{ L} \cdot \text{atm} \cdot \text{K}^{-1} \cdot \text{mol}^{-1})(473 \text{ K})}{\left(\dfrac{745 \text{ Torr}}{760 \text{ Torr} \cdot \text{atm}^{-1}}\right)(0.235 \text{ L})}$$

$$= 254 \text{ g} \cdot \text{mol}^{-1}$$

If the molecular formula is OsO_x, then the molar mass will be given by:

$$190.2 \text{ g} \cdot \text{mol}^{-1} + x(16.00 \text{ g} \cdot \text{mol}^{-1}) = 254 \text{ g} \cdot \text{mol}^{-1})$$

$$x = 3.99$$

The formula is OsO_4.

4.115 In this problem, the volume, pressure, and molar mass of the substance stay constant. In order to calculate the new mass with the same conditions, we can resort to using the ideal gas equation rearranged to group the constant terms on one side of the equation:

$$PV = nRT$$

$$PV = \frac{m}{M} RT$$

but M, P, and V are constants, so we can write

$$\frac{MPV}{R} = mT$$

Now we have two sets of conditions, 1 and 2, for which $\dfrac{MPV}{R}$ is constant

so we can set them equal:

$$m_1 T_1 = m_2 T_2$$

$$(46.2 \text{ g})(300. \text{ K}) = (m_2)(600. \text{ K}); \text{ therefore } m_2 = 23.1 \text{ g}$$

The mass of gas released must therefore be $46.2 \text{ g} - 23.1 \text{ g} = 23.1 \text{ g}$.

4.117 (a) volume of one atom = molar volume $\div$ Avogadro's number

$$2.38 \times 10^{-2} \text{ L} \cdot \text{mol}^{-1} \div 6.022 \times 10^{23} \text{ atoms} \cdot \text{mol}^{-1} = 3.95 \times 10^{-26} \text{ L} \cdot \text{mol}^{-1}$$

$$3.95 \times 10^{-26} \ \text{L·mol}^{-1} \left(\frac{1000 \ cm^3}{1 \ L} \right) \left(\frac{1.00 \times 10^{30} \ pm^3}{1 \ cm^3} \right)$$

$$= 3.95 \times 10^{7} \ \text{pm}^{3} \cdot \text{atom}^{-1}$$

$$3.95 \times 10^{7} \ \text{pm}^{3} = \frac{4}{3} \pi r^{3}; \ \ r = 211 \ \text{pm}$$

(b) The atomic radius of He is 128 pm (Appendix 2D).

The volume of the He atom, based upon this radius, is

$$V = \frac{4}{3} \pi r^3$$
$$= \frac{4}{3} \pi (128 \ \text{pm})^3$$
$$= 8.78 \times 10^{6} \ \text{pm}^3$$

(c) The difference in these values illustrates that there is no easy definition for the boundaries of an atom. The van der Waals value obtained from the correction for molar volume is considerably larger than the atomic radius, owing perhaps to longer range and weak interactions between atoms. One should also bear in mind that the value for the van der Waals b is a parameter used to obtain a good fit to a curve, and its interpretation is more complicated than a simple molar volume.

4.119 (a) $ClNO_2$

(b) $ClNO_2$

(c)

(d) trigonal planar

4.121 (a) Substitute the van der (vdW) parameters for ammonia as well as the given values of $n, R, P,$ and T into the vdW equation, then solve for V. Since the equation is cubic, solve graphically for the three roots or use an appropriate program such as Math Cad. Only one of the three roots is physically possible:

$$V^3 + n\left(\frac{RT + bP}{P}\right)V^2 + \left(\frac{n^2 a}{P}\right)V - \frac{n^3 ab}{P} = 0$$

$V^3 + (0.505 \text{ mol})$

$$\times \left(\frac{(0.08206 \text{ L} \cdot \text{atm} \cdot \text{K}^{-1} \cdot \text{mol}^{-1})(298 \text{ K}) + (3.707 \text{ L} \cdot \text{mol}^{-1})(95.0 \text{ atm})}{(95.0 \text{ atm})}\right)V^2$$

$$+ \left(\frac{(0.505 \text{ mol})^2 (4.225 \text{ L}^2 \cdot \text{atm} \cdot \text{mol}^{-2})}{(95.0 \text{ atm})}\right)V$$

$$- \frac{(0.505 \text{ mol})^3 (4.225 \text{ L}^2 \cdot \text{atm} \cdot \text{mol}^{-2})(3.707 \text{ L} \cdot \text{mol}^{-1})}{(95.0 \text{ atm})} = 0$$

$V^3 + (2.002 \text{ L})V^2 + (0.01134 \text{ L}^2)V - 0.02123 \text{ L}^3 = 0$

$V = -1.991$, or -0.1089, or 0.09789 L

but only the positive root is physically possible, so

$V = 0.0979$ L

(b) Compare the volume calculated in part (a) to that of an ideal gas under the same conditions.

$PV = nRT$

$$V = \frac{(0.505 \text{ mol})(0.08206 \text{ L} \cdot \text{atm} \cdot \text{K}^{-1} \cdot \text{mol}^{-1})(298 \text{ K})}{95.0 \text{ atm}} = 0.130 \text{ L}$$

$V_{\text{ideal}} = 0.130 \text{ L} > V_{\text{vdW}}, = 0.0979 \text{ L}$. Attractive forces dominate because the van der Waals, or "real", gas occupies less volume than the "ideal" gas. If the molecules are attracted to one another they will behave less independently, reducing the effective number of moles of gas.

CHAPTER 5

LIQUIDS AND SOLIDS

5.1 (a) London forces, dipole-dipole, hydrogen bonding; (b) London forces; (c) London forces, dipole-dipole, hydrogen bonding; (d) London forces, dipole-dipole.

5.3 Only (b) CH_3Cl, (c) CH_2Cl_2, and (d) $CHCl_3$ will have dipole-dipole interactions. The molecules CH_4 and CCl_4 do not have dipole moments.

5.5 (a) NaCl (801°C vs. −114.8°C) because it is an ionic compound as opposed to a molecular compound; (b) butanol (−90°C vs. −116°C) due to hydrogen bonding in butanol that is not possible in diethyl ether; (c) triiodomethane because it will have much stronger London dispersion forces (−82.2°C for trifluoromethane vs. 219°C for triiodomethane); (d) Methanol (-94 °C vs. -169 °C) because of the hydrogen bonding in methanol but not in ethylene.

5.7 (a) PF_3 and PBr_3 are both trigonal pyramidal and should have similar intermolecular forces, but PBr_3 has the greater number of electrons and should have the higher boiling point. The boiling point of PF_3 is −101.5°C and that of PBr_3 is 173.2 °C; (b) SO_2 is bent and has a dipole moment whereas CO_2 is linear and will be nonpolar. SO_2 should have the higher boiling point. SO_2 boils at −10°C, whereas CO_2

sublimes at −78°C.   (c) BF_3 and BCl_3 are both trigonal planar, so the choice of higher boiling point depends on the difference in total number of electrons. BCl_3 should have the higher boiling point (12.5°C vs. −99.9°C).

5.9   The interaction energies can be ordered based on the relationship the energy has to the distance separating the interacting species. Thus ion-ion interactions are the strongest and are directly proportional to the distance separating the two interacting species. Ion-dipole energies are inversely proportional to d^2, whereas dipole-dipole for constrained molecules (i.e., solid state) is inversely proportional to d^3. Dipole-dipole interactions where the molecules are free to rotate become comparable to induced dipole-induced dipole interactions, which are both inversely related to d^6. The order thus derived is: (b) dipole-induced dipole $\cong$ (c) dipole-dipole in the gas phase < (e) dipole-dipole in the solid phase < (a) ion-dipole < (d) ion-ion.

5.11   Only molecules with H attached to the electronegative atoms F, N, and O can hydrogen bond. Additionally, there must be lone pairs available for the H's to bond to. This is true only of (d) HNO_2.

5.13   II because the dipoles are aligned with oppositely charged ends closest to each other thereby maximizing dipole-dipole attractions.

5.15   The ionic radius of Al^{3+} is 53 pm and that of Be^{2+} is 27 pm. The ratio of energies will be given by

$$E_p \propto \left(\frac{-|Z|\mu}{d^2} \right)$$

$$E_{pAl^{3+}} \propto \left(\frac{-|Z|\mu}{d^2} \right) = \left(\frac{-|3|\mu}{(53+100)^2} \right)$$

$$E_{pBe^{2+}} \propto \left(\frac{-|Z|\mu}{d^2} \right) = \left(\frac{-|2|\mu}{(27+100)^2} \right)$$

The electric dipole moment of the water molecule (μ) will cancel:

$$\text{Ratio} \left(\frac{E_{pAl^{2+}}}{E_{pBe^{2+}}} \right) = \left(\frac{-|3|\mu/(53+100)^2}{-|2|\mu/(27+100)^2} \right) = \left(\frac{3(127)^2}{2(153)^2} \right) = 1.03$$

The attraction of the Al^{3+} ion will be greater than that of the Be^{2+} ion. Even though the Al^{3+} ion has a larger radius, its charge is higher than that of Be^{2+}, making the attraction greater over all.

5.17   (a) Xenon is larger, with more electrons, giving rise to larger London forces that increase the melting point.   (b) Hydrogen bonding in water causes the molecules to be held together more tightly than in diethyl ether. (c) Both molecules have the same molar mass, but pentane is a linear molecule compared to dimethylpropane, which is a compact, spherical molecule. The compactness of the dimethylpropane gives it a lower surface area. That means that the intermolecular attractive forces, which are of the same type (London forces) for both molecules, will have a larger effect for pentane.

5.19 $$F = \frac{-dE_P}{dr} = \frac{-d}{dr}\left(\frac{1}{r^6} \right) = -\left(\frac{-6}{r^7} \right) \propto \frac{1}{r^7}$$

5.21   (a) As intermolecular forces increase, boiling point will increase because more thermal energy is required to separate molecules apart; (b) viscosity will increase with the increase of intermolecular forces increase because stronger intermolecular forces will make molecules less mobile (or do not let molecules move past one another easily); (c) surface tension also increases as the increase of intermolecular forces because stronger intermolecular forces intend to pull molecules together and inward.

5.23 (a) *cis*-Dichloroethene is polar, whereas *trans*-dichloroethene, whose individual bond dipole moments cancel, is nonpolar. Therefore, *cis*-dichloroethene has the greater intermolecular forces and the greater surface tension. (b) Surface tension of liquids decreases with increasing temperature as a result of thermal motion as temperature rises. Increased thermal motion allows the molecules to more easily break away from each other, which manifests itself as decreased surface tension.

5.25 At 50°C all three compounds are liquids. C_6H_6 (nonpolar) $< C_6H_5SH$ (polar, but no hydrogen bonding) $< C_6H_5OH$ (polar and with hydrogen bonding). The viscosity will show the same ordering as the boiling points, which are 80°C for C_6H_6, 169°C for C_6H_5SH, 182° for C_6H_5OH.

5.27 CH_4, -162°C; CH_3CH_3, -88.5°C; $(CH_3)_2CHCH_2CH_3$, 28°C; $CH_3(CH_2)_3CH_3$, 36°C; CH_3OH, 64.5°C; CH_3CH_2OH, 78.3 °C; $CH_3CHOHCH_3$, 82.5°C; C_5H_9OH (cyclic, but not aromatic), 140°C; $C_6H_5CH_3OH$ (aromatic ring), 205°C; $OHCH_2CHOHCH_2OH$, 290°C

5.29 (a) hydrogen bonding; (b) London dispersion forces increase

5.31 Using $h = \dfrac{2\gamma}{gdr}$ we can calculate the height. For water:

$$r = \frac{1}{2} \, diameter = \frac{1}{2}\,(0.15 \text{ mm}) \left(\frac{1 \text{ m}}{1000 \text{ mm}} \right) = 7.5 \times 10^{-5} \text{ m}$$

$$d = 0.997 \text{ g} \cdot \text{cm}^{-3} \left(\frac{1 \text{ kg}}{1000 \text{ g}} \right) \left(\frac{10^6 \text{ cm}^3}{\text{m}^3} \right) = 9.97 \times 10^2 \text{ kg} \cdot \text{m}^{-3}$$

$$h = \frac{2(72.75 \times 10^{-3} \text{ N} \cdot \text{m}^{-1})}{(9.81 \text{ m} \cdot \text{s}^{-1})(9.97 \times 10^2 \text{ kg} \cdot \text{m}^{-3})(7.5 \times 10^{-5} \text{ m})} = 0.20 \text{ m or } 200 \text{ mm}$$

Remember that $1 \text{ N} = 1 \text{ kg} \cdot \text{m}^{-1} \cdot \text{s}^{-2}$

For ethanol:

$$d = 0.79 \text{ g} \cdot \text{cm}^{-3} \left(\frac{1 \text{ kg}}{1000 \text{ g}} \right) \left(\frac{10^6 \text{ cm}^3}{\text{m}^3} \right) = 7.9 \times 10^2 \text{ kg} \cdot \text{m}^{-3}$$

$$h = \frac{2(22.8 \times 10^{-3} \text{ N} \cdot \text{m}^{-1})}{(9.81 \text{ m} \cdot \text{s}^{-1})(7.9 \times 10^2 \text{ kg} \cdot \text{m}^{-3})(7.5 \times 10^{-5} \text{ m})} = 0.078 \text{ m or } 78 \text{ mm}$$

Water will rise to a higher level than ethanol. There are two opposing effects to consider. While the greater density of water, as compared to ethanol, acts against it rising as high, it has a much higher surface tension.

5.33 Glucose will be held in the solid by London forces, dipole-dipole interactions, and hydrogen bonds; benzophenone will be held in the solid by dipole-dipole interactions and London forces; methane will be held together by London forces only. London forces are strongest in benzophenone, but glucose can experience hydrogen bonding, which is a strong interaction and dominates intermolecular forces. Methane has few electrons so experiences only weak London forces. We would expect the melting points to increase in the order CH_4 (m.p. $= -182°C$)

< benzophenone (m.p. $= 48°C$) < glucose (m.p. $= 148 - 155°C$).

5.35 (a) network; (b) ionic; (c) molecular; (d) molecular; (e) network

5.37 Substance A: ionic; substance B: metallic; substance C: molecular solid

5.39 (a) At center: 1 center $\times$ 1 atom $\cdot$ center^{-1} = 1 atom; at 8 corners, 8 corners $\times$ $\frac{1}{8}$ atom $\cdot$ corner^{-1} = 1 atom; total = 2 atoms; (b) There are eight nearest neighbors, hence a coordination number of 8; (c) The direction along which atoms touch each other is the body diagonal of the unit cell. This body diagonal will be composed of four times the radius of the atom. In terms of the unit cell edge length a, the body diagonal will be $\sqrt{3}\, a$. The unit cell edge length will, therefore, be given by

$$4r = \sqrt{3}\, a \text{ or } a = \frac{4r}{\sqrt{3}} = \frac{4 \cdot (124 \text{ pm})}{\sqrt{3}} = 286 \text{ pm}$$

5.41 (a) a = length of side for a unit cell; for an fcc unit

cell, $a = \sqrt{8}\, r$ or $2\sqrt{2}\, r = 404$ pm.

$V = a^3 = (404 \text{ pm} \times 10^{-12} \text{ m} \cdot \text{pm}^{-1})^3 = 6.59 \times 10^{-29} \text{ m}^3 = 6.59 \times 10^{-23} \text{ cm}^3$.

Because for a fcc unit cell there are 4 atoms per unit cell, we have

$$\text{mass(g)} = 4 \text{ Al atoms} \times \frac{1 \text{ mol Al atoms}}{6.022 \times 10^{23} \text{ atoms} \cdot \text{mol}^{-1}} \times \frac{26.98 \text{ g}}{\text{mol Al atoms}}$$

$$= 1.79 \times 10^{-22} \text{ g}$$

$$d = \frac{1.79 \times 10^{-22} \text{ g}}{6.59 \times 10^{-23} \text{ cm}^3} = 2.72 \text{ g} \cdot \text{cm}^{-3}$$

(b) $a = \dfrac{4r}{\sqrt{3}} = \dfrac{4 \times 235 \text{ pm}}{\sqrt{3}} = 543 \text{ pm}$

$V = (543 \times 10^{-12} \text{ m})^3 = 1.60 \times 10^{-28} \text{ m}^3 = 1.60 \times 10^{-22} \text{ cm}^3$

There are 2 atoms per bcc unit cell:

$$\text{mass(g)} = 2 \text{ K atoms} \times \frac{1 \text{ mol K atoms}}{6.022 \times 10^{23} \text{ atoms} \cdot \text{mol}^{-1}} \times \frac{39.10 \text{ g}}{\text{mol K atoms}}$$

$$= 1.30 \times 10^{-22} \text{ g}$$

$$d = \frac{1.30 \times 10^{-22} \text{ g}}{1.60 \times 10^{-22} \text{ cm}^3} = 0.813 \text{ g} \cdot \text{cm}^{-3}$$

5.43 a = length of unit cell edge

$$V = \frac{\text{mass of unit cell}}{d}$$

(a)

$$V = a^3 = \dfrac{(1 \text{ unit cell})\left(\dfrac{195.09 \text{ g Pt}}{\text{mol Pt}}\right)\left(\dfrac{1 \text{ mol Pt}}{6.022 \times 10^{23} \text{ atoms Pt}}\right)\left(\dfrac{4 \text{ atoms}}{1 \text{ unit cell}}\right)}{21.450 \text{ g} \cdot \text{cm}^3}$$

$a = 3.92 \times 10^{-8}$ cm

Because for an fcc cell, a $= \sqrt{8}\, r$, $r = \dfrac{\sqrt{2}\,a}{4} = \dfrac{\sqrt{2}\,(3.92 \times 10^{-8} \text{ cm})}{4}$

$$= 1.39 \times 10^{-8} \text{ cm} = 139 \text{ pm}$$

(b) $V = a^3$

$$= \dfrac{(1 \text{ unit cell})\left(\dfrac{180.95 \text{ g Ta}}{1 \text{ mol Ta}}\right)\left(\dfrac{1 \text{ mol Ta}}{6.022 \times 10^{23} \text{ atoms Ta}}\right)\left(\dfrac{2 \text{ atoms}}{1 \text{ unit cell}}\right)}{16.654 \text{ g} \cdot \text{cm}^3}$$

$$= 3.61 \times 10^{-23} \text{ cm}^3$$

$a = 3.30 \times 10^{-8}$ cm

$r = \dfrac{\sqrt{3}\,a}{4} = \dfrac{\sqrt{3}\left(3.30 \times 10^{-8} \text{ cm}\right)}{4} = 1.43 \times 10^{-8} \text{ cm} = 143 \text{ pm}$

5.45 Let's assume that rhodium metal structure is *ccp* (*fcc*). In *ccp* structure, there are 4 atoms in each unit cell. The atomic mass (M) of rhodium is 102.91 g/mol. Therefore, the density of rhodium is:

$d = m/a^3$, m = mass of one unit cell = $4M/N_A$, a = the length of the side of a *fcc* unit cell $= 8^{1/2}r$

$$d = \dfrac{4M/N_A}{(8^{1/2}r)^3} = \dfrac{4M}{N_A(8^{1/2}r)^3}$$

$$= \dfrac{4 \text{ atoms} \times (102.91\text{g} \cdot mol^{-1})}{(6.022 \times 10^{23} \text{ atom} \cdot mol^{-1}) \times (8^{1/2} \times 1.34 \times 10^{-8} cm)^3} = 12.55 \text{ g·cm}^{-3}$$

Then we assume the rhodium metal structure is *bcc*. In bcc structure, there are 2 atoms in each unit cell, and a = the length of the side of a *bcc* unit cell $= 4r/3^{1/2}$. The density of rhodium is:

$$d = \dfrac{2M/N_A}{(4r/3^{1/2})^3} = \dfrac{2M}{N_A(4r/3^{1/2})^3}$$

$$= \frac{2 \; atoms \times (102.91 g \cdot mol^{-1})}{(6.022 \times 10^{23} \; atom \cdot mol^{-1}) \times (4 \times 1.34 \times 10^{-8} cm / 3^{1/2})^3} = 11.53 \; g \cdot cm^{-3}$$

The density value for the ccp structure is closer to 12. 42 $g \cdot cm^{-3}$.

Therefore, rhodium metal is close-packed structure.

5.47   (a) The volume of the unit cell is

$$V_{unit \; cell} = \left(543 \; pm \times \frac{10^{-12} \; m}{pm} \times \frac{100 \; cm}{m} \right)^3 = 1.601 \times 10^{-22} \; cm^3$$

mass in unit cell $= (1.601 \times 10^{-22} \; cm^3) \times (2.33 \; g \cdot cm^{-3}) = 3.73 \times 10^{-22} g$

(b)   The mass of a Si atom is

28.09 $g \cdot mol^{-1} \div (6.022 \times 10^{23})$ atoms $\cdot \; mol^{-1} = 4.665 \times 10^{-23} \; g \cdot atom^{-1}$

Therefore, there are $3.73 \times 10^{-22} g \div 4.665 \times 10^{-23} \; g \cdot atom^{-1} = 8$ atoms per

unit cell.

5.49   volume of a cylinder = base area x length = $\pi r^2 \; l$

volume of a triangular prism = base area x length = (bhl)/2 = $\sqrt{3}r^2 l$ for a

triangle inscribed in the centers of three touching cylinder bases

Each of the inscribed triangle's 60° angles accounts for 1/6 of the volume

of each of the three touching cylinders, or a total cylinder volume of (3/6)

$\pi r^2 \; l$. So the percent of the space occupied by the cylinders is (0.5) $\pi r^2 \; l$

*100/ $\sqrt{3}r^2 l = 0.5\pi/\sqrt{3} = 90.7\%$

5.51   (a) Indium arsenide has a (4,4)-coordination because each unit cell has 4

indium and 4 arsenic; (b) The formula of indium arsenide is InAs.

5.53   (a) There are eight chloride ions at the eight corners, giving a total of

8 corners $\times \frac{1}{8}$ atom $\cdot$ corner$^{-1} = 1 \; Cl^{-}$ ion

There is one Cs^{+} that lies at the center of the unit cell. All of this ion

belongs to the unit cell. The ratio is thus 1 : 1 for an empirical formula of

CsCl, with one formula unit per unit cell.

(b) The titanium atoms lie at the corners of the unit cell and at the body center: 8 corners $\times \frac{1}{8}$ atom $\cdot$ corner^{-1} + 1 at body center = 2 atoms per unit cell

Four Oxygen atoms lie on the faces of the unit cell and two lie completely within the unit cell, giving:

4 atoms in faces $\times \frac{1}{2}$ atom $\cdot$ face^{-1} + 2 atoms wholly within cell = 4 atoms

The ratio is thus two Ti per four O, or an empirical formula of TiO_2 with two formula units per unit cell (c) The Ti atoms are 6-coordinate and the O atoms are 3-coordinate.

5.55 (a) Each unit cell has one rhenium ($8 \times \frac{1}{8}$) and 3 oxide ion ($12 \times \frac{1}{4}$).

Therefore, rhenium oxide has a (6,2)-*coordination*. (b) The formula of rhenium oxide is ReO_3.

5.57 (a) ratio $= \dfrac{149 \text{ pm}}{133 \text{ pm}} = 1.12$, predict cesium-chloride structure with (8,8) coordination; however, rubidium fluoride actually adopts the rock-salt structure

(b) ratio $= \dfrac{72 \text{ pm}}{140 \text{ pm}} = 0.51$, predict rock-salt structure with (6,6) coordination

(c) ratio $= \dfrac{102 \text{ pm}}{196 \text{ pm}} = 0.520$, predict rock-salt structure with (6,6) coordination

5.59 (a) In the rock-salt structure, the unit cell edge length is equal to two times the radius of the cation plus two times the radius of the anion. Thus for CaO, $a =$ 2(100 pm) + 2(140 pm) = 480 pm. The volume of the unit cell will be given by (converting to cm^3 because density is normally given in terms of $g \cdot cm^{-3}$)

$$V = \left(480 \text{ pm} \times \frac{10^{-12} \text{ m}}{\text{pm}} \times \frac{100 \text{ cm}}{\text{m}} \right)^3 = 1.11 \times 10^{-22} \text{ cm}^3$$

There are four formula units in the unit cell, so the mass in the unit cell will be given by

$$\text{mass in unit cell} = \frac{\left(\dfrac{4 \text{ formula units}}{1 \text{ unit cell}} \right) \times \left(\dfrac{56.08 \text{ g CaO}}{1 \text{ mol CaO}} \right)}{6.022 \times 10^{23} \text{ molecules} \cdot \text{mol}^{-1}} = 3.725 \times 10^{-22} \text{ g}$$

The density will be given by the mass in the unit cell divided by the volume of the unit cell:

$$d = \frac{3.725 \times 10^{-22} \text{ g}}{1.11 \times 10^{-22} \text{ cm}^3} = 3.36 \, g \cdot cm^{-3}$$

(b) For a cesium chloride-like structure, it is the body diagonal that represents two times the radius of the cation and plus two times the radius of the anion. Thus the body diagonal for CsBr is equal to $2(170 \text{ pm}) + 2(196) = 732 \text{ pm}$.

For a cubic cell, the body diagonal $= \sqrt{3} \, a = 732 \text{ pm}$

$$a = 423 \text{ pm}$$

$$a^3 = V = \left(423 \text{ pm} \times \frac{10^{-12} \text{ m}}{\text{pm}} \times \frac{100 \text{ cm}}{\text{m}} \right)^3 = 7.57 \times 10^{-23} \text{ cm}^3$$

There is one formula unit of CsBr in the unit cell, so the mass in the unit cell will be given by

$$\text{mass in unit cell} = \frac{\left(\dfrac{1 \text{ formula units}}{1 \text{ unit cell}} \right) \left(\dfrac{212.82 \text{ g CsBr}}{1 \text{ mol CsBr}} \right)}{6.022 \times 10^{23} \text{ molecules} \cdot \text{mol}^{-1}} = 3.534 \times 10^{-22} \text{ g}$$

$$d = \frac{3.534 \times 10^{-22} \text{ g}}{7.57 \times 10^{-23} \text{ cm}^3} = 4.67 \, g \cdot cm^3$$

5.61 (a) the smallest possible rectangular unit cell is as following:

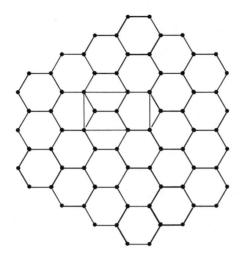

(b) There are 4 of carbon atoms in each unit cell

(c) The coordination number is 3 (C_3)

5.63 Each unit cell contains 4 RbI; Total of unit cells = $(6.022 \times 10^{23}$ RbI molecules/mole)/(4 RbI molecules/unit cell) = 1.505×10^{23} unit cells/mole; The volume of one unit cell = $(732.6 \text{ pm})^3$

Assume that the edge length of a cubic crystal of RbI that contains one mole RbI is X, then the volume of one mole RbI = X^3/mole =

$$\left(\frac{(732.6 \, pm)^3}{unit \, cell} \right) \times \left(\frac{1.505x10^{23} \, unit \, cell}{mole} \right);$$

$X = (732.6 \text{ pm}) \times \sqrt[3]{1.505x10^{23}} = 3.897 \times 10^{10} \text{ pm} = 3.897 \text{ cm}$

5.65 There are too many ways that these molecules can rotate and twist so that they do not remain rod-like. The molecular backbone when the molecule is stretched out is rod-like, but the molecules tend to curl up on themselves, destroying any possibility of long range order with neighboring molecules. This is partly due to the fact that the molecules have only single bonds that allow rotation about the bonds, so that each molecule can adopt many configurations. If multiple bonds are present, the bonding is more rigid.

5.67 Use of a nonpolar solvent such as hexane or benzene (etc.) in place of water should give rise to the formation of inverse micelles.

5.69 Because benzene is isotropic solvent and its viscosity is the same in every direction. However, a liquid crystal solvent is anisotropic solvent and its viscosity is smaller in the direction parallel to the long axis of the molecule than the "sideway" direction. Since methylbenzene is a small sphere molecule and its diffusion movement does not affected greatly whether solvent is isotropic or anisotropic.

5.71 (a) The NI_3 Lewis structure is:

, the molecular shape is trigonal pyramidal and it is a polar molecule. It can participate dipole-dipole interactions.

(b) The BI_3 Lewis structure is:

, the molecular shape is trigonal planar and it is a Non-polar molecule. It can not participate dipole-dipole interactions.

5.73 (a) Pentane and 2,2-dimethylpropane are isomers; both have the chemical formula C_5H_{12}. We will assume that 2,2-dimethylpropane is roughly spherical and that all the hydrogen atoms lie on this sphere. The surface area of a sphere is given by $A = 4\pi r^2$. For this particular sphere, $A = 4\pi(254 \text{ pm})^2 = 8.11 \times 10^5 \text{ pm}^2$. For pentane, the surface area of rectangular prism is

$2(295 \text{ pm} \times 766 \text{ pm}) +$
$2(295 \text{ pm} \times 254 \text{ pm}) + 2(254 \text{ pm} \times 766 \text{ pm}) = 9.91 \times 10^5 \text{ pm}^2$.

(b) Pentane has the larger surface area.

(c) The pentane should have the higher boiling point. It has a significantly larger surface area and should have stronger intermolecular forces between the molecules.

5.75 The unit cell for a cubic close-packed lattice is the fcc unit cell. For this cell, the relation between the radius of the atom r and the unit cell edge length a is

$$4r = \sqrt{2}\,a$$

$$a = \frac{4r}{\sqrt{2}}$$

The volume of the unit cell is given by

$$V = a^3 = \left(\frac{4r}{\sqrt{2}}\right)^3$$

If r is given in pm, then a conversion factor to cm is required:

$$V = a^3 = \left(\frac{4r}{\sqrt{2}} \times \frac{10^{-12}\ \text{m}}{\text{pm}} \times \frac{100\ \text{cm}}{\text{m}}\right)^3$$

Because there are four atoms per fcc unit cell, the mass in the unit cell will be given by

$$\text{mass} = \left(\frac{4\ \text{atoms}}{\text{unit cell}}\right)\left(\frac{M}{6.022 \times 10^{23}\ \text{atoms} \cdot \text{mol}^{-1}}\right)$$

The density will be given by

$$d = \frac{\text{mass of unit cell}}{\text{volume of unit cell}} = \frac{\left(\dfrac{4\ \text{atoms}}{\text{unit cell}}\right) \times \left(\dfrac{M}{6.022 \times 10^{23}\ \text{atoms} \cdot \text{mol}^{-1}}\right)}{\left(\dfrac{4\,r}{\sqrt{2}} \times \dfrac{10^{-12}\ \text{m}}{\text{pm}} \times \dfrac{100\ \text{cm}}{\text{m}}\right)^3}$$

$$= \frac{(2.936 \times 10^5)M}{r^3}$$

or

$$r = \sqrt[3]{\frac{(2.936 \times 10^5)M}{d}}$$

where M is the atomic mass in $g \cdot mol^{-1}$ and r is the radius in pm.

For the different gases we calculate the results given in the following table:

Gas	Density $(g \cdot cm^3)$	Molar mass $(g \cdot mol^{-1})$	Radius (pm)
Neon	1.20	20.18	170
Argon	1.40	39.95	203
Krypton	2.16	83.80	225
Xenon	2.83	131.30	239
Radon	4.4	222	246

5.77 There are two approaches to this problem. The information given does not specify the radius of the tungsten atom. This value can be looked up in Appendix 2D. We can calculate an answer, however, based simply upon the fact that the density is $19.3 \, g \cdot cm^{-3}$ for the bcc cell, by taking the ratio between the expected densities, based upon the assumption that the atomic radius of tungsten will be the same for both. The unit cell for a cubic close-packed lattice is the fcc unit cell. For this cell, the relation between the radius of the atom r and the unit cell edge length a is

$$4r = \sqrt{2}\,a$$
$$a = \frac{4r}{\sqrt{2}}$$

The volume of the unit cell is given by

$$V = a^3 = \left(\frac{4r}{\sqrt{2}}\right)^3$$

If r is given in pm, then a conversion factor to cm is required:

$$V = a^3 = \left(\frac{4r}{\sqrt{2}} \times \frac{10^{-12} \, m}{pm} \times \frac{100 \, cm}{m}\right)^3$$

Because there are four atoms per fcc unit cell, the mass in the unit cell is given by

$$\text{mass} = \left(\frac{4 \text{ atoms}}{\text{unit cell}}\right)\left(\frac{M}{6.022 \times 10^{23} \text{ atoms} \cdot \text{mol}^{-1}}\right)$$

The density is given by

$$d = \frac{\text{mass of unit cell}}{\text{volume of unit cell}} = \frac{\left(\dfrac{4 \text{ atoms}}{\text{unit cell}}\right)\left(\dfrac{M}{6.022 \times 10^{23} \text{ atoms} \cdot \text{mol}^{-1}}\right)}{\left(\dfrac{4r}{\sqrt{2}} \times \dfrac{10^{-12} \text{ m}}{\text{pm}} \times \dfrac{100 \text{ cm}}{\text{m}}\right)^3}$$

$$= \frac{(2.936 \times 10^5)M}{r^3}$$

or

$$r = \sqrt[3]{\frac{(2.936 \times 10^5)M}{d}}$$

where M is the atomic mass in $\text{g} \cdot \text{mol}^{-1}$ and r is the radius in pm.

Likewise, for a body-centered cubic lattice there will be two atoms per unit cell. For this cell, the relationship between the radius of the atom r and the unit cell edge length a is derived from the body diagonal of the cell, which is equal to 4 times the radius of the atom. The body diagonal is found from the Pythagorean theorem to be equal to the $\sqrt{3}\,a$.

$$4r = \sqrt{3}\,a$$

$$a = \frac{4r}{\sqrt{3}}$$

The volume of the unit cell is given by

$$V = a^3 = \left(\frac{4r}{\sqrt{3}}\right)^3$$

If r is given in pm, then a conversion factor to cm is required:

$$V = a^3 = \left(\frac{4r}{\sqrt{3}} \times \frac{10^{-12} \text{ m}}{\text{pm}} \times \frac{100 \text{ cm}}{\text{m}}\right)^3$$

Because there are two atoms per bcc unit cell, the mass in the unit cell will be given by

$$\text{mass} = \left(\frac{2 \text{ atoms}}{\text{unit cell}}\right) \times \left(\frac{M}{6.022 \times 10^{23} \text{ atoms} \cdot \text{mol}^{-1}}\right)$$

The density will be given by

$$d = \frac{\text{mass of unit cell}}{\text{volume of unit cell}} = \frac{\left(\dfrac{2 \text{ atoms}}{\text{unit cell}}\right)\left(\dfrac{M}{6.022 \times 10^{23} \text{ atoms} \cdot \text{mol}^{-1}}\right)}{\left(\dfrac{4\,r}{\sqrt{3}} \times \dfrac{10^{-12} \text{ m}}{\text{pm}} \times \dfrac{100 \text{ cm}}{\text{m}}\right)^3}$$

$$= \frac{\left(\dfrac{2 \text{ atoms}}{\text{unit cell}}\right)\left(\dfrac{M}{6.022 \times 10^{23} \text{ atoms} \cdot \text{mol}^{-1}}\right)}{(2.309 \times 10^{-10}\, r)^3}$$

$$= \frac{(2.698 \times 10^5)M}{r^3}$$

or

$$r = \sqrt[3]{\frac{(2.698 \times 10^5)M}{d}}$$

Setting these two expressions equal and cubing both sides, we obtain

$$\frac{(2.936 \times 10^5)M}{d_{\text{fcc}}} = \frac{(2.698 \times 10^5)M}{d_{\text{bcc}}}$$

The molar mass of tungsten M is the same for both ratios and will cancel from the equation.

$$\frac{(2.936 \times 10^5)}{d_{\text{fcc}}} = \frac{(2.698 \times 10^5)}{d_{\text{bcc}}}$$

Rearranging, we get

$$d_{\text{fcc}} = \frac{(2.936 \times 10^5)}{(2.698 \times 10^5)} d_{\text{bcc}}$$

$$= 1.088\, d_{\text{bcc}}$$

For W, $d_{\text{fcc}} = 1.088 \times 19.6 \text{ g} \cdot \text{cm}^3$

$$= 21.3 \text{ g} \cdot \text{cm}^{-3}$$

5.79 (a) The oxidation state of the titanium atoms must balance the charge on the oxide ions, O^{2-}. The presence of 1.18 O^{2-} ions means that the Ti present must have a charge to compensate the −2.36 charge on the oxide ions. The average oxidation state of Ti is thus +2.36.   (b) This is most easily solved by setting up a set of two equations in two unknowns. We know that the total charge on the titanium atoms present must equal 2.36, so if we multiply the charge on each type of titanium by the fraction of titanium present in that oxidation state and sum the values, we should get 2.36:

let x = fraction of Ti^{2+}, y = fraction of Ti^{3+}, then
2x + 3y = 2.36

Also, because we are assuming all the titanium is either +2 or +3, the fractions of each present must add up to 1:

x + y = 1

Solving these two equations simultaneously, we obtain y = 0.36, x = 0.64.

5.81 (a) True. If this is not the case, the unit cell will not match with other unit cells of the same type when stacked to form the entire lattice.

(b) False. Unit cells do not have to have atoms at the corners.

(c) True. In order for the unit cell to repeat properly, opposite faces must have the same composition.

(d) False. If one face is centered, the opposing face must be centered, but the other faces do not necessarily have to be centered.

5.83 (a) There is only one way to draw unit cells that will repeat to generate the entire lattice. The choice of unit cell is determined by conventions that are beyond the scope of this text (the smallest unit cell that indicates all of the symmetry present in the lattice is typically the one of choice). (b) There are two ways to draw the unit cells as shown below.

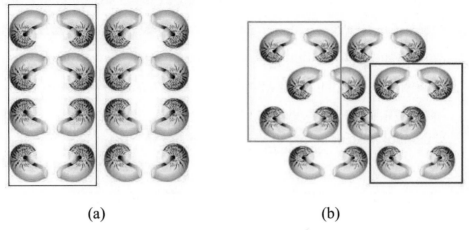

(a)   (b)

5.85 (a) The face diagonal the unit cell equals to 4r (r is the radius of the buckminsterfullerene). Therefore, $(4r)^2 = 2(142pm)^2$

Solve the equation: $4r = \sqrt{2}\,(142)$ pm; $r = 50.2$ pm

(b) The radius of K^+ is 133 pm. In close-packed cubic unit cell (face centered cubic unit cell), the tetrahedral holes is smaller than the octahedral holes. The sizes of tetrahedral holes can be calculated as following:

One unit cell can be divided into 8 subunit cells and each subunit cell contains one tetrahedral hole. The face diagonal of each subunit cell

$$f^2 = (\frac{a}{2})^2 + (\frac{a}{2})^2 \text{ (a is the edge length); the body diagonal } b^2 = f^2 + (\frac{a}{2})^2$$

$$b^2 = 2(\frac{a}{2})^2 + (\frac{a}{2})^2 = 3(\frac{a}{2})^2; \quad b = \sqrt{3}\,(\frac{a}{2}) = \sqrt{3}\,(\frac{142\,pm}{2}) = 123 \text{ pm}$$

The radius of the tetrahedral hole is: $r_{tetrahedral} + r_{buckminsterfullerene} = \dfrac{b}{2}$

$r_{tetrahedral} = \dfrac{123\,pm}{2} - 50.20$ pm $= 11.3$ pm which is not large enough to K^+ ion. Therefore, K^+ must take octahedral holes.

Each unit cell has 4 octahedral holes and 3 are occupied by the K^+. The percentage of holes filled are 75%.

5.87 Fused silica, also known as fused quartz, is predominantly SiO_2 with very few impurities. This glass is the most refractory, which means it can be used at the highest temperatures. Quartz vessels are routinely used for reactions that must be carried out at temperatures up to 1000°C. Vycor, which is 96% SiO_2, can be used at up to ca. 900°C, and normal borosilicate glasses at up to approximately 550°C. This is due to the fact that the glasses melt at lower temperatures, as the amount of materials other than SiO_2 increases. The borosilicate glasses (Pyrex or Kimax) are commonly used because the lower softening points allow them to be more easily molded and shaped into different types of glass objects, such as reaction flasks, beakers, and other types of laboratory and technical glassware.

5.89 For a bcc unit cell (see Example 5.3), d = $3^{3/2}$ (55.85 g/mol)/32(6.022x10^{23}/mole)x(1.24x10^{-8}cm)^3 = 7.90 g/cm^3
$(1.5$ cm)^3 x 7.90 g/cm^3 x (1 mole/55.85 g) =0.477 mole Fe
(1.00 atm x15.5 L) / (0.08206 L atm /K mol) (298K) = 0.634 mol O_2
Fe is LR, so (0.5)(0.477 mole)(159.70 g/mole) = 38.1 g Fe_2O_3 is theoretical yield, or maximum mass that can be produced.

5.91 For M, there is total of one atom in the unit cell from the corners;
cations: 8 corners $\times \frac{1}{8}$ atom $\cdot$ corner^{-1} + 6 faces + $\frac{1}{2}$ atom $\cdot$ face^{-1} = 4 atoms
anions: 8 tetrahedral holes $\times$ 1 atom $\cdot$ tetrahedral hole^{-1} = 8 atoms
The cation to anion ratio is thus 4 : 8 or 1 : 2; the empirical formula is MA_2.

5.93 The cesium chloride lattice is a simple cubic lattice of Cl$^-$ ions with a Cs$^+$ at the center of the unit cell (See 5.43). In the unit cell, there is a total of one Cl$^-$ ion and one Cs$^+$ ion. If the density is 3.988 g $\cdot$ cm^{-3}, then we can determine the volume and unit cell edge length. The molar mass of CsCl is 168.36 g $\cdot$ mol^{-1}.

$$3.988 \text{ g} \cdot \text{cm}^{-3} = \frac{\left(\dfrac{168.36 \text{ g} \cdot \text{mol}^{-1}}{6.022 \times 10^{23} \text{ formula units} \cdot \text{mol}^{-1}} \right)}{a^3}$$

$$a^3 = \frac{\left(\dfrac{168.36 \text{ g} \cdot \text{mol}^{-1}}{6.022 \times 10^{23} \text{ formula units} \cdot \text{mol}^{-1}} \right)}{3.988 \text{ g} \cdot \text{cm}^{-3}}$$

$$a = 4.12 \times 10^{-8} \text{ cm}$$
$$= 412 \text{ pm}$$

The volume of the unit cell is $(412 \text{ pm})^3 = 6.99 \times 10^7 \text{ pm}^3$.

We will determine the size of the Cs^+ and Cl^- ions from ionic radii given in Appendix 2D, but we can check these values against the unit cell dimensions. For this type of unit cell, the body diagonal will be equal to $2\,r(Cs^+) + 2\,r(Cl^-) = a\sqrt{3} = 714 \text{ pm}$. The sum of the ionic radii gives us $2\,(170 \text{ pm}) + 2\,(181 \text{ pm}) = 702 \text{ pm}$, which is in very good agreement.

Note that we cannot calculate the size of these ions independently from the unit cell data without more information, because this lattice is not close-packed. We will assume that the ions are spherical. The volume occupied in the unit cell will be

$$V_{Cs^+} = \frac{4}{3}\pi r^3 = \frac{4}{3}\pi(170 \text{ pm})^3 = 2.06 \times 10^7 \text{ pm}^3$$

$$V_{Cl^-} = \frac{4}{3}\pi r^3 = \frac{4}{3}\pi(181 \text{ pm})^3 = 2.48 \times 10^7 \text{ pm}^3$$

The total occupied volume in the cell is

$2.06 \times 10^7 \text{ pm}^3 + 2.48 \times 10^7 \text{ pm}^3 = 4.54 \times 10^7 \text{ pm}^3$. The empty space is $6.99 \times 10^7 \text{ pm}^3 - 4.54 \times 10^7 \text{ pm}^3 = 2.45 \times 10^7 \text{ pm}^3$. The percent empty space is $2.45 \times 10^7 \text{ pm}^3 \div 6.99 \times 10^7 \text{ pm}^3 \times 100 = 35\%$.

5.95 (a) In a simple cubic unit cell there is a total of one atom. The volume of the atom is given by $\frac{4}{3}\pi r^3$. The volume of the unit cell is given by a^3 and

$a = 2r$, so the volume of the unit cell is $8r^3$. The fraction of occupied space in the unit cell is given by

$$\frac{\frac{4}{3}\pi r^3}{8r^3} = \frac{\frac{4}{3}\pi}{8} = 0.52$$

52% of the space is occupied, so 48% of this unit cell would be empty.

(b) The percentage of empty space in an fcc unit cell is 26%, so the fcc cell is much more efficient at occupying the space available.

5.97 The Lewis structure of (HF)$_3$ chain is as follows:

$$\text{H}\!-\!\ddot{\underset{..}{\text{F}}}\text{:} \;\text{- - -}\; \text{H}\!-\!\ddot{\underset{..}{\text{F}}}\text{:} \;\text{- - -}\; \text{H}\!-\!\ddot{\underset{..}{\text{F}}}\text{:}$$

the bond angle is $\sim 180^\circ$

5.99 (a) This oxidation will increase the electrical conductivity of graphite because of the mobile HSO_4^- anions.

(b) This oxidation will increase the spacing, d, due to the HSO_4^-. Based on the Bragg equation: $2d\sin\theta = \lambda$, the angle of the x-ray beam will decrease (if the wavelength is the same).

5.101 The lattice layers from which constructive x-ray diffraction occurs are parallel. First draw perpendicular lines from the point of intersection of the top x-ray with the lattice plane to the lower x-ray for both the incident and diffracted rays. The x-rays are in phase and parallel at point A. If we want them to be still in phase and parallel when they exit the crystal, then they must still be in phase when they reach point C. In order for this to be true, the extra distance that the second beam travels with respect to the first must be equal to some integral number of wave-lengths. The total extra distance traveled, $A \rightarrow B \rightarrow C$, is equal to $2x$. From the diagram, we can see that the angle A-D-B must also be equal to θ. The angles θ and α sum to $90°$, as do the angles α and A-D-B. We can then write $\sin\theta = \dfrac{x}{d}$ and $x = d\sin\theta$. The total distance traveled is $2x = 2d\sin\theta$. So, for the two x-rays to be in phase as they exit the crystal, $2d \sin \theta$ must be equal to an integral number of wave-lengths.

5.103 The answer to this problem is obtained from Bragg's law: $\lambda = 2d \sin \theta$ where λ is the wavelength of radiation, d is the interplanar spacing, and θ is the angle of incidence of the x-ray beam. Here $d = 401.8$ pm and $\lambda = 71.07$ pm.

$71.07 \text{ pm} = 2(401.8 \text{ pm}) \sin \theta$

$\theta = 5.074°$

5.105 $3/2 \, kT = 1/2 \, mv^2$ $\lambda = h/mv$ $3kT = h^2/m\lambda^2$ $T = 633$ K

Chemistry Connections

5.107 (a) The Lewis structures of ethylammonium nitrate (EAN) and the formal charge of each atom are showing as follows:

(b) The hybridization schemes of C and N atoms are assigned in the Lewis structure of (a).

(c) The anion (NO_3^-). N^{5+} can be reduced to a lower oxidation state by a reducing agent.

(d) $CH_3CH_2NH_2 + HNO_3 \rightarrow CH_3CH_2NH_3^+ \, NO_3^-$ (acid-base)

(e) $n_{\text{ethylamine}} = \left(\dfrac{PV}{RT} \right) = \left(\dfrac{(0.960 \, atm)(2.00L)}{(0.08206 \, L \cdot amt \cdot K^{-1} \cdot mol^{-1})(296.2K)} \right) =$

0.0790 mole

$n_{\text{nitric acid}} = (0.250 \text{ L})(0.240\text{M}) = 0.0600 \text{ mole}$ (limiting reagent)

Theoretical yield $= 0.0600 \text{ mol } HNO_3$

$$x \left(\frac{(1 \, mole \, EAN)}{1 \, mole \, HNO_3} \right) \left(\frac{108.104 \, g EAN}{1 \, mole \, EAN} \right) = 6.49 \text{ g EAN}$$

Percentage yield $= \left(\dfrac{4.10}{6.49} \right) \times 100\% = 63.2\%$

(f) The forces hold EAN together in the solid contain ion-ion, hydrogen-bonding, and London forces. NaCl and NaBr only have ion-ion and London forces to hold them together (No H-bonding involved). However, the ion-ion forces in NaCl and NaBr are much stronger that those of EAN because of the smaller ion sizes of NaCl and NaBr.

(g) As the ion size increases, the distance between ions will increase and the ion-ion forces will decrease, and the melting point will decrease.

CHAPTER 6

INORGANIC MATERIALS

6.1 $E = \dfrac{hc}{\lambda} = \dfrac{(6.626 \times 10^{-34}\ J \cdot s)(2.998 \times 10^{8}\ m \cdot s^{-1})}{445 \times 10^{-9}\ m} = 4.47 \times 10^{-19}\ J$

6.3 $D_{Al} = \dfrac{m_1}{V} = \dfrac{26.98n}{V}$;

$D_{magnalium} = \dfrac{m_2}{V} = \dfrac{(26.89 \times 70\%n + 24.31 \times 30\%n)}{V} = \dfrac{26.18n}{V}$

$D = \dfrac{D_{magnalium}}{D_{Al}} = 0.9703$

6.5 The major impurity is carbon; it is removed by oxidation of carbon to CO_2, followed by capture of the CO_2 by base to form a slag.

6.7 Alloys are usually (1) harder and more brittle, and (2) poorer conductors of electricity than the metals from which they are made.

6.9 (a) The alloy is undoubtedly interstitial because the atomic radius of nitrogen is much smaller (74 pm vs. 124 pm) than that of iron. The rule of thumb is that the solute atom be less than 60% the solvent atom in radius in order for an interstitial alloy to form. That criterion is met here. (b) We expect that nitriding will make iron harder and stronger, with a lower electrical conductivity.

6.11 (a) These problems are most easily solved by assuming 100 g of substance. In 100 g of Ni-Cu alloy there will be 25 g Ni and 75 g Cu, corresponding to 0.43 mol Ni and 1.2 mol Cu. The atom ratio will be the same as mole ratio:

$$\frac{1.2\ mol\ Cu}{0.43\ mol\ Ni} = 2.8 \text{ Cu per Ni}$$

(b) Pewter, which is 7% Sb, 3% Cu, and 90% Sn, will contain 7 g Sb, 3 g Cu and 90 g Sn per 100 g of alloy. This will correspond to 0.06 mol Sb, 0.05 mol Cu, and 0.76 mol Sn per 100 g alloy. The atom ratio will be 15 Sn : 1.2 Sb : 1 Cu.

6.13 In diamond, carbon is sp^3-hybridized and forms a tetrahedral, three-dimensional network structure, which is extremely rigid. Graphite carbon is sp^2-hybridized and planar. Its application as a lubricant results from the fact that the two-dimensional sheets can "slide" across one another, thereby reducing friction. In graphite, the unhybridized p-electrons are free to move from one carbon atom to another, which results in its high electrical conductivity. In diamond, all electrons are localized in sp^3-hybridized C-C σ-bonds, so diamond is a poor conductor of electricity.

6.15 (a) Carbon atomic radius is 77 pm. The number of carbon atom on the 2.0 cm × 1.0 cm tape is:

$$\frac{2.0\ cm \times 1.0\ cm}{\pi(77 \times 10^{-10}\ cm)^2} = 1.1 \times 10^{16} \text{ C atoms}$$

(b) $\left(1.1 \times 10^{16} C\quad atoms\right)\left(\dfrac{1mol\quad C\quad atoms}{6.022 \times 10^{23} atoms \cdot mol^{-1}}\right) = 1.8 \times 10^{-8}\ mol$

$\left(1.8 \times 10^{-8}\ mol\right)\left(\dfrac{1nmol}{10^{-9}\ mol}\right) = 18 nmol$

6.17 Due to the greater polarizing ability of the smaller Mg^{2+} ion on the carbonate ion, we expect $MgCO_3$ to be less stable than $CaCO_3$. Therefore, we expect that the oxide formed is MgO and that the carbonate is $CaCO_3$. Furthermore, for $MgCO_3(s) \rightarrow MgO(s) + CO_2(g)$, ΔG_r° is roughly +50 kJ, and for $CaCO_3(s) \rightarrow CaO(s) + CO_2(g)$, ΔG_r° is roughly +130 kJ (as calculated from the data in Appendix 2A). Thus, the formation of MgO versus that of CaO is thermodynamically favored.

6.19   (a) The unit cell of moissanite is:

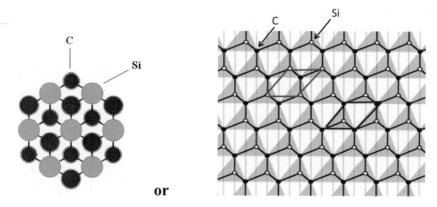

or

(b) $D = \dfrac{m}{V}$.   Although a Si atom is heavier than a carbon atom, a Si atom is larger than a carbon atom. This will result in more empty spaces between C and Si atoms during packing process, which makes moissanite less dense than diamond.

6.21   By viewing the unit cell from different directions, it is clear that it belongs to a hexagonal crystal system. (b) There are eight carbonate ions on edges, giving ¼ × 8 = 2 carbonate ions, plus there are four carbonate ions completely within the unit cell. The total number of carbonate ions in the unit cell is six. Calcium ions lie at the corners of the unit cell ($\dfrac{1}{8} \times 8 = 1$) as well as on the edges (¼ × 4) and within the cell (4). The total number of calcium ions in the cell is six, agreeing with the overall stoichiometry of calcite, $CaCO_3$.

6.23 $D = \dfrac{m}{V}, \quad m = DV, \quad V = \dfrac{m}{D}$

$V = \dfrac{m}{D} = \dfrac{19.28 \ g \cdot cm^{-3} \times 4.0 \ cm^3}{5.01 \ g \cdot cm^{-3}} = 15.4 \ cm^3$

6.25

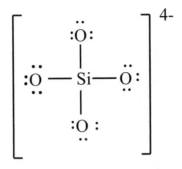

Formal charge: Si = 0, O = -1; Oxidation number: Si = +4, O = -2; this is an AX$_4$ VSEPR structure; therefore, the shape is tetrahedral.

6.27 The Si$_2$O$_7^{6-}$ ion is built from two SiO$_4^{4-}$ tetrahedral ions in which the silicate tetrahedral share one O atom. This is the only case in which one O is shared. (b) The pyroxenes, for example, jade, NaAl(SiO$_3$)$_2$, consist of chains of SiO$_4$ units in which two O atoms are shared by neighboring units. The repeating unit has the formula SiO$_3^{2-}$.

6.29 When glass is etched by HF, the following reaction happens:

$$SiO_2(s) + 6HF(aq) \rightarrow SiF_6^{2-}(aq) + 2H_3O^+(aq)$$

The oxidation number of silicon in the anion is +4.

6.31 The reaction of the following will occur:

$$2KAlSi_3O_8(s) + 2H_2O(l) + CO_2(g) \rightarrow$$
$$K_2CO_3(aq) + Al_2Si_2O_5(OH)_4(s) + 4SiO_2(s)$$

$$\text{Mass of clay} = \frac{2.0 \times 1000 \text{ g}}{278.34 \frac{g}{mol}} \times \left(\frac{1 \text{ mol clay}}{2 \text{ mol feldspar}}\right)\left(\frac{258.17 \text{ g}}{\text{mol clay}}\right) = 930 \text{ g}$$

6.33 Ionic fluorides react with water to liberate HF, which then reacts with the glass. Glass bottles used to store metal fluorides become brittle and may disintegrate upon standing on the shelf.

6.35 The iron ions impart a deep red color to the clay, which is not desirable for the manufacture of fine china; a white base is aesthetically more pleasing.

6.37 (a) The oxidation number of phosphorus in $Li_7P_3S_{11}$ is +5;
(b) The oxidation number of titanium in $BaTiO_3$ is +4.

6.39 The conductivity of a semiconductor increases with temperature as increasing numbers of electrons are promoted into the conduction band, whereas the conductivity of a metal will decrease as the motion of atoms will slow down the migration of electrons.

6.41 $\kappa = \dfrac{l}{A \times R}$; $\qquad R = \dfrac{l}{A \times \kappa}$

$R = 2.0 \text{ m} / (0.12 \times 10^{-6} \text{ m}^2 \times 60.7 \times 10^6 \text{ S·m}^{-1}) = 0.27 \text{ S}^{-1} = 0.27 \text{ } \Omega$

6.43 The diameter of silver is: 144 pm $\times$ 2 = 288 pm; for a 1.0 mm length of one-dimensional line, the total number of Ag atoms:

$$\dfrac{1.0 \text{ } mm \times \dfrac{1.0 \times 10^9 \text{ } pm}{1 \text{ } mm}}{288 \text{ } pm} = 3.5 \times 10^6 \text{ Ag atoms}$$

Silver has 2 states since 5s is half filled. So, $n = 1.8 \times 10^6$

6.45 (a) In and Ga; (b) P and Sb

6.47 Y: 8 atoms $\times \frac{1}{8}$ atom $\cdot$ corner^{-1} = 1 Y atom

Ba: 8 atoms $\times \frac{1}{4}$ atom $\cdot$ edge^{-1} = 2 Ba atoms

Cu: 3 Cu atoms completely inside unit cell = 3 Cu atoms

O: 10 atoms on faces $\times \frac{1}{2}$ atom $\cdot$ face^{-1} + 2 atoms completely inside unit

cell = 7 O atoms

Formula = $YBa_2Cu_3O_7$

6.49   In fluorescence, light absorbed by molecules is immediately emitted, whereas, in phosphorescence, molecules remain in an excited state for a period of time before emitting the absorbed light. In both phenomena, the energy of the emitted photon is lower than that of the absorbed photon (emitted photons have a longer wavelength).

6.51   Fluorescence dyes allow much smaller concentrations of biomolecules to be detected than normally possible.

6.53   Paramagnetic elements have at least one unpaired electron.

Sc = $3d^1 4s^2$ → Paramagnetic;   Ti = $4s^2 3d^2$ → Paramagnetic

V = $3d^3 4s^2$ → Paramagnetic;   Mn= $4s^2 3d^5$ → Paramagnetic

Fe = $3d^6 4s^2$ → Paramagnetic;   Co= $4s^2 3d^7$ → Paramagnetic

Ni= $4s^2 3d^8$ → Paramagnetic;   Y = $4d^1 5s^2$ → Paramagnetic

Zr = $4d^2 5s^2$ → Paramagnetic;   Nb = $4d^4 5s^1$ → Paramagnetic

Mo = $4d^5 5s^1$ → Paramagnetic;   Tc = $4d^5 5s^2$ → Paramagnetic

Ru= $4d^7 5s^1$ → Paramagnetic;   Rh= $4d^8 5s^1$ → Paramagnetic

Lu = $6s^2 4f^{14} 5d^1$ → Paramagnetic;   Hf = $6s^2 4f^{14} 5d^2$ → Paramagnetic

Ta = $6s^2 4f^{14} 5d^3$ → Paramagnetic;   W = $6s^2 4f^{14} 5d^4$ → Paramagnetic

Re = $6s^2 4f^{14} 5d^5$ → Paramagnetic;   Os = $6s^2 4f^{14} 5d^6$ → Paramagnetic

Ir = $6s^2 4f^{14} 5d^7$ → Paramagnetic;   Pt = $4f^{14} 5d^9 6s^1$ → Paramagnetic

Au = $4f^{14} 5d^{10} 6s^1$ → Paramagnetic

6.55   The compound is ferromagnetic below T_c because the magnetization is higher. Above the Curie temperature, the compound is simple paramagnet with randomly oriented spins, but below that temperature, the spins align and the magnetization increases.

6.57 The "top down" approach to manufacturing nanomaterials refers to physically assembling the nanoparticles into the form factor desired; on the other hand, the "bottom up" approach" utilizes specific intermolecular interactions to cause the nanomaterials to self-assemble.

6.59 Quantum dots florescence at 580.8 nm. With reference to the color chart, 580.8 nm radiation is yellow.

$$E_n = \frac{n^2 h^2}{8m_e r^2}$$

Energy Separation of the two lowest levels is: $n = 2$ & $n = 1$

$$\Delta E = \frac{4h^2}{8m_e r^2} - \frac{h^2}{8m_e r^2} = \frac{3h^2}{8m_e r^2} ; \quad \Delta E = \frac{hc}{\lambda}$$

Therefore, $r = \sqrt{\dfrac{3h\lambda}{8m_e c}} = \sqrt{\dfrac{3 \times 6.626 \times 10^{-34} J \cdot s \times 580.8 \times 10^{-9} m}{8 \times 9.109 \times 10^{-31} kg \times 2.998 \times 10^8 m \cdot s^{-1}}}$

$r = 7.269 \times 10^{-10}$ m $= 726.9$ pm

6.61 The uncertainty principle equation is given: $\Delta p \Delta x \geq \frac{1}{2} \hbar$

Where $\hbar = \dfrac{h}{2\pi} = \dfrac{6.62608 \times 10^{-34} J \cdot s}{2\pi} = 1.05457 \times 10^{-34}$ J $\cdot$ s

$\Delta p \Delta x \geq \frac{1}{2} (1.05457 \times 10^{-34}$ J$\cdot$s$) = 5.27285 \times 10^{-35}$ J $\cdot$ s

The minimum uncertainty occurs at the point where this relationship is an equality (i.e., using $=$ rather than $\geq$). The uncertainty in position will be taken as the 200 nm corresponding to the length of the box:

$\Delta p \Delta x = 5.27285 \times 10^{-35}$ J $\cdot$ s

$\Delta p = \dfrac{5.27285 \, kg \cdot m^2 \cdot s^{-1}}{200 \times 10^{-9} m} = 2.64 \times 10^{-28}$ kg $\cdot$ m $\cdot$ s^{-1} $= \Delta(mv) = m\Delta v$

The mass of an electron is 9.10939×10^{-31} kg, the uncertainty in velocity will be:

$$\Delta v = \frac{2.64 \times 10^{-28} \, kg \cdot m \cdot s^{-1}}{9.10939 \times 10^{-31} \, kg} = 290 \text{ m} \cdot \text{s}^{-1}$$

6.63 Given the expression of the cubic box is:

$$E_{xyz} = \frac{h^2}{8m_e L^2}(n_x^2 + n_y^2 + n_z^2)$$

Where E_{xyz} represents the energy of a given level having n values corresponding to x, y and z we can describe each new level by incrementing each n by one; any levels with the same energy will be degenerate. The three lowest energy levels will therefore be E_{111}, E_{211}, and E_{221}. From this we can arrive at expressions for the energy of each level:

$$E_{111} = \frac{h^2}{8m_e L^2}(1^2 + 1^2 + 1^2) = \frac{3h^2}{8m_e L^2}$$

Similarly, the other two energy levels are determined to be:

$$E_{211} = \frac{h^2}{8m_e L^2}(2^2 + 1^2 + 1^2) = \frac{3h^2}{4m_e L^2} \quad \text{and}$$

$$E_{221} = \frac{h^2}{8m_e L^2}(2^2 + 2^2 + 1^2) = \frac{9h^2}{8m_e L^2}$$

The 211, 121, and 112 levels are degenerate; that is $E_{211} = E_{121} = E_{112}$;

The 221, 122, and 212 levels are degenerate; that is $E_{221} = E_{122} = E_{212}$;

6.65 $SiC_{(s)} + 2NaOH_{(l)} + 2O_{2(g)} \rightarrow Na_2SiO_{3(s)} + H_2O_{(g)} + CO_{2(g)}$

6.67 (a) Molar Mass of $ZrO_2 = 123.2228$ g $\cdot mol^{-1}$

Number of moles of ZrO_2 in 15.0 Kg $= \dfrac{15.0 \times 10^3 \, g}{123.222 g \cdot mol^{-1}} = 122.$ mol

Total ZrO_2 molecules $= 6.022 \times 10^{23}$ mol$^{-1} \times 122.$ mol $= 7.35 \times 10^{25}$

Total number of electrons in Zr $= 40$;

Total number of electrons in 2 O $= 16$

Total number of electrons in one $ZrO_2 = 56$

Total Number of electrons in 15.0 Kg $ZrO_2 = 56 \times 7.35 \times 10^{25} = 4.12 \times 10^{27}$

Mass of an Electron $= 9.109 \times 10^{-31}$ Kg

Total Mass of electrons in 15.0 Kg $ZrO_2 = 4.12 \times 10^{27} \times 9.109 \times 10^{-31}$ Kg

$= 3.75 \times 10^{-3}$ Kg $= 3.75$ g

(b) Number of electrons = Number of protons

Mass of a Proton $= 1.673 \times 10^{-27}$ Kg

Total Mass of Protons $= 4.12 \times 10^{27} \times 1.673 \times 10^{-27}$ Kg $= 6.89$ Kg

6.69　When semiconductors are at very low temperatures, the valence band is fully occupied and conduction band is empty, so the conductivity is low. As temperature increases in vacuum, some electrons are promoted from valence band to the conduction band and cause increases in conductivity. However, when semiconductors are heated in oxygen, chemisorptions will happen and adsorbed oxygen on the surface will trap electrons forming oxygen negative ion, which will decrease the electron conductivity.

6.71　Plot (a) was obtained from an organic dye and plot (b) was obtained from a quantum dot suspension. For an organic dye solution, since the sizes of organic dye molecules are the same and the quantum levels (n) are well fixed so a distinct absorption spectrum will be obtained. For a quantum dot suspension, due to the diameter variation of quantum dots and variable quantum transition levels, the separation of energy levels will vary (as $1/r^2$). Therefore, the wavelength of light that can cause excitation will also vary. In addition, since the sizes of quantum dots vary only slightly, a continuous absorption spectrum is normally observed.

6.73　(a) C: sp^2;　B: sp^2; N: sp^2

(b) A tube with a diameter of about 1.3 nm will have a circumference of $2\pi r$ or πd. Thus, the circumference of a 1.3×10^{-9} m diameter nanotube will be $(1.3 \times 10^{-9} \text{ m})(\pi) = 4.1 \times 10^{-9}$ m.

To calculate the number of C_6 rings that will be strung together, we need to calculate the distance across the C_6 ring as shown by the arrow.

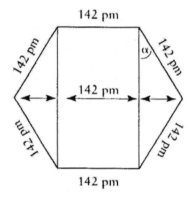

The total distance from one carbon to the opposite carbon on the ring will be given by $d = 142 \text{ pm} + 2(142 \text{ pm} \times \sin 30°) = 284$ pm.

One repeat unit will be one benzene ring plus one C—C bond,

$284 \text{ pm} + 142 \text{ pm} = 426 \text{ pm}$ or 4.26×10^{-10} m. There are thus about

$4.1 \times 10^{-9} \div 4.26 \times 10^{-10} = 10$ of units strung together around the smallest nanotube known.

(c) In C_{60} the carbon atoms are sp^2-hybridized and are nearly planar. However, the curvature of the molecule introduces some strain at the carbon atoms so that there is some tendency for some of the carbon atoms to undergo conversion to sp^3 hybridization. However, to make every carbon sp^3-hybridized would introduce much more strain on the carbon cage and, after a certain point, further addition of hydrogen becomes unfavorable.

(d) The spherical structures require the formation of five-member rings (see structure **10**, C_{60}). Boron nitride cannot form these rings because they would require high-energy boron-boron or nitrogen-nitrogen bonds.

(e) The unit cell described will contain a total of 4 B atoms and 4 N atoms. The volume of the cell is

$(361.5 \text{ pm})^3 = (3.615 \times 10^{-8} \text{ cm})^3 =$ or $4.724 \times 10^{-23} \text{ cm}^{-3}$. The mass in the unit cell will be

$$(4 \times 10.81 \text{ g} \cdot \text{mol}^{-1} + 4 \times 14.01 \text{ g} \cdot \text{mol}^{-1}) \div 6.022 \times 10^{23} \text{ mol}^{-1}$$
$$= 1.649 \times 10^{-22} \text{ g.}$$

$$d = \frac{1.649 \times 10^{-22} \text{ g}}{4.724 \times 10^{-23} \text{ cm}^3} = 3.491 \text{ g} \cdot \text{cm}^{-3}$$

(f) Because the density of cubic boron nitride is greater than that of hexagonal BN, we would expect the cubic form to be favored at high pressures, exactly as found for the cubic (diamond) and hexagonal (graphite) forms of carbon.

CHAPTER 7

THERMODYNAMICS: THE FIRST LAW

7.1   (a) isolated;   (b) closed;   (c) isolated;   (d) open;   (e) closed;

(f) open

7.3   (a) Work is given by $w = -P_{ext}\Delta V$. The applied external pressure is

known, but we must calculate the change in volume given the physical

dimensions of the pump and the distance, d, the piston in the pump

moves:

$$\Delta V = -\pi r^2 d = \pi(1.5 \text{ cm})^2 (20. \text{ cm})\left(\frac{1 \text{ L}}{1000 \text{ cm}^3}\right) = -0.14 \text{ L}$$

ΔV is negative because the air in the pump is compressed to a smaller

volume; work is then:

$$w = -(2.00 \text{ atm})(-0.14 \text{ L})\left(\frac{101.325 \text{ L}}{\text{L} \cdot \text{atm}}\right) = 28 \text{ J}$$

(b) Work on the air is positive by convention as work is done on the air, it

is compressed.

7.5   The change in internal energy ΔU is given simply by summing the two

energy terms involved in this process. We must be careful, however, that

the signs on the energy changes are appropriate. In this case, internal

energy will be added to the gas sample by heating and through

compression. Therefore the change in internal energy is:

$\Delta U = 524 \text{ kJ} + 340 \text{ kJ} = 864 \text{ kJ}$

7.7   (a) The internal energy increased by more than the amount of heat added.

Therefore, the extra energy must have come from work done on the

system.

(b) $w = \Delta U - q = 982 \text{ J} - 492 \text{ J} = +4.90 \times 10^2 \text{ J}.$

7.9 To get the entire internal energy change, we must sum the changes due to heat and work. In this problem, $q = +5500$ kJ. Work will be given by $w = -P_{ext}\Delta V$ because it is an expansion against a constant opposing pressure:

$$w = -\left(\frac{750 \text{ Torr}}{760 \text{ Torr}\cdot\text{atm}^{-1}}\right)\left(\frac{1846 \text{ mL} - 345 \text{ mL}}{1000 \text{ mL}\cdot\text{L}^{-1}}\right) = -1.48 \text{ L}\cdot\text{atm}$$

To convert to J we use the equivalency of the ideal gas constants:

$$w = -(1.48 \text{ L}\cdot\text{atm})\left(\frac{8.314 \text{ J}\cdot\text{K}^{-1}\cdot\text{mol}^{-1}}{0.082\,06 \text{ L}\cdot\text{atm}\cdot\text{K}^{-1}\cdot\text{mol}^{-1}}\right) = -1.50\times10^2 \text{ J}$$

$\Delta U = q + w = 5500 \text{ kJ} - 0.150 \text{ kJ} = 5500 \text{ kJ}$

The energy change due to the work term turns out to be negligible in this problem.

7.11 Using $\Delta U = q + w$, where $\Delta U = -2573$ kJ and $q = -947$ kJ,

- 2573 kJ $= -947$ kJ $+ w$.

Therefore, $w = -1626$ kJ.

1626 kJ of work can be done by the system on its surroundings.

7.13 (a) true if no work is done; (b) always true; (c) always false; (d) true only if $w = 0$ (in which case $\Delta U = q = 0$); (e) always true

7.15 (a) During melting heat is absorbed and q is positive.

Since the change is occurring at constant temperature $\Delta E = 0$.

Therefore work is done by the system and w is negative.

(b) During condensation heat is released and q is negative.

Since the change is occurring at constant temperature $\Delta E = 0$.

Therefore work is done on the system and w is positive.

7.17 (a) The heat change will be made up of two terms: one term to raise the temperature of the copper and the other to raise the temperature of the water:

$$q = (400.0 \text{ g})(4.18 \text{ J} \cdot (°C)^{-1} \cdot g^{-1})(100.0°C - 22.0°C)$$
$$+ (500.0 \text{ g})(0.38 \text{ J} \cdot (°C)^{-1} \cdot g^{-1})(100.0°C - 22.0°C)$$
$$= 1.30 \times 10^5 \text{ J} + 1.48 \times 10^4 \text{ J} = 1.45 \times 10^5 \text{ J} = 1.4 \times 10^2 \text{ kJ}$$

(b) The percentage of heat attributable to raising the temperature of water will be

$$\left(\frac{1.30 \times 10^5 \text{ J}}{1.45 \times 10^5 \text{ J}} \right)(100) = 90\%$$

7.19 heat lost by metal $= -$ heat gained by water

$$(20.0 \text{ g})(T_{final} - 100.0°C)(0.38 \text{ J} \cdot (°C)^{-1} \cdot g^{-1})$$
$$= -(50.7 \text{ g})(4.18 \text{ J} \cdot (°C)^{-1} \cdot g^{-1})(T_{final} - 22.0°C)$$
$$(T_{final} - 100.0°C)(7.6 \text{ J} \cdot (°C)^{-1}) = -(212 \text{ J} \cdot (°C)^{-1})(T_{final} - 22.0°C)$$
$$T_{final} - 100.0°C = -28(T_{final} - 22.0°C)$$
$$T_{final} + 28 \, T_{final} = 100.0°C + 616°C$$
$$29 \, T_{final} = 716°C$$
$$T_{final} = 25°C$$

7.21 $C_{cal} = \dfrac{22.5 \text{ kJ}}{23.97°C - 22.45°C} = 14.8 \text{ kJ} \cdot (°C)^{-1}$

7.23 (a) The irreversible work of expansion against a constant opposing pressure is given by

$$w = -P_{ex}\Delta V$$
$$w = -(1.00 \text{ atm})(6.52 \text{ L} - 4.29 \text{ L})$$
$$= -2.23 \text{ L} \cdot \text{atm}$$
$$= -2.23 \text{ L} \cdot \text{atm} \times 101.325 \text{ J} \cdot L^{-1} \cdot \text{atm}^{-1} = -226 \text{ J}$$

(b) An isothermal expansion will be given by

$$w = -nRT \frac{V_2}{V_1}$$

n is calculated from the ideal gas law:

$$n = \frac{PV}{RT} = \frac{(1.79 \text{ atm})(4.29 \text{ L})}{(0.082\ 06 \text{ L} \cdot \text{atm} \cdot \text{K}^{-1} \cdot \text{mol}^{-1})(305 \text{ K})} = 0.307 \text{ mol}$$

$$w = -(0.307 \text{ mol})(8.314 \text{ J} \cdot \text{K}^{-1} \cdot \text{mol}^{-1})(305 \text{ K}) \ln \frac{6.52}{4.29}$$

$$= -326 \text{ J}$$

Note that the work done is greater when the process is carried out reversibly.

7.25　NO_2. The heat capacity increases with molecular complexity—as more atoms are present in the molecule, there are more possible bond vibrations that can absorb added energy.

7.27　(a) The molar heat capacity of a monatomic ideal gas at constant pressure is $C_{P,m} = \frac{5}{2}R$. The heat released will be given by

$$q = \left(\frac{5.025 \text{ g}}{83.80 \text{ g} \cdot \text{mol}^{-1}} \right)(25.0°\text{C} - 97.6°\text{C})(20.8 \text{ J} \cdot \text{mol}^{-1} \cdot (°\text{C})^{-1}) = -90.6 \text{ J}$$

(b) Similarly, the molar heat capacity of a monatomic ideal gas at constant volume is $C_{V,m} = \frac{3}{2}R$. The heat released will be given by

$$q = \left(\frac{5.025 \text{ g}}{83.80 \text{ g} \cdot \text{mol}^{-1}} \right)(25.0°\text{C} - 97.6°\text{C})(12.5 \text{ J} \cdot \text{mol}^{-1} \cdot (°\text{C})^{-1}) = -54.4 \text{ J}$$

7.29　(a) HCN is a linear molecule. The contribution from molecular motions will be $5/2\ R$.

(b) C_2H_6 is a polyatomic, nonlinear molecule. The contribution from molecular motions will be $3R$.

(c) Ar is a monatomic ideal gas. The contribution from molecular motions to the heat capacity will be 3/2 R.

(d) HBr is a diatomic, linear molecule. The contribution from molecular motions will be 5/2 R.

7.31 The strategy here is to determine the amount of energy per photon and the amount of energy needed to heat the water. Dividing the latter by the former will give the number of photons needed. Energy per photon is given by:

$$E = \frac{hc}{\lambda} = \frac{(6.626 \times 10^{-34} \text{ J} \cdot \text{s})(2.9979 \times 10^{8} \text{ m} \cdot \text{s}^{-1})}{4.50 \times 10^{-3} \text{ m}} = 4.41 \times 10^{-23} \text{ J} \cdot \text{photon}^{-1}$$

The energy needed to heat the water is:

$$350 \text{ g } (4.184 \text{ J} \cdot \text{g}^{-1} \cdot {}^{\circ}\text{C}^{-1}) \, (100.0\,^{\circ}\text{C} - 25.0\,^{\circ}\text{C}) = 1.10 \times 10^{5} \text{ J}$$

The number of photons needed is therefore:

$$\frac{1.10 \times 10^{5} \text{ J}}{4.41 \times 10^{-23} \text{ J} \cdot \text{photon}^{-1}} = 2.49 \times 10^{27} \text{ photons}$$

7.33 (a) Using the estimation that $3R = C$:

$$3R = C = (0.392 \text{ J} \cdot \text{K}^{-1} \cdot \text{g}^{-1})(M)$$

$$M = \frac{3R}{0.392 \text{ J} \cdot \text{K}^{-1} \cdot \text{g}^{-1}} = 63.6 \text{ g mol}^{-1}$$

This molar mass indicates that the atomic solid is Cu(s)

(b) From Example 5.3 we find that the density of a substance which forms a face-centered cubic unit cell is given by: $d = \dfrac{4M}{8^{3/2} N_A r^{3}}$. Therefore, we expect the density of copper to be:

$$d = \frac{4 \, (63.55 \text{ g} \cdot \text{mol}^{-1})}{8^{3/2} (6.022 \times 10^{23} \text{ mol}^{-1})(1.28 \times 10^{-8} \text{ cm})^{3}} = 8.90 \text{ g} \cdot \text{cm}^{-3}$$

7.35 (a) $\Delta H_{vap} = \dfrac{4.76 \text{ kJ}}{0.579 \text{ mol}} = 8.22 \text{ kJ} \cdot \text{mol}^{-1}$

(b) $\Delta H_{vap} = \dfrac{21.2 \text{ kJ}}{\left(\dfrac{22.45 \text{ g}}{46.07 \text{ g} \cdot \text{mol}^{-1}} \right)} = 43.5 \text{ kJ} \cdot \text{mol}^{-1}$

7.37 This process is composed of two steps: melting the ice at $0°C$ and then raising the temperature of the liquid water from $0°C$ to $25°C$:

Step 1: $\Delta H = \left(\dfrac{80.0 \text{ g}}{18.02 \text{ g} \cdot \text{mol}^{-1}} \right) (6.01 \text{ kJ} \cdot \text{mol}^{-1}) = 26.7 \text{ kJ}$

Step 2: $\Delta H = (80.0 \text{ g})(4.18 \text{ J} \cdot (°C)^{-1} \cdot \text{g}^{-1})(20.0°C - 0.0°C) = 6.69 \text{ kJ}$

Total heat required $= 26.7 \text{ kJ} + 6.69 \text{ kJ} = 33.4 \text{ kJ}$

7.39 The heat gained by the water in the ice cube will be equal to the heat lost by the initial sample of hot water. The enthalpy change for the water in the ice cube will be composed of two terms: the heat to melt the ice at $0°C$ and the heat required to raise the ice from $0°C$ to the final temperature.

$\text{heat (ice cube)} = \left(\dfrac{50.0 \text{ g}}{18.02 \text{ g} \cdot \text{mol}^{-1}} \right) (6.01 \times 10^3 \text{ J} \cdot \text{mol}^{-1})$

$+ (50.0 \text{ g})(4.184 \text{ J} \cdot (°C)^{-1} \cdot \text{g}^{-1})(T_f - 0°)$

$= 1.67 \times 10^4 \text{ J} + (209 \text{ J} \cdot (°C)^{-1})(T_f - 0°)$

$\text{heat (water)} = (400 \text{ g})(4.184 \text{ J} \cdot (°C)^{-1} \cdot \text{g}^{-1})(T_f - 45°)$

$= (1.67 \times 10^3 \text{ J} \cdot (°C)^{-1})(T_f - 45°)$

Setting these equal:

$-(1.67 \times 10^3 \text{ J} \cdot (°C)^{-1})T_f + 7.5 \times 10^4 \text{ J} = 1.67 \times 10^4 \text{ J} + (209 \text{ J} \cdot (°C)^{-1})T_f$

Solving for T_f:

$T_f = \dfrac{5.8 \times 10^4 \text{ J}}{1.88 \times 10^3 \text{ J} \cdot (°C)^{-1}} = 31 °C$

7.41 Based on the $\Delta H_{fus} = 10.0$ kJ·mol^{-1} and $\Delta H_{vap} = 20.0$ kJ·mol^{-1} and a constant heating rate, melting should occur twice as fast as vaporization which is observed in (b), (c) and (d). Heating curve (d) can be eliminated because the slope for the solid, liquid and gas are all the same. Curve (c) can also be eliminated because the heating slope for the solid and liquid are the same (yet they have different heat capacities). This leaves (b) as the best match.

7.43 (a) $\Delta H = (1.25 \text{ mol})(+358.8 \text{ kJ} \cdot \text{mol}^{-1}) = 448$ kJ

(b) $\Delta H = \left(\dfrac{197 \text{ g C}}{12.01 \text{ g} \cdot \text{mol}^{-1} \text{ C}}\right)\left(\dfrac{358.8 \text{ kJ}}{4 \text{ mol C}}\right) = 1.47 \times 10^3$ kJ

(c) $\Delta H = 415 \text{ kJ} = (n_{CS_2})\left(\dfrac{358.8 \text{ kJ} \cdot \text{mol}^{-1}}{4 \text{ mol CS}_2}\right)$

$n_{CS_2} = 4.63$ mol CS$_2$ or $(4.63 \text{ mol})(76.13 \text{ g} \cdot \text{mol}^{-1}) = 352$ g CS$_2$

7.45 (a) $(12 \text{ ft} \times 12 \text{ ft} \times 8 \text{ ft})\left(\dfrac{30.48 \text{ cm}}{1 \text{ ft}}\right)^3 = 3.26 \times 10^7$ cm^3

The heat capacity of air is 1.01 J·(°C)$^{-1}$·mol^{-1} and the average molar mass of air is 28.97 g·mol^{-1} (see Table 4.1). The density of air can be calculated from the ideal gas law:

$$d = \dfrac{P}{MRT} = \dfrac{1.00 \text{ atm}}{(28.97 \text{ g} \cdot \text{mol}^{-1})(0.082 \, 06 \text{ L} \cdot \text{atm} \cdot \text{K}^{-1} \cdot \text{mol}^{-1})(277.6 \text{ K})}$$
$$d = 0.001 \, 52 \text{ g} \cdot \text{cm}^{-3}$$

$40°F = 4.44°C,\ 78°F = 25.55°C$
$\Delta T = 25.55°C - 4.44°C = 21.1°C$

The heat required is

$(3.26 \times 10^7 \text{ cm}^3)(0.001 \, 52 \text{ g} \cdot \text{cm}^{-3})(1.01 \text{ J} \cdot (°C)^{-1} \cdot \text{mol}^{-1})(21.1°C)$
$= 1.056 \times 10^6 \text{ J} = 1.056 \times 10^3 \text{ } kJ$

The mass of octane required to produce this much heat will be given by

$$\left(\frac{-1.056 \times 10^3 \text{ kJ}}{-5471 \text{ kJ} \cdot \text{mol}^{-1}} \right)(114.22 \text{ g} \cdot \text{mol}^{-1}) = 22.0465 \text{ g} = 2.2 \times 10^1 \text{ g}$$

(b) $$\Delta H = \left(\frac{(1.0 \text{ gal})(3.785 \times 10^3 \text{ mL} \cdot \text{gal}^{-1})(0.70 \text{ g} \cdot \text{mL}^{-1})}{114.22 \text{ g} \cdot \text{mol}^{-1}} \right)$$
$$\times \left(\frac{-10\,942 \text{ kJ}}{2 \text{ mol octane}} \right)$$
$$= -1.3 \times 10^5 \text{ kJ}$$

7.47 (a) $$\left(1250 \frac{\text{kJ}}{\text{hr}} \right)\left(1 \frac{\text{hr}}{\text{day}} \right)\left(150 \frac{\text{days}}{\text{year}} \right) = 1.9 \times 10^5 \frac{\text{kJ}}{\text{year}}$$

(b) $$\left(150 \frac{\text{trips}}{\text{year}} \right)\left(0.40 \frac{\text{gal.}}{\text{trip}} \right)\left(3.785 \frac{\text{L}}{\text{gal.}} \right)\left(1000 \frac{\text{mL}}{\text{L}} \right)\left(0.702 \frac{\text{g}}{\text{mL}} \right)$$
$$\left(\frac{1 \text{ mol}}{114.23 \text{ g}} \right)\left(5471 \frac{\text{kJ}}{\text{mol}} \right) = 7.6 \times 10^6 \frac{\text{kJ}}{\text{year}}$$

7.49 From $\Delta H = \Delta U + P\Delta V$ at constant pressure, or $\Delta U = \Delta H - P\Delta V$.

Because $w = -P\Delta V = +22 \text{ kJ}$, we get $-15 \text{ kJ} + 22 \text{ kJ} = \Delta U = +7 \text{ kJ}$.

7.51 Net production of 1 mole of gas:

$$\Delta n = +1 \text{ mol}$$
$$PV = nRT$$
$$P\Delta V = \Delta nRT$$
$$-P\Delta V = -\Delta nRT = -(1\text{mol})(8.314 \text{ J.K}^{-1}.\text{mol}^{-1})(298\text{K})$$
$$= -2.48 \text{ kJ}$$

$$\Delta U = q + w = \Delta H - P\Delta V = -318 \text{kJ} - 2.48 \text{kJ} = -320 \text{kJ}$$

7.53 To determine the enthalpy of the reaction we must start with a balanced chemical reaction and determine the limiting reagent:

$$2HCl(aq) + Zn(s) \rightarrow H_2(g) + ZnCl_2(aq) .$$

$$0.800 \text{ L} \cdot 0.500 \text{ M HCl} = 0.400 \text{ mol HCl}$$

$$\frac{8.5 \text{ g}}{65.37 \text{g} \cdot \text{mol}^{-1}} = 0.130 \text{ mol Zn}$$

Examining the reaction stoichiometry and the initial quantities of HCl and Zn, we note that Zn is the limiting reagent (0.260 mol of HCl is needed to completely react with 0.130 moles of Zn). The enthalpy of reaction may be obtained using tabulated enthalpies of formation:

$$\Delta H_r = -153.89 \frac{\text{kJ}}{\text{mol}} + 2\left(-167.16\frac{\text{kJ}}{\text{mol}}\right) - 2\left(-167.16\frac{\text{kJ}}{\text{mol}}\right) - 0$$

$$= -153.89 \frac{\text{kJ}}{\text{mol}}$$

This is the enthalpy per mole of Zinc consumed. Therefore, the energy released by the reaction of 8.5 g of Zinc is:

$$\left(-153.89\frac{\text{kJ}}{\text{mol}}\right)(0.130 \text{ mol}) = -20.0 \text{ kJ}$$

The change in the temperature of the water is then:

$$-20000 \text{ J} = \left(-4.184 \ \frac{\text{J}}{{}^{\circ}\text{C g}}\right)(800 \text{ g})\Delta T$$

$$\Delta T = 5.98 \, {}^{\circ}\text{C and } T_f = 25\,{}^{\circ}\text{C} + 5.98\,{}^{\circ}\text{C} = 31\,{}^{\circ}\text{C}$$

7.55 The enthalpy of reaction for the reaction

$$4 \ C_7H_5N_3O_6(s) + 21 \ O_2(g) \rightarrow 28 \ CO_2(g) + 10 \ H_2O(g) + 6 \ N_2(g)$$

may be found using enthalpies of formation:

$$28\left(-393.51\frac{\text{kJ}}{\text{mol}}\right) + 10\left(-241.82\frac{\text{kJ}}{\text{mol}}\right) - 4\left(-67\frac{\text{kJ}}{\text{mol}}\right) = -13168\frac{\text{kJ}}{\text{mol}}$$

This is the energy released per mole of reaction as written. One fourth of this amount of energy or $3292 \frac{\text{kJ}}{\text{mol}}$ will be *released* per mole of TNT consumed. The energy density in kJ per L may be found by dividing this amount of energy with the mass of one mole of TNT and then by multiplying with the density of TNT:

$$\frac{3292\,\frac{kJ}{mol}}{227.14\,\frac{g}{mol}}\left(1.65\,\frac{g}{cm^3}\right)\left(\frac{10^3\,cm^3}{1\,L}\right) = +23.9 \times 10^3\,\frac{kJ}{L}$$

7.57 (a) $CO(g) + H_2O(g) \rightarrow CO_2(g) + H_2(g)$

(b) Bomb calorimeter means volume is fixed. Therefore, $w = 0$.

$\Delta U = q + w$ Need to find q.   Use $q = C\Delta T$

First work out how much heat was released on burning 1.40 g of carbon monoxide.

$q_{reaction} + q_{calorimeter} = 0$

$q_{reaction} = -q_{calorimeter}$

$\qquad = -C\Delta T$

$\qquad = -(3.00\,kJ \cdot (°C)^{-1})(22.799\,°C - 22.113\,°C)$

$\qquad = -2.058\,kJ$

$MW\,(CO) = 12.0107\,g.mol^{-1} + 15.9994\,g.mol^{-1} = 28.0101\,g.mol^{-1}$

$moles\,(CO) = \dfrac{1.40\,g}{28.0101\,g.mol^{-1}} = 4.998 \times 10^{-2}$

For 1.00 mol $CO(g)$, $q = \dfrac{-2.058\,kJ}{4.998 \times 10^{-2}\,mol}$

$\qquad\qquad\qquad\qquad = -41.176\,kJ.mol^{-1} = -41.2\,kJ.mol^{-1}$

For the combustion of 1.00 mol $CO(g)$, $\Delta U = q = -41.2\,kJ.mol^{-1}$

7.59 The combustion reaction of diamond is reversed and added to the combustion reaction of graphite to give the desired reaction:

$C(gr) + O_2(g) \longrightarrow CO_2(g)$ $\Delta H° = -393.51\,kJ$

$CO_2(g) \longrightarrow C(dia) + O_2(g)$ $\Delta H° = +395.41\,kJ$

$C(gr) \longrightarrow C(dia)$ $\Delta H° = +1.90\,kJ$

7.61

$$3BaO(s) \rightarrow 2Ba(s) + \frac{3}{2}O_2(g) \qquad\qquad \Delta H° = 166 \text{ kJ}$$

$$2Al(s) + \frac{3}{2}O_2(g) \rightarrow Al_2O_3(s) \qquad\qquad \Delta H° = -1676 \text{ kJ}$$

$$\overline{3BaO(s) + 2Al(s) \rightarrow Al_2O_3(s) + 3Ba(s) \qquad \Delta H° = -15 \text{ kJ}}$$

7.63 First, write the balanced equations for the reaction given:

$$C_2H_2(g) + \tfrac{5}{2}O_2(g) \longrightarrow 2\,CO_2(g) + H_2O(l) \qquad \Delta H° = -1300 \text{ kJ}$$

$$C_2H_6(g) + \tfrac{7}{2}O_2(g) \longrightarrow 2\,CO_2(g) + 3\,H_2O(l) \qquad \Delta H° = -1560 \text{ kJ}$$

$$H_2(g) + \tfrac{1}{2}O_2(g) \longrightarrow H_2O(l) \qquad \Delta H° = -286 \text{ kJ}$$

The second equation is reversed and added to the first, plus 2x the third:

$$C_2H_2(g) + \tfrac{5}{2}O_2(g) \longrightarrow 2\,CO_2(g) + H_2O(l) \qquad \Delta H° = -1300 \text{ kJ}$$

$$2\,CO_2(g) + 3\,H_2O(l) \longrightarrow C_2H_6(g) + \tfrac{7}{2}O_2(g) \qquad \Delta H° = +1560 \text{ kJ}$$

$$2[H_2(g) + \tfrac{1}{2}O_2(g) \longrightarrow H_2O(l)] \qquad 2[\Delta H° = -286 \text{ kJ}]$$

$$C_2H_2(g) + 2\,H_2(g) \longrightarrow C_2H_6(g)$$
$$\Delta H° = -1300 \text{ kJ} \cdot \text{mol}^{-1} + 1560 \text{ kJ} \cdot \text{mol}^{-1} + 2(-286 \text{ kJ} \cdot \text{mol}^{-1})$$
$$= -312 \text{ kJ} \cdot \text{mol}^{-1}$$

7.65 The reaction enthalpy for this reaction is given by:

$$\Delta H° = 12\,(\Delta H°_f(H_2O,l))$$
$$- [4\,(\Delta H°_f(HNO_3,l)) + 5\,(\Delta H°_f(N_2H_4,l))]$$
$$= 12(-285.83 \text{ kJ} \cdot \text{mol}^{-1})$$
$$- [4(-174.10 \text{ kJ} \cdot \text{mol}^{-1}) + 5(+50.63 \text{ kJ} \cdot \text{mol}^{-1})]$$
$$= -2986.71 \text{ kJ} \cdot \text{mol}^{-1}$$

7.67

$$2\,NH_4Cl(s) \rightarrow 2\,NH_3(g) + 2\,HCl(g) \qquad \Delta H° = 2(176.0) \text{ kJ}$$

$$N_2(g) + 4\,H_2(g) + Cl_2(g) \rightarrow 2\,NH_4Cl(s) \qquad \Delta H° = -628.86 \text{ kJ}$$

$$\underline{2\,NH_3(g) \rightarrow N_2(g) + 3\,H_2(g) \qquad\qquad \Delta H° = 92.22 \text{ kJ}}$$

$$H_2(g) + Cl_2(g) \rightarrow 2\,HCl(g) \qquad \Delta H° = -184.6 \text{ kJ}$$

7.69

$2NH_4Br(s) \rightarrow 2NH_3(g) + 2HBr(g)$ $\Delta H^\circ = 2(+188.32 \text{ kJ})$

$2 NH_3(g) \rightarrow N_2(g) + 3 H_2(g)$ $\Delta H^\circ = +92.22 \text{ kJ}$

$N_2(g) + 4 H_2(g) + Br_2(l) \rightarrow 2 NH_4Br(s)$ $\Delta H^\circ = -541.66 \text{ kJ}$

$H_2(g) + Br_2(l) \rightarrow 2 HBr(g)$ $\Delta H^\circ = -72.80 \text{ kJ}$

7.71 From Appendix 2A, ΔH°_f (NO) $= +90.25 \text{ kJ}$

The reaction we want is

$$N_2(g) + \tfrac{5}{2} O_2(g) \longrightarrow N_2O_5(g)$$

Adding the first reaction to half of the second gives

$2 NO(g) + O_2(g) \longrightarrow 2 NO_2(g)$ $\Delta H^\circ = -114.1 \text{ kJ}$

$2 NO_2(g) + \tfrac{1}{2} O_2(g) \longrightarrow N_2O_5(g)$ $\Delta H^\circ = -55.1 \text{ kJ}$

$2 NO(g) + \tfrac{3}{2} O_2(g) \longrightarrow N_2O_5(g)$ -169.2 kJ

The enthalpy of this reaction equals the enthalpy of formation of N_2O_5 (g)

minus twice the enthalpy of formation of NO, so we can write

$-169.2 \text{ kJ} = \Delta H^\circ_f (N_2O_5) - 2(+90.25 \text{ kJ})$

$\Delta H^\circ_f (N_2O_5) = +11.3 \text{ kJ}$

7.73 The enthalpy of the reaction

$PCl_3(l) + Cl_2(g) \longrightarrow PCl_5(s)$ $\Delta H^\circ = -124 \text{ kJ}$

is $\Delta H^\circ_r = \Sigma \Delta H^\circ_f \text{ (products)} - \Sigma \Delta H^\circ_f \text{ (reactants)}$

$-124 \text{ kJ} = \Delta H^\circ_f (PCl_5, s) - \Delta H^\circ_f (PCl_3, l)$

Remember that the standard enthalpy of formation of $Cl_2(g)$ will be 0 by

definition because this is an element in its reference state. From the

Appendix we find that

$\Delta H^\circ_f (PCl_3, l) = -319.7 \text{ kJ} \cdot \text{mol}^{-1}$

$-124 \text{ kJ} = \Delta H^\circ_f (PCl_5, s) - (-319.7 \text{ kJ})$

$\Delta H^\circ_f (PCl_5, s) = -444 \text{ kJ} \cdot \text{mol}^{-1}$

7.75 (a) For $H_2O(l)$, we want to find the enthalpy of the reaction

$$H_2(g) + \tfrac{1}{2}O_2(g) \longrightarrow H_2O(l)$$

The enthalpy change can be estimated from bond enthalpies. We will need to put in $(1\ mol)(436\ kJ \cdot mol^{-1})$ to break the H—H bonds in $1\ mol\ H_2(g)$, $(\tfrac{1}{2}\ mol)\ (496\ kJ \cdot mol^{-1})$ to break the O—O bonds in $\tfrac{1}{2}\ mol\ O_2(g)$; we will get back $(2\ mol)\ (463\ kJ \cdot mol^{-1})$ for the formation of 2 mol O—H bonds. This will give $\Delta H = -242\ kJ \cdot mol^{-1}$. This value, however, will be to produce water in the gas phase. In order to get the value for the liquid, we will need to take into account the amount of heat given off when the gaseous water condenses to the liquid phase. This is $44.0\ kJ \cdot mol^{-1}$ at 298 K:

$$\Delta H^{\circ}_{f, water(l)} = \Delta H^{\circ}_{f, water(g)} - \Delta H^{\circ}_{vap} = -242\ kJ \cdot mol^{-1} - 44.0\ kJ \cdot mol^{-1}$$
$$= -286\ kJ \cdot mol^{-1}$$

(b) The calculation for methanol is done similarly:

$$C(gr) + 2\ H_2(g) + \tfrac{1}{2}O_2(g) \longrightarrow CH_3OH(l)$$

ΔH for individual bond contributions:

atomize 1 mol C(gr)	$(1\ mol)(717\ kJ \cdot mol^{-1})$
break 2 mol H—H bonds	$(2\ mol)(436\ kJ \cdot mol^{-1})$
break $\tfrac{1}{2}$ mol O_2 bonds	$(\tfrac{1}{2}\ mol)(496\ kJ \cdot mol^{-1})$
form 3 mol C—H bonds	$-(3\ mol)(412\ kJ \cdot mol^{-1})$
form 1 mol C—O bonds	$-(1\ mol)(360\ kJ \cdot mol^{-1})$
form 1 mol O—H bonds	$-(1\ mol)(463\ kJ \cdot mol^{-1})$

Total	-222kJ

$$\Delta H^{\circ}_{f, methanol(l)} = \Delta H^{\circ}_{f, methanol(g)} - \Delta H^{\circ}_{vap}$$
$$= -222\ kJ \cdot mol^{-1} - 35.3\ kJ \cdot mol^{-1}$$
$$= -257\ kJ \cdot mol^{-1}$$

(c) $6\,C(gr) + 3\,H_2(g) \longrightarrow C_6H_6(l)$

Without resonance, we do the calculation considering benzene to have three double and three single C—C bonds:

atomize:	6 mol C(gr)	$(6\ mol)(717\ kJ \cdot mol^{-1})$
break:	3 mol H—H bonds	$(3\ mol)(436\ kJ \cdot mol^{-1})$
form:	3 mol CRC bonds	$-(3\ mol)(612\ kJ \cdot mol^{-1})$
form:	3 mol C—C bonds	$-(3\ mol)(348\ kJ \cdot mol^{-1})$
form:	6 mol C—H bonds	$-(6\ mol)(412\ kJ \cdot mol^{-1})$

Total	+258 kJ

$$\Delta H^{\circ}_{f,\,benzene(l)} = \Delta H^{\circ}_{f,\,benzene(g)} - \Delta H^{\circ}_{vap} = +258\ kJ \cdot mol^{-1} - 30.8\ kJ \cdot mol^{-1}$$
$$= +227\ kJ \cdot mol^{-1}$$

(d) $6\,C(gr) + 3\,H_2(g) \longrightarrow C_6H_6(l)$

With resonance, we repeat the calculation considering benzene to have six resonance-stabilized C—C bonds:

atomize:	6 mol C(gr)	$(6\ mol)(717\ kJ \cdot mol^{-1})$
break:	3 mol H—H bonds	$(3\ mol)(436\ kJ \cdot mol^{-1})$
form:	6 mol C—C bonds, resonance	$-(6\ mol)(518\ kJ \cdot mol^{-1})$
form:	6 mol C—H bonds	$-(6\ mol)(412\ kJ \cdot mol^{-1})$

Total	+30 kJ

$$\Delta H^{\circ}_{f,\,benzene(l)} = \Delta H^{\circ}_{f,\,benzene(g)} - \Delta H^{\circ}_{vap} = +30\ kJ \cdot mol^{-1} - 30.8\ kJ \cdot mol^{-1}$$
$$= -1\ kJ \cdot mol^{-1}$$

7.77 For the reaction $Na_2O(s) \longrightarrow 2\,Na^+(g) + O^{2-}(g)$

$$\Delta H_L = 2\,\Delta H°_f(Na, g) + \Delta H°_f(O, g) + 2\,I_1(Na)$$
$$-E_{ea1}(O) - E_{ea2}(O) - \Delta H_f(Na_2O(s))$$

$$\Delta H_L = 2(107.32\ kJ \cdot mol^{-1}) + 249\ kJ \cdot mol^{-1} + 2(494\ kJ \cdot mol^{-1})$$
$$-141\ kJ \cdot mol^{-1} + 844\ kJ \cdot mol^{-1} + 409\ kJ \cdot mol^{-1}$$

$$\Delta H_L = 2564\ kJ \cdot mol^{-1}$$

7.79 (a) $\quad \Delta H_L = \Delta H°_f(Na, g) + \Delta H°_f(Cl, g) + I_1(Na)$
$$- E_{ea}\ of\ Cl - \Delta H_f(NaCl(s))$$

$$787\ kJ \cdot mol^{-1} = 108\ kJ \cdot mol^{-1} + 122\ kJ \cdot mol^{-1} + 494\ kJ \cdot mol^{-1}$$
$$- 349\ kJ \cdot mol^{-1} - \Delta H_f(NaCl(s))$$

$$\Delta H_f(NaCl(s)) = -412\ kJ \cdot mol^{-1}$$

(b) $\quad \Delta H_L = \Delta H°_f(K, g) + \Delta H°_f(Br, g) + I_1(K)$
$$- E_{ea}(Br) - \Delta H_f(KBr(s))$$

$$\Delta H_L = 89\ kJ \cdot mol^{-1} + 97\ kJ \cdot mol^{-1} + 418\ kJ \cdot mol^{-1}$$
$$- 325\ kJ \cdot mol^{-1} + 394\ kJ \cdot mol^{-1}$$
$$= 673\ kJ \cdot mol^{-1}$$

(c) $\Delta H_L = \Delta H°_f(Rb, g) + \Delta H°_f(F, g) + I_1(Rb) - E_{ea}(F) - \Delta H_f(RbF(s))$

$$774\ kJ \cdot mol^{-1} = \Delta H°_f(Rb, g) + 79\ kJ \cdot mol^{-1}$$
$$+ 402\ kJ \cdot mol^{-1} - 328\ kJ \cdot mol^{-1} + 558\ kJ \cdot mol^{-1}$$

$$\Delta H°_f(Rb, g) = 63\ kJ \cdot mol^{-1}$$

7.81 (a) break: 3 mol $C \equiv C$ bonds $3(837)\ kJ \cdot mol^{-1}$

       form: 6 mol $C=C$ bonds $-6(518)\ kJ \cdot mol^{-1}$

       Total $- 597\ kJ \cdot mol^{-1}$

(b) break: 4 mol $C{-}H$ bonds $4(412)\ kJ \cdot mol^{-1}$

                4 mol $Cl{-}Cl$ bonds $4(242)\ kJ \cdot mol^{-1}$

       form: 4 mol $C{-}Cl$ bonds $-4(338)\ kJ \cdot mol^{-1}$

4 mol H—Cl bonds $-4(431)$ kJ $\cdot$ mol^{-1}

___

Total -460 kJ $\cdot$ mol^{-1}

(c) The number and types of bonds on both sides of the equations are equal, so we expect the enthalpy of the reaction to be essentially 0.

7.83 (a) break: 1 mol N—N triple bonds $(1 \text{ mol})(944 \text{ kJ} \cdot \text{mol}^{-1})$

                  3 mol F—F bonds $(3 \text{ mol})(158 \text{ kJ} \cdot \text{mol}^{-1})$

        form: 6 mol N—F bonds $(6 \text{ mol})(-195 \text{ kJ} \cdot \text{mol}^{-1})$

___

Total $+248$ kJ $\cdot$ mol^{-1}

(b) break: 1 mol C=C bonds $(1 \text{ mol})(612 \text{ kJ} \cdot \text{mol}^{-1})$

                  1 mol O—H bonds $(1 \text{ mol})(463 \text{ kJ} \cdot \text{mol}^{-1})$

        form: 1 mol C—C bonds $-(1 \text{ mol})(348 \text{ kJ} \cdot \text{mol}^{-1})$

                  1 mol C—O bonds $-(1 \text{ mol})(360 \text{ kJ} \cdot \text{mol}^{-1})$

                  1 mol C—H bonds $-(1 \text{ mol})(412 \text{ kJ} \cdot \text{mol}^{-1})$

___

Total -45 kJ $\cdot$ mol^{-1}

(c) break: 1 mol C—H bonds $(1 \text{ mol})(412 \text{ kJ} \cdot \text{mol}^{-1})$

                  1 mol Cl—Cl bonds $(1 \text{ mol})(242 \text{ kJ} \cdot \text{mol}^{-1})$

        form: 1 mol C—Cl bonds $-(1 \text{ mol})(338 \text{ kJ} \cdot \text{mol}^{-1})$

                  1 mol H—Cl bonds $-(1 \text{ mol})(431 \text{ kJ} \cdot \text{mol}^{-1})$

___

Total -115 kJ $\cdot$ mol^{-1}

7.85 The value that we want is given simply by the difference between three isolated C=C bonds and three isolated C—C single bonds, versus six resonance-stabilized bonds:

3 C=C bonds + 3 C—C bonds = 3(348 kJ) + 3(612 kJ) = 2880 kJ

6 resonance-stabilized bonds = 6(518 kJ) = 3108 kJ

As can be seen, the six resonance-stabilized bonds are more stable by ca. 228 kJ.

7.87 (a) The enthalpy of vaporization is the enthalpy change associated with the conversion $C_6H_6(l) \longrightarrow C_6H_6(g)$ at constant pressure. The value at 298.2 K will be given by

$$\Delta H°_{\text{vaporization at 298 K}} = \Delta H°_f \ (C_6H_6, g) - \Delta H°_f \ (C_6H_6, l)$$
$$= 82.93 \text{ kJ} \cdot \text{mol}^{-1} - (49.0 \text{ kJ} \cdot \text{mol}^{-1})$$
$$= 33.93 \text{ kJ} \cdot \text{mol}^{-1}$$

(b) In order to take into account the difference in temperature, we need to use the heat capacities of the reactants and products in order to raise the temperature of the system to 353.2 K. We can rewrite the reactions as follows, to emphasize temperature, and then combine them according to Hess's law:

$C_6H_6(l)_{\text{at 298 K}} \longrightarrow C_6H_6(g)_{\text{at 298 K}} \quad \Delta H° = 33.93 \text{ kJ}$

$C_6H_6(l)_{\text{at 298 K}} \longrightarrow C_6H_6(l)_{\text{at 353.2 K}} \quad \Delta H° = (1 \text{ mol})(353.2 \text{ K} - 298.2 \text{ K})$
$$(136.1 \text{ J} \cdot \text{mol}^{-1} \cdot \text{K}^{-1})$$
$$= 7.48 \text{ kJ}$$

$C_6H_6(g)_{\text{at 298 K}} \longrightarrow C_6H_6(g)_{\text{at 353.2 K}} \quad \Delta H° = (1 \text{ mol})(353.2 \text{ K} - 298.2 \text{ K})$
$$(81.67 \text{ J} \cdot \text{mol}^{-1} \cdot \text{K}^{-1})$$
$$= 4.49 \text{ kJ}$$

To add these together to get the overall equation at 353.2 K, we must reverse the second equation:

$C_6H_6(l)_{\text{at 298 K}} \longrightarrow C_6H_6(g)_{\text{at 298 K}} \quad \Delta H° = 33.93 \text{ kJ}$

$$C_6H_6(l)_{at\ 353.2\ K} \longrightarrow C_6H_6(l)_{at\ 298\ K} \qquad \Delta H° = -7.48\ kJ$$

$$C_6H_6(g)_{at\ 298\ K} \longrightarrow C_6H_6(g)_{at\ 353.2\ K} \qquad \Delta H° = 4.49\ kJ$$

$$C_6H_6(l)_{at\ 353.2\ K} \longrightarrow C_6H_6(g)_{at\ 353.2\ K}$$
$$\Delta H° = 33.93\ kJ \cdot mol^{-1} - 7.48\ kJ \cdot mol^{-1} + 4.49\ kJ \cdot mol^{-1}$$
$$= 30.94\ kJ \cdot mol^{-1}$$

(c) The value in the table is $30.8\ kJ \cdot mol^{-1}$ for the enthalpy of vaporization of benzene. The value is close to that calculated as corrected by heat capacities. At least part of the error can be attributed to the fact that heat capacities are not strictly constant with temperature.

7.89 For the reaction: $A + 2B \rightarrow 3C + D$ the molar enthalpy of reaction at temperature 2 is given by:

$$\Delta H_{r,2}° = H_{m,2}°(\text{products}) - H_{m,2}°(\text{reactants})$$

$$= 3H_{m,2}°(C) + H_{m,2}°(D) - H_{m,2}°(A) - 2H_{m,2}°(B)$$
$$= 3[H_{m,1}°(C) + C_{p,m}(C)(T_2 - T_1)] + [H_{m,1}°(D) + C_{p,m}(D)(T_2 - T_1)]$$
$$- [H_{m,1}°(A) + C_{p,m}(A)(T_2 - T_1)] - 2[H_{m,1}°(B) + C_{p,m}(B)(T_2 - T_1)]$$
$$= 3H_{m,1}°(C) + H_{m,1}°(D) - H_{m,1}°(A) - 2H_{m,1}°(B)$$

$$+[3C_{p,m}(C) + C_{p,m}(D) - C_{p,m}(A) - 2C_{p,m}(B)](T_2 - T_1)$$
$$= \Delta H_{r,1}° + [3C_{p,m}(C) + C_{p,m}(D) - C_{p,m}(A) - 2C_{p,m}(B)](T_2 - T_1)$$

Finally, $\Delta H_{r,2}° = \Delta H_{r,1}° + \Delta C_p(T_2 - T_1)$, which is Kirchhoff's law.

7.91 This process involves five separate steps: (1) raising the temperature of the ice from −5.042 °C to 0.00 °C. (2) melting the ice at 0.00°C, (3) raising the temperature of the liquid water from 0.00°C to 100.00°C, (4) vaporizing the water at 100.00°C, and (5) raising the temperature of the water vapor from 100.00°C to 150.35°C.

Step 1:

$$\Delta H = (42.30\ g)(2.03\ J \cdot (°C)^{-1} \cdot g^{-1})(0.00\ °C - (-5.042°C)) = 0.433\ kJ$$

Step 2: $\Delta H = \left(\dfrac{42.30 \text{ g}}{18.02 \text{ g} \cdot \text{mol}^{-1}} \right)(6.01 \text{ kJ} \cdot \text{mol}^{-1}) = 14.1 \text{ kJ}$

Step 3:

$\Delta H = (42.30 \text{ g})(4.18 \text{ J} \cdot (^\circ \text{C})^{-1} \cdot \text{g}^{-1})(100.00 \text{ }^\circ\text{C} - 0.00 ^\circ\text{C}) = 17.7 \text{ kJ}$

Step 4: $\Delta H = \left(\dfrac{42.30 \text{ g}}{18.02 \text{ g} \cdot \text{mol}^{-1}} \right)(40.7 \text{ kJ} \cdot \text{mol}^{-1}) = 95.5 \text{ kJ}$

Step 5:

$\Delta H = (42.30 \text{ g})(2.01 \text{ J} \cdot (^\circ\text{C})^{-1} \cdot \text{g}^{-1})(150.35 ^\circ\text{C} - 100.00 ^\circ\text{C}) = 4.3 \text{ kJ}$

The total heat required

$= 0.4 \text{ kJ} + 14.1 \text{ kJ} + 17.7 \text{ kJ} + 95.5 \text{ kJ} + 4.3 \text{ kJ}$

$= 132.0 \text{ kJ}$

7.93 (a)

$$2H_2S(g) + 3O_2(g) \rightarrow 2SO_2(g) + 2H_2O(l)$$
$$\underline{4H_2S(g) + 2SO_2(g) \rightarrow 6S(s) + 4H_2O(l)}$$
$$6H_2S(g) + 3O_2(g) \rightarrow 6S(s) + 6H_2O(l)$$

or

$$2H_2S(g) + O_2(g) \rightarrow 2S(s) + 2H_2O(l)$$

(b)

$$\frac{60.0 \times 10^3 \text{g}}{32.065 \text{ g. mol}^{-1}} = 1871.2 \text{ mol} = 1.87 \times 10^3 \text{ mol}$$

$\Delta H^\circ (\text{reaction}) = \Sigma \Delta H_f^\circ (\text{products}) - \Sigma \Delta H_f^\circ (\text{reactants})$

$= 2(-2.85.83 \text{ kJ. mol}^{-1}) - 2(-20.63 \text{kJ. mol}^{-1})$

$= -571.66 \text{ kJ. mol}^{-1} + 41.26 \text{ kJ. mol}^{-1}$

$= -530.40 \text{ kJ. mol}^{-1}$

Since -530.40 kJ.mol^{-1} reaction produces 2 moles of S(s), then the

enthalpy change associated with the production of 60.0 kg of S(s) would

be $-4.96 \times 10^5 \text{kJ}$

(c) Highly exothermic reaction and the reactor would need to be cooled.

7.95 Appendix 2A provides us with the heat of formation of $I_2(g)$ at 298K ($+62.44 \text{ kJ} \cdot \text{mol}^{-1}$) and the heat capacities of $I_2(g)$ ($36.90 \text{ J} \cdot \text{K}^{-1} \cdot \text{mol}^{-1}$) and $I_2(s)$ ($54.44 \text{ J} \cdot \text{K}^{-1} \cdot \text{mol}^{-1}$). We can calculate the $\Delta H_{sub}{}^0$ at 298K:

$$I_2(s) \longrightarrow I_2(g) \qquad\qquad \Delta H_{sub}{}^0 = +62.44 \text{ kJ} \cdot \text{mol}^{-1}$$

We can calculate the enthalpy of fusion from the relationship

$$\Delta H_{sub}{}^0 = \Delta H_{fus}{}^0 + \Delta H_{vap}{}^0$$

but these values need to be at the same temperature. To correct the value for the fact that we want all the numbers for 298K, we need to alter the heat of vaporization, using the heat capacities for liquid and gaseous iodine.

$$I_2(l) \text{ at } 184.3°C \longrightarrow I_2(g) \text{ at } 184.3°C \qquad \Delta H_{vap}{}^0 = +41.96 \text{ kJ} \cdot \text{mol}^{-1}$$

From Section 6.22, we find the following relationship

$$\Delta H_{r,2}{}^0 = \Delta H_{r,1}{}^0 + \Delta C_{P,m}{}^0 (T_2 - T_1)$$

$$\Delta H_{vap, 298K}{}^0 = \Delta H_{vap, 475.5K}{}^0 + (C_{P,m}{}^0(I_2,g) - C_{P,m}{}^0(I_2,l)) (T_2 - T_1)$$

$$\Delta H_{vap, 298K}{}^0 = +41.96 \text{ kJ} \cdot \text{mol}^{-1}$$
$$+ (36.90 \text{ J} \cdot \text{K}^{-1} \cdot \text{mol}^{-1} - 80.7 \text{ J} \cdot \text{K}^{-1} \cdot \text{mol}^{-1})(298K - 475.5K)$$
$$= +49.73 \text{ kJ} \cdot \text{mol}^{-1}$$

So, at 298K:

$$+62.44 \text{ kJ} \cdot \text{mol}^{-1} = \Delta H_{fus}{}^0 + 49.73 \text{ kJ} \cdot \text{mol}^{-1}$$
$$\Delta H_{fus}{}^0 = +12.71 \text{ kJ} \cdot \text{mol}^{-1}$$

7.97 First, we'll work out how much heat was required to raise the temperature of the water sample: $q = gC\Delta T = (150 \text{ g}) (4.18 \text{ J.°C}^{-1}.\text{g}^{-1})(5.00°C) = 3135 \text{ J}$ Since both the water and ice samples were at 0.00° C and the water sample took 0.5 h to get to 5.00° C, we can estimate that the ice took 10.0 h to melt. If 3135 J of heat were transferred in 0.5 h, then the amount of heat transferred in 10.0 h is,

$$\frac{10.0\,h}{0.5\,h} \times 3135\,J = 62.7\,kJ$$

(for 150 g sample).

$$\frac{150\,g}{18.0\,g.\,mol^{-1}} = 8.33\,mol\,H_2O \qquad \frac{62.7\,kJ}{8.33\,mol} = 7.53\,kJ.\,mol^{-1}$$

Which given the assumptions is fairly close to 6.01 kJ.mol^{-1}.

7.99 (a) $\quad C_6H_5NH_2\,(l) + \frac{31}{4}\,O_2\,(g) \longrightarrow 6\,CO_2\,(g) + \frac{7}{2}\,H_2O(l) + \frac{1}{2}\,N_2\,(g)$

(b)

$$m_{CO_2} = \left(\frac{0.1754\text{ g aniline}}{93.12\text{ g} \cdot mol^{-1}\text{ anline}}\right)\left(\frac{6\text{ mol }CO_2}{1\text{ mol aniline}}\right)(28.01\text{ g} \cdot mol^{-1}\,CO_2)$$

$$= 0.4873\text{ g }CO_2\,(g)$$

$$m_{H_2O} = \left(\frac{0.1754\text{ g aniline}}{93.12\text{ g} \cdot mol^{-1}\text{ aniline}}\right)\left(\frac{3.5\text{ mol }H_2O}{1\text{ mol aniline}}\right)(18.02\text{ g} \cdot mol^{-1}\,H_2O)$$

$$= 0.1188\text{ g }H_2O(l)$$

$$m_{N_2} = \left(\frac{0.1754\text{ g aniline}}{93.12\text{ g} \cdot mol^{-1}\text{ aniline}}\right)\left(\frac{0.5\text{ mol }N_2}{1\text{ mol aniline}}\right)(28.02\text{ g} \cdot mol^{-1}\,N_2)$$

$$= 0.026\,39\text{ g }N_2\,(g)$$

(c)

$$n_{O_2} = \left(\frac{0.1754\text{ g aniline}}{93.12\text{ g} \cdot mol^{-1}\text{ aniline}}\right)\left(\frac{\frac{31}{4}\text{ mol }O_2}{1\text{ mol aniline}}\right)$$

$$= 0.014\,60\text{ g }O_2\,(g)$$

$$P = \frac{nRT}{V} = \frac{(0.014\,60\text{ mol }O_2)(0.082\,06\text{ L} \cdot atm \cdot K^{-1} \cdot mol^{-1})(296K)}{0.355\text{ L}}$$

$$= 0.999\text{ atm}$$

7.101 (a) The reaction enthalpy is obtained by Hess's law:

$$\Delta H°_r = \Delta H°_f\,(CO,\,g) - \Delta H°_f\,(H_2O,\,g)$$

$$\Delta H°_r = (1)(-110.53\text{ kJ} \cdot mol^{-1}) - (1)(-241.82\text{ kJ} \cdot mol^{-1})$$

$$\Delta H°_r = +131.29\text{ kJ} \cdot mol^{-1}$$

endothermic

(b) The number of moles of H_2 produced is obtained from the ideal gas

law: $n = \dfrac{PV}{RT} = \dfrac{\left(\dfrac{500\ \text{Torr}}{760\ \text{Torr} \cdot \text{atm}^{-1}}\right)(200\ \text{L})}{(0.082\ 06\ \text{L} \cdot \text{atm} \cdot \text{K}^{-1} \cdot \text{mol}^{-1})(338\ \text{K})} = 4.74\ \text{mol}$

The enthalpy change accompanying the production of this amount of

hydrogen will be given by: $\Delta H = (4.74\ \text{mol})(131.29\ \text{kJ} \cdot \text{mol}^{-1}) = 623\ \text{kJ}$

7.103 (a) The number of moles burned may be obtained by taking the difference
in the number of moles of gas present in the tank before and after the drive
using the ideal gas equation:

$n_1 - n_2 = \dfrac{P_1 V}{RT} - \dfrac{P_2 V}{RT} = (P_1 - P_2)\left(\dfrac{V}{RT}\right)$

$= (16.0\ \text{atm} - 4.0\ \text{atm})\left(\dfrac{30.0\ \text{L}}{(0.0820574\ \text{L} \cdot \text{atm} \cdot \text{K}^{-1} \cdot \text{mol}^{-1})(298\ \text{K})}\right)$

$= 14.7\ \text{mol}$

(b) From a table of enthalpies of combustion, the enthalpy of combustion
of H_2 is found to be $-286\ \text{kJ} \cdot \text{mol}^{-1}$. The energy change is, therefore,

$(14.7\ \text{mol})(-286\ \text{kJ} \cdot \text{mol}^{-1}) = -4.20 \times 10^3\ \text{kJ}$.

7.105 (a) First we must balance the chemical reaction:

$C_6H_6(l) + \tfrac{15}{2} O_2(g) \longrightarrow 6\ CO_2(g) + 3\ H_2O(l)$

For 1 mol $C_6H_6(l)$ burned, the change in the number of moles of gas is

$(6.00 - 7.50)\ \text{mol} = -1.50\ \text{mol} = \Delta n$

$w = -P\Delta V = -P\left(\dfrac{\Delta n RT}{P}\right) = -\Delta n RT$

$w = -(-1.50\ \text{mol})(8.314\ \text{J} \cdot \text{K}^{-1} \cdot \text{mol}^{-1})(298\ \text{K}) = +3.716 \times 10^3\ \text{J} = +3.72\ \text{kJ}$

(b) $\Delta H_c = 6(-393.51\ \text{kJ} \cdot \text{mol}^{-1}) + 3(-285.83\ \text{kJ} \cdot \text{mol}^{-1}) - (+49.0\ \text{kJ} \cdot \text{mol}^{-1})$
$= -3267.5\ \text{kJ}$

(c) $\Delta U° = \Delta H° + w = (-3267.5 + 3.72)\ \text{kJ} = -3263.8\ \text{kJ}$

7.107

(a)　(1)　　(2)　(3)

(b) From bond enthalpies, each step is identical, as the number and types of bonds broken and formed are the same:

break:	1 mol C=C bonds	612 kJ
	1 mol H—H bonds	436 kJ
form:	1 mol C—C bonds	−348 kJ
	2 mol C—H bonds	2(−412 kJ)

Total:		−124 kJ

The total energy change should be equal to the sum of the three steps or $3(-124 \text{ kJ}) = -372 \text{ kJ}$.

(c) The Hess's law calculation using standard enthalpies of formation is easily performed on the composite reaction:

$$C_6H_6(l) + 3 H_2(g) \longrightarrow C_6H_{12}(l)$$
$$\Delta H°_r = \Sigma \Delta H°_f (\text{products}) - \Sigma \Delta H°_f (\text{reactants})$$

$$= \Delta H°_f (\text{cyclohexane}) - \Delta H°_f (\text{benzene})$$
$$= -156.4 \text{ kJ} \cdot \text{mol}^{-1} - (+49.0 \text{ kJ} \cdot \text{mol}^{-1})$$
$$= -205.4 \text{ kJ} \cdot \text{mol}^{-1}$$

(d) The hydrogenation of benzene is much less exothermic than predicted by bond enthalpy estimations. Part of this difference can be due to the inherent inaccuracy of using average values, but the difference is so large that this cannot be the complete explanation. As may be expected, the resonance energy of benzene makes it more stable than would be expected by treating it as a set of three isolated double and three isolated single bonds. The difference in these two values [$-205 \text{ kJ} - (-372 \text{ kJ}) = 167 \text{ kJ}$] is a measure of how much more stable benzene is than the Kekulé structure would predict.

7.109 (a) The combustion reaction is

$$C_{60}(s) + 60\,O_2(g) \longrightarrow 60\,CO_2(g)$$

The enthalpy of formation of $C_{60}(s)$ will be given by

$$\Delta H^\circ_c = 60\,\Delta H^\circ_f\,(CO_2,g) - \Delta H^\circ_f\,(C_{60},s)$$
$$-25\,937\text{ kJ} = 60\text{ mol} \times (-393.51\text{ kJ} \cdot \text{mol}^{-1}) - \Delta H^\circ_f\,(C_{60},s)$$
$$\Delta H^\circ_f\,(C_{60},s) = +2326\text{ kJ} \cdot \text{mol}^{-1}$$

(b) The bond enthalpy calculation is

$60\,C(gr) \longrightarrow 60\,C(g)$	$(60)(+717\text{ kJ} \cdot \text{mol}^{-1)})$
Form 60 mol C—C bonds	$-60(348\text{ kJ} \cdot \text{mol}^{-1})$
Form 30 mol C=C bonds	$-30(612\text{ kJ} \cdot \text{mol}^{-1})$
$C_{60}(g) \longrightarrow C_{60}(s)$	-233 kJ
$60\,C(gr) \longrightarrow C_{60}(s)$	$+3547\text{ kJ}$

(c) From the experimental data, the enthalpy of formation of C_{60} shows that it is *more* stable by $(3547\text{ kJ} - 2326\text{ kJ}) = 1221\text{ kJ}$ than predicted by the isolated bond model.

(d) $1221\text{ kJ} \div 60 = 20\text{ kJ}$ per carbon atom

(e) $150\text{ kJ} \div 6 = 25\text{ kJ}$ per carbon atom

(f) Although the comparison of the stabilization of benzene with that of C_{60} should be treated with caution, it does appear that there is slightly less stabilization per carbon atom in C_{60} than in benzene. This fits with expectations, as the C_{60} molecule is forced by its geometry to be curved. This means that the overlap of the *p*-orbitals, which gives rise to the delocalization that results in resonance, will not be as favorable as in the planar benzene molecule. Another perspective on this is obtained by noting that the C atoms in C_{60} are forced to be partially sp^3 hybridized because they cannot be rigorously planar as required by sp^2 hybridization.

7.111 The balanced combustion reactions are

$$C_6H_3(NO_2)_3(s) + \tfrac{15}{4}O_2(g) \longrightarrow 6\,CO_2(g) + \tfrac{3}{2}H_2O(l) + \tfrac{3}{2}N_2(g)$$

$$C_6H_3(NH_2)_3(s) + \tfrac{33}{4}O_2(g) \longrightarrow 6\,CO_2(g) + \tfrac{9}{2}H_2O(l) + \tfrac{3}{2}N_2(g)$$

Because the fundamental structures of the two molecules are the same, we need only look at the differences between the two, which in this case are concerned with the groups attached to nitrogen. From the combustion equations we can see that the differences are (1) the consumption of $\tfrac{18}{4}$ more moles of $O_2(g)$ and (2) the production of three more moles of $H_2O(l)$ for the combustion of aniline. Because the ΔH°_f of $O_2(g)$ is 0, the net difference will be the production of 3 more moles of $H_2O(l)$ or $3 \times (-285.83\ \text{kJ} \cdot \text{mol}^{-1}) = -857.49\ \text{kJ}$.

7.113 (a)
$$V_{init} = \frac{nRT}{P} = \frac{(0.060\ \text{mol})(0.0820578\ \text{L} \cdot \text{atm} \cdot \text{K}^{-1} \cdot \text{mol}^{-1})(298.15\ \text{K})}{1.00\ \text{atm}}$$
$$= 1.5\ \text{L}$$

(b) The combustion reaction is: $2\,SO_2(g) + O_2(g) \rightarrow 2\,SO_3(g)$. If equal molar amounts of SO_2 and O_2 are mixed, as in this case, SO_2 is the limiting reagent.

(c) The total number of moles remaining in the container will be: 0.030 mol $SO_3(g)$ + 0.015 mol $O_2(g)$ = 0.045 mol of gas at the end of the reaction. The final volume will, therefore, be:

$$V_f = \frac{nRT}{P} = \frac{(0.045\ \text{mol})(0.0820578\ \text{L} \cdot \text{atm} \cdot \text{K}^{-1} \cdot \text{mol}^{-1})(298.15\ \text{K})}{1.00\ \text{atm}}$$
$$= 1.1\ \text{L}$$

(d) $\Delta V = 1.1\ \text{L} - 1.5\ \text{L} = -0.4\ \text{L}$

$w = -P\Delta V = (1.00\ \text{atm})(-0.4\ \text{L})(101.325\ \text{J} \cdot \text{L}^{-1} \cdot \text{atm}^{-1})$

$\qquad = 40\ \text{J}$ of work done on the system (work is positive)

(e) The enthalpy of reaction may be found using standard enthalpies of formation and the balanced equation given above:

$$\Delta H_r = 2(-395.72 \text{ kJ} \cdot \text{mol}^{-1}) - 2(-296.83 \text{ kJ} \cdot \text{mol}^{-1}) = -197.78 \text{ kJ} \cdot \text{mol}^{-1}.$$

If 0.030 mol of SO_2 are consumed, then enthalpy change is:

$$(0.030 \text{ mol SO}_2)\left(\frac{-197.78 \text{ kJ}}{2 \text{ mol SO}_2}\right) = -2966.7 \text{ kJ} = -3.0 \text{ kJ (2 sf)} = -3000 \text{ J}.$$

(f) $\Delta U_r = q + w = -3000 \text{ J} + 40 \text{ J} = -2960 \text{ J}$

7.115 (a)

(b) $CH_4(g) + 2O_2(g) \rightarrow CO_2(g) + 2H_2O(g)$

$$\Delta H_c^o = (4 \times 412) + (2 \times 496) + (2 \times -743) + (4 \times -463)$$

$$= +2640 - 3338 = -698 \text{ kJ.mol}^{-1}$$

$H_3C\text{-}O\text{-}CH_3(g) + 3O_2(g) \rightarrow 2CO_2(g) + 3H_2O(g)$

$$\Delta H_c^o = (2 \times 360) + (6 \times 412) + (3 \times 496) + (4 \times -743) + (6 \times -463)$$

$$= -1070 \text{ kJ.mol}^{-1}$$

Given: $CH_3CH_2OH(l) \rightarrow CH_3CH_2OH(g) \quad \Delta H_{vap}^o = 43.5 \text{ kJ mol}^{-1}$

$CH_3CH_2OH(g) + 3O_2(g) \rightarrow 2CO_2(g) + 3H_2O(g)$

$$\Delta H_c^o = (360) + (463) + (348) + (5 \times 412) + (3 \times 496) + (4 \times -743) + (6 \times -463)$$

$$= -1031 \text{ kJ.mol}^{-1}$$

For the burning of 1 mole of $CH_3CH_2OH(l)$, $\Delta H_c^o = 43.5 - 1031 = -988 \text{ kJ.mol}^{-1}$.

The burning of 1 mole of dimethyl ether releases the most heat.

(c) $\dfrac{-890\,\text{kJ. mol}^{-1}}{16.01\text{g. mol}^{-1}} = -55.6\,\text{kJ. g}^{-1}$

$\dfrac{-1368\,\text{kJ. mol}^{-1}}{46.02\,\text{g. mol}^{-1}} = -29.73\,\text{kJ. g}^{-1}$

$\dfrac{-5471\,\text{kJ. mol}^{-1}}{114.08\,\text{g. mol}^{-1}} = -47.96\,\text{kJ. g}^{-1}$

Methane as it releases the most heat per gram.

(d) mass of 10.00 L octane $= \left(0.70\,\text{g. mol}^{-1}\right)\left(10{,}000\,\text{mL}\right) = 7000\,\text{g}$

heat released $= \left(7000\,\text{g}\right)\left(-55.6\,\text{kJ. g}^{-1}\right) = -3.892 \times 10^5\,\text{kJ}$

moles of methane gas $= \dfrac{3.892 \times 10^5\,\text{kJ}}{890\,\text{kJ. mol}^{-1}} = 437.3\,\text{mol}$

$V = \dfrac{nRT}{P} = \dfrac{\left(437.3\,\text{mol}\right)\left(8.20574 \times 10^{-2}\ \text{L.atm.K}^{-1}.\text{mol}^{-1}\right)\left(298\ \text{K}\right)}{10.00\ \text{atm}}$

$= 1069\,\text{L} = 1.1 \times 10^3\,\text{L}$

(e) methane gas, $-890\,\text{kJ. mol}^{-1}\,CO_2$ (less CO_2)

ethanol liquid, $-684\,\text{kJ. mol}^{-1}\,CO_2$ (more CO_2)

octane liquid, $-684\,\text{kJ. mol}^{-1}\,CO_2$ (more CO_2)

CHAPTER 8

THERMODYNAMICS: THE SECOND AND THIRD LAWS

8.1 Assume the transfer of energy as heat is reversible.

(a) rate of entropy generation $= \dfrac{\Delta S_{surr}}{time} = -\dfrac{q_{rev}}{time \cdot T}$

$$= -\dfrac{\text{rate of heat generation}}{T}$$

$$= \dfrac{-(100. \, \text{J} \cdot \text{s}^{-1})}{293 \, \text{K}} = 0.341 \, \text{J} \cdot \text{K}^{-1} \cdot \text{s}^{-1}$$

(b) $\Delta S_{day} = (0.341 \, \text{J} \cdot \text{K}^{-1} \cdot \text{s}^{-1})(60 \, \text{sec} \cdot \text{min}^{-1})(60 \, \text{min} \cdot \text{hr}^{-1})(24 \, \text{hr} \cdot \text{day}^{-1})$

$$= 29.5 \, \text{kJ} \cdot \text{K}^{-1} \cdot \text{day}^{-1}$$

(c) Less, because in the equation $\Delta S = \dfrac{-\Delta H}{T}$, if T is larger, ΔS is smaller.

8.3 (a) $\Delta S = \dfrac{q_{rev}}{T} = \dfrac{65 \, \text{J}}{298 \, \text{K}} = 0.22 \, \text{J} \cdot \text{K}^{-1}$

(b) $\Delta S = \dfrac{65 \, \text{J}}{373 \, \text{K}} = 0.17 \, \text{J} \cdot \text{K}^{-1}$

(c) Since T is in the denominator the entropy change is smaller at higher temperatures. The same amount of heat has a greater effect on entropy changes when transferred at lower temperatures.

8.5 (a) The relationship to use is $dS = \dfrac{dq}{T}$. At constant pressure, we can

$$dq = n \, C_P \, dT :$$

substitute $dS = \dfrac{n \, C_p \, dT}{T}$

Upon integration, this gives $\Delta S = n \, Cp \, \ln \dfrac{T_2}{T_1}$. The answer is calculated by

simply plugging in the known quantities. Remember that for an ideal monatomic gas

$C_P = \frac{5}{2} R:$

$$\Delta S = (1.00 \text{ mol}) \left(\frac{5}{2} \times 8.314 \text{ J} \cdot \text{K}^{-1} \cdot \text{mol}^{-1} \right) \ln \frac{431.0 \text{ K}}{310.8 \text{ K}} = 6.80 \text{ J} \cdot \text{K}^{-1}$$

(b) A similar analysis using C_V gives $\Delta S = n \, C_V \ln \frac{T_2}{T_1}$, where C_V for a monatomic ideal gas is $\frac{3}{2} R$:

$$\Delta S = (1.00 \text{ mol}) \left(\frac{3}{2} \times 8.314 \text{ J} \cdot \text{K}^{-1} \cdot \text{mol}^{-1} \right) \ln \frac{431.0 \text{ K}}{310.8 \text{ K}} = 4.08 \text{ J} \cdot \text{K}^{-1}$$

8.7   Assume the process is isothermal and reversible and use the relationship $dS = \dfrac{dq}{T}$. Because the process is isothermal, $\Delta U = 0$ and hence $q = -w$, where $w = -PdV$. Making this substitution, we obtain

$$dS = \frac{P \, dV}{T} = \frac{nRT}{TV} dV = \frac{nR}{V} dV$$

$$\therefore \Delta S = nR \ln \frac{V_2}{V_1}$$

Substituting the known quantities, we obtain

$$\Delta S = (5.25 \text{ mol}) (8.314 \text{ J} \cdot \text{K}^{-1} \cdot \text{mol}^{-1}) \ln \frac{34.058 \text{ L}}{24.252 \text{ L}}$$

$$= 14.8 \text{ J} \cdot \text{K}^{-1}$$

8.9

$$\Delta S = nR \ln \frac{V_2}{V_1}$$

Since P is the inverse of V for ideal gases we can write:

$$\Delta S = nR \ln \frac{P_1}{P_2}$$

$$= (1.50 \text{ mol})(8.314 \text{ J.K}^{-1}.\text{mol}^{-1}) \ln \frac{15.0 \text{ atm}}{0.500 \text{ atm}}$$

$$= 42.4 \ \text{J.K}^{-1}$$

8.11 First calculate the decrease in entropy resulting from the decrease in volume. Then calculate the increase in entropy resulting from the increase in temperature. Then add these to get the net entropy change. Assume ideal behavior and 1 mol N_2 gas.

$$\Delta S = nR \ln \frac{V_2}{V_1}$$

$$= (1.00 \text{ mol})(8.314 \text{ J.K}^{-1}.\text{mol}^{-1}) \ln \frac{0.500 \text{ L}}{3.00 \text{ L}}$$

$$= -14.897 \text{ J.K}^{-1} \quad (-14.9 \text{ J.K}^{-1} \text{ using 3 sig. fig.})$$

$$\Delta S = nR \ln \frac{T_2}{T_1}$$

$$= (1.00 \text{ mol})(8.314 \text{ J.K}^{-1}.\text{mol}^{-1}) \ln \frac{301.25 \text{ K}}{291.65 \text{ K}}$$

$$= 0.269 \text{ J.K}^{-1}$$

Net change in entropy, $\Delta S_{net} = (-14.897 + 0.269) \text{ J.K}^{-1} = -14.6 \text{ J.K}^{-1}$

(Rounding off at the end using 3 sig. fig.)

8.13 Assume the transfer of energy as heat is reversible. The change in entropy for each block (block 1 and block 2) are: $\Delta S_1 = \frac{q_1}{T_1}$ and $\Delta S_2 = \frac{q_2}{T_2}$ energy is transferred, then the change in entropy for the system is:

$$\Delta S = \Delta S_1 + \Delta S_2 = \frac{q_1}{T_1} + \frac{q_2}{T_2}$$

If 1 J of energy is transferred from block 2 to block 1,

$q_1 = +1 \text{ J}, q_2 = -1 \text{ J}$, and $\Delta S = \frac{1}{T_1} - \frac{1}{T_2}$.

For the transfer of heat from block 2 to block 1 to be spontaneous, ΔS must be positive and, therefore, T_2 would have to be greater than T_1.

8.15 (a) Heat enters house interior and entropy of the house interior increases. Heat leaves house exterior and entropy of the house exterior decreases.

(b) $T_{interior} > T_{exterior}$ with $\Delta S_{(interior)} = \frac{q}{T_{interior}}$ and $\Delta S_{(exterior)} = \frac{q}{T_{exterior}}$

$\therefore \Delta S_{interior} < \Delta S_{exterior}$

8.17 (a) $\Delta S° = \dfrac{q}{T} = \dfrac{\Delta H°}{T} = \dfrac{1.00\ \text{mol} \times (-6.01\ \text{kJ} \cdot \text{mol}^{-1})}{273.2\ \text{K}} = -22.0\ \text{J} \cdot \text{K}^{-1}$

(b) $\Delta S = \dfrac{q}{T} = \dfrac{\Delta H}{T} = \dfrac{\dfrac{50.0\ \text{g}}{46.07\ \text{g} \cdot \text{mol}^{-1}} \times 43.5\ \text{kJ} \cdot \text{mol}^{-1}}{351.5\ \text{K}} = +134\ \text{J} \cdot \text{K}^{-1}$

8.19 (a) The boiling point of a liquid may be obtained from the relationship

$\Delta S_{vap} = \dfrac{\Delta H_{vap}}{T_B}$, or $T_B = \dfrac{\Delta H_{vap}}{\Delta S_{vap}}$. This relationship should be rigorously true

if we have the actual enthalpy and entropy of vaporization. The data in the

Appendix, however, are for 298 K. Thus, calculation of $\Delta H°_{vap}$ or $\Delta S°_{vap}$,

using the enthalpy and entropy differences between the gas and liquid

forms at 298 K, give a good approximation of these quantities but the

values are not exact. For ethanal(l) $\longrightarrow$ ethanal(g), the data in the

appendix give

$\Delta H_{vap} \cong -166.19\ \text{kJ} \cdot \text{mol}^{-1} - (-192.30)\ \text{kJ} \cdot \text{mol}^{-1} = 26.11\ \text{kJ} \cdot \text{mol}^{-1}$

$\Delta S_{vap} \cong 250.3\ \text{J} \cdot \text{K}^{-1} \cdot \text{mol}^{-1} - 160.2\ \text{J} \cdot \text{K}^{-1} \cdot \text{mol}^{-1} = 90.1\ \text{J} \cdot \text{K}^{-1} \cdot \text{mol}^{-1}$

$T_B = \dfrac{26.11 \times 10^3\ \text{J} \cdot \text{mol}^{-1}}{90.1\ \text{J} \cdot \text{K}^{-1} \cdot \text{mol}^{-1}} = 290.\ \text{K}$

(b) The boiling point of ethanal is 20.8°C or 293.9 K.

(c) These numbers are in very good agreement.

(d) Differences arise partly because the enthalpy and entropy of

vaporization are slightly different from the values calculated at 298 K, but

the boiling point of ethanal is not 298 K.

8.21 (a) Trouton's rule indicates that the entropy of vaporization for a number

of organic liquids is approximately $85\ \text{J} \cdot \text{K}^{-1} \cdot \text{mol}^{-1}$. Using this

information and the relationship

$T_B = \dfrac{\Delta H°_{vap}}{\Delta S°_{vap}} = \dfrac{21.51 \times 10^3\ \text{J} \cdot \text{mol}^{-1}}{85\ \text{J} \cdot \text{K}^{-1} \cdot \text{mol}^{-1}} = 253\ \text{K}.$

(b) The experimental boiling point of dimethyl ether is 248 K, which is in reasonably close agreement. The value 85 J.K^{-1}.mol^{-1} is an average value for the entropy of vaporization of organic liquids and therefore deviations are expected when using this average value for individual organic liquids.

8.23 (a) The value can be estimated from

$$\Delta H°_{vap} = T \Delta S°_{vap}$$
$$\Delta H°_{vap} = (353 \text{ K})(85 \text{ J} \cdot \text{mol}^{-1} \cdot \text{K}^{-1})$$
$$= +30. \text{ kJ} \cdot \text{mol}^{-1}$$

(b) $\Delta S°_{surr} = -\dfrac{\Delta H°_{system}}{T}$

$$\Delta S°_{surr} = -\left(\frac{10 \text{ g}}{78.11 \text{ g} \cdot \text{mol}^{-1}}\right)\left(\frac{30 \text{ kJ} \cdot \text{mol}^{-1}}{353 \text{ K}}\right) = -11 \text{ J} \cdot \text{K}^{-1}$$

8.25 COF_2. COF_2 and BF_3 are both trigonal planar molecules, but it would be possible for the molecule to be disordered with the fluorine and oxygen atoms occupying the same locations. Because all the groups attached to boron are identical, such disorder is not possible.

8.27 There are six orientations of an SO_2F_2 molecule as shown below:

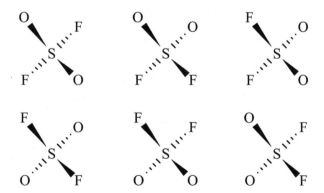

The Boltzmann expression for one mole of SO_2F_2 molecules having six possible orientations is

$$S = k \ln 6^{6.02 \times 10^{23}} = (1.38 \times 10^{-23} \text{ J} \cdot \text{K}^{-1}) \ln 6^{6.02 \times 10^{23}}$$
$$S = 14.9 \text{ J} \cdot \text{K}^{-1}$$

8.29 (a) HBr(g), because Br is more massive and contains more elementary particles than F in HF; (b) NH$_3$(g), because it has greater complexity, being a molecule rather than a single atom; (c) I$_2$ (l), because molecules in liquids are more randomly oriented than molecules in solids; (d) 1.0 mol Ar(g) at 1.00 atm, because it will occupy a larger volume than 1.0 mol of Ar(g) at 2.00 atm.

8.31 It is easy to order H$_2$O in its various phases because entropy will increase when going from a solid to a liquid to a gas. The main question concerns where to place C(s, diamond) in this order, and that will essentially become a question of whether C(s, diamond) should have more or less entropy than H$_2$O(s), because we would automatically expect C(s, diamond) to have less entropy than any liquid. Because water is a molecular substance held together in the solid phase by weak hydrogen bonds, and in C(s, diamond) the carbon is more rigidly held in place and will have less entropy.

In summary, C(s, diamond) < H$_2$O(s) < H$_2$O(l) < H$_2$O(g).

8.33 (a) Iodine is expected to have higher entropy due to its larger mass and consequently a larger number of fundamental particles.

(b) When we consider the two structures, it is clear that 1-pentene will have more flexibility in its framework than cyclopentane, which will be comparatively rigid. Therefore, we predict 1-pentene to have a higher entropy.

(c) Ethene (or ethylene) is a gas and polyethylene is a solid, so we automatically expect ethene to have a higher entropy. Also, for the same mass, a sample of ethene will be composed of many small molecules, whereas polyethylene will be made up of fewer but larger molecules.

8.35 (a) Entropy should decrease because the number of moles of gas is less on the product side of the reaction.

(b) Entropy should increase because the dissolution of the solid copper phosphate will increase the randomness of the copper and phosphate ions.

(c) Entropy should decrease as the total number of moles decreases.

8.37 $\Delta S_B < \Delta S_C < \Delta S_A$. The change in entropy for container A is greater than that for container B or C due to the greater number of particles. The change in entropy in container C is greater than that of container B because of the disorder due to the vibrational motion of the molecules in container C.

8.39 (a) $H_2(g) + \frac{1}{2}O_2(g) \longrightarrow H_2O(l)$

$$\Delta S^\circ_f = \Sigma\, S^\circ_{\text{products}} - \Sigma\, S^\circ_{\text{reactants}}$$
$$= 69.91\ \text{J}\cdot\text{K}^{-1}\cdot\text{mol}^{-1} - [130.68\ \text{J}\cdot\text{K}^{-1}\cdot\text{mol}^{-1} + \tfrac{1}{2}(205.14\ \text{J}\cdot\text{K}^{-1}\cdot\text{mol}^{-1})]$$
$$= -163.34\ \text{J}\cdot\text{K}^{-1}\cdot\text{mol}^{-1}$$

The entropy change is negative because the number of moles of gas has decreased by 1.5. Note that the absolute entropies of the elements are not 0, and that the entropy change for the reaction in which a compound is formed from the elements is also not 0.

(b) $CO(g) + \frac{1}{2}O_2(g) \longrightarrow CO_2(g)$

$$\Delta S^\circ_r = 213.74\ \text{J}\cdot\text{K}^{-1}\cdot\text{mol}^{-1}$$
$$- [197.67\ \text{J}\cdot\text{K}^{-1}\cdot\text{mol}^{-1} + \tfrac{1}{2}(205.14\ \text{J}\cdot\text{K}^{-1}\cdot\text{mol}^{-1})]$$
$$= -86.5\ \text{J}\cdot\text{K}^{-1}\cdot\text{mol}^{-1}$$

The entropy change is negative because the number of moles of gas has decreased by 0.5.

(c) $CaCO_3(s) \longrightarrow CaO(s) + CO_2(g)$

$$\Delta S^\circ_r = 39.75\ \text{J}\cdot\text{K}^{-1}\cdot\text{mol}^{-1} + 213.74\ \text{J}\cdot\text{K}^{-1}\cdot\text{mol}^{-1} - [+92.9\ \text{J}\cdot\text{K}^{-1}\cdot\text{mol}^{-1}]$$
$$= +160.6\ \text{J}\cdot\text{K}^{-1}\cdot\text{mol}^{-1}$$

The entropy change is positive because the number of moles of gas has increased by 1.

(d) $4 \, KClO_3 \, (s) \longrightarrow 3 \, KClO_4 \, (s) + KCl(s)$

$$\Delta S^\circ_r = 3(151.0 \, J \cdot K^{-1} \cdot mol^{-1}) + 82.59 \, J \cdot K^{-1} \cdot mol^{-1}$$
$$-[4(143.1 \, J \cdot K^{-1} \cdot mol^{-1})]$$
$$= -36.8 \, J \cdot K^{-1} \cdot mol^{-1}$$

It is not immediately obvious, but the four moles of solid products are more ordered than the four moles of solid reactants.

8.41 $dS = \dfrac{dq_{rev}}{T} = \dfrac{C_{P,m} dT}{T}$ and, therefore, $\Delta S = \displaystyle\int_{T_1}^{T_2} \dfrac{C_{P,m}}{T} \, dT.$

If $C_{P,m} = a + bT + c/T^2$, then

$$\Delta S = \int_{T_1}^{T_2} \frac{a + bT + c/T^2}{T} \, dT$$

$$= \int_{T_1}^{T_2} \left(\frac{a}{T} + b + \frac{c}{T^3} \right) dT = \left(a \ln(T) + bT - \frac{c}{2T^2} \right) \Bigg|_{T_1}^{T_2}$$

$$= a \ln\left(\frac{T_2}{T_1} \right) + b(T_2 - T_1) - \frac{c}{2} \left(\frac{1}{T_2^2} - \frac{1}{T_1^2} \right)$$

ΔS(true) for heating graphite from 298 K to 400 K is

$$= (16.86 \, J \cdot K^{-1} \cdot mol^{-1}) \ln\left(\frac{400 \, K}{298 \, K} \right)$$

$$+ (0.00477 \, J \cdot K^{-2} \cdot mol^{-1})(400 \, K - 298 \, K)$$

$$- \frac{(-8.54 \times 10^5 \, J \cdot K \cdot mol^{-1})}{2} \left(\frac{1}{(400 \, K)^2} - \frac{1}{(298 \, K)^2} \right)$$

$$= 3.31 \, J \cdot K^{-1} \cdot mol^{-1}$$

If we assume a constant heat capacity at the mean temperature of 350 K:

$$C_{P,m} = (16.86 \text{ J} \cdot \text{K}^{-1} \cdot \text{mol}^{-1}) + (0.00477 \text{ J} \cdot \text{K}^{-2} \cdot \text{mol}^{-1})(350 \text{ K})$$

$$+ \frac{(-8.54 \times 10^5 \text{ J} \cdot \text{K} \cdot \text{mol}^{-1})}{(350 \text{ K})^2}$$

$$= 11.6 \text{ J} \cdot \text{K}^{-1} \cdot \text{mol}^{-1}$$

and

$$\Delta S(\text{mean}) = C_{P,m} \ln \frac{T_2}{T_1} = (11.6 \text{ J} \cdot \text{K}^{-1} \cdot \text{mol}^{-1}) \ln\left(\frac{400 \text{ K}}{298 \text{ K}}\right) = 3.41 \text{ J} \cdot \text{K}^{-1} \cdot \text{mol}^{-1}$$

Not integrating the heat capacity leads to roughly a 3.0% error in ΔS.

8.43 The entropy of vaporization of water at 85 °C may be carried out through a series of three reversible steps. Namely, reversibly heating the reactants to 100 °C, carrying out the phase change at this temperature, and finally cooling the products back to 85 °C. The sum of the ΔS's for these three steps will be equivalent to vaporizing water at 85 °C in one irreversible step.

Step 1, heating the reactants to 100 °C:

$$\Delta S_1 = C_{P,m} \ln\left(\frac{T_2}{T_1}\right) = (75.3 \text{ J} \cdot \text{K}^{-1} \cdot \text{mol}^{-1}) \ln\left(\frac{373 \text{ K}}{358 \text{ K}}\right) = 3.09 \text{ J} \cdot \text{K}^{-1} \cdot \text{mol}^{-1}$$

Step 2, the entropy of vaporization of H_2O at 100 °C is 109.0 $\text{J} \cdot \text{K}^{-1} \cdot \text{mol}^{-1}$

Step 3, cooling the products to 85 °C:

$$\Delta S_3 = C_{P,m} \ln\left(\frac{T_2}{T_1}\right) = (33.6 \text{ J} \cdot \text{K}^{-1} \cdot \text{mol}^{-1}) \ln\left(\frac{358 \text{ K}}{373 \text{ K}}\right) = -1.38 \text{ J} \cdot \text{K}^{-1} \cdot \text{mol}^{-1}$$

Therefore, the molar entropy of vaporization is H_2O at 85 °C is:

$$\Delta S_{\text{v,m}} = \Delta S_1 + \Delta S_2 + \Delta S_3 = 111 \text{ J} \cdot \text{K}^{-1} \cdot \text{mol}^{-1}.$$

8.45 Assume no change in volume. To calculate the change in entropy for the hot and cold water, the amount of energy which flows from one to the other must first be calculated:

$$q_c = -q_h$$
$$= n_c \cdot C_{p,m}(H_2O) \cdot (T_f - T_{i,c}) = -n_h \cdot C_{p,m}(H_2O) \cdot (T_f - T_{i,h})$$

where n_c and n_h are the moles of cold and hot water, respectively, and $T_{i,c}$ and $T_{i,h}$ are the initial temperatures of the cold and hot water, respectively. Dividing both sides by $C_{p,m}(H_2O)$ we obtain:

$$n_c \cdot (T_f - T_{i,c}) = -n_h \cdot (T_f - T_{i,h}).$$

The moles of hot and cold water are:

$$n_c = \frac{50 \text{ g}}{18.015 \text{ g} \cdot \text{mol}^{-1}} = 2.78 \text{ mol and } n_h = \frac{65 \text{ g}}{18.015 \text{ g} \cdot \text{mol}^{-1}} = 3.61 \text{ mol}$$

and the final temperature T_f is therefore:

$$T_f = \frac{(2.78 \text{ mol} \cdot 293.15 \text{ K}) + (3.61 \text{ mol} \cdot 323.15 \text{ K})}{(2.78 \text{ mol} + 3.61 \text{ mol})} = 310.1 \text{ K}.$$

With T_f, we can calculate ΔS for the hot and cold water, and the total ΔS for the entire system:

$$\Delta S_c = n_c \cdot C_{p,m} \ln\left(\frac{T_f}{T_{i,c}}\right) = (2.78 \text{ mol})(75.3 \text{ J} \cdot \text{K}^{-1} \cdot \text{mol}^{-1})\ln\left(\frac{310.1 \text{ K}}{293 \text{ K}}\right)$$

$$= +11.9 \text{ J} \cdot \text{K}^{-1}$$

$$\Delta S_h = n_h \cdot C_{p,m} \ln\left(\frac{T_f}{T_{i,h}}\right) = (3.61 \text{ mol})(75.3 \text{ J} \cdot \text{K}^{-1} \cdot \text{mol}^{-1})\ln\left(\frac{310.1 \text{ K}}{323 \text{ K}}\right)$$

$$= -11.1 \text{ J} \cdot \text{K}^{-1}$$

Therefore, ΔS_{tot} is :

$$\Delta S_{tot} = \Delta S_c + \Delta S_h = +0.8 \text{ J} \cdot \text{K}^{-1}.$$

8.47 (a) The change in entropy will be given by

$$\Delta S_{surr} = \frac{-\Delta H_{system}}{T} = \frac{-1.00 \text{ mol} \times 8.2 \times 10^3 \text{ J} \cdot \text{mol}^{-1}}{111.7 \text{ K}} = -73 \text{ J} \cdot \text{K}^{-1}$$

$$\Delta S_{system} = \frac{\Delta H_{system}}{T} = \frac{1.00 \text{ mol} \times 8.2 \times 10^3 \text{ J} \cdot \text{mol}^{-1}}{111.7 \text{ K}} = +73 \text{ J} \cdot \text{K}^{-1}$$

(b) $\Delta S_{surr} = \frac{-\Delta H_{system}}{T} = \frac{-1.00 \text{ mol} \times 4.60 \times 10^3 \text{ J} \cdot \text{mol}^{-1}}{158.7 \text{ K}} = -29.0 \text{ J} \cdot \text{K}^{-1}$

$$\Delta S_{system} = \frac{\Delta H_{system}}{T} = \frac{1.00 \text{ mol} \times 4.60 \times 10^3 \text{ J} \cdot \text{mol}^{-1}}{158.7 \text{ K}} = +29.0 \text{ J} \cdot \text{K}^{-1}$$

(c) $\Delta S_{surr} = \dfrac{-\Delta H_{system}}{T} = \dfrac{-(1.00 \text{ mol} \times -4.60 \times 10^3 \text{ J} \cdot \text{mol}^{-1})}{158.7 \text{ K}}$

$= +29.0 \text{ J} \cdot \text{K}^{-1}$

$\Delta S_{system} = \dfrac{\Delta H_{system}}{T} = \dfrac{1.00 \text{ mol} \times -4.60 \times 10^3 \text{ J} \cdot \text{mol}^{-1}}{158.7 \text{ K}} = -29.0 \text{ J} \cdot \text{K}^{-1}$

8.49 (a) The total entropy change is given by $\Delta S_{tot} = \Delta S_{surr} + \Delta S$. ΔS for an isothermal, reversible process is calculated from

$\Delta S = \dfrac{q_{rev}}{T} = \dfrac{-w_{rev}}{T} = nR \ln \dfrac{V_2}{V_1}$. To do the calculation we need the value of

n, which is obtained by use of the ideal gas law:

$(4.95 \text{ atm})(1.67 \text{ L}) = n(0.082\ 06 \text{ L} \cdot \text{atm} \cdot \text{K}^{-1} \cdot \text{mol}^{-1})(323 \text{ K})$; $n = 0.312$ mol.

$\Delta S = (0.312 \text{ mol})(8.314 \text{ J} \cdot \text{K}^{-1} \cdot \text{mol}^{-1}) \ln \dfrac{7.33 \text{ L}}{1.67 \text{ L}} = +3.84 \text{ J} \cdot \text{K}^{-1}$. Because the

process is reversible, $\Delta S_{tot} = 0$, so $\Delta S_{surr} = -\Delta S = -3.84 \text{ J} \cdot \text{K}^{-1}$.

(b) For the irreversible process, ΔS is the same, $+3.84 \text{ J} \cdot \text{K}^{-1}$. No work is done in free expansion (see Section 6.6) so $w = 0$. Because $\Delta U = 0$, it follows that $q = 0$. Therefore, no heat is transferred into the surroundings, and their entropy is unchanged: $\Delta S_{surr} = 0$. The total change in entropy is therefore $\Delta S_{tot} = +3.84 \text{ J} \cdot \text{K}^{-1}$.

8.51 Spontaneous change means $\Delta S_{tot} > 0$.

A vapor to liquid change is condensation and it releases heat. However as shown in the diagram, the temperature of the system does not change. Therefore the released heat must have left system to the surroundings. Since heat flows from hot to cold, the temperature of the surroundings (T_{surr}) must be lower than the temperature of the system (T_{sys}), and since

$\Delta S_{sys} = \dfrac{q_{rev}}{T_{sys}}$ and $\Delta S_{surr} = \dfrac{q_{rev}}{T_{surr}}$, then $\Delta S_{tot} > 0$.

8.53 Exothermic reactions tend to be spontaneous because the result is an increase in the entropy of the surroundings. Using the mathematical relationship $\Delta G_r = \Delta H_r - T\Delta S_r$, it is clear that if ΔH_r is large and negative compared to ΔS_r, then the reaction will generally be spontaneous.

8.55 (a) $\Delta H°_r = 3(-824.2 \text{ kJ} \cdot \text{mol}^{-1}) - [2(-1118.4 \text{ kJ} \cdot \text{mol}^{-1})]$

$= -235.8 \text{ kJ} \cdot \text{mol}^{-1}$

$\Delta S°_r = 3(87.40 \text{ J} \cdot \text{K}^{-1} \cdot \text{mol}^{-1}) - [2(146.4 \text{ J} \cdot \text{K}^{-1} \cdot \text{mol}^{-1})$

$+ \frac{1}{2}(205.14 \text{ J} \cdot \text{K}^{-1} \cdot \text{mol}^{-1})]$

$= -133.17 \text{ J} \cdot \text{K}^{-1} \cdot \text{mol}^{-1}$

$\Delta G°_r = 3(-742.2 \text{ kJ} \cdot \text{mol}^{-1}) - [2(-1015.4 \text{ kJ} \cdot \text{mol}^{-1})]$

$= -195.8 \text{ kJ} \cdot \text{mol}^{-1}$

ΔG_r may also be calculated from $\Delta H°_r$ and $\Delta S°_r$ (the numbers calculated differ slightly from the two methods due to rounding differences):

$\Delta G°_r = \Delta H°_r - T\Delta S°_r$

$= -235.8 \text{ kJ} \cdot \text{mol}^{-1} - (298 \text{ K})(-133.17 \text{ J} \cdot \text{K}^{-1} \cdot \text{mol}^{-1})/(100 \text{ J} \cdot \text{kJ}^{-1})$

$= -196.1 \text{ kJ} \cdot \text{mol}^{-1}$

(b) $\Delta H°_r = -1208.09 \text{ kJ} \cdot \text{mol}^{-1} - [-1219.6 \text{ kJ} \cdot \text{mol}^{-1}]$

$= 11.5 \text{ kJ} \cdot \text{mol}^{-1}$

$\Delta S°_r = -80.8 \text{ J} \cdot \text{K}^{-1} \cdot \text{mol}^{-1} - [68.87 \text{ J} \cdot \text{K}^{-1} \cdot \text{mol}^{-1}]$

$= -149.7 \text{ J} \cdot \text{K}^{-1} \cdot \text{mol}^{-1}$

$\Delta G°_r = -1111.15 \text{ kJ} \cdot \text{mol}^{-1} - [-1167.3 \text{ kJ} \cdot \text{mol}^{-1}]$

$= +56.2 \text{ kJ} \cdot \text{mol}^{-1}$

or

$\Delta G°_r = \Delta H°_r - T\Delta S°_r$

$= 11.5 \text{ kJ} \cdot \text{mol}^{-1} - (298 \text{ K})(-149.7 \text{ J} \cdot \text{K}^{-1} \cdot \text{mol}^{-1})/(1000 \text{ J} \cdot \text{kJ}^{-1})$

$= 56.1 \text{kJ} \cdot \text{mol}^{-1}$

(c) $\Delta H°_r = 9.16 \text{ kJ} \cdot \text{mol}^{-1} - [2(33.18 \text{ kJ} \cdot \text{mol}^{-1})] = -57.20 \text{ kJ} \cdot \text{mol}^{-1}$

$\Delta S°_r = 304.29 \text{ J} \cdot \text{K}^{-1} \cdot \text{mol}^{-1} - [2(240.06 \text{ J} \cdot \text{K}^{-1} \cdot \text{mol}^{-1})]$

$= -175.83 \text{ J} \cdot \text{K}^{-1} \cdot \text{mol}^{-1}$

$\Delta G°_r = 97.89 \text{ kJ} \cdot \text{mol}^{-1} - [2(51.31 \text{ kJ} \cdot \text{mol}^{-1})] = -4.73 \text{ kJ} \cdot \text{mol}^{-1}$

or

$\Delta G°_r = \Delta H°_r - T\Delta S°_r$

$= -57.2 \text{ kJ} \cdot \text{mol}^{-1} - (298 \text{ K})(-175.83 \text{ J} \cdot \text{K}^{-1} \cdot \text{mol}^{-1})/(1000 \text{ J} \cdot \text{kJ}^{-1})$

$= -4.80 \text{ kJ} \cdot \text{mol}^{-1}$

8.57 (a) $\frac{1}{2}N_2(g) + \frac{3}{2}H_2(g) \longrightarrow NH_3(g)$

$\Delta H°_r = \Delta H°_f(NH_3) = -46.11 \text{ kJ} \cdot \text{mol}^{-1}$

$\Delta S°_r = S°_m(NH_3, g) - [\frac{1}{2}S°_m(N_2, g) + \frac{3}{2}S°_m(H_2, g)]$

$= 192.45 \text{ J} \cdot \text{K}^{-1} \cdot \text{mol}^{-1} - [\frac{1}{2}(191.61 \text{ J} \cdot \text{K}^{-1} \cdot \text{mol}^{-1})$

$+ \frac{3}{2}(130.68 \text{ J} \cdot \text{K}^{-1} \cdot \text{mol}^{-1})]$

$= -99.38 \text{ J} \cdot \text{K}^{-1} \cdot \text{mol}^{-1}$

$\Delta G°_r = -46.11 \text{ kJ} \cdot \text{mol}^{-1} - (298 \text{ K})(-99.38 \text{ J} \cdot \text{K}^{-1} \cdot \text{mol}^{-1})/(1000 \text{ J} \cdot \text{kJ}^{-1})$

$= -16.49 \text{ kJ} \cdot \text{mol}^{-1}$

$S°_m(NH_3) = 192.45 \text{ J} \cdot \text{K}^{-1} \cdot \text{mol}^{-1}$

$\Delta S°_f(NH_3)$ is negative because several gas molecules combine to form 1

NH_3 molecule.

(b) $H_2(g) + \frac{1}{2}O_2(g) \longrightarrow H_2O(g)$

$\Delta H°_r = \Delta H°_f(H_2O, g) = -241.82 \text{ kJ} \cdot \text{mol}^{-1}$

$\Delta S°_r = S°_m(H_2O, g) - [S°_m(H_2, g) + \frac{1}{2}S°_m(O_2, g)]$

$= 188.83 \text{ J} \cdot \text{K}^{-1} \cdot \text{mol}^{-1} - [130.68 \text{ J} \cdot \text{K}^{-1} \cdot \text{mol}^{-1} + \frac{1}{2}(205.14 \text{ J} \cdot \text{K}^{-1} \cdot \text{mol}^{-1})]$

$= -44.42 \text{ J} \cdot \text{K}^{-1} \cdot \text{mol}^{-1}$

$\Delta G°_r = -241.82 \text{ kJ} \cdot \text{mol}^{-1} - (298 \text{ K})(-44.42 \text{ J} \cdot \text{K}^{-1} \cdot \text{mol}^{-1})/(1000 \text{ J} \cdot \text{kJ}^{-1})$

$= -228.58 \text{ kJ} \cdot \text{mol}^{-1}$

$S°_m(H_2O, g) = 188.83 \text{ J} \cdot \text{K}^{-1} \cdot \text{mol}^{-1}$

$\Delta S°_f(H_2O, g)$ is a negative number because there is a reduction in the

number of gas molecules in the reaction when $S°_m$ is positive.

(c) $C(s, graphite) + \frac{1}{2}O_2(g) \longrightarrow CO(g)$

$\Delta H°_r = \Delta H°_f(CO, g) = -110.53 \text{ kJ} \cdot \text{mol}^{-1}$

$\Delta S°_r = S°_m(CO, g) - [S°_m(C, s) + \frac{1}{2}S°_m(O_2, g)]$

$\qquad = 197.67 \text{ J} \cdot \text{K}^{-1} \cdot \text{mol}^{-1} - [5.740 \text{ J} \cdot \text{K}^{-1} \cdot \text{mol}^{-1} + \frac{1}{2}(205.14 \text{ J} \cdot \text{K}^{-1} \cdot \text{mol}^{-1})]$

$\qquad = +89.36 \text{ J} \cdot \text{K}^{-1} \cdot \text{mol}^{-1}$

$\Delta G°_r = -110.53 \text{ kJ} \cdot \text{mol}^{-1} - (298 \text{ K})(89.36 \text{ J} \cdot \text{K}^{-1} \cdot \text{mol}^{-1})/(1000 \text{ J} \cdot \text{kJ}^{-1})$

$\qquad = -137.2 \text{ kJ} \cdot \text{mol}^{-1}$

$S°_m(CO, g) = 197.67 \text{ J} \cdot \text{K}^{-1} \cdot \text{mol}^{-1}$

The $S°_m(CO, g)$ is larger than $\Delta S°_f(CO, g)$ because in the formation reaction the number of moles of gas is reduced.

(d) $\frac{1}{2}N_2(g) + O_2(g) \longrightarrow NO_2(g)$

$\Delta H°_r = \Delta H°_f(NO_2) = +33.18 \text{ kJ} \cdot \text{mol}^{-1}$

$\Delta S°_r = S°_m(NO_2, g) - [\frac{1}{2}S°_m(N_2, g) + S°_m(O_2, g)]$

$\qquad = 240.06 \text{ J} \cdot \text{K}^{-1} \cdot \text{mol}^{-1} - [\frac{1}{2}(191.61 \text{ J} \cdot \text{K}^{-1} \cdot \text{mol}^{-1}) + 205.14 \text{ J} \cdot \text{K}^{-1} \cdot \text{mol}^{-1}]$

$\qquad = -60.89 \text{ J} \cdot \text{K}^{-1} \cdot \text{mol}^{-1}$

$\Delta G°_r = 33.18 \text{ kJ} \cdot \text{mol}^{-1} - (298 \text{ K})(-60.89 \text{ J} \cdot \text{K}^{-1} \cdot \text{mol}^{-1})/(1000 \text{ J} \cdot \text{kJ}^{-1})$

$\qquad = +51.33 \text{ kJ} \cdot \text{mol}^{-1}$

$S°_m(NO_2, g) = 240.06 \text{ J} \cdot \text{K}^{-1} \cdot \text{mol}^{-1}$

The $\Delta S°_f(NO_2, g)$ is somewhat negative due to the reduction in the number of gas molecules during the reaction. For all of these, the important point to gain is that the $S°_m$ value of a compound is not the same as the $\Delta S°_f$ for the formation of that compound. $\Delta S°_f$ is often negative because one is bringing together a number of elements to form that compound.

8.59 Use the relationship $\Delta G°_r = \Sigma \Delta G°_f(\text{products}) - \Sigma \Delta G°_f(\text{reactants})$:

(a) $\Delta G°_r = 2\Delta G°_f(SO_3, g) - [2\Delta G°_f(SO_2, g)]$

$\qquad = 2(-371.06 \text{ kJ} \cdot \text{mol}^{-1}) - [2(-300.19 \text{ kJ} \cdot \text{mol}^{-1})]$

$\qquad = -141.74 \text{ kJ} \cdot \text{mol}^{-1}$

The reaction is spontaneous.

(b) $\Delta G^{\circ}{}_r = \Delta G^{\circ}{}_f(CaO, s) + \Delta G^{\circ}{}_f(CO_2, g) - \Delta G^{\circ}{}_f(CaCO_3, s)$

$\qquad = (-604.03 \text{ kJ} \cdot \text{mol}^{-1}) + (-394.36 \text{ kJ} \cdot \text{mol}^{-1})$

$\qquad\quad - (-1128.8 \text{ kJ} \cdot \text{mol}^{-1})$

$\qquad = +130.4 \text{ kJ} \cdot \text{mol}^{-1}$

The reaction is not spontaneous.

(c) $\Delta G^{\circ}{}_r = 16\Delta G^{\circ}{}_f(CO_2, g) + 18\Delta G^{\circ}{}_f(H_2O, l) - [2\Delta G^{\circ}{}_f(C_8H_{18}, l)]$

$\qquad = 16(-394.36 \text{ kJ} \cdot \text{mol}^{-1}) + 18(-237.13 \text{ kJ} \cdot \text{mol}^{-1})$

$\qquad\quad - [2(6.4 \text{ kJ} \cdot \text{mol}^{-1})]$

$\qquad = -10\,590.9 \text{ kJ} \cdot \text{mol}^{-1}$

The reaction is spontaneous.

8.61 The standard free energies of formation of the compounds are: (a) $PCl_5(g)$, $-305.0 \text{ kJ} \cdot \text{mol}^{-1}$; (b) $HCN(g)$, $+124.7 \text{ kJ} \cdot \text{mol}^{-1}$; (c) $NO(g)$, $+86.55 \text{ kJ} \cdot \text{mol}^{-1}$; (d) $SO_2(g)$, $-300.19 \text{ kJ} \cdot \text{mol}^{-1}$. Those compounds with a positive free energy of formation are unstable with respect to the elements. Thus (a) and (d) are thermodynamically stable.

8.63 To understand what happens to $\Delta G^{\circ}{}_r$ as temperature is raised, we use the relationship $\Delta G^{\circ}{}_r = \Delta H^{\circ}{}_r - T\Delta S^{\circ}{}_r$. From this it is clear that the free energy of the reaction becomes less favorable (more positive) as temperature increases, only if $\Delta S^{\circ}{}_r$ is a negative number. Therefore, we need only to find out whether the standard entropy of formation of the compound is a negative number. Assuming constant standard pressure this is calculated for each compound as follows:

(a) $P(s) + \frac{5}{2}Cl_2(g) \longrightarrow PCl_5(g)$

$\Delta S^{\circ}{}_r = S^{\circ}{}_m(PCl_5, g) - [S^{\circ}{}_m(P, s) + \frac{5}{2}S^{\circ}{}_m(Cl_2, g)]$

$\qquad = 364.6 \text{ J} \cdot \text{K}^{-1} \cdot \text{mol}^{-1} - [41.09 \text{ J} \cdot \text{K}^{-1} \cdot \text{mol}^{-1} + \frac{5}{2}(223.07 \text{ J} \cdot \text{K}^{-1} \cdot \text{mol}^{-1})]$

$\qquad = -234.2 \text{ J} \cdot \text{K}^{-1} \cdot \text{mol}^{-1}$

The compound is less stable at higher temperatures.

(b) $C(s), \text{graphite} + \frac{1}{2}N_2(g) + \frac{1}{2}H_2(g) \longrightarrow HCN(g)$

$$\Delta S^\circ_r = S^\circ_m(HCN, g) - [S^\circ_m(C, s) + \tfrac{1}{2} S^\circ_m(N_2, g) + \tfrac{1}{2} S^\circ_m(H_2, g)]$$

$$= 201.78\ J \cdot K^{-1} \cdot mol^{-1} - [5.740\ J \cdot K^{-1} \cdot mol^{-1}$$

$$+ \tfrac{1}{2}(191.61\ J \cdot K^{-1} \cdot mol^{-1}) + \tfrac{1}{2}(130.68\ J \cdot K^{-1} \cdot mol^{-1})]$$

$$= +34.90\ J \cdot K^{-1} \cdot mol^{-1}$$

HCN(g) is more stable at higher T.

(c) $\tfrac{1}{2} N_2(g) + \tfrac{1}{2} O_2(g) \longrightarrow NO(g)$

$$\Delta S^\circ_r = S^\circ_m(NO, g) - [\tfrac{1}{2} S^\circ_m(N_2, g) + \tfrac{1}{2} S^\circ_m(O_2, g)]$$

$$= 210.76\ J \cdot K^{-1} \cdot mol^{-1} - [\tfrac{1}{2}(191.61\ J \cdot K^{-1} \cdot mol^{-1})$$

$$+ \tfrac{1}{2}(205.14\ J \cdot K^{-1} \cdot mol^{-1})]$$

$$= +12.38\ J \cdot K^{-1} \cdot mol^{-1}$$

NO(g) is more stable as T increases.

(d) $S(s) + O_2(g) \longrightarrow SO_2(g)$

$$\Delta S^\circ_r = S^\circ(SO_2, g) - [S^\circ(S, s) + S^\circ(O_2, g)]$$

$$= 248.22\ J \cdot K^{-1} \cdot mol^{-1} - [31.80\ J \cdot K^{-1} \cdot mol^{-1}$$

$$+ 205.14\ J \cdot K^{-1} \cdot mol^{-1}]$$

$$= +11.28\ J \cdot K^{-1} \cdot mol^{-1}$$

$SO_2(g)$ is more stable as T increases.

8.65 (a) $2\ H_2O_2(l) \longrightarrow 2\ H_2O(l) + O_2(g)$

$$\Delta S^\circ_r = 2S^\circ_m(H_2O, l) + S^\circ_m(O_2, g) - 2S^\circ_m(H_2O_2, l)$$

$$= 2(69.91\ J \cdot K^{-1} \cdot mol^{-1}) + 205.14\ J \cdot K^{-1} \cdot mol^{-1}$$

$$- 2(109.6\ J \cdot K^{-1} \cdot mol^{-1})$$

$$= +125.8\ J \cdot K^{-1} \cdot mol^{-1}$$

$$\Delta H^\circ_r = 2\Delta H^\circ_f(H_2O, aq) - 2\Delta H^\circ_f(H_2O_2, l)$$

$$= 2(-285.83\ kJ \cdot mol^{-1}) - 2(-187.78\ kJ \cdot mol^{-1})$$

$$= -196.10\ kJ \cdot mol^{-1}$$

$$\Delta G^\circ_r = 2\Delta G^\circ_f(H_2O, aq) - 2\Delta G^\circ_f(H_2O_2, l)$$

$$= 2(-237.13\ kJ \cdot mol^{-1}) - 2(-120.35\ kJ \cdot mol^{-1})$$

$$= -233.56\ kJ \cdot mol^{-1}$$

ΔG°_r can also be calculated from ΔS°_r and ΔH°_r using the relationship:

$$\Delta G^\circ_r = \Delta H^\circ_r - T\Delta S^\circ_r$$
$$= -196.1\,\text{kJ}\cdot\text{mol}^{-1} - (298\,\text{K})(+125.8\,\text{J}\cdot\text{K}^{-1}\cdot\text{mol}^{-1})/(1000\,\text{J}\cdot\text{kJ}^{-1})$$
$$= -233.6\,\text{kJ}\cdot\text{mol}^{-1}$$

(b) $2\,F_2(g) + 2\,H_2O(l) \longrightarrow 4\,HF(aq) + O_2(g)$

$$\Delta S^\circ_r = 4S^\circ_m(HF,aq) + S^\circ_m(O_2,g) - [2S^\circ_m(F_2,g) + 2S^\circ_m(H_2O,l)]$$
$$= 4(88.7\,\text{J}\cdot\text{K}^{-1}\cdot\text{mol}^{-1}) + 205.14\,\text{J}\cdot\text{K}^{-1}\cdot\text{mol}^{-1}$$
$$- [2(202.78\,\text{J}\cdot\text{K}^{-1}\cdot\text{mol}^{-1}) + 2(69.91\,\text{J}\cdot\text{K}^{-1}\cdot\text{mol}^{-1})]$$
$$= +14.6\,\text{J}\cdot\text{K}^{-1}\cdot\text{mol}^{-1}$$

$$\Delta H^\circ_r = 4\Delta H^\circ_f(HF,aq) - 2\Delta H^\circ_f(H_2O,l)$$
$$= 4(-330.08\,\text{kJ}\cdot\text{mol}^{-1}) - 2(-285.83\,\text{kJ}\cdot\text{mol}^{-1})$$
$$= -748.66\,\text{kJ}\cdot\text{mol}^{-1}$$

$$\Delta G^\circ_r = 4\Delta G^\circ_f(HF,aq) - 2\Delta G^\circ_f(H_2O,l)$$
$$= 4(-296.82\,\text{kJ}\cdot\text{mol}^{-1}) - 2(-237.13\,\text{kJ}\cdot\text{mol}^{-1})$$
$$= -713.02\,\text{kJ}\cdot\text{mol}^{-1}$$

ΔG°_r can also be calculated from ΔS°_r *and* ΔH°_r using the relationship:

$$\Delta G^\circ_r = \Delta H^\circ_r - T\Delta S^\circ_r$$
$$= -748.66\,\text{kJ}\cdot\text{mol}^{-1} - (298\,\text{K})(14.6\,\text{J}\cdot\text{K}^{-1}\cdot\text{mol}^{-1})/(1000\,\text{J}\cdot\text{kJ}^{-1})$$
$$= -753.01\,\text{kJ}\cdot\text{mol}^{-1}$$

8.67 In order to find ΔG°_r at a temperature other than 298 K, we must first calculate ΔH°_r and ΔS°_r and then use the relationship

$\Delta G^\circ_r = \Delta H^\circ_r + T\Delta S^\circ_r$ to calculate ΔG°_r.

(a) In order to determine the range over which the reaction will be spontaneous, we consider the relative signs of ΔH°_r and ΔS°_r and their effect on ΔG°_r.

$$\Delta H^\circ_r = 2\Delta H^\circ_f(BF_3, g) + 3\Delta H^\circ_f(H_2O, l) - [\Delta H^\circ_f(B_2O_3, s)$$
$$+ 6\Delta H^\circ_f(HF, g)]$$
$$= 2(-1137.0 \text{ kJ} \cdot \text{mol}^{-1}) + 3(-285.83 \text{ kJ} \cdot \text{mol}^{-1})$$
$$- [(-1272.8 \text{ kJ} \cdot \text{mol}^{-1}) + 6(-271.1 \text{ kJ} \cdot \text{mol}^{-1})]$$
$$= -232.1 \text{ kJ} \cdot \text{mol}^{-1}$$
$$\Delta S^\circ_r = 2S^\circ_m(BF_3, g) + 3S^\circ_m(H_2O, l)$$
$$- [S^\circ_m(B_2O_3, S) + 6S^\circ_m(HF, g)]$$
$$= 2(254.12 \text{ J} \cdot \text{K}^{-1} \cdot \text{mol}^{-1}) + 3(69.91 \text{ J} \cdot \text{K}^{-1} \cdot \text{mol}^{-1})$$
$$- [53.97 \text{ J} \cdot \text{K}^{-1} \cdot \text{mol}^{-1} + 6(173.78 \text{ J} \cdot \text{K}^{-1} \cdot \text{mol}^{-1})]$$
$$= -378.68 \text{ J} \cdot \text{K}^{-1} \cdot \text{mol}^{-1}$$
$$\Delta G^\circ_r = -232.1 \text{ J} \cdot \text{K}^{-1} \cdot \text{mol}^{-1} - (353 \text{ K})(-378.68 \text{ J} \cdot \text{K}^{-1} \cdot \text{mol}^{-1})/(1000 \text{ J} \cdot \text{kJ}^{-1})$$
$$= -98.42 \text{ kJ} \cdot \text{mol}^{-1}$$

Because ΔH°_r is negative and ΔS°_r is also negative, we expect the reaction to be spontaneous at low temperatures, where the term $T\Delta S^\circ_r$ will be less than ΔH°_r. To find the temperature of the cutoff, we calculate the temperature at which $\Delta G^\circ_r = 0$. For this reaction, that temperature is

$$\Delta G^\circ_r = 0 = -232.1 \text{ kJ} \cdot \text{mol}^{-1} - (T)(-378.68 \text{ J} \cdot \text{K}^{-1} \cdot \text{mol}^{-1})/(1000 \text{ J} \cdot \text{kJ}^{-1})$$
$$T = 612.9 \text{ K}$$

The reaction should be spontaneous below 612.9 K.

(b) $$\Delta H^\circ_r = \Delta H^\circ_f(CaCl_2, aq) + \Delta H^\circ_f(C_2H_2, g) - [\Delta H^\circ_f(CaC_2, s)$$
$$+ 2\Delta H^\circ_f(HCl, aq)]$$
$$= (-877.1 \text{ kJ} \cdot \text{mol}^{-1}) + 226.73 \text{ kJ} \cdot \text{mol}^{-1}$$
$$- [(-59.8 \text{ kJ} \cdot \text{mol}^{-1}) + 2(-167.16 \text{ kJ} \cdot \text{mol}^{-1})]$$
$$= -256.3 \text{ kJ} \cdot \text{mol}^{-1}$$
$$\Delta S^\circ_r = S^\circ_m(CaCl_2, aq) + S^\circ_m(C_2H_2, g)$$
$$- [S^\circ_m(CaC_2, s) + 2S^\circ_m(HCl, aq)]$$
$$= 59.8 \text{ J} \cdot \text{K}^{-1} \cdot \text{mol}^{-1} + 200.94 \text{ J} \cdot \text{K}^{-1} \cdot \text{mol}^{-1}$$
$$- [69.96 \text{ J} \cdot \text{K}^{-1} \cdot \text{mol}^{-1} + 2(56.5 \text{ J} \cdot \text{K}^{-1} \cdot \text{mol}^{-1})]$$
$$= +77.8 \text{ J} \cdot \text{K}^{-1} \cdot \text{mol}^{-1}$$
$$\Delta G^\circ_r = -256.2 \text{ kJ} \cdot \text{mol}^{-1} - (353 \text{ K})(+77.8 \text{ J} \cdot \text{K}^{-1} \cdot \text{mol}^{-1})/(1000 \text{ J} \cdot \text{kJ}^{-1})$$
$$= -283.7 \text{ kJ} \cdot \text{mol}^{-1}$$

Because ΔH°_r is negative and ΔS°_r is positive, the reaction will be spontaneous at all temperatures.

(c)

$\Delta H^\circ_r = \Delta H^\circ_f(C\ (s),\ diamond) = +1.895\ kJ \cdot mol^{-1}$

$\Delta S^\circ_r = S^\circ_m(C\ (s),\ diamond) - S^\circ_m(C\ (s),\ graphite)$

$\quad = +2.377\ J \cdot K^{-1} \cdot mol^{-1} - 5.740\ J \cdot K^{-1} \cdot mol^{-1}$

$\quad = -3.363\ J \cdot K^{-1}$

$\Delta G^\circ_r = +1.895\ kJ \cdot mol^{-1} - (353\ K)(-3.363\ J \cdot K^{-1} \cdot mol^{-1})/(1000\ J \cdot kJ^{-1})$

$\quad = +3.082\ kJ \cdot mol^{-1}$

Because ΔH°_r is positive and ΔS°_r is negative, the reaction will be nonspontaneous at all temperatures. Note: This calculation is for atmospheric pressure. Diamond can be produced from graphite at elevated pressures and high temperatures.

8.69 Assuming standard state conditions, ΔG_r° for $C_2H_4(g) + H_2O(g) \rightarrow$ $CH_3CH_2OH(l)$ is:

$\Delta G^\circ_r = (-174.8\ kJ \cdot mol^{-1}) - (68.15\ kJ \cdot mol^{-1}) - (-228.57\ kJ \cdot mol^{-1})$

$\quad = -14.4\ kJ \cdot mol^{-1}$ Negative ΔG°_r indicates a spontaneous reaction.

ΔG_r° for $C_2H_6(g) + H_2O(g) \rightarrow CH_3CH_2OH(l) + H_2(g)$ is:

$\Delta G^\circ_r = (-174.8\ kJ \cdot mol^{-1}) - (-32.82\ kJ \cdot mol^{-1}) - (-228.57\ kJ \cdot mol^{-1})$

$\quad = +86.6\ kJ \cdot mol^{-1}$ Positive ΔG°_r indicates a nonspontaneous reaction.
Reaction A is spontaneous, but reaction B is not spontaneous at any temperature.

8.71 The reactions (2) and (3) in this problem are used to drive the regeneration of ATP. For every three moles of NADH that undergo reaction:

$3\ NADH(aq) \rightarrow 3NAD^+(aq) + 3H^+(aq) + 6e^- \qquad \Delta G_r = 3(-158.2\ kJ)$

$\frac{3}{2}O_2(g) + 6H^+(aq) + 6e^- \rightarrow 3H_2O(l) \qquad \Delta G_r = 3(-61.9\ kJ)$

$3\ NADH(aq) + \frac{3}{2}O_2(g) + 3H^+(aq) \rightarrow 3H_2O(l) + 3NAD^+(aq)$

$\qquad\qquad\qquad\qquad\qquad\qquad \Delta G_r = -660.3\ kJ$

Therefore, for every three moles of NADH, 660.3 kJ / 30.5 kJ = 21.6 moles of ATP would be regenerated if all of the free energy released were used to drive reaction (1).

8.73 (a) 1-propanol (C_3H_8O) and 2-propanone (C_3H_6O) have similar numbers of electrons so that we would expect the molar entropies to be similar. Because 1-propanol exhibits hydrogen bonding, however, we might expect the liquid phase to be more ordered than for 2-propanone. This is observed. The standard molar entropy for 2-propanone is 200 $J \cdot K^{-1} \cdot mol^{-1}$ while that of 1-propanol is 193 $J \cdot K^{-1} \cdot mol^{-1}$.

(b) In the gas phase, hydrogen bonding will not be important because the molecules are too far apart, so the standard molar entropies should be more similar.

8.75 (a) $\Delta G^{\circ} < 0$

(b) ΔH° unable to tell

(c) ΔS° unable to tell

(d) $\Delta S_{totalr} > 0$

8.77 For the cis compound there will be 12 different orientations:

For the trans compound there will only be 3 different orientations.

Comparing the Boltzmann entropy calculations for the cis and trans forms:

cis:

$S = k \ln 12^{6.02 \times 10^{23}} = (1.38 \times 10^{-23} \ J \cdot K^{-1}) \ln 12^{6.02 \times 10^{23}}$

$S = 20.6 \ J \cdot K^{-1}$

trans:

$S = k \ln 3^{6.02 \times 10^{23}} = (1.38 \times 10^{-23} \ J \cdot K^{-1}) \ln 3^{6.02 \times 10^{23}}$

$S = 9.13 \ J \cdot K^{-1}$

The cis form should have the higher residual entropy.

8.79 (a) $W = 3$

(b) $W = 12$

(c)   Initially one of the three atom systems had two atoms in higher energy states. In part (b) the system will be at equilibrium when each three atom system has one quantum of energy. Therefore energy will flow from the system with two quanta to the system with none.

8.81   According to Trouton's rule, the entropy of vaporization of an organic liquid is a constant of approximately 85 $J \cdot mol^{-1} \cdot K^{-1}$. The relationship between entropy of fusion, enthalpy of fusion, and melting point is given by $\Delta S^{\circ}_{fus} = \dfrac{\Delta H^{\circ}_{fus}}{T_{fus}}$.

For Pb : $\Delta S^{\circ}_{fus} = \dfrac{5100 \text{ J}}{600 \text{ K}} = 8.50 \text{ J} \cdot K^{-1}$

For Hg : $\Delta S^{\circ}_{fus} = \dfrac{2290 \text{ J}}{234 \text{ K}} = 9.79 \text{ J} \cdot K^{-1}$

For Na : $\Delta S^{\circ}_{fus} = \dfrac{2640 \text{ J}}{371 \text{ K}} = 7.12 \text{ J} \cdot K^{-1}$

These numbers are reasonably close but clearly much smaller than the value associated with Trouton's rule.

8.83   This is best answered by considering the reaction that interconverts the two compounds

$$4 \text{ Fe}_3\text{O}_4(s) + \text{O}_2(g) \longrightarrow 6 \text{ Fe}_2\text{O}_3(s)$$

We calculate ΔG°_r using data from Appendix 2A:

$\Delta G^{\circ}_r = 6\Delta G^{\circ}_f(\text{Fe}_2\text{O}_3, s) - [4\Delta G^{\circ}_f(\text{Fe}_3\text{O}_4, s)]$

$\Delta G^{\circ}_r = 6(-742.2 \text{ kJ} \cdot mol^{-1}) - [4(-1015.4 \text{ kJ} \cdot mol^{-1})]$

$\qquad = -391.6 \text{ kJ} \cdot mol^{-1}$

Because ΔG°_r is negative, the process is spontaneous at 25°C.

Therefore, Fe_2O_3 is thermodynamically more stable.

8.85 (a) $H_2(g) + Br_2(g) \rightarrow 2\, HBr(g)$

Use thermodynamic data at 25 °C from the appendix, and the relationship

$\Delta G°_r = \Sigma\, \Delta G°_f \text{(products)} - \Sigma\, \Delta G°_r \text{(reactants)}$:

$$\Delta G°_r = 2\Delta G°_f(HBr, g) - [\Delta G°_f(H_2, g) + \Delta G°_f(Br_2, g)]$$
$$= 2(-53.45 \text{ kJ} \cdot \text{mol}^{-1}) - [0 + (3.11 \text{ kJ} \cdot \text{mol}^{-1})]$$
$$= -110.0 \text{ kJ} \cdot \text{mol}^{-1}$$

(b)   STP: 0°C or 273.15K and 1 atm

moles of H_2 gas:

$$n = \frac{P \cdot V}{R \cdot T} = \frac{1. \text{ atm} \cdot 0.120 \text{ L}}{\left(0.0820574 \text{ L} \cdot \text{atm} \cdot \text{K}^{-1} \cdot \text{mol}^{-1}\right) \cdot 273.15 \text{ K}} = 5.35 \times 10^{-3} \text{ mol}$$

A stoichiometric amount of bromine gas is $5.35 \times 10^{-3} \text{ mol}$.

Moles of HBr(g) formed $= 2(5.35 \times 10^{-3} \text{ mol}) = 1.07 \times 10^{-2} \text{ mol}$

Molar concentration of the resulting hydrobromic acid,

$$M = \frac{n}{V} = \frac{1.07 \times 10^{-2}}{0.150 \text{ L}} = 7.14 \times 10^{-2} \text{ mol} \cdot \text{L}^{-1}$$

8.87 (a) Because the enthalpy change for dissolution is positive, the entropy change of the surroundings must be a negative number

$\left(\Delta S°_{surr} = -\dfrac{\Delta H°_{system}}{T} \right)$. Because spontaneous processes are accompanied

by an increase in entropy, the change in enthalpy does not favor the dissolution process. (b) In order for the process to be spontaneous (because it occurs readily, we know it is spontaneous), the entropy change of the system must be positive. (c) Positional disorder is dominant. (d) Because the surroundings participate in the solution process only as a source of heat, the entropy change of the surroundings is primarily a result of the dispersal of thermal motion. (e) The driving force for the dissolution is the dispersal of matter, resulting in an overall positive ΔS.

8.89 (a)(i) $\Delta G_r^o = \sum \Delta G_f^o \left(products \right) - \sum \Delta G_f^o \left(reactants \right) = -120.35 \, \text{kJ.mol}^{-1}$

(ii) $\Delta G_r^o = 2(-120.35) - 2(-237.13) = +233.56 \, \text{kJ.mol}^{-1}$

Method (i) releases more energy per mole of O_2.

(b) Method (i) has the more negative standard Gibbs free energy.

(c) No.

8.91 The entries all correspond to aqueous ions. The fact that they are negative is due to the reference point that has been established. Because ions cannot actually be separated and measured independently, a reference point that defines $S^\circ_m (H^+, aq) = 0$ has been established. This definition is then used to calculate the standard entropies for the other ions. The fact that they are negative will arise in part because the solvated ion $M(H_2O)_x^{n+}$ will be more ordered than the isolated ion and solvent molecules $(M^{n+} + x \, H_2O)$.

8.93 (a) In order to calculate the free energy at different temperatures, we need to know ΔH° and ΔS° for the process: $H_2O(l) \longrightarrow H_2O(g)$

$\Delta H^\circ_r = \Delta H^\circ_f (H_2O, g) - \Delta H^\circ_f (H_2O, l)$
$= (-241.82 \, \text{kJ} \cdot \text{mol}^{-1}) - [-285.83 \, \text{kJ} \cdot \text{mol}^{-1}]$
$= 44.01 \, \text{kJ} \cdot \text{mol}^{-1}$

$\Delta S^\circ_r = S^\circ_m (H_2O, g) - S^\circ_m (H_2O, l)$
$= 188.83 \, \text{J} \cdot \text{K}^{-1} \cdot \text{mol}^{-1} - [69.91 \, \text{J} \cdot \text{K}^{-1} \cdot \text{mol}^{-1}]$
$= 118.92 \, \text{J} \cdot \text{K}^{-1} \cdot \text{mol}^{-1}$

$\Delta G^\circ_r = \Delta H^\circ_r - T \Delta S^\circ_r$
$= 44.01 \, \text{kJ} \cdot \text{mol}^{-1} - T(118.92 \, \text{J} \cdot \text{K}^{-1} \cdot \text{mol}^{-1})/(1000 \, \text{J} \cdot \text{kJ}^{-1})$

$T(K)$	$\Delta G^\circ_r (kJ)$
298	8.57 kJ
373	-0.35 kJ
423	-6.29 kJ

The reaction goes from being nonspontaneous near room temperature to being spontaneous above 100°C.

(b) The value at 100°C should be exactly 0, because this is the normal boiling point of water.

(c) The discrepancy arises because the enthalpy and entropy values calculated from the tables are not rigorously constant with temperature. Better values would be obtained using the actual enthalpy and entropy of vaporization measured at the boiling point.

8.95

8.97 $p_{down} = p_{up}$

By inspection of the equations in exercise 8.96,

$e^{-\Delta E/kT} = 1$ for $p_{down} = p_{up}$. $p_{down} = p_{up}$

$$\frac{1}{1+1} = \frac{1}{1+1}$$

Since ΔE is the energy difference between the low energy and high energy spin states it is a positive value. Since ΔE is a positive value, $e^{-\Delta E/kT}$ will tend 1 as T → ∞.

8.99 (a) Since standard molar entropies increase with temperature (more translational, vibrational and rotational motion), the $-T\Delta S_m^o$ term becomes more negative at higher temperatures.

(b) For gases with translational motion, their increase in standard molar entropy, on heating, is much larger than for solids and liquids.

Therefore their $-T\Delta S_m^o$ term is more negative.

8.101 The dehydrogenation of cyclohexane to benzene follows the following equation:

$$C_6H_{12}(l) \longrightarrow C_6H_6(l) + 3\,H_2(g)$$

We can confirm that this process is nonspontaneous by calculating the ΔG°_r for the process, using data in Appendix 2A:

$$\Delta G^\circ_r = \Delta G^\circ_f(C_6H_6, l) - \Delta G^\circ_f(C_6H_{12}, l)$$
$$= 124.3\text{ kJ}\cdot\text{mol}^{-1} - 26.7\text{ kJ}\cdot\text{mol}^{-1}$$
$$= +97.6\text{ kJ}\cdot\text{mol}^{-1}$$

The reaction of ethane with hydrogen can be examined similarly:

$$C_2H_2(g) + H_2(g) \longrightarrow C_2H_6(g)$$
$$\Delta G^\circ_r = \Delta G^\circ_f(C_2H_6, g) - \Delta G^\circ_f(C_2H_{12}, g)$$
$$= (-32.82\text{ kJ}\cdot\text{mol}^{-1}) - 68.15\text{ kJ}\cdot\text{mol}^{-1}$$
$$= -100.97\text{ kJ}\cdot\text{mol}^{-1}$$

We can now combine these two reactions so that $C_2H_2(g)$ accepts the hydrogen that is formed in the dehydrogenation reaction:

$$C_6H_{12}(l) \longrightarrow C_6H_6(l) + 3\,H_2(g)$$
$$+ 3[C_2H_2(g) + H_2(g) \longrightarrow C_2H_6(g)]$$

$$\Delta G^\circ_r = +97.6\text{ kJ}\cdot\text{mol}^{-1}$$
$$\Delta G^\circ_r = 3(-100.97\text{ kJ}\cdot\text{mol}^{-1})$$

$$C_6H_{12}(l) + 3\,C_2H_2(g) \longrightarrow C_6H_6(l) + 3\,C_2H_6(g)$$
$$\Delta G^\circ_r = -205.13\text{ kJ}\cdot\text{mol}^{-1}$$

We can see that by combining these two reactions, the overall process becomes spontaneous. Essentially, we are using the energy of the favorable reaction to drive the nonfavorable process.

8.103 (a)

| *cis*-2-Butene | *trans*-2-butene | 2-Methylpropene |
| (1) | (2) | (3) |

(b) For the three reactions, the calculation of $\Delta G°$, $\Delta H°$, and $\Delta S°$ are as follows:

$$\Delta G°_r = \Delta G°_f(2) - \Delta G°_f(1)$$
$$= 62.97 \text{ kJ} \cdot \text{mol}^{-1} - 65.86 \text{ kJ} \cdot \text{mol}^{-1}$$
$$= -2.89 \text{ kJ} \cdot \text{mol}^{-1}$$
$$\Delta H°_r = \Delta H°_f(2) - \Delta H°_f(1)$$
$$= (-11.17 \text{ kJ} \cdot \text{mol}^{-1}) - (-6.99 \text{ kJ} \cdot \text{mol}^{-1})$$
$$= -4.18 \text{ kJ} \cdot \text{mol}^{-1}$$
$$\Delta G°_r = \Delta H°_r - T\Delta S°_r$$
$$-2.89 \text{ kJ} \cdot \text{mol}^{-1} = -4.18 \text{ kJ} \cdot \text{mol}^{-1} - (298 \text{ K})(\Delta S°_r)/(1000 \text{ J} \cdot \text{kJ}^{-1})$$
$$\Delta S°_r = -4.33 \text{ J} \cdot \text{K}^{-1} \cdot \text{mol}^{-1}$$

$$\Delta G°_r = \Delta G°_f(3) - \Delta G°_f(1)$$
$$= 58.07 \text{ kJ} \cdot \text{mol}^{-1} - 65.86 \text{ kJ} \cdot \text{mol}^{-1}$$
$$= -7.79 \text{ kJ} \cdot \text{mol}^{-1}$$
$$\Delta H°_f = \Delta H°_f(3) - \Delta H°_f(1)$$
$$= (-16.90 \text{ kJ} \cdot \text{mol}^{-1}) - (-6.99 \text{ kJ} \cdot \text{mol}^{-1})$$
$$= -9.91 \text{ kJ} \cdot \text{mol}^{-1}$$
$$\Delta G°_r = \Delta H°_r - T\Delta S°_r$$
$$-7.79 \text{ kJ} \cdot \text{mol}^{-1} = -9.91 \text{ kJ} \cdot \text{mol}^{-1} - (298 \text{ K})(\Delta S°_r)/(1000 \text{ J} \cdot \text{kJ}^{-1})$$
$$\Delta S°_r = -7.11 \text{ J} \cdot \text{K}^{-1} \cdot \text{mol}^{-1}$$

$$\Delta G^{\circ}_{\text{r}} = \Delta G^{\circ}_{\text{f}}(3) - \Delta G^{\circ}_{\text{f}}(2)$$
$$= 58.07 \text{ kJ} \cdot \text{mol}^{-1} - 62.97 \text{ kJ} \cdot \text{mol}^{-1}$$
$$= -4.90 \text{ kJ} \cdot \text{mol}^{-1}$$
$$\Delta H^{\circ}_{\text{r}} = \Delta H^{\circ}_{\text{f}}(3) - \Delta H^{\circ}_{\text{f}}(2)$$
$$= (-16.90 \text{ kJ} \cdot \text{mol}^{-1}) - (-11.17 \text{ kJ} \cdot \text{mol}^{-1})$$
$$= -5.73 \text{ kJ} \cdot \text{mol}^{-1}$$
$$\Delta G^{\circ}_{\text{r}} = \Delta H^{\circ}_{\text{r}} - T\Delta S^{\circ}_{\text{r}}$$
$$-4.90 \text{ kJ} \cdot \text{mol}^{-1} = -5.73 \text{ kJ} \cdot \text{mol}^{-1} - (298 \text{ K})(\Delta S^{\circ}_{\text{r}})/(1000 \text{ J} \cdot \text{kJ}^{-1})$$
$$\Delta S^{\circ}_{\text{r}} = -2.78 \text{ J} \cdot \text{K}^{-1} \cdot \text{mol}^{-1}$$

(c) The most stable of the three compounds is 2-methylpropene.

(d) Because ΔS° is also equal to the difference in the S°_{m} values for the compounds, we can examine those values to place the three compounds in order of their relative absolute entropies.
The ordering is $S^{\circ}_{\text{m}}(1) > S^{\circ}_{\text{m}}(2) > S^{\circ}_{\text{m}}(3)$.

8.105 (a) $2\text{NaN}_3(s) \rightarrow 3\text{N}_2(g) + 2\text{Na}(g)$

(b) ΔS^{o}_{r} is positive because a gas is produced (from a solid).

(c) $\overset{+}{\text{Na}}\overset{-}{\text{N}_3} = \overset{-}{\text{N}} = \overset{-}{\text{N}}$ $\quad\quad \overset{0}{N} \equiv \overset{0}{N}$ $\quad$ Overall the nitrogen is oxidized.

(d) $\Delta S^{o}_{r} = \sum S^{o}_{m}(\text{products}) - \sum S^{o}_{m}(\text{reactants})$

$$= \left[3(191.61) + 2(153.71)\right] - \left[2(96.9))\right]$$

$$= \left[574.83 + 307.42\right] - \left[193.8\right]$$

$$= +688.45 \text{J.K}^{-1}.\text{mol}^{-1} = +688.4 \text{J.K}^{-1}.\text{mol}^{-1}$$

(e) $\Delta H^{o}_{r} = \sum \Delta H^{o}_{f}(\text{products}) - \sum \Delta H^{o}_{f}(\text{reactants})$

$$= \left[3(0) + 2(107.32)\right] - \left[2(21.7)\right]$$

$$= 214.64 - 43.4 = 171.24 = 171.2 \text{ kJ.mol}^{-1}$$

$$\Delta G^{o}_{r} = \Delta H^{o}_{r} - T\Delta S^{o}_{r} = (171.2 \times 10^{3} \text{J.mol}^{-1}) - 298\text{K}(688.4 \text{ J.k}^{-1}.\text{mol}^{-1})$$
$$= -3.39 \times 10^{4} \text{J.mol}^{-1}$$

(f) yes $\quad\quad$ (g) ΔH^{o}_{r} and ΔS^{o}_{r} are both positive. Therefore lowered.

CHAPTER 9
PHYSICAL EQUILIBRIA

9.1 In a 1.0 L vessel at 20°C, there will be 17.5 Torr of water vapor. The ideal gas law can be used to calculate the mass of water present:

$PV = nRT$

let $m = $ mass of water

$$\left(\frac{17.5 \text{ Torr}}{760 \text{ Torr} \cdot \text{atm}^{-1}}\right)(1.0 \text{ L}) = \left(\frac{m}{18.02 \text{ g} \cdot \text{mol}^{-1}}\right)(0.082\,06 \text{ L} \cdot \text{atm} \cdot \text{K}^{-1} \cdot \text{mol}^{-1})(293 \text{ K})$$

$$m = \frac{(17.5 \text{ Torr})(1.0 \text{ L})(18.02 \text{ g} \cdot \text{mol}^{-1})}{(760 \text{ Torr} \cdot \text{atm}^{-1})(0.082\,06 \text{ L} \cdot \text{atm} \cdot \text{K}^{-1} \cdot \text{mol}^{-1})(293 \text{ K})}$$

$$m = 0.017 \text{ g}$$

9.3 (a) 87 °C (b) 113 °C

9.5 Remember that at the normal boiling of a material its vapor pressure will be 1.00 atm (101.325 kPa). Since we know the vapor pressure at -28 °C (245.15 K) is 17.0 kPa and the $\Delta H^{\circ}_{vap} = 23.77$ kJ $\cdot$ mol^{-1} we can use the

relationship $\ln \dfrac{P_2}{P_1} = \dfrac{\Delta H^{\circ}_{vap}}{R}\left[\dfrac{1}{T_1} - \dfrac{1}{T_2}\right]$ to calculate the temperature at

which the vapor pressure will be 101.325 kPa (be careful with your units):

$$\ln \frac{101.325 kPa}{17.0 kPa} = \frac{23.77 kJ \cdot mol^{-1}}{8.314\,47 \times 10^{-3} kJ \cdot K^{-1} \cdot mol^{-1}}\left(\frac{1}{245.15 K} - \frac{1}{T_2}\right);$$

solving for T_2, we get the normal boiling point of BCl_3 to be 289 K, or 16 °C.

9.7   (a) The quantities $\Delta H°_{vap}$ and $\Delta S°_{vap}$ can be calculated using the

relationship $\ln P = -\dfrac{\Delta H°_{vap}}{R} \cdot \dfrac{1}{T} + \dfrac{\Delta S°_{vap}}{R}$

Because we have two temperatures with corresponding vapor pressures,

we can set up two equations with two unknowns and solve for $\Delta H°_{vap}$ and

$\Delta S°_{vap}$. If the equation is used as is, P must be expressed in atm, which is

the standard reference state. Remember that the value used for P is really

activity, which for pressure is P divided by the reference state of 1 atm, so

that the quantity inside the ln term is dimensionless.

$$\left(8.314 \text{ J} \cdot \text{K}^{-1} \cdot \text{mol}^{-1}\right) \ln\left(\frac{58 \text{ Torr}}{760 \text{ Torr}}\right) = -\frac{\Delta H°_{vap}}{250.4 \text{ K}} + \Delta S°_{vap}$$

$$\left(8.314 \text{ J} \cdot \text{K}^{-1} \cdot \text{mol}^{-1}\right) \ln\left(\frac{512 \text{ Torr}}{760 \text{ Torr}}\right) = -\frac{\Delta H°_{vap}}{298.2 \text{ K}} + \Delta S°_{vap}$$

which give, upon combining terms,

$$-21.39 \text{ J} \cdot \text{K}^{-1} \cdot \text{mol}^{-1} = \left(-0.003\ 994 \text{ K}^{-1}\right)\Delta H°_{vap} + \Delta S°_{vap}$$

$$-3.284 \text{ J} \cdot \text{K}^{-1} \cdot \text{mol}^{-1} = \left(-0.003\ 353 \text{ K}^{-1}\right)\Delta H°_{vap} + \Delta S°_{vap}$$

Subtracting one equation from the other will eliminate the $\Delta S°_{vap}$ term and

allow us to solve for $\Delta H°_{vap}$:

$$-18.11 \text{ J} \cdot \text{K}^{-1} \cdot \text{mol}^{-1} = \left(-0.000\ 641 \text{ K}^{-1}\right)\Delta H°_{vap}$$

$$\Delta H°_{vap} = +28.3 \text{ kJ} \cdot \text{mol}^{-1}$$

(b) We can then use $\Delta H°_{vap}$ to calculate $\Delta S°_{vap}$ using either of the two

equations:

$$-21.39 \text{ J} \cdot \text{K}^{-1} \cdot \text{mol}^{-1} = -0.003\ 994 \text{ K}^{-1}\ (28\ 200 \text{ J} \cdot \text{mol}^{-1}) + \Delta S°_{vap}$$

$$\Delta S°_{vap} = 91.2 \text{ J} \cdot \text{K}^{-1} \cdot \text{mol}^{-1}$$

$$-3.284 \text{ J} \cdot \text{K}^{-1} \cdot \text{mol}^{-1} = -0.003\ 353 \text{ K}^{-1}\ (28\ 200 \text{ J} \cdot \text{mol}^{-1}) + \Delta S°_{vap}$$

$$\Delta S°_{vap} = 91.3 \text{ J} \cdot \text{K}^{-1} \cdot \text{mol}^{-1}$$

(c) The $\Delta G°_{vap}$ is calculated using $\Delta G°_r = \Delta H°_r - T\Delta S°_r$

$$\Delta G°_r = +28.3 \text{ kJ} \cdot \text{mol}^{-1} - (298 \text{ K})(91.2 \text{ J} \cdot \text{K}^{-1} \cdot \text{mol}^{-1})/(1000 \text{ J} \cdot \text{kJ}^{-1})$$

$$\Delta G°_r = +1.1 \text{ kJ} \cdot \text{mol}^{-1}$$

(d) The boiling point can be calculated using one of several methods. The easiest to use is the one developed in the last chapter:

$$\Delta G^\circ{}_{vap} = \Delta H^\circ{}_{vap} - T_B \, \Delta S^\circ{}_{vap} = 0$$

$$\Delta H^\circ{}_{vap} = T_B \, \Delta S^\circ{}_{vap} \text{ or } T_B = \frac{\Delta H^\circ{}_{vap}}{\Delta S^\circ{}_{vap}}$$

$$T_B = \frac{\left(28.2 \text{ kJ} \cdot \text{mol}^{-1}\right)\left(1000 \text{ J} \cdot \text{kJ}^{-1}\right)}{91.2 \text{ J} \cdot \text{K}^{-1} \cdot \text{mol}^{-1}} = 309 \text{ K or } 36°C$$

Alternatively, we could use the relationship $\ln \dfrac{P_2}{P_1} = -\dfrac{\Delta H^\circ{}_{vap}}{R}\left[\dfrac{1}{T_2} - \dfrac{1}{T_1}\right].$

Here we would substitute, in one of the known vapor pressure points, the value of the enthalpy of vaporization and the condition that $P = 1$ atm at the normal boiling point.

9.9 (a) The quantities $\Delta H^\circ{}_{vap}$ and $\Delta S^\circ{}_{vap}$ can be calculated, using the relationship

$$\ln P = -\frac{\Delta H^\circ{}_{vap}}{R} \cdot \frac{1}{T} + \frac{\Delta S^\circ{}_{vap}}{R}$$

Because we have two temperatures with corresponding vapor pressures (we know that the vapor pressure = 1 atm at the boiling point), we can set up two equations with two unknowns and solve for $\Delta H^\circ{}_{vap}$ and $\Delta S^\circ{}_{vap}$. If the equation is used as is, P must be expressed in atm which is the standard reference state. Remember that the value used for P is really activity which, for pressure, is P divided by the reference state of 1 atm, so that the quantity inside the ln term is dimensionless.

$$\left(8.314 \text{ J} \cdot \text{K}^{-1} \cdot \text{mol}^{-1}\right)\ln\left(1\right) = -\frac{\Delta H^\circ{}_{vap}}{315.58 \text{ K}} + \Delta S^\circ{}_{vap}$$

$$\left(8.314 \text{ J} \cdot \text{K}^{-1} \cdot \text{mol}^{-1}\right)\ln\left(\frac{140 \text{ Torr}}{760 \text{ Torr}}\right) = -\frac{\Delta H^\circ{}_{vap}}{273.2 \text{ K}} + \Delta S^\circ{}_{vap}$$

which give, upon combining terms,

$$0 \text{ J} \cdot \text{K}^{-1} \cdot \text{mol}^{-1} = \left(-0.003\,169 \text{ K}^{-1}\right)\Delta H^{\circ}{}_{\text{vap}} + \Delta S^{\circ}{}_{\text{vap}}$$

$$-14.06 \text{ J} \cdot \text{K}^{-1} \cdot \text{mol}^{-1} = \left(-0.003\,660 \text{ K}^{-1}\right)\Delta H^{\circ}{}_{\text{vap}} + \Delta S^{\circ}{}_{\text{vap}}$$

Subtracting one equation from the other will eliminate the $\Delta S^{\circ}{}_{\text{vap}}$ term and allow us to solve for $\Delta H^{\circ}{}_{\text{vap}}$ of iodomethane.

$$-14.06 \text{ J} \cdot \text{K}^{-1} \cdot \text{mol}^{-1} = \left(-0.000\,491 \text{ K}^{-1}\right)\Delta H^{\circ}{}_{\text{vap}}$$

$$\Delta H^{\circ}{}_{\text{vap}} = +28.6 \text{ kJ} \cdot \text{mol}^{-1}$$

(b) We can then use $\Delta H^{\circ}{}_{\text{vap}}$ to calculate $\Delta S^{\circ}{}_{\text{vap}}$ using either of the two equations:

$$0 = \left(-0.003\,169 \text{ K}^{-1}\right)(28\,600 \text{ J} \cdot \text{mol}^{-1}) + \Delta S^{\circ}{}_{\text{vap}}$$

$$\Delta S^{\circ}{}_{\text{vap}} = 90.6 \text{ J} \cdot \text{K}^{-1} \cdot \text{mol}^{-1}$$

$$-14.06 \text{ J} \cdot \text{K}^{-1} \cdot \text{mol}^{-1} = \left(-0.003\,660 \text{ K}^{-1}\right)(28\,600 \text{ J} \cdot \text{mol}^{-1}) + \Delta S^{\circ}{}_{\text{vap}}$$

$$\Delta S^{\circ}{}_{\text{vap}} = 90.6 \text{ J} \cdot \text{K}^{-1} \cdot \text{mol}^{-1}$$

(c) The vapor pressure of CH_3I at another temperature is calculated using

$$\ln\frac{P_2}{P_1} = -\frac{\Delta H^{\circ}{}_{\text{vap}}}{R}\left[\frac{1}{T_2} - \frac{1}{T_1}\right]$$

We need to insert the calculated value of the enthalpy of vaporization and one of the known vapor pressure points:

$$\ln\left(\frac{P_{\text{at }25.0^{\circ}\text{C}}}{1 \text{ atm}}\right) = -\frac{28\,600 \text{ J} \cdot \text{mol}^{-1}}{8.314 \text{ J} \cdot \text{K}^{-1} \cdot \text{mol}^{-1}}\left[\frac{1}{298.2 \text{ K}} - \frac{1}{315.58 \text{ K}}\right]$$

$$P_{\text{at }25.0^{\circ}\text{C}} = 0.53 \text{ atm or } 4.0 \times 10^2 \text{ Torr}$$

9.11 Table 7.3 contains the enthalpy of vaporization and the boiling point of methanol (at which the vapor pressure $= 1$ atm). Using this data and the equation

$$\ln\frac{P_2}{P_1} = -\frac{\Delta H^{\circ}{}_{\text{vap}}}{R}\left[\frac{1}{T_2} - \frac{1}{T_1}\right]$$

$$\ln\left(\frac{P_{25.0^{\circ}\text{C}}}{1}\right) = -\frac{35\,300 \text{ J} \cdot \text{mol}^{-1}}{8.314 \text{ J} \cdot \text{K}^{-1} \cdot \text{mol}^{-1}}\left[\frac{1}{298.2 \text{ K}} - \frac{1}{337.2 \text{ K}}\right]$$

$P_{25.0°C} = 0.19$ atm or 1.5×10^2 Torr

9.13   (a) vapor;   (b) liquid;   (c) vapor

9.15   (a) 2.4 K;   (b) about 10 atm;   (c) 5.5 K;   (d) no

9.17   (a) At the lower pressure triple point, liquid helium-I and -II are in equilibrium with helium gas; at the higher pressure triple point, liquid helium-I and -II are in equilibrium with solid helium.   (b) helium-I

9.19   The pressure increase would bring CO_2 into the solid region.

9.21   (a) KCl is an ionic solid, so water would be the best choice;   (b) CCl_4 is non-polar, so the best choice is benzene;   (c) CH_3COOH is polar, so water is the better choice.

9.23   (a) hydrophilic, because NH_2 is polar, and has a lone pair and H atoms that can participate in hydrogen bonding to water molecules;
(b) hydrophobic, because the CH_3 group is not very polar;
(c) hydrophobic, because the Br group is not very polar;
(d) hydrophilic, because the carboxylic acid group has lone pairs on oxygen and an acidic proton that can participate in hydrogen bonding to water molecules.

9.25   (a) The solubility of $O_2(g)$ in water is 1.3×10^{-3} mol·L^{-1}·atm^{-1}.

$$\text{solubility at 50 kPa} = \left(\frac{50 \text{ kPa}}{101.325 \text{ kPa} \cdot \text{atm}^{-1}} \right) \left(1.3 \times 10^{-3} \text{ mol} \cdot \text{L}^{-1} \cdot \text{atm}^{-1} \right)$$
$$= 6.4 \times 10^{-4} \text{ mol} \cdot \text{L}^{-1}$$

(b) The solubility of $CO_2 (g)$ in water is 2.3×10^{-2} mol$\cdot$L$^{-1}\cdot$atm^{-1}.

$$\text{solubility at 500 Torr} = \left(\frac{500 \text{ Torr}}{760 \text{ Torr}\cdot\text{atm}^{-1}} \right)\left(2.3 \times 10^{-2} \text{ mol}\cdot\text{L}^{-1}\cdot\text{atm}^{-1} \right)$$

$$= 1.5 \times 10^{-2} \text{ mol}\cdot\text{L}^{-1}$$

(c) solubility =

$$\left(0.10 \text{ atm} \right)\left(2.3 \times 10^{-2} \text{ mol}\cdot\text{L}^{-1}\cdot\text{atm}^{-1} \right) = 2.3 \times 10^{-3} \text{ mol}\cdot\text{L}^{-1}.$$

9.27 (a) $\left(4 \text{ mg}\cdot\text{L}^{-1} \right)\left(1000 \text{ mL}\cdot\text{L}^{-1} \right)\left(1.00 \text{ g}\cdot\text{mL}^{-1} \right)\left(1 \text{ kg}/1000 \text{ g} \right)$

$= 4 \text{ mg}\cdot\text{L}^{-1}$ or 4 ppm

(b) The solubility of $O_2 (g)$ in water is 1.3×10^{-3} mol$\cdot$L$^{-1}\cdot$atm^{-1} which

can be converted to parts per million as follows:

In 1.00 L (corresponding to 1 kg) of solution there will be

1.3×10^{-3} mol O_2

$\left(1.3 \times 10^{-3} \text{mol}\cdot\text{kg}^{-1}\cdot\text{atm}^{-1} O_2 \right)\left(32.00 \text{ g}\cdot\text{mol}^{-1} O_2 \right)\left(10^3 \text{ mg}\cdot\text{g}^{-1} \right)$

$= 42 \text{ mg}\cdot\text{kg}^{-1}\cdot\text{atm}^{-1}$ or 42 ppm$\cdot$atm^{-1}

$4 \text{ mg} \div 41 \text{ mg}\cdot\text{kg}^{-1}\cdot\text{atm}^{-1} = 0.1 \text{ atm}$

(c) $P = \dfrac{0.1 \text{ atm}}{0.21} = 0.5 \text{ atm}$

9.29 (a) By Henry's law, the concentration of CO_2 in solution will double.

(b) No change in the equilibrium will occur; the partial pressure of CO_2

is unchanged and the concentration is unchanged.

9.31 The solubility of CO_2 can be determined using Henry's law:

$$S_{CO_2} = k_H \times P = \left(2.3 \times 10^{-2} \text{ mol}\cdot\text{L}^{-1}\cdot\text{atm}^{-1} \right)\left(3.60 \text{ atm} \right)$$

$$n_{CO_2} = S \times 0.420 \text{ L} = 3.48 \times 10^{-2} \text{ mol } CO_2$$

This is equal to 1.5 g of CO_2.

9.33   (a) Because it is exothermic, the enthalpy change must be negative;

(b) $Li_2SO_4(s) \rightarrow 2\ Li^+(aq) + SO_4^{2-}(aq) + heat$;   (c) Given that

$\Delta H_L + \Delta H_{hydration} = \Delta H$ of solution, the enthalpy of hydration should be

larger. If the lattice energy were greater, the overall process would be

endothermic.

9.35   To answer this question we must first determine the molar enthalpies of

solution and multiply this by the number of moles of solid dissolved to get

the actual amount of heat released.

(a) $\Delta H^{\circ}_{sol} = 3.9\ kJ \cdot mol^{-1}$

$$\Delta H = \left(\frac{10.0\ g\ NaCl}{58.44\ g \cdot mol^{-1}}\right)(3.9\ kJ \cdot mol^{-1}) = 0.67\ kJ\ or\ 6.7 \times 10^2\ J$$

(b) $\Delta H^{\circ}_{sol} = -7.5\ kJ \cdot mol^{-1}$

$$\Delta H = \left(\frac{10.0\ g\ NaI}{149.89\ g \cdot mol^{-1}}\right)(-7.5\ kJ \cdot mol^{-1}) = -0.50\ kJ = -5.0 \times 10^2\ J$$

(c) $\Delta H^{\circ}_{sol} = -329\ kJ \cdot mol^{-1}$

$$\Delta H = \left(\frac{10.0\ g\ AlCl_3}{133.33\ g \cdot mol^{-1}}\right)(-329\ kJ \cdot mol^{-1}) = -24.7\ kJ$$

(d) $\Delta H^{\circ}_{sol} = 6.6\ kJ \cdot mol^{-1}$

$$\Delta H = \left(\frac{10.0\ g\ NH_4NO_3}{80.05\ g \cdot mol^{-1}}\right)(6.6\ kJ \cdot mol^{-1}) = +0.82\ kJ = +8.2 \times 10^2\ J$$

9.37   (a) $m_{NaCl} = \dfrac{\left(\dfrac{25.0\ g\ NaCl}{58.44\ g \cdot mol^{-1}}\right)}{0.5000\ kg} = 0.856\ m$

(b) $m_{NaOH} = \dfrac{\left(\dfrac{mass\ NaOH}{40.00\ g \cdot mol^{-1}}\right)}{0.345\ kg} = 0.18\ m$

mass NaOH $= 2.5\ g$

(c) $m_{NaCl} = \dfrac{\left(\dfrac{0.978 \text{ g urea}}{60.06 \text{ g} \cdot \text{mol}^{-1}} \right)}{(285 \text{ mL})(1 \text{ g} \cdot \text{mL}^{-1})(10^{-3} \text{ kg} \cdot \text{g}^{-1})} = 0.0571 \, m$

9.39   (a) 1 kg of 5.00% K_3PO_4 will contain 50.0 g K_3PO_4 and 950.0 g H_2O.

$$\dfrac{\left(\dfrac{50.0 \text{ g } K_3PO_4}{212.27 \text{ g} \cdot \text{mol}^{-1} K_3PO_4} \right)}{0.950 \text{ kg}} = 0.248 \, m$$

(b) The mass of 1.00 L of solution will be 1043 g, which will contain

$1043 \text{ g} \times 0.0500 = 52.2 \text{ g } K_3PO_4$.

$$\dfrac{\left(\dfrac{52.2 \text{ g } K_3PO_4}{212.27 \text{ g} \cdot \text{mol}^{-1} K_3PO_4} \right)}{1.00 \text{ L}} = 0.246 \text{ M}$$

9.41   (a) If x_{MgCl_2} is 0.0120, then there are 0.0120 mol $MgCl_2$ for every 0.9880

mol H_2O. The mass of water will be

$18.02 \text{ g} \cdot \text{mol}^{-1} \times 0.9880 \text{ mol} = 17.80 \text{ g or } 0.01780 \text{ kg}.$

$$m_{Cl^-} = \dfrac{\left(\dfrac{2 \text{ mol } Cl^-}{1 \text{ mol } MgCl_2} \right)(0.0120 \text{ mol } MgCl_2)}{0.01780 \text{ kg solvent}} = 1.35 \, m$$

(b) $m_{NaOH} = \dfrac{\left(\dfrac{6.75 \text{ g } NaOH}{40.00 \text{ g} \cdot \text{mol}^{-1}} \right)}{0.325 \text{ kg solvent}} = 0.519 \, m$

(c) 1.000 L of 15.00 M HCl(aq) will contain 15.00 mol with a mass of

$15.00 \times 36.46 \text{ g} \cdot \text{mol}^{-1} = 546.9 \text{ g}.$ The density of the 1.000 L of solution

is $1.0745 \text{ g} \cdot \text{cm}^{-3}$ so the total mass in the solution is 1074.5 g. This leaves

$1074.5 \text{ g} - 546.9 \text{ g} = 527.6 \text{ g}$ as water.

$\dfrac{15.00 \text{ mol HCl}}{0.5276 \text{ kg solvent}} = 28.43 \, m$

9.43 (a) Molar mass of $CaCl_2 \cdot 6\,H_2O = 219.08$ g·mol^{-1}, which consists of 110.98 g $CaCl_2$ and 108.10 g of water.

$$m_{CaCl_2} = \frac{x \text{ mol } CaCl_2 \cdot 6\,H_2O}{0.500 \text{ kg} + x\,(6 \times 0.018\,02 \text{ kg } H_2O)}$$

Note: 18.02 g $H_2O = 0.018\,02$ kg $H_2O = 1.000$ mol H_2O

$x =$ number of moles of $CaCl_2 \cdot 6\,H_2O$ needed to prepare a solution of molality m_{CaCl_2}, in which each mole of $CaCl_2 \cdot 6\,H_2O$ produces 6(0.018 02 kg) of water as solvent (assuming we begin with 0.250 kg H_2O).

For a 0.125 m solution of $CaCl_2 \cdot 6\,H_2O$,

$$0.125 \ m = \frac{x}{0.500 \text{ kg} + x(6)(0.01802 \text{ kg } H_2O)}$$

$x = 0.0625$ mol $+ x(0.0135$ mol$)$

$x - 0.0135x = 0.0625$ mol

$0.986 \ x = 0.0625$ mol

$x = 0.0634$ mol $CaCl_2 \cdot 6\,H_2O$

$$\therefore (0.0634 \text{ mol } CaCl_2 \cdot 6\,H_2O)\left(\frac{219.08 \text{ g } CaCl_2 \cdot 6\,H_2O}{1 \text{ mol } CaCl_2 \cdot 6\,H_2O}\right) = 13.9 \text{ g } CaCl_2 \cdot 6\,H_2O$$

(b) Molar mass of $NiSO_4 \cdot 6\,H_2O = 262.86$ g·mol^{-1}, which consists of 154.77 g $NiSO_4$ and 108.09 g H_2O.

$$m_{NiSO_4} = \frac{x \text{ mol } NiSO_4 \cdot 6\,H_2O}{0.500 \text{ kg} + x\,(6 \times 0.01802 \text{ kg } H_2O)}$$

where $x =$ number of moles of $NiSO_4 \cdot 6H_2O$ needed to prepare a solution of molality m_{NiSO_4}, in which each mole of $NiSO_4 \cdot 6\,H_2O$ produces 6(0.018 02 kg) of water as solvent. Assuming we begin with 0.500 kg H_2O, for a 0.22 m solution of $NiSO_4 \cdot 6\ H_2O$,

$$0.22\ m = \frac{x}{0.500\ \text{kg} + x(6)(0.01802\ \text{kg H}_2)}$$

$$x = 0.11\ \text{mol} + x(0.0238\ \text{mol})$$

$$x - 0.0238\ x = 0.11\ \text{mol}$$

$$0.976\ x = 0.11\ \text{mol}$$

$$x = 0.11\ \text{mol NiSO}_4 \cdot 6\ \text{H}_2\text{O}$$

$$\therefore (0.11\ \text{mol NiSO}_4 \cdot 6\ \text{H}_2\text{O}) \left(\frac{262.86\ \text{g NiSO}_4 \cdot 6\ \text{H}_2\text{O}}{1\ \text{mol NiSO}_4 \cdot 6\ \text{H}_2\text{O}} \right) = 29\ \text{g NiSO}_4 \cdot 6\ \text{H}_2\text{O}$$

9.45 The free energy of the solvent in the NaCl solution will always be lower than the free energy of pure water; if given enough time, all of the water from the "pure water" beaker will become part of the NaCl solution, leaving an empty beaker. An equilibrium will be established between the solvent and the water vapor.

9.47 (a) $P = x_{\text{solvent}} \times P_{\text{pure solvent}}$

At 100°C, the normal boiling point of water, the vapor pressure of water is 1.00 atm. If the mole fraction of sucrose is 0.100, then the mole fraction of water is 0.900:

$P = 0.900 \times 1.000\ \text{atm} = 0.900\ \text{atm or 684 Torr}$

(b) First, the molality must be converted to mole fraction. If the molality is $0.100\ \text{mol} \cdot \text{kg}^{-1}$, then there will be 0.100 mol sucrose per 1000 g of water.

$$x_{\text{H}_2\text{O}} = \frac{n_{\text{H}_2\text{O}}}{n_{\text{H}_2\text{O}} + n_{\text{sucrose}}} = \frac{\dfrac{1000\ \text{g}}{18.02\ \text{g} \cdot \text{mol}^{-1}}}{\dfrac{1000\ \text{g}}{18.02\ \text{g} \cdot \text{mol}^{-1}} + 0.100\ \text{mol}} = 0.998$$

$P = 0.998 \times 1.000\ \text{atm} = 0.998\ \text{atm or 758 Torr}$

9.49 The mole fraction of benzene in the mixture can be calculated using Raoult's law :

$$P_{\text{benzene soln}} = x_{\text{benzene}} \times P_{\text{pure benzene}}; \text{ so } x_{\text{benzene}} = \frac{P_{\text{benzene soln}}}{P_{\text{pure benzene}}}$$

$$x_{\text{benzene}} = \frac{75.0 \text{ Torr}}{94.6 \text{ Torr}} = 0.793$$

Since the sum of the mole fractions of solute and solvent must equal 1 and since we know the moles of benzene that are present calculation of the moles of solute added is straightforward:

$$x_{\text{unknown}} = 1 - x_{\text{benzene}} = 1 - 0.793 = 0.207$$

$$x_{\text{unknown}} = \frac{n_{\text{unknown}}}{n_{\text{unknown}} + n_{\text{benzene}}} = \frac{n_{\text{unknown}}}{n_{\text{unknown}} + 0.300} = 0.207$$

Solving this gives $n_{\text{unknown}} = 7.80 \times 10^{-2}$ mol.

9.51 (a) From the relationship $P = x_{\text{solvent}} \left(P_{\text{pure solvent}} \right)$ we can calculate the mole fraction of the solvent:

$$94.8 \text{ Torr} = x_{\text{solvent}} \left(100.0 \text{ Torr} \right)$$

$$x_{\text{solvent}} = 0.948$$

The mole fraction of the unknown compound will be $1.000 - 0.948 = 0.052$.

(b) The molar mass can be calculated by using the definition of mole fraction for either the solvent or solute. In this case, the math is slightly easier if the definition of mole fraction of the solvent is used:

$$x_{\text{solvent}} = \frac{n_{\text{solvent}}}{n_{\text{unknown}} + n_{\text{solvent}}}$$

$$0.948 = \frac{\dfrac{100 \text{ g}}{78.11 \text{ g} \cdot \text{mol}^{-1}}}{\dfrac{100 \text{ g}}{78.11 \text{ g} \cdot \text{mol}^{-1}} + \dfrac{8.05 \text{ g}}{M_{\text{unknown}}}}$$

$$M_{\text{unknown}} = \frac{8.05 \text{ g}}{\left[\left(\dfrac{100 \text{ g}}{78.11 \text{ g} \cdot \text{mol}^{-1}} \right) \Big/ 0.948 \right] - \left(\dfrac{100 \text{ g}}{78.11 \text{ g} \cdot \text{mol}^{-1}} \right)} = 115 \text{ g} \cdot \text{mol}^{-1}$$

9.53 (a) Pure water has a vapor pressure of 760.00 Torr at 100°C. The mole fraction of the solution can be determined from $P = x_{solvent} \left(P_{pure\ solvent} \right)$

$751 \text{ Torr} = x_{solvent} \left(760.00 \text{ Torr} \right)$

$x_{solvent} = 0.988$

The mole fraction needs to be converted to molality:

$x_{solvent} = 0.988 = \dfrac{n_{H_2O}}{n_{H_2O} + n_{solute}}$

Because the absolute amount of solution is not important, we can assume that the total number of moles = 1.00.

$0.988 = \dfrac{n_{H_2O}}{1.00 \text{ mol}}$

$n_{H_2O} = 0.988 \text{ mol}; n_{solute} = 0.012 \text{ mol}$

$\text{molality} = \dfrac{0.012 \text{ mol}}{\left(\dfrac{0.988 \text{ mol} \left(18.02 \text{ g} \cdot \text{mol}^{-1} \right)}{1000 \text{ g} \cdot \text{kg}^{-1}} \right)} = 0.67 \text{ mol} \cdot \text{kg}^{-1}$

Knowing the mole fraction, one can calculate ΔT_b:

$\Delta T_b = k_b m$

$\Delta T_b = 0.51 \text{ K} \cdot \text{mol} \cdot \text{kg}^{-1} \left(0.67 \text{ mol} \cdot \text{kg}^{-1} \right) = 0.34 \text{ K or } 0.34°C$

Boiling point $= 100.00°C + 0.34°C = 100.34°C$

(b) The procedure is the same as in (a). Pure benzene has a vapor pressure of 760.00 Torr at 80.1°C. The mole fraction of the solution can be determined from

$P = x_{solvent} \left(P_{pure\ solvent} \right)$

$740 \text{ Torr} = x_{solvent} \left(760.00 \text{ Torr} \right)$

$x_{solvent} = 0.974$

The mole fraction needs to be converted to molality:

$x_{solvent} = 0.974 = \dfrac{n_{benzene}}{n_{benzene} + n_{solute}}$

Because the absolute amount of solution is not important, we can assume that the total number of moles = 1.00:

$$0.974 = \frac{n_{benzene}}{1.00 \text{ mol}}$$

$$n_{benzene} = 0.974 \text{ mol}; \ n_{solute} = 0.026 \text{ mol}$$

$$\text{molality} = \frac{0.026 \text{ mol}}{\left(\frac{0.974 \text{ mol}\left(78.11 \text{ g} \cdot \text{mol}^{-1}\right)}{1000 \text{ g} \cdot \text{kg}^{-1}}\right)} = 0.34 \text{ mol} \cdot \text{kg}^{-1}$$

Knowing the mole fraction, one can calculate ΔT_b:

$$\Delta T_b = k_b m$$
$$\Delta T_b = 2.53 \text{ K} \cdot \text{kg} \cdot \text{mol}^{-1} \left(0.34 \text{ mol} \cdot \text{kg}^{-1}\right) = 0.86 \text{ K or } 0.86°C$$
$$\text{Boiling point} = 80.1°C + 0.86°C = 81.0°C$$

9.55 $\Delta T_f = 179.8°C - 176.9°C = 2.9°C$ or 2.9 K

$$\Delta T_f = k_f m$$
$$2.9 \text{ K} = (39.7 \text{ K} \cdot \text{kg} \cdot \text{mol}^{-1})m$$
$$2.9 \text{ K} = (39.7 \text{ K} \cdot \text{kg} \cdot \text{mol}^{-1})\left(\frac{\left(\frac{1.14 \text{ g}}{M_{unknown}}\right)}{0.100 \text{ kg}}\right)$$

$$\frac{\left(0.100 \text{ kg}\right)\left(2.9 \text{ K}\right)}{39.7 \text{ K} \cdot \text{kg} \cdot \text{mol}^{-1}} = \frac{1.14 \text{ g}}{M_{unknown}}$$

$$M_{unknown} = \frac{\left(1.14 \text{ g}\right)\left(39.7 \text{ K} \cdot \text{kg} \cdot \text{mol}^{-1}\right)}{\left(0.100 \text{ kg}\right)\left(2.9 \text{ K}\right)} = 1.6 \times 10^2 \text{ g} \cdot \text{mol}^{-1}$$

9.57 (a) A 1.00% aqueous solution of NaCl will contain 1.00 g of NaCl for 99.0 g of water. To use the freezing point depression equation, we need the molality of the solution:

$$\text{molality} = \frac{\left(\frac{1.00 \text{ g}}{58.44 \text{ g} \cdot \text{mol}^{-1}}\right)}{0.0990 \text{ kg}} = 0.173 \text{ mol} \cdot \text{kg}^{-1}$$

$$\Delta T_f = i k_f m$$

$$\Delta T_f = i (1.86 \text{ K} \cdot \text{kg} \cdot \text{mol}^{-1})(0.173 \text{ mol} \cdot \text{kg}^{-1}) = 0.593 \text{ K}$$

$$i = 1.84$$

(b) molality of all solute species (undissociated NaCl(aq) plus Na^+

(aq) + Cl^- (aq)) $= 1.84 \times 0.173 \text{ mol} \cdot \text{kg}^{-1} = 0.318 \text{ mol} \cdot \text{kg}^{-1}$

(c) If all the NaCl had dissociated, the total molality in solution would

have been $0.346 \text{ mol} \cdot \text{kg}^{-1}$, giving an i value equal to 2. If no dissociation

had taken place, the molality in solution would have equaled 0.173

$\text{mol} \cdot \text{kg}^{-1}$.

$$NaCl(aq) \rightleftharpoons Na^+(aq) + Cl^-(aq)$$

$$\left(0.173 \text{ mol} \cdot \text{kg}^{-1} - x\right) \qquad x \qquad x$$

$$0.173 \text{ mol} \cdot \text{kg}^{-1} - x + x + x = 0.318 \text{ mol} \cdot \text{kg}^{-1}$$

$$0.173 \text{ mol} \cdot \text{kg}^{-1} + x = 0.318 \text{ mol} \cdot \text{kg}^{-1}$$

$$x = 0.145 \text{ mol} \cdot \text{kg}^{-1}$$

$$\% \text{ dissociation} = \left(\frac{0.145 \text{ mol} \cdot \text{kg}^{-1}}{0.173 \text{ mol} \cdot \text{kg}^{-1}} \right)(100) = 83.8\%$$

9.59 The compound that freezes at the lower temperature will have the largest

$\left|\Delta T_f\right|$ (assuming that $i = 1$, since we are told they are both molecular

compounds). Since the same mass of each compound is present in the same

amount of solvent for each solution, the compound with the lower molar

mass will have the larger number of moles and therefore the larger $\left|\Delta T_f\right|$

(since the compound with the largest molar amount present will have the

largest amount of solute particles present). Since compound A caused the

water to freeze at a lower T than compound B, compound A must have the

lower molar mass. Compound B must therefore have the greater molar

mass.

9.61 For an electrolyte that dissociates into two ions, the van't Hoff i factor will be 1 plus the degree of dissociation, in this case 0.075. This can be readily seen for the general case MX. Let A = initial concentration of MX (if none is dissociated) and let Y = the concentration of MX that subsequently dissociates:

$$MX(aq) \rightleftharpoons M^{n+}(aq) + X^{n-}(aq)$$

$$A - Y Y Y$$

The total concentration of solute species is $(A - Y) + Y + Y = A + Y$

The value of i will then be equal to A + Y or 1.075.

The freezing point change is then easy to calculate:

$$\Delta T_f = i k_f m$$
$$= 1.075 \left(1.86 \text{ K} \cdot \text{kg} \cdot \text{mol}^{-1}\right)\left(0.10 \text{ mol} \cdot \text{kg}^{-1}\right)$$
$$= 0.20$$

Freezing point of the solution will be $0.00°C - 0.20°C = -0.20°C$.

9.63 (a) $\Pi = iRT(\text{molarity})$
$$= 1\left(0.082\,06 \text{ L} \cdot \text{atm} \cdot \text{K}^{-1} \cdot \text{mol}^{-1}\right)\left(293 \text{ K}\right)\left(0.010 \text{ mol} \cdot \text{L}^{-1}\right)$$
$$= 0.24 \text{ atm or } 1.8 \times 10^2 \text{ Torr}$$

(b) Because HCl is a strong acid, it should dissociate into two ions, H^+ and Cl^-, so $i = 2$.

$$\Pi = 2\left(0.082\,06 \text{ L} \cdot \text{atm} \cdot \text{K}^{-1} \cdot \text{mol}^{-1}\right)\left(293 \text{ K}\right)\left(1.0 \text{ mol} \cdot \text{L}^{-1}\right)$$
$$= 48 \text{ atm}$$

(c) $CaCl_2$ should dissociate into 3 ions in solution, therefore $i = 3$.

$$\Pi = 3\left(0.082\,06 \text{ L} \cdot \text{atm} \cdot \text{K}^{-1} \cdot \text{mol}^{-1}\right)\left(293 \text{ K}\right)\left(0.010 \text{ mol} \cdot \text{L}^{-1}\right)$$
$$= 0.72 \text{ atm or } 5.5 \times 10^2 \text{ Torr}$$

9.65 (a) Mannitol has a molar mass of $180.15 \text{ g} \cdot \text{mol}^{-1} = 180.15 \text{ mg} \cdot \text{mmol}^{-1}$

Remembering that $\text{molarity} = \text{mol} \cdot \text{L}^{-1} = \text{mmol} \cdot \text{mL}^{-1}$ we get

$$\frac{180 \text{ mg mannitol}}{1 \text{ mL}} \times \frac{1 \text{ mmol mannitol}}{180.15 \text{ mg mannitol}} = 1.0 \text{ mmol} \cdot \text{mL}^{-1} = 1.0 \text{ M}$$

(b) Assuming an ambient temperature of 25 °C (298 K),

$$\Pi = iRT \,(\text{molarity})$$
$$= 1 \left(0.082\,06\ \text{L} \cdot \text{atm} \cdot \text{K}^{-1} \cdot \text{mol}^{-1}\right)\left(298\ \text{K}\right)\left(1.0\ \text{mol} \cdot \text{L}^{-1}\right)$$
$$= 24\ \text{atm}$$

9.67 The polypeptide is a nonelectrolye, so $i = 1$.

$$\Pi = iRT \,(\text{molarity})$$

$$\Pi = \frac{3.74\ \text{Torr}}{760\ \text{Torr} \cdot \text{atm}^{-1}} = 1 \left(0.082\,06\ \text{L} \cdot \text{atm} \cdot \text{K}^{-1} \cdot \text{mol}^{-1}\right)\left(300\ \text{K}\right)\left(\frac{\left(\dfrac{0.40\ \text{g}}{M_{\text{unknown}}}\right)}{1.0\ \text{L}}\right)$$

$$M_{\text{unknown}} = \frac{\left(0.082\,06\ \text{L} \cdot \text{atm} \cdot \text{K}^{-1} \cdot \text{mol}^{-1}\right)\left(300\ \text{K}\right)\left(0.40\ \text{g}\right)\left(760\ \text{Torr} \cdot \text{atm}^{-1}\right)}{\left(3.74\ \text{Torr}\right)\left(1.0\ \text{L}\right)}$$

$$= 2.0 \times 10^{3}\ \text{g} \cdot \text{mol}^{-1}$$

9.69 We assume the polymer to be a nonelectrolyte, so $i = 1$.

$$\Pi = iRT \,(\text{molarity})$$

$$\Pi = \frac{6.3\ \text{Torr}}{760\ \text{Torr} \cdot \text{atm}^{-1}} = 1 \left(0.08206\ \text{L} \cdot \text{atm} \cdot \text{K}^{-1} \cdot \text{mol}^{-1}\right)\left(293\ \text{K}\right)\left(\frac{\left(\dfrac{0.20\ \text{g}}{M_{\text{unknown}}}\right)}{0.100\ \text{L}}\right)$$

$$M_{\text{unknown}} = \frac{\left(0.08206\ \text{L} \cdot \text{atm} \cdot \text{K}^{-1} \cdot \text{mol}^{-1}\right)\left(293\ \text{K}\right)\left(0.20\ \text{g}\right)\left(760\ \text{Torr} \cdot \text{atm}^{-1}\right)}{\left(6.3\ \text{Torr}\right)\left(0.100\ \text{L}\right)}$$

$$= 5.8 \times 10^{3}\ \text{g} \cdot \text{mol}^{-1}$$

9.71 (a) $C_{12}H_{22}O_{11}$ should be a nonelectrolyte, so $i = 1$.

$$\Pi = iRT \left(\text{molarity}\right)$$
$$= 1\left(0.082\,06 \text{ L} \cdot \text{atm} \cdot \text{K}^{-1} \cdot \text{mol}^{-1}\right)\left(293 \text{ K}\right)\left(0.050 \text{ mol} \cdot \text{L}^{-1}\right)$$
$$= 1.2 \text{ atm}$$

(b) NaCl dissociates to give 2 ions in solution, so $i = 2$.

$$\Pi = iRT \left(\text{molarity}\right)$$
$$= 2\left(0.082\,06 \text{ L} \cdot \text{atm} \cdot \text{K}^{-1} \cdot \text{mol}^{-1}\right)\left(293 \text{ K}\right)\left(0.0010 \text{ mol} \cdot \text{L}^{-1}\right)$$
$$= 0.048 \text{ atm or 36 Torr}$$

(c) AgCN dissociates in solution to give two ions (Ag^+ and CN^-),

so $i = 2$.

$$\Pi = iRT \left(\text{molarity}\right)$$

We must assume that the AgCN does not significantly affect either the volume or density of the solution, which is reasonable given the very small amount of it that dissolves.

$$\Pi = 2\left(0.082\,06 \text{ L} \cdot \text{atm} \cdot \text{K}^{-1} \cdot \text{mol}^{-1}\right)\left(293 \text{ K}\right)\left(\dfrac{\left(\dfrac{2.3 \times 10^{-5} \text{ g}}{133.89 \text{ g} \cdot \text{mol}^{-1}}\right)}{\left(\dfrac{100 \text{ g H}_2\text{O}}{1 \text{ g} \cdot \text{cm}^{-3} \text{ H}_2\text{O} \times 1000 \text{ cm}^3 \cdot \text{L}^{-1}}\right)}\right)$$

$$= 8.3 \times 10^{-5} \text{ atm}$$

9.73 $$\Pi = n\dfrac{RT}{V} = \dfrac{m}{M}\dfrac{RT}{V}$$

where m is the mass of unknown compound.

$$M = \dfrac{mRT}{V\Pi}$$

$$M = \dfrac{(0.166 \text{ g})(0.082\,06 \text{ L} \cdot \text{atm} \cdot \text{K}^{-1} \cdot \text{mol}^{-1})(293 \text{ K})}{(0.010 \text{ L})\left(\dfrac{1.2 \text{ Torr}}{760 \text{ Torr} \cdot \text{atm}^{-1}}\right)} = 2.5 \times 10^5 \text{ g} \cdot \text{mol}^{-1}$$

9.75 (a) To determine the vapor pressure of the solution, we need to know the mole fraction of each component.

$$x_{benzene} = \frac{1.50 \text{ mol}}{1.50 \text{ mol} + 0.50 \text{ mol}} = 0.75$$

$$x_{toluene} = 1 - x_{benzene} = 0.25$$

$$P_{total} = 0.75\,(94.6 \text{ Torr}) + 0.25\,(29.1 \text{ Torr}) = 78.2 \text{ Torr}$$

The vapor phase composition will be given by

$$x_{benzene\ in\ vapor\ phase} = \frac{P_{benzene}}{P_{total}} = \frac{0.75\,(94.6 \text{ Torr})}{78.2 \text{ Torr}} = 0.91$$

$$x_{toluene\ in\ vapor\ phase} = 1 - 0.91 = 0.09$$

The vapor is richer in the more volatile benzene, as expected.

(b) The procedure is the same as in (a) but the number of moles of each component must be calculated first:

$$n_{benzene} = \frac{15.0 \text{ g}}{78.11 \text{ g} \cdot \text{mol}^{-1}} = 0.192$$

$$n_{toluene} = \frac{65.3 \text{ g}}{92.14 \text{ g} \cdot \text{mol}^{-1}} = 0.709$$

$$x_{benzene} = \frac{0.192 \text{ mol}}{0.192 \text{ mol} + 0.709 \text{ mol}} = 0.213$$

$$x_{toluene} = 1 - x_{benzene} = 0.787$$

$$P_{total} = 0.213\,(94.6 \text{ Torr}) + 0.787\,(29.1 \text{ Torr}) = 43.0 \text{ Torr}$$

The vapor phase composition will be given by

$$x_{benzene\ in\ vapor\ phase} = \frac{P_{benzene}}{P_{total}} = \frac{0.213\,(94.6 \text{ Torr})}{43.0 \text{ Torr}} = 0.469$$

$$x_{toluene\ in\ vapor\ phase} = 1 - 0.469 = 0.531$$

9.77 To calculate this quantity, we must first find the mole fraction of each that will be present in the mixture. This value is obtained from the relationship

$$P_{\text{total}} = \left[x_{\text{1,1-dichloroethane}} \left(P_{\text{pure 1,1-dichloroethane}} \right) \right]$$
$$+ \left[x_{\text{1,1-dichlorotetrafluoroethane}} \left(P_{\text{pure 1,1-dichlorotetrafluoroethane}} \right) \right]$$

$$157 \text{ Torr} = \left[x_{\text{1,1-dichloroethane}} \left(228 \text{ Torr} \right) \right] + \left[x_{\text{1,1-dichlorotetrafluoroethane}} \left(79 \text{ Torr} \right) \right]$$

$$157 \text{ Torr} = \left[x_{\text{1,1-dichloroethane}} \left(228 \text{ Torr} \right) \right] + \left[\left(1 - x_{\text{1,1-dichloroethane}} \right) \left(79 \text{ Torr} \right) \right]$$

$$157 \text{ Torr} = 79 \text{ Torr} + \left[x_{\text{1,1-dichloroethane}} \left(228 \text{ Torr} \right) - x_{\text{1,1-dichloroethane}} \left(79 \text{ Torr} \right) \right]$$

$$78 \text{ Torr} = x_{\text{1,1-dichloroethane}} \left(149 \text{ Torr} \right)$$

$$x_{\text{1,1-dichloroethane}} = 0.52$$

$$x_{\text{1,1-dichlorotetrafluoroethane}} = 1 - 0.52 = 0.48$$

To calculate the number of grams of 1,1-dichloroethane, we use the definition of mole fraction. Mathematically, it is simpler to use the $x_{\text{1,1-dichlorotetrafluoroethane}}$ definition:

$$n_{\text{1,1-dichlorotetrafluoroethane}} = \frac{100.0 \text{ g}}{170.92 \text{ g} \cdot \text{mol}^{-1}} = 0.5851 \text{ mol}$$

$$x_{\text{1,1-dichlorotetrafluoroethane}} = \frac{0.5851 \text{ mol}}{0.5851 \text{ mol} + \dfrac{m}{98.95 \text{ g} \cdot \text{mol}^{-1}}} = 0.48$$

$$0.5851 \text{ mol} = 0.48 \left[0.5851 \text{ mol} + \frac{m}{98.95 \text{ g} \cdot \text{mol}^{-1}} \right]$$

$$0.5851 \text{ mol} - \left[0.48 \left(0.5851 \text{ mol} \right) \right] = 0.48 \left[\frac{m}{98.95 \text{ g} \cdot \text{mol}^{-1}} \right]$$

$$0.3042 \text{ mol} = \left(0.004\,851 \text{ mol} \cdot \text{g}^{-1} \right) m$$

$$m = 63 \text{ g}$$

9.79 Raoult's Law applies to the vapor pressure of the mixture, so positive deviation means that the vapor pressure is higher than expected for an ideal solution. Negative deviation means that the vapor pressure is lower than expected for an ideal solution. Negative deviation will occur when the interactions between the different molecules are somewhat stronger than the interactions between molecules of the same kind   (a) For methanol and ethanol, we expect the types of intermolecular attractions in the mixture to be similar to those in the component liquids, so that an ideal solution is predicted.   (b) For HF and H_2O, the possibility of intermolecular hydrogen bonding between water and HF would suggest that negative deviation would be observed, which is the case. HF and H_2O form an azeotrope that boils at $111°C$, a temperature higher than the boiling point of either HF $(19.4°C)$ or water.   (c) Because hexane is nonpolar and water is polar with hydrogen bonding, we would expect a mixture of these two to exhibit positive deviation (the interactions between the different molecules would be weaker than the intermolecular forces between like molecules). Hexane and water do form an azeotrope that boils at $61.6°C$, a temperature below the boiling point of either hexane or water.

9.81 A *foam* is a colloid formed by suspending a gas in a liquid or a solid matrix, while a *sol* is a suspension of a solid in a liquid. Some examples of foams are styrofoam and soapsuds; examples of sols are muddy water and mayonnaise.

9.83 Colloids will reflect or scatter light while true solutions do not (this is known as the Tyndall effect).

9.85 Water is a polar molecule and as a result will orient itself differently around cations and anions, aligning its dipole in such a way as to present the most favorable interaction possible:

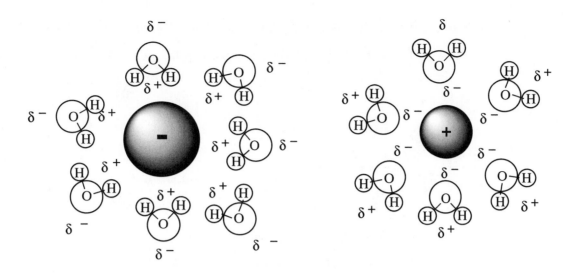

9.87 (a) stronger; (b) low; (c) high; (d) weaker; (e) weak, low; (f) low; (g) strong, high

9.89 (a, b) Viscosity and surface tension decrease with increasing temperature; at high temperatures the molecules readily move away from their neighbors because of increased kinetic energy. (c, d) Evaporation rate and vapor pressure increase with increasing temperature because the kinetic energy of the molecules increase with temperature, and the molecules are more likely to escape into the gas phase.

9.91 (a) Butane (a nonpolar molecule) will dissolve in the nonpolar solvent (tetrachloromethane).

(b) Calcium chloride will dissolve in the polar solvent, water, since the solvation of the ions formed will be aided by water's dipole moment:

9.93 (a) $\left(\dfrac{25.0 \text{ Torr}}{31.83 \text{ Torr}}\right)(100) = 78.5\%$; (b) At 25°C the vapor pressure of water is only 23.76 Torr, so some of the water vapor in the air would condense as dew or fog.

9.95 $\Delta T = k_f m$; also $m = \dfrac{n_{\text{solute}}}{\text{mass}_{\text{solvent(kg)}}}$ and $n_{\text{solute}} = \dfrac{\text{mass}_{\text{solute}}}{M_{\text{solute}}}$

so

$$\Delta T = k_f \dfrac{\text{mass}_{\text{solute}}}{M_{\text{solute}}\left(\text{mass}_{\text{solvent(kg)}}\right)}$$

Solving for M_{solute},

$$M_{\text{solute}} = \dfrac{k_f\left(\text{mass}_{\text{solute}}\right)}{\Delta T_f\left(\text{mass}_{\text{solvent(kg)}}\right)}$$

(a) If mass$_{\text{solute}}$ appears greater, M_{solute} appears greater than actual molar mass, as mass$_{\text{solute}}$ occurs in the numerator above. Also, the ΔT measured will be smaller because less solute will actually be dissolved. This has the same effect as increasing the apparent M_{solute}.

(b) Because the true mass$_{\text{solvent}} = d \times V$, if d_{solvent} is less than $1.00 \text{ g} \cdot \text{cm}^{-3}$, then the true mass$_{\text{solvent}}$ will be less than the assumed mass. M_{solute} is inversely proportional to mass$_{solvent}$, so an artificially high mass$_{solvent}$ will lead to an artificially low M$_{solute}$.

(c) If true freezing point is higher than the recorded freezing point, true $\Delta T <$ assumed ΔT, or assumed $\Delta T >$ true ΔT, and M_{solute} appears less than actual M_{solute}, as ΔT occurs in the denominator.

(d) If not all solute dissolved, the true mass$_{solute} <$ assumed mass$_{solute}$ or assumed mass$_{solute} >$ true mass$_{solute}$, and M_{solute} appears greater than the actual M_{solute}, as mass$_{solute}$ occurs in the numerator.

9.97 The water Coleridge referred to was seawater. The boards shrank due to osmosis (a net movement of water from the cells of the wood to the saline water). You can't drink seawater: osmosis would cause a net flow of water from the cells of the body to the saline-enriched surround solution and the cells would die.

9.99 (a) The 5.22 cm or 52.2 mm rise for an aqueous solution must be converted to Torr or mmHg in order to be expressed into consistent units.

$$52.2 \text{ mm} \left(\frac{0.998 \text{ g} \cdot \text{cm}^{-3}}{13.6 \text{ g} \cdot \text{cm}^{-3}} \right) = 3.83 \text{ mmHg or } 3.83 \text{ Torr}$$

The molar mass can be calculated using the osmotic pressure equation:

$$\Pi = iRT \times \text{molarity}$$

Assume that the protein is a nonelectrolyte with $i = 1$ and that the amount of protein added does not significantly affect the volume of the solution.

$$\Pi = 1 \left(0.082\,06 \text{ L} \cdot \text{atm} \cdot \text{K}^{-1} \cdot \text{mol}^{-1} \right) \left(293 \text{ K} \right) \left(\frac{\left(\frac{0.010 \text{ g}}{M_{protein}} \right)}{0.010 \text{ L}} \right) = \frac{3.83 \text{ Torr}}{760 \text{ Torr} \cdot \text{atm}^{-1}}$$

$$M_{protein} = \frac{1 \left(0.082\,06 \text{ L} \cdot \text{atm} \cdot \text{K}^{-1} \cdot \text{mol}^{-1} \right) \left(293 \text{ K} \right) \left(0.010 \text{ g} \right) \left(760 \text{ Torr} \cdot \text{atm}^{-1} \right)}{\left(0.010 \text{ L} \right) \left(3.83 \text{ Torr} \right)}$$

$$M_{\text{protein}} = 4.8 \times 10^3 \text{ g} \cdot \text{mol}^{-1}$$

(b) The freezing point can be calculated using the relationship $\Delta T_f = ik_f m$:

$$\Delta T_f = 1\left(1.86 \text{ K} \cdot \text{kg} \cdot \text{mol}^{-1}\right) \left(\frac{\left(\dfrac{0.010 \text{ g}}{4.8 \times 10^3 \text{ g} \cdot \text{mol}^{-1}} \right)}{\left(\dfrac{10 \text{ mL} \times 1.00 \text{ g} \cdot \text{mL}^{-1}}{1000 \text{ g} \cdot \text{kg}^{-1}} \right)} \right)$$

$$= 3.9 \times 10^{-4} \text{ K or } 3.9 \times 10^{-4} \, ^\circ\text{C}$$

The freezing point will be $0.00^\circ\text{C} - 3.9 \times 10^{-4}\,^\circ\text{C} = -3.9 \times 10^{-4}\,^\circ\text{C}$.

(c) The freezing point change is so small that it cannot be measured accurately, so osmotic pressure would be the preferred method for measuring the molecular weight.

9.101 $\Delta T = 77.19^\circ\text{C} - 76.54^\circ\text{C} = 0.65^\circ \; 0.65 \text{ K}$

$$\Delta T_b = ik_b m = ik_b \left(\frac{\left(\dfrac{m_{\text{solute}}}{M_{\text{solute}}} \right)}{\text{kg solvent}} \right)$$

$$M_{\text{solute}} = \frac{ik_b m_{\text{solute}}}{(\text{kg solvent})(\Delta T_b)} = \frac{(1)(4.95 \text{ K} \cdot \text{kg} \cdot \text{mol}^{-1})(0.30 \text{ g})}{(0.0300 \text{ kg})(0.65 \text{ K})}$$

$$= 76 \text{ g} \cdot \text{mol}^{-1}$$

9.103 (a) The data in Appendix 2A can be used to calculate the change in enthalpy and entropy for the vaporization of methanol:

$$\text{CH}_3\text{OH(l)} \rightleftharpoons \text{CH}_3\text{OH(g)}$$

$$\Delta H^\circ_{\text{vap}} = \Delta H^\circ_f(\text{CH}_3\text{OH, g}) - \Delta H^\circ_f(\text{CH}_3\text{OH, l})$$
$$= (-200.66 \text{ kJ} \cdot \text{mol}^{-1}) - (-238.86 \text{ kJ} \cdot \text{mol}^{-1})$$
$$= 38.20 \text{ kJ} \cdot \text{mol}^{-1}$$

$$\Delta S^{\circ}{}_{\text{vap}} = S^{\circ}{}_{\text{m}}(CH_3OH(g)) - S^{\circ}{}_{\text{m}}(CH_3OH(l))$$

$$= 239.81 \text{ J} \cdot \text{K}^{-1} \cdot \text{mol}^{-1} - 126.8 \text{ J} \cdot \text{K}^{-1} \cdot \text{mol}^{-1}$$

$$= 113.0 \text{ J} \cdot \text{K}^{-1} \cdot \text{mol}^{-1}$$

To derive the general equation, we start with the expression that $\Delta G^{\circ}{}_{\text{vap}} = -RT \ln P$, where P is the vapor pressure of the solvent. Because $\Delta G^{\circ}{}_{\text{vap}} = \Delta H^{\circ}{}_{\text{vap}} - T\Delta S^{\circ}{}_{\text{vap}}$, this is the relationship to use to determine the temperature dependence of $\ln P$:

$$\Delta H^{\circ}{}_{\text{vap}} - T\Delta S^{\circ}{}_{\text{vap}} = -RT \ln P$$

This equation can be rearranged to give

$$\ln P = -\frac{\Delta H^{\circ}{}_{\text{vap}}}{R} \cdot \frac{1}{T} + \frac{\Delta S^{\circ}{}_{\text{vap}}}{R}$$

To create an equation specific to methanol, we can plug in the actual values of R, $\Delta H^{\circ}{}_{\text{vap}}$, and $\Delta S^{\circ}{}_{\text{vap}}$:

$$\ln P = -\frac{38\,200 \text{ J} \cdot \text{mol}^{-1}}{8.314 \text{ J} \cdot \text{K}^{-1} \cdot \text{mol}^{-1}} \cdot \frac{1}{T} + \frac{113.0 \text{ J} \cdot \text{K}^{-1} \cdot \text{mol}^{-1}}{8.314 \text{ J} \cdot \text{K}^{-1} \cdot \text{mol}^{-1}}$$

$$= -\frac{4595 \text{ K}}{T} + 13.59$$

(b) The relationship to plot is $\ln P$ versus $\dfrac{1}{T}$. This should result in a straight line whose slope is $-\dfrac{\Delta H^{\circ}{}_{\text{vap}}}{R}$ and whose intercept is $\dfrac{\Delta S^{\circ}{}_{\text{vap}}}{R}$. The pressure must be given in atm for this relationship, because atm is the standard state condition.

(c) Because we have already determined the equation, it is easiest to calculate the vapor by inserting the value of 0.0°C or 273.2 K :

$$\ln P = -\frac{4595 \text{ K}}{T} + 13.59 = -\frac{4595 \text{ K}}{273.2 \text{ K}} + 13.59 = -16.82 + 13.59 = -3.23$$

$P = 0.040$ atm or 30. Torr

(d) As in (c) we can use the equation to find the point where the vapor pressure of methanol = 1 atm.

$$\ln P = \ln 1 = 0 = -\frac{4595 \text{ K}}{T} + 13.59$$

$T = 338.1$ K

9.105 (a) $\Delta T_f = 1.72$ K $= ik_f m = \left(1.86 \text{ K} \cdot \text{kg} \cdot \text{mol}^{-1}\right)\left(\dfrac{n_{aa}}{0.95 \text{ kg}}\right)$

where n_{aa} are the moles of acetic acid in solution.
Assuming $i = 1$:

$$n_{aa} = \frac{\left(1.72 \text{ K}\right)\left(0.95 \text{ kg}\right)}{\left(1.86 \text{ K} \cdot \text{kg} \cdot \text{mol}^{-1}\right)} = 0.878 \text{ mol}$$

Experimentally, the molar mass of acetic acid is, therefore:

$$\frac{50 \text{ g}}{0.878 \text{ mol}} = 56.9 \text{ g} \cdot \text{mol}^{-1}$$

This experimental molar mass of acetic acid is less than the
known molecular mass of the acid (60.0 g $\cdot$ mol^{-1}) indicating that
the acid is dissociating in solution giving an $i > 1$.

(b)   As in part (a) the experimental molar mass is first found:

$$\Delta T_f = 2.32 \text{ K} = ik_f m = (5.12 \text{ K} \cdot kg \cdot mol^{-1})\left(\frac{n_{aa}}{0.95 \text{ kg}}\right)$$

where n_{aa} are the moles of acetic acid in solution.

Assuming $i = 1$: $n_{aa} = \dfrac{(2.32 \text{ K})(0.95 \text{ kg})}{(5.12 \text{ K} \cdot \text{kg} \cdot \text{mol}^{-1})} = 0.430$ mol.

Experimentally, the molar mass of acetic acid is, therefore:

$$\frac{50 \text{ g}}{0.430 \text{ mol}} = 116 \text{ g} \cdot \text{mol}^{-1}.$$

This experimental molar mass of acetic acid is significantly higher than
the known molar mass of the acid (60.0 g·mol^{-1}) indicating that the acid is
dissociating in solution giving an $i < 1$. A van't Hoff factor less than 1
indicates that acetic acid is *dimerizing* in the benzene, or acetic acid
molecules are aggregating together in solution.

9.107 (a) The plot of the data is shown below. On this plot, the slope $= -\dfrac{\Delta H°_{vap}}{R}$

and the intercept $= \dfrac{\Delta S°_{vap}}{R}$.

Temp. (K)	T^{-1} (K^{-1})	Vapor pressure (Torr)	V.P. (atm)	ln P
190	0.005 26	3.2	0.0042	−5.47
228	0.004 38	68	0.089	−2.41
250	0.004 00	240	0.316	−1.15
273	0.003 66	672	0.884	−0.123

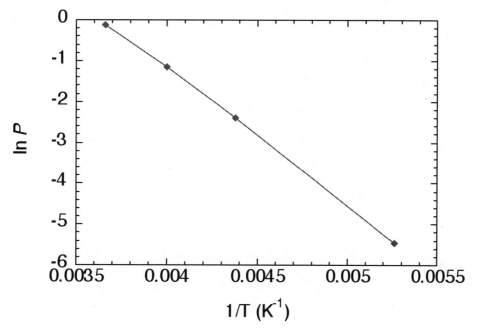

From the curve fitting program:

$y = -3358.714x + 12.247$

(b) $-\dfrac{\Delta H°_{vap}}{R} = -3359$

$\Delta H°_{vap} = (3359)(8.314 \text{ J} \cdot \text{K}^{-1} \cdot \text{mol}^{-1}) = 28 \text{ kJ} \cdot \text{mol}^{-1}$

(c) $\dfrac{\Delta S°_{vap}}{R} = 12.25$

$\Delta S°_{vap} = (12.25)(8.314 \text{ J} \cdot \text{K}^{-1} \cdot \text{mol}^{-1}) = 1.0 \times 10^2 \text{ J} \cdot \text{K}^{-1} \cdot \text{mol}^{-1}$

(d) The normal boiling point will be the temperature at which the vapor pressure = 1 atm or at which the ln 1 = 0. This will occur when $T^{-1} = 0.0036$ or $T = 2.7 \times 10^2$ K.

(e) This is most easily done by using an equation derived from $\Delta H°_{vap}$ and $\Delta S°_{vap}$:

$$\ln P = -\frac{\Delta H°_{vap}}{R} \cdot \frac{1}{T} + \frac{\Delta S°_{vap}}{R}$$

$$\ln \frac{15 \text{ Torr}}{760 \text{ Torr}} = -\frac{28\,000 \text{ J} \cdot \text{mol}^{-1}}{8.314 \text{ J} \cdot \text{K}^{-1} \cdot \text{mol}^{-1}} \cdot \frac{1}{T} + \frac{100 \text{ J} \cdot \text{K}^{-1} \cdot \text{mol}^{-1}}{8.314 \text{ J} \cdot \text{K}^{-1} \cdot \text{mol}^{-1}}$$

$$T = 2.1 \times 10^2 \text{ K}$$

9.109 The critical temperatures are

Compound T_C (°C)

Compound	T_C (°C)
CH_4	−82.1
C_2H_6	32.2
C_3H_8	96.8
C_4H_{10}	152

The critical temperatures increase with increasing mass, showing the influence of the stronger London forces.

9.111 (a) If sufficient chloroform and acetone are available, the pressures in the flasks will be the equilibrium vapor pressures at that temperature. We can calculate these amounts using the ideal gas equation:

$$\left(\frac{195 \text{ Torr}}{760 \text{ Torr} \cdot \text{atm}^{-1}} \right)(1.00 \text{ L}) =$$

$$\left(\frac{m_{chloroform}}{119.37 \text{ g} \cdot \text{mol}^{-1} \text{ chloroform}} \right)(0.08206 \text{ L} \cdot \text{atm} \cdot \text{K}^{-1} \cdot \text{mol}^{-1})(298 \text{ K})$$

$$m_{chloroform} = 1.25 \text{ g}$$

$$\left(\frac{225 \text{ Torr}}{760 \text{ Torr} \cdot \text{atm}^{-1}} \right)(1.00 \text{ L}) =$$

$$\left(\frac{m_{acetone}}{58.08 \text{ g} \cdot \text{mol}^{-1} \text{ acetone}} \right)(0.08206 \text{ L} \cdot \text{atm} \cdot \text{K}^{-1} \cdot \text{mol}^{-1})(298 \text{ K})$$

$m_{acetone} = 0.703 \text{ g}$

In both cases, sufficient compound is available to achieve the vapor pressure; flask A will have a pressure of 195 Torr and flask B will have a pressure of 222 Torr.

(b) When the stopcock is opened, some chloroform will move into flask B and acetone will move into flask A to restore the equilibrium vapor pressure. Additionally, however, some acetone vapor will dissolve in the liquid chloroform and vice versa. Ultimately the system will reach an equilibrium state in which the compositions of the liquid phases in both flasks are the same and the gas phase composition is uniform. The gas phase and liquid phase compositions will be established by Raoult's law. It is conceptually most convenient for this calculation to start by putting all the material into one liquid phase. Such a solution would have the following composition:

$$\chi_{acetone} = \frac{\dfrac{35.0 \text{ g}}{58.08 \text{ g} \cdot \text{mol}^{-1} \text{ acetone}}}{\left(\dfrac{35.0 \text{ g}}{58.08 \text{ g} \cdot \text{mol}^{-1} \text{ acetone}} \right) + \left(\dfrac{35.0 \text{ g}}{119.37 \text{ g} \cdot \text{mol}^{-1} \text{ chloroform}} \right)}$$

$\chi_{acetone} = 0.67$

$$\chi_{chloroform} = \frac{\dfrac{35.0 \text{ g}}{119.37 \text{ g} \cdot \text{mol}^{-1} \text{ chloroform}}}{\left(\dfrac{35.0 \text{ g}}{58.08 \text{ g} \cdot \text{mol}^{-1} \text{ acetone}} \right) + \left(\dfrac{35.0 \text{ g}}{119.37 \text{ g} \cdot \text{mol}^{-1} \text{ chloroform}} \right)} \text{ or } 1 - \chi_{acetone}$$

$\chi_{chloroform} = 0.33$

This gives the composition of the liquid phase. The composition of the gas phase will be determined from the pressures of the gases:

$$P_{\text{acetone, gas}} = \chi_{\text{acetone, liquid}} \cdot P^\circ_{\text{acetone}} = (0.67)(222 \text{ Torr}) = 149 \text{ Torr}$$

$$P_{\text{chloroform, gas}} = \chi_{\text{chloroform, liquid}} \cdot P^\circ_{\text{chloroform}} = (0.33)(195 \text{ Torr}) = 64 \text{ Torr}$$

$$\chi_{\text{acetone, gas}} = \frac{P_{\text{acetone}}}{P_{\text{acetone}} + P_{\text{chloroform}}}$$

$$\chi_{\text{acetone, gas}} = \frac{149 \text{ Torr}}{149 \text{ Torr} + 64 \text{ Torr}} = 0.70$$

$$\chi_{\text{chloroform, gas}} = 1 - \chi_{\text{acetone, gas}} = 0.30$$

The gas phase composition will, therefore, be slightly richer in acetone than in chloroform. The total pressure in the flask will be 213 Torr.

(c) The solution shows negative deviation from Raoult's law. This means that the molecules of acetone and chloroform attract each other slightly more than molecules of the same kind. Under such circumstances, the vapor pressure is lower than expected from the ideal calculation. This will give rise to a high-boiling azeotrope. The gas phase composition will also be slightly different from that calculated from the ideal state, but whether acetone or chloroform would be richer in the gas phase depends on which side of the azeotrope the composition lies. Because we are not given the composition of the azeotrope, we cannot state which way the values will vary.

9.113 The vapor pressure is more sensitive if ΔH_{vap} is small. The fact that ΔH_{vap} is small indicates that it takes little energy to volatilize the sample, which means that the intermolecular forces are weaker. Hence we expect the ratio P_2/P_1 in the Clausius-Clapeyron equation to be larger.

9.115 The initial information tells us that the detector is more responsive to A than to B. Under these conditions, the response is

$5.44 \text{ cm}^2 \div 0.52 \text{ mg A} = 11 \text{ cm}^2 \cdot \text{mg}^{-1}$ A whereas the response for B is

$8.72 \text{ cm}^2 \div 2.30 \text{ mg B} = 3.79 \text{ cm}^2 \cdot \text{mg}^{-1}$ B. The detector is thus

$11 \text{ cm}^2 \cdot \text{mg}^{-1} \div 3.79 \text{ cm}^2 \cdot \text{mg}^{-1} = 2.9$ times more sensitive for A than B.

Because the conditions are different for determining the unknown amount of A present, we cannot use the area ratios directly to determine the quantity of A. Instead, we use the standard B as a reference.

(X = mg A)

$$\frac{3.52 \text{ cm}^2}{X} \div \frac{7.58 \text{ cm}^2}{2.00 \text{ mg (B)}} = 2.9$$

$$X = \frac{3.52 \text{ cm}^2}{2.9} \left(\frac{2.00 \text{ mg}}{7.58 \text{ cm}^2} \right)$$

$$X = 0.32 \text{ mg A}$$

9.117

$$\Pi = iRT \left(\text{molarity} \right)$$

Assuming $i = 1$

$$7.7 \text{ atm} = \left(0.0821 \text{ L} \cdot \text{atm} \cdot \text{K}^{-1} \cdot \text{mol}^{-1} \right) \left(310 \text{ K} \right) (M)$$

$$M = 0.303 \text{ mol} \cdot \text{L}^{-1}$$

If 0.500 L of solution is needed:

$$\left(0.303 \text{ mol} \cdot \text{L}^{-1} \right) \left(0.500 \text{ L} \right) = 0.151 \text{ mol of glucose}$$

Molecular mass of glucose is $81.158 \text{ g} \cdot \text{mol}^{-1}$

$$\left(0.151 \text{ mol} \right) \left(81.158 \text{ g} \cdot \text{mol}^{-1} \right) = 27.3 \text{ g}$$

9.119 First, calculate the weight % of acetic acid in a solution with $x_{\text{acetic acid}} = 0.15$:

If you have 1 mol total of a solution with $x_{\text{acetic acid}} = 0.15$ and $x_{\text{H}_2\text{O}} = 0.85$, you would have 0.15 mol of acetic acid ($60.053 \text{ g} \cdot \text{mol}^{-1}$) in 0.85 mol of H_2O ($18.015 \text{ g} \cdot \text{mol}^{-1}$). The weight % of acetic acid in such a solution is:

$$\left(0.15 \text{ mol} \right) (60.053 \text{ g} \cdot \text{mol}^{-1}) = 9.01 \text{ g of acetic acid}$$

$$\left(0.85 \text{ mol} \right) (18.015 \text{ g} \cdot \text{mol}^{-1}) = 15.3 \text{ g of H}_2\text{O}$$

$$\frac{9.01 \text{ g}}{9.01 \text{ g} + 15.3 \text{ g}} \left(100\% \right) = 37.0\% \text{ acetic acid by weight}$$

If a solution with $x_{\text{acetic acid}} = 0.15$ is 37.0% acetic acid by weight, 30 g of such a solution will contain:

$$\left(30 \text{ g}\right)\left(0.370\right) = 11.1 \text{ g acetic acid}$$

$$\frac{11.1 \text{ g}}{60.053 \text{ g} \cdot \text{mol}^{-1}} = 0.185 \text{ mol of acetic acid}$$

given that acetic acid is a monoprotic acid which will react in a 1:1 fashion with NaOH, 0.185 mol of NaOH will completely react with 0.185 mol of acetic acid. The volume of a 0.010 M solution of NaOH needed to completely consume the acetic acid is: $\dfrac{0.185 \text{ mol}}{0.010 \text{ mol} \cdot \text{L}^{-1}} = 18.5 \text{ L}$

9.121 The non-polar chains of both the surfactant and pentanol will interact to form a hydrophobic region with the heads of the two molecules pointing away from this region towards the aqueous solution. To prevent the heads of the shorter pentanol molecules from winding up in the hydrophobic region, the layered structure might be comprised of a water region, a surfactant layer (heads pointing toward the water) a pentanol layer (with tails pointing toward the hydrophobic tails of the surfactant) and back to an aqueous region.

9.123 (a) The hybridization of the carbon atoms in citric acid are:

(b) yes; all four —OH groups present can participate in hydrogen bonding; the remaining three oxygens can accept H-bonds.

(c) since citric acid can hydrogen bond with itself it should have large intermolecular forces and should be a solid. The ability to both donate and accept hydrogen bonds should also make it soluble in water.

(d) 100.0 g of a 0.9% NaCl solution will contain 0.9 g of NaCl (MM = 58.44 g/mol), which is 1.54×10^{-2} mol of NaCl. Assuming that NaCl completely dissociates, the moles of solute present is two times this amount or 3.08×10^{-2} mol of solute in 100.0 mL. Therefore, the concentration of normal saline will be 0.3 M.

(e) To make a 500.0 mL sports drink isotonic, you need it to have a total molarity of 0.3 M (from part d). This means that the amount of solute needed is: 0.3 $M \times 0.5000$ L = 0.15 mol total solute. Addition of 1.0 g of NaCl to the 500 mL sports drink will account for 0.034 mol of solute (twice the moles of NaCl added), leaving 0.13 mol of glucose being needed to achieve isotonicity. Since glucose has a molar mass of 180 g/mol, this means that in addition to the 1.0 g of NaCl, 2×10^{1} g of glucose needs to be added to the 500.0 mL sports drink.

(f) 300.0 mL of a 1.00% boric acid = 300.0 g of solution (assuming the density of the solution is 1.00 g cm^{-3}), meaning there are 3.00 g of B(OH)$_3$ or 4.85×10^{-2} mol of B(OH)$_3$ in the solution (since $MM_{B(OH)_3} = 61.81$ g·mol^{-1}). To be isotonic, 300.0 mL should have 0.3 M $\times$ 0.3000 L = 0.09 mol total solute. Subtracting out the amount of boric acid present means that 0.04 mol of solute needed to be added to achieve isotonicity; since 1 mol of NaCl provides 2 mol of solute, 0.02 mol, or 1 g, of NaCl must be added to the boric acid solution.

CHAPTER 10
CHEMICAL EQUILIBRIA

10.1 (a) False. Equilibrium is dynamic. At equilibrium, the concentrations of reactants and products will not change, but the reaction will continue to proceed in both directions.

(b) False. Equilibrium reactions are affected by the presence of both products and reactants.

(c) False. The value of the equilibrium constant is not affected by the amounts of reactants or products added as long as the temperature is constant.

(d) True.

10.3

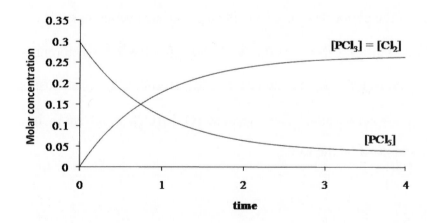

10.5 (a) Flask 3 represents the point of reaction equilibrium.

(b) % decomposition = $(6/11) \times 100 = 54.5\%$

(c) Decomposition: $X_2 \ \xrightleftharpoons{\hspace{1em}} \ 2X$

$$K = \frac{(P_X)^2}{P_{X_2}} = \frac{[(\text{mole fraction of } P_X) P_i]^2}{(\text{mole fraction of } P_{X_2}) P_i} = \frac{[\left(\frac{12}{17}\right) \times 0.1]^2}{\left(\frac{5}{17}\right) \times 0.1} = 0.16$$

10.7 (a) $CH_4(g) + 2\,O_2(g) \leftrightarrow CO_2(g) + 2\,H_2O(g)$

$$K_C = \frac{[CO_2][H_2O]^2}{[CH_4][O_2]^2}$$

(b) $I_2(g) + 5\,F_2(g) \leftrightarrow 2\,IF_5(g)$

$$K_C = \frac{[IF_5]^2}{[I_2][F_2]^5}$$

(c) $2\,NO_2(g) + F_2(g) \leftrightarrow 2\,FNO_2(g)$

$$K_C = \frac{[FNO_2]^2}{[F_2][NO_2]^2}$$

10.9 (a) Because the volume is the same, the number of moles of O_2 is larger in the second experiment. (b) Because K_C is a constant and the denominator is larger in the second case, the numerator must also be larger; so the concentration of O_2 is larger in the second case.

(c) Although $[O_2]^3/[O_3]^2$ is the same, $[O_2]/[O_3]$ will be different, a result seen by solving for K_C in each case. (d) Because K_C is a constant, $[O_2]^3/[O_3]^2$ is the same. (e) Because $[O_2]^3/[O_3]^2$ is the same, its reciprocal must be the same.

10.11 $K = \dfrac{P_{HI}^2}{P_{H_2} P_{I_2}}$

for condition 1, $K = \dfrac{(0.0137)^2}{(6.47 \times 10^{-3})(0.594 \times 10^{-3})} = 48.8$

for condition 2, $K = \dfrac{(0.0169)^2}{(3.84 \times 10^{-3})(1.52 \times 10^{-3})} = 48.9$

for condition 1, $K = \dfrac{(0.0100)^2}{(1.43 \times 10^{-3})(1.43 \times 10^{-3})} = 48.9$

10.13 (a) $\dfrac{1}{P_{BCl_3}{}^2}$

(b) $[H_3PO_4]^4 [H_2S]^{10}$

(c) $\dfrac{P_{BrF_3}{}^2}{P_{Br_2} P_{F_2}{}^3}$

10.15 The free energy at a specific set of conditions is given by

$\Delta G_r = \Delta G°_r + RT \ln Q$

$\Delta G_r = -RT \ln K + RT \ln Q$

$\Delta G_r = -RT \ln K + RT \ln \dfrac{[I]^2}{[I_2]}$

$\quad = -(8.314 \text{ J} \cdot \text{K}^{-1} \cdot \text{mol}^{-1})(1200 \text{ K}) \ln (6.8)$

$\qquad\qquad + (8.314 \text{ J} \cdot \text{K}^{-1} \cdot \text{mol}^{-1})(1200 \text{ K}) \ln \dfrac{(0.98)^2}{(0.13)}$

$\quad = 8.3 \times 10^{-1} \text{ kJ} \cdot \text{mol}^{-1}$

Because ΔG_r is positive, the reaction will be spontaneous to produce I_2.

10.17 The free energy at a specific set of conditions is given by

$\Delta G_r = \Delta G°_r + RT \ln Q$

$\Delta G_r = -RT \ln K + RT \ln Q$

$\Delta G_r = -RT \ln K + RT \ln \dfrac{[NH_3]^2}{[N_2][H_2]^3}$

$\quad = -(8.314 \text{ J} \cdot \text{K}^{-1} \cdot \text{mol}^{-1})(400 \text{ K}) \ln (41)$

$\qquad\qquad + (8.314 \text{ J} \cdot \text{K}^{-1} \cdot \text{mol}^{-1})(400 \text{ K}) \ln \dfrac{(21)^2}{(4.2)(1.8)^3}$

$\quad = -27 \text{ kJ} \cdot \text{mol}^{-1}.$

Because ΔG_r is negative, the reaction will proceed to form products.

10.19 To answer these questions, we will first calculate ΔG° for each reaction and then use that value in the expression $\Delta G^\circ = -RT \ln K$.

(a) $2\,H_2(g) + O_2(g) \longrightarrow 2\,H_2O(g)$

$\Delta G^\circ_r = 2(\Delta G^\circ_f(H_2O, g) = 2(-228.57\,\text{kJ}\cdot\text{mol}^{-1}) = -457.14\,\text{kJ}$

$\Delta G^\circ_r = -RT \ln K$ or

$$\ln K = -\frac{\Delta G^\circ_p}{RT}$$

$$\ln K = -\frac{-457\,140\,\text{J}}{(8.314\,\text{J}\cdot\text{K}^{-1}\cdot\text{mol}^{-1})(298.15\,\text{K})} = +184.42$$

$K = 1 \times 10^{80}$

(b) $2\,CO(g) + O_2(g) \longrightarrow 2\,CO_2(g)$

$\Delta G^\circ_r = 2 \times \Delta G^\circ_f(CO_2, g) - [2 \times \Delta G^\circ_f(CO, g)]$

$\qquad = 2(-394.36\,\text{kJ}\cdot\text{mol}^{-1}) - [2(-137.17\,\text{kJ}\cdot\text{mol}^{-1})]$

$\qquad = -514.38\,\text{kJ}$

$$\ln K = -\frac{-514\,380\,\text{J}}{(8.314\,\text{J}\cdot\text{K}^{-1}\cdot\text{mol}^{-1})(298.15\,\text{K})} = +207.5$$

$K = 1 \times 10^{90}$

(c) $CaCO_3(s) \longrightarrow CaO(s) + CO_2(g)$

$\Delta G^\circ_r = \Delta G^\circ_f(CaO, s) + \Delta G^\circ_f(CO_2, g) - [\Delta G^\circ_f(CaCO_3(s)]$

$\qquad = (-604.03\,\text{kJ}\cdot\text{mol}^{-1}) + (-394.36\,\text{kJ}\cdot\text{mol}^{-1}) - [-1128.8\,\text{kJ}\cdot\text{mol}^{-1}]$

$\qquad = +130.4\,\text{kJ}$

$$\ln K = -\frac{+130\,400\,\text{J}}{(8.314\,\text{J}\cdot\text{K}^{-1}\cdot\text{mol}^{-1})(298\,\text{K})} = -52.6$$

$K = 1 \times 10^{-23}$

10.21 (a) $\Delta G^\circ_r = -RT \ln K$

$\qquad\qquad = -(8.314\,\text{J}\cdot\text{K}^{-1}\cdot\text{mol}^{-1})(1200\,\text{K})\ln 6.8 = -19\,\text{kJ}\cdot\text{mol}^{-1}$

(b) $\Delta G^\circ_r = -RT \ln K = -(8.314\,\text{J}\cdot\text{K}^{-1}\cdot\text{mol}^{-1})(298\,\text{K})\ln 1.1 \times 10^{-12}$

$\qquad\qquad = +68\,\text{kJ}\cdot\text{mol}^{-1}$

10.23 First we must calculate K for the reaction, which can be done using data from Appendix 2A:

$$\Delta G^\circ_r = 2 \times \Delta G^\circ_f(NO, g) = 2 \times 86.55 \, kJ \cdot mol^{-1} = 173.1 \, kJ \cdot mol^{-1}$$

$$\Delta G^\circ = -RT \ln K$$

$$\ln K = -\frac{\Delta G^\circ}{RT}$$

$$\ln K = -\frac{+173\,100 \, J \cdot mol^{-1}}{(8.314 \, J \cdot K^{-1} \cdot mol^{-1})(298 \, K)} = -69.9$$

$$K = 4 \times 10^{-31}$$

Because $Q > K$, the reaction will tend to proceed to produce reactants.

10.25 $2 \, BrCl(g) = Br_2(g) + Cl_2(g)$ $K = 377$ (1)

$H_2(g) + Cl_2(g) = 2 \, HCl(g)$ $K = 4.0 \times 10^{31}$ (2)

(1) + (2): $2 \, BrCl(g) + H_2(g) = Br_2(g) + 2 \, HCl(g)$

K for this reaction $= K_1 \times K_2 = 377 \times 4.0 \times 10^{31} = 1.5 \times 10^{34}$.

10.27 $H_2(g) + I_2(g) \rightarrow 2 \, HI(g)$ $K_C = 160$

$$K_C = \frac{[HI]^2}{[H_2][I_2]}$$

$$160 = \frac{(2.21 \times 10^{-3})^2}{[H_2](1.46 \times 10^{-3})}$$

$$[H_2] = \frac{(2.21 \times 10^{-3})^2}{(160)(1.46 \times 10^{-3})}$$

$$[H_2] = 2.1 \times 10^{-5} \, mol \cdot L^{-1}$$

10.29 $K = \dfrac{P_{PCl_3} P_{Cl_2}}{P_{PCl_5}}$

$$25 = \frac{P_{PCl_3}(5.43)}{1.18}$$

$$P_{PCl_3} = \frac{(25)(1.18)}{5.43} = 5.4$$

$$P_{PCl_3} = 5.4 \text{ bar}$$

10.31 (a) $K = \dfrac{p_{HI}^2}{p_{H_2} \cdot p_{I_2}} = 160$

$$Q = \frac{(0.10)^2}{(0.20)(0.10)} = 0.50$$

(b) $Q \neq K$, $\therefore$ the system is not at equilibrium.

(c) Because $Q < K$, more products will be formed.

10.33 (a) $K_C = \dfrac{[SO_3]^2}{[SO_2]^2[O_2]} = 1.7 \times 10^6$

$$Q_C = \frac{\left[\dfrac{1.0 \times 10^{-4}}{0.500}\right]^2}{\left[\dfrac{1.20 \times 10^{-3}}{0.500}\right]^2 \left[\dfrac{5.0 \times 10^{-4}}{0.500}\right]} = 6.9$$

(b) Because $Q_C < K_C$, more products will tend to form, which will result in the formation of more SO_3.

10.35 (a) $K = \dfrac{P_{NO}^2 P_{Cl_2}}{P_{NOCl}^2} = 1.8 \times 10^{-2}$

$$K = \left(\frac{T}{12.027 \text{ K}}\right)^{\Delta n} K_C$$

$$K_C = \left(\frac{12.027 \text{ K}}{T}\right)^{\Delta n} K = \left(\frac{12.027 \text{ K}}{500 \text{ K}}\right)^{(3-2)} 1.8 \times 10^{-2} = 4.3 \times 10^{-4}$$

(b) $K = P_{CO_2} = 167$

$$K_C = \left(\frac{12.027 \text{ K}}{T}\right)^{\Delta n} \quad K = \left(\frac{12.027 \text{ K}}{1073 \text{ K}}\right)^{(1)} \quad 167 = 1.87$$

10.37 For the reaction written as $N_2(g) + 3 H_2(g) \rightarrow 2 NH_3(g)$ Eq. 1,

$$K = \frac{P_{NH_3}^{\ 2}}{P_{N_2} P_{H_2}^{\ 3}} = 41.$$

(a) For the reaction written as $2 NH_3(g) \rightarrow N_2(g) + 3 H_2(g)$ Eq. 2,

$$K = \frac{P_{N_2} P_{H_2}^{\ 3}}{P_{NH_3}^{\ 2}}.$$

This is $\dfrac{1}{K_{Eq.\ 1}} = \dfrac{1}{41} = 0.024.$

(b) For the reaction written as $\frac{1}{2} N_2(g) + \frac{3}{2} H_2(g) \rightarrow NH_3(g)$ Eq. 3,

$$K_{Eq.3} = \frac{P_{NH_3}}{P_{N_2}^{\ 1/2} P_{H_2}^{\ 3/2}} = \sqrt{K_{Eq.1}} = \sqrt{41} = 6.4.$$

Note that Eq. 3 = $\frac{1}{2}$ Eq. 1 and thus $K_{Eq.3} = K_{Eq.1}^{\ 1/2}$.

(c) For the reaction written as $2 N_2(g) + 6 H_2(g) \rightarrow 4 NH_3(g)$ Eq. 4,

$$K_{Eq.4} = \frac{P_{NH_3}^{\ 4}}{P_{N_2}^{\ 2} P_{H_2}^{\ 6}} = K_{Eq.1}^{\ 2} = 41^2 = 1.7 \times 10^3.$$

Note that Eq. 4 = 2 Eq. 1 and thus $K_{Eq.3} = K_{Eq.1}^{\ 2}$.

10.39 $\dfrac{1.90 \text{ g HI}}{127.91 \text{ g} \cdot \text{mol}^{-1}} = 0.0148$ mol HI

$2 HI(g) \longrightarrow H_2(g) + I_2$

$0.0172 \text{ mol} - 2x \quad x \quad x$

$$0.0148 \text{ mol} = 0.0172 \text{ mol} - 2x$$

$$x = 0.0012 \text{ mol}$$

$$K_C = \frac{\left[\dfrac{0.0012}{2.00}\right]\left[\dfrac{0.0012}{2.00}\right]}{\left[\dfrac{0.0148}{2.00}\right]^2} = \frac{\left[\dfrac{0.0012}{2.00}\right]^2}{\left[\dfrac{0.0148}{2.00}\right]^2} = 0.0066 \text{ or } 6.6 \times 10^{-3}$$

$$K = (RT)^{\Delta n} K_c = K_c \quad (\text{since } \Delta n = 0)$$

10.41 $\dfrac{25.0 \text{ g } NH_4(NH_2CO_2)}{78.07 \text{ g} \cdot \text{mol}^{-1} \; NH_4(NH_2CO_2)} = 0.320 \text{ mol } NH_4(NH_2CO_2)$

$\dfrac{0.0174 \text{ g } CO_2}{44.01 \text{ g} \cdot \text{mol}^{-1} CO_2} = 3.95 \times 10^{-4} \text{ mol } CO_2$

2 mol NH_3 are formed per mol of CO_2, so mol $NH_3 = 2 \times 3.95 \times 10^{-4}$

$\qquad = 7.90 \times 10^{-4}.$

$$K_C = [NH_3]^2[CO_2] = \left(\frac{7.90 \times 10^{-4}}{0.250}\right)^2 \left(\frac{3.95 \times 10^{-4}}{0.250}\right) = 1.58 \times 10^{-8}$$

10.43 (c) $K = x^2/(1.0 - 2x)^2$

10.45 (a) The balanced equation is $Cl_2(g) \rightarrow 2\,Cl(g)$.

The initial concentration of $Cl_2(g)$ is $\dfrac{0.0020 \text{ mol } Cl_2}{2.0 \text{ L}} = 0.0010 \text{ mol} \cdot L^{-1}.$

Concentration $(\text{mol} \cdot L^{-1})$	$Cl_2(g)$	$\rightarrow$	$2\,Cl(g)$
initial	0.0010		0
change	$-x$		$+2x$
equilibrium	$0.0010 - x$		$+2x$

$$K_C = \frac{[\text{Cl}]^2}{[\text{Cl}_2]} = \frac{(2x)^2}{(0.0010 - x)} = 1.2 \times 10^{-7}$$

$$4x^2 = (1.2 \times 10^{-7})(0.0010 - x)$$

$$4x^2 + (1.2 \times 10^{-7})x - (1.2 \times 10^{-10}) = 0$$

$$x = \frac{-(1.2 \times 10^{-7}) \pm \sqrt{(1.2 \times 10^{-7})^2 - 4(4)(-1.2 \times 10^{-10})}}{2 \times 4}$$

$$x = \frac{-(1.2 \times 10^{-7}) \pm 4.4 \times 10^{-5}}{8}$$

$$x = -5.5 \times 10^{-6} \text{ or } +5.5 \times 10^{-6}$$

The negative answer is not meaningful, so we choose

$x = +5.5 \times 10^{-6} \text{ mol} \cdot \text{L}^{-1}$. The concentration of Cl_2 is essentially

unchanged because $0.0010 - 5.5 \times 10^{-6} \cong 0.0010$. The concentration of

Cl atoms is $2 \times (5.5 \times 10^{-6}) = 1.1 \times 10^{-5} \text{ mol} \cdot \text{L}^{-1}$.

(b) The balanced equation is: $\text{F}_2(g) \rightarrow 2\,\text{F}(g)$.

The problem is worked in an identical fashion to (a) but the equilibrium

constant is now 1.2×10^{-4}.

The initial concentration of $\text{F}_2(g)$ is $\dfrac{0.0020 \text{ mol } \text{F}_2}{2.0 \text{ L}} = 0.0010 \text{ mol} \cdot \text{L}^{-1}$.

Concentration $(\text{mol} \cdot \text{L}^{-1})$	$\text{F}_2(g)$	$\rightarrow$	$2\,\text{F}(g)$
initial	0.0010		0
change	$-x$		$+2x$
equilibrium	$0.0010 - x$		$+2x$

$$K_C = \frac{[\text{F}]^2}{[\text{F}_2]} = \frac{(2x)^2}{(0.0010 - x)} = 1.2 \times 10^{-4}$$

$$4x^2 = (1.2 \times 10^{-4})(0.0010 - x)$$

$$4x^2 + (1.2 \times 10^{-4})x - (1.2 \times 10^{-7}) = 0$$

$$x = \frac{-(1.2 \times 10^{-4}) \pm \sqrt{(1.2 \times 10^{-4})^2 - 4(4)(-1.2 \times 10^{-7})}}{2.4}$$

$$x = \frac{-(1.2 \times 10^{-4}) \pm 1.4 \times 10^{-3}}{8}$$

$$x = -1.9 \times 10^{-4} \text{ or } +1.6 \times 10^{-4}$$

The negative answer is not meaningful, so we choose

$x = +1.6 \times 10^{-4} \text{ mol} \cdot \text{L}^{-1}$. The concentration of F_2 is

$0.0010 - 1.6 \times 10^{-4} = 8 \times 10^{-4} \text{ mol} \cdot \text{L}^{-1}$. The concentration of F atoms is

$2 \times (1.6 \times 10^{-4}) = 3.2 \times 10^{-4} \text{ mol} \cdot \text{L}^{-1}$.

(c) Cl_2 is more stable. This can be seen even without the aid of the calculation from the larger equilibrium constant for the dissociation for F_2 compared to Cl_2.

10.47 (a) Concentration of PCl_5 initially

$$= \frac{\left(\dfrac{1.0 \text{ g PCl}_5}{208.22 \text{ g} \cdot \text{mol}^{-1} \text{ PCl}_5}\right)}{0.250 \text{ L}} = 0.019 \text{ mol} \cdot \text{L}^{-1}.$$

Concentration $(\text{mol} \cdot \text{L}^{-1})$	$PCl_5(g)$	$\rightarrow$	$PCl_3(g)$	$+$	Cl_2
initial	0.019		0		0
change	$-x$		$+x$		$+x$
final	$0.019 - x$		$+x$		$+x$

$$K_C = \frac{[PCl_2][Cl_2]}{[PCl_5]} = \frac{(x)(x)}{(0.019 - x)} = \frac{x^2}{(0.019 - x)}$$

$$\frac{x^2}{(0.019 - x)} = 1.1 \times 10^{-2}$$

$$x^2 = (1.1 \times 10^{-2})(0.019 - x)$$

$$x^2 + (1.1 \times 10^{-2})x - 2.1 \times 10^{-4} = 0$$

$$x = \frac{-(1.1 \times 10^{-2}) \pm \sqrt{(1.1 \times 10^{-2})^2 - (4)(1)(-2.1 \times 10^{-4})}}{2.1}$$

$$x = \frac{-(1.1 \times 10^{-2}) \pm 0.031}{2.1} = +0.010 \text{ or } -0.021$$

The negative root is not meaningful, so we choose $x = 0.010 \text{ mol} \cdot \text{L}^{-1}$.

$[PCl_3] = [Cl_2] = 0.010 \text{ mol} \cdot \text{L}^{-1}$; $[PCl_5] = 0.009 \text{ mol} \cdot \text{L}^{-1}$.

(b) The percentage decomposition is given by

$$\frac{0.010}{0.019} \times 100 = 53\%.$$

10.49 Starting concentration of $NH_3 = \dfrac{0.400 \text{ mol}}{2.00 \text{ L}} = 0.200 \text{ mol} \cdot L^{-1}$.

Concentration (mol·L^{-1})	NH$_4$HS(s) $\rightarrow$	NH$_3$(g)	+	H$_2$S(g)
initial	—	0.200		0
change	—	+x		+x
final	—	0.200 + x		+x

$$K_C = [NH_3][H_2S] = (0.200 + x)(x)$$
$$1.6 \times 10^{-4} = (0.200 + x)(x)$$
$$x^2 + 0.200x - 1.6 \times 10^{-4} = 0$$
$$x = \frac{-(+0.200) \pm \sqrt{(+0.200)^2 - (4)(1)(-1.6 \times 10^{-4})}}{2.1}$$
$$x = \frac{-0.200 \pm 0.2016}{2.1} = +0.0008 \text{ or } -0.2008$$

The negative root is not meaningful, so we choose $x = 8 \times 10^{-4} \text{ mol} \cdot L^{-1}$

(note that in order to get this number we have had to ignore our normal

significant figure conventions).

$$[NH_3] = +0.200 \text{ mol} \cdot L^{-1} + 8 \times 10^{-4} \text{ mol} \cdot L^{-1} = 0.200 \text{ mol} \cdot L^{-1}$$

$$[H_2S] = 8 \times 10^{-4} \text{ mol} \cdot L^{-1}$$

Alternatively, we could have assumed that

$x \ll 0.2$, the $0.200x = 1.6 \times 10^{-4}$, $x = 8.0 \times 10^{-4}$.

10.51 The initial concentrations of N_2 and O_2 are equal at $0.114 \text{ mol} \cdot L^{-1}$

because the vessel has a volume of 1.00 L.

Concentrations (mol·L^{-1})	N$_2$(g)	+	O$_2$(g)	$\rightarrow$	2 NO(g)
initial	0.114		0.114		0
change	−x		−x		+2x

final $\qquad$ $0.114 - x$ $\qquad$ $0.114 - x$ $\qquad$ $+2x$

$$K_C = \frac{[NO]^2}{[N_2][O_2]} = \frac{(2x)^2}{(0.114 - x)(0.114 - x)} = \frac{(2x)^2}{(0.114 - x)^2}$$

$$1.00 \times 10^{-5} = \frac{(2x)^2}{(0.114 - x)^2}$$

$$\sqrt{1.00 \times 10^{-5}} = \sqrt{\frac{(2x)^2}{(0.114 - x)^2}}$$

$$3.16 \times 10^{-3} = \frac{(2x)}{(0.114 - x)}$$

$$2x = (3.16 \times 10^{-3})(0.114 - x)$$

$$2.003\,16\,x = 3.60 \times 10^{-4}$$

$$x = 1.8 \times 10^{-4}$$

$[NO] = 2x = 2 \times 1.8 \times 10^{-4} = 3.6 \times 10^{-4}\ mol \cdot L^{-1}$; the concentrations of N_2

and O_2 remain essentially unchanged at $0.114\ mol \cdot L^{-1}$.

10.53 The initial concentrations of H_2 and I_2 are

$$[H_2] = \frac{0.400\ mol}{3.00\ L} = 0.133\ mol \cdot L^{-1}; [I_2] = \frac{1.60\ mol}{3.00\ L} = 0.533\ mol \cdot L^{-1}$$

Concentrations $(mol \cdot L^{-1})$	$H_2(g)$	+	$I_2(g)$	$\rightarrow$	$2\,HI(g)$
initial	0.133		0.533		0
change	$-x$		$-x$		$+2x$
final	$0.133 - x$		$0.533 - x$		$+2x$

At equilibrium, 60.0% of the H_2 had reacted, so 40.0% of the H_2

remains:

$$(0.400)(0.133\ mol \cdot L^{-1}) = 0.133\ mol \cdot L^{-1} - x$$

$$x = 0.133\ mol \cdot L^{-1} - (0.400)(0.133\ mol \cdot L^{-1})$$

$$x = 0.080\ mol \cdot L^{-1}$$

At equilibrium: $[H_2] = 0.133\ mol \cdot L^{-1} - 0.080\ mol \cdot L^{-1} = 0.053\ mol \cdot L^{-1}$

$$[I_2] = 0.533 \text{ mol} \cdot L^{-1} - 0.080 \text{ mol} \cdot L^{-1} = 0.453 \text{ mol} \cdot L^{-1}$$

$$[HI] = 2 \times 0.080 \text{ mol} \cdot L^{-1} = 0.16 \text{ mol} \cdot L^{-1}$$

$$K_C = \frac{[HI]^2}{[H_2][I_2]} = \frac{0.16^2}{(0.053)(0.453)} = 1.1$$

10.55 Initial concentrations of CO and O_2 are given by

$$[CO] = \frac{\left(\dfrac{0.28 \text{ g CO}}{28.01 \text{ g} \cdot \text{mol}^{-1} \text{CO}}\right)}{2.0 \text{ L}} = 5.0 \times 10^{-3} \text{ mol} \cdot L^{-1}$$

$$[O_2] = \frac{\left(\dfrac{0.032 \text{ g } O_2}{32.00 \text{ g} \cdot \text{mol}^{-1} \text{ } O_2}\right)}{2.0 \text{ L}} = 5.0 \times 10^{-4} \text{ mol} \cdot L^{-1}.$$

Concentration (mol·L⁻¹)	2 CO(g)	+	O_2(g)	→	2 CO_2(g)
initial	5.0×10^{-3}		5.0×10^{-4}		0
change	$-2x$		$-x$		$+2x$
final	$5.0 \times 10^{-3} - 2x$		$5.0 \times 10^{-4} - x$		$+2x$

$$K_C = \frac{[CO_2]^2}{[CO]^2 [O_2]}$$

$$= \frac{(2x)^2}{(5.0 \times 10^{-3} - 2x)^2 (5.0 \times 10^{-4} - x)}$$

$$= \frac{4x^2}{(4x^2 - 0.020x + 2.5 \times 10^{-5})(5.0 \times 10^{-4} - x)}$$

$$0.66 = \frac{4x^2}{-4x^3 + 0.022x^2 - 3.5 \times 10^{-5}x + 1.25 \times 10^{-8}}$$

$$4x^2 = (0.66)(-4x^3 + 0.022x^2 - 3.5 \times 10^{-5}x + 1.25 \times 10^{-8})$$

$$6.06x^2 = -4x^3 + 0.022x^2 - 3.5 \times 10^{-5}x + 1.25 \times 10^{-8}$$

$$0 = -4x^3 - 6.04x^2 - 3.5 \times 10^{-5}x + 1.25 \times 10^{-8}$$

$$x = 4.3 \times 10^{-5}$$

$$[CO_2] = 8.6 \times 10^{-5} \text{ mol} \cdot L^{-1}; [CO] = 4.9 \times 10^{-3} \text{ mol} \cdot L^{-1}$$

$$[O_2] = 4.6 \times 10^{-4} \text{ mol} \cdot L^{-1}$$

10.57 $K_C = \dfrac{[BrCl]^2}{[Cl_2][Br_2]}$

$0.031 = \dfrac{(0.145)^2}{(0.495)[Br_2]}$

$[Br_2] = \dfrac{(0.145)^2}{(0.495)(0.031)} = 1.4 \text{ mol} \cdot \text{L}^{-1}$

10.59 We can calculate changes according to the reaction stoichiometry:

Amount (mol)	$CO(g)$	+	$3 H_2(g)$	$\rightarrow$	$CH_4(g)$	+	$H_2O(g)$
initial	2.00		3.00		0		0
change	$-x$		$-3x$		$+x$		$+x$
final	$2.00 - x$		$3.00 - 3x$		0.478		$+x$

According to the stoichiometry, 0.478 mol = x; therefore, at equilibrium, there are 2.00 mol – 0.478 mol = 1.52 mol CO, 3.00 – 3(0.478 mol) = 1.57 mol H_2, and 0.478 mol H_2O. To employ the equilibrium expression, we need either concentrations or pressures; because K_C is given, we will choose to express these as concentrations. This calculation is easy because $V = 10.0$ L:

$[CO] = 0.152 \text{ mol} \cdot \text{L}^{-1}$; $[H_2] = 0.157 \text{ mol} \cdot \text{L}^{-1}$; $[CH_4] = 0.0478 \text{ mol} \cdot \text{L}^{-1}$; $[H_2O] = 0.0478 \text{ mol} \cdot \text{L}^{-1}$

$K_C = \dfrac{[CH_4][H_2O]}{[CO][H_2]^3} = \dfrac{(0.0478)(0.0478)}{(0.152)(0.157)^3} = 3.88$

10.61 First, we calculate the initial concentrations of each species:

$[SO_2] = [NO] = \dfrac{0.100 \text{ mol}}{5.00 \text{ L}} = 0.0200 \text{ mol} \cdot \text{L}^{-1}$;

$[NO_2] = \dfrac{0.200 \text{ mol}}{5.00 \text{ L}} = 0.0400 \text{ mol} \cdot \text{L}^{-1}$; $[SO_3] = \dfrac{0.150 \text{ mol}}{5.00 \text{ L}} = 0.0300 \text{ mol} \cdot \text{L}^{-1}$

We can use these values to calculate Q in order to see which direction the reactions will go:

$$Q = \frac{(0.0200)(0.0300)}{(0.0200)(0.0400)} = 0.75$$

Because $Q < K_C$, the reaction will proceed to produce more products.

Concentration $(\text{mol} \cdot \text{L}^{-1})$

	$SO_2(g)$	+	$NO_2(g)$	$\rightarrow$	$NO(g)$	+	$SO_3(g)$
initial	0.0200		0.0400		0.0200		0.0300
change	$-x$		$-x$		$+x$		$+x$
final	$0.0200 - x$		$0.0400 - x$		$0.0200 + x$		$0.0300 + x$

$$K_C = \frac{[NO][SO_3]}{[SO_2][NO_2]}$$

$$85.0 = \frac{(0.0200 + x)(0.0300 + x)}{(0.0200 - x)(0.0400 - x)}$$

$$= \frac{x^2 + 0.0500\,x + 0.000\,600}{x^2 - 0.0600\,x + 0.000\,800}$$

$$85.0\,(x^2 - 0.0600\,x + 0.000\,800) = x^2 + 0.0500\,x + 0.000\,600$$

$$85.0\,x^2 - 5.10\,x + 0.0680 = x^2 + 0.0500\,x + 0.000\,600$$

$$84.0\,x^2 - 5.15\,x + 0.0674 = 0$$

$$x = \frac{-(-5.15) \pm \sqrt{(-5.15)^2 - (4)(84.0)(0.0674)}}{(2)(84.0)} = \frac{+5.15 \pm 1.97}{168}$$

$$x = +0.0424 \quad \text{or} \quad +0.0189$$

The root 0.0424 is not meaningful because it is larger than the concentration of NO_2. The root of choice is therefore 0.0189.

At equilibrium:

$$[SO_2] = 0.0200 \text{ mol} \cdot \text{L}^{-1} - 0.0189 \text{ mol} \cdot \text{L}^{-1} = 0.0011 \text{ mol} \cdot \text{L}^{-1}$$
$$[NO_2] = 0.0400 \text{ mol} \cdot \text{L}^{-1} - 0.0189 \text{ mol} \cdot \text{L}^{-1} = 0.0211 \text{ mol} \cdot \text{L}^{-1}$$
$$[NO] = 0.0200 \text{ mol} \cdot \text{L}^{-1} + 0.0189 \text{ mol} \cdot \text{L}^{-1} = 0.0389 \text{ mol} \cdot \text{L}^{-1}$$
$$[SO_3] = 0.0300 \text{ mol} \cdot \text{L}^{-1} + 0.0189 \text{ mol} \cdot \text{L}^{-1} = 0.0489 \text{ mol} \cdot \text{L}^{-1}$$

To check, we can put these numbers back into the equilibrium constant expression:

$$K_C = \frac{[NO][SO_3]}{[SO_2][NO_2]}$$

$$\frac{(0.0389)(0.0489)}{(0.0011)(0.0211)} = 82.0$$

Compared to $K_C = 85.0$, this is reasonably good agreement given the nature of the calculation. We can check to see, by trial and error, if a better answer could be obtained. Because the K_C value is low for the concentrations we calculated, we can choose to alter x slightly so that this ratio becomes larger. If we let $x = 0.0190$, the concentrations of NO and SO$_3$ are increased to 0.0390 and 0.0490, and the concentrations of SO$_2$ and NO$_2$ are decreased to 0.0010 and 0.0200 (the stoichiometry of the reaction is maintained by calculating the concentrations in this fashion). Then the quotient becomes 91.0, which is further from the value for K_C than the original answer. So, although the agreement is not the best with the numbers we obtained, it is the best possible, given the limitation on the number of significant figures we are allowed to use in the calculation.

10.63 (a) The initial concentrations are

$$[PCl_5] = \frac{1.50 \text{ mol}}{0.500 \text{ L}} = 3.00 \text{ mol} \cdot L^{-1}; [PCl_3] = \frac{3.00 \text{ mol}}{0.500 \text{ L}} = 6.00 \text{ mol} \cdot L^{-1};$$

$$[Cl_2] = \frac{0.500 \text{ mol}}{0.500 \text{ L}} = 1.00 \text{ mol} \cdot L^{-1}.$$

First calculate Q:

$$Q = \frac{[PCl_5]}{[PCl_3][Cl_2]} = \frac{3.00}{(6.00)(1.00)} = 0.500$$

Because $Q \neq K$, the reaction is not at equilibrium.

(b) Because $Q < K_C$, the reaction will proceed to form products.

(c)

Concentrations (mol $\cdot$ L^{-1})	PCl$_3$(g)	+	Cl$_2$(g)	$\rightarrow$	PCl$_5$(g)
initial	6.00		1.00		3.00

change $\qquad\qquad\qquad -x \qquad\qquad -x \qquad\qquad +x$

final $\qquad\qquad\qquad 6.00 - x \qquad 1.00 - x \qquad 3.00 + x$

$$K_C = \frac{[PCl_5]}{[PCl_3][Cl_2]}$$

$$0.56 = \frac{3.00 + x}{(6.00 - x)(1.00 - x)} = \frac{3.00 + x}{x^2 - 7x + 6.00}$$

$$(0.56)(x^2 - 7x + 6.00) = 3.00 + x$$

$$0.56x^2 - 3.92x + 3.36 = 3.00 + x$$

$$0.56x^2 - 4.92x + 0.36 = 0$$

$$x = \frac{-(-4.92) \pm \sqrt{(-4.92)^2 - (4)(0.56)(0.36)}}{(2)(0.56)} = \frac{+4.92 \pm 4.48}{1.12}$$

$x = 9.2$ or 0.07

Because the root 9.2 is larger than the amount of PCl_3 or Cl_2 available, it is physically meaningless and can be discarded. Thus, $x = 0.071 \ mol \cdot L^{-1}$, giving

$$[PCl_5] = 3.00 \ mol \cdot L^{-1} + 0.07 \ mol \cdot L^{-1} = 3.07 \ mol \cdot L^{-1}$$

$$[PCl_3] = 6.00 \ mol \cdot L^{-1} - 0.07 \ mol \cdot L^{-1} = 5.93 \ mol \cdot L^{-1}$$

$$[Cl_2] = 1.00 \ mol \cdot L^{-1} - 0.07 \ mol \cdot L^{-1} = 0.93 \ mol \cdot L^{-1}$$

The number can be checked by substituting them back into the equilibrium constant expression:

$$K_C = \frac{[PCl_5]}{[PCl_3][Cl_2]}$$

$$\frac{(3.07)}{(5.93)(0.93)} \overset{?}{=} 0.56$$

$$0.56 \overset{?}{=} 0.56$$

10.65 Pressures (bar) $\qquad$ 2 HCl(g) $\qquad \rightarrow \qquad$ H$_2$(g) $\quad + \quad$ Cl$_2$(g)

initial $\qquad\qquad\qquad\qquad 0.22 \qquad\qquad\qquad\quad 0 \qquad\qquad\quad 0$

change $\qquad\qquad\qquad\quad -2x \qquad\qquad\qquad +x \qquad\qquad +x$

final $\qquad\qquad\qquad\quad 0.22 - 2x \qquad\qquad +x \qquad\qquad +x$

$$K = \frac{P_{H_2} P_{Cl_2}}{P_{HCl}^{\,2}}$$

$$3.2 \times 10^{-34} = \frac{(x)(x)}{(0.22 - 2x)^2}$$

Because the equilibrium constant is small, assume that $x \ll 0.22$:

$$3.2 \times 10^{-34} = \frac{x^2}{(0.22)^2}$$

$$x^2 = (3.2 \times 10^{-34})(0.22)^2$$

$$x = \sqrt{(3.2 \times 10^{-34})(0.22)^2}$$

$$x = \pm 3.9 \times 10^{-18}$$

The negative root is not physically meaningful and can be discarded. x is small compared to 0.22, so the initial assumption was valid. The pressures at equilibrium are

$$P_{HCl} = 0.22 \text{ bar}; \; P_{H_2} = P_{Cl_2} = 3.9 \times 10^{-18} \text{ bar}$$

The values can be checked by substituting them into the equilibrium expression:

$$\frac{(3.9 \times 10^{-18})(3.9 \times 10^{-18})}{(0.22)^2} \overset{?}{=} 3.2 \times 10^{-34}$$

$$3.1 \times 10^{-34} \overset{\checkmark}{=} 3.2 \times 10^{-34}$$

$$P_{HCl} = 0.22 \text{ bar}; \; P_{H_2} = P_{Cl_2} = 3.9 \times 10^{-18} \text{ bar}$$

The numbers agree very well for a calculation of this type.

10.67 (a) To determine on which side of the equilibrium position the conditions lie, we will calculate Q:

$$[CO] = \frac{0.342 \text{ mol}}{3.00 \text{ L}} = 0.114 \text{ mol} \cdot L^{-1}; \; [H_2] = \frac{0.215 \text{ mol}}{3.00 \text{ L}} = 0.0717 \text{ mol} \cdot L^{-1};$$

$$[CH_3OH] = \frac{0.125 \text{ mol}}{3.00 \text{ L}} = 0.0417 \text{ mol} \cdot L^{-1}$$

$$Q = \frac{[CH_3OH]}{[CO][H_2]^2} = \frac{0.0417}{(0.114)(0.0717)^2} = 71.1 \times 10^3$$

Because $Q > K_C$, the reaction will proceed to produce more of the reactants, which means that the concentration of methanol will decrease.

(b)

Concentrations $(\text{mol} \cdot \text{L}^{-1})$	$CO(g) +$	$2\,H_2(g)$	$\rightarrow$	$CH_3OH(g)$
initial	0.114	0.0717		0.0417
change	$+x$	$+2x$		$-x$
final	$0.0114 + x$	$0.0717 + 2x$		$0.0417 - x$

$$K_C = \frac{0.0417 - x}{(0.0114 + x)(0.0717 + 2x)^2} = 1.1 \times 10^{-2}$$

$$
\begin{aligned}
0.0417 - x &= (1.1 \times 10^{-2})(0.114 + x)(0.0717 + 2x)^2 \\
&= (1.2 \times 10^{-3} + 1.1 \times 10^{-2}x)(4x^2 + 0.287x + 5.14 \times 10^{-3}) \\
&= 4.4 \times 10^{-2}x^3 + 8.0 \times 10^{-3}x^2 + 4.0 \times 10^{-4}x + 6.2 \times 10^{-6} \\
0 &= 4.4 \times 10^{-2}x^3 + 8.0 \times 10^{-3}x^2 + 1.00x - 0.0417
\end{aligned}
$$

This equation can be solved approximately, simply by inspection: It is clear that the x term will be very much larger than the x^3 and the x^2 terms because their coefficients are very small compared to 1.00. This leads to a prediction that $x = 0.0417\,\text{mol} \cdot \text{L}^{-1}$ to within the accuracy of the data. Essentially all of the CH_3OH will react, so that

$[CO] = 0.114\,\text{mol} \cdot \text{L}^{-1} + 0.0417\,\text{mol} \cdot \text{L}^{-1} = 0.156\,\text{mol} \cdot \text{L}^{-1}$;

$[H_2] = 0.0717\,\text{mol} \cdot \text{L}^{-1} + 2\,(0.0417\,\text{mol} \cdot \text{L}^{-1}) = 0.155\,\text{mol} \cdot \text{L}^{-1}$. The mathematical situation is odd in that clearly a $[CH_3OH] = 0$ will not satisfy the equilibrium constant. Knowing that the methanol concentration is very small compared to the CO and H_2 concentrations, we can now back-calculate to get a concentration value that will satisfy the equilibrium expression:

$$K_C = \frac{y}{(0.156)(0.155)^2} = 1.1 \times 10^{-2}$$

$$y = (1.1 \times 10^{-2})(0.156)(0.155)^2 = 2.6 \times 10^{-4}$$

Alternatively, the cubic equation can be solved with the aid of a graphing calculator like the one supplied on the CD accompanying this book.

10.69 Since reactants are strongly favored, it is easier to push the reaction as far to the left as possible then start from new initial conditions.

Pressures (bar)	$2\,HCl(g)$	$\rightarrow$	$H_2(g)$	$+$	$Cl_2(g)$
original	2.0		1.0		3.0
new initial	4.0		0		2.0
change	$-2x$		$+x$		$+x$
final	$4.0 - 2x$		$+x$		$2.0 + x$

$$K = \frac{P_{H_2} P_{Cl_2}}{P_{HCl}^{\;2}}$$

$$3.2 \times 10^{-34} = \frac{(x)(2.0 + x)}{(4.0 - 2x)^2} \approx \frac{(x)(2.0)}{(4.0)^2} = \frac{x}{16}$$

$$x = 2.6 \times 10^{-33}\,\text{bar}$$

$$V = \frac{nRT}{p} = \frac{(1\ \text{mole})(8.314 \times 10^{-2}\,\text{L bar K}^{-1}\,\text{mol}^{-1})(298\text{K})}{2.6 \times 10^{-33}\,\text{bar}}$$

$$= 9.7 \times 10^{33}\,\text{L}$$

10.71 (a) According to Le Chatelier's principle, an increase in the partial pressure of CO_2 will result in creation of reactants, which will decrease the H_2 partial pressure.

(b) According to Le Chatelier's principle, if the CO pressure is reduced, the reaction will shift to form more CO, which will decrease the pressure of CO_2.

(c) According to Le Chatelier's principle, if the concentration of CO is increased, the reaction will proceed to form more products, which will result in a higher pressure of H_2.

(d) The equilibrium constant for the reaction is unchanged because it is unaffected by any change in concentration.

10.73 (a) According to Le Chatelier's principle, increasing the concentration of NO will cause the reaction to form reactants in order to reduce the concentration of NO; the amount of water will decrease.

(b) For the same reason as in (a), the amount of O_2 will increase.

(c) According to Le Chatelier's principle, removing water will cause the reaction to shift toward products, resulting in the formation of more NO.

(d) According to Le Chatelier's principle, removing a reactant will cause the reaction to shift in the direction to replace the removed substance; the amount of NH_3 should increase.

(e) According to Le Chatelier's principle, adding ammonia will shift the reaction to the right, but the equilibrium constant, which is a constant, will not be affected.

(f) According to Le Chatelier's principle, removing NO will cause the formation of more products; the amount of NH_3 will decrease.

(g) According to Le Chatelier's principle, adding reactants will promote the formation of products; the amount of oxygen will decrease.

10.75 As per Le Chatelier's principle, whether increasing the pressure on a reaction will affect the distribution of species within an equilibrium mixture of gases depends largely upon the difference in the number of moles of gases between the reactant and product sides of the equation. If there is a net increase in the amount of gas, then applying pressure will shift the reaction toward reactants in order to remove the stress applied by increasing the pressure. Similarly, if there is a net decrease in the amount of gas, applying pressure will cause the formation of products. If the number of moles of gas is the same on the product and reactant side, then changing the pressure will have little or no effect on the equilibrium distribution of species present. Using this information, we can apply it to the specific reactions given. The answers are: (a) reactants; (b) reactants; (c) reactants; (d) no change (there is the same number of moles of gas on both sides of the equation); (e) reactants.

10.77 (a) If the pressure of NO (a product) is increased, the reaction will shift to form more reactants; the pressure of NH_3 should increase.

(b) If the pressure of NH_3 (a reactant) is decreased, then the reaction will shift to form more reactants; the pressure of O_2 should increase.

10.79 If a reaction is exothermic, raising the temperature will tend to shift the reaction toward reactants, whereas if the reaction is endothermic, a shift toward products will be observed. For the specific examples given, (a) and (b) are endothermic (the values for (b) can be calculated, but we know that it requires energy to break an X—X bond, so those processes will all be endothermic) and raising the temperature should favor the formation of products; (c) and (d) are exothermic and raising the temperature should favor the formation of reactants.

10.81 First, we calculate the equilibrium constant for the conditions given.

$$K = \frac{(23.72)^2}{(3.11)(1.64)^3} = 41.0,$$ which corresponds to the reaction written as

$$N_2(g) + 3\,H_2(g) \rightarrow 2\,NH_3(g)$$

We then set up the table of anticipated changes upon introduction of the nitrogen:

	$N_2(g)$	$+$	$3\,H_2(g)$	$\rightarrow$	$2\,NH_3(g)$
initial	4.68 bar		1.64 bar		23.72 bar
change	$-x$		$-3x$		$+2x$
total	$4.68 - x$		$1.64 - 3x$		$23.72 + 2x$

$$41.0 = \frac{(23.72 + 2x)^2}{(4.68 - x)(1.64 - 3x)^3}$$

The equation can be solved using a graphing calculator, other computer software, or by trial and error. The solution is $x = 0.0656$ and the pressures of gases are $P_{N_2} = 4.61$ bar, $P_{H_2} = 1.44$ bar, $P_{NH_3} = 23.85$ bar.

10.83 (a) $\Delta G_r^o = -RT \ln K; \quad \ln K = -\dfrac{\Delta G_r^o}{RT}$

$\Delta G_r^o = (-191.28 \text{ kJ} \cdot \text{mol}^{-1}) - (-198.08 \text{ kJ} \cdot \text{mol}^{-1}) = 6.8 \text{ kJ} \cdot \text{mol}^{-1}$

$\ln K = -\dfrac{6800 J \cdot mol^{-1}}{(8.314 J \cdot K^{-1} \cdot mol^{-1})(900 K)} = -0.909;$

$K_C = 0.403$

Since K_C is quite small, there will not be large amount of hydrogen gas produced when the reaction reaches to equilibrium. To increase the amount of hydrogen gas production, temperature needs to be increased.

(b) 5.20×10^3 g C $\times \dfrac{1 mol\ C}{12.011 g\ C} = 433.$ mol C

125 g $H_2O \times \dfrac{1 mol\ H_2O}{18.016 g\ H_2O} = 6.94$ mol H_2O

H_2O is limiting. Conc. of $H_2O = 6.94$ mol/10.0 L $= 0.694$ mol·L^{-1}

Concentration (mol·L^{-1})	H_2O (g)	CO(g)	H_2(g)
initial	0.694	0	0
change	$-x$	$+x$	$+x$
final	0.694 - x	$+x$	$+x$

$K_C = \dfrac{[CO][H_2]}{[H_2O]} = \dfrac{(x)(x)}{(0.694 - x)} = 0.403$

$x^2 = 0.280 - 0.403x; \quad x^2 + 0.403x - 0.280 = 0$

Solve the quadratic equation: $x = +0.364$ or -0.766

The negative root is not meaningful, so we choose $x = 0.364$ mol · L^{-1}

$[CO] = [H_2] = 0.364$ mol · L^{-1}

$[H_2O] = 0.694 - 0.364 = 0.330$ mol · L^{-1}.

10.85 Even though numbers are given, we do not need to do a calculation to answer this qualitative question. Because the equilibrium constant for the formation of ammonia is smaller at the higher temperature, raising the temperature will favor the formation of reactants. Less ammonia will be present at higher temperature, assuming no other changes occur to the system (i.e., the volume does not change, no reactants or products are added or removed from the container, etc.).

10.87 To answer this question we must calculate Q:

$$Q = \frac{[NH_3]^2}{[N_2][H_2]^3} = \frac{(0.500)^2}{(3.00)(2.00)^3} = 0.0104$$

Because $Q \neq K$, the system is not at equilibrium, and because $Q < K$, the reaction will proceed to produce more products.

10.89 Because we want the equilibrium constant at two temperatures, we will need to calculate $\Delta H°_r$ and $\Delta S°_r$ for each reaction:

(a) $NH_4Cl \rightarrow NH_3(g) + HCl(g)$

$\Delta H°_r = \Delta H°_f(NH_3, g) + \Delta H°_f(HCl, g) - \Delta H°_f(NH_4Cl, s)$

$\Delta H°_r = (-46.11 \text{ kJ} \cdot \text{mol}^{-1}) + (-92.31 \text{ kJ} \cdot \text{mol}^{-1}) - (-314.43 \text{ kJ} \cdot \text{mol}^{-1})$

$\Delta H°_r = 176.01 \text{ kJ} \cdot \text{mol}^{-1}$

$\Delta S°_r = S°(NH_3, g) + S°(HCl, g) - S°(NH_4Cl, s)$

$\Delta S°_r = 192.45 \text{ J} \cdot \text{K}^{-1} \cdot \text{mol}^{-1} + 186.91 \text{ J} \cdot \text{K}^{-1} \cdot \text{mol}^{-1} - 94.6 \text{ J} \cdot \text{K}^{-1} \cdot \text{mol}^{-1}$

$\Delta S°_r = 284.8 \text{ J} \cdot \text{K}^{-1} \cdot \text{mol}^{-1}$

$\Delta G°_r = \Delta H°_r - T\Delta S°_r$

At 298 K:

$\Delta G°_{r(298 \text{ K})} = 176.01 \text{ kJ} - (298 \text{ K})(284.8 \text{ J} \cdot \text{K}^{-1})/(1000 \text{ J} \cdot \text{kJ}^{-1})$

$= 91.14 \text{ kJ} \cdot \text{mol}^{-1}$

$\Delta G°_{r(298 \text{ K})} = -RT \ln K$

$$\ln K = -\frac{\Delta G°_{r(298\,K)}}{RT}$$

$$= -\frac{91140\,\text{J}}{(8.314\,\text{J} \cdot \text{K}^{-1})(298\,\text{K})} = -36.8$$

$$K = 1 \times 10^{-16}$$

At 423 K:

$$\Delta G°_{r(423\,K)} = 176.01\,\text{kJ} - (423\,\text{K})(284.8\,\text{J} \cdot \text{K}^{-1})/(1000\,\text{J} \cdot \text{kJ}^{-1})$$

$$= 55.54\,\text{kJ} \cdot \text{mol}^{-1}$$

$$\Delta G°_{r(423\,K)} = -RT \ln K$$

$$\ln K = -\frac{\Delta G°_{r(423\,K)}}{RT}$$

$$= -\frac{55\,540\,\text{J}}{(8.314\,\text{J} \cdot \text{K}^{-1})(423\,\text{K})} = -15.8$$

$$K = 1 \times 10^{-7}$$

(b) $H_2(g) + D_2O(l) \rightarrow D_2(g) + H_2O(l)$

$$\Delta H°_r = \Delta H°_f(H_2O, l) - [\Delta H°_f(D_2O, l)]$$

$$\Delta H°_r = (-285.83\,\text{kJ} \cdot \text{mol}^{-1}) - [-294.60\,\text{kJ} \cdot \text{mol}^{-1}]$$

$$\Delta H°_r = 8.77\,\text{kJ} \cdot \text{mol}^{-1}$$

$$\Delta S°_r = S°(D_2, g) + S°(H_2O, l) - [S°(H_2, g) + S°(D_2O, l)]$$

$$\Delta S°_r = 144.96\,\text{J} \cdot \text{K}^{-1} \cdot \text{mol}^{-1} + 69.91\,\text{J} \cdot \text{K}^{-1} \cdot \text{mol}^{-1}$$

$$- [130.68\,\text{J} \cdot \text{K}^{-1} \cdot \text{mol}^{-1} + 75.94\,\text{J} \cdot \text{K}^{-1} \cdot \text{mol}^{-1}]$$

$$\Delta S°_r = 8.25\,\text{J} \cdot \text{K}^{-1} \cdot \text{mol}^{-1}$$

At 298 K:

$$\Delta G°_{r(298\,K)} = 8.77\,\text{kJ} \cdot \text{mol}^{-1} - (298\,\text{K})(8.25\,\text{J} \cdot \text{K}^{-1} \cdot \text{mol}^{-1})/(1000\,\text{J} \cdot \text{kJ}^{-1})$$

$$= 6.31\,\text{kJ} \cdot \text{mol}^{-1}$$

$$\Delta G°_{r(298\,K)} = -RT \ln K$$

$$\ln K = -\frac{\Delta G°_{r(298\,K)}}{RT}$$

$$= -\frac{6310\,\text{J}}{(8.314\,\text{J} \cdot \text{K}^{-1})(298\,\text{K})} = -2.55$$

$K = 7.8 \times 10^{-2}$

At 423 K:

$$\Delta G^\circ_{r(423\,K)} = 8.77 \text{ kJ} \cdot \text{mol}^{-1} - (423 \text{ K})(8.25 \text{ J} \cdot \text{K}^{-1} \cdot \text{mol}^{-1})/(1000 \text{ J} \cdot \text{kJ}^{-1})$$
$$= 5.28 \text{ kJ} \cdot \text{mol}^{-1}$$

$$\ln K = -\frac{5280 \text{ J}}{(8.314 \text{ J} \cdot \text{K}^{-1})(423 \text{ K})} = -1.50$$

$K = 0.22$

10.91 $\quad K = (RT)^{\Delta n} K_c$

$$\ln \frac{K_2}{K_1} = \frac{\Delta H_r^\circ}{R} \left(\frac{1}{T_1} - \frac{1}{T_2} \right)$$

$$\ln \left(\frac{(RT_2)^{\Delta n} K_{c2}}{(RT_1)^{\Delta n} K_{c1}} \right) = \frac{\Delta H_r^\circ}{R} \left(\frac{1}{T_1} - \frac{1}{T_2} \right)$$

$$\Delta n \ln \left(\frac{T_2}{T_1} \right) + \ln \left(\frac{K_{c2}}{K_{c1}} \right) = \frac{\Delta H_r^\circ}{R} \left(\frac{1}{T_1} - \frac{1}{T_2} \right)$$

$$\ln \left(\frac{K_{c2}}{K_{c1}} \right) = \frac{\Delta H_r^\circ}{R} \left(\frac{1}{T_1} - \frac{1}{T_2} \right) - \Delta n \ln \left(\frac{T_2}{T_1} \right)$$

10.93 The change increased the concentration of X(g) and decreased the concentration of X_2(g) in Flask 2. Increasing the temperature (a) would increase the formation of X(g) because the reaction is endothermic. Adding X(g) atoms or decreasing the volume (increasing pressure) would increase the formation of X_2(g), which did not occur. Adding a catalyst would not favor the formation of either X(g) or X_2(g). Also, the value of the equilibrium constant is larger in Flask 2 than in Flask 1, which is consistent with an increase in temperature of an endothermic reaction.

10.95 (a) According to Le Chatelier's principle, adding a product should cause a shift in the equilibrium toward the reactants side of the equation.

(b) Because there are equal numbers of moles of gas on both sides of the equation, there will be little or no effect on compressing the system.

(c) If the amount of CO_2 is increased, this will cause the reaction to shift toward the formation of products.

(d) Because the reaction is endothermic, raising the temperature will favor the formation of products.

(e) If the amount of $C_6H_{12}O_6$ is removed, this will cause the reaction to shift toward the formation of products.

(f) Because water is a liquid, it is by definition present at unit concentration, so changing the amount of water will not affect the reaction. As long as the glucose solution is dilute, its concentration can be considered unchanged.

(g) Decreasing the concentration of a reactant will favor the production of more reactants.

10.97 (a) $2A(g) \rightarrow B(g) + 2C(g)$

(b) $K = \dfrac{P_B(P_C)^2}{(P_A)^2} = \dfrac{(5/100)(10/100)^2}{(18/100)^2} = 1.54 \times 10^{-2}$

Note: The partial pressure units are changed to bar

(c) $K = (RT)^{\Delta n}K_c; \quad K_c = (RT)^{-\Delta n}K = (RT)^{-1}K = K/RT =$

$$\frac{1.54 \times 10^{-2}}{(8.3145 \times 10^{-2} \text{ L} \cdot \text{bar} \cdot K^{-1} \cdot \text{mol}^{-1})(298.15 \text{ K})} = 6.21 \times 10^{-4}$$

10.99 (a) In order to solve this problem, we will manipulate the equations with the known K's so that we can combine them to give the desired overall reaction.

First, reverse equation (1) and multiply it by $\frac{1}{2}$:

$$H_2(g) + \tfrac{1}{2}O_2(g) \rightarrow H_2O(g) \quad K_4 = \frac{1}{(1.6 \times 10^{-11})^{1/2}} \tag{4}$$

Multiply equation (2) by $\frac{1}{2}$ also:

$$CO_2(g) \rightarrow CO(g) + \tfrac{1}{2}O_2(g) \quad K_5 = (1.3 \times 10^{-10})^{1/2} \tag{5}$$

Adding equations (4) and (5) gives the desired reaction. The resultant equilibrium constant will be the product of the K's for (4) and (5):

$$CO_2(g) + H_2(g) \rightarrow H_2O(g) + CO(g) \quad K_5 = \left(\frac{1.3 \times 10^{-10}}{1.6 \times 10^{-11}}\right)^{1/2} = 2.9 \tag{3}$$

(b) To obtain the K value for a net equation from two (or more) others, the K's are multiplied, but $\Delta G°_r$'s are added:

$$2\,H_2O(g) \rightarrow 2\,H_2(g) + O_2(g) \tag{1}$$

$$\begin{aligned}
\Delta G°_r &= -RT \ln K \\
&= -(8.314\,\text{J} \cdot \text{K}^{-1} \cdot \text{mol}^{-1})(1565\,\text{K}) \ln 1.6 \times 10^{-11} \\
&= +3.2 \times 10^2\,\text{kJ} \cdot \text{mol}^{-1}
\end{aligned}$$

$$2\,CO_2(g) \rightarrow 2\,CO(g) + O_2(g) \tag{2}$$

$$\begin{aligned}
\Delta G°_r &= -RT \ln K \\
&= -(8.314\,\text{J} \cdot \text{K}^{-1} \cdot \text{mol}^{-1})(1565\,\text{K}) \ln 1.3 \times 10^{-10} \\
&= +3.0 \times 10^2\,\text{kJ} \cdot \text{mol}^{-1}
\end{aligned}$$

The corresponding values for (4) and (5) are

$$\Delta G°_{r(4)} = -\tfrac{1}{2}(3.2 \times 10^2\,\text{kJ}) = -1.6 \times 10^2\,\text{kJ}$$

$$\Delta G°_{r(5)} = \tfrac{1}{2}(3.0 \times 10^2\,\text{kJ}) = 1.5 \times 10^2\,\text{kJ}.$$

Summing these two values will give

$$\Delta G°_{r(3)} = -1.6 \times 10^2\,\text{kJ} + 1.5 \times 10^2\,\text{kJ} = -10\,\text{kJ} \cdot \text{mol}^{-1}$$

$$\ln K = -\frac{\Delta G°_r}{RT} = -\frac{-10\,000\,\text{J} \cdot \text{mol}^{-1}}{(8.314\,\text{J} \cdot \text{K}^{-1} \cdot \text{mol}^{-1})(1565\,\text{K})} = +0.8$$

$$K = 2.$$

This value is in reasonable agreement with the one obtained in (a), given the problems in significant figures and due to rounding errors.

10.101 (a) The initial concentration of $NO_2(g)$ is 0.020 $mol \cdot L^{-1}$

Concentration $(mol \cdot L^{-1})$	$2\,NO_2(g)$	$\rightarrow$	$N_2O_4(g)$
initial	0.020		0
change	-2x		+x
equilibrium	0.020 – 2x		+x

$$K_C = \frac{[N_2O_4]}{[NO_2]^2} = \frac{x}{(0.020-2x)^2} = \frac{1}{6.1\times10^{-3}} = 1.6\times10^2$$

Note: $(0.020-2x)^2 = 4.0\times10^{-4} - 0.08x + 4x^2$

$x = 0.064 - 13x + 6.4\times10^2 x^2$

$6.4\times10^2 x^2 - 14x + 0.064 = 0$

Solve the quadratic equation: $x = +0.0065$ or $+0.015$

It is obvious that $x = +0.015$ is not meaningful, because a negative NO_2 concentration will be obtained at equilibrium:

$0.020 - 2x = 0.020 - 2(0.015) = -0.01 mol \cdot L^{-1}$

so we choose $x = 0.0065\ mol \cdot L^{-1}$

$[N_2O_4] = 0.0065\ mol \cdot L^{-1}$

$[NO_2] = 0.020 - 2x = 0.020 - 2(0.0065) = 7.0\times10^{-3}\ mol \cdot L^{-1}$

(b) If the volume is decreased by half ($V = 500.$ mL). The initial concentration of NO_2 would be 0.040 $mol \cdot L^{-1}$. By following the same procedure in (a):

Concentration $(mol \cdot L^{-1})$	$2\,NO_2(g)$	$\rightarrow$	$N_2O_4(g)$
initial	0.040		0
change	-2x		+x
equilibrium	0.040 – 2x		+x

$$K_C = \frac{[N_2O_4]}{[NO_2]^2} = \frac{x}{(0.040-2x)^2} = \frac{1}{6.1\times10^{-3}} = 1.6\times10^2$$

Note: $(0.040-2x)^2 = 1.6\times10^{-3} - 0.16x + 4x^2$

$x = 0.26 - 26x + 6.4\times10^2 x^2$

$6.4 \times 10^2 x^2 - 27 x + 0.26 = 0$

Solve the quadratic equation: $x = +0.015$ or $+0.027$

It is obvious that $x = +0.027$ is not meaningful, because a negative NO_2 concentration will be obtained at equilibrium:

$0.040 - 2x = 0.040 - 2(0.027) = -0.014 \; mol \cdot L^{-1}$

so we choose $x = 0.015 \; mol \cdot L^{-1}$

$[N_2O_4] = 0.015 \; mol \cdot L^{-1}$

$[NO_2] = 0.040 - 2x = 0.040 - 2(0.015) = 0.010 \; mol \cdot L^{-1}$

10.103

Pressure	$PCl_5(g)$	$\rightarrow$	$PCl_3(g)$	$+$	$Cl_2(g)$
initial	n				
change	$-n\alpha$		$+n\alpha$		$+n\alpha$
final	$n(1-\alpha)$		$+n\alpha$		$+n\alpha$

$$K = \frac{(n\alpha)(n\alpha)}{n(1-\alpha)} = \frac{n^2\alpha^2}{n(1-\alpha)} = \frac{n\alpha^2}{(1-\alpha)}$$

$$P = n(1-\alpha) + n\alpha + n\alpha = n(1+\alpha)$$

$$n = \frac{P}{1+\alpha}$$

$$K = \frac{n\alpha^2}{(1-\alpha)} = \left(\frac{P}{1+\alpha}\right)\left(\frac{\alpha^2}{1-\alpha}\right) = \frac{P\alpha^2}{1-\alpha^2}$$

$$(1-\alpha^2)K = P\alpha^2$$

$$K - K\alpha^2 = P\alpha^2$$

$$K = P\alpha^2 + K\alpha^2$$

$$\alpha^2 = \frac{K}{P+K}$$

$$\alpha = \sqrt{\frac{K}{P+K}}$$

(a) For the specific conditions $K = 4.96$ and $P = 0.50$ bar,

$$\alpha = \sqrt{\frac{4.96}{0.50+4.96}} = 0.953 \,.$$

(b) For the specific conditions $K = 4.96$ and $P = 1.00$ bar,

$$\alpha = \sqrt{\frac{4.96}{1.00 + 4.96}} = 0.912 \,.$$

10.105 (a) If $K = 1.00$, then $\Delta G°$ must be equal to 0 ($\Delta G° = -RT \ln K$).

(b) This can be calculated by determining the values for $\Delta H°$ and $\Delta S°$ at 25°C.

$$\Delta H° = -393.51 \text{ kJ} \cdot \text{mol}^{-1} - [(-110.53 \text{ kJ} \cdot \text{mol}^{-1}) + (-241.82 \text{ kJ} \cdot \text{mol}^{-1})]$$
$$= -41.16 \text{ kJ} \cdot \text{mol}^{-1}$$

$$\Delta S° = 130.68 \text{ J} \cdot \text{K}^{-1} \cdot \text{mol}^{-1} + 213.74 \text{ J} \cdot \text{K}^{-1} \cdot \text{mol}^{-1}$$
$$- [197.67 \text{ J} \cdot \text{K}^{-1} \cdot \text{mol}^{-1} + 188.83 \text{ J} \cdot \text{K}^{-1} \cdot \text{mol}^{-1}]$$
$$= -42.08 \text{ J} \cdot \text{K}^{-1} \cdot \text{mol}^{-1}$$

$$\Delta G = (-41.16 \text{ kJ} \cdot \text{mol}^{-1})(1000 \text{ J} \cdot \text{kJ}^{-1}) - T(-42.08 \text{ J} \cdot \text{K}^{-1} \cdot \text{mol}^{-1}) = 0$$
$$T = 978 \ K \text{ (or 705°C)}$$

(c)

	$CO(g) +$	$H_2O(g)$	$\rightarrow CO_2(g) +$	$H_2(g)$
	10.00 bar	10.00 bar	5.00 bar	5.00 bar
change	$-x$	$-x$	$+x$	$+x$
net	$10.00 - x$	$10.00 - x$	$5.00 + x$	$5.00 + x$

$x = 2.50$ bar

All pressures are equal to 7.50 bar.

(d) First, check Q to determine the direction of the reaction:

$$Q = \frac{(10.00)(5.00)}{(6.00)(4.00)} = 2.08$$

Because Q is greater than 1, the reaction will shift to produce reactants.

	$CO(g) +$	$H_2O(g)$	$\rightarrow$	$CO_2(g) +$	$H_2(g)$
	6.00 bar	4.00 bar		10.00 bar	5.00 bar
change	$+x$	$+x$		$-x$	$-x$
net	$6.00 + x$	$4.00 + x$		$10.00 - x$	$5.00 - x$

$$\frac{(10.00 - x)(5.00 - x)}{(6.00 + x)(4.00 + x)} = 1$$

$$(10.00 - x)(5.00 - x) = (6.00 + x)(4.00 + x)$$

$$x^2 - 15.00\,x + 50.0 = x^2 + 10.00\,x + 24.0$$

$$25.00\,x = 26.0$$

$$x = 1.04 \text{ bar}$$

$P_{CO(g)} = 7.04$ bar; $P_{H_2O(g)} = 5.04$ bar; $P_{CO_2(g)} = 8.96$ bar; $P_{H_2(g)} = 3.96$ bar

10.107 (a) These values are easily calculated from the relationship $\Delta G° = -RT$ ln K. For the atomic species, the free energy of the reaction will be $\frac{1}{2}$ of this value because the equilibrium reactions are for the formation of two moles of halogen atoms.

The results:

Halogen	Bond Dissociation Energy $(kJ \cdot mol^{-1})$	$\Delta G°$ $(kJ \cdot mol^{-1})$
fluorine	146	19.1
chlorine	230	47.9
bromine	181	42.8
iodine	139	5.6

(b)

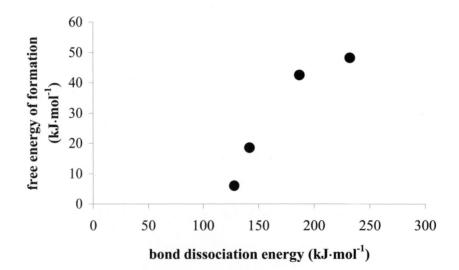

There is a correlation between the bond dissociation energy and the free energy of formation of the atomic species, but the relationship is clearly not linear.

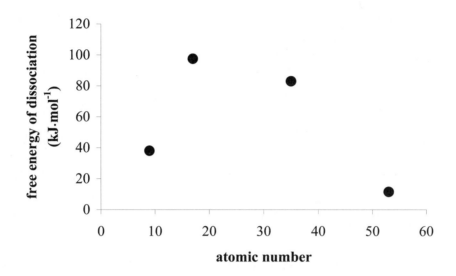

For the heavier three halogens, there is a trend to decreasing free energy of formation of the atoms as the element becomes heavier, but fluorine is anomalous. The F—F bond energy is lower than expected, owing to repulsions of the lone pairs of electrons on the adjacent F atoms because the F—F bond distance is so short.

10.109 (a) Using the thermodynamic data in Appendix 2A:

$Br_2(g) \rightarrow 2\,Br(g)$

$\Delta G° = 2(82.40\ kJ \cdot mol^{-1}) - 3.11\ kJ \cdot mol^{-1} = 161.69\ kJ \cdot mol^{-1}$

$K = e^{-\Delta G°/RT} = 4.5 \times 10^{-29}$

For equilibrium constant calculations, this is reasonably good agreement with the value obtained from part (a), especially if one considers that $\Delta H°$ will not be perfectly constant over so large a temperature range.

(b) We will use data from Appendix 2A to calculate the vapor pressure of bromine:

$Br_2(l) \rightarrow Br_2(g)$

$\Delta G° = 3.11\,kJ \cdot mol^{-1}$

$K = e^{-\Delta G°/RT} = 0.285$

The vapor pressure of bromine will, therefore, be 0.285 bar or 0.289 atm. Remember that because the standard state for the thermodynamic quantities is 1 bar, the values in K will be derived in bar as well.

(c) $4.5 \times 10^{-29} = \dfrac{P_{Br(g)}^{\,2}}{P_{Br_2(l)}} = \dfrac{P_{Br(g)}^{\,2}}{0.285\ \text{bar}}$

$P_{Br(g)}^{\,2} = 3.6 \times 10^{-15}$ bar or 3.6×10^{-15} atm

(d) Use the ideal gas law:

$PV = nRT$

$(0.289\ \text{atm})V = (0.0100\ \text{mol})(0.082\,06\ L \cdot atm \cdot K^{-1} \cdot mol^{-1})(298\ K)$

$V = 0.846$ L or 846 mL

10.111 (a) Using the thermodynamic data in Appendix 2A:

$2\,H_2S(g) + SO_2(g) \leftrightarrow 3\,S(s) \quad + 2\,H_2O(g)$

$\Delta G° = 2(\text{-228.57 kJ·mol}^{-1}) - [2(\text{-33.56 kJ·mol}^{-1}) + (\text{-300.19 kJ·mol}^{-1})]$

$\qquad = \text{- 89.83 kJ·mol}^{-1}$

$\ln K = -\dfrac{\Delta G^0}{RT} = -\dfrac{-89.83 \times 10^3\ J \cdot mol^{-1}}{(8.314 J \cdot K^{-1} \cdot mol^{-1})(298K)} = 36.26$

$K = 5.6 \times 10^{15}$

Part (b) The effect of each change to the equilibrium is:

(a) S is solid and already present in excess in the original equilibrium. Adding more solid will not affect the equilibrium so the amount of H_2O will not change.

(b) According to Le Chatelier's principle, increasing the concentration of H_2S will cause the reaction to form more products in order to reduce the concentration of H_2S; some of the SO_2 will be consumed. Therefore, the amount of SO_2 will decrease.

(c) Removing a product will shift the reaction toward the formation of more products; the amount of SO_2 should decrease.

(d) Removing the reactant SO_2 will cause the reaction to shift toward the production of more reactants; the amount of S should decrease since it will be used in the production of more reactants.

(e) The equilibrium constant will be unaffected by changes in the concentrations of any of the species present.

(f) Decreasing volume (or increase pressure) will cause the reaction to shift to the side which has less number of gas molecules. In this specific case, the reaction will shift toward the product side. The amount of SO_2 should decrease.

(g) Based on the expression in (a), increasing temperature will decrease equilibrium constant, shifting reaction to the reactant side. The amount of SO_2 should increase.

10.113 The equilibrium reaction across the membrane is

$$Na^+ \text{ (outside of the cell)} \longleftrightarrow Na^+ \text{(inside the cell)}$$

The equilibrium constant $K_c = \dfrac{[Na^+]_{\text{inside the cell}}}{[Na^+]_{\text{outside the cell}}} = \dfrac{10\,mM}{140\,mM} = 7.1 \times 10^{-2}$

$\Delta G_r^\circ = -RT \ln K = -(8.314 \text{ J} \cdot \text{K}^{-1} \cdot \text{mol}^{-1})(310.15 \text{ K}) \ln(7.1 \times 10^{-2})$
$\qquad = 6.8 \times 10^3 \text{ J} = 6.8 \text{ kJ}$

A positive ΔG_r° means that the reaction above is not favorable to the right side (the Na^+ inside the cell).

Note: $K = K_c$ in this problem so K_c is directly used in ΔG_r° calculation.

10.115 (a) $K_C = \dfrac{[NO_2]^2}{[N_2O_4]} = \dfrac{(2.13)^2}{0.405} = 11.2$

(b) If NO_2 is added, the equilibrium will shift to produce more N_2O_4. The amount of NO_2 will be greater than initially present, but less than the $3.13 \text{ mol} \cdot L^{-1}$ present immediately upon making the addition. K_C will not be affected.

(c)

Concentrations (mol·L⁻¹)	N_2O_4	$\rightarrow$	$2\,NO_2$
initial	0.405		3.13
change	$+x$		$-2x$
final	$0.405 + x$		$3.13 - 2x$

$$11.2 = \frac{(3.13 - 2x)^2}{0.405 + x}$$

$$(11.2)(0.405 + x) = (3.13 - 2x)^2$$

$$11.2x + 4.536 = 4x^2 - 12.52x + 9.797$$

$$4x^2 - 23.7x + 5.26 = 0$$

$$x = \frac{-(-23.7) \pm \sqrt{(-23.7)^2 - (4)(4)(5.26)}}{(2)(4)} = \frac{23.7 \pm 21.9}{8}$$

$$x = 5.70 \text{ or } 0.23$$

At equilibrium, $[N_2O_4] = 0.405\ \text{mol·L}^{-1} + 0.23\ \text{mol·L}^{-1} = 0.64\ \text{mol·L}^{-1}$

$[NO_2] = 3.13\ \text{mol·L}^{-1} - 2(0.23\ \text{mol·L}^{-1}) = 2.67\ \text{mol·L}^{-1}$.

These concentrations are consistent with the predictions in (b).

10.117 To find the vapor pressure, we first calculate $\Delta G°$ for the conversion of the liquid to the gas at 298 K, using the free energies of formation found in the appendix:

$$\begin{aligned}
\Delta G°_{H_2O(l) \to H_2O(g)} &= \Delta G°_{f(H_2O(g))} - \Delta G°_{f(H_2O(l))} \\
&= (-228.57\ \text{kJ·mol}^{-1}) - [-237.13\ \text{kJ·mol}^{-1}] \\
&= 8.56\ \text{kJ·mol}^{-1}
\end{aligned}$$

$$\begin{aligned}
\Delta G°_{D_2O(l) \to D_2O(g)} &= \Delta G°_{f(D_2O(g))} - \Delta G°_{f(D_2O(l))} \\
&= (-234.54\ \text{kJ·mol}^{-1}) - [-243.44\ \text{kJ·mol}^{-1}] \\
&= 8.90\ \text{kJ·mol}^{-1}
\end{aligned}$$

The equilibrium constant for these processes is the vapor pressure of the liquid:

$$K = P_{H_2O} \text{ or } K = P_{D_2O}$$

Using $\Delta G° = -RT \ln K$, we can calculate the desired values.

For H_2O:

$$\ln K = -\frac{\Delta G^\circ}{RT} = -\frac{8560 \text{ J} \cdot \text{mol}^{-1}}{(8.314 \text{ J} \cdot \text{K}^{-1} \cdot \text{mol}^{-1})(298 \text{ K})} = -3.45$$

$$K = 0.032 \text{ bar}$$

$$0.032 \text{ bar} \times \frac{1 \text{ atm}}{1.013\,25 \text{ bar}} \times \frac{760 \text{ Torr}}{1 \text{ atm}} = 24 \text{ Torr}$$

For D_2O:

$$\ln K = -\frac{\Delta G^\circ}{RT} = -\frac{8900 \text{ J} \cdot \text{mol}^{-1}}{(8.314 \text{ J} \cdot \text{K}^{-1} \cdot \text{mol}^{-1})(298 \text{ K})} = -3.59$$

$$K = 0.028 \text{ bar}$$

$$0.028 \text{ bar} \times \frac{1 \text{ atm}}{1.013\,25 \text{ bar}} \times \frac{760 \text{ Torr}}{1 \text{ atm}} = 21 \text{ Torr}$$

The answer is that D has a lower zero point energy than H. This makes the D_2O — D_2O "hydrogen bond" stronger than the H_2O — H_2O hydrogen bond. Because the hydrogen bond is stronger, the intermolecular forces are stronger, the vapor pressure is lower, and the boiling point is higher. Potential energy curves for the O—H and O—D bonds as a function of distance:

ΔE = energy required to break O—H or O—D bond

10.119 First, we derive a general relationship that relates ΔG°_f values to nonstandard state conditions. We do this by returning to the fundamental definition of ΔG and ΔG°:

$$\Delta G^\circ_f = \Sigma G^\circ_{m(products)} - \Sigma G^\circ_{m(reactants)}$$

Similarly, for conditions other than standard state, we can write

$$\Delta G_f = \Sigma G_{m(products)} - \Sigma G_{m(reactants)}.$$

Because $G_m = G^\circ_m + RT \ln Q$,

$$\Delta G_f = \Sigma (G^\circ_m + RT \ln P_i)_{(products)} - \Sigma (G^\circ_m + RT \ln P_i)_{(reactants)}.$$

But because all reactants or products in the same system refer to the same standard state,

$$\Delta G_f = \Sigma G^\circ_{m(products)} - \Sigma G^\circ_{m(reactants)} + \Delta n(RT \ln Q_{(reactants)})$$

$$\Delta G_{f} = \Delta G^{\circ}{}_{f} + \Delta n(RT \ln Q_{(reactants)}).$$

The value ΔG_{f}, which is the nonstandard value for the conditions of 1 atm,

$1 \, mol \cdot L^{-1}$, etc. becomes the new standard value.

(a) $\frac{1}{2} H_2(g) + \frac{1}{2} I_2(g) \rightarrow HI(g)$, $\Delta n = 1 \, mol - (\frac{1}{2} \, mol + \frac{1}{2} \, mol) = 0$

$1 \, atm = 1.013 \, 25 \, bar$

$\Delta G_{f} = 1.70 \, kJ \cdot mol^{-1}$
$\quad + (0)(8.314 \, J \cdot K^{-1} \cdot mol^{-1})(298 \, K)(\ln 1.013 \, 25)/(1000 \, J \cdot kJ^{-1})$
$\quad = 1.70 \, kJ \cdot mol^{-1}$

(b) $C(s) + \frac{1}{2} O_2(g) \rightarrow CO(g)$, $\Delta n = 1 \, mol - \frac{1}{2} \, mol = \frac{1}{2} \, mol$

$\Delta G_{f} = -137.17 \, kJ \cdot mol^{-1}$
$\quad + (\frac{1}{2})(8.314 \, J \cdot K^{-1} . mol^{-1})(298 \, K)(\ln 1.013 \, 25)/(1000 \, J \cdot kJ^{-1})$
$\quad = -137.15 \, kJ \cdot mol^{-1}$

(c) $C(s) + \frac{1}{2} N_2(g) + \frac{1}{2} H_2(g) \rightarrow HCN(g)$, $\Delta n = 1 \, mol - (\frac{1}{2} \, mol + \frac{1}{2} \, mol)$
$\qquad\qquad\qquad = 0$

$$1 \, Torr \times \frac{1 \, atm}{760 \, Torr} \times \frac{1.013 \, 25 \, bar}{1 \, atm} = 1.333 \times 10^{-3} \, bar$$

$\Delta G_{f} = 124.7 \, kJ \cdot mol^{-1}$
$\quad + (0)(8.314 \, J \cdot K^{-1} \cdot mol^{-1})(298 \, K)(\ln 1.333 \times 10^{-3})/(1000 \, J \cdot kJ^{-1})$
$\quad = 124.7 \, kJ \cdot mol^{-1}$

(d) $C(s) + 2 H_2(g) \rightarrow CH_4(g)$, $\Delta n = 1 \, mol - 2 \, mol = -1 \, mol$

$1 \, Pa = 10^{-5} \, bar$

$\Delta G_{f} = -50.72 \, kJ \cdot mol^{-1}$
$\quad + (-1)(8.314 \, J \cdot K^{-1} \cdot mol^{-1})(298 \, K)(\ln 10^{-5})/(1000 \, J \cdot kJ^{-1})$
$\quad = -22.2 \, kJ \cdot mol^{-1}$

10.121

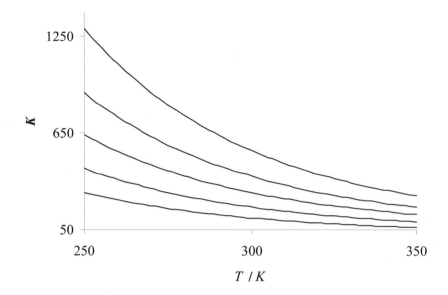

10.123 (a) (i) Increasing the amount of a reactant will push the equilibrium toward the products. More NO_2 will form. (ii) Removing a product will pull the equilibrium toward products. More NO_2 will form. (iii) Increasing total pressure by adding an inert gas not change the relative partial pressures. There will be no change in the amount of NO_2. (b) Since there are two moles of gas on both sides of the reaction, the volume cancels out of the equilibrium constant expression and it is possible to use moles directly for each component.

At equilibrium, SO_3 = 0.245 moles – 0.240 moles = 0.005 moles, SO_2 = 0.240 moles and NO_2 = 0.240 moles since the reaction is 1:1:1:1. The original number of moles of NO is set to x.

$K = 6.0 \times 10^3 = (0.240)(0.240)/(0.005)(x - 0.240) = (11.52)/(x - 0.240)$

$x = (11.52/6.0 \times 10^3) + 0.240$

$x = 0.242$ moles of NO

CHAPTER 11

ACIDS AND BASES

11.1 (a) $CH_3NH_3^+$ (b) $NH_2NH_3^+$ (c) H_2CO_3 (d) CO_3^{2-}

(e) $C_6H_5O^-$ (f) $CH_3CO_2^-$

11.3 For all parts (a) to (e), H_2O and H_3O^+ form a conjugate acid-base pair in which H_2O is the base and H_3O is the acid.

(a) $H_2SO_4(aq) + H_2O(l) \rightleftharpoons H_3O^+(aq) + HSO_4^-(aq)$

H_2SO_4 and $HSO_4^-(aq)$ form a conjugate acid-base pair in which

H_2SO_4 is the acid and $HSO_4^-(aq)$ is the base.

(b) $C_6H_5NH_3^+(aq) + H_2O(l) \rightleftharpoons H_3O^+(aq) + C_6H_5NH_2(aq)$

$C_6H_5NH_3^+$ and $C_6H_5NH_2(aq)$ form a conjugate acid-base pair in which

$C_6H_5NH_3^+$ is the acid and $C_6H_5NH_2(aq)$ is the base.

(c) $H_2PO_4^-(aq) + H_2O(l) \rightleftharpoons H_3O^+(aq) + HPO_4^{2-}(aq)$

$H_2PO_4^-(aq)$ and $HPO_4^{2-}(aq)$ form a conjugate acid-base pair in which

$H_2PO_4^-(aq)$ is the acid and $HPO_4^{2-}(aq)$ is the base.

(d) $HCOOH(aq) + H_2O(l) \rightleftharpoons H_3O^+(aq) + HCO_2^-(aq)$

$HCOOH(aq)$ and $HCO_2^-(aq)$ form a conjugate acid-base pair in which

$HCOOH(aq)$ is the acid and $HCO_2^-(aq)$ is the base.

(e) $NH_2NH_3^+(aq) + H_2O(l) \rightleftharpoons H_3O^+(aq) + NH_2NH_2(aq)$

$NH_2NH_3^+(aq)$ and $NH_2NH_2(aq)$ form a conjugate acid-base pair in

which $NH_2NH_3^+(aq)$ is the acid and $NH_2NH_2(aq)$ is the base.

11.5 (a) Brønsted acid: HNO_3

Brønsted base: $HPO_4{}^{2-}$

(b) conjugate base to HNO_3: $NO_3{}^-$

conjugate acid to $HPO_4{}^{2-}$: $H_2PO_4{}^-$

11.7 (a) $HClO_3$ (chloric acid); conjugate base: $ClO_3{}^-$

(b) HNO_2 (nitrous acid); conjugate base: $NO_2{}^-$

11.9 (a) proton transferred from $NH_4{}^+$ to H_2O, $NH_4{}^+$ (acid), H_2O (base);

(b) proton transferred from $NH_4{}^+$ to I^-, $NH_4{}^+$ (acid), I^- (base);

(c) no proton transferred

(d) proton transferred from $NH_4{}^+$ to $NH_2{}^-$, $NH_4{}^+$ (acid), $NH_2{}^-$ (base)

11.11 (a) $HCO_3{}^-$ as an acid: $HCO_3{}^-(aq) + H_2O(l) \rightleftharpoons H_3O^+(aq) + CO_3{}^{2-}(aq)$,

$HCO_3{}^-$ (acid) and $CO_3{}^{2-}$ (base), H_2O (base) and H_3O^+ (acid);

$HCO_3{}^-$ as a base: $H_2O(l) + HCO_3{}^-(aq) \rightleftharpoons H_2CO_3(aq) + OH^-(aq)$,

$HCO_3{}^-$ (base) and H_2CO_3 (acid), H_2O (acid) and OH^- (base)

(b) $HPO_4{}^{2-}$ as an acid: $HPO_4{}^{2-}(aq) + H_2O(l) \rightleftharpoons H_3O^+(aq) + PO_4{}^{3-}(aq)$,

$HPO_4{}^{2-}$ (acid) and $PO_4{}^{3-}$ (base), H_2O (base) and H_3O^+ (acid);

$HPO_4{}^{2-}$ as a base: $HPO_4{}^{2-}(aq) + H_2O(l) \rightleftharpoons H_2PO_4{}^-(aq) + OH^-(aq)$,

$HPO_4{}^{2-}$ (base) and $H_2PO_4{}^-$ (acid), H_2O (acid) and OH^- (base)

11.13 (a) Two protons can be accepted (one on each N);

(b)

or

(c) Each of the two nitro- groups will show amphiprotic behavior in aqueous solution (can either accept a proton and donate a proton).

11.15 The Lewis structures of (a) to (e) are as follows:

(a) Lewis base   (b) Lewis acid   (c) Lewis acid   (d) Lewis base   (e) Lewis base

11.17 (a)

Lewis acid   Lewis base   product

(b)

Lewis acid   Lewis base   product

11.19 (a) basic; (b) acidic; (c) amphoteric; (d) basic

11.21 HCl is a strong acid. If concentration decreases to 12% of its initial value (i.e. $0.12 \times [HCl]_0$), the pH change $(\Delta pH) = -\log(\dfrac{0.12 \times [HCl]_0}{[HCl]_0}) = 0.92$

(i.e., pH will increase by 0.92).

11.23 In each case, use $K_w = [H_3O^+][OH^-] = 1.0 \times 10^{-14}$, then

$$[OH^-] = \frac{K_w}{[H_3O^+]} = \frac{1.0 \times 10^{-14}}{[H_3O^+]}.$$

(a) $[OH^-] = \dfrac{1.0 \times 10^{-14}}{0.02} = 5.0 \times 10^{-13} \text{ mol} \cdot \text{L}^{-1}$

(b) $[OH^-] = \dfrac{1.0 \times 10^{-14}}{1.0 \times 10^{-5}} = 1.0 \times 10^{-9} \text{ mol} \cdot \text{L}^{-1}$

(c) $[OH^-] = \dfrac{1.0 \times 10^{-14}}{3.1 \times 10^{-3}} = 3.2 \times 10^{-12} \text{ mol} \cdot \text{L}^{-1}$

11.25 (a) $K_w = 2.1 \times 10^{-14} = [H_3O^+][OH^-] = x^2$, where $x = [H_3O^+] = [OH^-]$

$x = \sqrt{2.1 \times 10^{-14}} = 1.4 \times 10^{-7} \text{ mol} \cdot \text{L}^{-1}$

$pH = -\log[H_3O^+] = 6.80$

(b) $[OH^-] = [H_3O^+] = 1.4 \times 10^{-7} \text{ mol} \cdot \text{L}^{-1}$

11.27 Because $Ba(OH)_2$ is a strong base,

$Ba(OH)_2 (aq) \longrightarrow Ba^{2+} (aq) + 2\, OH^- (aq)$, 100%.

Then

$[Ba(OH)_2]_0 = [Ba^{2+}], [OH^-] = 2 \times [Ba(OH)_2]_0$, where $[Ba(OH)_2]_0 =$ nominal concentration of $Ba(OH)_2$.

$$\text{moles of Ba(OH)}_2 = \frac{0.43 \ g}{171.36 \ g \cdot mol^{-1}} = 2.5 \times 10^{-3} \ \text{mol}$$

$$[\text{Ba(OH)}_2]_0 = \frac{2.5 \times 10^{-3} \ mol}{0.100 \ L} = 2.5 \times 10^{-2} \ \text{mol} \cdot \text{L}^{-1} = [\text{Ba}^{2+}]$$

$$[\text{OH}^-] = 2 \times [\text{Ba(OH)}_2]_0 = 5.0 \times 10^{-2} \ \text{mol} \cdot \text{L}^{-1}$$

$$[\text{H}_3\text{O}^+] = \frac{K_W}{[OH^-]} = \frac{1.0 \times 10^{-14}}{5.0 \times 10^{-2}} = 2.0 \times 10^{-13} \ \text{mol} \cdot \text{L}^{-1}$$

11.29 (a) The pH of the desired solution = $-\log(0.025) = 1.6$

(b) The actual pH = $-\log\left(\dfrac{200.0 \ mL \times 0.025 \ M}{250.0}\right) = 1.7$

11.31 Because $\text{pH} = -\log[\text{H}_3\text{O}^+]$, $\log[\text{H}_3\text{O}^+] = -\text{pH}$. Taking the antilogs of both sides gives $[\text{H}_3\text{O}^+] = 10^{-\text{pH}} \ \text{mol} \cdot \text{L}^{-1}$.

(a) $[\text{H}_3\text{O}^+] = 10^{-3.3} = 5 \times 10^{-4} \ \text{mol} \cdot \text{L}^{-1}$

(b) $[\text{H}_3\text{O}^+] = 10^{-6.7} \ \text{mol} \cdot \text{L}^{-1} = 2 \times 10^{-7} \ \text{mol} \cdot \text{L}^{-1}$

(c) $[\text{H}_3\text{O}^+] = 10^{-4.4} \ \text{mol} \cdot \text{L}^{-1} = 4 \times 10^{-5} \ \text{mol} \cdot \text{L}^{-1}$

(d) $[\text{H}_3\text{O}^+] = 10^{-5.3} \ \text{mol} \cdot \text{L}^{-1} = 5 \times 10^{-6} \ \text{mol} \cdot \text{L}^{-1}$

11.33 (a) $[\text{HNO}_3] = [\text{H}_3\text{O}^+] = 0.0146 \ \text{mol} \cdot \text{L}^{-1}$

$\text{pH} = -\log(0.0146) = 1.84, \ \text{pOH} = 14.00 - (-1.84) = 12.16$

(b) $[\text{HCl}] = [\text{H}_3\text{O}^+] = 0.11 \ \text{mol} \cdot \text{L}^{-1}$

$\text{pH} = -\log(0.11) = 0.96, \ \text{pOH} = 14.00 - 0.96 = 13.04$

(c) $[\text{OH}^-] = 2 \times [\text{Ba(OH)}_2] = 2 \times 0.0092 \ \text{M} = 0.018 \ \text{mol} \cdot \text{L}^{-1}$

$\text{pOH} = -\log(0.018) = 1.74, \ \text{pH} = 14.00 - 1.74 = 12.26$

(d) $[\text{KOH}]_0 = [\text{OH}^-]$

$$[\text{OH}^-] = \left(\frac{2.00 \ mL}{500 \ mL}\right) \times (0.175 \ \text{mol} \cdot \text{L}^{-1}) = 7.0 \times 10^{-4} \ \text{mol} \cdot \text{L}^{-1}$$

$$\text{pOH} = -\log(7.0 \times 10^{-4}) = 3.15, \ \text{pH} = 14.00 - 3.15 = 10.85$$

(e) $[\text{NaOH}]_0 = [\text{OH}^-]$

$$\text{number of moles of NaOH} = \frac{0.0136 \text{ g}}{40.00 \text{ g} \cdot \text{mol}^{-1}} = 3.40 \times 10^{-4} \text{ mol}$$

$$[\text{NaOH}]_0 = \frac{3.40 \times 10^{-4} \text{ mol}}{0.350 \text{ L}} = 9.71 \times 10^{-4} \text{ mol} \cdot \text{L}^{-1} = [\text{OH}^-]$$

$$\text{pOH} = -\log(9.71 \times 10^{-4}) = 3.01, \ \text{pH} = 14.00 - 3.01 = 10.99$$

(f) $[\text{HBr}]_0 = [\text{H}_3\text{O}^+]$

$$[\text{H}_3\text{O}^+] = \left(\frac{75.0 \text{ mL}}{500 \text{ mL}}\right) \times (3.5 \times 10^{-4} \text{ mol} \cdot \text{L}^{-1}) = 5.2 \times 10^{-5} \text{ mol} \cdot \text{L}^{-1}$$

$$\text{pH} = -\log(5.3 \times 10^{-5}) = 4.28, \ \text{pOH} = 14.00 - 4.28 = 9.72$$

11.35 (a) pH = 13.25, $[\text{H}]^+ = 10^{-13.25} = 5.6 \times 10^{-14} \text{ mol} \cdot \text{L}^{-1}$, $[\text{OH}^-] = 0.18$ mol $\cdot$ L^{-1}

(b) $[\text{OH}^-] = 0.18 \times \left(\frac{500.0 \text{ mL}}{5.00 \text{ mL}}\right) = 18$ mol $\cdot$ L^{-1}

(c) The reaction is $\text{Na}_2\text{O(s)} + \text{H}_2\text{O(l)} \rightarrow 2 \text{ NaOH(aq)}$

The mass of Na_2O is 18 mol $\cdot$ L$^{-1} \times (0.200 \text{ L}) \times \left(\frac{1 \text{ mol Na}_2\text{O}}{2 \text{ mol NaOH}}\right)$

$\times \left(\frac{61.98 \text{ g Na}_2\text{O}}{1 \text{ mol Na}_2\text{O}}\right) = 110$ grams Na_2O.

11.37 $\text{p}K_{\text{a1}} = -\log K_{\text{a1}}$; therefore, after taking antilogs, $K_{\text{a1}} = 10^{-\text{p}K_{\text{a1}}}$.

Acid	$\text{p}K_{\text{a1}}$	K_{a1}
(a) H_3PO_4	2.12	7.6×10^{-3}
(b) H_3PO_3	2.00	1.0×10^{-2}
(c) H_2SeO_3	2.46	3.5×10^{-3}
(d) HSeO_4	1.92	1.2×10^{-2}

(e) The larger K_{a1}, the stronger the acid; therefore

$$H_2SeO_3 < H_3PO_4 < H_3PO_3 < HSeO_4^-.$$

11.39 (a) $HClO_2(aq) + H_2O(l) \rightleftharpoons H_3O^+(aq) + ClO_2^-(aq)$

$$K_a = \frac{[H_3O^+][ClO_2^-]}{[HClO_2]}$$

$ClO_2^-(aq) + H_2O(l) \rightleftharpoons HClO_2(aq) + OH^-(aq)$

$$K_b = \frac{[HClO_2][OH^-]}{[ClO_2^-]}$$

(b) $HCN(aq) + H_2O(l) \rightleftharpoons H_3O^+(aq) + CN^-(aq)$

$$K_a = \frac{[H_3O^+][CN^-]}{[HCN]}$$

$CN^-(aq) + H_2O(l) \rightleftharpoons HCN(aq) + OH^-(aq)$

$$K_b = \frac{[HCN][OH^-]}{[CN^-]}$$

(c) $C_6H_5OH(aq) + H_2O(l) \rightleftharpoons H_3O^+(aq) + C_6H_5O^-(aq)$

$$K_a = \frac{[H_3O^+][C_6H_5O^-]}{[C_6H_5OH]}$$

$C_6H_5O^-(aq) + H_2O(l) \rightleftharpoons C_6H_5OH(aq) + OH^-(aq)$

$$K_b = \frac{[C_6H_5OH][OH^-]}{[C_6H_5O^-]}$$

11.41 Decreasing pK_a will correspond to increasing acid strength because

$pK_a = -\log K_a$. The pK_a values (given in parentheses) determine the

following ordering:

$(CH_3)_2NH_2^+ \; (14.00 - 3.27 = 10.73) < {}^+NH_3OH \; (14.00 - 7.97 = 6.03)$

$< HNO_2 \; (3.37) < HClO_2 \; (2.00).$

Remember that the pK_a for the conjugate acid of a weak base will be given by $pK_a + pK_b = 14$.

11.43 Decreasing pK_b will correspond to increasing base strength because $pK_b = -\log K_b$. The pK_b values (given in parentheses) determine the following ordering:

$$F^- \ (14.00 - 3.45 = 10.55) < CH_3COO^- \ (14.00 - 4.75 = 9.25)$$
$$< C_5H_5N \ (8.75) \ll NH_3 \ (4.75).$$

Remember that the pK_b for the conjugate base of a weak acid will be given by $pK_a + pK_b = 14$.

11.45 Any acid whose conjugate base lies above water in Table 10.3 will be a strong acid; that is, the conjugate base of the acid will be a weaker base than water, and so water will accept the H^+ preferentially. Based on this information, we obtain the following analysis: (a) $HClO_3$, strong; (b) H_2S, weak; (c) HSO_4^-, weak (Note: Even though H_2SO_4 is a strong acid, HSO_4^- is a weak acid. Its conjugate base is SO_4^{2-}.); (d) $CH_3NH_3^+$, weak acid; (e) HCO_3^-, weak; (f) HNO_3, strong; (g) CH_4, weak.

11.47 The larger the K_a, the stronger the corresponding acid. 2,4,6-Trichlorophenol is the stronger acid because the chlorine atoms have a greater electron-withdrawing power than the hydrogen atoms present in the unsubstituted phenol.

11.49 The larger the pK_a of an acid, the stronger the corresponding conjugate base; hence, the order is aniline < ammonia < methylamine < ethylamine.

Although we should not draw conclusions from such a small data set, we might suggest the possibility that

(1) arylamines < ammonia < alkylamines

(2) methyl < ethyl < etc.

(Arylamines are amines in which the nitrogen of the amine is attached to a benzene ring.)

11.51 For oxoacids, the greater the number of highly electronegative O atoms attached to the central atom, the stronger the acid. This effect is related to the increased oxidation number of the central atom as the number of O atoms increases. Therefore, HIO_3 is the stronger acid, with the lower pK_a.

11.53 (a) HCl is the stronger acid, because its bond strength is much weaker than the bond in HF, and bond strength is the dominant factor in determining the strength of binary acids.

(b) $HClO_2$ is stronger; there is one more O atom attached to the Cl atom in $HClO_2$ than in $HClO$. The additional O in $HClO_2$ helps to pull the electron of the H atom out of the H—O bond. The oxidation state of Cl is higher in $HClO_2$ than in $HClO$.

(c) $HClO_2$ is stronger; Cl has a greater electronegativity than Br, making the H—O bond $HClO_2$ more polar than in $HBrO_2$.

(d) $HClO_4$ is stronger; Cl has a greater electronegativity than P.

(e) HNO_3 is stronger. The explanation is the same as that for part (b). HNO_3 has one more O atom.

(f) H_2CO_3 is stronger; C has greater electronegativity than Ge. See part (c).

11.55 (a) The $—CCl_3$ group that is bonded to the carboxyl group, $—COOH$, in trichloroacetic acid, is more electron withdrawing than the $—CH_3$ group in acetic acid. Thus, trichloroacetic acid is the stronger acid.

(b) The $—CH_3$ group in acetic acid has electron-donating properties, which means that it is less electron withdrawing than the $—H$ attached to the carboxyl group in formic acid, $HCOOH$. Thus, formic acid is a slightly stronger acid than acetic acid. However, it is not nearly as strong as trichloroacetic acid. The order is $CCl_3COOH \gg HCOOH > CH_3COOH$.

11.57 (a) Concentration

$(mol \cdot L^{-1})$	CH_3COOH	$+ H_2O$	$\leftrightarrow$	H_3O^+	$+ CH_3CO_2^-$
initial	0.20	—		0	0
change	$-x$	—		$+x$	$+x$
equilibrium	$0.20 - x$	—		x	x

$$K_a = 1.8 \times 10^{-5} = \frac{[H_3O^+][CH_3CO_2^-]}{[CH_3COOH]} = \frac{x^2}{0.20 - x} \approx \frac{x^2}{0.20}$$

$x = [H_3O^+] = 1.9 \times 10^{-3} \ mol \cdot L^{-1}$

$pH = -\log(1.9 \times 10^{-3}) = 2.72; \ pOH = 14.00 - 2.72 = 11.28$

$\%$ deprotonation $= (\dfrac{1.9 \times 10^{-3}}{0.20}) \times 100 = 0.95\%$

(b) The equilibrium table for (b) is similar to that for (a).

$$K_a = 3.0 \times 10^{-1} = \frac{[H_3O^+][CCl_3CO_2^-]}{[CCl_3COOH]} = \frac{x^2}{0.20 - x}$$

$x^2 + 0.30 \, x - 0.060 = 0; \quad x = 0.14, \, -0.44$

The negative root is not possible and can be eliminated.

$x = [H_3O^+] = 0.14 \ mol \cdot L^{-1},$

$pH = -\log(0.14) = 0.85; \ pOH = 14.00 - 0.85 = 13.15$

$\%$ deprotonation $= (\dfrac{0.14}{0.20}) \times 100 = 70\%$

(c)   The equilibrium table for (c) is similar to that for (a).

$$K_a = 1.8 \times 10^{-4} = \frac{[H_3O^+][HCO_2^-]}{[HCOOH]} = \frac{x^2}{0.20 - x} \approx \frac{x^2}{0.20}$$

$$x = [H_3O^+] = 6.0 \times 10^{-3} \text{ mol} \cdot L^{-1}$$

$$pH = -\log(6.0 \times 10^{-3}) = 2.22; \quad pOH = 14.00 - 2.72 = 11.78$$

$$\% \text{ deprotonation} = (\frac{6.0 \times 10^{-3}}{0.20}) \times 100 = 3.0\%$$

(d) Acidity increases when the hydrogen atoms in the methyl group of acetic acid are replaced by atoms that have a higher electronegativity, such as chlorine.

11.59   (a) $HClO_2 + H_2O \rightarrow H_3O^+ + ClO_2^-$

$$[H_3O^+] = [ClO_2^-] = 10^{-pH} = 10^{-1.2} = 0.06 \text{ mol} \cdot L^{-1}$$

$$K_a = \frac{[H_3O^+][ClO_2^-]}{[HClO_2]} = \frac{(0.06)^2}{0.10 - 0.06} = 0.09 \text{ (1 sf)}$$

$$pK_a = -\log(0.09) = 1.0$$

(b) $C_3H_7NH_2 + H_2O \rightarrow C_3H_7NH_3^+ + OH^-$

$$pOH = 14.00 - 11.86 = 2.14$$

$$[C_3H_7NH_3^+] = [OH^-] = 10^{-2.14} = 7.2 \times 10^{-3} \text{ mol} \cdot L^{-1}$$

$$K_b = \frac{[C_3H_7NH_3^+][OH^-]}{[C_3H_7NH_2]} = \frac{(7.2 \times 10^{-3})^2}{0.10 - 7.2 \times 10^{-3}} = 5.6 \times 10^{-4}$$

$$pK_b = -\log(5.6 \times 10^{-4}) = 3.25$$

11.61   (a) Concentration

(mol·L^{-1})	H$_2$O	+	NH$_3$	$\rightleftharpoons$	NH$_4^+$	+	OH$^-$
initial	—		0.057		0		0
change	—		$-x$		$+x$		$+x$
equilibrium	—		$0.057 - x$		x		x

$$K_b = \frac{[NH_4^+][OH^-]}{[NH_3]} = \frac{x \cdot x}{0.057 - x} \approx \frac{x^2}{0.057} = 1.8 \times 10^{-5}$$

$$x = [OH^-] = \sqrt{0.057 \times 1.8 \times 10^{-5}} = 1.0 \times 10^{-3} \ mol \cdot L^{-1}$$

$$pOH = -\log(1.0 \times 10^{-3}) = 3.00, \ pH = 14.00 - 3.00 = 11.00$$

$$percentage \ protonation = \frac{1.0 \times 10^{-3}}{0.057} \times 100\% = 1.8\%$$

(b) Concentration

$(mol \cdot L^{-1})$	NH_2OH	+	H_2O	$\rightleftharpoons$	$^+NH_3OH$	+	OH^-
initial	0.162		—		0		0
change	$-x$		—		$+x$		$+x$
equilibrium	$0.162 - x$		—		x		x

$$K_b = 1.1 \times 10^{-8} = \frac{x^2}{0.162 - x} \approx \frac{x^2}{0.162}$$

$$x = [OH^-] = 4.2 \times 10^{-5} \ mol \cdot L^{-1}$$

$$pOH = -\log(4.2 \times 10^{-5}) = 4.38, \ pH = 14.00 - 4.38 = 9.62$$

$$percentage \ protonation = \frac{4.2 \times 10^{-5}}{0.162} \times 100\% = 0.026\%$$

(c) Concentration

$(mol \cdot L^{-1})$	$(CH_3)_3N$	+	H_2O	$\rightleftharpoons$	$(CH_3)_3NH^+$	+	OH^-
initial	0.35		—		0		0
change	$-x$		—		$+x$		$+x$
equilibrium	$0.35 - x$		—		$+x$		$+x$

$$6.5 \times 10^{-5} = \frac{x^2}{0.35 - x}$$

Assume $x \ll 0.35$

Then $x = 4.8 \times 10^{-3} \ mol \cdot L^{-1}$

$[OH^-] = 4.8 \times 10^{-3} \ mol \cdot L^{-1}$

$pOH = -\log(4.8 \times 10^{-3}) = 2.32, \ pH = 14.00 - 2.32 = 11.68$

$$percentage \ protonation = \frac{4.8 \times 10^{-3}}{0.35} \times 100\% = 1.4\%$$

(d) $pK_b = 14.00 - pK_a = 14.00 - 8.21 = 5.79, \ K_b = 1.6 \times 10^{-6}$

$$\text{codeine} + H_2O \rightleftharpoons \text{codeineH}^+ + OH^-$$

$$K_b = 1.6 \times 10^{-6} = \frac{x^2}{0.073 - x} \approx \frac{x^2}{0.073}$$

$$x = [OH^-] = 1.1 \times 10^{-4} \text{ mol} \cdot L^{-1}$$

$$pOH = -\log(1.1 \times 10^{-4}) = 3.96, \quad pH = 14.00 - 3.96 = 10.04$$

$$\text{percentage protonation} = \frac{1.1 \times 10^{-4}}{0.0073} \times 100\% = 2.5\%$$

11.63 (a) $pH = 4.60$, $[H_3O^+] = 10^{-pH} = 10^{-4.60} = 2.5 \times 10^{-5} \text{ mol} \cdot L^{-1}$

Let x = nominal concentration of HClO, then

Concentration

$(\text{mol} \cdot L^{-1})$	HClO	+	H_2O	$\rightleftharpoons$	H_3O^+	+	ClO^-
nominal	x		—		0		0
equilibrium	$x - 2.5 \times 10^{-5}$		—		2.5×10^{-5}		2.5×10^{-5}

$$K_a = 3.0 \times 10^{-8} = \frac{(2.5 \times 10^{-5})^2}{x - 2.5 \times 10^{-5}}$$

Solve for x; $x = \dfrac{(2.5 \times 10^{-5})^2 + (2.5 \times 10^{-5})(3.0 \times 10^{-8})}{3.0 \times 10^{-8}}$

$$= 2.1 \times 10^{-2} \text{ mol} \cdot L^{-1} = 0.021 \text{ mol} \cdot L^{-1}$$

(b) $pOH = 14.00 - pH = 14.00 - 10.20 = 3.80$

$[OH^-] = 10^{-pOH} = 10^{-3.80} = 1.6 \times 10^{-4}$

Let x = nominal concentration of NH_2NH_2, then

Concentration

$(\text{mol} \cdot L^{-1})$	NH_2NH_2	+	H_2O	$\rightleftharpoons$	$NH_2NH_3^+$	+	OH^-
nominal	x		—		0		0
equilibrium	$x - 1.6 \times 10^{-4}$		—		1.6×10^{-4}		1.6×10^{-4}

$$K_b = 1.7 \times 10^{-6} = \frac{(1.6 \times 10^{-4})^2}{x - 1.6 \times 10^{-4}}$$

Solve for x; $x = 1.5 \times 10^{-2} \text{ mol} \cdot L^{-1}$

11.65 Concentration

(mol·L^{-1})	C$_6$H$_5$COOH	+ H$_2$O	$\rightleftharpoons$	H$_3$O$^+$	+ C$_6$H$_5$CO$_2$$^-$
initial	0.110	—		0	0
change	$-x$	—		$+x$	$+x$
equilibrium	$0.110 - x$	—		x	x

$$H_2O + octylamine \rightleftharpoons octylamineH^+ + OH^-$$

$$x = 0.024 \times 0.110 \; mol \cdot L^{-1} = [H_3O^+] = [C_6H_5CO_2{}^-]$$

$$K_a = \frac{[H_3O^+][C_6H_5COO^-]}{[C_6H_5COOH]} = \frac{(0.024 \times 0.110)^2}{(1 - 0.024) \times 0.110} = 6.5 \times 10^{-5}$$

$$pH = -\log(2.6 \times 10^{-3}) = 2.58$$

11.67 The change in the concentration of octylamine is

$x = 0.067 \times 0.10 = 0.0067 \; mol \cdot L^{-1}$. Thus the equilibrium table is

(mol·L^{-1})	H$_2$O	+ octylamine	$\rightleftharpoons$	octylamineH$^+$	+ OH$^-$
initial	–	0.100		0	0
change	–	$-x$		$+x$	$+x$
equilibrium	–	$0.100 - x$		x	x

$$x = 0.067 \times 0.100 \; mol \cdot L^{-1} = [OH^-] = [octylamine\,H^+]$$

$$K_b = \frac{[OH^-][octylamine\,H^+]}{[octylamine]} = \frac{(0.067 \times 0.100)^2}{(1 - 0.067) \times 0.100} = 4.8 \times 10^{-5}$$

$$pH = 14 - \left[-\log(6.7 \times 10^{-3})\right] = 11.83$$

11.69 $K = \dfrac{[NO_2{}^-][NH_4{}^+]}{[HNO_2][NH_3]} = \dfrac{K_a(HNO_2) \times K_b(NH_3)}{K_w}$

$$= \frac{(4.3 \times 10^{-4})(1.8 \times 10^{-5})}{1.0 \times 10^{-14}} = 7.7 \times 10^5$$

This means that the reaction is in favor of the products.

11.71 (a) less than 7, $NH_4{}^+(aq) + H_2O(l) \rightleftharpoons H_3O^+(aq) + NH_3(aq)$

(b) greater than 7, $H_2O(l) + CO_3{}^{2-}(aq) \rightleftharpoons HCO_3{}^-(aq) + OH^-(aq)$

(c) greater than 7, $H_2O(l) + F^-(aq) \rightleftharpoons HF(aq) + OH^-(aq)$

(d) neutral

(e) less than 7,

$Al(H_2O)_6^{3+}(aq) + H_2O(l) \rightleftharpoons H_3O^+(aq) + Al(H_2O)_5OH^{2+}(aq)$

(f) less than 7,

$Cu(H_2O)_6^{2+}(aq) + H_2O(l) \rightleftharpoons H_3O^+(aq) + Cu(H_2O)_5OH^+(aq)$

11.73 The rank in increasing solution pH is:

(c) < (a) < (b) < (d)

Justification: (d) is a weak base, pH will be the highest (pH = 8.97)
(c) is a weak acid but the concentration high (pH = 2.72, see Exercise
11.57(a)) which gives this solution the lowest pH; (a) is a strong acid, but
the concentration is low ($pH = -\log(1.0 \times 10^{-5}) = 5.00$); (b)
is a conjugated acid of CH_3NH_2 ($K_b = 3.6 \times 10^{-4}$) with a $K_a = 2.8 \times 10^{-11}$,
which is a very weak acid (pH = 5.63).

11.75 (a) $K_b = \dfrac{K_w}{K_a} = \dfrac{1.00 \times 10^{-14}}{1.8 \times 10^{-5}} = 5.6 \times 10^{-10}$

Concentration

$(mol \cdot L^{-1})\ CH_3CO_2^-(aq) + H_2O(l) \rightleftharpoons HCH_3CO_2(aq) + OH^-(aq)$

initial	0.63	—	0	0
change	$-x$	—	$+x$	$+x$
equilibrium	$0.63 - x$	—	x	x

$K_b = \dfrac{[HCH_3CO_2][OH^-]}{[CH_3CO_2^-]} = 5.6 \times 10^{-10} = \dfrac{x^2}{0.63 - x} \approx \dfrac{x^2}{0.63}$

$x = 1.9 \times 10^{-5} = [OH^-]$, $pOH = -\log(1.9 \times 10^{-5}) = 4.72$

$pH = 14.00 - pOH = 14.00 - 4.72 = 9.28$

(b) $K_a = \dfrac{K_w}{K_b} = \dfrac{1.00 \times 10^{-14}}{1.8 \times 10^{-15}} = 5.6 \times 10^{-10}$

Concentration (mol·L^{-1}) $NH_4^+(aq) + H_2O(l) \rightleftharpoons H_3O^+(aq) + NH_3(aq)$

initial		0.19	—	0	0
change		$-x$	—	$+x$	$+x$
equilibrium		$0.19 - x$	—	x	x

$$K_a = \frac{[H_3O^+][NH_3]}{[NH_4Cl]} = 5.6 \times 10^{-10} = \frac{x^2}{0.19 - x} \approx \frac{x^2}{0.19}$$

$$x = 1.0 \times 10^{-5} \text{ mol·L}^{-1} = [H_3O^+]$$

$$pH = -\log(1.0 \times 10^{-5}) = 5.00$$

(c) Concentration

$(\text{mol·L}^{-1})\ Al(H_2O)_6^{3+}(aq) + H_2O(l) \rightleftharpoons H_3O^+(aq) + Al(H_2O)_5OH^{2+}(aq)$

initial	0.055	—	0	0
change	$-x$	—	$+x$	$+x$
equilibrium	$0.055 - x$	—	x	x

$$K_a = \frac{[H_3O^+][Al(H_2O)_5OH^{2+}]}{[Al(H_2O)_6^{3+}]} = 1.4 \times 10^{-5} = \frac{x^2}{0.055 - x} \approx \frac{x^2}{0.055}$$

$$x = 8.8 \times 10^{-4} \text{ mol·L}^{-1} = [H_3O^+]$$

$$pH = -\log(8.8 \times 10^{-4}) = 3.06$$

(d) Concentration

$(\text{mol·L}^{-1})\quad H_2O(l) + CN^-(aq) \rightleftharpoons HCN(aq) + OH^-(aq)$

initial	—	0.065	0	0
change	—	$-x$	$+x$	$+x$
equilibrium	—	$0.065 - x$	x	x

$$K_b = \frac{K_w}{K_a} = \frac{1.00 \times 10^{-14}}{4.9 \times 10^{-10}} = 2.0 \times 10^{-5} = \frac{[HCN][OH^-]}{[CN^-]} = \frac{x^2}{0.065 - x} \approx \frac{x^2}{0.065}$$

$$x = [OH^-] = 1.1 \times 10^{-3} \text{ mol·L}^{-1}$$

$$pOH = -\log(1.1 \times 10^{-3}) = 2.96,\ pH = 11.04$$

11.77 Concentration

$(mol \cdot L^{-1})$	$CH_3NH_3^+(aq)$	+ $H_2O(l)$	$\rightleftharpoons$ $H_3O^+(aq)$	+ $CH_3NH_2(aq)$
initial	0.510	—	0	0
change	$-x$	—	$+x$	$+x$
equilibrium	$0.510 - x$	—	x	x

$$K_a = \frac{[H_3O^+][CH_3NH_2]}{[CH_3NH_3^+]} = 2.8 \times 10^{-11} = \frac{x^2}{0.510 - x} \approx \frac{x^2}{0.510}$$

$$x = 3.8 \times 10^{-6} \ mol \cdot L^{-1} = [H_3O^+]$$

$$pH = -\log(3.8 \times 10^{-6}) = 5.42$$

11.79 pH of the NaA is 10.35, $[H^+] = 10^{-10.35}$, $[OH^-] = 10^{-3.65} = 2.2 \times 10^{-4}$

Hydrolysis reaction: $A^-(aq) + H_2O(l) \rightarrow HA(aq) + OH^-(aq)$

$$K_b = \frac{[HA][OH^-]}{[A^-]} = \frac{(2.2 \times 10^{-4})^2}{(0.010)} = 4.84 \times 10^{-6}$$

$K_a = K_w/K_b = 2.1 \times 10^{-9}$. The formula of the acid is HBrO.

11.81 (a) $\dfrac{0.020 \ mol \cdot L^{-1} \ NaCH_3CO_2 \times 0.150 \ L}{0.500 \ L} = 0.00600 \ mol \cdot L^{-1}$

Concentration

$(mol \cdot L^{-1})$	$H_2O(l)$	+ $CH_3CO_2^-(aq)$	$\rightleftharpoons$ $CH_3COOH(aq)$	+ $OH^-(aq)$
initial	—	0.00600	0	0
change	—	$-x$	$+x$	$+x$
equilibrium	—	$0.00600 - x$	x	x

$$K_b = \frac{K_w}{K_a} = \frac{1.00 \times 10^{-14}}{1.8 \times 10^{-5}} = 5.6 \times 10^{-10} = \frac{[CH_3COOH][OH^-]}{[CH_3CO_2^-]}$$

$$5.6 \times 10^{-10} = \frac{x^2}{0.00600 - x} \approx \frac{x^2}{0.00600}$$

$$x = 1.8 \times 10^{-6} \ mol \cdot L^{-1} = [CH_3COOH]$$

(b) $\left(\dfrac{2.16 \text{ g NH}_4\text{Br}}{400 \text{ mL}}\right)\left(\dfrac{1 \text{ mL}}{10^{-3} \text{ L}}\right)\left(\dfrac{1 \text{ mol NH}_4\text{Br}}{97.95 \text{ g NH}_4\text{Br}}\right)$

$= 0.0551 \,(\text{mol NH}_4\text{Br}) \cdot \text{L}^{-1}$

Concentration

(mol·L^{-1})	NH$_4^+$(aq)	+	H$_2$O(l)	$\rightleftharpoons$	H$_3$O$^+$(aq)	+	NH$_3$(aq)
initial	0.0551		—		0		0
change	$-x$		—		$+x$		$+x$
equilibrium	$0.0551 - x$		—		x		x

$K_a = \dfrac{K_w}{K_b} = \dfrac{1.00 \times 10^{-14}}{1.8 \times 10^{-5}} = 5.6 \times 10^{-10} = \dfrac{[\text{NH}_3][\text{H}_3\text{O}^+]}{[\text{NH}_4^+]}$

$5.6 \times 10^{-10} = \dfrac{x^2}{0.0551 - x} \approx \dfrac{x^2}{0.0551}$

$x = 5.5 \times 10^{-6} \text{ mol} \cdot \text{L}^{-1} = [\text{H}_3\text{O}^+]$ and $\text{pH} = -\log(5.5 \times 10^{-6}) = 5.26$

11.83 $\text{pH} = \frac{1}{2}(\text{p}K_{a_1} + \text{p}K_{a_2}) = \frac{1}{2}(2.340 + 9.886) = 6.113$

11.85 The initial concentrations of HSO$_4^-$ and H$_3$O$^+$ are both 0.15 mol·L^{-1} as a result of the complete ionization of H$_2$SO$_4$ in the first step. The second ionization is incomplete.

Concentration

(mol·L^{-1})	HSO$_4^-$	+	H$_2$O	$\rightleftharpoons$	H$_3$O$^+$	+	SO$_4^{2-}$
initial	0.15		—		0.15		0
change	$-x$		—		$+x$		$+x$
equilibrium	$0.15 - x$		—		$0.15 + x$		x

$$K_{a2} = 1.2 \times 10^{-2} = \frac{[H_3O^+][SO_4{}^{2-}]}{[HSO_4{}^-]} = \frac{(0.15 + x)(x)}{0.15 - x}$$

$$x^2 + 0.162x - 1.8 \times 10^{-3} = 0$$

$$x = \frac{-0.162 + \sqrt{(0.162)^2 + (4)(1.8 \times 10^{-3})}}{2} = 0.0104 \text{ mol} \cdot L^{-1}$$

$$[H_3O^+] = 0.15 + x = (0.15 + 0.0104) \text{ mol} \cdot L^{-1} = 0.16 \text{ mol} \cdot L^{-1}$$

$$pH = -\log(0.16) = 0.80$$

11.87 (a) Because $K_{a2} \ll K_{a1}$, the second ionization can be ignored.

Concentration

(mol·L^{-1})	H$_2$CO$_3$	+	H$_2$O	⇌	H$_3$O$^+$	+	HCO$_3{}^-$
initial	0.010		—		0		0
change	$-x$		—		$+x$		$+x$
equilibrium	0.010 − x		—		x		x

$$K_{a1} = \frac{[H_3O^+][HCO_3{}^-]}{[H_2CO_3]} = \frac{x^2}{0.010 - x} \approx \frac{x^2}{0.010} = 4.3 \times 10^{-7}$$

$$x = [H_3O^+] = 6.6 \times 10^{-5} \text{ mol} \cdot L^{-1}$$

$$pH = -\log(6.6 \times 10^{-5}) = 4.18$$

(b) Because $K_{a2} \ll K_{a1}$, the second ionization can be ignored.

Concentration

(mol·L^{-1})	(COOH)$_2$	+	H$_2$O	⇌	H$_3$O$^+$	+	(COOH)CO$_2{}^-$
initial	0.10		—		0		0
change	$-x$		—		$+x$		$+x$
equilibrium	0.10 − x		—		x		x

$$K_{a1} = 5.9 \times 10^{-2} = \frac{[H_3O^+][(COOH)CO_2{}^-]}{[(COOH)_2]} = \frac{x^2}{0.10 - x}$$

$$x^2 + 5.9 \times 10^{-2} x - 5.9 \times 10^{-3} = 0$$

$$x = \frac{-5.9 \times 10^{-2} + \sqrt{(5.9 \times 10^{-2})^2 + (4)(5.9 \times 10^{-3})}}{2} = 0.053 \text{ mol} \cdot L^{-1}$$

$$pH = -\log(0.053) = 1.28$$

(c) Because $K_{a2} \ll K_{a1}$, the second ionization can be ignored.

Concentration $(mol \cdot L^{-1})$ $\quad H_2S \quad + \quad H_2O \quad \rightleftharpoons \quad H_3O^+ \quad + \quad HS^-$

equilibrium $\qquad\qquad\qquad 0.20 - x \qquad\quad - \qquad\qquad x \qquad\quad x$

$K_{a1} = 1.3 \times 10^{-7} = \dfrac{[H_3O^+][HS^-]}{[H_2S]} = \dfrac{x^2}{0.20 - x} \approx \dfrac{x^2}{0.20}$

$x = [H_3O^+] = 1.6 \times 10^{-4} \; mol \cdot L^{-1}$

$pH = -\log(1.6 \times 10^{-4}) = 3.80$

11.89 (a) The pH is given by $pH = \frac{1}{2}(pK_{a1} + pK_{a2})$. From Table 11.10, we find

$K_{a1} = 1.5 \times 10^{-2} \quad pK_{a1} = 1.82$

$K_{a2} = 1.2 \times 10^{-7} \quad pK_{a2} = 6.92$

$pH = \frac{1}{2}(1.82 + 6.92) = 4.37$

(b) The pH of a salt solution of a polyprotic acid is independent of the concentration of the salt; therefore, $pH = 4.37$.

11.91 (a) The pH is given by $pH = \frac{1}{2}(pK_{a1} + pK_{a2})$. For the monosodium salt, the pertinent values are pK_{a1} and pK_{a2} :

$pH = \frac{1}{2}(3.14 + 5.95) = 4.55$

(b) For the disodium salt, the pertinent values are pK_{a2} and pK_{a3} :

$pH = \frac{1}{2}(5.95 + 6.39) = 6.17$

11.93 The equilibrium reactions of interest are

$H_2CO_3(aq) + H_2O(l) \rightleftharpoons H_3O^+(aq) + HCO_3^-(aq) \quad K_{a1} = 4.3 \times 10^{-7}$

$HCO_3^-(aq) + H_2O(l) \rightleftharpoons H_3O^+(aq) + CO_3^{2-}(aq) \quad K_{a2} = 5.6 \times 10^{-11}.$

Because the second ionization constant is much smaller than the first, we can assume that the first step dominates:

Concentration

$$(\text{mol} \cdot \text{L}^{-1}) \quad H_2CO_3(aq) + H_2O(l) \rightleftharpoons H_3O^+(aq) + HCO_3^-(aq)$$

initial	0.0456	—	0	0
change	$-x$	—	$+x$	$+x$
final	$0.0456 - x$	—	$+x$	$+x$

$$K_{a1} = \frac{[H_3O^+][HCO_3^-]}{[H_2CO_3]}$$

$$4.3 \times 10^{-7} = \frac{(x)(x)}{0.0456 - x} = \frac{x^2}{0.0456 - x}$$

Assume that $x \ll 0.0456$.

Then $x^2 = (4.3 \times 10^{-7})(0.0456)$.

$x = 1.4 \times 10^{-4}$

Because $x < 1\%$ of 0.0456, the assumption was valid.

$x = [H_3O^+] = [HCO_3^-] = 1.4 \times 10^{-4} \text{ mol} \cdot \text{L}^{-1}$

This means that the concentration of H_2CO_3 is

$0.0455 \text{ mol} \cdot \text{L}^{-1} - 0.00014 \text{ mol} \cdot \text{L}^{-1} = 0.0455 \text{ mol} \cdot \text{L}^{-1}$.

We can then use the other equilibria to determine the remaining concentrations:

$$K_{a2} = \frac{[H_3O^+][CO_3^{2-}]}{[HCO_3^-]}$$

$$5.6 \times 10^{-11} = \frac{(1.4 \times 10^{-4})[CO_3^{2-}]}{(1.4 \times 10^{-4})}$$

$[CO_3^{2-}] = 5.6 \times 10^{-11} \text{ mol} \cdot \text{L}^{-1}$

Because $5.6 \times 10^{-11} \ll 1.4 \times 10^{-4}$, the initial assumption that the first ionization would dominate is valid.

To calculate $[OH^-]$, we use the K_w relationship:

$$K_w = [H_3O^+][OH^-]$$

$$[OH^-] = \frac{K_w}{[H_3O^+]} = \frac{1.00 \times 10^{-14}}{1.4 \times 10^{-4}} = 7.1 \times 10^{-11} \text{ mol} \cdot L^{-1}$$

In summary, $[H_2CO_3] = 0.0455 \text{ mol} \cdot L^{-1}$, $[H_3O^+] = [HCO_3^-]$
$$= 1.4 \times 10^{-4} \text{ mol} \cdot L^{-1},$$

$[CO_3^{2-}] = 5.6 \times 10^{-11} \text{ mol} \cdot L^{-1}$, $[OH^-] = 7.1 \times 10^{-11} \text{ mol} \cdot L^{-1}$.

11.95 The equilibrium reactions of interest are now the base forms of the carbonic acid equilibria, so K_b values should be calculated for the following changes:

$$CO_3^{2-}(aq) + H_2O(l) \rightleftharpoons HCO_3^-(aq) + OH^-(aq)$$

$$K_{b1} = \frac{K_w}{K_{a2}} = \frac{1.00 \times 10^{-14}}{5.6 \times 10^{-11}} = 1.8 \times 10^{-4}$$

$$HCO_3^-(aq) + H_2O(l) \rightleftharpoons H_2CO_3(aq) + OH^-(aq)$$

$$K_{b2} = \frac{K_w}{K_{a1}} = \frac{1.00 \times 10^{-14}}{4.3 \times 10^{-7}} = 2.3 \times 10^{-8}$$

Because the second hydrolysis constant is much smaller than the first, we can assume that the first step dominates:

Concentration

$(\text{mol} \cdot L^{-1})$	$CO_3^{2-}(aq)$	$+$	$H_2O(l)$	$\rightleftharpoons$	$HCO_3^-(aq)$	$+$	$OH^-(aq)$
initial	0.0456		—		0		0
change	$-x$		—		$+x$		$+x$
final	$0.0456 - x$		—		$+x$		$+x$

$$K_{b1} = \frac{[HCO_3^-][OH^-]}{[CO_3^{2-}]}$$

$$1.8 \times 10^{-4} = \frac{(x)(x)}{0.0456 - x} = \frac{x^2}{[0.0456 - x]}$$

Assume that $x \ll 0.0456$.

Then $x^2 = (1.8 \times 10^{-4})(0.0456)$.

$x = 2.9 \times 10^{-3}$

Because $x > 5\%$ of 0.0456, the assumption was not valid and the full expression should be solved using the quadratic equation:

$x^2 + 1.8 \times 10^{-4} x - (1.8 \times 10^{-4})(0.0456) = 0$

Solving using the quadratic equation gives $x = 0.0028 \text{ mol} \cdot \text{L}^{-1}$.

$x = [\text{HCO}_3{}^-] = [\text{OH}^-] = 0.0028 \text{ mol} \cdot \text{L}^{-1}$

Therefore, $[\text{CO}_3{}^{2-}] = 0.0456 \text{ mol} \cdot \text{L}^{-1} - 0.0028 \text{ mol} \cdot \text{L}^{-1} = 0.0428 \text{ mol} \cdot \text{L}^{-1}$.

We can then use the other equilibria to determine the remaining concentrations:

$$K_{b2} = \frac{[\text{H}_2\text{CO}_3][\text{OH}^-]}{[\text{HCO}_3{}^-]}$$

$$2.3 \times 10^{-8} = \frac{[\text{H}_2\text{CO}_3](0.0028)}{(0.0028)}$$

$[\text{H}_2\text{CO}_3] = 2.3 \times 10^{-8} \text{ mol} \cdot \text{L}^{-1}$

Because $2.3 \times 10^{-8} \ll 0.0028$, the initial assumption that the first hydrolysis would dominate is valid. To calculate $[\text{H}_3\text{O}^+]$, we use the K_w relationship:

$K_w = [\text{H}_3\text{O}^+][\text{OH}^-]$

$$[\text{H}_3\text{O}^+] = \frac{K_w}{[\text{OH}^-]} = \frac{1.00 \times 10^{-14}}{0.0028} = 3.6 \times 10^{-12} \text{ mol} \cdot \text{L}^{-1}$$

In summary, $[\text{H}_2\text{CO}_3] = 2.3 \times 10^{-8} \text{ mol} \cdot \text{L}^{-1}$, $[\text{OH}^-] = [\text{HCO}_3{}^-] =$ 0.0028 $\text{mol} \cdot \text{L}^{-1}$, $[\text{CO}_3{}^{2-}] = 0.0428 \text{ mol} \cdot \text{L}^{-1}$, $[\text{H}_3\text{O}^+] = 3.6 \times 10^{-12} \text{ mol} \cdot \text{L}^{-1}$.

11.97 (a) Phosphorous acid: The two pK_a values are 2.00 and 6.59. Because pH $= 6.30$ lies between pK_{a1} and pK_{a2}, the dominant form will be the singly deprotonated HA^- ion.

(b) Oxalic acid: The two pK_a values are 1.23 and 4.19. Because pH = 6.30 lies above pK_{a2}, the species present in largest concentration will be the doubly deprotonated A^{2-} ion.

(c) Hydrosulfuric acid: The two pK_a values are 6.89 and 14.15. Because pH = 6.30 lies below both pK_a values, the species present in highest concentrations will be the fully protonated H_2A form.

11.99 The equilibria present in solution are

$$H_2SO_3(aq) + H_2O(l) \rightleftharpoons H_3O^+(aq) + HSO_3^-(aq) \quad K_{a1} = 1.5 \times 10^{-2}$$
$$HSO_3^-(aq) + H_2O(l) \rightleftharpoons H_3O^+(aq) + SO_3^{2-}(aq) \quad K_{a2} = 1.2 \times 10^{-7}.$$

The calculation of the desired concentrations follows exactly after the method derived in Equation 25, substituting H_2SO_3 for H_2CO_3, HSO_3^- for HCO_3^-, and SO_3^{2-} for CO_3^{2-}. First, calculate the quantity f (at pH = 5.50 $[H_3O^+] = 10^{-5.5} = 3.2 \times 10^{-6}$ mol $\cdot$ L^{-1}):

$$f = [H_3O^+]^2 + [H_3O^+]K_{a1} + K_{a1}K_{a2}$$
$$= (3.2 \times 10^{-6})^2 + (3.2 \times 10^{-6})(1.5 \times 10^{-2}) + (1.5 \times 10^{-2})(1.2 \times 10^{-7})$$
$$= 5.0 \times 10^{-8}$$

The fractions of the species present are then given by

$$\alpha(H_2SO_3) = \frac{[H_3O^+]}{f} = \frac{(3.2 \times 10^{-6})^2}{5.0 \times 10^{-8}} = 2.1 \times 10^{-4}$$

$$\alpha(HSO_3^-) = \frac{[H_3O^+]K_{a1}}{f} = \frac{(3.2 \times 10^{-6})(1.5 \times 10^{-2})}{5.0 \times 10^{-8}} = 0.96$$

$$\alpha(SO_3^{2-}) = \frac{K_{a1}K_{a2}}{f} = \frac{(1.5 \times 10^{-2})(1.2 \times 10^{-7})}{5.0 \times 10^{-8}} = 0.036$$

Thus, in a 0.150 mol $\cdot$ L^{-1} solution at pH 5.50, the dominant species will be HSO_3^- with a concentration of $(0.150$ mol $\cdot$ L$^{-1})(0.96) = 0.14$ mol $\cdot$ L^{-1}.

The concentration of H_2SO_3 will be $(2.1 \times 10^{-4})(0.150$ mol $\cdot$ L$^{-1}) = 3.2 \times 10^{-5}$ mol $\cdot$ L^{-1} and the concentration of SO_3^{2-} will be

$(0.036)(0.150 \text{ mol} \cdot L^{-1}) = 0.0054 \text{ mol} \cdot L^{-1}$.

11.101 (a) Concentration

$(\text{mol} \cdot L^{-1})$	$B(OH)_3$	$+$	$2 H_2O$	$\rightleftharpoons$	H_3O^+	$+$	$B(OH)_4^-$
initial	1.0×10^{-4}		—		0		0
change	$-x$		—		$+x$		$+x$
equilibrium	$1.0 \times 10^{-4} - x$		—		x		x

$$K_a = 7.2 \times 10^{-10} = \frac{[H_3O^+][B(OH)_4^-]}{[B(OH)_3]} = \frac{x^2}{1.0 \times 10^{-4} - x} \approx \frac{x^2}{1.0 \times 10^{-4}}$$

$x = [H_3O^+] = 2.7 \times 10^{-7} \text{ mol} \cdot L^{-1}$

$pH = -\log(2.7 \times 10^{-7}) = 6.57$

Note: This value of $[H_3O^+]$ is not much different from the value for pure water, $1.0 \times 10^{-7} \text{ mol} \cdot L^{-1}$; therefore, it is at the lower limit of safely ignoring the contribution to $[H_3O^+]$ from the autoprotolysis of water. The exercise should be solved by simultaneously considering both equilibria. Concentration

$(\text{mol} \cdot L^{-1})$	$B(OH)_3$	$+$	$2 H_2O$	$\rightleftharpoons$	H_3O^+	$+$	$B(OH)_4^-$
equilibrium	$1.0 \times 10^{-4} - x$		—		x		y

Concentration $(\text{mol} \cdot L^{-1})$	$2 H_2O$	$\rightleftharpoons$	H_3O^+	$+$	OH^-
equilibrium	—		x		z

Because there are now two contributions to $[H_3O^+]$, $[H_3O^+]$ is no longer equal to $[B(OH)_4^-]$, nor is it equal to $[OH^-]$, as in pure water. To avoid a cubic equation, x will again be ignored relative to $1.0 \times 10^{-4} \text{ mol} \cdot L^{-1}$. This approximation is justified by the approximate calculation above, and because K_a is very small relative to 1.0×10^{-4}. Let $a = $ initial concentration of $B(OH)_3$, then

$$K_a = 7.2 \times 10^{-10} = \frac{xy}{a-x} \approx \frac{xy}{a} \text{ or } y = \frac{aK_a}{x}$$

$$K_w = 1.0 \times 10^{-14} = xz.$$

Electroneutrality requires

$x = y + z$ or $z = x - y$; hence, $K_w = xz = x(x - y)$.

Substituting for y from above:

$$x \times \left(x - \frac{aK_a}{x}\right) = K_w$$

$$x^2 - aK_a = K_w$$

$$x^2 = K_w + aK_a$$

$$x = \sqrt{K_w + aK_a} = \sqrt{1.0 \times 10^{-14} + 1.0 \times 10^{-4} \times 7.2 \times 10^{-10}}$$

$$x = 2.9 \times 10^{-7} \text{ mol} \cdot \text{L}^{-1} = [H_3O^+]$$

$$pH = -\log(2.9 \times 10^{-7}) = 6.54$$

This value is slightly, but measurably, different from the value 6.57 obtained by ignoring the contribution to $[H_3O^+]$ from water.

(b) In this case, the second ionization can safely be ignored; $K_{a2} \ll K_{a1}$.

Concentration

$(\text{mol} \cdot \text{L}^{-1})$	H_3PO_4	$+$	H_2O	$\rightleftharpoons$	H_3O^+	$+$	$H_2PO_4^-$
initial	0.015		—		0		0
change	$-x$		—		$+x$		$+x$
equilibrium	$0.015 - x$		—		x		x

$$K_{a1} = 7.6 \times 10^{-3} = \frac{x^2}{0.015 - x}$$

$$x^2 + 7.6 \times 10^{-3}x - 1.14 \times 10^{-4} = 0$$

$$x = [H_3O^+] = \frac{-7.6 \times 10^{-3} + \sqrt{(7.6 \times 10^{-3})^2 + 4.56 \times 10^{-4}}}{2}$$

$$= 7.5 \times 10^{-3} \text{ mol} \cdot \text{L}^{-1}$$

$$pH = -\log(7.5 \times 10^{-3}) = 2.12$$

(c) In this case, the second ionization can safely be ignored; $K_{a2} \ll K_{a1}$.

Concentration

(mol·L^{-1})	H$_2$SO$_3$	+	H$_2$O	$\rightleftharpoons$	H$_3$O$^+$	+	HSO$_3^-$
initial	0.1		—		0		0
change	$-x$		—		$+x$		$+x$
equilibrium	$0.1-x$		—		x		x

$$K_{a1} = 1.5 \times 10^{-2} = \frac{x^2}{0.10 - x}$$

$$x^2 + 1.5 \times 10^{-2} x - 1.5 \times 10^{-3} = 0$$

$$x = [H_3O^+] = \frac{-1.5 \times 10^{-2} + \sqrt{(1.5 \times 10^{-2})^2 + 6.0 \times 10^{-3}}}{2} = 0.032 \text{ mol} \cdot L^{-1}$$

$$pH = -\log(0.032) = 1.49$$

11.103 The three equilibria involved are

$$H_3PO_4(aq) \rightleftharpoons H_2PO_4^-(aq) + H_3O^+(aq), \quad K_{a_1} = 7.6 \times 10^{-3} = \frac{[H_3O^+][H_2PO_4^-]}{[H_3PO_4]}$$

$$H_2PO_4^-(aq) \rightleftharpoons HPO_4^{2-}(aq) + H_3O^+(aq), \quad K_{a_2} = 6.2 \times 10^{-8} = \frac{[H_3O^+][HPO_4^{2-}]}{[H_2PO_4^-]}$$

$$HPO_4^{2-}(aq) \rightleftharpoons PO_4^{3-}(aq) + H_3O^+(aq), \quad K_{a_3} = 2.1 \times 10^{-13} = \frac{[H_3O^+][PO_4^{3-}]}{[HPO_4^{2-}]}$$

We also know that the combined concentration of all the phosphate species is

$$[H_3PO_4] + [H_2PO_4^-] + [HPO_4^{2-}] + [PO_4^{3-}] = 1.5 \times 10^{-2}$$

and the hydronium ion concentration is

$$[H] = 10^{-pH} = 10^{-2.25} = 5.62 \times 10^{-3} \text{ mol} \cdot L^{-1}.$$

At this point it is a matter of solving this set of simultaneous equations to obtain the concentrations of the phosphate containing species. We start by dividing both sides of the equilibrium constant expressions above by the given hydronium ion concentration to obtain three ratios:

$$1.35 = \frac{[H_2PO_4^-]}{[H_3PO_4]}, \quad 1.10 \times 10^{-5} = \frac{[HPO_4^{2-}]}{[H_2PO_4^-]}, \quad \text{and} \quad 3.74 \times 10^{-11} = \frac{[PO_4^{3-}]}{[HPO_4^{2-}]}$$

Through rearrangement and substitution of these three ratios, we can obtain the following expressions:

$[H_2PO_4^-] = 1.35 \cdot [H_3PO_4]$,

$[HPO_4^{2-}] = 1.10 \times 10^{-5} \cdot [H_2PO_4^-] = 1.10 \times 10^{-5} \cdot 1.35 \cdot [H_3PO_4]$

$\qquad = 1.48 \times 10^{-5} \cdot [H_3PO_4]$, and

$[PO_4^{3-}] = 3.74 \times 10^{-11} \cdot [HPO_4^{2-}] = 3.74 \times 10^{-11} \cdot 1.48 \times 10^{-5} \cdot [H_3PO_4]$

$\qquad = 5.54 \times 10^{-16} \cdot [H_3PO_4]$

Substituting these expressions back into the sum:

$[H_3PO_4] + [H_2PO_4^-] + [HPO_4^{2-}] + [PO_4^{3-}] =$

$= [H_3PO_4] + (1.35 \cdot [H_3PO_4]) + (1.48 \times 10^{-5} \cdot [H_3PO_4]) + (5.54 \times 10^{-16} \cdot [H_3PO_4])$

$= 1.5 \times 10^{-2}$

we find:

$[H_3PO_4] = 6.4 \times 10^{-3} \ mol \cdot L^{-1}$,

$[H_2PO_4^-] = 1.35 \cdot [H_3PO_4] = 8.6 \times 10^{-3} \ mol \cdot L^{-1}$,

$[HPO_4^{2-}] = 1.48 \times 10^{-5} \cdot [H_3PO_4] = 9.5 \times 10^{-8} \ mol \cdot L^{-1}$, and

$[PO_4^{3-}] = 5.54 \times 10^{-16} \cdot [H_3PO_4] = 3.5 \times 10^{-18} \ mol \cdot L^{-1}$.

11.105 We can use the relationship derived in the text,

$[H_3O^+]^2 - [HA]_{initial}[H_3O^+] - K_w = 0$, in which HA is any strong acid.

$[H_3O^+]^2 - (6.55 \times 10^{-7})[H_3O^+] - (1.00 \times 10^{-14}) = 0$

Solving using the quadratic equation gives

$[H_3O^+] = 6.70 \times 10^{-7}$, pH = 6.174.

This value is slightly lower than the value calculated, based on the acid

concentration alone (pH $= -\log(6.55 \times 10^{-7}) = 6.184$).

11.107 We can use the relationship derived in the text,

$[H_3O^+]^2 + [B]_{initial}[H_3O^+] - K_w = 0$, in which B is any strong base.

$[H_3O^+]^2 + (9.78 \times 10^{-8})[H_3O^+] - (1.00 \times 10^{-14}) = 0$

Solving using the quadratic equation gives

$[H_3O^+] = 6.24 \times 10^{-8}$, pH = 7.205.

This value is higher than the value calculated, based on the base concentration alone $(\text{pOH} = -\log(9.78 \times 10^{-8}) = 7.009)$.

11.109 In general, when $[\text{H}_3\text{O}^+] \leq 10^{-6}$ (that is, when the pH lies between 6 and 7), the autoprotolysis of water must be taken into account. We can use Equation 18b:

$$x^3 + K_a x^2 - (K_w + K_a \cdot [\text{HA}]_{\text{initial}})x - K_w \cdot K_a = 0, \text{ where } x = [\text{H}_3\text{O}^+]$$

To solve the $[\text{HA}]_{\text{initial}}$, you substitute the values of $[\text{H}_3\text{O}^+] = 1.0 \times 10^{-6}$, $K_w = 1.00 \times 10^{-14}$, and $K_a = 1.8 \times 10^{-5}$ into this equation and then solve for $[\text{HA}]_{\text{initial}}$. $[\text{HA}]_{\text{initial}} = 1.0 \times 10^{-6}$ M.

That is, when acetic acid concentration is 1.0×10^{-6} M or lower, autoprotolysis of water must be taken into account.

11.111 (a) In the absence of a significant effect due to the autoprotolysis of water, the pH values of the 1.00×10^{-4} M and 1.00×10^{-6} M HBrO solutions can be calculated as described earlier.

For 1.00×10^{-4} mol$\cdot$L^{-1} :

Concentration

$(\text{mol} \cdot \text{L}^{-1})$ HBrO(aq)	+	H$_2$O(l)	$\rightleftharpoons$	H$_3$O$^+$(aq)	+	BrO$^-$(aq)
initial	1.00×10^{-4}		—	0		0
change	$-x$		—	$+x$		$+x$
final	$1.00 \times 10^{-4} - x$		—	$+x$		$+x$

$$K_a = \frac{[\text{H}_3\text{O}^+][\text{BrO}^-]}{[\text{HBrO}]}$$

$$2.0 \times 10^{-9} = \frac{(x)(x)}{1.00 \times 10^{-4} - x} = \frac{x^2}{[1.00 \times 10^{-4} - x]}$$

Assume $x \ll 1.00 \times 10^{-4}$.

$$x^2 = (2.0 \times 10^{-9})(1.00 \times 10^{-4})$$

$x = 4.5 \times 10^{-7}$

Because $x < 1\%$ of 1.00×10^{-4}, the assumption was valid. Given this value, the pH is then calculated to be $-\log(4.5 \times 10^{-7}) = 6.35$.

For $1.00 \times 10^{-6} \ \text{mol} \cdot \text{L}^{-1}$:

Concentration

$(\text{mol} \cdot \text{L}^{-1})$	HBrO(aq)	+	H$_2$O(l) $\rightleftharpoons$ H$_3$O$^+$(aq)	+	BrO$^-$(aq)
initial	1.00×10^{-6}	—	0		0
change	$-x$	—	$+x$		$+x$
final	$1.00 \times 10^{-6} - x$	—	$+x$		$+x$

$$K_a = \frac{[\text{H}_3\text{O}^+][\text{BrO}^-]}{[\text{HBrO}]}$$

$$2.0 \times 10^{-9} = \frac{(x)(x)}{1.00 \times 10^{-6} - x} = \frac{x^2}{[1.00 \times 10^{-6} - x]}$$

Assume $x \ll 1.00 \times 10^{-6}$.

$$x^2 = (2.0 \times 10^{-9})(1.00 \times 10^{-6})$$
$$x = 4.5 \times 10^{-8}$$

x is 4.5% of 1.00×10^{-6}, so the assumption is less acceptable. The pH is calculated to be $-\log(4.5 \times 10^{-8}) = 7.35$. Because this predicts a basic solution, it is not reasonable.

(b) To calculate the value taking into account the autoprotolysis of water, we can use Equation 22:

$$x^3 + K_a x^2 - (K_w + K_a \cdot [\text{HA}]_{\text{initial}})x - K_w \cdot K_a = 0, \text{ where } x = [\text{H}_3\text{O}^+]$$

To solve the expression, you substitute the values of $K_w = 1.00 \times 10^{-14}$, the initial concentration of acid, and $K_a = 2.0 \times 10^{-9}$ into this equation and then solve the expression either by trial and error or, preferably, using a graphing calculator such as the one found on the CD accompanying this text.

Alternatively, you can use a computer program designed to solve simultaneous equations. Because the unknowns include $[H_3O^+]$, $[OH^-]$, $[HBrO]$, and $[BrO^-]$, you will need four equations. As seen in the text, pertinent equations are

$$K_a = \frac{[H_3O^+][BrO^-]}{[HBrO]}$$

$$K_w = [H_3O^+][OH^-]$$

$$[H_3O^+] = [OH^-] + [BrO^-].$$

$$[HBrO]_{initial} = [HBrO] + [BrO^-]$$

Both methods should produce the same result.

The values obtained are

$[H_3O^+] = 4.6 \times 10^{-7}$ mol·L^{-1}, pH $= 6.34$ (compare to 6.35 obtained in (a))
$[BrO^-] = 4.4 \times 10^{-7}$ mol·L^{-1}
$[HBrO] \cong 1.0 \times 10^{-5}$ mol·L^{-1}
$[OH^-] = 2.2 \times 10^{-8}$ mol·L^{-1}.

Similarly, for $[HBrO]_{initial} = 1.00 \times 10^{-6}$,

$[H_3O^+] = 1.1 \times 10^{-7}$ mol·L^{-1}, pH $= 6.96$ (compare to 7.35 obtained in (a))
$[BrO^-] = 1.8 \times 10^{-8}$ mol·L^{-1}
$[HBrO] \cong 9.8 \times 10^{-7}$ mol·L^{-1}
$[OH^-] = 9.1 \times 10^{-8}$ mol·L^{-1}

Note that for the more concentrated solution, the effect of the autoprotolysis of water is very small. Notice also that the less concentrated solution is more acidic, due to the autoprotolysis of water, than would be predicted if this effect were not operating.

11.113 (a) In the absence of a significant effect due to the autoprotolysis of water, the pH values of the 8.50×10^{-5} M and 7.37×10^{-6} M HCN solutions can be calculated as described earlier.

For 8.50×10^{-5} mol·L^{-1} :

Concentration

$(mol \cdot L^{-1})$	HCN(aq)	+	$H_2O(l)$	$\rightleftharpoons$	$H_3O^+(aq)$	+	$CN^-(aq)$
initial	8.50×10^{-5}		—		0		0
change	$-x$		—		$+x$		$+x$
final	$8.50 \times 10^{-5} - x$		—		$+x$		$+x$

$$K_a = \frac{[H_3O^+][CN^-]}{[HCN]}$$

$$4.9 \times 10^{-10} = \frac{(x)(x)}{8.5 \times 10^{-5} - x} = \frac{x^2}{[8.5 \times 10^{-5} - x]}$$

Assume $x \ll 8.5 \times 10^{-5}$.

$$x^2 = (4.9 \times 10^{-10})(8.5 \times 10^{-5})$$

$$x = 2.0 \times 10^{-7}$$

Because $x < 1\%$ of 8.50×10^{-5}, the assumption was valid. Given this value, the pH is then calculated to be $-\log(2.0 \times 10^{-7}) = 6.69$.

For 7.37×10^{-6} $mol \cdot L^{-1}$:

Concentration

$(mol \cdot L^-)$	HCN(aq)	+	$H_2O(l)$	$\rightleftharpoons$	$H_3O^+(aq)$	+	$CN^-(aq)$
initial	7.37×10^{-6}		—		0		0
change	$-x$		—		$+x$		$+x$
final	$7.37 \times 10^{-6} - x$		—		$+x$		$+x$

$$K_a = \frac{[H_3O^+][CN^-]}{[HCN]}$$

$$4.9 \times 10^{-10} = \frac{(x)(x)}{7.37 \times 10^{-6} - x} = \frac{x^2}{[7.37 \times 10^{-6} - x]}$$

Assume $x \ll 7.37 \times 10^{-6}$.

$$x^2 = (4.9 \times 10^{-10})(7.37 \times 10^{-6})$$

$$x = 6.0 \times 10^{-8}$$

x is $< 1\%$ of 7.37×10^{-6}, so the assumption is still reasonable. The pH is then calculated to be $-\log(6.0 \times 10^{-8}) = 7.22$. This answer is not reasonable because we know HCN is an acid.

(b) To calculate the value, taking into account the autoprotolysis of water, we can use Equation 21:

$$x^3 + K_a x^2 - (K_w + K_a \cdot [HA]_{initial})x - K_w \cdot K_a = 0, \text{ where } x = [H_3O^+]$$

To solve the expression, you substitute the values of $K_w = 1.00 \times 10^{-14}$, the initial concentration of acid, and $K_a = 4.9 \times 10^{-10}$ into this equation and then solve the expression either by trial and error or, preferably, using a graphing calculator such as the one found on the CD accompanying this text.

Alternatively, you can use a computer program designed to solve simultaneous equations. Because the unknowns include $[H_3O^+]$, $[OH^-]$, $[HBrO]$, and $[BrO^-]$, you will need four equations. As seen in the text, the pertinent equations are

$$K_a = \frac{[H_3O^+][CN^-]}{[HCN]}$$

$$K_w = [H_3O^+][OH^-]$$

$$[H_3O^+] = [OH^-] + [CN^-]$$

$$[HCN]_{initial} = [HCN] + [CN^-].$$

Both methods should produce the same result.

For $[HCN] = 8.5 \times 10^{-5} \text{ mol} \cdot L^{-1}$, the values obtained are

$[H_3O^+] = 2.3 \times 10^{-7} \text{ mol} \cdot L^{-1}$, pH $= 6.64$ (compare to 6.69 obtained in (a))

$[CN^-] = 1.8 \times 10^{-7} \text{ mol} \cdot L^{-1}$

$[HCN] \cong 8.5 \times 10^{-5} \text{ mol} \cdot L^{-1}$

$[OH^-] = 4.4 \times 10^{-8} \text{ mol} \cdot L^{-1}$.

Similarly, for $[HCN]_{initial} = 7.37 \times 10^{-6}$,

$[H_3O^+] = 1.2 \times 10^{-7} \text{ mol} \cdot L^{-1}$, pH $= 6.92$ (compare to 7.22 obtained in (a))

$[CN^-] = 3.1 \times 10^{-8} \text{ mol} \cdot L^{-1}$

$[HCN] \cong 7.3 \times 10^{-6} \text{ mol} \cdot L^{-1}$

$[OH^-] = 8.6 \times 10^{-8} \text{ mol} \cdot L^{-1}$.

Note that for the more concentrated solution, the effect of the autoprotolysis of water is smaller. Notice also that the less concentrated solution is more acidic, due to the autoprotolysis of water, than would be predicted if this effect were not operating.

11.115 (a) Acid (1) is a strong acid because it totally dissociates.

(b) Acid (3) has the strongest base because it is the weakest acid.

(c) Acid (3) has the largest pK_a. Since $pK_a = -\log K_a$. The smallest K_a value will result the largest pK_a.

11.117 The apparent motion of hydronium and hydroxide ions is not as dependent on the diffusion of individual ions as other ions; they are formed and reformed as protons are transferred from and to water molecules in solution. Autoprotolysis allows rapid proton transfer between water molecules.

11.119 (a) We begin by finding the empirical formula of the compound:

$$C: \quad \frac{0.942 \text{ g CO}_2}{44.011 \text{ g} \cdot \text{mol}^{-1}} = 0.0214 \text{ mol CO}_2 \quad \therefore \quad 0.214 \text{ mol C}$$

$$(0.214 \text{ mol C})(12.011 \text{ g} \cdot \text{mol}^{1}) = 0.257 \text{ g C}$$

$$H: \quad \frac{0.0964 \text{ g H}_2\text{O}}{18.0158 \text{ g} \cdot \text{mol}^{-1}} = 0.00535 \text{ mol H}_2\text{O} \quad \therefore \quad 0.0107 \text{ mol H}$$

$$(0.0107 \text{ mol H})(1.008 \text{ g} \cdot \text{mol}^{-1}) = 0.0108 \text{ g H}$$

$$Na: \quad \frac{0.246 \text{ g H}_2\text{O}}{22.99 \text{ g} \cdot \text{mol}^{-1}} = 0.0107 \text{ mol Na}$$

$$(0.0107 \text{ mol Na})(22.99 \text{ g} \cdot \text{mol}^{-1}) = 0.0246 \text{ g Na}$$

$$O: \quad \text{mass of O} = 1.200 \text{ g} - 0.257 \text{ g} - 0.0108 \text{ g} - 0.0246 \text{ g} = 0.686 \text{ g of O}$$

$$\frac{0.686 \text{ g O}}{16.00 \text{ g} \cdot \text{mol}^{-1}} = 0.0429 \text{ mol O}$$

Dividing through by 0.0107 moles, we find the empirical formula to be: C_2HNaO_4.

A molar mass of 112.02 g·mol^{-1} indicates that this is also the molecular formula.

(b)

(c) The dissolved substance is sodium oxalate, it is capable of gaining or losing a proton and, therefore, amphiprotic. $\text{pH} = \frac{1}{2}\left(\text{p}K_{a_1} + \text{p}K_{a2}\right) = 2.7$.

11.121 The Lewis structure of boric acid is

boric acid   conjugate base

There is no resonance in boric acid (there is no double bond). $B(OH)_3$ is a very weak acid because it does not have a conjugate system to delocalize the electrons on the oxygen to weaken the O-H bond.
(b) In that reaction, boric acid acts as a Lewis acid because it accepts electron pairs (OH$^-$) from water. But, its acidity is not due to its dissociation.

11.123 (a) Nitrous acid will act like a strong acid because its conjugate base, the nitrite ion, is a weaker base than the acetate ion. The presence of acetate ions will shift the proton transfer equilibrium of nitrous acid toward the products (NO_2^- and H_3O^+) by consuming H_3O^+, which will increase the apparent K_a of nitrous acid. Carbonic acid is a weaker acid than acetic acid, so no shift in equilibrium occurs. (b) Ammonia will act like a strong base because its conjugate acid, the ammonium ion, is a weaker acid than acetic acid. The presence of acetic acid will shift the proton transfer equilibrium of ammonia with water toward the products (NH_4^+ and OH^-) by consuming OH^-, which will increase the apparent K_b of ammonia.

11.125 We wish to calculate K_a for the reaction

$$HF(aq) + H_2O(l) \rightleftharpoons H_3O^+(aq) + F^-(aq).$$

This equation is equivalent to

$$HF(aq) \rightleftharpoons H^+(aq) + F^-(aq).$$

This latter writing of the expression is simpler for the purpose of the thermodynamic calculations.

The $\Delta G°$ value for this reaction is easily calculated from the free energies given in the appendix:

$$= (-278.79 \text{ kJ} \cdot \text{mol}^{-1}) - (-296.82 \text{ kJ} \cdot \text{mol}^{-1}) = 18.03 \text{ kJ} \cdot \text{mol}^{-1}$$
$$- RT \ln K$$
$$K = e^{-\Delta G°/RT}$$
$$K = e^{-(18030 \text{ J} \cdot \text{mol}^{-1})/[8.314 \text{ J} \cdot \text{K}^{-1} \cdot \text{mol}^{-1})(298 \text{ K})]} = 6.9 \times 10^{-4}$$

11.127 (a) $D_2O + D_2O \leftrightarrow D_3O^+ + OD^-$

(b) $K_{hw} = [D_3O^+][OD^-] = 1.35 \times 10^{-15}$; $pK_{hw} = -\log K_{hw} = 14.870$

(c) $[D_3O^+] = [OD^-] = \sqrt{1.35 \times 10^{-15}} = 3.67 \times 10^{-8} \text{ mol} \cdot \text{L}^{-1}$

(d) $pD = -\log(3.67 \times 10^{-8}) = 7.435 = pOD$

(e) $pD + pOD = pK_{hw} = 14.870$

11.129

$$CH_3CH(OH)COOH(aq) + H_2O(l) \rightleftharpoons H_3O^+(aq) + CH_3CH(OH)COO^-(aq)$$
Concentration (mol $\cdot$ L^{-1})

initial	1.00	—	0	0
change	$-x$	—	$+x$	$+x$
equilibrium	$1.00 - x$	—	x	x

$$K_a = \frac{[H_3O^+][CH_3CH(OH)COO^-]}{[CH_3CH(OH)COOH]} = \frac{x^2}{1.00\text{-}x} = 8.4 \times 10^{-4}$$

$x = 0.029$

Percent deprotonation $= \dfrac{0.029}{1.00}(100\%) = 2.9\%$ deprotonated.

To calculate T_f for this solution we need the concentration in moles of solute per kilogram of solvent. Assuming a density of $1 \text{ g} \cdot \text{cm}^{-1}$, 1.00 L of solution will weigh 1000 g. One liter of solution will contain 1 mole (90.08 g) of lactic acid. Therefore, one liter of solution will contain 1000 g $- 90.08$ g $= 910$ g or 0.910 kg of solvent.

The molarity of the solution is $\dfrac{1.029 \text{ mol solute}}{0.910 \text{ kg solvent}} = 1.13 \, m$.

$$\Delta T_f = \left(1.86 \text{ K} \cdot \text{kg} \cdot \text{mol}^{-1}\right)\left(1.13 \text{ mol} \cdot \text{Kg}^{-1}\right) = 2.1 \text{ K}$$

The temperature at which the solution will freeze is 271 K.

11.131 $T' = 30°C = 303 \text{ K}, T = 20°C = 293 \text{ K}$

$$\ln\left(\frac{K_a'}{K_a}\right) = \frac{\Delta H°}{R}\left(\frac{1}{T} - \frac{1}{T'}\right) = \frac{\Delta H°}{R}\left(\frac{T'-T}{TT'}\right)$$

This is the van't Hoff equation.

$$\ln\left(\frac{1.768 \times 10^{-4}}{1.765 \times 10^{-4}}\right) = \frac{\Delta H°}{8.314 \text{ J} \cdot \text{K}^{-1} \cdot \text{mol}^{-1}}\left(\frac{303 \text{ K} - 293 \text{ K}}{293 \text{ K} \times 303 \text{ K}}\right)$$

$$\Delta H° = \frac{\ln\left(\frac{1.768 \times 10^{-4}}{1.765 \times 10^{-4}}\right)\left(8.314 \text{ J} \cdot \text{K}^{-1} \cdot \text{mol}^{-1}\right)}{\left(\frac{303 \text{ K} - 293 \text{ K}}{293 \text{ K} \times 303 \text{ K}}\right)} = 1.3 \times 10^2 \text{ J} \cdot \text{mol}^{-1}$$

11.133 The greater difference in acidity constants (K_{a1} and K_{a2}) of adipic acid than those of oxalic acid is due to the structural differences of these two acids. Adipic acid has four CH_2 groups between the two carboxylic groups and this will cause a greater electron delocalization which stabilizes the carboxylic anions. As a result, the two protons on the adipic acid can be easily deprotonated almost simultaneously. This is why the two pK_a values are so close. However, oxalic acid is lack of CH_2 groups between the two carboxylic groups and has a weaker electron delocalization ability of anions and results a large separation of two pK_a values.

11.135 (a) The H^+ in lactic acid can interact with HbO_2^- to produce HHb and release more oxygen molecules, which means that the concentration of HbO_2^- will be lower in the tissues.

(b) Since more Hb^- will return to the lung, more oxygen will be bound to Hb^- to produce HbO_2^-, which will increase the concentration of HbO_2^-.

11.137 (a) Given that CO_2 will react with water to form carbonic acid, H_2CO_3, it only remains to determine the concentration of H_3O^+ due to the deprotonation of H_2CO_3.

$$[CO_2] = (2.3 \times 10^{-2}\ mol \cdot L^{-1} \cdot atm^{-1})(3.04 \times 10^{-4}\ atm) = 7.0 \times 10^{-6}\ mol \cdot L^{-1}$$

	$H_2CO_3(aq)$	$+$	$H_2O(l)$	$\rightleftharpoons$	$H_3O^+(aq)$	$+$	$HCO_3^-(aq)$
initial	7.0×10^{-6}		—		0		0
change	$-x$		—		$+x$		$+x$
equilibrium	$7.0 \times 10^{-6} - x$		—		x		x

$$K_a = 10^{-pKa} = 10^{-6.37} = 4.27 \times 10^{-7} = \left(\frac{x^2}{7.0 \times 10^{-6}} \right)$$

$$x = [H_3O^+] = 1.73 \times 10^{-6}$$

$$pH = -\log(1.73 \times 10^{-6}) = 5.75$$

(b) The equilibria involved are:

$$H_2CO_3(aq) + H_2O(l) \rightleftharpoons HCO_3^-(aq) + H_3O^+ \qquad K_{a1} = 4.3 \times 10^{-7}$$

$$HCO_3^-(aq) + H_2O(l) \rightleftharpoons CO_3^{2-}(aq) + H_3O^+ \qquad K_{a2} = 5.6 \times 10^{-11}$$

Given these two equilibria, two equations can be written:

$$K_{a1} = \frac{\left[HCO_3^-\right]\left[H_3O^+\right]}{\left[H_2CO_3\right]}$$

$$K_{a2} = \frac{\left[CO_3^{2-}\right]\left[H_3O^+\right]}{\left[HCO_3^-\right]}$$

Also, enough information is provided to construct two other equations:

$$\left[H_2CO_3\right] + \left[HCO_3^-\right] + \left[CO_3^{2-}\right] = 4.5 \times 10^{-3}$$

$$\left[H_3O^+\right] = 10^{-4.8} = 1.6 \times 10^{-5}$$

Solving this set of simultaneous equations for the three unknown concentrations, one finds:

$[H_2CO_3] = 4.4 \times 10^{-3}$ mol $\cdot$ L^{-1}

$[HCO_3^-] = 1.2 \times 10^{-4}$ mol $\cdot$ L^{-1}

$[CO_3^{2-}] = 4.2 \times 10^{-10}$ mol $\cdot$ L^{-1}

(c) Assuming all sulfur is converted to SO_2, the amount of SO_2 produced is

$$\frac{\left(1.00 \times 10^3 \text{ kg}\right)\left(0.025\right)\left(1000 \text{ g} \cdot \text{kg}^{-1}\right)}{32.07 \text{ g} \cdot \text{mol}^{-1}} = 780 \text{ mol of S}$$

Therefore, $(780 \text{ mol})\left(64.07 \text{ g} \cdot \text{mol}^{-1}\right) = 50,000$ g, or 50 kg, of SO_2 is produced

(d) To determine the pH, we first must calculate the volume of water in which this 50 kg of SO_2 is dissolved:

$$V = (2 \text{ cm})(2.6 \text{ Km})\left(\frac{1000 \text{ m}}{1 \text{ Km}}\right)^2\left(\frac{100 \text{ cm}}{1 \text{ m}}\right)^2\left(\frac{1 \text{ L}}{1000 \text{ cm}^3}\right) = 5.2 \times 10^7 \text{ L}$$

The concentration of SO_2(aq) is then

$(780 \text{ mol})/\left(5.2 \times 10^7 \text{ L}\right) = 1.5 \times 10^{-5}$ M . We can assume that upon solution the SO_2(aq) is converted to sulfurous acid and the pH is calculated as described earlier:

Concentration (mol $\cdot$ L^{-1})	H_2SO_3(aq)	+	H_2O(l)	$\rightleftharpoons$	H_3O^+(aq)	+	HSO_3^-(aq)
initial	1.50×10^{-5}		—		0		0
change	$-x$		—		$+x$		$+x$

| final | $1.50 \times 10^{-5} - x$ | — | $+x$ | $+x$ |

$$K_a = \frac{[H_3O^+][HSO_3^-]}{[H_2SO_3]}$$

$$1.55 \times 10^{-2} = \frac{(x)(x)}{1.5 \times 10^{-5} - x}$$

Employing the quadratic formula we find $x = [H_3O^+] = 1.5 \times 10^{-5}$, giving a pH of $pH = \log(1.5 \times 10^{-5}) = 4.82$.

(e) When dissolved in water, SO_3 will form sulfuric acid:

$$SO_3(aq) + 2 H_2O(l) \rightarrow H_3O^+(aq) + HSO_4^-(aq)$$

The fist deprotonation of sulfuric acid is complete, the concentration of hydronium ion due to the second deprotonation of sulfuric acid may be found as described earlier:

Concentration

$(mol \cdot L^{-1})$	$HSO_4^-(aq)$	+	$H_2O(l)$	$\rightleftharpoons$	$H_3O^+(aq)$	+	$SO_4^{2-}(aq)$
initial	1.50×10^{-5}		—		1.50×10^{-5}		0
change	$-x$		—		$+x$		$+x$
final	$1.50 \times 10^{-5} - x$		—		$1.50 \times 10^{-5} - x$		$+x$

$$K_a = \frac{[H_3O^+][SO_4^{2-}]}{[HSO_4^-]}$$

$$1.2 \times 10^{-2} = \frac{(1.5 \times 10^{-5} + x)(x)}{1.5 \times 10^{-5} - x}$$

Employing the quadratic formula we find $x = [H_3O^+] = 1.495 \times 10^{-5}$, and the total hydronium ion concentration to be $1.495 \times 10^{-5} + 1.5 \times 10^{-5} = 3.0 \times 10^{-5}$. Therefore, $pH = -\log(3.0 \times 10^{-5}) = 4.52$.

(f) Because the process is only 90% efficient, to remove 50.0 kg of SO_2 one must supply enough $CaCO_3$ to remove 55.6 kg of SO_2 (50.0 kg is 90% of 55.6 kg). The moles of SO_2 to be removed is given by

$$\frac{55600 \text{ g}}{64.06 \text{ g mol}^{-1}} = 868 \text{ mol } SO_2.$$

1 mole of SO_2 is consumed by 1 mole of $CaCO_3$. Therefore,

$$(868 \text{ mol CaCO}_3)(100.09 \text{ g mol}^{-1}) = 86.9 \text{ kg.}$$

CHAPTER 12

AQUEOUS EQUILIBRIA

12.1 (a) When solid sodium acetate is added to an acetic acid solution, the concentration of H_3O^+ decreases because the equilibrium

$$HC_2H_3O_2(aq) + H_2O(l) \rightleftharpoons H_3O^+(aq) + C_2H_3O_2^-(aq)$$

shifts to the left to relieve the stress imposed by the increase of $[C_2H_3O_2^-]$ (Le Chatelier's principle).

(b) When HCl is added to a benzoic acid solution, the percentage of benzoic acid that is deprotonated decreases because the equilibrium

$$C_6H_5COOH(aq) + H_2O(l) \rightleftharpoons H_3O^+(aq) + C_6H_5CO_2^-(aq)$$

shifts to the left to relieve the stress imposed by the increased $[H_3O^+]$ (Le Chatelier's principle).

(c) When solid NH_4Cl is added to an ammonia solution, the concentration of OH^- decreases because the equilibrium

$$NH_3(aq) + H_2O(l) \rightleftharpoons NH_4^+(aq) + OH^-(aq)$$ shifts to the left to relieve the stress imposed by the increased $[NH_4^+]$ (Le Chatelier's principle).

Because $[OH^-]$ decreases, $[H_3O^+]$ increases and pH decreases.

12.3 (a) $K_a = \dfrac{[H_3O^+][A^-]}{[HA]}$; $pK_a = pH - \log \dfrac{[A^-]}{[HA]}$. If $[A^-] = [HA]$, then

$pK_a = pH$.

$pH = pK_a = 3.08$, $K_a = 8.3 \times 10^{-4}$

(b) Let $x = [\text{lactate ion}] = [L^{-1}]$ and $y = [H_3O^+]$

Concentration

(mol·L^{-1})	HL(aq)	+	H$_2$O(l)	$\rightleftharpoons$	H$_3$O$^+$(aq)	+	L$^-$(aq)
initial	$2x$		—		—		x
change	$-y$		—		$+y$		$+y$
equilibrium	$2x - y$		—		y		$y + x$

$$K_a = \frac{[H_3O^+][L^-]}{[HL]} = \frac{(y)(y+x)}{(2x-y)} \cong \frac{(y)(x)}{(2x)} = 8.3 \times 10^{-4}$$

$$y = 2(8.3 \times 10^{-4}) \cong 1.7 \times 10^{-3} \text{ mol·L}^{-1} \cong [H_3O^+]$$

$$\text{pH} \approx 2.77$$

12.5 (a) The reaction is $HCN(aq) + H_2O(l) \rightleftharpoons H_3O^+(aq) + CN^-(aq)$.

Concentration

(mol·L^{-1})	HCN(aq) +	H$_2$O(l)	$\rightleftharpoons$	H$_3$O$^+$(aq) +	CN$^-$(aq)
initial	0.075	—		0	0.060
change	$-x$	—		$+x$	$+x$
equilibrium	$0.075 - x$	—		x	$0.060 + x$

$$K_a = \frac{[H_3O^+][CN^-]}{[HCN]} = \frac{(x)(0.060 + x)}{(0.075 - x)} = 4.9 \times 10^{-10}$$

$$x = [[H_3O^+] \approx 6.1 \times 10^{-10} \text{ mol·L}^{-1}$$

(b) 0.30 M NaCl will have no effect:

Concentration

(mol·L^{-1})	NH$_2$NH$_2$(aq) +	H$_2$O(l)	$\rightleftharpoons$	NH$_2$NH$_3$$^+$(aq) +	OH$^-$(aq)
initial	0.20	0		0	
change	$-x$	—		$+x$	$+x$
equilibrium	$0.20 - x$	—		x	x

$$K_b = \frac{[NH_2NH_3^+][OH^-]}{[NH_2NH_2]}$$

$$1.7 \times 10^{-6} = \frac{[OH^-]^2}{0.20}$$

$$[OH^-] = \sqrt{3.4 \times 10^{-7}} = 5.8 \times 10^{-4} \ mol \cdot L^{-1}$$

$$[H_3O^+] = \frac{K_w}{[OH^-]} = \frac{1.00 \times 10^{-14}}{5.8 \times 10^{-4}} = 1.7 \times 10^{-11} \ mol \cdot L^{-1}$$

(c) Setup is similar to part (a).

$$K_a = 4.9 \times 10^{-10} = \frac{[H_3O^+][CN^-]}{[HCN]} = \frac{(x)(0.030 + x)}{(0.015 - x)}$$

$$4.9 \times 10^{-10} \approx \frac{[H_3O^+][0.030]}{(0.015)}$$

$$[H_3O^+] = 2.5 \times 10^{-10} \ mol \cdot L^{-1}$$

(d) When the concentrations of a weak base and its conjugate acid are equal, the pOH equals the pK_b. Therefore, the pOH of hydrazine = pK_b = 5.77, and

$$pH = 14.00 - pOH = 14.00 - 5.77 = 8.23$$

$$[H_3O^+] = 10^{-8.23} = 5.9 \times 10^{-9} \ mol \cdot L^{-1}$$

12.7 In each case, the equilibrium involved is

$$HSO_4^-(aq) + H_2O(l) \rightleftharpoons H_3O^+(aq) + SO_4^{2-}(aq).$$

$HSO_4^-(aq)$ and $SO_4^{2-}(aq)$ are conjugate acid and base; therefore, the pH calculation is most easily performed with the Henderson-Hasselbalch equation:

$$pH = pK_a + \log\left(\frac{[base]}{[acid]}\right) = pK_a + \log\left(\frac{[SO_4^{2-}]}{[HSO_4^-]}\right)$$

(a) $pH = 1.92 + \log\left(\frac{0.25 \ mol \cdot L^{-1}}{0.5 \ mol \cdot L^{-1}}\right) = 1.62, \quad pOH = 14.00 - 1.62 = 12.38$

(b) $pH = 1.92 + \log\left(\frac{0.10 \ mol \cdot L^{-1}}{0.50 \ mol \cdot L^{-1}}\right) = 1.22, \quad pOH = 12.78$

(c) $pH = pK_a = 1.92$, $pOH = 12.08$

See solution to Exercise 11.3.

12.9 (a) $HCN(aq) + H_2O(l) \rightleftharpoons H_3O^+(aq) + CN^-(aq)$

total volume $= 100$ mL $= 0.100$ L

moles of HCN $= 0.0300$ L $\times 0.050$ mol $\cdot$ L^{-1} $= 1.5 \times 10^{-3}$ mol HCN

moles of NaCN $= 0.0700$ L $\times 0.030$ mol $\cdot$ L^{-1} $= 2.1 \times 10^{-3}$ mol NaCN

initial $[HCN]_0 = \dfrac{1.5 \times 10^{-3} \text{ mol}}{0.100 \text{ L}} = 1.5 \times 10^{-2}$ mol $\cdot$ L^{-1}

initial $[CN^-]_0 = \dfrac{2.1 \times 10^{-3} \text{ mol}}{0.100 \text{ L}} = 2.1 \times 10^{-2}$ mol $\cdot$ L^{-1}

Concentration

(mol $\cdot$ L^{-1})	HCN(aq)	+	H$_2$O(l)	$\rightleftharpoons$	H$_3$O$^+$(aq)	+	CN$^-$(aq)
initial	1.5×10^{-2}		—		0		2.1×10^{-2}
change	$-x$		—		$+x$		$+x$
equilibrium	$1.5 \times 10^{-2} - x$		—		x		$2.1 \times 10^{-2} + x$

$K_a = \dfrac{[H_3O^+][CN^-]}{[HCN]} = \dfrac{(x)(2.1 \times 10^{-2} + x)}{(1.5 \times 10^{-2} - x)} \approx \dfrac{(x)(2.1 \times 10^{-2})}{(1.5 \times 10^{-2})} = 4.9 \times 10^{-10}$

$x \approx 3.5 \times 10^{-10}$ mol $\cdot$ L^{-1} $\approx [H_3O^+]$

$pH = -\log[H_3O^+] = -\log(3.5 \times 10^{-10}) = 9.46$

(b) The solution here is the same as for part (a), except for the initial concentrations:

$[HCN]_0 = \dfrac{0.0400 \text{ L} \times 0.030 \text{ mol} \cdot \text{L}^{-1}}{0.100 \text{ L}} = 1.2 \times 10^{-2}$ mol $\cdot$ L^{-1}

$[CN^-]_0 = \dfrac{0.0600 \text{ L} \times 0.050 \text{ mol} \cdot \text{L}^{-1}}{0.100 \text{ L}} = 3.0 \times 10^{-2}$ mol $\cdot$ L^{-1}

$K_a = 4.9 \times 10^{-10} = \dfrac{(x)(3.0 \times 10^{-2})}{(1.2 \times 10^{-2})}$

$$x = [H_3O^+] = 2.0 \times 10^{-10} \text{ mol} \cdot \text{L}^{-1}$$

$$pH = -\log(2.0 \times 10^{-10}) = 9.71$$

(c) $[HCN]_0 = [NaCN]_0$ after mixing; therefore,

$$K_a = 4.9 \times 10^{-10} = \frac{(x)[NaCN]_0}{[HCN]_0} = x = [H_3O^+]$$

$$pH = pK_a = -\log(4.9 \times 10^{-10}) = 9.31$$

12.11 (a) $pH = pK_a + \log\left(\dfrac{[CH_3CO_2^-]}{[CH_3COOH]}\right)$ (see Exercise 11.23)

$$pH = pK_a + \log\left(\frac{0.100}{0.100}\right) = 4.75 \text{ (initial pH)}$$

final pH: $(0.0100 \text{ L})(0.950 \text{ mol} \cdot \text{L}^{-1}) = 9.50 \times 10^{-3}$ mol NaOH (strong

base) produces 9.50×10^{-3} mol $CH_3CO_2^-$ from CH_3COOH

$0.100 \text{ mol} \cdot \text{L}^{-1} \times 0.100 \text{ L} = 1.00 \times 10^{-2}$ mol CH_3COOH initially

$0.100 \text{ mol} \cdot \text{L}^{-1} \times 0.100 \text{ L} = 1.00 \times 10^{-2}$ mol $CH_3CO_2^-$ initially

After adding NaOH:

$$[CH_3COOH] = \frac{(1.00 \times 10^{-2} - 9.5 \times 10^{-3}) \text{ mol}}{0.110 \text{ L}} = 5 \times 10^{-3} \text{ mol} \cdot \text{L}^{-1}$$

$$[CH_3CO_2^-] = \frac{(1.00 \times 10^{-2} + 9.5 \times 10^{-3}) \text{ mol}}{0.110 \text{ L}} = 1.77 \times 10^{-1} \text{ mol} \cdot \text{L}^{-1}$$

$$pH = 4.75 + \log\left(\frac{1.77 \times 10^{-1} \text{ mol} \cdot \text{L}^{-1}}{5 \times 10^{-3} \text{ mol} \cdot \text{L}^{-1}}\right) = 4.75 + 1.5 = 6.3$$

(b) $(0.0200 \text{ L})(0.100 \text{ mol} \cdot \text{L}^{-1}) = 2.00 \times 10^{-3}$ mol HNO_3 (strong acid)

produces

2.00×10^{-3} mol CH_3COOH from $CH_3CO_2^-$.

After adding HNO_3 [see part (a) of this exercise]:

$$[CH_3COOH] = \frac{(1.00 \times 10^{-2} + 2.00 \times 10^{-3})\ mol}{0.120\ L} = 1.00 \times 10^{-1}\ mol \cdot L^{-1}$$

$$[CH_3CO_2^{\ -}] = \frac{(1.00 \times 10^{-2} - 2.00 \times 10^{-3})\ mol}{0.120\ L} = 6.7 \times 10^{-2}\ mol \cdot L^{-1}$$

$$pH = 4.75 + \log\left(\frac{6.7 \times 10^{-2}\ mol \cdot L^{-1}}{1.00 \times 10^{-1}\ mol \cdot L^{-1}}\right) = 4.75 - 0.17 = 4.58$$

$$\Delta pH = -0.17$$

12.13 $\left(\dfrac{0.356\ g\ NaF}{0.050\ L}\right)\left(\dfrac{1\ mol\ NaF}{41.99\ g\ NaF}\right) = 0.17\ mol \cdot L^{-1}$

Concentration

$(mol \cdot L^{-1})$	$HF(aq)$	$+$	$H_2O(l)$	$\rightleftharpoons$	$H_3O^+(aq)$	$+$	$F^-(aq)$
initial	0.40		—		0		0.17
change	$-x$		—		$+x$		$+x$
equilibrium	$0.40 - x$		—		x		$0.17 + x$

$$K_a = \frac{[H_3O^+][F^-]}{[HF]} = \frac{(x)(0.17 + x)}{(0.40 - x)} \approx \frac{(x)(0.17)}{(0.40)} = 3.5 \times 10^{-4}$$

$$x \cong 8.2 \times 10^{-4}\ mol \cdot L^{-1} \cong [H_3O^+]$$

$$pH = -\log[H_3O^+] = -\log(8.2 \times 10^{-4}) = 3.09$$

$$change\ in\ pH = 3.09 - 1.93 = 1.16$$

12.15 In a solution containing $HClO(aq)$ and $ClO^-(aq)$, the following equilibrium occurs:

$$HClO(aq) + H_2O(l) \rightleftharpoons H_3O^+(aq) + ClO^-(aq)$$

The ratio $[ClO^-]/[HClO]$ is related to pH, as given by the Henderson-Hasselbalch equation: $pH = pK_a + \log\left(\dfrac{[ClO^-]}{[HClO]}\right)$, or

$$\log\left(\frac{[ClO^-]}{[HClO]}\right) = pH - pK_a = 6.50 - 7.53 = -1.03$$

$$\frac{[ClO^-]}{[HClO]} = 9.3 \times 10^{-2}$$

12.17 The rule of thumb we use is that the effective range of a buffer is roughly within plus or minus one pH unit of the pK_a of the acid. Therefore:

(a) $pK_a = 3.08$; pH range 2–4

(b) $pK_a = 4.19$; pH range 3–5

(c) $pK_{a3} = 12.68$; pH range 11.5–13.5

(d) $pK_{a2} = 7.21$; pH range 6–8

(e) $pK_b = 7.97$, $pK_a = 6.03$; pH range 5–7

12.19 Choose a buffer system in which the conjugate acid has a pK_a close to the desired pH. Therefore:

(a) $HClO_2$ and $NaClO_2$, $pK_a = 2.00$

(b) NaH_2PO_4 and Na_2HPO_4, $pK_{a2} = 7.21$

(c) $CH_2ClCOOH$ and $NaCH_2ClCO_2$, $pK_a = 2.85$

(d) Na_2HPO_4 and Na_3PO_4, $pK_a = 12.68$

12.21 (a) $HCO_3^-(aq) + H_2O(l) \rightarrow CO_3^{2-}(aq) + H_3O^+(aq)$

$$K_{a2} = \frac{[H_3O^+][CO_3^{2-}]}{[HCO_3^-]}, \quad pK_{a2} = 10.25$$

$$pH = pK_{a2} + \log\left(\frac{[CO_3^{2-}]}{[HCO_3^-]}\right)$$

$$\log\left(\frac{[CO_3^{2-}]}{[HCO_3^-]}\right) = pH - pK_{a2} = 11.0 - 10.25 = 0.75$$

$$\frac{[CO_3^{2-}]}{[HCO_3^-]} = 5.6$$

(b) $[CO_3^{2-}] = 5.6 \times [HCO_3^-] = 5.6 \times 0.100 \text{ mol} \cdot L^{-1} = 0.56 \text{ mol} \cdot L^{-1}$

moles of $CO_3{}^{2-}$ = moles of K_2CO_3 = $0.56 \text{ mol} \cdot L^{-1} \times 1 \text{ L} = 0.56 \text{ mol}$

mass of K_2CO_3 = $0.56 \text{ mol} \times \left(\dfrac{138.21 \text{ g } K_2CO_3}{1 \text{ mol } K_2CO_3} \right) = 77 \text{ g } K_2CO_3$

(c) $[HCO_3{}^-] = \dfrac{[CO_3{}^{2-}]}{5.6} = \dfrac{0.100 \text{ mol} \cdot L^{-1}}{5.6} = 1.8 \times 10^{-2} \text{ mol} \cdot L^{-1}$

moles of $HCO_3{}^-$ = moles of $KHCO_3$ = $1.8 \times 10^{-2} \text{ mol} \cdot L^{-1} \times 1 \text{ L}$
$$= 1.8 \times 10^{-2} \text{ mol}$$

mass $KHCO_3$ = $1.8 \times 10^{-2} \text{ mol} \times 100.12 \text{ g} \cdot \text{mol}^{-1} = 1.8 \text{ g } KHCO_3$

(d) $[CO_3{}^{2-}] = 5.6 \times [HCO_3{}^-]$

moles of

$HCO_3{}^-$ = moles $KHCO_3$ = $0.100 \text{ mol} \cdot L^{-1} \times 0.100 \text{ L} = 1.00 \times 10^{-2} \text{ mol}$

Because the final total volume is the same for both

$KHCO_3$ and K_2CO_3, the number of moles of K_2CO_3 required is

$5.6 \times 1.00 \times 10^{-2} \text{ mol} = 5.6 \times 10^{-2} \text{ mol}$.

Thus,

volume of K_2CO_3 solution $= \dfrac{5.6 \times 10^{-2} \text{ mol}}{0.200 \text{ mol} \cdot L^{-1}} = 0.28 \text{ L} = 2.8 \times 10^2 \text{ mL}$.

12.23 (a) The titration curve is as follows:

The initial pH was 12.00; the final pH was 2.54 (when 30.0 mL of HCl was added); pH at stoichiometric point was 7.00

(b) 10.0 mL; (c) 5.00 mL

12.25 $HCl(aq) + NaOH(aq) \longrightarrow H_2O(l) + Na^+(aq) + Cl^-(aq)$

(a) $V_{HCl} = (\frac{1}{2})(25.0 \text{ mL})\left(\dfrac{10^{-3} \text{ L}}{1 \text{ mL}}\right)\left(\dfrac{0.110 \text{ mol NaOH}}{1 \text{ L}}\right)$

$\left(\dfrac{1 \text{ mol HCl}}{1 \text{ mol NaOH}}\right)\left(\dfrac{1 \text{ L HCl}}{0.150 \text{ mol HCl}}\right)$

$= 9.17 \times 10^{-3} \text{ L HCl}$

(b) $2 \times 9.17 \times 10^{-3} \text{ L} = 0.0183 \text{ L}$

(c) volume $= (0.0250 + 0.0183) \text{ L} = 0.0433 \text{ L}$

$[Na^+] = (0.0250 \text{ L})\left(\dfrac{0.110 \text{ mol NaOH}}{1 \text{ L}}\right)\left(\dfrac{1 \text{ mol Na}^+}{1 \text{ mol NaOH}}\right)\left(\dfrac{1}{0.0433 \text{ L}}\right)$

$= 0.0635 \text{ mol} \cdot \text{L}^{-1}$

(d) number of moles of H_3O^+ (from acid) $= (0.0200 \text{ L})\left(\dfrac{0.150 \text{ mol}}{1 \text{ L}}\right)$

$= 3.00 \times 10^{-3} \text{ mol } H_3O^+$

number of moles of OH^- (from base) $= (0.0250 \text{ L})\left(\dfrac{0.110 \text{ mol Na}^+}{1 \text{ L}}\right)$

$= 2.75 \times 10^{-3} \text{ mol } OH^-$

excess $H_3O^+ = (3.00 - 2.75) \times 10^{-3} \text{ mol} = 2.5 \times 10^{-4} \text{ mol } H_3O^+$

$[H_3O^+] = \dfrac{2.5 \times 10^{-4} \text{ mol}}{0.0450 \text{ L}} = 5.6 \times 10^{-3} \text{ mol} \cdot \text{L}^{-1}$

$pH = -\log(5.6 \times 10^{-3}) = 2.25$

12.27 The moles of OH^- added are equivalent to the number of moles of HA present:

$(0.350 \text{ mol} \cdot \text{L}^{-1})(0.052 \text{ L}) = 0.0182 \text{ mol } OH^-$

Therefore, 0.0182 mol HA were present in solution

(a) molar mass of HA: $\dfrac{4.25 \text{ g}}{0.0182 \text{ mol}} = 234 \text{ g} \cdot \text{mol}^{-1}$

(b) pKa = pH at half-way titration = 3.82

12.29 (a) $pOH = -\log(0.110) = 0.959$, $pH = 14.00 - 0.959 = 13.04$

(b) initial moles of OH^- (from base) $= (0.0250 \text{ L})\left(\dfrac{0.110 \text{ mol}}{1 \text{ L}}\right)$

$$= 2.75 \times 10^{-3} \text{ mol OH}^-$$

moles of H_3O^+ added $= (0.0050 \text{ L})\left(\dfrac{0.150 \text{ mol}}{1 \text{ L}}\right) = 7.5 \times 10^{-4} \text{ mol H}_3O^+$

excess $OH^- = (2.75 - 0.75) \times 10^{-3} \text{ mol} = 2.00 \times 10^{-3} \text{ mol OH}^-$

$[OH^-] = \dfrac{2.00 \times 10^{-3} \text{ mol}}{0.0300 \text{ L}} = 0.0667 \text{ mol} \cdot L^{-1}$

$pOH = -\log(0.0667) = 1.176$, $pH = 14.00 - 1.176 = 12.82$

(c) moles of H_3O^+ added $= 2 \times 7.5 \times 10^{-4} \text{ mol} = 1.50 \times 10^{-3} \text{ mol H}_3O^+$

excess $OH^- = (2.75 - 1.50) \times 10^{-3} \text{ mol} = 1.25 \times 10^{-3} \text{ mol OH}^-$

$[OH^-] = \dfrac{1.25 \times 10^{-3} \text{ mol}}{0.0350 \text{ L}} = 0.0357 \text{ mol} \cdot L^{-1}$

$pOH = -\log(0.0357) = 1.447$, $pH = 14.00 - 1.447 = 12.55$

(d) $pH = 7.00$

$$V_{HCl} = (2.75 \times 10^{-3} \text{ mol NaOH})\left(\dfrac{1 \text{ mol HCl}}{1 \text{ mol NaOH}}\right)\left(\dfrac{1 \text{ L HCl}}{0.150 \text{ mol HCl}}\right)$$

$$= 0.0183 \text{ L}$$

(e) $[H_3O^+] = (0.0050 \text{ L})\left(\dfrac{0.150 \text{ mol}}{1 \text{ L}}\right)\left(\dfrac{1}{(0.0250 + 0.0183 + 0.0050) \text{ L}}\right)$

$$= 0.016 \text{ mol} \cdot L^{-1}$$

$pH = -\log(0.016) = 1.80$

(f) $[H_3O^+] = \left(\dfrac{0.010 \text{ L}}{0.0533 \text{ L}}\right)\left(\dfrac{0.150 \text{ mol}}{1 \text{ L}}\right) = 0.028 \text{ mol} \cdot L^{-1}$

$pH = -\log(0.028) = 1.55$

12.31 mass of pure NaOH $= (0.0342 \text{ L HCl})\left(\dfrac{0.0695 \text{ mol HCl}}{1 \text{ L HCl}}\right)$

$$\left(\frac{1 \text{ mol NaOH}}{1 \text{ mol HCl}}\right)\left(\frac{40.00 \text{ g NaOH}}{1 \text{ mol NaOH}}\right)\left(\frac{300 \text{ mL}}{25.0 \text{ mL}}\right)$$

$$= 1.14 \text{ g}$$

$$\text{percent purity} = \frac{1.14 \text{ g}}{1.436 \text{ g}} \times 100\% = 79.4$$

12.33 (a) Concentration

$$(\text{mol} \cdot \text{L}^{-1})\ CH_3COOH(aq) + H_2O(l) \rightleftharpoons H_3O^+(aq) + CH_3CO_2^-(aq)$$

initial	0.10	—	0	0
change	$-x$	—	$+x$	$+x$
equilibrium	$0.10 - x$	—	x	x

$$K_a = \frac{[H_3O^+][CH_3CO_2^-]}{[CH_3COOH]} = 1.8 \times 10^{-5} = \frac{x^2}{0.10 - x} \approx \frac{x^2}{0.10}$$

$$x^2 = 1.8 \times 10^{-6}$$

$$x = 1.3 \times 10^{-3} \text{ mol} \cdot \text{L}^{-1} = [H_3O^+]$$

$$\text{initial pH} = -\log(1.3 \times 10^{-3}) = 2.89$$

(b)   moles of $CH_3COOH = (0.0250 \text{ L})(0.10 \text{ M})$

$$= 2.50 \times 10^{-3} \text{ mol } CH_3COOH$$

moles of $NaOH = (0.0100 \text{ L})(0.10 \text{ M}) = 1.0 \times 10^{-3} \text{ mol } OH^-$

After neutralization,

$$\frac{1.50 \times 10^{-3} \text{ mol } CH_3COOH}{0.0350 \text{ L}} = 4.29 \times 10^{-2} \text{ mol} \cdot \text{L}^{-1} \ CH_3COOH$$

$$\frac{1.0 \times 10^{-3} \text{ mol } CH_3CO_2^-}{0.0350 \text{ L}} = 2.86 \times 10^{-2} \text{ mol} \cdot \text{L}^{-1} \ CH_3CO_2^-$$

Then consider equilibrium, $K_a = \dfrac{[H_3O^+][CH_3CO_2^-]}{[CH_3COOH]}$

Concentration

$$(\text{mol} \cdot \text{L}^{-1}) \quad CH_3COOH(aq) + H_2O(l) \rightleftharpoons H_3O^+(aq) + CH_3CO_2^-(aq)$$

initial	4.29×10^{-2}	—	0	2.86×10^{-2}
change	$-x$	—	$+x$	$+x$

equilibrium $\quad 4.29 \times 10^{-2} - x \qquad\qquad - \qquad\qquad x \qquad 2.86 \times 10^{-2} + x$

$$1.8 \times 10^{-5} = \frac{(x)(x + 2.86 \times 10^{-2})}{(4.29 \times 10^{-2} - x)} ; \text{ assume } +x \text{ and } -x \text{ negligible.}$$

$[H_3O^+] = x = 2.7 \times 10^{-5} \text{ mol} \cdot L^{-1}$ and $pH = -\log(2.7 \times 10^{-5}) = 4.56$

(c) Because acid and base concentrations are equal, their volumes are equal at the stoichiometric point. Therefore, 25.0 mL NaOH is required to reach the stoichiometric point, and 12.5 mL NaOH is required to reach the half stoichiometric point.

(d) At the half stoichiometric point, $pH = pK_a = 4.75$.

(e) 25.0 mL; see part (c)

(f) The pH is that of 0.050 M $NaCH_3CO_2$.

Concentration

(mol $\cdot$ L^{-1})	$H_2O(l)$ +	$CH_3CO_2^-$(aq)	$\rightleftharpoons CH_3COOH$(aq) +	OH^-(aq)
initial	—	0.050	0	0
change	—	$-x$	$+x$	$+x$
equilibrium	—	$0.050 - x$	x	x

$$K_b = \frac{K_w}{K_a} = \frac{1.00 \times 10^{-14}}{1.8 \times 10^{-5}} = 5.6 \times 10^{-10} = \frac{x^2}{0.050 - x} \approx \frac{x^2}{0.050}$$

$x^2 = 2.8 \times 10^{-11}$

$x = 5.3 \times 10^{-6} \text{ mol} \cdot L^{-1} = [OH^-]$

$pOH = 5.28, pH = 14.00 - 5.28 = 8.72$

12.35 (a) $K_b = \dfrac{[NH_4^+][OH^-]}{[NH_3]} = 1.8 \times 10^{-5}$

Concentration

(mol $\cdot$ L^{-1})	$H_2O(l)$ +	NH_3(aq)	$\rightleftharpoons$	NH_4^+(aq) +	OH^-(aq)
initial	—	0.15		0	0
change	—	$-x$		$+x$	$+x$
equilibrium	—	$0.15 - x$		x	x

$$1.8 \times 10^{-5} = \frac{x^2}{0.15 - x} \approx \frac{x^2}{0.15}$$

$$[OH^-] = x = 1.6 \times 10^{-3} \text{ mol} \cdot L^{-1}$$

$$pOH = 2.80, \text{ initial pH} = 14.00 - 2.80 = 11.20$$

(b) initial moles of

$$NH_3 = (0.0150 \text{ L})(0.15 \text{ mol} \cdot L^{-1}) = 2.3 \times 10^{-3} \text{ mol NH}_3$$

$$\text{moles of HCl} = (0.0150 \text{ L})(0.10 \text{ mol} \cdot L^{-1}) = 1.5 \times 10^{-3} \text{ mol HCl}$$

$$\frac{(2.3 \times 10^{-3} - 1.5 \times 10^{-3}) \text{ mol NH}_3}{0.0300 \text{ L}} = 2.7 \times 10^{-2} \text{ mol} \cdot L^{-1} \text{ NH}_3$$

$$\frac{1.5 \times 10^{-3} \text{ mol HCl}}{0.0300 \text{ L}} = 5.0 \times 10^{-2} \text{ mol} \cdot L^{-1} \text{ HCl} \approx 5.0 \times 10^{-2} \text{ mol} \cdot L^{-1} \text{ NH}_4^+$$

Then consider the equilibrium:

Concentration

$(\text{mol} \cdot L^{-1})$	$H_2O(l)$	$+$	$NH_3(aq)$	$\rightleftharpoons$	$NH_4^+(aq)$	$+$	$OH^-(aq)$
initial	—		2.7×10^{-2}		5.0×10^{-2}		0
change	—		$-x$		$+x$		$+x$
equilibrium	—		$2.7 \times 10^{-2} - x$		$5.0 \times 10^{-2} + x$		x

$$K_b = \frac{[NH_4^+][OH^-]}{[NH_3]} = 1.8 \times 10^{-5}$$

$$= \frac{(x)(5.0 \times 10^{-2} + x)}{(2.7 \times 10^{-2} - x)}; \text{ assume that } +x \text{ and } -x \text{ are negligible}$$

$$[OH^-] = x = 9.7 \times 10^{-6} \text{ mol} \cdot L^{-1} \text{ and pOH} = 5.01$$

Therefore, pH $= 14.00 - 5.01 = 8.99$.

(c) At the stoichiometric point, moles of NH_3 = moles of HCl.

$$\text{volume HCl added} = \frac{(0.15 \text{ mol} \cdot L^{-1} \text{ NH}_3)(0.0150 \text{ L})}{0.10 \text{ mol} \cdot L^{-1} \text{ HCl}} = 0.022 \text{ L HCl}$$

Therefore, halfway to the stoichiometric point, volume HCl added

$= 22/2 = 11 \text{ mL}$.

(d) At half stoichiometric point, $pOH = pK_b$ and $pOH = 4.75$.

Therefore, pH $= 14.00 - 4.75 = 9.25$.

(e) 22 mL; see part (c)

(f) $NH_4^+(aq) + H_2O(l) \rightleftharpoons H_3O^+(aq) + NH_3(aq)$

The initial moles of NH_3 have now been converted to moles of NH_4^+ in a

$(15 + 22.5 = 37.5)$ mL volume:

$$[NH_4^+] = \frac{2.25 \times 10^{-3} \text{ mol}}{0.0375 \text{ L}} = 0.060 \text{ mol} \cdot L^{-1}$$

$$K_a = \frac{K_w}{K_b} = \frac{1.00 \times 10^{-14}}{1.8 \times 10^{-5}} = 5.6 \times 10^{-10}$$

Concentration

$(mol \cdot L^{-1})$	$NH_4^+(aq)$	+	$H_2O(l)$	$\rightleftharpoons$	$H_3O^+(aq)$	+	$NH_3(aq)$
initial	0.060		—		0		0
change	$-x$		—		$+x$		$+x$
equilibrium	$0.060 - x$		—		x		x

$$K_a = 5.6 \times 10^{-10} = \frac{x^2}{0.060 - x} \approx \frac{x^2}{0.060}$$

$$x = [H_3O^+] = 5.8 \times 10^{-6} \text{ mol} \cdot L^{-1}$$

$$pH = -\log(5.8 \times 10^{-6}) = 5.24$$

(g) Methyl red will be appropriate for this titration.

12.37 (a) The acid is a weak acid because the initial pH of the acid $=5.0$ and the stoichiometric point occurs at pH $=10$ (the anion is a base).

(b) pH $=5.0$, $[H_3O^+] = 1 \times 10^{-5}$ mol $\cdot L^{-1}$

(c) At halfway titration, pH $= pK_a$. It takes 5.0 mL base to reach halfway, the pH $= 7.5 = pK_a$ of the acid. $K_a = 10^{-7.5} = 3 \times 10^{-8}$.

(d) $HA(aq) + H_2O(l) \rightarrow H_3O^+(aq) + A^-(aq)$

When HA self-dissociation reaches to an equilibrium:

$[H_3O^+] = [A^-] = 1 \times 10^{-5}$ mol $\cdot L^{-1}$; $[HA]_{eq} = [HA]_{initial} - (1 \times 10^{-5})$

$$K_a = 3 \times 10^{-8} = \frac{[H_3O^+][A^-]}{[HA]_{eq}} = \frac{(1 \times 10^{-5})^2}{[HA]_{initial} - (1 \times 10^{-5})}$$

Since K_a is small, the dissociation level of HA will be very small.

Therefore, $[HA]_{initial} - (1 \times 10^{-5}) \approx [HA]_{initial}$.

Solve for the equation: $[HA]_{initial} = 3 \times 10^{-3}$ mol $\cdot$ L^{-1}

(e) It takes 10.0 mL base to reach to stoichiometric point. $[HA]_{initial} = 3.16 \times 10^{-3}$ mol $\cdot$ L^{-1}. If the volume of acid = 25.0 mL, then

$$[B] = \frac{[HA]V_{HA}}{V_{base}} = \frac{(3 \times 10^{-3})(25\ mL)}{(10.0\ mL)} = 8 \times 10^{-3}\ M.$$

(f) Phenolphthalein will be an appropriate indicator.

12.39 At the stoichiometric point, the volume of solution will have doubled; therefore, the concentration of $CH_3CO_2^-$ will be 0.10 M. The equilibrium is

Concentration

(mol $\cdot$ L^{-1})	CH_3CO_2 (aq) + H_2O(l) $\rightleftharpoons$	HCH_3CO_2 (aq) +	OH^- (aq)
initial	0.10	—	0	0
change	$-x$	—	$+x$	$+x$
equilibrium	$0.10 - x$	—	x	x

$$K_b = \frac{K_w}{K_a} = \frac{1.00 \times 10^{-14}}{1.8 \times 10^{-5}} = 5.6 \times 10^{-10}$$

$$K_b = \frac{[HCH_3CO_2][OH^-]}{[CH_3CO_2^-]} = \frac{x^2}{0.10 - x} \approx \frac{x^2}{0.10} = 5.6 \times 10^{-10}$$

$x = 7.5 \times 10^{-6}$ mol $\cdot$ L^{-1} = $[OH^-]$

$pOH = -\log(7.5 \times 10^{-6}) = 5.12$, $pH = 14.00 - 5.12 = 8.88$

From Table 11.3, we see that this pH value lies within the range for phenolphthalein and thymol blue; the others would not.

12.41 Exercise 11.33: thymol blue or phenolphthalein; Exercise 11.35: methyl red or bromocresol green.

12.43 (a) To reach the first stoichiometric point, we must add enough solution to neutralize one H^+ on the H_3AsO_4. To do this, we will require

0.0750 L $\times$ 0.137 mol $\cdot$ L^{-1} = 0.0103 mol OH$^-$. The volume of base required will be given by the number of moles of base required divided by the concentration of base solution:

$$\frac{0.0750 \text{ L} \times 0.137 \text{ mol} \cdot \text{L}^{-1}}{0.275 \text{ mol} \cdot \text{L}^{-1}} = 0.0374 \text{ L or } 37.4 \text{ mL}$$

(b) and (c) To reach the second stoichiometric point will require double the amount calculated in (a), or 74.8 mL, and the third stoichiometric point will be reached with three times the amount added in (a), or 112 mL.

12.45 (a) The base HPO_3^{2-} is the fully deprotonated form of phosphorous acid H_3PO_3 (the remaining H attached to P is not acidic). It will require an equal number of moles of HNO_3 to react with HPO_3^{2-} in order to reach the first stoichiometric point (formation of $H_2PO_3^-$). The value will be

given by $\dfrac{0.0355 \text{ L} \times 0.158 \text{ mol} \cdot \text{L}^{-1}}{0.255 \text{ mol} \cdot \text{L}^{-1}} = 0.0220 \text{ L or } 22.0 \text{ mL}$

(b) To reach the second stoichiometric point would require double the amount of solution calculated in (a), or 44.0 mL.

12.47 (a)

Point (a): initial point, point (b): first halfway point, point (c): first stoichiometric point, point (d): second halfway point, point (e) second stoichiometric point

Note: (i) The initial pH is higher than points b and c because we assume that the initial $[H^+]$ from the first dissociation only (actually second dissociation contributes also); (ii) due to the close pK_a values of thiosulfuric acid, only one obvious titration curve can be observed. Two protons will be titrated at the same time for this acid.

(b) First stoichiometric point: (C) = 5.0 mL; second stoichiometric point: (E) = 10. mL

(c) First stoichiometric point: $pH = \frac{1}{2}(pK_{a_1} + pK_{a_2}) = \frac{1}{2}(0.6+1.74) = 1.2$; second stoichiometric point:

$$S_2O_3^{2-} + H_2O \rightleftharpoons HS_2O_3^- + OH^-$$

$$K_{b_1} = \frac{K_W}{K_{a_2}} = 5.56 \times 10^{-13}, \; [S_2O_3^{2-}]_{initial} = (5.0 \times 0.01)/15.0 = 0.0033 \text{ M}$$

$$K_{b_1} = \frac{x^2}{0.0033 - x} = 5.56 \times 10^{-13}, \; x = [OH^-] = 4.30 \times 10^{-8} < [OH^-]_{water}$$

Therefore, pH = 7.0.

12.49 (a) This value is calculated as described in Example 11.12. First, we calculate the molarity of the starting phosphorous acid solution:

$$\frac{0.122 \text{ g}}{81.99 \text{ g} \cdot \text{mol}^{-1}} \bigg/ 0.0500 \text{ L} = 0.0298 \text{ mol} \cdot \text{L}^{-1}. \text{ We then use the first acid}$$

dissociation of phosphorous acid as the dominant equilibrium. The K_{a1} is 1.0×10^{-2}. Let H_2P represent the fully-protonated phosphorus acid.

Concentration

(mol·L⁻¹)	$H_2P(aq)$ + $H_2O(l)$	$\rightleftharpoons$ $HP^-(aq)$ +	$H_3O^+(aq)$
initial	0.0298	— 0	0
change	−x	— +x	+x
final	0.0298 − x	— +x	+x

$$K_a = \frac{[H_3O^+][HP^-]}{[H_2P]} = 1.0 \times 10^{-2}$$

$$1.0 \times 10^{-2} = \frac{x \cdot x}{0.0298 - x} = \frac{x^2}{0.0298 - x}$$

If we assume $x \ll 0.0298$, then the equation becomes

$$x^2 = (1.0 \times 10^{-2})(0.0298) = 2.98 \times 10^{-4}$$

$$x = 1.73 \times 10^{-2}.$$

Because this value is more than 10% of 0.0300, the full quadratic solution should be undertaken. The equation is

$$x^2 = (1.0 \times 10^{-2})(0.0298 - x) \text{ or}$$

$$x^2 + (1.0 \times 10^{-2} x) - (2.98 \times 10^{-4}) = 0.$$

Using the quadratic formula, we obtain $x = 0.013$.

pH $= 1.89$

(b) First, carry out the reaction between phosphorous acid and the strong base to completion:

$$H_2P(aq) + OH^-(aq) \longrightarrow HP^-(aq) + H_2O(l)$$

moles of $H_2P = (0.0298 \text{ mol} \cdot L^{-1})(0.0500 \text{ L}) = 1.49 \times 10^{-3} \text{ mol}$

moles of $OH^- = (0.00500 \text{ L})(0.175 \text{ mol} \cdot L^{-1}) = 8.75 \times 10^{-4} \text{ mol}$

8.75×10^{-4} mol OH^- will react completely with 1.49×10^{-3} mol H_2P to give 8.75×10^{-4} mol HP^-, with 6.15×10^{-4} moles of H_2P remaining.

$$[H_2P] = \frac{6.15 \times 10^{-4} \text{ mol}}{0.0550 \text{ L}} = 0.0112 \text{ mol} \cdot L^{-1}$$

$$[HP^-] = \frac{8.75 \times 10^{-4} \text{ mol}}{0.0550 \text{ L}} = 0.0159 \text{ mol} \cdot L^{-1}$$

Concentration

(mol$\cdot$L^{-1})	H$_2$P(aq)	+	H$_2$O(l)	$\rightleftharpoons$	HP$^-$(aq)	+	H$_3$O$^+$(aq)
initial	0.0112		—		0.0159		0
change	$-x$		—		$+x$		$+x$
final	$0.0112 - x$		—		$0.0159 + x$		$+x$

The calculation is performed as in part (a):

$$1.0 \times 10^{-2} = \frac{(0.0159 + x)x}{0.0112 - x}$$

$$x = 3.75 \times 10^{-3}$$

$$\text{pH} = 2.42$$

(c) moles of $H_2P = 1.49 \times 10^{-3}$

moles of $OH^- = 8.75 \times 10^{-4} + 8.75 \times 10^{-4} = 1.75 \times 10^{-3}$

Following the reaction between H_2P and OH^-, 0 mol of H_2P remain and 2.60×10^{-4} mol OH^- remain. $[OH^-] = 4.33 \times 10^{-3}$ M

1.49×10^{-3} mol of HP^- remain, $[HP^-] = \dfrac{1.49 \times 10^{-3} \text{ mol}}{0.060 \text{ L}} = 2.48 \times 10^{-2}$ M.

Since 4.33×10^{-3} M OH^- will produce 4.33×10^{-3} M P^{2-},

the equilibrium $[HP^-] = 2.48 \times 10^{-2} - 4.33 \times 10^{-3} = 2.05 \times 10^{-2}$ M

Concentration

$(\text{mol} \cdot L^{-1})$	HP^-	+	H_2O	$\rightleftharpoons$	P^{2-}	+	H_3O^+
initial	2.05×10^{-2}		—		4.33×10^{-3}		0
change	$-x$		$-x$		$+x$		$+x$
final	$2.05 \times 10^{-2} - x$		—		$4.33 \times 10^{-3} + x$		x

The calculation is performed as in part (a):

$$k_{a2} = 2.60 \times 10^{-7} = \frac{x(4.33 \times 10^{-3} + x)}{\left(2.05 \times 10^{-2} - x\right)}$$

Assuming x is small, we find $x = 1.23 \times 10^{-6}$,

$$\text{pH} = -\log(1.23 \times 10^{-6}) = 5.91.$$

12.51 (a) The reaction of the base Na_2HPO_4 with the strong acid will be taken to completion first:

$$HPO_4^{2-}(aq) + H_3O^+(aq) \longrightarrow H_2PO_4^- + H_2O(l)$$

Initially, moles of

$$HPO_4^{2-} = \text{moles of } H_3O^+ = 0.0500 \text{ L} \times 0.275 \text{ mol} \cdot L^{-1} = 0.0138 \text{ mol}$$

Because this reaction proceeds with no excess base or acid, we are dealing with a solution that can be viewed as being composed of $H_2PO_4^-$. The problem then becomes one of estimating the pH of this solution, which can be done from the relationship

$pH = \frac{1}{2}(pK_{a1} + pK_{a2})$

$pH = \frac{1}{2}(2.12 + 7.21) = 4.66$.

(b) This reaction proceeds as in (a), but there is more strong acid available, so the excess acid will react with $H_2PO_4^-$ to produce H_3PO_4. Addition of the first 50.0 mL of acid solution will convert all the HPO_4^{2-} into $H_2PO_4^-$. The additional 25.0 mL of the strong acid will react with $H_2PO_4^-$:

$H_2PO_4^-(aq) + H_3O^+(aq) \longrightarrow H_3PO_4(aq) + H_2O(l)$

0.0138 mol $H_2PO_4^-$ will react with 0.006 88 mol H_3O^+ to give 0.0069 mol H_3PO_4, with 0.069 mol $H_2PO_4^-$ in excess. The concentrations will be

$[H_3PO_4] = [H_2PO_4^-] = \dfrac{0.0069 \text{ mol}}{0.125 \text{ L}} = 0.055$. The appropriate relationship

to use is then

Concentration

$(mol \cdot L^{-1})$ $\quad H_3PO_4(aq) \quad + \quad H_2O(l) \quad \rightleftharpoons \quad H_2PO_4^-(aq) \quad + \quad H_3O^+(aq)$

initial	0.055	—	0.055	0
change	$-x$	—	$+x$	$+x$
final	$0.055 - x$	—	$0.055 + x$	$+x$

$K_{a1} = \dfrac{[H_2PO_4^-][H_3O^+]}{[H_3PO_4]}$

Because the equilibrium constant is not small compared to 0.055, the full quadratic solution must be calculated:

$x^2 + 0.055x = 7.6 \times 10^{-3}(0.055 - x)$

$x^2 + 0.063x - 4.2 \times 10^{-4} = 0$

$x = 1.6 \times 10^{-3}$

$pH = -\log(1.6 \times 10^{-3}) = 2.80$

(c) The reaction of Na_2HPO_4 with strong acid goes only halfway to completion. 0.275 mol of HPO_4^{2-} will react with

$(0.025 \text{ L} \times 0.275 \text{ mol} \cdot \text{L}^{-1}) = 6.9 \times 10^{-3}$ mol HCl to produce 6.9×10^{-3} mol

$H_2PO_4^{-}$ and leave 6.9×10^{-3} HPO_4^{2-} unreacted.

6.9×10^{-3} mol $\div$ 0.075 L $= 0.092$ mol $\cdot \text{L}^{-1}$

Concentration

$(\text{mol} \cdot \text{L}^{-1})$	$H_2PO_4^{-}(aq)$	$+$ $H_2O(l)$	$\rightleftharpoons$ $H_2PO_4^{2-}(aq)$	$+$ $H_3O^{+}(aq)$
intial	0.092	—	0.092	0
change	$-x$	—	$+x$	$+x$
final	$0.092 - x$	—	$0.092 + x$	$+x$

$K_{a2} = 6.2 \times 10^{-8} = \dfrac{[HPO_4^{2-}][H_3O^{+}]}{[H_2PO_4^{-}]}$

$\dfrac{[0.092 + x][H_3O^{+}]}{[0.092 - x]} = 6.2 \times 10^{-8}$

assuming $x \ll$ than 0.092

$x = [H_3O^{+}] = 6.2 \times 10^{-8}$

$pH = -\log(6.2 \times 10^{-8}) = 7.21$

12.53 (a) The solubility equilibrium is $AgBr(s) \rightleftharpoons Ag^{+}(aq) + Br^{-}(aq)$.

$[Ag^{+}] = [Br^{-}] = 8.8 \times 10^{-7}$ mol $\cdot \text{L}^{-1} = S =$ solubility

$K_{sp} = [Ag^{+}][Br^{-}] = (8.8 \times 10^{-7})(8.8 \times 10^{-7}) = 7.7 \times 10^{-13}$

(b) The solubility equilibrium is $PbCrO_4(s) \rightleftharpoons Pb^{2+}(aq) + CrO_4^{2-}(aq)$.

$[Pb^{2+}] = 1.3 \times 10^{-7}$ mol $\cdot \text{L}^{-1} = S$, $[CrO_4^{2-}] = 1.3 \times 10^{-7}$ mol $\cdot \text{L}^{-1} = S$

$K_{sp} = [Pb^{2+}][CrO_4^{2-}] = (1.3 \times 10^{-7})(1.3 \times 10^{-7}) = 1.7 \times 10^{-14}$

(c) The solubility equilibrium is $Ba(OH)_2(s) \rightleftharpoons Ba^{2+}(aq) + 2\,OH^{-}(aq)$.

$$[Ba^{2+}] = 0.11 \ mol \cdot L^{-1} = S, \quad [OH^-] = 0.22 \ mol \cdot L^{-1} = 2S$$

$$K_{sp} = [Ba^{2+}][OH^-]^2 = (0.11)(0.22)^2 = 5.3 \times 10^{-3}$$

(d) The solubility equilibrium is $MgF_2(s) \rightleftharpoons Mg^{2+}(aq) + 2 \ F^-(aq)$.

$$[Mg^{2+}] = 1.2 \times 10^{-3} \ mol \cdot L^{-1} = S, \quad [F^-] = 2.4 \times 10^{-3} \ mol \cdot L^{-1} = 2S$$

$$K_{sp} = [Mg^{2+}][F^-]^2 = (1.2 \times 10^{-3})(2.4 \times 10^{-3})^2 = 6.9 \times 10^{-9}$$

12.55 $Tl_2CrO_4(s) \rightleftharpoons 2 \ Tl^+(aq) + CrO_4^{\ 2-}(aq)$

$$[CrO_4^{\ 2-}] = S = 6.3 \times 10^{-5} \ mol \cdot L^{-1}$$
$$[Tl^+] = 2S = 2(6.3 \times 10^{-5}) \ mol \cdot L^{-1}$$
$$K_{sp} = [Tl^+]^2[CrO_4^{\ 2-}] = (2S)^2 \times (S)$$
$$K_{sp} = [2(6.3 \times 10^{-5})]^2 \times (6.3 \times 10^{-5}) = 1.0 \times 10^{-12}$$

12.57 (a) Equilibrium equation: $BiI_3(s) \leftrightarrow Bi^{3+}(aq) \ + 3 \ I^-(aq)$
$$K_{sp} = 7.71 \times 10^{-19} = [Bi][I^-]^3 = (S)(3S)^3$$
$$S = 1.30 \times 10^{-5} \ mol \cdot L^{-1}$$

(b) Equilibrium equation: $CuCl(s) \leftrightarrow Cu^+(aq) \ + \ Cl^-(aq)$
$$K_{sp} = 1.0 \times 10^{-6} = [Cu^+][Cl] = S^2$$
$$S = 1.0 \times 10^{-3} \ mol \cdot L^{-1}$$

(c) Equilibrium equation: $CaCO_3(s) \rightleftharpoons Ca^{2+}(aq) + CO_3^{\ 2-}(aq)$

$$K_{sp} = [Ca^{2+}][CO_3^{\ 2-}] = S \times S = S^2 = 8.7 \times 10^{-9}$$
$$S = 9.3 \times 10^{-5} \ mol \cdot L^{-1}$$

12.59 (a) Concentration $(mol \cdot L^{-1})$ $AgCl(s) \rightleftharpoons Ag^+(aq) \ + \ Cl^-(aq)$

initial	—	0	0.20
change	—	$+S$	$+S$
equilibrium	—	S	$S + 0.20$

$$K_{sp} = [Ag^+][Cl^-] = (S) \times (S + 0.20) = 1.6 \times 10^{-10}$$

Assume S in $(S + 0.20)$ is negligible, so $0.20 \ S = 1.6 \times 10^{-10}$

$S = 8.0 \times 10^{-10} \text{ mol} \cdot \text{L}^{-1} = [\text{Ag}^+] = $ molar solubility of AgCl in

0.15 M NaCl

(b) Concentration (mol $\cdot$ L^{-1}) Hg$_2$Cl$_2$(s) $\rightleftharpoons$ Hg$_2^{2+}$(aq) + 2 Cl$^-$(aq)

initial	—	0	0.150
change	—	+S	+2S
equilibrium	—	S	0.150 + 2S

$K_{sp} = [\text{Hg}_2^{2+}][\text{Cl}^-]^2 = (S) \times (2S + 0.150)^2 = 2.6 \times 10^{-18}$

Assume 2S in (2S + 0.150) is negligible, so $(0.150)^2 S = 2.6 \times 10^{-18}$.

$S = 1.2 \times 10^{-16} \text{ mol} \cdot \text{L}^{-1} = [\text{Hg}_2^{2+}] = $ molar solubility of Hg$_2$Cl$_2$ in

0.150 M NaCl.

(c) Concentration (mol $\cdot$ L^{-1}) PbCl$_2$(s) $\rightleftharpoons$ Pb^{2+}(aq) + 2 Cl$^-$(aq)

initial	—	0	2 × 0.025 = 0.05
change	—	+S	+S
equilibrium	—	S	S + 0.05

$K_{sp} = [\text{Pb}^{2+}][\text{Cl}^-]^2 = S \times (2S + 0.05)^2 = 1.6 \times 10^{-5}$

S may not be negligible relative to 0.05, so the full cubic form may be required. We do it both ways:

For $S^3 + 0.20\, S^2 + (0.0025 \times 10^{-2}\, S) - (1.6 \times 10^{-5}) = 0$, the solution by standard methods is $S = 4.6 \times 10^{-3} \text{ mol} \cdot \text{L}^{-1}$.

If S had been neglected, the answer would be S = 6.4×10^{-3}.

(d) Concentration (mol $\cdot$ L^{-1}) Fe(OH)$_2$(s) $\rightleftharpoons$ Fe^{2+}(aq) + 2 OH$^-$(aq)

initial	—	2.5×10^{-3}	0
change	—	+S	+2S
equilibrium	—	$2.5 \times 10^{-3} + S$	2S

$K_{sp} = [\text{Fe}^{2+}][\text{OH}^-]^2 = (S + 2.5 \times 10^{-3}) \times (2S)^2 = 1.6 \times 10^{-14}$

Assume S in $(S + 2.5 \times 10^{-3})$ is negligible, so

$4S^2 \times (2.5 \times 10^{-3}) = 1.6 \times 10^{-1}$.

$$S^2 = 1.6 \times 10^{-12}$$

$$S = 1.3 \times 10^{-6} \text{ mol} \cdot \text{L}^{-1} = \text{molar solubility of Fe(OH)}_2 \text{ in } 2.5 \times 10^{-3} \text{ M FeCl}_2.$$

12.61 (a) $pH = 7.0; [OH^-] = 1.0 \times 10^{-7} \text{ mol} \cdot \text{L}^{-1}$

$$Al^{3+}(aq) + 3 \, OH^-(aq) \rightleftharpoons Al(OH)_3(s)$$

$$[Al^{3+}][OH^-]^3 = K_{sp} = 1.0 \times 10^{-33}$$

$$S \times (10^{-7})^3 = 1.0 \times 10^{-33}$$

$$S = \frac{1.0 \times 10^{-33}}{1 \times 10^{-21}} = 1.0 \times 10^{-12} \text{ mol} \cdot \text{L}^{-1} = [Al^{3+}]$$

$= \text{molar solubility of Al(OH)}_3 \text{ at } pH = 7.0$

(b) $pH = 4.5; pOH = 9.5; [OH^-] = 3.2 \times 10^{-10} \text{ mol} \cdot \text{L}^{-1}$

$$[Al^{3+}][OH^-]^3 = K_{sp} = 1.0 \times 10^{-33}$$

$$S \times (3.2 \times 10^{-10})^3 = 1.0 \times 10^{-33}$$

$$S = \frac{1.0 \times 10^{-33}}{3.3 \times 10^{-29}} = 3.1 \times 10^{-5} \text{ mol} \cdot \text{L}^{-1} = [Al^{3+}]$$

$= \text{molar solubility of Al(OH)}_3 \text{ at } pH = 4.5$

(c) $pH = 7.0; [OH^-] = 1.0 \times 10^{-7} \text{ mol} \cdot \text{L}^{-1}$

$$Zn^{2+}(aq) + 2 \, OH^-(aq) \rightleftharpoons Zn(OH)_2(s)$$

$$[Zn^{2+}][OH^-]^2 = K_{sp} = 2.0 \times 10^{-17}$$

$$S \times (1.0 \times 10^{-7})^2 = 2.0 \times 10^{-17}$$

$$S = \frac{2.0 \times 10^{-17}}{1.0 \times 10^{-14}} = 2.0 \times 10^{-3} \text{ mol} \cdot \text{L}^{-1} = [Zn^{2+}]$$

$= \text{molar solubility of Zn(OH)}_2 \text{ at } pH = 7.0$

(d) $pH = 6.0; pOH = 8.0; [OH^-] = 1.0 \times 10^{-8} \text{ mol} \cdot \text{L}^{-1}$

$$[Zn^{2+}][OH^-]^2 = 2.0 \times 10^{-17} = K_{sp}$$

$$S \times (1.0 \times 10^{-8})^2 = 2.0 \times 10^{-17}$$

$$S = \frac{2.0 \times 10^{-17}}{1.0 \times 10^{-16}} = 2.0 \times 10^{-1} = 0.20 \text{ mol} \cdot \text{L}^{-1} = [Zn^{2+}]$$

$= \text{molar solubility of Zn(OH)}_2 \text{ at } pH = 6.0$

12.63 (a) $Ag^+(aq) + Cl^-(aq) \rightleftharpoons AgCl(s)$

Concentration $(mol \cdot L^{-1})$	Ag^+	Cl^-
initial	0	1.0×10^{-5}
change	$+x$	0
equilibrium	x	1.0×10^{-5}

$K_{sp} = [Ag^+][Cl^-] = 1.6 \times 10^{-10} = (x)(1.0 \times 10^{-5})$

$x = [Ag^+] = 1.6 \times 10^{-5} \text{ mol} \cdot L^{-1}$

(b) mass $AgNO_3$

$$= \left(\frac{1.6 \times 10^{-5} \text{ mol } AgNO_3}{1 \text{ L}} \right)(0.100 \text{ L})\left(\frac{169.88 \text{ g } AgNO_3}{1 \text{ mol } AgNO_3} \right)\left(\frac{1 \mu g}{10^{-6} \text{ g}} \right)$$

$= 2.7 \times 10^2 \ \mu g \ AgNO_3$

12.65 (a) $Ni^{2+}(aq) + 2 OH^-(aq) \rightleftharpoons Ni(OH)_2(s)$

Concentration $(mol \cdot L^{-1})$	Ni^{2+}	OH^-
initial	0.060	0
change	0	$+x$
equilibrium	0.060	x

$K_{sp} = [Ni^{2+}][OH^-]^2 = 6.5 \times 10^{-18} = (0.060)(x)^2$

$[OH^-] = x = 1.0 \times 10^{-8} \text{ mol} \cdot L^{-1}$

$pOH = -\log(1.0 \times 10^{-8}) = 8.00, \ pH = 14.00 - 8.00 = 6.00$

(b) A similar setup for $[Ni^{2+}] = 0.030$ M.

gives $x = 1.5 \times 10^{-8}$

$pOH = -\log(1.5 \times 10^{-8}) = 7.82$

$pH = 14.00 - 7.82 = 6.18$

12.67 $\left(\dfrac{1 \text{ mL}}{20 \text{ drops}} \right) \times 1 \text{ drop} = 0.05 \text{ mL} = 0.05 \times 10^{-3} \text{ L} = 5 \times 10^{-5} \text{ L}$

and $(5 \times 10^{-5} \text{ L})(0.010 \text{ mol} \cdot L^{-1}) = 5 \times 10^{-7} \text{ mol NaCl} = 5 \times 10^{-7} \text{ mol } Cl^-$

(a) $Ag^+(aq) + Cl^-(aq) \rightleftharpoons AgCl(s)$, $[Ag^+][Cl^-] = K_{sp}$

$$Q_{sp} = \left[\frac{(0.010 \text{ L})(0.0040 \text{ mol} \cdot \text{L}^{-1})}{0.010 \text{ L}}\right]\left[\frac{5 \times 10^{-7} \text{ mol}}{0.010 \text{ L}}\right] = 2 \times 10^{-7}$$

Will precipitate because $Q_{sp}(2 \times 10^{-7}) > K_{sp}(1.6 \times 10^{-10})$.

(b) $Pb^{2+}(aq) + 2 Cl^-(aq) \rightleftharpoons PbCl_2(s)$, $[Pb^{2+}][Cl^-]^2 = K_{sp}$

$$Q_{sp} = \left[\frac{(0.0100 \text{ L})(0.0040 \text{ mol} \cdot \text{L}^{-1})}{0.010 \text{ L}}\right]\left[\frac{5 \times 10^{-7} \text{ mol}}{0.010 \text{ L}}\right]^2 = 1 \times 10^{-11}$$

Will not precipitate because $Q_{sp}(1 \times 10^{-11}) < K_{sp}(1.6 \times 10^{-5})$.

12.69 (a) $K_{sp}[Ni(OH)_2] < K_{sp}[Mg(OH)_2] < K_{sp}[Ca(OH)_2]$

This is the order for the solubility products of these hydroxides. Thus, the order of precipitation is (first to last) $Ni(OH)_2$, $Mg(OH)_2$, $Ca(OH)_2$.

(b) $K_{sp}[Ni(OH)_2] = 6.5 \times 10^{-18} = [Ni^{2+}][OH^-]^2$

$$[OH^-]^2 = \frac{6.5 \times 10^{-18}}{0.0010} = 6.5 \times 10^{-15}$$
$$[OH^-] = 8.1 \times 10^{-8}$$
$$pOH = -\log[OH^-] = 7.09 \quad pH \approx 7$$
$$K_{sp}[Mg(OH)_2] = 1.1 \times 10^{-11} = [Mg^{2+}][OH^-]^2$$
$$[OH^-] = \sqrt{\frac{1.1 \times 10^{-11}}{0.0010}} = 1.0 \times 10^{-4}$$
$$pOH = -\log(1.0 \times 10^{-4}) = 4.00 \quad pH = 14.00 - 4.00 = 10.00, pH \approx 10$$
$$K_{sp}[Ca(OH)^2] = 5.5 \times 10^{-6} = [Ca^{2+}][OH^-]^2$$
$$[OH^-] = \sqrt{\frac{5.5 \times 10^{-6}}{0.0010}} = 7.4 \times 10^{-2}$$
$$pOH = -\log(7.4 \times 10^{-2}) = 1.13 \quad pH = 14.00 - 1.13 = 12.87, pH \approx 13$$

12.71 The K_{sp} values are: MgF_2 6.4×10^{-9}

                                 BaF_2 1.7×10^{-6}

$$MgCO_3 \quad 1.0 \times 10^{-5}$$

$$BaCO_3 \quad 8.1 \times 10^{-9}$$

The difference in these numbers suggests that there is a greater solubility difference between the carbonates, and thus this anion should give a better separation. Because different numbers of ions are involved, it is instructive to convert the K_{sp} values into molar solubility. For the fluorides the reaction is:

$$MF_2(s) \rightleftharpoons M^{2+}(aq) + 2\,F^-(aq)$$

Change $\qquad +x \qquad +2x$

$$K_{sp} = x(2x)^2$$

Solving this for MgF_2 gives 0.0012 M and for BaF_2 gives 0.0075 M.

For the carbonates:

$$MCO_3(s) \rightleftharpoons M^{2+}(aq) + CO_3{}^{2-}(aq)$$

$$\qquad\qquad +x \qquad +x$$

$$K_{sp} = x^2$$

Solving this for $MgCO_3$ gives 0.0032 M and for $BaCO_3$ gives 9.0×10^{-5} M. Clearly, the solubility difference is greatest between the two carbonates, and $CO_3{}^{2-}$ is the better choice of anion.

12.73 $Cu(IO_3)_2$ ($K_{sp} = 1.4 \times 10^{-7}$) is more soluble than $Pb(IO_3)_2$ ($K_{sp} = 2.6 \times 10^{-13}$), so Cu^{2+} will remain in solution until essentially all the $Pb(IO_3)_2$ has precipitated. Thus, we expect very little Pb^{2+} to be the left in solution by the time we reach the point at which $Cu(IO_3)_2$ begins to precipitate.

The concentration of $IO_3{}^-$ at which Cu^{2+} begins to precipitate will be given by
$$K_{sp} = [Cu^{2+}][IO_3{}^-]^2 = 1.4 \times 10^{-7} = [0.0010][IO_3{}^-]^2$$
$$[IO_3{}^-] = 0.012 \ mol \cdot L^{-1}.$$

The concentration of Pb in solution when the $[IO_3^-] = 0.012$ mol·L^{-1} is given by

$$K_{sp} = [Pb^{2+}][IO_3^-]^2 = 2.6 \times 10^{-13} = [Pb^{2+}][0.012]^2$$
$$[Pb^{2+}] = 1.8 \times 10^{-9} \text{ mol·}L^{-1}.$$

12.75 $CaF_2(s) \rightleftharpoons Ca^{2+}(aq) + 2 F^-(aq)$ $K_{sp} = 4.0 \times 10^{-11}$

   $F^-(aq) + H_2O(l) \rightleftharpoons HF(aq) + OH^-(aq)$ $K_b(F^-) = 2.9 \times 10^{-11}$

(a) Multiply the second equilibrium equation by 2 and add to the first equilibrium:

$CaF_2(s) + 2 H_2O(l) \rightleftharpoons Ca^{2+}(aq) + 2 HF(aq) + 2 OH^-(aq)$
$K = K_w \cdot K_b^2 = (4.0 \times 10^{-11})(2.9 \times 10^{-11})^2 = 3.4 \times 10^{-32}$

(b) The calculation of K_{sp} is complicated by the fact that the anion of the salt is part of a weak base-acid pair. If we wish to solve the equation algebraically, then we need to consider which equilibrium is the dominant one at pH = 7.0, for which $[H_3O^+] = 1 \times 10^{-7}$.

To determine whether F^- or HF is the dominant species at this pH (if either), consider the base hydrolysis reaction:

$F^-(aq) + H_2O(l) \rightleftharpoons HF(aq) + OH^-(aq)$ $K_b = 2.9 \times 10^{-11}$

$$K_b = \frac{[HF][OH^-]}{[F^-]}$$

$$2.9 \times 10^{-11} = \frac{[HF][1 \times 10^{-7}]}{[F^-]}$$

$$\frac{[HF]}{[F^-]} = \frac{2.9 \times 10^{-11}}{1 \times 10^{-7}} = 3 \times 10^{-4}$$

Given that the ratio of HF to F^- is on the order of 10^{-4} to 1, the F^- species is still dominant. The appropriate equation to use is thus the original one for the K_{sp} of $CaF_2(s)$:

$$CaF_2(s) \rightleftharpoons Ca^{2+}(aq) + 2F^-(aq) \qquad K_{sp} = 4.0 \times 10^{-11}$$

$$K_{sp} = [Ca^{2+}][F^-]^2$$

$$4.0 \times 10^{-11} = x(2x)^2 = 4x^3$$

$$x = 2.0 \times 10^{-4}$$

molar solubility $= 2.0 \times 10^{-4}$ mol$\cdot$L^{-1}

(c) At

pH $= 3.0$, $[H_3O^+] = 1 \times 10^{-3}$ mol$\cdot$L^{-1} and $[OH^-] = 1 \times 10^{-11}$ mol$\cdot$L^{-1}.

Under these conditions,

$$K_b = \frac{[HF][OH^-]}{[F^-]}$$

$$2.9 \times 10^{-11} = \frac{[HF][1 \times 10^{-11}]}{[F^-]}$$

$$\frac{[HF]}{[F^-]} = \frac{2.9 \times 10^{-11}}{1 \times 10^{-11}} = 3$$

$$[HF] = 3[F^-].$$

As can be seen, at pH 3.0 the amounts of F^- and HF are comparable, so the protonation of F^- to form HF cannot be ignored. The relation $2[Ca^{2+}] = [F^-] + [HF]$ is required by the mass balance as imposed by the stoichiometry of the dissolution equilibrium:

$$2[Ca^{2+}] = [F^-] + 3[F^-]$$

$$2[Ca^{2+}] = 4[F^-]$$

$$[Ca^{2+}] = 2[F^-]$$

Using this with K_{sp} relationship:

$$(2[F^-])[F^-]^2 = 4.0 \times 10^{-11}$$

$$2[F^-]^3 = 4.0 \times 10^{-11}$$

$$[F^-] = 2 \times 10^{-4}$$

$$[Ca^{2+}] = (2)(2 \times 10^{-4}) = 4 \times 10^{-4} \text{mol} \cdot \text{L}^{-1}$$

The solubility is about double that at pH $= 7.0$.

12.77 $AgBr(s) \rightleftharpoons Ag^+(aq) + Br^-(aq)$ $K_{sp} = 7.7 \times 10^{-13}$

$Ag^+(aq) + 2\,CN^-(aq) \rightleftharpoons Ag(CN)_2^{\,-}(aq)$ $K_f = 5.6 \times 10^8$

$AgBr(s) + 2\,CN^-(aq) \rightleftharpoons Ag(CN)_2^{\,-}(aq) + Br^-(aq)$ $K = 4.3 \times 10^{-4}$

Hence, $K = \dfrac{[Ag(CN)_2^{\,-}][Br^-]}{[CN^-]^2} = 4.3 \times 10^{-4}$

Concentration

(mol·L^{-1})	$AgBr(s)$	$+\ 2\,CN^-(aq)$	$\rightleftharpoons$	$Ag(CN)_2^{\,-}(aq)$	$+$	$Br^-(aq)$
initial	—	0.10		0		0
change	—	$-2S$		$+S$		$+S$
equilibrium	—	$0.10 - 2S$		S		S

$\dfrac{[Ag(CN)_2^{\,-}][Br^-]}{[CN^-]^2} = \dfrac{S^2}{(0.10 - 2S)^2} = 4.3 \times 10^{-4}$

$\dfrac{S}{0.10 - 2S} = \sqrt{4.3 \times 10^{-4}} = 2.1 \times 10^{-2}$

$S = (2.1 \times 10^{-3}) - (4.2 \times 10^{-2}\,S)$

$1.042S = 2.1 \times 10^{-3}$

$S = 2.0 \times 10^{-3}\ \text{mol·L}^{-1} = \text{molar solubility of AgBr}$

12.79 The two salts can be distinguished by their solubility in NH_3. The equilibria that are pertinent are:

$AgCl(s) + 2\,NH_3(aq) \rightleftharpoons Ag(NH_3)_2^{\,+}(aq) + Cl^-(aq)$
$K = K_{sp} \cdot K_f = 2.6 \times 10^{-3}$

$AgI(s) + 2\,NH_3(aq) \rightleftharpoons Ag(NH_3)_2^{\,+}(aq) + I^-(aq)$

$K = K_{sp} \cdot K_f = 2.4 \times 10^{-9}$

For example, let's consider the solubility of these two salts in 1.00 M NH_3 solution:

For AgCl $K = \dfrac{[Ag(NH_3)_2^{\,+}][Cl^-]}{[NH_3]^2} = 2.6 \times 10^{-3}$

Concentration

(mol·L^{-1})	AgCl(s)	+	2 HN$_3$(aq)	$\rightleftharpoons$	Ag(NH$_3$)$_2{}^+$(aq)	+	Cl$^-$(aq)
initial	—		1.00		0		0
change	—		$-2x$		$+x$		$+x$
final	—		$1.00 - 2x$		$+x$		$+x$

$$K = \frac{[\text{Ag(NH}_3)_2{}^+][\text{Cl}^-]}{[\text{NH}_3]^2} = 2.6 \times 10^{-3}$$

$$2.6 \times 10^{-3} = \frac{[x][x]}{[1.00 - 2x]^2} = \frac{x^2}{[1.00 - 2x]^2}$$

$$0.051 = \frac{x}{1.00 - 2x}$$

$$x = 0.046$$

0.046 mol AgCl will dissolve in 1.00 L of aqueous solution. The molar mass of AgCl is 143.32 g·mol^{-1}; this corresponds to

0.046 mol·L$^{-1} \times 142.32$ g·mol$^{-1} = 6.5$ g·L^{-1}.

For AgI, the same calculation gives $x = 4.9 \times 10^{-5}$ mol·L^{-1}. The molar mass of AgI is 234.77 g·mol^{-1}, giving a solubility of

4.9×10^{-5} mol·L$^{-1} \times 234.77$ g·mol$^{-1} = 0.023$ g·L^{-1}.

Thus, we could treat a 0.10 g sample of the compound with 20.0 mL of 1.00 M NH$_3$. The AgCl would all dissolve, whereas practically none of the AgI would.

Note: AgI is also slightly yellow in color, whereas AgCl is white, so an initial distinction could be made based on the color of the sample.

12.81 In order to use qualitative analyses, the sample must first be dissolved. This can be accomplished by digesting the sample with concentrated HNO$_3$ and then diluting the resulting solution. HCl or H$_2$SO$_4$ could not be used because some of the metal compounds formed would be insoluble, whereas all of the nitrates would dissolve. Once the sample is dissolved and diluted, an aqueous solution containing chloride ions can be introduced. This should precipitate the Ag^+ as AgCl but would leave the bismuth and nickel in solution, as long as the solution was acidic. The remaining solution can then be treated with H_2S. In acidic solution, Bi_2S_3 will precipitate but NiS will not. Once the Bi_2S_3 has been precipitated, the pH of the solution can be raised by addition of base. Once this is done, NiS should precipitate.

12.83 The relation to use is $pH = pK_a + \log\dfrac{[\text{Base form}]}{[\text{Acid form}]}$. The K_a value for acetic acid is 1.8×10^{-5} and $pK_a = 4.74$. Because we are adding acid, the pH will fall upon the addition, and we want the final pH to be no more than 0.20 pH units different from the initial pH, or 4.54.

$$4.54 = 4.74 + \log\frac{[\text{Base form}]}{[\text{Acid form}]}$$

$$-0.20 = \log\frac{[\text{Base form}]}{[\text{Acid form}]}$$

$$\frac{[\text{Base form}]}{[\text{Acid form}]} = 0.63$$

We want the concentration of the base form to be 0.63 times that of the acid form. We do not know the initial number of moles of base or acid forms, but we know that the two amounts were equal. Let $C =$ initial number of moles of acetic acid and the initial number of moles of sodium acetate. The number of moles of H_3O^+ to be added (in the form of HCl(aq)) is $0.001\,00 \text{ L} \times 6.00 \text{ mol} \cdot \text{L}^{-1} = 0.006\,00$ mol. The total final volume will be 0.1010 L.

$$\frac{\dfrac{C - 0.006\,00}{0.1010\ \text{L}}}{\dfrac{C + 0.006\,00}{0.1010\ \text{L}}} = 0.63$$

$$\frac{C - 0.006\,00}{C + 0.006\,00} = 0.63$$

$C - 0.006\,00 = 0.63(C + 0.006\,00)$

$C - 0.006\,00 = 0.63\,C + 0.003\,78$

$0.37\,C = 0.009\,78$

$C = 0.026$

The initial buffer solution must contain at least 0.026 mol acetic acid and 0.026 mol sodium acetate. The concentration of the initial solution will then be $0.026\ \text{mol} \div 0.100\ \text{L} = 0.260\ \text{M}$ in both acetic acid and sodium acetate.

12.85 (a) $\text{pH} = \text{pK}_{a1} + \log\dfrac{[HOOCCH_2CO_2^-]}{[HOOCCH_2COOH]}$;

when $[HOOCCH_2COOH] = [HOOCCH_2COO^-]$, $\text{pH} = \text{pK}_{a1} = 2.8$

(b) Similar to (a), when $[HOOCCH_2COO^-] = [^-OOCCH_2COO^-]$,

$\text{pH} = \text{pK}_{a2} = 5.7$

(c) At pH = 4.2, $HOOCCH_2COO^-$ is the predominant species. The reason is that between pH > 2.8 and pH < 5.7, $HOOCCH_2COO^-$ is the major species in the solution from the results in (a) and (b).

12.87 Let novocaine = N; $N(aq) + H_2O(l) \rightleftharpoons HN^+(aq) + OH^-(aq)$.

$$K_b = \frac{[HN^+][OH^-]}{[N]}$$

$\text{p}K_a = \text{p}K_w - \text{p}K_b = 14.00 - 5.05 = 8.95$

$\text{pH} = \text{p}K_a + \log\left(\dfrac{[N]}{[HN^+]}\right)$

$\log\left(\dfrac{[N]}{[HN^+]}\right) = \text{pH} - \text{p}K_a = 7.4 - 8.95 = -1.55$

Therefore, the ratio of the concentrations of novocaine and its conjugate acid is $[N]/[HN^+] = 10^{-1.55} = 2.8 \times 10^{-2}$.

12.89 In addition to the reaction corresponding to the dissolution of $PbF_2(s)$:

$PbF_2(s) \leftrightarrow Pb^{2+}(aq) + 2\ F^-(aq)$ $\qquad K = 3.7 \times 10^{-8}$

The buffer will provide a source of $H_3O^+(aq)$ ions, which will allow the reaction

$H_3O^+(aq) + F^- \leftrightarrow HF(aq) + H_2O(l)$ $\quad K = 1/(K_a(HF)) = 1/(3.5 \times 10^{-4}) = 2.86 \times 10^3$.

These two coupled reactions give two equilibrium expressions that must be simultaneously satisfied:

$$[F^-]^2[Pb^{2+}] = 1.7 \times 10^{-6} \quad \text{and} \quad \frac{[HF]}{[H_3O^+][F^-]} = 2.86 \times 10^{-3}$$

Given that all fluoride ions come from $PbF_2(s)$ and wind up as either $F^-(aq)$ or $HF(aq)$, and that for every 1 mole of $Pb^{2+}(aq)$ generated 2 moles of $F^-(aq)$ are also produced, we can write a third equation that relates the concentration of the fluoride-containing species to the concentration of dissolved barium:

$[Pb^{2+}] = \frac{1}{2}([F^-] + [HF])$

In the end, the concentration of $Pb^{2+}(aq)$ will be equal to the solubility of $PbF_2(s)$. To determine the equilibrium concentration of $Pb^{2+}(aq)$, we first determine $[H_3O^+]$, which is fixed by the buffer system, and then use the three simultaneous equations above to solve for $[Pb^{2+}]_{eq}$.

The buffer determines the equilibrium concentration of $H_3O^+(aq)$. The initial concentrations of $H_3O^+(aq)$ and $NaCH_3CO_2(aq)$ are

$$[H_3O^+]_i = \frac{(0.055\ L)(0.15\ mol \cdot L^{-1})}{0.10\ L} = 0.0825\ M \quad \text{and}$$

$$[CH_3CO_2^-]_i = \frac{(0.045\ L)(0.65\ mol \cdot L^{-1})}{0.10\ L} = 0.293\ M.$$

To determine their equilibrium concentrations we solve using the familiar method:

Concentration

| (mol·L⁻¹) | CH₃COOH(aq) | + | H₂O(l) | ⇌ | H₃O⁺(aq) | + | CH₃CO₂⁻(aq) |

(mol·L^{-1})	CH_3COOH(aq)	+	H_2O(l)	$\rightleftharpoons$	H_3O^+(aq)	+	$CH_3CO_2^-$(aq)
initial	0		—		0.0825		0.293
change	$+x$		—		$-x$		$-x$
equilibrium	x		—		$0.0825 - x$		$0.292 - x$

$$K_a = 1.8 \times 10^{-5} = \frac{[H_3O^+][CH_3CO_2^-]}{[CH_3COOH]} = \frac{(0.0825 - x)(0.292 - x)}{(x)}$$

Rearranging this expression we obtain $0.024072 - 0.3745\,x + x^2 = 0$.

Using the quadratic formula, we find $x = 0.082493$, and $[H_3O^+] = 7.1 \times 10^{-6}$ M.

With this equilibrium concentration of H_3O^+(aq), we revisit the three simultaneous equations from above, namely

$$[F^-]^2[Pb] = 3.7 \times 10^{-8};$$

$$[Pb^{2+}] = \tfrac{1}{2}\big([F^-] + [HF]\big); \text{ and}$$

$$\frac{[HF]}{[H_3O^+][F^-]} = 2.86 \times 10^{-3}.$$

Due to the presence of the buffer, $[H_3O^+] = 7.1 \times 10^{-6}$, and this last equation simplifies to

$$\frac{[HF]}{[F^-]} = 2.03 \times 10^{-2}.$$

Rearranging these three simultaneous equations we find that

$$[Pb^{2+}] = \frac{3.7 \times 10^{-8}}{[F^-]^2}, \quad [HF] = [F^-] \times \big(2.03 \times 10^{-2}\big), \quad \text{and}$$

$$[Pb^{2+}] = \tfrac{1}{2}\big([F^-] + [HF]\big)$$

$$\frac{3.7 \times 10^{-8}}{[F^-]^2} = \tfrac{1}{2}\Big[[F^-] + \big([F^-] \times \big(2.03 \times 10^{-2}\big)\big)\Big]. \quad \text{Solving this expression for } [F^-]:$$

$$[F^-] = \sqrt[3]{\frac{3.7 \times 10^{-8}}{0.5203}} = 4.14 \times 10^{-3}.$$

The equilibrium concentration of Pb^{2+}(aq) is then:

$$[Pb^{2+}] = \frac{3.7 \times 10^{-8}}{[F^-]^2} = \frac{3.7 \times 10^{-8}}{(4.14 \times 10^{-3})^2} = 2.16 \times 10^{-3} \text{ M}$$

Therefore, the solubility of PbF_2(s) is 2.2×10^{-3} M

12.91 If the two pK_a values of a diprotic acid is far apart (≥ 1000), there will be two stoichiometric points observed. If the two pK_a values of a diprotic acid is close (<500) (due to the electron delocalization of anions), like those of tartaric acid, there will be one stoichiometric point observed (i.e., two protons are titrated at about the same time). The titration curve in this case is not very steep due to the overlap titration of the first proton and the second proton.

12.93 The K_{sp} values from Tables 12.4 are $Cu^{2+}, 1.3 \times 10^{-36}$; $Co^{2+}, 5 \times 10^{-22}$; $Cd^{2+}, 4 \times 10^{-29}$. All the salts have the same expression for K_{sp}, $K_{sp} = [M^{2+}][S^{2-}]$, so the compound with the smallest K_{sp} will precipitate first, in this case CuS.

12.95 (a) $CO_3{}^{2-}$(aq) $+ 2\,Ag^+$(aq) $\rightarrow Ag_2CO_3$(s)

Moles of Ag^+ = (0.0362 L Ag^+)(0.110 mol/L) = 3.98×10^{-3} mol Ag^+

Moles of $CO_3{}^{2-}$ = 3.98×10^{-3} mol $Ag^+ \times \dfrac{1\ mol\ CO_3^{2-}}{2\ mol\ Ag^+}$

= 1.99×10^{-3} mol $CO_3{}^{2-}$

$$[CO_3{}^{2-}] = \frac{1.99 \times 10^{-3}\ mol\ CO_3^{2-}}{0.0250\ L} = 0.0796\ mol \cdot L^{-1}\ CO_3{}^{2-}$$

(b) You can use excel to make the titration plot. Before stoichiometric point, $[CO_3{}^{2-}]$ is dominant, and the concentration of Ag^+ is very low.

$$[CO_3{}^{2-}] = \frac{(V_{CO_3^{2-}} M_{CO_3^{2-}}) - (V_{Ag^+} M_{Ag^+})/2}{V_{Total}}, \text{ and } [Ag^+] = \sqrt{\frac{K_{sp}}{[CO_3^{2-}]}}\ ; \text{ At}$$

stoichiometric point, $[Ag^+] = \sqrt[3]{\dfrac{K_{sp}}{4}}$; After stoichiometric point, $[Ag^+]$ is

dominant, and the extra $[Ag^+] = \dfrac{(V_{Ag^+} M_{Ag^+})/2 - (V_{CO_3^{2-}} M_{CO_3^{2-}})}{V_{Total}}$. The K_{sp}

value of Ag_2CO_3 is 6.2×10^{-12}. The plot is as follows:

At stoichiometric point, $pAg^+ = 3.94$

12.97 (a) $SO_2(g) + H_2O(l) \rightleftharpoons H_2SO_3(aq)$

(b) $(0.050 \times 1.0 \times 10^{-4}) - (0.0302 \times 1.5 \times 10^{-4}) = 5 \times 10^{-7}$ mol NaOH

(c)

$$5 \times 10^{-7} \text{ mol NaOH} \times \left(\frac{1 \text{ mol } H_2SO_3}{2 \text{ mol NaOH}}\right)\left(\frac{1 \text{ mol } SO_2}{1 \text{ mol } H_2SO_3}\right)$$

$$= 2.5 \times 10^{-7} \text{ mol } SO_2$$

Mass of $SO_2 = (2.5 \times 10^{-7} \text{ mol}) \times (64.06 \text{ g/mol}) = 1.6 \times 10^{-5}$ g SO_2

Moles of air: PV=nRT,

$$(\frac{753 \text{ Torr}}{760 \text{ Torr/atm}})\left(3.0\,L \cdot h^{-1} \times 2.5\,h\right) = n(0.08206\,L \cdot atm \cdot mol^{-1} \cdot K^{-1})(295\,K)$$

$n = 0.31$ mole air

Mass of air (assuming 78% N_2, 22% O_2 to simplify the calculation):

$0.31 \times (0.78) \times (28.0 \text{ g } N_2/\text{mol } N_2) + 0.31 \times (0.22) \times (32.0 \text{ g } O_2/\text{mole})$

$= 9.0$ g

Concentration of SO_2 in the air (in parts per million (ppm)):

$\left(\dfrac{1.6 \times 10^{-5}}{9.0} \right) \times (1.0 \times 10^{-6}) = 1.8$ ppm

12.99 The K_{sp} value for PbF_2 obtained from Table 11.5 is 3.7×10^{-8}. Using this value, the ΔG° of the dissolution reaction can be obtained from

$\Delta G^\circ = -RT \ln K$.

$\Delta G^\circ = -(8.314 \text{ J} \cdot \text{K}^{-1} \cdot \text{mol}^{-1})(298.2 \text{ K}) \ln (3.7 \times 10^{-8})$

$\Delta G^\circ = +42.43 \text{ kJ} \cdot \text{mol}^{-1}$

From the Appendices we find that $\Delta G^\circ_f (F^-, aq) = -278.79 \text{ kJ} \cdot \text{mol}^{-1}$ and $\Delta G^\circ_f (Pb^{2+}, aq) = -24.43 \text{ kJ} \cdot \text{mol}^{-1}$.

$\Delta G^\circ = +42.43 \text{ kJ} \cdot \text{mol}^{-1} = \Delta G^\circ_f (Pb^{2+}, aq) + \Delta G^\circ_f (F^-, aq) - \Delta G^\circ_f (PbF_2, s)$

$+42.43 \text{ kJ} \cdot \text{mol}^{-1} = (-24.43 \text{ kJ} \cdot \text{mol}^{-1}) + (-278.79 \text{ kJ} \cdot \text{mol}^{-1})$

$- \Delta G^\circ_f (PbF_2, s)$

$\Delta G^\circ_f (PbF_2, s) = -345.65 \text{ kJ} \cdot \text{mol}^{-1}$

12.101 (a) The amount of CO_2 present at equilibrium may be found using the equilibrium expression for the reaction of interest:

$H_3O^+ (aq) + HCO_3^- \rightleftharpoons 2 \, H_2O(l) + CO_2(aq)$

$K = 7.9 \times 10^{-7} = \dfrac{[CO_2]}{[H_3O^+][HCO_3^-]}$

Solving for $[CO_2]$:

$[CO_2] = (7.9 \times 10^{-7})[H_3O^+][HCO_3^-]$

Given: $[H_3O^+] = 10^{-6.1} = 8 \times 10^{-7}$ M, and $[HCO_3^-] = 5.5 \ \mu\text{mol} \cdot \text{L}^{-1} = 5.5 \times 10^{-6}$ M

$[CO_2] = (7.9 \times 10^{-7})(8 \times 10^{-7})(5.5 \times 10^{-6}) = 3 \times 10^{-18} \text{mol} \cdot \text{L}^{-1}$

In 1.0 L of solution there will be 3×10^{-18} mol of $CO_2(aq)$.

(b) Adding 0.65×10^{-6} mol of $H_3O^+(aq)$ to the equilibrium system in (a) will give an initial $[H_3O^+]$ of 1.44×10^{-6} M. To determine the equilibrium concentration of $H_3O^+(aq)$ we set up the familiar problem:

Concentration

$$(\text{mol}\cdot\text{L}^{-1}) \quad H_3O^+(aq) \;+\; HCO_3^-(l) \;\rightleftharpoons\; 2\,H_2O(l) \;+\; CO_2(aq)$$

initial	1.44×10^{-6}	5.5×10^{-6}	—	3.45×10^{-18}
change	$-x$	$-x$	—	$+x$
equilibrium	$1.44 \times 10^{-6} - x$	$5.5 \times 10^{-6} - x$	—	$3.45 \times 10^{-18} + x$

$$K = 7.9 \times 10^{-7} = \frac{3.45 \times 10^{-18} + x}{\left(1.44 \times 10^{-6} - x\right)\left(5.5 \times 10^{-6} + x\right)}$$

Rearranging to obtain a polynomial in x:

$$2.83 \times 10^{-18} \;-\; x \;+\; 7.9 \times 10^{-7} x^2 = 0$$

Using the quadratic forumla one finds:

$$x = 2.825 \times 10^{-18}$$

Giving: $[H_3O^+] = 1.4 \times 10^{-6}$ and pH $= 5.8$, the same pH as the initial solution.

12.103 (a) pH = pKa + log ([base]/[acid]) = 4.75 + log(0.300/0.200) = 4.93

(b) 6.0 = 4.75 + log{(0.300 L × 0.300 M + x)/(0.300 L × 0.200 M − x)}

= 4.75 + log{(0.0900 mol + x)/(0.0600 mol − x)}

log{(0.0900 mol + x)/(0.0600 mol − x)} = 1.25;

x = mol of OH$^-$ = 0.0520 mol = mole of NaOH

mass of NaOH = (0.0520 mol) × (40.01 g/mol) = 2.08 g NaOH

12.105 (a) Equilibrium equation: $Fe(OH)_3(s) \leftrightarrow Fe^{3+}(aq) \;+\; 3\,OH^-(aq)$

$K_{sp} = 2.0 \times 10^{-39} = [Fe^{3+}][OH^-]^3 = (S)(3S)^3 = 27(S)^4$

$S = 9.3 \times 10^{-11}$ mol$\cdot$L^{-1} (the OH$^-$ from water is ignored)

(b) $[OH^-]_{\text{iron hydroxide}} = 3S = 2.8 \times 10^{-10}$ M; pOH = 9.55; pH = 4.45

(c) Results in (b) is not reasonable because the OH$^-$ from water is ignored.
$[OH^-]_{\text{water}} = 1.0 \times 10^{-7}$ M;   Actually, $[OH^-] = 1.0 \times 10^{-7}$ M in a saturated solution of $Fe(OH)_3$ because the $[OH^-]$ from $Fe(OH)_3$ is much smaller than the $[OH^-]$ from water; pH = 7.0.

(d) $Fe(OH)_3(s) \leftrightarrow Fe^{3+}(aq) \;+\; 3\,OH^-(aq)$ .... (1)

   $2\,H_2O(l) \leftrightarrow H_3O^+(aq) \;+\; OH^-(aq)$ .... (2)

$3 \times (2) - (1)$: $Fe^{3+}(aq) \;+\; 6\,H_2O(l) \leftrightarrow 3\,H_3O^+(aq) \;+\; Fe(OH)_3(s)$

$K = (K_w)^3/K_{sp} = (1.0 \times 10^{-14})^3/(2.0 \times 10^{-39}) = 5 \times 10^{-4}$

(e) 10.0 g of $Fe_2(SO_4)_3 \times (1.0 \text{ mol}/399.9\text{g}) = 0.025$ mol $Fe_2(SO_4)_3$

Mol of $Fe^{3+} = 0.025 \times 2 = 0.050$ mol;

when pH is maintained at 8.0, $[H_3O^+] = 1.0 \times 10^{-8}$, $[OH^-] = 1.0 \times 10^{-6}$;

$Q = [Fe^{3+}][OH^-]^3 = (0.050)(1.0 \times 10^{-6})^3 = 5.0 \times 10^{-20} >> K_{sp}$, all of the Fe^{3+}

will be precipitated as long as pH is maintained. The moles of $Fe(OH)_3$ will be the same as that of Fe^{3+} (0.050 mol).

Mass of $Fe(OH)_3 = 0.050$ mol $\times$ (106.9 g/mol) = 5.3 grams.

(f) Moles of $[Cl^-]$ in 25.0 mL NaCl sample:

(24.72 g NaCl) $\times$ (1.0 mol NaCl/58.44 g) $\times$ (25.00 mL/1000 mL) $\times$

(1 mol Cl^-/1 mol NaCl) = 0.01057 mol Cl^-

The moles of Cl^- left in solution:

(0.604 g AgCl) $\times$ (1.0 mole AgCl/143.32 g) (1 mol Cl^-/1 mol AgCl)

$= 4.21 \times 10^{-3}$ mol

Percentage of Cl^- removed: $[(0.01057 - 4.21 \times 10^{-3})/0.01057] \times 100\%$

= 60.1%

CHAPTER 13
ELECTROCHEMISTRY

13.1 (a) Cr reduced from 6+ to 3+; C oxidized from 2- to 1-;

(b) $C_2H_5OH(aq) \rightarrow C_2H_4O(aq) + 2\,H^+(aq) + 2\,e^-$;

(c) $Cr_2O_7^{2-}(aq) + 14\,H^+(aq) + 6\,e^- \rightarrow 2\,Cr^{3+}(aq) + 7\,H_2O(l)$;

(d) $8\,H^+(aq) + Cr_2O_7^{2-}(aq) + 3\,C_2H_5OH(aq) \rightarrow$
     $2\,Cr^{3+}(aq) + 3\,C_2H_4O(aq) + 7\,H_2O(l)$

13.3 In each case, first obtain the balanced half-reactions. Multiply the oxidation and reduction half-reactions by appropriate factors that will result in the same number of electrons being present in both half-reactions. Then add the half-reactions, canceling electrons in the process, to obtain the balanced equation for the whole reaction. Check to see that the final equation is balanced.

(a) $4[Cl_2(g) + 2\,e^- \rightarrow 2\,Cl^-(aq)]$

$1[S_2O_3^{2-}(aq) + 5\,H_2O(l) \rightarrow 2\,SO_4^{2-}(aq) + 10\,H^+(aq) + 8\,e^-]$
$4\,Cl_2(g) + S_2O_3^{2-}(aq) + 5\,H_2O(l) + 8\,e^- \rightarrow$
  $8\,Cl^-(aq) + 2\,SO_4^{2-}(aq) + 10\,H^+(aq) + 8\,e^-$
$4\,Cl_2(g) + S_2O_3^{2-}(aq) + 5\,H_2O(l) \rightarrow 8\,Cl^-(aq) + 2\,SO_4^{2-}(aq) + 10\,H^+(aq)$
Cl_2 is the oxidizing agent and $S_2O_3^{2-}$ is the reducing agent.

(b) $2[MnO_4^-(aq) + 8\,H^+(aq) + 5\,e^- \rightarrow Mn^{2+}(aq) + 4\,H_2O(l)]$

$5[H_2SO_3(aq) + H_2O(l) \rightarrow HSO_4^-(aq) + 3\,H^+(aq) + 2\,e^-]$
$2\,MnO_4^-(aq) + 16\,H^+(aq) + 5\,H_2SO_3(aq) + 5\,H_2O(l) + 10\,e^- \rightarrow$
    $2\,Mn^{2+}(aq) + 8\,H_2O(l) + 5\,HSO_4^-(aq) + 15\,H^+(aq) + 10\,e^-$
$2\,MnO_4^-(aq) + H^+(aq) + 5\,H_2SO_3(aq) \rightarrow$
    $2\,Mn^{2+}(aq) + 3\,H_2O(l) + 5\,HSO_4^-(aq)$

MnO_4^- is the oxidizing agent and H_2SO_3 is the reducing agent.

(c) $Cl_2(g) + 2\,e^- \rightarrow 2\,Cl^-(aq)$

$H_2S(aq) \rightarrow S(s) + 2\,H^+(aq) + 2\,e^-$

$Cl_2(g) + H_2S(aq) + 2\,e^- \rightarrow 2\,Cl^-(aq) + S(s) + 2\,H^+(aq) + 2\,e^-$

$Cl_2(g) + H_2S(aq) \rightarrow 2\,Cl^-(aq) + S(s) + 2\,H^+(aq)$

Cl_2 is the oxidizing agent and H_2S is the reducing agent.

(d) $Cl_2(g) + 2\,e^- \rightarrow 2\,Cl^-(aq)$

$2\,H_2O(l) + Cl_2(g) \rightarrow 2\,HOCl(aq) + 2\,H^+(aq) + 2\,e^-$

$2\,H_2O(l) + 2\,Cl_2(g) + 2\,e^- \rightarrow 2\,HOCl(aq) + 2\,H^+(aq) + 2\,Cl^-(aq) + 2\,e^-$

or $H_2O(l) + Cl_2(g) \rightarrow HOCl(aq) + H^+(aq) + Cl^-(aq)$

Cl_2 is both the oxidizing and the reducing agent.

13.5 (a) $O_3(g) \rightarrow O_2(g)$

$O_3(g) \rightarrow O_2(g) + H_2O(l)$ (balances O's)

$2\,H_2O(l) + O_3(g) \rightarrow O_2(g) + H_2O(l) + 2\,OH^-(aq)$ (balances H's)

$H_2O(l) + O_3(g) \rightarrow O_2(g) + 2\,OH^-(aq)$ (cancels H_2O)

$H_2O(l) + O_3(g) + 2\,e^- \rightarrow O_2(g) + 2\,OH^-(aq)$ (balances charge);

$Br^-(aq) \rightarrow BrO_3^-(aq)$

$3\,H_2O(l) + Br^-(aq) \rightarrow BrO_3^-(aq)$ (balances O's)

$6\,OH^-(aq) + 3\,H_2O(l) + Br^-(aq) \rightarrow BrO_3^-(aq) + 6\,H_2O(l)$ (balances H's)

$6\,OH^-(aq) + 3\,H_2O(l) + Br^-(aq) \rightarrow BrO_3^-(aq) + 6\,H_2O(l) + 6\,e^-$

(balances charge)

Combining half-reactions yields

$3[H_2O(l) + O_3(g) + 2\,e^- \rightarrow O_2(g) + 2\,OH^-(aq)]$

$6\,OH^-(aq) + 3\,H_2O(l) + Br^-(aq) \rightarrow BrO_3^-(aq) + 6\,H_2O(l) + 6\,e^-$

$6\,H_2O(l) + 3\,O_3(g) + 6\,OH^-(aq) + Br^-(aq) + 6\,e^- \rightarrow$

$\qquad\qquad 3\,O_2(g) + 6\,OH^-(aq) + BrO_3^-(aq) + 6\,H_2O(l) + 6\,e^-$

and $3\,O_3(g) + Br^-(aq) \rightarrow 3\,O_2(g) + BrO_3^-(aq)$

O_3 is the oxidizing agent and Br^- is the reducing agent.

(b) $Br_2(l) + 2 e^- \rightarrow 2 Br^-(aq)$ (balanced reduction half-reaction)

$Br_2(l) + 6 H_2O(l) \rightarrow 2 BrO_3^-(aq)$ (O's balanced); then

$Br_2(l) + 6 H_2O(l) + 12 OH^-(aq) \rightarrow 2 BrO_3^-(aq) + 12 H_2O(l)$ (H's balanced);
and $Br_2(l) + 12 OH^-(aq) \rightarrow 2 BrO_3^-(aq) + 6 H_2O(l) + 10 e^-$ (electrons

balanced)

Combining half-reactions yields

$5[Br_2(l) + 2 e^- \rightarrow 2 Br^-(aq)]$
$1[Br_2(l) + 12 OH^-(aq) \rightarrow 2 BrO_3^-(aq) + 6 H_2O(l) + 10 e^-]$
$6 Br_2(l) + 12 OH^-(aq) + 10 e^- \rightarrow$
$\quad 10 Br^-(aq) + 2 BrO_3^-(aq) + 6 H_2O(l) + 10 e^-$
$6 Br_2(l) + 12 OH^-(aq) \rightarrow 10 Br^-(aq) + 2 BrO_3^-(aq) + 6 H_2O(l)$

Dividing by 2 gives

$3 Br_2(l) + 6 OH^-(aq) \rightarrow 5 Br^-(aq) + BrO_3^-(aq) + 3 H_2O(l)$

Br_2 is both the oxidizing agent and the reducing agent.

(c) $Cr^{3+}(aq) + 4 H_2O(l) \rightarrow CrO_4^{2-}(aq)$ (O's balanced); then

$Cr^{3+}(aq) + 4 H_2O(l) + 8 OH^-(aq) \rightarrow CrO_4^{2-}(aq) + 8 H_2O(l)$ (H's

balanced); and

$Cr^{3+}(aq) + 8 OH^-(aq) \rightarrow CrO_4^{2-}(aq) + 4 H_2O(l) + 3 e^-$ (charge

balanced)

$MnO_2(s) \rightarrow Mn^{2+}(aq) + 2H_2O(l)$; then

$MnO_2(s) + 4 H_2O(l) \rightarrow Mn^{2+}(aq) + 2 H_2O(l) + 4 OH^-(aq)$ (H's

balanced); and

$MnO_2(s) + 2 H_2O(l) + 2 e^- \rightarrow Mn^{2+}(aq) + 4 OH^-(aq)$ (charge

balanced)

Combining half-reactions yields

$$2[Cr^{3+}(aq) + 8 OH^-(aq) \rightarrow CrO_4^{2-}(aq) + 4 H_2O(l) + 3 e^-]$$

$$3[MnO_2(s) + 2 H_2O(l) + 2 e^- \rightarrow Mn^{2+}(aq) + 4 OH^-(aq)]$$

$$2 Cr^{3+}(aq) + 16 OH^-(aq) + 3 MnO_2(s) + 6 H_2O(l) + 6 e^- \rightarrow$$
$$2 CrO_4^{2-}(aq) + 8 H_2O(l) + 3 Mn^{2+}(aq) + 12 OH^-(aq) + 6 e^-$$

$$2 Cr^{3+}(aq) + 4 OH^-(aq) + 3 MnO_2(s) \rightarrow$$
$$2 CrO_4^{2-}(aq) + 2 H_2O(l) + 3 Mn^{2+}(aq)$$

Cr^{3+} is the reducing agent and MnO_2 is the oxidizing agent.

(d) $3[P_4(s) + 8 OH^-(aq) \rightarrow 4 H_2PO_2^-(aq) + 4 e^-]$

$$P_4(s) + 12 H_2O(l) + 12 e^- \rightarrow 4 PH_3(g) + 12 OH^-(aq)$$

$$4 P_4(s) + 12 H_2O(l) + 24 OH^-(aq) + 12 e^- \rightarrow$$
$$12 H_2PO_2^-(aq) + 4 PH_3(g) + 12 OH^-(aq) + 12 e^-$$

$$4 P_4(s) + 12 H_2O(l) + 12 OH^-(aq) \rightarrow 12 H_2PO_2^-(aq) + 4 PH_3(g)$$

or $P_4(s) + 3 H_2O(l) + 3 OH^-(aq) \rightarrow 3 H_2PO_2^-(aq) + PH_3(g)$

$P_4(s)$ is both the oxidizing and the reducing agent.

13.7 $P_4S_3(aq) \rightarrow H_3PO_4(aq) + SO_4^{2-}(aq)$

For the oxidation of P_4S_3, both the P and S atoms are oxidized. The assignment of oxidation states to the P and S atoms is complicated by the presence of P—P bonds in the molecule, which leads to non-integral values. As long as we are consistent in our assignments, the end result should be the same. We will assume that S in P_4S_3 is 2− and, therefore, loses 8 electrons on going to S^{6+} in the sulfate ion. Because P_4S_3 is a neutral molecule and, if S has an oxidation number of −2, then each phosphorus atom will have an oxidation number of +1.5. Phosphorus in phosphoric acid has an oxidation number of +5. so each P atom of P_4S_3 must lose 3.5 electrons. The total number of electrons lost is $(4 \times 3.5) + (3 \times 8) = 38$.

$$P_4S_3(aq) \rightarrow 4 H_3PO_4(aq) + 3 SO_4^{2-}(aq) + 38 e^-$$

We balance the charge by adding H^+ in an acidic solution:

$P_4S_3(aq) \rightarrow 4\,H_3PO_4(aq) + 3\,SO_4{}^{2-}(aq) + 44\,H^+(aq) + 38\,e^-$

The final balance is achieved by adding water to provide the oxygen and hydrogen atoms:

$P_4S_3(aq) + 28\,H_2O(l) \rightarrow 4\,H_3PO_4(aq) + 3\,SO_4{}^{2-}(aq) + 44\,H^+(aq) + 38\,e^-$

The other half-reaction is simpler.

$NO_3{}^-(aq) \rightarrow NO(g)$

N has an oxidation number of $+5$ in the nitrate ion and $+2$ in nitric oxide. Each nitrogen atom gains three electrons in the course of the reaction.

$NO_3{}^-(aq) + 3\,e^- \rightarrow NO(g)$

Charge balance is again achieved by adding H^+:

$NO_3{}^-(aq) + 4\,H^+(aq) + 3\,e^- \rightarrow NO(g)$

The number of hydrogen and oxygen atoms is completed by the addition of water:

$NO_3{}^-(aq) + 4\,H^+(aq) + 3\,e^- \rightarrow NO(aq) + 2\,H_2O(l)$

Combining the two half-reactions gives

$38\,[NO_3{}^-(aq) + 4\,H^+(aq) + 3\,e^- \rightarrow NO(g) + 2\,H_2O(l)]$
$+3\,[P_4S_3(aq) + 28\,H_2O(l) \rightarrow$
$\qquad 4\,H_3PO_4(aq) + 3\,SO_4{}^{2-}(aq) + 44\,H^+(aq) + 38\,e^-]$
$3\,P_4S_3(aq) + 38\,NO_3{}^-(aq) + 20\,H^+(aq) + 8\,H_2O(l) \rightarrow$
$\qquad 12\,H_3PO_4(aq) + 9\,SO_4{}^{2-}(aq) + 38\,NO(g)$

13.9 (a) $\quad \Delta G_r{}^\circ = -nFE^\circ = -2(9.6485 \times 10^4\,\text{C. mol}^{-1})(1.08\,\text{J. C}^{-1})$
$$= -2.08 \times 10^5\,\text{J. mol}^{-1}$$

(b) $\quad \Delta G_r{}^\circ = -nFE^\circ = -6(9.6485 \times 10^4\,\text{C. mol}^{-1})(-1.29\,\text{J. C}^{-1})$
$$= 7.47 \times 10^5\,\text{J. mol}^{-1}$$

13.11 (a) $Ag^+(aq) + e^- \rightarrow Ag(s) \quad E^\circ(\text{cathode}) = +0.80\,\text{V}$

$Ni^{2+}(aq) + 2\,e^- \rightarrow Ni(s) \quad E^\circ(\text{anode}) = -0.23\,\text{V}$

Reversing the anode half-reaction yields

$$Ni(s) \rightarrow Ni^{2+}(aq) + 2\ e^{-}$$

and the cell reaction is, upon addition of the half-reactions,

$$2\ Ag^{+}(aq) + Ni(s) \rightarrow 2\ Ag(s) + Ni^{2+}(aq) \qquad E°_{cell} = +0.80\ V - (-0.23)\ V$$
$$= +1.03\ V$$

(b) $2\ H^{+}(aq) + 2\ e^{-} \rightarrow H_{2}(g) \qquad E°(anode) = 0.00\ V$

$Cl_{2}(g) + 2\ e^{-} \rightarrow 2\ Cl^{-}(aq) \qquad E°(cathode) = +1.36\ V$

Therefore, at the anode, after reversal,

$$H_{2}(g) \rightarrow 2\ H^{+}(aq) + 2\ e^{-}$$

and, the cell reaction is, upon addition of the half-reactions,

$$Cl_{2}(g) + H_{2}(g) \rightarrow 2\ H^{+}(aq) + 2\ Cl^{-}(aq) \qquad E°_{cell} = +1.36\ V - 0.00\ V$$
$$= +1.36\ V$$

(c) $Cu^{2+}(aq) + 2\ e^{-} \rightarrow Cu(s) \qquad E°(anode) = +0.34\ V$

$Ce^{4+}(aq) + e^{-} \rightarrow Ce^{3+}(aq) \qquad E°(cathode) = +1.61\ V$

Therefore, at the anode, after reversal,

$$Cu(s) \rightarrow Cu^{2+}(aq) + 2\ e^{-}$$

and, the cell reaction is, upon addition of the half-reactions,

$$2\ Ce^{4+}(aq) + Cu(s) \rightarrow Cu^{2+}(aq) + 2\ Ce^{3+}(aq) \qquad E°_{cell} = 1.61\ V - (0.34\ V)$$
$$= +1.27\ V$$

(d) $O_{2}(g) + 2\ H_{2}O(l) + 4\ e^{-} \rightarrow 4\ OH^{-}(aq) \qquad E°(cathode) = 0.40\ V$

$O_{2}(g) + 4\ H^{+}(aq) + 4\ e^{-} \rightarrow 2\ H_{2}O(l) \qquad E°(anode) = 1.23\ V$

Reversing the anode half-reaction yields

$$2\ H_{2}O(l) \rightarrow O_{2}(g) + 4\ H^{+}(aq) + 4\ e^{-}$$

and the cell reaction is, upon addition of the half-reactions,

$$4\ H_{2}O(l) \rightarrow 4\ H^{+}(aq) + 4\ OH^{-}(aq) \qquad E°_{cell} = 0.40\ V - 1.23\ V = -0.83\ V$$

or, $H_{2}O(l) \rightarrow H^{+}(aq) + OH^{-}(aq)$

Note: This balanced equation corresponds to the cell notation given. The spontaneous process is the reverse of this reaction.

(e) $Sn^{4+}(aq) + 2\ e^{-} \rightarrow Sn^{2+}(aq) \qquad\qquad\qquad E°(anode) = +0.15\ V$

$$Hg_2Cl_2(s) + 2\ e^- \rightarrow 2\ Hg(l) + 2\ Cl^-(aq) \qquad E°(cathode) = +0.27\ V$$

Therefore, at the anode, after reversal,

$$Sn^{2+}(aq) \rightarrow Sn^{4+}(aq) + 2\ e^-$$

and the cell reaction is, upon addition of the half-reactions,

$$Sn^{2+}(aq) + Hg_2Cl_2(s) \rightarrow 2\ Hg(l) + 2\ Cl^-(aq) + Sn^{4+}(aq)$$

$$E°_{cell} = 0.27\ V - 0.15\ V = 0.12\ V$$

13.13 (a) $Ni^{2+}(aq) + 2\ e^- \rightarrow Ni(s)$ $\qquad E°(cathode) = -0.23\ V$

$$Zn^{2+}(aq) + 2\ e^- \rightarrow Zn(s) \qquad E°(anode) = -0.76\ V$$

Reversing the anode reaction yields

$$Zn(s) \rightarrow Zn^{2+}(aq) + 2\ e^- \quad \text{(at anode); then, upon addition,}$$

$$Ni^{2+}(aq) + Zn(s) \rightarrow Ni(s) + Zn^{2+}(aq) \quad \text{(overall cell)}$$

$$E°_{cell} = -0.23\ V - (-0.76\ V) = +0.53\ V$$

and $Zn(s)\,|\,Zn^{2+}(aq)\,\|\,Ni^{2+}(aq)\,|\,Ni(s)$

(b) $2[Ce^{4+}(aq) + e^- \rightarrow Ce^{3+}(aq)]$ $\qquad E°(cathode) = +1.61\ V$

$$I_2(s) + 2\ e^- \rightarrow 2\ I^-(aq) \qquad E°(anode) = +0.54\ V$$

Reversing the anode reaction yields

$$2\ I^-(aq) \rightarrow 2\ e^- + I_2(s) \quad \text{(at anode); then, upon addition,}$$

$$2\ I^-(aq) + 2\ Ce^{4+}(aq) \rightarrow 2\ Ce^{3+}(aq) + I_2(s) \quad \text{(overall cell)}$$

$$E°_{cell} = +1.61\ V - 0.54\ V = +1.07\ V$$

and $Pt(s)\,|\,I^-(aq)\,|\,I_2(s)\,\|\,Ce^{4+}(aq),\ Ce^{3+}(aq)\,|\,Pt(s)$

An inert electrode such as Pt is necessary when both oxidized and reduced species are in the same solution.

(c) $Cl_2(g) + 2\ e^- \rightarrow 2\ Cl^-(aq)$ $\ E°(cathode) = +1.36\ V$

$$2\ H^+(aq) + 2\ e^- \rightarrow H_2(g) \ \ E°(anode) = 0.00\ V$$

Reversing the anode reaction yields

$$H_2(g) \rightarrow 2\ H^+(aq) + 2\ e^- \quad \text{(at anode); then, upon addition,}$$

$$H_2(g) + Cl_2(g) \rightarrow 2\ HCl(aq) \quad \text{(overall cell)} \quad E°_{cell} = +1.36\ V - 0.00\ V$$
$$= +1.36\ V$$

and $Pt(s)\,|\,H_2(g)\,|\,H^+(aq)\,||\,Cl^-(aq)\,|\,Cl_2(g)\,|\,Pt(s)$

An inert electrode such as Pt is necessary for gas/ion electrode reactions.

(d)　$3[Au^+(aq) + e^- \rightarrow Au(s)]$　$E°(\text{cathode}) = +1.69\ V$

$Au^{3+}(aq) + 3\ e^- \rightarrow Au(s)$　$E°(\text{anode}) = +1.40\ V$

Reversing the anode reaction yields

$Au(s) \rightarrow Au^{3+}(aq) + 3\ e^-$　then, upon addition (anode),

$$3\ Au^+(aq) \rightarrow$$
$$2\ Au(s) + Au^{3+}(aq) \quad \text{(overall cell)} \quad E°_{cell} = +1.69\ V - 1.40\ V$$
$$= +0.29\ V$$

and $Au(s)\,|\,Au^{3+}(aq)\,||\,Au^+(aq)\,|\,Au(s)$

13.15　(a)　$Ag^+(aq) + e^- \rightarrow Ag(s)$　$E°(\text{cathode}) = +0.80\ V$

$AgBr(s) + e^- \rightarrow Ag(s) + Br^-(aq)$　$E°(\text{anode}) = +0.07\ V$

Reversing the anode reaction yields

$Ag(s) + Br^-(aq) \rightarrow AgBr(s) + e^-$　then, upon addition,

$$Ag^+(aq) + Br^-(aq) \rightarrow AgBr(s) \text{ (overall cell)}\quad E°_{cell} = +0.80\ V - 0.07\ V$$
$$= +0.73\ V$$

This is the direction of the spontaneous standard cell reaction that could be used to study the reverse of the given solubility equilibrium. A cell diagram for this favorable process is

$Ag(s)\,|AgBr(s)\,|\,Br^-(aq)\,||\,Ag^+(aq)\,|\,Ag(s)$

(b) To conform to the notation of this chapter, the neutralization is rewritten as

$$H^+(aq) + OH^- \rightarrow H_2O(l)$$

$O_2(g) + 4\ H^+(aq) + 4\ e^- \rightarrow 2\ H_2O(l)$　$E°(\text{cathode}) = +1.23\ V$

$O_2(g) + 2\ H_2O(l) + 4\ e^- \rightarrow 4\ OH^-(aq)$　$E°(\text{anode}) = +0.40\ V$

Reversing the anode reaction yields

$4 \text{ OH}^-(\text{aq}) \rightarrow O_2(\text{g}) + 2 \text{ H}_2O(\text{l}) + 4 \text{ e}^-$; then, upon addition,

$4 \text{ H}^+(\text{aq}) + 4 \text{ OH}^-(\text{aq}) \rightarrow 4 \text{ H}_2O(\text{l})$

or $\text{H}^+(\text{aq}) + \text{OH}^-(\text{aq}) \rightarrow \text{H}_2O(\text{l})$ (overall cell) $E° = +1.23 \text{ V} - 0.40 \text{ V}$
$$= +0.83 \text{ V}$$

and $\text{Pt(s)} |O_2(\text{g})| \text{OH}^-(\text{aq}) \| \text{H}^+(\text{aq}) |O_2(\text{g})| \text{Pt(s)}$

(c) $\text{Cd(OH)}_2(\text{s}) + 2 \text{ e}^- \rightarrow \text{Cd(s)} + 2 \text{ OH}^-(\text{aq})$ $E°(\text{anode}) = -0.81 \text{ V}$

$\text{Ni(OH)}_3(\text{s}) + \text{e}^- \rightarrow \text{Ni(OH)}_2(\text{s}) + \text{OH}^-(\text{aq})$ $E°(\text{cathode}) = +0.49 \text{ V}$

Reversing the anode reaction and multiplying the cathode reaction by 2 yields

$\text{Cd(s)} + 2 \text{ OH}^-(\text{aq}) \rightarrow \text{Cd(OH)}_2(\text{s}) + 2 \text{ e}^-$

$2 \text{ Ni(OH)}_3 + 2 \text{ e}^- \rightarrow 2 \text{ Ni(OH)}_2(\text{s}) + 2 \text{ OH}^-(\text{aq})$ then, upon addition,

$2 \text{ Ni(OH)}_2(\text{s}) + \text{Cd(s)} \rightarrow \text{Cd(OH)}_2(\text{s}) + 2 \text{ Ni(OH)}_2(\text{s})$

overall cell $E° = +1.30 \text{ V}$

and $\text{Cd(s)} | \text{Cd(OH)}_2(\text{s}) | \text{KOH(aq)} \| \text{Ni(OH)}_3(\text{s}) | \text{Ni(OH)}_2(\text{s}) | \text{Ni(s)}$

13.17 (a) $\text{MnO}_4^-(\text{aq}) + 8 \text{ H}^+(\text{aq}) + 5 \text{ e}^- \rightarrow \text{Mn}^{2+}(\text{aq}) + 4 \text{ H}_2O(\text{l})$ (cathode half-reaction)

$5[\text{Fe}^{2+}(\text{aq}) \rightarrow \text{Fe}^{3+}(\text{aq}) + \text{e}^-]$ (anode half-reaction)

(b) Reversing the anode reaction and adding the two equations yields

$\text{MnO}_4^-(\text{aq}) + 5 \text{ Fe}^{2+}(\text{aq}) + 8 \text{ H}^+(\text{aq}) \rightarrow \text{Mn}^{2+}(\text{aq}) + 5 \text{ Fe}^{3+}(\text{aq}) + 4 \text{ H}_2O(\text{l})$

The cell diagram is

$\text{Pt(s)} | \text{Fe}^{3+}(\text{aq}), \text{Fe}^{2+}(\text{aq}) \| \text{H}^+(\text{aq}), \text{MnO}_4^-(\text{aq}), \text{Mn}^{2+}(\text{aq}) | \text{Pt(s)}$

13.19 A galvanic cell has a positive potential difference; therefore, identify as cathode and anode the electrodes that make $E°$ (cell) positive upon calculating

$E°(\text{cell}) = E°(\text{cathode}) - E°(\text{anode})$

There are only two possibilities: If your first guess gives a negative $E°$ (cell), switch your identification.

(a) $Cu^{2+}(aq) + 2 e^- \rightarrow Cu(s)$ $E°(\text{cathode}) = +0.34$ V

$Cr^{3+}(aq) + e^- \rightarrow Cr^{2+}(aq)$ $E°(\text{anode}) = -0.41$ V
$E°(\text{cell}) = +0.34$ V $- (-0.41$ V$) = +0.75$ V

(b) $AgCl(s) + e^- \rightarrow Ag(s) + Cl^-(aq)$ $E°(\text{cathode}) = +0.22$ V

$AgI(s) + e^- \rightarrow Ag(s) + I^-(aq)$ $E°(\text{anode}) = -0.15$ V
$E°(\text{cell}) = +0.22$ V $- (-0.15$ V$) = +0.37$ V

(c) $Hg_2^{2+}(aq) + 2 e^- \rightarrow 2 Hg(l)$ $E°(\text{cathode}) = +0.79$ V

$Hg_2Cl_2(s) + 2 e^- \rightarrow 2 Hg(l) + 2 Cl^-(aq)$ $E°(\text{anode}) = +0.27$ V
$E°(\text{cell}) = +0.79$ V $- (+0.27$ V$) = +0.52$ V

(d) $Pb^{4+}(aq) + 2 e^- \rightarrow Pb^{2+}(aq)$ $E°(\text{cathode}) = +1.67$ V

$Sn^{4+}(aq) + 2 e^- \rightarrow Sn^{2+}(aq)$ $E°(\text{anode}) = +0.15$ V
$E°(\text{cell}) = +1.67$ V $- (+0.15$ V$) = +1.52$ V

13.21 In each case, determine the cathode and anode half-reactions corresponding to the reactions *as written*. Look up the standard reduction potentials for these half-reactions and then calculate $E°_{\text{cell}} = E°(\text{cathode}) - E°(\text{anode})$. If $E°_{\text{cell}}$ is positive, the reaction is spontaneous under standard conditions.

(a) $E°_{\text{cell}} = E°(\text{cathode}) - E°(\text{anode}) = +0.96$ V $- (+0.79$ V$) = +0.17$ V.

Therefore, spontaneous galvanic cell:

$Hg(l) \mid Hg_2^{2+}(aq) \parallel NO_3^-(aq), H^+(aq) \mid NO(g) \mid Pt(s)$
$\Delta G°_r = -nFE° = -(6)(9.65 \times 10^4 \text{ C} \cdot \text{mol}^{-1})(+0.17 \text{ J} \cdot \text{C}^{-1}) = -98 \text{ kJ} \cdot \text{mol}^{-1}$

(b) $E°_{\text{cell}} = E°(\text{cathode}) - E°(\text{anode}) = +0.92$ V $- (+1.09$ V$) = -0.17$ V

Therefore, not spontaneous.

(c) $E°_{\text{cell}} = E°(\text{cathode}) - E°(\text{anode}) = +1.33$ V $- (+0.97$ V$) = +0.36$ V

Therefore, spontaneous galvanic cell.

$$Pt(s) \mid Pu^{3+}(aq), Pu^{4+}(aq) \parallel Cr_2O_7^{2-}(aq), Cr^{3+}(aq), H^+(aq) \mid Pt(s)$$

$$\Delta G_r^\circ = -nFE^\circ = -(6)(9.65 \times 10^4 \text{ C} \cdot \text{mol}^{-1})(0.36 \text{ J} \cdot \text{C}^{-1}) = -208 \text{ kJ} \cdot \text{mol}^{-1}$$

13.23 The cell, as written $Cu(s) \mid Cu^{2+}(aq) \parallel M^{2+}(aq) \mid M(s)$, makes the Cu/Cu^{2+} electrode the anode, because this is where oxidation is occurring; the M^{2+}/M electrode is the cathode. The calculation is

$$E^\circ = E^\circ(\text{cathode}) - E^\circ(\text{anode})$$

$$-0.689 \text{ V} = E^\circ(\text{cathode}) - (+0.34 \text{ V})$$

$$E^\circ(\text{cathode}) = -0.349 \text{ V}$$

13.25 Refer to Appendix 2B. The more negative (less positive) the standard reduction potential, the stronger is the metal as a reducing agent.

(a) $Cu < Fe < Zn < Cr$

(b) $Mg < Na < K < Li$

(c) $V < Ti < Al < U$

(d) $Au < Ag < Sn < Ni$

13.27 The appropriate half-reactions are:

$$U^{4+} + e^- \rightarrow U^{3+} \qquad E^\circ = -0.61 \qquad \text{(A)}$$

$$U^{3+} + 3 e^- \rightarrow U \qquad E^\circ = -1.79 \qquad \text{(B)}$$

(A) and (B) add to give the desired half-reaction (C):

$$U^{4+} + 4 e^- \rightarrow U \qquad\qquad E^\circ = ? \qquad \text{(C)}$$

In order to calculate the potential of a *half-reaction*, we need to convert the E° values into ΔG° values:

$$\Delta G^\circ(A) = -nFE^\circ(A) = -1F(-0.61 \text{ V})$$
$$\Delta G^\circ(B) = -nFE^\circ(B) = -3F(-1.79 \text{ V})$$
$$\Delta G^\circ(C) = -nFE^\circ(C) = -4FE^\circ(C)$$
$$\Delta G^\circ(C) = \Delta G^\circ(A) + \Delta G^\circ(B)$$
$$-4FE^\circ(C) = -1F(-0.61 \text{ V}) + [-3F(-1.79 \text{ V})]$$

The constant F will cancel from both sides, leaving:

$-4E°(C) = -1(-0.61 \text{ V}) - 3(-1.79 \text{ V})$

$E°(C) = -[0.61 \text{ V} + 5.37 \text{ V}]/4 = -1.50 \text{ V}$

13.29 In each case, identify the couple with the more positive reduction potential. This will be the couple at which reduction occurs, and therefore which contains the oxidizing agent. The other couple contains the reducing agent.

(a) Co^{2+}/Co $E° = -0.28 \text{ V}$, Co^{2+} is the oxidizing agent (cathode)

Ti^{3+}/Ti^{2+} $E° = -0.37 \text{ V}$, Ti^{2+} is the reducing agent (anode)

$Pt(s) \mid Ti^{2+}(aq), Ti^{3+}(aq) \parallel Co^{2+}(aq) \mid Co(s)$

$E°_{cell} = E°(\text{cathode}) - E°(\text{anode}) = -0.28 \text{ V} - (-0.37 \text{ V}) = +0.09 \text{ V}$

(b) U^{3+}/U $E° = -1.79 \text{ V}$, U^{3+} is the oxidizing agent (cathode)

La^{3+}/La $E° = -2.52 \text{ V}$, La is the reducing agent (anode)

$La(s) \mid La^{3+}(aq) \parallel U^{3+}(aq) \mid U(s)$

$E°_{cell} = -1.79 \text{ V} - (-2.52 \text{ V}) = +0.73 \text{ V}$

(c) Fe^{3+}/Fe^{2+} $E° = +0.77 \text{ V}$, Fe^{3+} is the oxdizing agent (cathode)

H^+/H_2 $E° = 0.00 \text{ V}$, H_2 is the reducing agent (anode)

$Pt(s) \mid H_2(g) \mid H^+(aq) \parallel Fe^{2+}(aq), Fe^{3+}(aq) \mid Pt(s)$

$E°_{cell} = +0.77 \text{ V} - 0.00 \text{ V} = +0.77 \text{ V}$

(d) $O_3/O_2,OH^-$ $E° = +1.24 \text{ V}$, O_3 is the oxidizing agent (cathode)

Ag^+/Ag $E° = +0.80 \text{ V}$, Ag is the reducing agent (anode)

$Ag(s) \mid Ag^+(aq) \parallel OH^-(aq) \mid O_3(g), O_2(g) \mid Pt(s)$

$E°_{cell} = +1.24 \text{ V} - 0.80 \text{ V} = +0.44 \text{ V}$

13.31 (a) $E°(Cl_2, Cl^-) = +1.36 \text{ V}$ (cathode)

$E°(Br_2, Br^-) = +1.09 \text{ V}$ (anode)

Because $E°(Cl_2, Cl^-) > E°(Br_2, Br^-)$ the reaction favors products.

$E°_{cell} = +1.36 \text{ V} - 1.09 \text{ V} = +0.27 \text{ V}$

$Cl_2(g)$ is the oxidizing agent.

(b) $E°(Ce^{4+}/Ce^{3+}) = +1.61$ V   (anode)

$E°(MnO_4^-/Mn^{2+}) = +1.51$ V   (cathode)

Because $E°(Ce^{4+}/Ce^{3+}) > E°(MnO_4^-/Mn^{2+})$, the reaction does not favor products.

(c) $E°(Pb^{4+}/Pb^{2+}) = +1.67$ V   (anode)

$E°(Pb^{2+}/Pb) = -0.13$ V   (cathode)

Because $E°(Pb^{4+}/Pb^{2+}) > E°(Pb^{2+}/Pb)$, the reaction does not favor products.

(d) $E°(NO_3^-/NO_2/H^+) = +0.80$ V   (cathode)

$E°(Zn^{2+}/Zn) = -0.76$ V   (anode)

Because $E°(NO_3^-/NO_2/H^+) > E°(Zn^{2+}/Zn)$, the reaction favors products.

$E°_{cell} = +0.80$ V $- (-0.76$ V$) = +1.56$ V

NO_3^- is the oxidizing agent.

13.33 (a) We want $E°$ for $Tl^{3+}(aq) + 3$ e$^- \rightarrow Tl(s)$, $n = 3$.

This reaction is the reverse of the formation reaction:

$Tl(s) \rightarrow Tl^{3+}(aq) + 3$ e$^-$

Therefore, for the Tl^{3+}/Tl couple, $\Delta G°_r = -215$ kJ$\cdot$mol^{-1}.

$\Delta G°_r = -nFE° = -215$ kJ$\cdot$mol^{-1}

$E° = \dfrac{\Delta G°_r}{-nF} = \dfrac{-2.15 \times 10^5 \text{ J}\cdot\text{mol}^{-1}}{-3 \times 9.65 \times 10^4 \text{ C}\cdot\text{mol}^{-1}} = +0.743$ V

(b) Using the potential from part (a) and the potential from Appendix 2B for the reduction of Tl^+, we can decide whether or not Tl^+ will disproportionate in solution. The equation of interest is

3 Tl$^+(aq) \rightarrow 2$ Tl$(s) +$ Tl$^{3+}(aq)$.

The half-reactions to combine are:

$Tl^+ \rightarrow Tl^{3+} + 2\,e^-$ (1)

$Tl^+ + e^- \rightarrow Tl$ (2)

The potential for reaction (1) must be obtained using the $\Delta G°$ values of the two known half-reactions:

$Tl \rightarrow Tl^{3+} + 3\,e^-$ $\Delta G° = +215 \text{ kJ} \cdot \text{mol}^{-1}$

$Tl^+ + e^- \rightarrow Tl$

$\begin{aligned}\Delta G° &= -nFE° \\ &= -1(9.65 \times 10^4 \text{ J} \cdot \text{V}^{-1} \cdot \text{mol}^{-1})(-0.34 \text{ V})/(1000 \text{ J} \cdot \text{kJ}^{-1}) \\ &= +33 \text{ kJ} \cdot \text{mol}^{-1}\end{aligned}$

$\Delta G°$ for the combined half-reaction $Tl^+ \rightarrow Tl^{3+} + 2\,e^-$ is the sum of these two numbers:

$+215 \text{ kJ} \cdot \text{mol}^{-1} + 33 \text{ kJ} \cdot \text{mol}^{-1} = +248 \text{ kJ} \cdot \text{mol}^{-1}$

We can now combine this with the reduction process for Tl^+ to get the desired equation:

$\begin{aligned}Tl^+ &\rightarrow Tl^{3+} + 2\,e^- & \Delta G° &= +248 \text{ kJ} \cdot \text{mol}^{-1} \\ +2(Tl^+ + e^- &\rightarrow Tl) & \Delta G° &= 2(+33 \text{ kJ} \cdot \text{mol}^{-1}) \\ 3\,Tl^+ &\rightarrow 2\,Tl + Tl^{3+} & \Delta G° &= +248 \text{ kJ} \cdot \text{mol}^{-1} + 2(+33 \text{ kJ} \cdot \text{mol}^{-1}) \\ & & &= +314 \text{ kJ} \cdot \text{mol}^{-1}\end{aligned}$

Since $\Delta G°$ is large and positive, $K \ll 1$. This disproportionation reaction does not favor products.

13.35 (a) $Ti^{2+}(aq) + 2\,e^- \rightarrow Ti(s)$ $E°(\text{cathode}) = -1.63 \text{ V}$

$Mn^{2+}(aq) + 2\,e^- \rightarrow Mn(s)$ $E°(\text{anode}) = -1.18 \text{ V}$

Note: These equations represent the cathode and anode half-reactions for the overall reaction as written. The spontaneous direction of this reaction under standard conditions is the opposite of that given.

$E°_{\text{cell}} = E°(\text{cathode}) - E°(\text{anode}) = -1.63 \text{ V} - (-1.18 \text{ V}) = -0.45 \text{ V}$, and

$$\ln K = \frac{nFE°}{RT}. \qquad \text{At } 25°C \quad \ln K = \frac{nE°}{0.02569 \text{ V}}.$$

$$\therefore \quad \ln K = \frac{(2)(-0.45 \text{ V})}{0.02569 \text{ V}} = -35 \quad \text{and} \quad K = 6 \times 10^{-16}.$$

(b) $In^{3+}(aq) + 2 e^- \rightarrow In^{2+}(aq)$ $E°(\text{cathode}) = -0.49 \text{ V}$

$U^{4+}(aq) + e^- \rightarrow U^{3+}(aq)$ $E°(\text{anode}) = -0.61 \text{ V}$

$E°_{\text{cell}} = E°(\text{cathode}) - E°(\text{anode}) = -0.49 \text{ V} - (-0.61 \text{ V}) = +0.12 \text{ V}$, and

at $25°C \quad \ln K = \frac{(2)(+0.12 \text{ V})}{0.02569 \text{ V}} = +9.3.$

$$\therefore \quad K = 1 \times 10^4.$$

13.37 (a) $Pb^{4+}(aq) + 2 e^- \rightarrow Pb^{2+}(aq)$ $E°(\text{cathode}) = +1.67 \text{ V}$

$Sn^{2+}(aq) \rightarrow Sn^{4+}(aq) + 2e^-$ $E°(\text{anode}) = +0.15 \text{ V}$

$Pb^{4+}(aq) + Sn^{2+}(aq) \rightarrow Pb^{2+}(aq) + Sn^{4+}(aq)$ $E°_{\text{cell}}$
$\qquad = 1.67 \text{ V} - (0.15 \text{ V}) = +1.52 \text{ V}$

Then, $E = E° - \left(\dfrac{0.025\ 693 \text{ V}}{n}\right) \ln Q; 1.33 \text{ V} = 1.52 \text{ V} - \left(\dfrac{0.025\ 693 \text{ V}}{2}\right) \ln Q$

$\quad \ln Q = \dfrac{1.52 \text{ V} - 1.33 \text{ V}}{0.0129 \text{ V}} = \dfrac{0.19 \text{ V}}{0.0129 \text{ V}} = 15 \quad Q = 10^6$

(b) $2[Cr_2O_7^{2-}(aq) + 14 H^+(aq) + 6 e^- \rightarrow$
$\qquad 2 Cr^{3+}(aq) + 7 H_2O(l)]$ $E°(\text{cathode}) = 1.33 \text{ V}$

$3[2 H_2O(l) \rightarrow O_2(g) + 4 H^+(aq) + 4 e^-]$ $E°(\text{anode}) = +1.23 \text{ V}$

$2 Cr_2O_7^{2-}(aq) + 16 H^+(aq) \rightarrow 4 Cr^{3+}(aq) + 8 H_2O(l) + 3 O_2(g)$ $E°_{\text{cell}}$
$\qquad = 0.10 \text{ V}$

Then, $E = E° - \left(\dfrac{0.0257 \text{ V}}{n}\right) \ln Q; \ 0.10 \text{ V} = +0.10 \text{ V} - \left(\dfrac{0.0257 \text{ V}}{12}\right) \ln Q$

$\ln Q = 0.00 \quad Q = 1.0$

13.39 (a) $Cu^{2+}(aq, 0.010 \text{ M}) + 2 e^- \rightarrow Cu(s)$ (cathode)

$Cu^{2+}(aq, 0.0010 \text{ M}) + 2 e^- \rightarrow Cu(s)$ (anode)

$$\text{Cu}^{2+}(\text{aq}, 0.010 \text{ M}) \rightarrow \text{Cu}^{2+}(\text{aq}, 0.0010 \text{ M}), \, n = 2$$

$$E°_{\text{cell}} = E°(\text{cathode}) - E°(\text{anode}) = 0 \text{ V}$$

$$E_{\text{cell}} = E°_{\text{cell}} - \left(\frac{RT}{nF}\right) \ln Q = -\left(\frac{0.025\,693 \text{ V}}{2}\right) \ln Q \text{ at } 25°\text{C}$$

$$E_{\text{cell}} = -\left(\frac{0.025\,693 \text{ V}}{2}\right) \ln \left(\frac{0.0010 \text{ M}}{0.010 \text{ M}}\right) = +0.030 \text{ V}$$

(b) at $\text{pH} = 3.0$, $[\text{H}^+] = 1 \times 10^{-3}$ M

at $\text{pH} = 4.0$, $[\text{H}^+] = 1 \times 10^{-4}$ M

Cell reaction is $\text{H}^+(\text{aq}, 1 \times 10^{-3} \text{ M}) \rightarrow \text{H}^+(\text{aq}, 1 \times 10^{-4} \text{ M}), \, n = 1$

$$E°_{\text{cell}} = 0 \text{ V} \quad E_{\text{cell}} = E°_{\text{cell}} - \left(\frac{RT}{nF}\right) \ln Q = -\left(\frac{0.025\,693 \text{ V}}{1}\right) \ln \left(\frac{1 \times 10^{-4}}{1 \times 10^{-3}}\right)$$

$$= +6 \times 10^{-2} \text{ V}$$

13.41 In each case, $E°_{\text{cell}} = E°(\text{cathode}) - E°(\text{anode})$. Recall that the values for $E°$ at the electrodes refer to the electrode potential for the half-reaction written as a reduction reaction. In balancing the cell reaction, the half-reaction at the anode is reversed. However, this does not reverse the sign of electrode potential used at the anode, because the value always refers to the reduction potential.

(a) $2 \text{ H}^+(\text{aq}, 1.0 \text{ M}) + 2 \text{ e}^- \rightarrow \text{H}_2(\text{g}, 1 \text{ atm}) \quad E°(\text{cathode}) = 0.00 \text{ V}$

$\text{H}_2(\text{g}, 1 \text{ atm}) \rightarrow 2 \text{ H}^+(\text{aq}, 0.075 \text{ M}) + 2 \text{ e}^- \quad E°(\text{anode}) = 0.00 \text{ V}$

$2 \text{ H}^+(\text{aq}, 1.0 \text{ M}) + \text{H}_2(\text{g}, 1 \text{ atm}) \rightarrow 2 \text{ H}^+(\text{aq}, 0.075 \text{ M}) + \text{H}_2(\text{g}, 1 \text{ atm})$

$E°_{\text{cell}} = 0.00 \text{ V}$

Then, $E = E° - \left(\dfrac{0.025\,693 \text{ V}}{n}\right) \ln \left(\dfrac{[\text{H}^+, 0.075 \text{ M}]^2 P_{\text{H}_2}}{[\text{H}^+, 1.0 \text{ M}]^2 P_{\text{H}_2}}\right)$

$E = 0.00 \text{ V} - \left(\dfrac{0.025\,693 \text{ V}}{2}\right) \ln \left(\dfrac{(0.075 \text{ M})^2 \times 1 \text{ atm}}{(1.0 \text{ M})^2 \times 1 \text{ atm}}\right)$

$E = -0.0129 \text{ V} \ln (0.075)^2 = +0.067 \text{ V}$

(b) $\text{Ni}^{2+}(\text{aq}) + 2 \text{ e}^- \rightarrow \text{Ni(s)} \quad E°(\text{cathode}) = -0.23 \text{ V}$

$Zn(s) \rightarrow Zn^{2+}(aq) + 2\ e^-\quad E°(\text{anode}) = -0.76\ V$

$Ni^{2+}(aq) + Zn(s) \rightarrow Ni(s) + Zn^{2+}(aq)\ E°_{\text{cell}} = +0.53\ V$

Then, $E = E° - \left(\dfrac{0.025\ 693\ V}{n}\right)\ln\left(\dfrac{[Zn^{2+}]}{[Ni^{2+}]}\right)$

$E = 0.53\ V - \left(\dfrac{0.025\ 693\ V}{2}\right)\ln\left(\dfrac{0.37}{0.059}\right) = 0.53\ V - 0.02\ V = 0.51\ V$

(c) $2\ H^+(aq) + 2\ e^- \rightarrow H_2(g)\ E°(\text{cathode}) = 0.00\ V$

$2\ Cl^-(aq) \rightarrow Cl_2(g) + 2\ e^-\ E°(\text{anode}) = +1.36\ V$

$2\ H^+(aq) + 2\ Cl^-(aq) \rightarrow H_2(g) + Cl_2(g)\ E°_{\text{cell}} = -1.36\ V$

Then,

$E = E° - \left(\dfrac{0.025\ 693\ V}{n}\right)\ln\left(\dfrac{P_{H_2}P_{Cl_2}}{[H^+]^2[Cl^-]^2}\right)$

$E = -1.36\ V - \left(\dfrac{0.025\ 693\ V}{2}\right)\ln\left(\dfrac{\left(\dfrac{125}{760}\right)\left(\dfrac{250}{760}\right)}{(0.85)^2\ (1.0)^2}(1.01325)^2\right)$

$E = -1.36\ V + 0.03\ V$

$\quad = -1.33\ V$

(d) $Sn^{4+}(aq, 0.867\ M) + 2\ e^- \rightarrow Sn^{2+}(aq, 0.55\ M)$

$\qquad E°(\text{cathode}) = +0.15\ V$

$Sn(s) \rightarrow Sn^{2+}(aq, 0.277\ M) + 2\ e^-\ E°(\text{anode}) = -0.14\ V$

$Sn^{4+}(aq, 0.867\ M) + Sn(s) \rightarrow Sn^{2+}(aq, 0.55) + Sn^{2+}(aq, 0.277\ M)$

$E°_{\text{cell}} = 0.29\ V$

$E = E° - \left(\dfrac{0.025\ 693\ V}{2}\right)\ln\left(\dfrac{(0.55)(0.277)}{(0.867)}\right)$

$E = 0.29\ V + 0.02\ V = 0.31\ V$

13.43 In each case, obtain the balanced equation for the cell reaction from the half-cell reactions at the electrodes, by reversing the reduction equation for the half-reaction at the anode, multiplying the half-reaction equations by an appropriate factor to balance the number of electrons, and then adding the half-reactions. Calculate $E°_{cell} = E°(cathode) - E°(anode)$. Then write the Nernst equation for the cell reaction and solve for the unknown.

(a) $Hg_2Cl_2(s) + 2 e^- \rightarrow 2 Hg(l) + 2 Cl^-(aq)$ $E°(cathode) = +0.27$ V

$H_2(g) \rightarrow 2 H^+(aq) + 2 e^-$ $E°(anode) = 0.00$ V

$H_2(g) + Hg_2Cl_2(s) \rightarrow 2 H^+(aq) + 2 Hg(l) + 2 Cl^-(aq)$ $E°_{cell} = +0.27$ V

$$E = E° - \left(\frac{0.025\ 693\ \text{V}}{n}\right) \ln\left(\frac{[H^+]^2[Cl^-]^2}{[H_2]}\right)$$

$$0.33\ \text{V} = 0.27\ \text{V} - \left(\frac{0.025\ 693\ \text{V}}{2}\right) \ln\left(\frac{[H^+]^2(1)^2}{(1)}\right)$$

$$= 0.27\ \text{V} - (0.0129\ \text{V}) \ln [H^+]^2$$

$$0.06\ \text{V} = -0.0257\ \text{V} \ln [H^+] = -0.0257\ \text{V} \times (2.303 \log [H^+])$$

$$pH = \frac{0.06\ \text{V}}{(2.303)\ (0.025\ 693\ \text{V})} = 1.0$$

(b)

$$2[MnO_4^-(aq) + 8 H^+(aq) + 5 e^- \rightarrow$$
$$Mn^{2+}(aq) + 4 H_2O(l)] \quad E°(cathode) = +1.51\ \text{V}$$

$$5[2 Cl^-(aq) \rightarrow Cl_2(g) + 2 e^-] \quad E°(anode) = +1.36\ \text{V}$$

$$2 MnO_4^-(aq) + 16 H^+(aq) + 10 Cl^-(aq) \rightarrow$$
$$5 Cl_2(g) + 2 Mn^{2+}(aq) + 8 H_2O(l)\ E°_{cell} = +0.15\ \text{V}$$

$$E = E° - \left(\frac{0.0257\ \text{V}}{n}\right) \ln\left(\frac{[Cl_2]^5[Mn^{2+}]^2}{[MnO_4^-]^2[H^+]^{16}[Cl^-]^{10}}\right)$$

$$-0.30 \text{ V} = +0.15 \text{ V} - \left(\frac{0.0257 \text{ V}}{10}\right) \ln\left(\frac{(1)^5 (0.10)^2}{(0.010)^2 \, (1\times10^{-4})^{16} (\text{Cl}^-)^{10}}\right)$$

$$-0.45 \text{ V} = -(0.002\,5693 \text{ V}) \log\left(\frac{1\times10^{-2}}{(1\times10^{-4})\,(1\times10^{-64})\,[\text{Cl}^-]^{10}}\right)$$

$$= -0.002\,5693 \text{ V}\left[\ln(1\times10^{66}) + \ln\left(\frac{1}{[\text{Cl}^-]^{10}}\right)\right]$$

$$= -0.390 \text{ V} + (0.0025\,693 \text{ V})\ln[\text{Cl}^-]^{10}$$

$$-0.0594 \text{ V} = 0.002\,5693 \text{ V}\,\ln[\text{Cl}^-]^{10}$$

$$= (0.025\,693 \text{ V})\ln[\text{Cl}^-]$$

$$\ln[\text{Cl}^-] = \frac{-0.06 \text{ V}}{0.025\,693 \text{ V}} = -2$$

$$[\text{Cl}^-] = 10^{-1} \text{ mol} \cdot \text{L}^{-1}$$

13.45 Since the reduction potential of tin(II) is negative relative to the S.H.E., we will assume the tin electrode to be the anode such that the standard cell potential would be positive. Then we can use the Nernst equation to solve for the hydrogen ion activity in order to calculate the pH. The cell reaction is $\text{Sn}(s) + 2 \text{ H}^+ \rightarrow \text{Sn}^{2+} + \text{H}_2(g)$.

$$E = E^\circ - \left(\frac{0.025\,693 \text{ V}}{n}\right) \ln\left(\frac{[\text{Sn}^{2+}]p_{\text{H}_2}}{[\text{H}^+]^2}\right)$$

$$0.061 \text{ V} = 0.14 \text{ V} - \left(\frac{0.025\,693 \text{ V}}{2}\right)\ln\left(\frac{[0.015][1]}{[\text{H}^+]^2}\right)$$

$$= 0.14 \text{ V} - (0.01284 \text{ V})(\ln(0.015) - \ln[\text{H}^+]^2)$$

$$\frac{0.079 \text{ V}}{0.01284 \text{ V}} = \ln(0.015) - 2\ln[\text{H}^+]$$

$$-10.352 = 2\ln[\text{H}^+]$$

$$\ln[\text{H}^+] = -5.176, \quad [\text{H}^+] = 5.650\times10^{-3}$$

$$\text{pH} = -\log(5.650\times10^{-3}) = 2.25$$

13.47 To calculate this value, we need to determine the $E°$ value for the solubility reaction:

$Hg_2Cl_2(s) \rightarrow Hg_2^{2+}(aq) + 2\ Cl^-(aq)$ $\quad E° = ?$

The relationship $\Delta G° = -nRT \ln K = -nFE°$ can be used to calculate the value of K_{sp}.

The equations that will add to give the net equation we want are

$Hg_2Cl_2(s) + 2\ e^- \rightarrow 2\ Hg(l) + 2\ Cl^-(aq)$ $\qquad E° = +0.27$ V

$2\ Hg(l) \rightarrow Hg_2^{2+}(aq) + 2\ e^-$ $\qquad\qquad\qquad E° = 0.79$ V

Notice that the second equation is reversed from the reduction reaction given in the Appendix, and consequently the $E°$ value is changed in sign. Adding these two equations together gives the desired net reaction, and summing the $E°$ values will give the $E°$ value for that process:

$E° = (+0.29$ V$) + (-0.79$ V$) = -0.50$ V

$$\ln K_{sp} = \frac{nFE°}{RT} = \frac{(2)(9.65 \times 10^4 \text{ C} \cdot \text{mol}^{-1})(-0.50 \text{ V})}{(8.314 \text{ J} \cdot \text{K}^{-1} \cdot \text{mol}^{-1})(298 \text{ K})} = -38.95$$

$K_{sp} = 1.2 \times 10^{-17}$

(b) This value is a factor of 10 greater than the value in Table 11.6 (1.3×10^{-18}).

13.49 This cell uses two silver electrodes, so $E° = 0$ and E is determined by the ratio of $[Ag^+]_{anode}$ to $[Ag^+]_{cathode}$. Since $[Ag^+]_{anode} < [Ag^+]_{cathode}$, the ratio is less than 1 and $E > 0$, so the cell can do work because $\Delta G = w_{max} = -nFE$.

$$E = -\left(\frac{0.025\ 693 \text{ V}}{n}\right) \ln \left(\frac{[Ag^+]_{anode}}{[Ag^+]_{cathode}}\right) = -\left(\frac{0.025\ 693 \text{ V}}{1}\right) \ln \left(\frac{5.0 \times 10^{-3}}{0.15}\right)$$

$= 0.0874$ V

$\Delta G = w_{max} = -nFE = -(1 \text{ mol})(96\ 485 \text{ J V}^{-1}\text{mol}^{-1})(0.0874 \text{ V})$

$= -8.4$ kJ

Therefore, the maximum work that the cell can perform is 8.4 kJ per mole of Ag.

13.51 For the standard calomel electrode, $E° = +0.27$ V. If this were set equal to 0, all other potentials would also be decreased by 0.27 V.   (a) Therefore, the standard hydrogen electrode's standard reduction potential would be 0.00 V – 0.27 V or –0.27 V.   (b) The standard reduction potential for Cu^{2+}/Cu would be 0.34 V – 0.27 V or +0.07 V.

13.53 The strategy is to consider the possible competing cathode and anode reactions. At the cathode, choose the reduction reaction with the most positive (least negative) standard reduction potential ($E°$ value). At the anode, choose the oxidation reaction with the least positive (most negative) standard reduction potential ($E°$ value, as given in the table). Then calculate $E°_{cell} = E°(\text{cathode}) - E°(\text{anode})$. The negative of this value is the minimum potential that must be supplied.

(a) cathode: Ni^{2+} (aq) + 2 e$^-$ → Ni(s) $E° = -0.23$ V

(rather than 2 H$_2$O(l) + 2 e$^-$ → H$_2$(g) + 2 OH$^-$(aq) $E° = -0.83$ V)

(b) anode: 2 H$_2$O(l) → O$_2$(g) + 4 H$^+$(aq) + 4 e$^-$ $E° = +1.23$ V

(the SO_4^{2-} ion will not oxidize)

(c) $E°_{cell} = E°(\text{cathode}) - E°(\text{anode}) = -0.23$ V – (+1.23 V) = –1.46 V

Therefore E (supplied) must be > +1.46 V (1.46 V is the minimum).

13.55 In each case, compare the reduction potential of the ion to the reduction potential of water ($E° = -0.42$ V) and choose the process with the least negative $E°$ value.

(a) Mn^{2+}(aq) + 2 e$^-$ → Mn(s) $E° = -1.18$ V

(b) Al^{3+}(aq) + 3 e$^-$ → Al(s) $E° = -1.66$ V

The reactions in (a) and (b) evolve hydrogen rather than yield a metallic deposit because water is reduced, according to

2 H$_2$O(l) + 2 e$^-$ → H$_2$(g) + 2 OH$^-$ (aq) ($E° = -0.42$ V, at pH = 7)

(c) $Ni^{2+}(aq) + 2\,e^- \rightarrow Ni(s)$ $E° = -0.23$ V

(d) $Au^{3+}(aq) + 3\,e^- \rightarrow Au(s)$ $E° = +1.69$ V

In (c) and (d) the metal ion will be reduced.

13.57 4500 C $\div 9.65 \times 10^4$ C·F^{-1} $= 0.047$ F $= 0.047$ mol e$^-$

(a) $(0.047 \div 3)$ mol Bi^{3+} $+ 0.047$ mol e$^-$ $\rightarrow$
      $(0.047 \div 3)$ mol Bi, or 0.016 mol Bi $= 3.3$ g

(b) 0.047 mol H$^+$ $+ 0.047$ mol e$^-$ $\rightarrow (0.047 \div 2)$ mol H$_2$

0.024 mol H$_2 \times 24.45$ L·mol^{-1} (at 298 K) $= 0.59$ L

(c) $(0.047 \div 3)$ mol Co^{3+} $+ 0.047$ mol e$^-$ $\rightarrow$
      $(0.047 \div 3)$ mol Co $= 0.016$ mol Co or 0.94 g

13.59 (a) $Ag^+(aq) + e^- \rightarrow Ag(s)$

$$\text{time} = (1.50 \text{ g Ag})\left(\frac{1 \text{ mol Ag}}{107.98 \text{ g Ag}}\right)\left(\frac{1 \text{ mol e}^-}{1 \text{ mol Ag}}\right)$$

$$\left(\frac{9.65 \times 10^4 \text{ C}}{1 \text{ mol e}^-}\right)\left(\frac{1 \text{ A·s}}{1 \text{ C}}\right)\left(\frac{1}{0.0136}\right) = 9.9 \times 10^4 \text{ s or 27 h}$$

(b) $Cu^{2+}(aq) + 2\,e^- \rightarrow Cu(s)$

$$\text{mass Cu} = (9.9 \times 10^4 \text{ s})(0.0136 \text{ A})\left(\frac{1 \text{ C}}{1 \text{ A·s}}\right)\left(\frac{1 \text{ mol e}^-}{9.65 \times 10^4 \text{ C}}\right)$$

$$\left(\frac{0.50 \text{ mol Cu}}{1 \text{ mol e}^-}\right)\left(\frac{63.5 \text{ g Cu}}{1 \text{ mol Cu}}\right) = 0.44 \text{ g Cu}$$

13.61 (a) $Cr(VI) + 6\,e^- \rightarrow Cr(s)$

$$\text{current} = \frac{\text{charge}}{\text{time}}$$

$$= \frac{2.5 \text{ g Cr} \left(\dfrac{1 \text{ mol Cr}}{52.00 \text{ g Cr}} \right) \left(\dfrac{6 \text{ mol e}^-}{1 \text{ mol Cr}} \right) \left(\dfrac{9.65 \times 10^4 \text{ C}}{1 \text{ mol e}^-} \right)}{12 \text{ h} \times 3600 \text{ s} \cdot \text{h}^{-1}}$$

$$= 0.64 \text{ C} \cdot \text{s}^{-1} = 0.64 \text{ A}$$

(b) $Na^+ + e^- \rightarrow Na(s)$

$$\text{current} = \frac{2.5 \text{ g Na} \left(\dfrac{1 \text{ mol Na}}{22.99 \text{ g Na}} \right) \left(\dfrac{1 \text{ mol e}^-}{1 \text{ mol Na}} \right) \left(\dfrac{9.65 \times 10^4 \text{ C}}{1 \text{ mol e}^-} \right)}{12 \text{ h} \times 3600 \text{ s} \cdot \text{h}^{-1}}$$

$$= 0.24 \text{ C} \cdot \text{s}^{-1} = 0.24 \text{ A}$$

13.63 $Ru^{n+}(aq) + n\,e^- \rightarrow Ru(s)$; solve for n

$$\text{moles of Ru} = (0.0310 \text{ g Ru}) \left(\frac{1 \text{ mol}}{101.07 \text{ g Ru}} \right) = 3.07 \times 10^{-4} \text{ mol}$$

$$\text{total charge} = (500 \text{ s})(120 \text{ mA}) \left(\frac{10^{-3} \text{ A}}{1 \text{ mA}} \right) \left(\frac{1 \text{ C} \cdot \text{s}^{-1}}{1 \text{ A}} \right) = 60 \text{ C}$$

$$\text{moles of e}^- = (60 \text{ C}) \left(\frac{1 \text{ mol e}^-}{96\,500 \text{ C}} \right) = 6.2 \times 10^{-4} \text{ mol e}^-$$

$$n = \frac{6.2 \times 10^{-4} \text{ mol e}^-}{3.07 \times 10^{-4} \text{ mol}} = \frac{2 \text{ mol charge}}{1 \text{ mol}}$$

Therefore, oxidation number of Ru^{2+} is +2.

13.65 $Hf^{n+} + n\,e^- \rightarrow Hf(s)$; solve for n.

$$\text{charge consumed} = 15.0 \text{ C} \cdot \text{s}^{-1} \times 2.00 \text{ h} \times 3600 \text{ s} \cdot \text{h}^{-1} = 1.08 \times 10^5 \text{ C}$$

$$\text{moles of charge consumed} = (1.08 \times 10^5 \text{ C}) \left(\frac{1 \text{ mol e}^-}{9.65 \times 10^4 \text{ C}} \right) = 1.12 \text{ mol e}^-$$

$$\text{moles of Hf plated} = (50.0 \text{ g Hf}) \left(\frac{1 \text{ mol Hf}}{178.49 \text{ g Hf}} \right) = 0.280 \text{ mol Hf}$$

Then, $n = \dfrac{1.12 \text{ mol e}^-}{0.280 \text{ mol Hf}} = 4.00 \text{ mol e}^- / \text{mol Hf}$

Therefore, the oxidation number is 4, that is, Hf^{4+}.

13.67 (a) $2Cu_{(aq)}^{2+} + 4e^- \rightarrow 2Cu(s)$ $\qquad\qquad\qquad E° = 0.34 \text{ V}$

$\dfrac{2H_2O(l) \rightarrow O_2(g) + 4H^+_{(aq)} + 4e^- \qquad\qquad E° = -0.82 \text{ V (pH7)}}{2Cu_{(aq)}^{2+} + 2H_2O(l) \rightarrow 2Cu(s) + O_2(g) + 4H^+_{(aq)} \qquad E° = -0.48 \text{ V}}$

$\therefore H^+_{(aq)}$ or hydronium ions are generated at the anode.

(b) $\qquad Q = nF$ and $Q = It$

$\therefore n = \dfrac{It}{F} = \dfrac{(0.120A)(1.08 \times 10^5 s)}{9.6485 \times 10^4 C. \text{mol}^{-1}} = 0.134 \text{ mol of electrons}$

Since 4 moles of electrons are needed in the balanced reaction to generate 4 moles of H_3O^+ there is a 1:1 ratio. Since 0.134 mol of electrons were supplied a maximum of 0.134 mol H_3O^+ can be generated.

(c)

$$[H_3O^+] = \dfrac{0.134 \text{ mol}}{0.2000 \text{ L}}$$

$$= 0.67 \text{ mol. L}^{-1}$$

$$pH = -\log[H_3O^+] = 0.17$$

13.69 (a) Assuming alkaline conditions:

oxidation/anode: $\qquad 6OH^-(aq) + 2Al(s) \rightarrow 2Al(OH)_3(aq) + 6e-$

reduction/cathode: $\qquad 3H_2O(l) + 1\dfrac{1}{2}O_2(g) + 6e- \rightarrow 6OH^-(aq)$

overall: $\qquad 3H_2O(l) + 2Al(s) + 1\dfrac{1}{2}O_2(g) \rightarrow 2Al(OH)_3(aq)$

(b) $\qquad E°_{cell} = E°_{cathode} - E°_{anode}$

$$= (+0.40 \text{ V}) - (-1.66 \text{ V})$$

$$= +2.06 \text{ V}$$

13.71 Assuming all the energy comes from reduction of oxygen focuses attention on this half reaction:

$$O_2(g) + 4\,H^+(aq) + 4\,e^- \rightarrow 2\,H_2O(l) \qquad E° = +1.23\ V$$

Body conditions are far from standard state values, so the actual value of E would be reduced by about 0.5 V if we take pH, p_{O_2} and T into account. However, we are only estimating an average current to one significant digit, so $E = 1.23 \pm 0.5\ V \approx 1\ V$ is adequate. With these approximations in mind, we can calculate the current.

$$It = nF = \frac{\Delta G}{-E} \qquad \text{or}$$

$$I = \frac{\Delta G}{-Et} = \frac{(-10 \times 10^6\ J)}{-(1\ V)(24\ h)(3600\ s \cdot h^{-1})} \cdot \frac{1\ V \cdot C}{1\ J} = 115\ A \approx 100\ A$$

13.73 (a) The electrolyte is KOH(aq)/HgO(s), which will have the consistency of a moist paste.

(b) The oxidizing agent is HgO(s).

(c) $HgO(s) + Zn(s) \rightarrow Hg(l) + ZnO(s)$

13.75 (a)
$$Pb_{(s)} + HSO^-_{4(aq)} \rightarrow PbSO_4(s) + H^+_{(aq)} + 2e^-$$

$$\underline{PbO_2(s) + 3\,H^+_{(aq)} + HSO_4^-_{(aq)} + 2e^- \rightarrow PbSO_4(s) + 2\,H_2O(l)}$$

$$Pb(s) + PbO_2(s) + 2\,HSO_4^-_{(aq)} + 2H^+_{(aq)} \rightarrow 2PbSO_4(s) + 2\,H_2O(l)$$

(b) pH increases

Amount of PbO_2(s) decreases.

Total amount of lead in the battery stays the same.

13.77 See Table 13.1   (a) KOH(aq)   (b) In the charging process, the cell reaction is the reverse of what occurs in discharge. Therefore, at the anode, $2\,Ni(OH)_2(s) + 2\,OH^-(aq) \rightarrow 2\,Ni(OH)_3 + 2\,e^-$.

13.79 $Fe^{3+}(aq) + 3\,e^- \rightarrow Fe(s)$ $E° = -0.04$ V

$Cr^{3+}(aq) + 3\,e^- \rightarrow Cr(s)$ $E° = -0.74$ V

$Fe^{2+}(aq) + 2\,e^- \rightarrow Fe(s)$ $E° = -0.44$ V

$Cr^{2+}(aq) + 2\,e^- \rightarrow Cr(s)$ $E° = -0.91$ V

Comparison of the reduction potentials shows that Cr is more easily oxidized than Fe, so the presence of Cr retards the rusting of Fe. At the position of the scratch, the gap is filled with oxidation products of Cr, thereby preventing contact of air and water with the iron.

13.81 (a) $n_{e^-} = n_{Ag^+} = \dfrac{It}{F} = \dfrac{(3.5\text{ A})(395.0\text{ s})}{(96\ 485\text{ C} \cdot \text{mol}^{-1})} = 1.43 \times 10^{-2}$ mol Ag

1.43×10^{-2} mol Ag $\left(\dfrac{107.87\text{ g Ag}}{\text{mol Ag}} \right) = 1.55$ g Ag

$\dfrac{1.55\text{ g}}{2.69\text{ g}} \times 100 = 57.4\%$ Ag

(b) 2.69 g-1.55 g=1.14 g X

Since the salt is 1:1 Ag:X, the molar mass of X is

$\dfrac{1.14\text{ g}}{1.43 \times 10^{-2}\text{ mol}} = 79.7$ g $\cdot$ mol^{-1}

This molar mass is closest to bromine, so the formula is AgBr.

13.83 (a) $Fe_2O_3 \cdot H_2O$ (b) H_2O and O_2 jointly oxidize iron.   (c) Water is more highly conducting if it contains dissolved ions, so the rate of rusting is increased.

13.85 (a) aluminum or magnesium; both are below titanium in the electrochemical series.

(b) cost, availability, and toxicity of products in the environment

(c) $Cu^{2+} + 2\,e^- \rightarrow Cu(s)$ $E° = +0.34$ V

$$Cu^+ + e^- \rightarrow Cu(s) \qquad E° = +0.52 \text{ V}$$

$$Fe^{3+} + 3 e^- \rightarrow Fe(s) \qquad E° = -0.04 \text{ V}$$

$$Fe^{2+} + 2 e^- \rightarrow Fe(s) \qquad E° = -0.44 \text{ V}$$

Fe could act as the anode of an electrochemical cell if Cu^{2+} or Cu^+ were present; therefore, it could be oxidized at the point of contact. Water with dissolved ions would act as the electrolyte.

13.87 (a) E

       (b)   both

       (c)   neither

       (d) $E°$

       (e) E

13.89   Al, Zn, Fe, Co, Ni, Cu, Ag, Au

13.91 $2[Zn^{2+}(aq) + 2 e^- \rightarrow Zn(s)] \qquad E°(\text{cathode}) = -0.76 \text{ V}$

$$M(s) \rightarrow M^{4+}(aq) + 4 e^- \qquad\qquad E°(\text{anode}) = x$$

$$M(s) + 2 Zn^{2+}(aq) \rightarrow 2 Zn(s) + M^{4+}(aq) \quad E°_{cell} = 0.16 \text{ V}$$

$$E°_{cell} = E°(\text{cathode}) - E°(\text{anode})$$

$$+0.16 \text{ V} = -0.76 \text{ V} - (x)$$

$$x = -0.92 \text{ V} = E°(M^{4+}/M)$$

13.93   The strategy is to find the $E°$ value for the solubility reaction and then find appropriate half-reactions that add to give that solubility reaction. One of these half-reactions is our unknown, the other is obtained from Appendix 2B:

$$Cu(IO_3)_2(s) + 2 e^- \rightarrow Cu(s) + 2 IO_3^-(aq) \qquad E° = ? \text{ (A)}$$

$$Cu(s) \qquad \rightarrow Cu^{2+}(aq) + 2 e^- \qquad\qquad\qquad E° = -0.34 \text{ V} \quad \text{(B)}$$

$$Cu(IO_3)_2 \, (s) + \qquad \rightarrow Cu^{2+} \, (aq) + 2 \, IO_3^- \, (aq) \qquad E^\circ = \frac{RT \ln K_{sp}}{nF} \quad (C)$$

$$E^\circ = \frac{RT \ln K_{sp}}{nF}$$

$$= \frac{(8.314 \, J \cdot K^{-1} \cdot mol^{-1})(298.2 \, K) \ln (1.4 \times 10^{-7})}{2(9.65 \times 10^4 \, C \cdot mol^{-1})}$$

$$= -0.20 \, V$$

$$-0.20 \, V = E^\circ(A) + (-0.34 \, V)$$

$$E^\circ(A) = +0.14 \, V$$

13.95 A negatively charged electrolyte flows from the cathode to the anode.

13.97 (a) In acidic solution, the relevant reactions are

$$O_2 + 4 \, H^+ + 4 \, e^- \rightarrow 2 \, H_2O \qquad\qquad E^\circ = +1.23 \, V$$

$$Ag \rightarrow Ag^+ + e^- \qquad\qquad E^\circ = -0.80 \, V$$

Overall reaction:

$$O_2(g) + 4 \, H^+(aq) + 4 \, Ag(s) \rightarrow 4 \, Ag^+(aq) + 2 \, H_2O(l) \qquad E^\circ = +0.43 \, V$$

Because the potential is positive, the reaction should be spontaneous and would be expected to occur. We should also consider the conditions; because air is only 20.95% O_2, the potential may be different from that calculated for standard conditions. If air is the source of oxygen, then it will be present at $0.2095 \times 1.013\,25 \, bar = 0.2123 \, bar$.

$$E = E^\circ - \frac{0.0592}{4} \log \frac{[Ag^+]^4}{P_{O_2}[H^+]^4}$$

$$= +0.43 \, V - \frac{0.0592}{4} \log \frac{[1.0]^4}{(0.2123)[1.0]^4}$$

$$= +0.43 \, V - \frac{0.0592}{4} \log \frac{1}{0.2123}$$

$$= +0.43 \, V - 0.010 \, V$$

$$= +0.42 \, V$$

The potential is still positive and the reaction is expected to be spontaneous.

(b) In basic solution, the relevant reactions are

$$O_2 + 2\,H_2O + 4\,e^- \rightarrow 4\,OH^- \qquad\qquad E° = +0.40\,\text{V}$$

$$Ag \rightarrow Ag^+ + e^- \qquad\qquad E° = -0.80\,\text{V}$$

Overall reaction:

$$O_2(g) + 2\,H_2O(l) + 4\,Ag(s) \rightarrow 4\,Ag^+(aq) + 4\,OH^-(aq)\ E° = -0.40\,\text{V}$$

This process as written is nonspontaneous and is not predicted to occur. However, AgOH forms an insoluble precipitate, changing the nature of the reaction. The K_{sp} value for AgOH is 1.5×10^{-8}. We use the Nernst equation to calculate the potential under these conditions:

$$E = E° - \frac{0.0592}{4}\log\frac{[Ag^+]^4\,[OH^-]^4}{P_{O_2}}$$

$$= -0.40\,\text{V} - \frac{0.0592}{4}\log\frac{K_{sp}{}^4}{P_{O_2}}$$

$$= -0.40\,\text{V} - \frac{0.0592}{4}\log\frac{(1.5\times10^{-8})^4}{0.2132}$$

$$= -0.40\,\text{V} + 0.45\,\text{V}$$

$$= +0.05\,\text{V}$$

Under these conditions, the potential is slightly positive and the oxidation should be spontaneous.

13.99 (a) Reduction takes place at the electrode with the higher concentration, which would be the chromium electrode in contact with the 1.0 M CrCl$_3$.

    (b)   Yes   (c)   Yes   (d)   No

13.101 (a) $M_{Ag^+}V_{Ag^+} = M_{I^-}V_{I^-}$

$$M_{Ag^+} = \frac{M_{I^-}V_{I^-}}{V_{Ag^+}} = \frac{(0.015\ \text{M})(16.7\ \text{mL})}{(25.0\ \text{mL})}$$

$$= 1.0 \times 10^{-2}\ \text{M}$$

(b) We can find $[Ag^+]$ by using the Nernst equation appropriately,

$E = E° - \left(\dfrac{0.025\,693\ \text{V}}{n} \right) \ln \left(\dfrac{1}{[\text{Ag}^+]} \right)$. Since the standard reduction

potential of silver(I) is +0.80 V, it will be the reduction half reaction

versus the S.H.E., so $[\text{Ag}^+]$ appears in the denominator of Q. In addition,

$n = 1$ and $E° = 0.080\ \text{V}$.

$$0.325\ \text{V} = 0.80\ \text{V} - \left(\dfrac{0.025\,693\ \text{V}}{1} \right) \ln \left(\dfrac{1}{[\text{Ag}^+]} \right)$$

$$-0.475\ \text{V} = (-2.567 \times 10^{-2}\ \text{V})(\ln 1 - \ln[\text{Ag}^+])$$

$$-18.50 = \ln[\text{Ag}^+]$$

$$[\text{Ag}^+] = 9.23 \times 10^{-9}\ \text{M}$$

Recalling that $K_{sp} = [\text{Ag}^+][\text{I}^-]$, and assuming $[\text{Ag}^+] = [\text{I}^-]$ at the

stoichiometric point of the titration,

$$K_{sp} = [\text{Ag}^+][\text{I}^-] = (9.23 \times 10^{-9})^2 = 8.5 \times 10^{-17}$$

13.103 $F_2(g) + 2\,e^- \rightarrow 2\,F^-(aq)\quad E°(\text{cathode}) = +2.87\ \text{V}$

$2\,HF(aq) \rightarrow F_2(g) + 2\,H^+(aq) + 2\,e^-\quad E°(\text{anode}) = +3.03\ \text{V}$

$2\,HF(aq) \rightarrow 2\,H^+(aq) + 2\,F^-(aq)$

$E°_{\text{cell}} = E°(\text{cathode}) - E°(\text{anode}) = +2.87\ \text{V} - (+3.03\ \text{V}) = -0.16\ \text{V}$

For the above reaction, $K = \dfrac{[\text{H}^+]^2[\text{F}^-]^2}{[\text{HF}]^2}$ and $\ln K = \dfrac{nFE°}{RT}$

at $25°\text{C} = \dfrac{nE°}{0.025\,69\ \text{V}} = \dfrac{(2)(-0.16\ \text{V})}{0.025\,69\ \text{V}} = -12$

$K = 10^{-5}$

$K_a = \sqrt{K} = \sqrt{10^{-5}} = 10^{-3}$

13.105(a) $\Delta G_{r1}° = -nFE_1° = \Delta H_{r1}° - T_1 \Delta S_{r1}°$ where r_1 represents the reaction at T_1.

$\Delta G_{r2}° = -nFE_2° = \Delta H_{r2}° - T_2 \Delta S_{r2}°$ where r_2 represents the reaction at T_2.

On subtracting the first reaction from the second, we obtain

$-nFE_2^\circ + nFE_1^\circ = \Delta H_{r2}^\circ - T_2\Delta S_{r2}^\circ - [\Delta H_{r1}^\circ - T_1\Delta S_{r1}^\circ].$

Since $\Delta H_{r1}^\circ = \Delta H_{r2}^\circ$ we obtain,

$nFE_1^\circ - nFE_2^\circ = - T_2\Delta S_{r2}^\circ + T_1\Delta S_{r1}^\circ$

which can be rewritten as $-nFE_1^\circ + nFE_2^\circ = + T_2\Delta S_{r2}^\circ - T_1\Delta S_{r1}^\circ.$

Since $\Delta S_{r1}^\circ = \Delta S_{r2}^\circ = \Delta S_r^\circ$ we obtain,

$-nFE_1^\circ + nFE_2^\circ = \Delta S_r^\circ(T_2 - T_1)$

$nFE_2^\circ = nFE_1^\circ + \Delta S_r^\circ(T_2 - T_1)$

$E_2^\circ = E_1^\circ + \Delta S_r^\circ(T_2 - T_1)/nF$

(b) The redox reaction is $2H_2(g) + O_2(g) \rightarrow 2H_2O(l)$.

Using the standard potentials at 25°C to calculate E_1°,

$O_2(g) + 4H^+(aq) + 4e^- \rightarrow 2H_2O(l)$	$E^\circ = +1.23$ V
$2H_2(g) \rightarrow 4H^+(aq) + 4e^-$	$E^\circ = 0$ V
$2H_2(g) + O_2(g) \rightarrow 2H_2O(l)$	$E_1^\circ = +1.23$ V

$\Delta S_r^\circ = \sum S_{m,\text{ products}}^\circ - \sum S_{m,\text{ reactants}}^\circ$

$= 2(69.91 \text{ J.K}^{-1}.\text{mol}^{-1}) - [2(130.68 \text{ J.K}^{-1}.\text{mol}^{-1}) + (205.14 \text{ J.K}^{-1}.\text{mol}^{-1})]$

$= 139.82 \text{ J.K}^{-1}.\text{mol}^{-1} - 466.50 \text{ J.K}^{-1}.\text{mol}^{-1}$

$= -326.68 \text{ J.K}^{-1}.\text{mol}^{-1}$

$E_{r,\, 80°C}^\circ = E_{r,\, 25°C}^\circ + \Delta E_r^\circ$

$\qquad = E_{r,\, 25°C}^\circ + \dfrac{\Delta S_r^\circ}{nF}\Delta T$

$\qquad = +1.23 \text{ V} + \dfrac{-326.68 \text{ J.K}^{-1}.\text{mol}^{-1}}{4(9.6485 \times 10^4 \text{ J.V}^{-1}.\text{mol}^{-1})}(353.15 \text{ K} - 298.15 \text{ K})$

$\qquad = +1.23 \text{ V} + \dfrac{-326.68 \text{ J.K}^{-1}.\text{mol}^{-1}}{3.8594 \times 10^5 \text{ J.V}^{-1}.\text{mol}^{-1}}(55.0 \text{ K})$

$\qquad = +1.23 \text{ V} - 4.65 \times 10^{-2} \text{ V}$

$\qquad = +1.18 \text{ V}$

13.107 The wording of this exercise suggests that K^+ ions participate in an electrolyte concentration cell reaction. Therefore, $E^\circ_{cell} = 0.00$ V, because the two half cells would be identical under standard conditions.

Then,

$$E = E^\circ - \left(\frac{0.0257\ V}{n}\right) \ln\left(\frac{[K_{out}^+]}{[K_{in}^+]}\right) = 0.00\ V - \left(\frac{0.0257\ V}{1}\right)\ln\left(\frac{1}{30}\right)$$

$$= +0.09\ V$$

and $E = 0.00\ V - \left(\frac{0.0257\ V}{1}\right)\ln\left(\frac{1}{20}\right) = +0.08\ V$

The range of potentials is 0.08 V to 0.09 V.

13.109 (a) $Ag^+ + e^- \longrightarrow Ag \qquad\qquad E^\circ = +0.80$ V

$I_2 + 2\,e^- \longrightarrow 2\,I^- \qquad\qquad\qquad E^\circ = +0.54$ V

For $\quad 2\,Ag^+ + 2\,I^- \longrightarrow Ag + I_2,\ \ E^\circ = +0.80\ V - 0.54\ V = +0.26$ V

The process should be spontaneous.

(b) The formation of AgI precipitate means that the concentration of Ag^+ ions is never high enough to achieve the conditions necessary for the redox reaction to take place. The solubility product K_{sp} limits the concentrations in solution, so that the actual redox potential is not the value calculated, which represents the values when $[Ag^+] = 1\,M$ and $[I^-] = 1\,M.$ If we use the concentrations established by the solubility equilibrium and the Nernst equation, we can calculate the actual redox potential:

$$E = E^\circ - \frac{RT}{nF}\ln Q$$

where, in this case, $\quad Q = \dfrac{1}{K_{sp}^2}\quad$ for the reaction as written

$K_{sp} = 1.5 \times 10^{-16}$ for AgI

$$E = +0.26 \text{ V} - \frac{(8.314 \text{ J} \cdot \text{K}^{-1} \cdot \text{mol}^{-1})(298 \text{ K})}{(2)(96\ 485 \text{ J} \cdot \text{V}^{-1} \cdot \text{mol}^{-1})} \ln\left(\frac{1}{(1.5 \times 10^{-16})^2}\right)$$

$$= +0.26 \text{ V} - \frac{(8.314 \text{ J} \cdot \text{K}^{-1} \cdot \text{mol}^{-1})(298 \text{ K})}{(96\ 485 \text{ J} \cdot \text{V}^{-1} \cdot \text{mol}^{-1})} \ln\left(\frac{1}{(1.5 \times 10^{-16})}\right)$$

$$= +0.26 \text{ V} - 0.94 \text{ V}$$

$$= -0.68 \text{ V}$$

The fact that the concentrations of Ag^+ and I^- are limited in solution means that the redox potential for a spontaneous reaction is never achieved.

13.111 Buffer system. $HA \rightarrow H^+ + A^-$

$$Q = \frac{(H^+)(A^-)}{(HA)}$$

Note: (H^+), as opposed to $[H^+]$, indicates a nonequilibrium molarity.

Because in a buffer system $(A) \approx (HA)$, we can write

$$Q = (H^+)$$

$$E_{cell} = E°_{cell} - \frac{RT}{nF} \ln (H^+)$$

$$0.060 \text{ V} = E°_{cell} - \left(\frac{0.025\ 693}{1}\right)(2.303)(\log(H^+))$$

Because $\log(H^+) = -[-\log(H^+)] = -\text{pH}$, we have

$$0.060 \text{ V} = E°_{cell} - 0.0592 \times (-\text{pH})$$

$$0.060 \text{ V} = E°_{cell} + 0.0592 \times \text{pH}$$

$$0.060 \text{ V} = E°_{cell} + 0.0592 \times 9.40$$

$$0.060 \text{ V} = E°_{cell} + 0.556 \text{ V}$$

$$E° = 0.060 \text{ V} - 0.556 \text{ V} = -0.496 \text{ V}$$

Similarly, $0.22 \text{ V} = -0.496 \text{ V} + 0.0592 \text{ V} \times \text{pH}$

$$\text{pH} = \frac{0.22 \text{ V} + 0.496 \text{ V}}{0.0592 \text{ V}} = 12$$

13.113 Using $\Delta G^\circ = -nFE^\circ$ and $\Delta G^\circ = -RT \ln K$, one obtains the relationship

$E^\circ = \dfrac{RT}{nF} \ln K$.

Must have $E^\circ = 0$ when pH $= 7$.

pH $= 7$ when concentration of H^+ and OH^- are equal to 1.0×10^{-7} mol.L^{-1}.

When $K = 1$, then $E^\circ = 0$.

Therefore at pH $= 7$: $\quad K = \dfrac{[H^+]}{[OH^-]} = \dfrac{1.0 \times 10^{-7} \text{ mol.L}^{-1}}{1.0 \times 10^{-7} \text{ mol.L}^{-1}} = 1 \quad$ and $\quad E^\circ = 0$

At pH $= 1$: $E^\circ = \dfrac{RT}{nF} \ln K = 0.025693 \text{ V } \ln \dfrac{1.0 \text{ mol.L}^{-1}}{1.0 \times 10^{-14} \text{ mol.L}^{-1}} = +0.828$ V

At pH $= 14$: $E^\circ = \dfrac{RT}{nF} \ln K = 0.025693 \text{ V } \ln \dfrac{1.0 \times 10^{-14} \text{ mol.L}^{-1}}{1.0 \text{ mol.L}^{-1}} = -0.828$V

13.115 (a) The more dilute electrolyte in a concentration cell is always the anode. The anode generates the electrons (in this case $Ag(s) \rightarrow Ag^+(aq) + 1e-$) and the concentration of $Ag^+(aq)$ increases.

In all chemical reactions the species that increases in concentration is the product. Therefore in the Nernst equation, for a concentration cell, the more dilute electrolyte is always the product.

$E = E^\circ - \dfrac{RT}{nF} \ln \dfrac{[Ag^+]_{\text{anode}}}{[Ag^+]_{\text{cathode}}}$

A plot of cell voltage, E, versus $\ln [Ag^+]_{\text{anode}}$ would be a linear increase with positive slope because a larger difference in concentration results in a larger cell voltage. In a concentration cell the larger the concentration difference between the anode and cathode the higher the cell voltage. In other words, a decreasing value for the numerator, $[Ag^+]_{\text{anode}}$, makes the ln term increasingly negative and therefore the term, $-\dfrac{RT}{nF} \ln \dfrac{[Ag^+]_{\text{anode}}}{[Ag^+]_{\text{cathode}}}$,

increasingly positive.

(b) $E = E^\circ - \dfrac{RT}{nF} \ln Q = E^\circ - \dfrac{0.025693 \text{ V}}{n} \ln Q$

For this redox reaction $n = 1$.

The slope is 0.025693 V, and this value corresponds to the terms $\dfrac{RT}{nF}$.

What is being done here is a rearrangement of the Nernst equation to

$$E = -\frac{0.025693\text{ V}}{n} \ln Q + E°,$$ resulting in a straight line plot of the form

$y = mx + c$ where the slope, m, is $\dfrac{RT}{nF}$ and the y-intercept, c, is $E°$.

(c) y-intercept is $E°$, and for all concentration cells $E° = 0$.

(d) Provided we know the values for R, T and n, the answer is yes since,

$$\frac{RT}{nF} = 0.025693\text{ V}.$$

13.117 The strategy for working this problem is to create a set of equations that will add to the desired equilibrium reaction:

$$HBrO(aq) + H_2O(l) \rightarrow H_3O^+(aq) + BrO^-(aq)$$

From Appendix 2B, we find

$$2\text{ HBrO} + 2\text{ H}^+ + 2\text{ e}^- \rightarrow Br_2 + 2\text{ H}_2O \qquad E° = +1.60\text{ V}$$

$$BrO^- + H_2O + 2\text{ e}^- \rightarrow Br^- + 2\text{ OH}^- \qquad E° = +0.76$$

On examination of these equations, it is clear that we will also need a half-reaction that, when combined with the two above, will eliminate

Br_2 and Br^-. The obvious choice is

$$Br_2 + 2\text{ e}^- \rightarrow 2\text{ Br}^- \qquad E° = +1.09$$

We combine these by adding twice the reverse reaction to the other two:

$$2\text{ HBrO} + 2\text{ H}^+ + 2\text{ e}^- \rightarrow Br_2 + 2\text{ H}_2O$$
$$2(Br^- + 2\text{ OH}^- \rightarrow BrO^- + H_2O + 2\text{ e}^-)$$
$$Br_2 + 2\text{ e}^- \rightarrow 2\text{ Br}^-$$

$$2\text{ HBrO} + 2\text{ H}^+ + 4\text{ OH}^- \rightarrow 2\text{ BrO}^- + 4\text{ H}_2O$$

Caution: We must be careful here in adding the $E°$ values—we have created essentially a new half-reaction by summing these reactions, which requires that we convert to ΔG values. Whenever one sums more than two half-reactions, it is necessary to convert to the ΔG values using $\Delta G° = nFE°$, in order to work the problem:

$$2\,HBrO + 2\,H^+ + 2\,e^- \rightarrow Br_2 + 2\,H_2O$$
$$\Delta G° = -2(9.65 \times 10^4\ C \cdot mol^{-1})(+1.60\ V) = -309\ kJ \cdot mol^{-1}$$

$$BrO^- + H_2O + 2\,e^- \rightarrow Br^- + 2\,OH^-$$
$$\Delta G° = -2(9.65 \times 10^4\ C \cdot mol^{-1})(+0.76\ V) = -147\ kJ \cdot mol^{-1}$$

$$Br_2^- + 2\,e^- \rightarrow 2\,Br^-$$
$$\Delta G° = -2(9.65 \times 10^4\ C \cdot mol^{-1})(+1.09\ V) = -210\ kJ \cdot mol^{-1}$$

For $2\,HBrO + 2\,H^+ + 4\,OH^- \rightarrow 2\,BrO^- + 4\,H_2O$
$$\Delta G° = -309\ kJ + 2(+147\ kJ) - 210\ kJ = -225\ kJ \cdot mol^{-1}$$

We now see that we will need to eliminate OH^- from the left side of the equation. This can be done in one of two ways: we can use the K_w value for the autoprotolysis of water or, equivalently, we can use appropriate half-reactions that sum to the autoprotolysis of water. The appropriate half-reactions are

$$2\,H_2O \rightarrow O_2 + 4\,H^+ + 4\,e^- \qquad E° = -1.23\ V$$

$$O_2 + 2\,H_2O + 4\,e^- \rightarrow 4\,OH^- \qquad E° = +0.40\ V$$

These sum to give

$$4\,H_2O \rightarrow 4\,H^+ + 4\,OH^- \qquad E° = -0.83\ V$$

This is a $4\,e^-$ reaction. Alternatively, one can write the $1\,e^-$ process that will have the same $E°$ value.

$$H_2O \rightarrow H^+ + OH^- \qquad E° = -0.83\ V$$

$$\Delta G° = -(1)(9.65 \times 10^4\ C \cdot mol^{-1})(-0.83\ V) = +80\ kJ \cdot mol^{-1}$$

$$2\,HBrO + 2\,H^+ + 4\,OH^- \rightarrow 2\,BrO^- + 4\,H_2O$$
$$\Delta G° = -225\ kJ \cdot mol^{-1}$$

$$4(H_2O \rightarrow H^+ + OH^-) \qquad 4(\Delta G° = +80\ kJ \cdot mol^{-1})$$

$$2\,HBrO \rightarrow 2\,H^+ + 2\,BrO^- \qquad \Delta G° = -225\ kJ \cdot mol^{-1} + 4(+80\ kJ \cdot mol^{-1})$$
$$= +95\ kJ \cdot mol^{-1}$$

The desired reaction is half of this, for which $\Delta G° = +48 \text{ kJ} \cdot \text{mol}^{-1}$.

Using $\Delta G° = -RT \ln K$, we obtain $K = 4 \times 10^{-9}$, which is in reasonable agreement for this type of calculation with the value of 2×10^{-9} given in Table 10.1.

13.119 In order to determine the current applied, we need to find the number of moles of electrons transferred. The electrolysis of water to produce gaseous oxygen and hydrogen,

$2 \text{ H}_2\text{O(l)} \rightarrow \text{O}_2(\text{g}) + 2 \text{ H}_2(\text{g})$, transfers 4 moles of electrons for each mole of oxygen gas produced: $4 \text{ OH}^-(\text{aq}) \rightarrow \text{O}_2(\text{g}) + 2 \text{ H}_2\text{O(l)} + 4 \text{ e}^-$. We can determine the number of moles of oxygen from its volume, partial pressure, and temperature.

$$n_{\text{O}_2} = \frac{p_{\text{O}_2}V}{RT} = \frac{(p_{tot} - p_{\text{H}_2\text{O}})V}{RT}$$

$$= \frac{(722 \text{ Torr} - 19.83 \text{ Torr})(25.0 \text{ mL})}{(0.08206 \text{ L} \cdot \text{atm} \cdot \text{K}^{-1} \cdot \text{mol}^{-1})(295 \text{ K})} \cdot \frac{1 \text{ L}}{1000 \text{ mL}} \cdot \frac{1 \text{ atm}}{760 \text{ Torr}}$$

$$= 9.542 \times 10^{-4} \text{ mol O}_2 \text{ produced}$$

$$n_{e^-} = n_{\text{O}_2} \times \frac{4 \text{ mol e}^-}{\text{mol O}_2} = 9.542 \times 10^{-4} \text{ mol O}_2 \times \frac{4 \text{ mol e}^-}{\text{mol O}_2}$$

$$= 3.817 \times 10^{-3} \text{ mol e}^-$$

$It = nF$

$$I = \frac{nF}{t} = \frac{(3.817 \times 10^{-3} \text{ mol e}^-)}{(30.0 \text{ min})(60 \text{ s} \cdot \text{min}^{-1})} \cdot \left(96 \text{ }485 \frac{\text{C}}{\text{mol e}^-}\right)$$

$$= 0.205 \text{ A}$$

13.121 (a) (i) $2\text{H}^+(\text{aq}) + 2\text{e}^- \rightarrow \text{H}_2(\text{g})$ $\qquad E° = 0$

$$E^\oplus = E° - \frac{RT}{nF} \ln Q$$

$$E^\oplus = E° - \frac{0.025693 \text{ V}}{2} \ln \frac{[1]}{\left[10^{-7}\right]^2}$$

$E^{\oplus} = 0 - 0.41 \text{ V} = \text{-0.41 V}$

(ii) $NO_3^-(aq) + 4H^+(aq) + 3e^- \rightarrow NO(g) + 2H_2O \qquad E° = +0.96 \, V$

$$E^{\oplus} = E° - \frac{RT}{nF} \ln Q$$

$$E^{\oplus} = E° - \frac{0.025693 \, \text{V}}{3} \ln \frac{[1][1]^2}{[1]\left[10^{-7}\right]^4}$$

$E^{\oplus} = +0.96 \text{ V} - 0.55 \text{ V} = +0.41 \text{ V}$

(b) $NAD^+(aq) + H^+(aq) + 2e^- \rightarrow NADH(aq) \qquad E° = \text{-0.099 V}$

$$E^{\oplus} = E° - \frac{RT}{nF} \ln Q$$

$$E^{\oplus} = E° - \frac{0.025693 \, \text{V}}{n} \ln \frac{\left[NADH\right]}{\left[NAD^+\right]\left[H^+\right]}$$

$$E^{\oplus} = E° - \frac{0.025693 \, \text{V}}{2} \ln \frac{[1]}{[1]\left[10^{-7}\right]}$$

$$E^{\oplus} = -0.099V - \frac{0.025693 \, \text{V}}{2} \ln 10^7$$

$E^{\oplus} = \text{-0.099 V} - 0.207 \text{ V} = \text{-0.31 V}$

(c)

(d)

(e)

$$\text{NADH(aq)} \rightarrow \text{NAD}^+(\text{aq}) + \text{H}^+(\text{aq}) + 2e^- \qquad\qquad E^\oplus = +0.31\,\text{V}$$

$$\text{pyruvate} + 2\text{H}^+(\text{aq}) + 2\text{e}^- \rightarrow \text{lactate} \qquad\qquad E^\oplus = -0.190\,\text{V}$$

$$\text{NADH(aq)} + \text{pyruvate} + \text{H}^+(\text{aq}) \rightarrow \text{NAD}^+(\text{aq}) + \text{lactate} \quad E^\oplus = +0.12\,\text{V}$$

$$E^\oplus = E^\circ - \frac{RT}{nF}\ln Q$$

$$E^\circ = E^\oplus + \frac{RT}{nF}\ln Q$$

$$E^\circ = E^\oplus + \frac{0.025693\,\text{V}}{n}\ln\frac{\big[\text{lactate}\big]\big[\text{NAD}^+(\text{aq})\big]}{\big[\text{NADH(aq)}\big]\big[\text{pyruvate}\big]\big[\text{H}^+(\text{aq})\big]}$$

$$E^\circ = +0.12\,\text{V} + \frac{0.025693\,\text{V}}{2}\ln\frac{[1][1]}{[1][1]\big[10^{-7}\big]}$$

$$E^\circ = +0.12\,\text{V} + 0.0128465\,\text{V}\,\ln 10^7$$

$$E^\circ = +0.12\,\text{V} + 0.207\,\text{V}$$

$$E^\circ = +0.33\,\text{V}$$

(f) $\Delta G_r^\circ = -nFE^\circ$

$$\Delta G_r^\circ = -2(9.64853\times10^4\,\text{C.mol}^{-1})(+0.33\,\text{V})$$

$$\Delta G_r^\circ = -63.7\,\text{kJ.mol}^{-1} = -64\,\text{kJ.mol}^{-1}$$

(g) $\Delta G_r^\circ = -RT\ln K$ and $\quad \Delta G_r^\circ = -nFE^\circ$

Therefore, $\quad \ln K = \dfrac{nFE^\circ}{RT}$

$$K = e^{\frac{nFE^\circ}{RT}}$$

$$K = e\Big(\frac{(+63700\,\text{J.mol}^{-1})}{(8.314\,\text{J.K}^{-1}.\text{mol}^{-1})(298.15\,\text{K})}\Big)$$

$$K = e^{25.69}$$

$$K = +1.43\times10^{11}$$

CHAPTER 14

CHEMICAL KINETICS

14.1 (a) $\text{rate}(N_2) = \text{rate}(H_2) \times \left(\dfrac{1 \text{ mol } N_2}{3 \text{ mol } H_2} \right) = \dfrac{1}{3} \times \text{rate}(H_2)$

(b) $\text{rate}(NH_3) = \text{rate}(H_2) \times \left(\dfrac{2 \text{ mol } NH_3}{3 \text{ mol } H_2} \right) = \dfrac{2}{3} \times \text{rate}(H_2)$

(c) $\text{rate}(NH_3) = \text{rate}(N_2) \times \left(\dfrac{2 \text{ mol } NH_3}{1 \text{ mol } N_2} \right) = 2 \times \text{rate}(N_2)$

14.3 (a) The rate of formation of dichromate ions =

$$\left(\frac{0.14 \text{ mol } Cr_2O_7^{2-}}{L \cdot s} \right) \left(\frac{2 \text{ mol } CrO_4^{2-}}{1 \text{ mol } Cr_2O_7^{2-}} \right) = 0.28 \text{ mol} \cdot L \cdot s^{-1}$$

(b) $0.14 \text{ mol} \cdot L^{-1} \cdot s^{-1} \div 1 = 0.14 \text{ mol} \cdot L^{-1} \cdot s^{-1}$

14.5 (a) $\text{rate of formation of } O_2 = \left(6.5 \times 10^{-3} \dfrac{\text{mol } NO_2}{L \cdot s} \right) \times \left(\dfrac{1 \text{ mol } O_2}{2 \text{ mol } NO_2} \right)$

$$= 3.3 \times 10^{-3} \text{ (mol } O_2) \cdot L^{-1} \cdot s^{-1}$$

(b) $6.5 \times 10^{-3} \text{ mol} \cdot L^{-1} \cdot s^{-1} \div 2 = 3.3 \times 10^{-3} \text{ mol} \cdot L^{-1} \cdot s^{-1}$

14.7 (a) and (c)

Note that the curves for the $[I_2]$ and $[H_2]$ are identical and only the $[I_2]$ curve is shown.

(b) The rates at individual points are given by the slopes of the lines tangent to the points in question. If these are determined graphically, there may be some variation from the numbers given below.

time, s	rate, $\text{mmol} \cdot L^{-1} \cdot s^{-1}$
0	20.0060
1000	20.003
2000	20.000 98
3000	20.000 61
4000	20.000 40
5000	20.000 31

14.9 For A $\longrightarrow$ products, rate $= (\text{mol A}) \cdot L^{-1} \cdot s^{-1}$

(a) $[(\text{mol A}) \cdot L^{-1} \cdot s^{-1}] = k_0 [A]^0 = k_0$, so units of k_0 are $(\text{mol A}) \cdot L^{-1} \cdot s^{-1}$ (same as the units for the rate, in this case)

(b) $[(\text{mol A}) \cdot L^{-1} \cdot s^{-1}] = k_1 [A]$, so units of k_1 are $\dfrac{(\text{mol A}) \cdot L^{-1} \cdot s^{-1}}{(\text{mol A}) \cdot L^{-1}} = s^{-1}$

(c) $[(\text{mol A}) \cdot L^{-1} \cdot s^{-1}] = k_1 [A]^2$, so units of k_1 are $\dfrac{(\text{mol A}) \cdot L^{-1} \cdot s^{-1}}{\left[(\text{mol A}) \cdot L^{-1} \right]^2}$

$= L \cdot (\text{mol A})^{-1} \cdot s^{-1}$

14.11 From the units of the rate constant, k, it follows that the reaction is first order, thus rate $= k[N_2O_5]$.

$$[N_2O_5] = \left(\frac{3.45 \text{ g } N_2O_5}{0.750 \text{ L}} \right) \left(\frac{1 \text{ mol } N_2O_5}{108.02 \text{ g } N_2O_5} \right) = 0.0426 \text{ mol} \cdot L^{-1}$$

rate $= 5.2 \times 10^{-3} \text{ s}^{-1} \times 0.0426 \text{ mol} \cdot L^{-1} = 2.2 \times 10^{-4} \text{ (mol } N_2O_5) \cdot L^{-1} \cdot s^{-1}$

14.13 (a) From the units of the rate constant, it follows that the reaction is second order; therefore, $\text{rate} = k[\text{H}_2][\text{I}_2]$

$$= (0.063 \text{ L} \cdot \text{mol}^{-1} \cdot \text{s}^{-1}) \left(\frac{0.52 \text{ g H}_2}{0.750 \text{ L}} \right) \left(\frac{1 \text{ mol H}_2}{2.016 \text{ g H}_2} \right) \left(\frac{0.19 \text{ g I}_2}{0.750 \text{ L}} \right) \left(\frac{1 \text{ mol I}_2}{253.8 \text{ g I}_2} \right)$$

$$= 2.2 \times 10^{-5} \text{ mol} \cdot \text{L}^{-1} \cdot \text{s}^{-1}$$

(b) $\text{rate(new)} = k \times 2 \times [\text{H}_2]_{\text{initial}}[\text{I}_2] = 2 \times \text{rate(initial)}$, so, by a factor of 2

14.15 Because the rate increased in direct proportion to the concentrations of both reactants, the rate is first order in both reactants.

$$\text{rate} = k[\text{CH}_3\text{Br}][\text{OH}^-]$$

14.17 (a) Use experiments 1 and 4 to show that [C] is independent of the rate. Use experiments 2 and 4 to solve for the order with respect to A:

$$\frac{20^a}{10} = \frac{4}{2} \qquad\qquad 2^a = 2 \qquad\qquad \text{Therefore } a = 1.$$

Use experiments 2 and 3 to solve for the order with respect to B:

$$\frac{200^b}{100} = \frac{16}{4} \qquad\qquad 2^b = 4 \qquad\qquad \text{Therefore } b = 2. \qquad \text{Overall order} = 3$$

(b) $\text{rate} = k[\text{A}][\text{B}]^2$

(c) Using experiment 1:

$$2.0 \text{ mmol L}^{-1} \text{ s}^{-1} = k(10 \text{ mmol L}^{-1}) (100 \text{ mmol L}^{-1})^2$$

$$k = 2.0 \times 10^{-5} \text{ L}^2 \text{ mmol}^{-2} \text{ s}^{-1}$$

(d) $\text{rate} = k[\text{A}][\text{B}]^2$

$$= (2.0 \times 10^{-5} \text{ L}^2 \text{ mmol}^{-2} \text{ s}^{-1})(4.62 \text{ mmol L}^{-1})(0.177 \text{ mmol L}^{-1})^2$$

$$= 2.9 \times 10^{-6} \text{ mmol L}^{-1} \text{ s}^{-1}$$

14.19 (a) Doubling the concentration of A (experiments 1 and 2) doubles the rate; therefore, the reaction is first order in A. Increasing the concentration of B by the ratio 3.02/1.25 (experiments 2 and 3) increases the rate by $(3.02/1.25)^2$; hence, the reaction is second order in B. Tripling the concentration of C (experiments 3 and 4) increases the rate by $3^2 = 9$; thus, the reaction is second order in C. Therefore, rate $= k[A][B]^2[C]^2$.

(b) overall order $= 5$

(c) $k = \dfrac{\text{rate}}{[A][B]^2[C]^2}$

Using the data from experiment 4, we get

$$k = \left(\frac{0.457\ \text{mol}}{L \cdot s}\right)\left(\frac{L}{1.25 \times 10^{-3}\ \text{mol}}\right)\left(\frac{L}{3.02 \times 10^{-3}\ \text{mol}}\right)^2 \left(\frac{L}{3.75 \times 10^{-3}\ \text{mol}}\right)^2$$

$$= 2.85 \times 10^{12}\ L^4 \cdot \text{mol}^{-4} \cdot s^{-1}$$

From experiment 3, we get

$$k = \left(\frac{5.08 \times 10^{-2}\ \text{mol}}{L \cdot s}\right)\left(\frac{L}{1.25 \times 10^{-3}\ \text{mol}}\right)\left(\frac{L}{3.02 \times 10^{-3}\ \text{mol}}\right)^2$$

$$\times \left(\frac{L}{1.25 \times 10^{-3}\ \text{mol}}\right)^2$$

$$= 2.85 \times 10^{12}\ L^4 \cdot \text{mol}^{-4} \cdot s^{-1}\ (\text{Checks!})$$

(d) rate $= \left(\dfrac{2.85 \times 10^{12}\ L^4}{\text{mol}^4 \cdot s}\right)\left(\dfrac{3.01 \times 10^{-3}\ \text{mol}}{L}\right)\left(\dfrac{1.00 \times 10^{-3}\ \text{mol}}{L}\right)^2$

$$\times \left(\frac{1.15 \times 10^{-3}\ \text{mol}}{L}\right)^2$$

$$= 1.13 \times 10^{-2}\ \text{mol} \cdot L^{-1} \cdot s^{-1}$$

14.21

$$\ln\frac{[A]_t}{[A]_0} = -kt$$

$$\ln\frac{[A]_t}{[A]_0} = -(7.6 \times 10^{-3}\ \text{min}^{-1})(300\ \text{min})$$

$$\frac{[A]_t}{[A]_0} = e^{-2.28}$$

$[A]_t = 0.10[A]_0$ \hspace{2cm} 10% of the initial drug concentration remains in the body.

Mass of drug remaining in the body, $A_t = (0.10)(20\ \text{mg}) = 2.0\ \text{mg}$

14.23 (a) $k = \dfrac{0.693}{t_{1/2}} = \dfrac{0.693}{1000\ \text{s}} = 6.93 \times 10^{-4}\ \text{s}^{-1}$

(b) We use $\ln\left(\dfrac{[A]_0}{[A]_t}\right) = kt$ and solve for k.

$$k = \frac{\ln\left([A]_0 / [A]_t\right)}{t} = \frac{\ln\left(\dfrac{0.67\ \text{mol}\cdot\text{L}^{-1}}{0.53\ \text{mol}\cdot\text{L}^{-1}}\right)}{25\ \text{s}} = 9.4 \times 10^{-3}\ \text{s}^{-1}$$

(c) $[A]_t = \left(\dfrac{0.153\ \text{mol A}}{\text{L}}\right) - \left[\left(\dfrac{2\ \text{mol A}}{1\ \text{mol B}}\right)\left(\dfrac{0.034\ \text{mol B}}{\text{L}}\right)\right]$

$\qquad = 0.085\ (\text{mol A})\cdot\text{L}^{-1}$

$$k = \frac{\ln\left(\dfrac{0.153\ \text{mol}\cdot\text{L}^{-1}}{0.085\ \text{mol}\cdot\text{L}^{-1}}\right)}{115\ \text{s}} = 5.1 \times 10^{-3}\ \text{s}^{-1}$$

14.25 (a) $t_{1/2} = \dfrac{0.693}{k} = \left(\dfrac{0.693\ \text{s}}{3.7 \times 10^{-5}}\right)\left(\dfrac{1\ \text{min}}{60\ \text{s}}\right)\left(\dfrac{1\ \text{h}}{60\ \text{min}}\right) = 5.2\ \text{h}$

(b) $[A]_t = [A]_0\ e^{-kt}$

$t = 3.5\ \text{h} \times 3600\ \text{s}\cdot\text{h}^{-1} = 1.3 \times 10^{4}\ \text{s}$

$[N_2O_5] = 0.0567\ \text{mol}\cdot\text{L}^{-1} \times e^{-(3.7 \times 10^{-5}\ \text{s}^{-1})(1.3 \times 10^{4}\ \text{s})} = 3.5 \times 10^{-2}\ \text{mol}\cdot\text{L}^{-1}$

(c) Solve for t from $\ln\left(\dfrac{[A]_0}{[A]_t}\right) = kt$, which gives

$$t = \frac{\ln\left(\dfrac{[A]_0}{[A]_t}\right)}{k} = \frac{\ln\left(\dfrac{[N_2O_5]_0}{[N_2O_5]_t}\right)}{k} = \frac{\ln\left(\dfrac{0.0567}{0.0135}\right)}{3.7 \times 10^{-5}\ \text{s}^{-1}} = 3.9 \times 10^{4}\ \text{s}$$

$\qquad = (3.9 \times 10^{4}\ \text{s})\left(\dfrac{1\ \text{min}}{60\ \text{s}}\right) = 6.5 \times 10^{2}\ \text{min}$

14.27 (a) $\dfrac{[A]}{[A]_0} = \dfrac{1}{8} = \left(\dfrac{1}{2}\right)^3$ so the time elapsed is 3 half-lives.

$t = 3 \times 355\ \text{s} = 1065\ \text{s}$

(b) $\dfrac{[A]}{[A]_0} = \dfrac{1}{4} = \left(\dfrac{1}{2}\right)^2$; so the time elapsed is 2 half-lives.

$t = 2 \times 355 \text{ s} = 710 \text{ s}$

(c) Because 15% is not a multiple of $\frac{1}{2}$, we cannot work directly from the half-life. But $k = 0.693/t_{1/2}$

so $k = \dfrac{0.693}{355 \text{ s}} = 1.95 \times 10^{-3} \text{ s}^{-1}$

$t = \dfrac{\ln\left(\dfrac{[A]_0}{[A]_t}\right)}{k} = \dfrac{\ln\left(\dfrac{1}{0.15}\right)}{1.95 \times 10^{-3} \text{ s}^{-1}} = 9.7 \times 10^2 \text{ s}$

(d) $t = \dfrac{\ln\dfrac{[A]_0}{\frac{1}{9}[A]_0}}{k} = \dfrac{\ln 9}{1.95 \times 10^{-3} \text{ s}^{-1}} = 1.1 \times 10^3 \text{ s}$

14.29 (a) We first calculate the concentration of A at 3.0 min.

$[A]_t = [A]_0 - \left(\dfrac{1 \text{ mol A}}{3 \text{ mol B}}\right) \times [B]_t$

$= 0.015 \text{ mol} \cdot \text{L}^{-1} - \left(\dfrac{1 \text{ mol A}}{3 \text{ mol B}}\right) \times 0.018 \text{ (mol B)} \cdot \text{L}^{-1}$

$= 0.009 \text{ mol} \cdot \text{L}^{-1}$

The rate constant is then determined from the first-order integrated rate law.

$k = \dfrac{\ln\left(\dfrac{[A]_0}{[A]_t}\right)}{t} = \dfrac{\ln\left(\dfrac{0.015}{0.009}\right)}{3.0 \text{ min}} = 0.17 \text{ min}^{-1}$

(b) $[A]_t = 0.015 \text{ mol} \cdot \text{L}^{-1} - \left(\dfrac{1 \text{ mol A}}{3 \text{ mol B}}\right) \times 0.030 \text{ (mol B)} \cdot \text{L}^{-1}$

$= 0.005 \text{ mol} \cdot \text{L}^{-1}$

$$t = \frac{\ln\left(\dfrac{[A]_0}{[A]_t}\right)}{k} = \frac{\ln\left(\dfrac{0.015}{0.005}\right)}{0.17 \text{ min}^{-1}} = 6.5 \text{ min}$$

additional time = 6.5 min − 3.0 min = 3.5 min

14.31 (a)

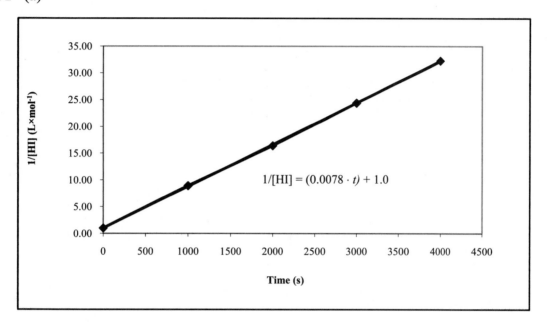

$$\frac{1}{[A]_t} = \frac{1 + [A]_0 \, kt}{[A]_0} = \frac{1}{[A]_0} + kt$$

Thus, if the reaction is second order, a plot of 1/[HI] against time should give a straight line of slope k. As can be seen from the graph, the data fit the equation for a second-order reaction quite well. The slope is determined by a least squares fit of the data by the graphing program.
(b) (i). The rate constant for the rate law for the loss of HI is simply the slope of the best fit line, 7.8×10^{-3} L · mol^{-1} · s^{-1}. (ii). Since two moles of HI are consumed per mole of reaction, the rate constant for the unique rate law is half the slope or 3.9×10^{-3} L · mol^{-1} · s^{-1}.

14.33 Obtain an expression for the half-life of a 2nd order reaction.

$$[A]_t = \frac{[A]_0}{1 + [A]_0 kt}$$

$$\frac{[A]_{t_{1/2}}}{[A]_0} = \frac{1}{2} = \frac{1}{1 + [A]_0 kt_{1/2}}$$

Therefore, $1 + [A]_0 kt_{1/2} = 2$, or $[A]_0 kt_{1/2} = 1$, or

$$t_{1/2} = \frac{1}{k[A]_0} \text{ and } k = \frac{1}{t_{1/2}[A]_0}$$

It is also convenient to solve for t. Take reciprocals:

$$\frac{1}{[A]_t} = \frac{1}{[A]_0} + kt$$

giving

$$t = \frac{\dfrac{1}{[A]_t} - \dfrac{1}{[A]_0}}{k}$$

(a) $k = \dfrac{1}{t_{1/2}[A]_0} = \dfrac{1}{(50.5 \text{ s})(0.84 \text{ mol} \cdot \text{L}^{-1})} = 0.024 \text{ L} \cdot \text{mol}^{-1} \cdot \text{s}^{-1}$

$$t = \frac{\dfrac{1}{[A]} - \dfrac{1}{[A]_0}}{k} = \frac{\dfrac{16}{[A]_0} - \dfrac{1}{[A]_0}}{k} = \frac{15}{k[A]_0}$$

$$= \frac{15}{(0.024 \text{ L} \cdot \text{mol}^{-1} \cdot \text{s}^{-1})(0.84 \text{ mol} \cdot \text{L}^{-1})} = 7.4 \times 10^2 \text{ s}$$

(b) $t = \dfrac{\dfrac{4}{[A]_0} - \dfrac{1}{[A]_0}}{k} = \dfrac{3}{k[A]_0}$

$$= \frac{3}{(0.024 \text{ L} \cdot \text{mol}^{-1} \cdot \text{s}^{-1})(0.84 \text{ mol} \cdot \text{L}^{-1})} = 1.5 \times 10^2 \text{ s}$$

(c) $t = \dfrac{\dfrac{5}{[A]_0} - \dfrac{1}{[A]_0}}{k} = \dfrac{4}{k[A]_0}$

$$= \frac{4}{(0.024 \text{ L} \cdot \text{mol}^{-1} \cdot \text{s}^{-1})(0.84 \text{ mol} \cdot \text{L}^{-1})} = 2.0 \times 10^2 \text{ s}$$

14.35 (a)
$$t_{1/2} = \frac{0.693}{k} = \frac{0.693}{2.81 \times 10^{-3} \text{ min}^{-1}} = 247 \text{ min}$$

(b)
$$t = \frac{\ln\left(\frac{[SO_2Cl_2]_0}{[SO_2Cl_2]_t}\right)}{k} = \frac{\ln 10}{2.81 \times 10^{-3} \text{ min}^{-1}} = 819 \text{ min}$$

(c) $[A]_t = [A]_0 \, e^{-kt}$

Because the vessel is sealed, masses and concentrations are proportional, and we write

$$(\text{mass left})_t = (\text{mass})_0 \, e^{-kt}$$

$$= 14.0 \text{ g} \times e^{-(2.81 \times 10^{-3} \text{ min}^{-1} \times 60 \text{ min} \cdot \text{h}^{-1} \times 1.5 \text{ h})}$$

$$= 10.9 \text{ g}$$

Note: Knowledge of the volume of the vessel is not required. However, we could have converted mass to concentration, solved for the new concentration at 1.5 h. and finally converted back to the new (remaining) mass. But this is not necessary

14.37 (a)
$$t = \frac{\frac{1}{[A]} - \frac{1}{[A]_0}}{k} = \frac{\frac{1 \text{ L}}{0.080 \text{ mol}} - \frac{1 \text{ L}}{0.10 \text{ mol}}}{0.015 \text{ L} \cdot \text{mol}^{-1} \cdot \text{min}^{-1}} = 1.7 \times 10^2 \text{ min}$$

(b)
$$[A] = \frac{0.15 \text{ mol A}}{L} - \left[\left(\frac{0.19 \text{ mol B}}{L}\right)\left(\frac{1 \text{ mol A}}{2 \text{ mol B}}\right)\right]$$

$$= 0.055(\text{mol A}) \cdot L^{-1} = 0.37[A]_0$$

$$t = \frac{\frac{1}{[A]_t} - \frac{1}{[A]_0}}{k}$$

$$= \frac{\frac{1}{0.055 \text{ mol} \cdot L^{-1}} - \frac{1}{0.15 \text{ mol} \cdot L^{-1}}}{0.0035 \text{ L} \cdot \text{mol}^{-1} \cdot \text{min}^{-1}}$$

$$= 3.3 \times 10^3 \text{ min}$$

14.39 $rate = -\dfrac{1}{a}\dfrac{d[A]}{dt} = k[A]$

$\dfrac{d[A]}{[A]} = -\, akt\,\mathrm{d}$

integrate from $[A]_0$ at $t \neq 0$ to $[A]$ at t :

$\displaystyle\iint_{[A]_0}^{[A]_t} \dfrac{d[A]}{[A]} = -\, akt\,'\mathrm{d}$

$\ln\dfrac{[A]_t}{[A]_0} = -akt,$ and $[A]_t = [A]_0\exp(-akt)$

For the half-life. At $t_{1/2}$, $[A]_t = \frac{1}{2}[A]_0$. Therefore:

$\ln\dfrac{[A]_0}{[A]_t} = \ln 2 = akt_{1/2},$ and

$t_{1/2} = \dfrac{\ln 2}{ak}$

14.41 Given: $\dfrac{d[A]}{dt} = -k[A]^3$, we can derive an expression for the amount of time needed for the inital concentration of A, $[A]_0$, to decrease by 1/2. Begin by obtaining the integrated rate law for a third-order reaction by separation of variables:

$\displaystyle\int_{[A]_0}^{[A]_t}[A]^{-3}d[A] = \int_0^t -k\,dt = -\dfrac{1}{2}\Big[[A]_t^{-2} - [A]_0^{-2}\Big] = -kt$

To obtain an expression for the half-life, let $[A]_t = \frac{1}{2}[A]_0$ and $t = t_{1/2}$:

$-\dfrac{1}{2}\Big[(\tfrac{1}{2}[A]_0)^{-2} - [A]_0^{-2}\Big] = -kt_{1/2}$

solving for the half-life:

$t_{1/2} = \dfrac{3}{2k[A]_0^2}$

14.43 $2AC + B \rightarrow A_2B + 2C$

Intermediate is AB.

14.45 The overall reaction is $CH_2=CHCOOH + HCl \rightarrow ClCH_2CH_2COOH$. The intermediates include chloride ion, $CH_2=CHC(OH)_2^+$ and $ClCH_2CHC(OH)_2$.

14.47 (a) Overall reaction: $O_3 + O \rightarrow 2O_2$

(b)

		Rate Law	Molecularity
	Step 1	rate $= k[O_3][NO]$	bimolecular
	Step 2	rate $= k[NO_2][O]$	bimolecular

(c) Reaction intermediate: NO_2

(d) Catalyst: NO

14.49 The first elementary reaction is the rate-controlling step, because it is the slow step. The second elementary reaction is fast and does not affect the overall reaction order, which is second order as a result of the fact that the rate-controlling step is bimolecular.

rate $= k[NO][Br_2]$

14.51 If mechanism (a) were correct, the rate law would be rate $= k[NO_2][CO]$. But this expression does not agree with the experimental result and can be eliminated as a possibility. Mechanism (b) has rate $= k[NO_2]^2$ from the slow step. Step 2 does not influence the overall rate, but it is necessary to achieve the correct overall reaction; thus this mechanism agrees with the experimental data. Mechanism (c) is not correct, which can be seen from the rate expression for the slow step, rate $= k[NO_3][CO]$. [CO] cannot be eliminated from this expression to yield the experimental result, which does not contain [CO].

14.53 (a) True;   (b) False. At equilibrium, the *rates* of the forward and reverse reactions are equal, *not the rate constants*.   (c) False. Increasing the concentration of a reactant causes the rate to increase by providing more reacting molecules. It does not affect the rate constant of the reaction.

14.55 The overall rate of formation of A is rate $= -k[A] + k'[B]$. The first term accounts for the forward reaction and is negative as this reaction reduces [A]. The second term, which is positive, accounts for the back reaction which increases [A]. Given the 1:1 stoichiometry of the reaction, if no B was present at the beginning of the reaction, [A] and [B] at any time are related by the equation: $[A] + [B] = [A]_o$ where $[A]_o$ is the initial concentration of A. Therefore, the rate law may be written:

$$\frac{d[A]}{dt} = -k[A] + k'([A]_o - [A]) = -(k + k')[A] + k'[A]_o$$

The solution of this first-order differential equation is:

$$[A] = \frac{k' + ke^{-(k'+k)t}}{k' + k}[A]_o$$

As $t \to \infty$ the exponential term in the numerator goes to zero and the concentrations reach their equilibrium values given by:

$$[A]_{eq} = \frac{k'[A]_o}{k' + k} \quad \text{and} \quad [B]_{eq} = [A]_o - [A]_\infty = \frac{k[A]_o}{k + k'}$$

taking the ratio of products over reactants we see that:

$$\frac{[B]_{eq}}{[A]_{eq}} = \frac{k}{k'} = K \quad \text{where } K \text{ is the equilibrium constant for the reaction.}$$

14.57

(a) Given the Arrhenius equation, $\ln k = \ln A - E_a/RT$, we see that the slope of the best fit line to the data $(-3.27 \times 10^4 \text{ K})$ is E_a/R and the y-intercept (35.0) is $\ln A$. Therefore,

$E_a = (3.27 \times 10^4 \text{ K})(8.31 \times 10^{-3} \text{ kJ} \cdot \text{mol}^{-1} \cdot \text{K}^{-1}) = 2.72 \times 10^2 \text{ kJ} \cdot \text{mol}^{-1}$.

(b) At 600 °C (or 873 K), the rate constant is:

$$\ln(k) = \left(-3.27 \times 10^4 \text{ K}\right)\frac{1}{873 \text{ K}} + 35.0 = -2.46$$

$k = 0.088$

14.59 We use $\ln\left(\dfrac{k'}{k}\right) = \dfrac{E_a}{R}\left(\dfrac{1}{T} - \dfrac{1}{T'}\right) = \dfrac{E_a}{R}\left(\dfrac{T'-T}{T'T}\right)$

$$\ln\left(\frac{k'}{k}\right) = \ln\left(\frac{0.87 \text{ s}^{-1}}{0.76 \text{ s}^{-1}}\right)$$

$$= \left(\frac{E_a}{8.31 \times 10^{-3} \text{ kJ} \cdot \text{K}^{-1} \cdot \text{mol}^{-1}}\right)\left(\frac{1030 \text{ K} - 1000 \text{ K}}{1030 \text{ K} \times 1000 \text{ K}}\right)$$

$$E_a = \frac{(8.31 \times 10^{-3} \text{ kJ} \cdot \text{K}^{-1} \cdot \text{mol}^{-1})(1000 \text{ K})(1030 \text{ K})}{(1030 \text{ K} - 1000 \text{ K})}\ln\left(\frac{0.87 \text{ s}^{-1}}{0.76 \text{ s}^{-1}}\right)$$

$= 39 \text{ kJ} \cdot \text{mol}^{-1}$

14.61 We use $\ln\left(\dfrac{k'}{k}\right) = \dfrac{E_a}{R} = \left(\dfrac{1}{T} - \dfrac{1}{T'}\right) = \dfrac{E_a}{R}\left(\dfrac{T'-T}{TT'}\right)$

$k' = $ rate constant at 700°C, $T' = (700 + 273) \text{ K} = 973 \text{ K}$

$$\ln\left(\frac{k'}{k}\right) = \left(\frac{315 \text{ kJ} \cdot \text{mol}^{-1}}{8.314 \times 10^{-3} \text{ kJ} \cdot \text{mol}^{-1}}\right)\left(\frac{973 \text{ K} - 1073 \text{ K}}{973 \text{ K} \times 1073 \text{ K}}\right)$$

$= -3.63;\ \dfrac{k'}{k} = 0.026$

$k' = 0.026 \times 9.7 \times 10^{10} \text{ L} \cdot \text{mol}^{-1} \cdot \text{s}^{-1} = 2.5 \times 10^9 \text{ L} \cdot \text{mol}^{-1} \cdot \text{s}^{-1}$

14.63 $\ln\left(\dfrac{k'}{k}\right) = \dfrac{E_a}{R}\left(\dfrac{1}{T} - \dfrac{1}{T'}\right) = \dfrac{E_a}{R}\left(\dfrac{T'-T}{TT'}\right)$

$$= \left(\dfrac{103\ \text{kJ}\cdot\text{mol}^{-1}}{8.314\times10^{-3}\ \text{kJ}\cdot\text{K}^{-1}\cdot\text{mol}^{-1}}\right)\left(\dfrac{323\ \text{K} - 318\ \text{K}}{318\ \text{K}\times323\ \text{K}}\right) = 0.60$$

$\dfrac{k'}{k} = 1.8$

$k' = 1.8\times5.1\times10^{-4}\,\text{s}^{-1} = 9.2\times10^{-4}\,\text{s}^{-1}$

14.65 (a) The equilibrium constant will be given by the ratio of the rate constant of the forward reaction to the rate constant of the reverse reaction:

$$K = \dfrac{k}{k'} = \dfrac{265\ \text{L}\cdot\text{mol}^{-1}\cdot\text{min}^{-1}}{392\ \text{L}\cdot\text{mol}^{-1}\cdot\text{min}^{-1}} = 0.676$$

(b) The reaction profile corresponds to a plot similar to that shown in Fig. 13.31a. The reaction is endothermic—the reverse reaction has a lower activation barrier than the forward reaction.

(c) Raising the temperature will increase the rate constant of the reaction with the higher activation barrier more than it will the rate constant of the reaction with the lower energy barrier. We expect the rate of the forward reaction to go up substantially more than for the reverse reaction in this case. k will increase more than k' and consequently the equilibrium constant K will increase. This is consistent with Le Chatelier's principle.

14.67 (a) cat = catalyzed, uncat = uncatalyzed $\quad E_{a,\text{cat}} = \frac{75}{125}\,E_{a,\text{uncat}} = 0.60\,E_{a,\text{uncat}}$

$$\dfrac{\text{rate(cat)}}{\text{rate(uncat)}} = \dfrac{k_{\text{cat}}}{k_{\text{uncat}}} = \dfrac{A e^{-E_{a,\text{cat}}/RT}}{A e^{-E_{a,\text{uncat}}/RT}} = \dfrac{e^{-(0.60)E_{a,\text{uncat}}/RT}}{e^{-E_{a,\text{uncat}}/RT}}$$

$$= e^{(-0.60+1.00)E_{a,\text{uncat}}/RT} = e^{(0.40)E_{a,\text{uncat}}/RT}$$

$$= e^{[(0.40)(125\ \text{kJ}\cdot\text{mol}^{-1})/(8.314\times10^{-3}\ \text{kJ}\cdot\text{K}^{-1}\cdot\text{mol}^{-1}\times298\ \text{K})]} = 6\times10^{8}$$

(b) The last step of the calculation in (a) is repeated with $T = 350$ K.

$$e^{[(0.40)(125\ \text{kJ}\cdot\text{mol}^{-1})/(8.314\times10^{-3}\ \text{kJ}\cdot\text{K}^{-1}\cdot\text{mol}^{-1}\times350\ \text{K})]} = 3\times10^{7}$$

The rate enhancement is lower at higher temperatures.

14.69 cat = catalyzed

$$\frac{\text{rate(cat)}}{\text{rate(uncat)}} = \frac{k_{\text{cat}}}{k_{\text{uncat}}} = 1000 = \frac{Ae^{-E_{a,\text{cat}}/RT}}{Ae^{-E_a/RT}} = \frac{e^{-E_{a,\text{cat}}/RT}}{e^{-E_a/RT}}$$

$\ln 1000$

$$E_{a,\text{cat}} = E_a - RT \ln 1000$$
$$= 98 \text{ kJ} \cdot \text{mol}^{-1} - (8.31 \times 10^{-3} \text{ kJ} \cdot \text{K}^{-1} \cdot \text{mol}^{-1})(298 \text{ K})(\ln 1000)$$
$$= 81 \text{ kJ} \cdot \text{mol}^{-1}$$

14.71 The overall reaction is $RCN + H_2O \longrightarrow RCONH_2$.

$RC(=N^-)OH$ and $RC(=NH)OH$ are intermediates.

The hydroxide ion serves as a catalyst for the reaction.

14.73 (a) False. A catalyst increases the rate of both the forward and reverse reactions by providing a completely different pathway. (b) True, although a catalyst may be poisoned and lose activity.

(c) False. There is a completely different pathway provided for the reaction in the presence of a catalyst.

(d) False. The position of the equilibrium is unaffected by the presence of a catalyst.

14.75 (a) To obtain the Michaelis-Menten rate equation, we will begin by employing the steady-state approximation, setting the rate of change in the concentration of the ES intermediate equal to zero:

$$\frac{d[ES]}{dt} = k_1[E][S] - k'_1[ES] - k_2[ES] = 0.$$

Rearranging gives: $[E][S] = \left(\frac{k_2 + k'_1}{k_1}\right)[ES] = K_M[ES].$

The total bound and unbound enzyme concentration, $[E]_0$, is given by:

$[E]_0 = [E] + [ES]$, and, therefore, $[E] = [ES] - [E]_0$.

Substituting this expression for [E] into the preceding equation, we obtain:

$([ES] - [E]_0)[S] = K_M[ES].$

Rearranging to obtain [ES] gives: $[ES] = \dfrac{[E]_0[S]}{K_M + [S]}$.

From the mechanism, the rate of appearance of the product is given by $rate = k_2[ES]$. Substituting the preceding equation for [ES], we obtain:

$$\text{Rate} = \dfrac{k_2[E]_0[S]}{K_M + [S]},$$

the Michaelis-Menten rate equation, which can be rearranged to obtain:

$$\dfrac{1}{\text{rate}} = \dfrac{K_M}{k_2[E]_0[S]} + \dfrac{1}{k_2[E]_0}.$$

If one plots $\dfrac{1}{\text{rate}}$ versus $\dfrac{1}{[S]}$, the slope will be $\dfrac{K_M}{k_2[E]_0}$ and the y-intercept

will be $\dfrac{1}{k_2[E]_0}$.

(b)

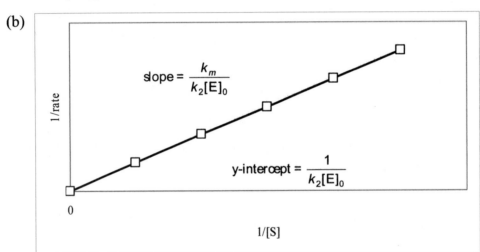

14.77

    (a) rate $= k[NO]^2$, bimolecular

    (b) rate $= k[Cl_2]$, unimolecular

    (c) rate $= k[NO_2]^2$, bimolecular

    (d) b and c

14.79 (a) $CH_3CH = CHCH_2Br$ because the positive charge on the reactive

intermediate is on a primary carbon atom. This also agrees with the result that at high temperatures this is the predominant product as there is sufficient energy to overcome the larger activation energy.

(b) Kinetic control predominates at low temperatures. The reaction pathway with the lower activation energy will predominate at low temperatures because the lower activation energy barrier results in a larger rate constant and therefore a faster reaction.

14.81 Given, $K = 326$ and $k_{attach} = 7.4 \times 10^7 \text{ L.mol}^{-1}.\text{s}^{-1}$

Since $K = \dfrac{k_{forward}}{k_{reverse}} = \dfrac{k_{attach}}{k_{loss}}$

Then $k_{loss} = \dfrac{k_{attach}}{K} = \dfrac{7.4 \times 10^7 \text{ L. mol}^{-1}.\text{s}^{-1}}{326} = 2.3 \times 10^5 \text{ L.mol}^{-1}.\text{s}^{-1}$

14.83 (a) $t_{1/2} = 5$ s since the flask contains 12 atoms at t = 0 and 6 atoms at t = 5s.

(b) Since $t_{1/2} = 5 \text{ s} = \dfrac{0.693}{k}$ then $k = 0.139 \text{ s}^{-1}$

First order, $\ln[A] = \ln[A]_0 - kt$

Then at 8 s, $\ln[A] = \ln 12 - (0.139 \text{ s}^{-1})(8 \text{ s}) = 1.37$

$A = 3.94 \approx 4$ There should be four molecules at t = 8 s.

14.85 (a) $CH_3CHO \rightarrow CH_3 + CHO$ unimolecular

$[CH_3\ldots\ldots\ldots\ldots\ldots CHO]^{\ddagger}$

(b) $2I \rightarrow I_2$ termolecular

$[I\ldots\ldots\ldots I\ldots\ldots\ldots Ar]^{\ddagger}$

(c) $O_2 + NO \rightarrow NO_2 + O$ bimolecular

$[\,O\ldots\ldots\ldots O\ldots\ldots NO]^{\ddagger}$

14.87 $x = $ amount of original sample $= 25.0$ mg

$n = $ number of half-lives

$\left(\dfrac{1}{2}\right)^n \times x = $ amount remaining

$$\frac{10.9}{12.3} = 0.886 \text{ half-lives}$$

$$\left(\frac{1}{2}\right)^{0.886} \times 25.0 \text{ mg} = 13.5 \text{ mg}$$

14.89 The anticipated rate for mechanism (i) is: rate = $k[C_{12}H_{22}O_{11}]$, while the expected rate for mechanism (ii) is: rate = $k[C_{12}H_{22}O_{11}][H_2O]$. The rate for mechanism (ii) will be pseudo-first-order in dilute solutions of sucrose because the concentration of water will not change. Therefore, in dilute solutions kinetic data can not be used to distinguish between the two mechanisms. However, in a highly concentrated solution of sucrose, the concentration of water will change during the course of the reaction. As a result, if mechanism (ii) is correct the kinetics will display a first-order dependence on the concentration of H_2O while mechanism (i) predicts that the rate of the reaction is independent of $[H_2O]$.

14.91 (a) The objective is to reproduce the observed rate law. If step 2 is the slow step, if step 1 is a rapid equilibrium, and if step 3 is fast also, then our proposed rate law will be rate = $k_2[N_2O_2][H_2]$. Consider the equilibrium

of Step 1: $\quad k_1[NO]^2 = k'_1[N_2O_2]$

$$[N_2O_2] = \frac{k_1}{k'_1}[NO]^2$$

Substituting in our proposed rate law, we have

$$\text{rate} = k_2\frac{k_1}{k'_1}[NO]^2[H_2] = k[NO]^2[H_2] \text{ where } k = k_2\frac{k_1}{k'_1}$$

The assumptions made above reproduce the observed rate law; therefore, step 2 is the slow step.

(b)

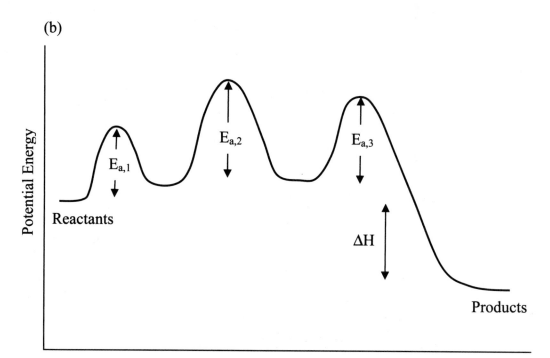

Note: The dips that represent the formation of the intermediate N_2O_2 and N_2O will not be at the same energy, but we have no information to determine which should be lower.

14.93 $\dfrac{\text{rate at 28°C}}{\text{rate at 5°C}} = \dfrac{k'}{k} = \dfrac{t}{t'} = \dfrac{48\ \text{h}}{4\ \text{h}}$

We use $\ln\left(\dfrac{k'}{k}\right) = \dfrac{E_a}{R}\left(\dfrac{1}{T} - \dfrac{1}{T'}\right)$ and solve for E_a.

$$E_a = \dfrac{R\ln\left(\dfrac{k'}{k}\right)}{\left(\dfrac{1}{T} - \dfrac{1}{T'}\right)} = \dfrac{(8.314 \times 10^{-3}\ \text{kJ} \cdot \text{K}^{-1} \cdot \text{mol}^{-1})\ln\left(\dfrac{48}{4}\right)}{\left(\dfrac{1}{278\ \text{K}} - \dfrac{1}{301\ \text{K}}\right)} = 75\ \text{kJ} \cdot \text{mol}^{-1}$$

14.95 $\text{rate} = \dfrac{k_2[E]_0[S]}{K_m + [S]}$

$$1.21 \times 10^{-3}\ \text{mol.\,L}^{-1}.\ \text{s}^{-1} = \dfrac{k_2[E]_0(0.156\ \text{mol.\,L}^{-1})}{(0.038\ \text{mol.\,L}^{-1}) + (0.156\ \text{mol.\,L}^{-1})}$$

$$1.5 \times 10^{-3} \text{ mol. L}^{-1}. \text{s}^{-1} = k_2[\text{E}]_0 = \text{max rate}$$

14.97

    (a) three steps

    (b) first step

    (c) third step

    (d) two

    (e) none

14.99 (a) ClO is the reaction intermediate; Cl is the catalyst.

    (b) Cl, ClO, O, O_2

    (c) Step 1 and step 2 are propagating.

    (d) $\text{Cl} + \text{Cl} \longrightarrow \text{Cl}_2$

14.101 For a third-order reaction,

$$t_{1/2} \propto \frac{1}{[A_0]^2} \text{ or } t_{1/2} = \frac{\text{constant}}{[A_0]^2}$$

(a) The time necessary for the concentration to fall to one-half of the initial concentration is one half-life:

$$\text{first half-life} = t_1 = t_{1/2} = \frac{\text{constant}}{[A_0]^2}$$

(b) This time, $t_{1/4}$, is two half-lives, but because of different starting concentrations, the half-lives are not the same:

$$\text{second half-life} = t_2 = \frac{\text{constant}}{(\frac{1}{2}[A_0])^2} = \frac{4(\text{constant})}{[A_0]^2} = 4t_1$$

$$\text{total time} = t_1 + t_2 = t_1 + 4t_1 = 5t_1 = t_{1/4}$$

(c) This time, $t_{1/16}$, is four half-lives; again, the half-lives are not the same:

$$\text{third half-life} = t_3 = \frac{\text{constant}}{(\frac{1}{4}[A_0])^2} = \frac{16(\text{constant})}{[A_0]^2} = 16t_1$$

$$\text{fourth half-life} = t_4 = \frac{\text{constant}}{(\frac{1}{8}[A_0])^2} = \frac{64(\text{constant})}{[A_0]^2} = 64t_1$$

$$\text{total time} = t_1 + t_2 + t_3 + t_4 = t_1 + 4t_1 + 16t_1 + 64t_1 = 85t_1 = t_{1/16}$$

If t_1 is known, the times $t_{1/4}$ and $t_{1/16}$ can be calculated easily.

14.103 The following plots are linear: (b) (c) (d) (f) (g)

14.105 (a) The overall reaction is: $OCl^- + I^- \longrightarrow OI^- + Cl^-$

(b) The rate law will be based upon the slow step of the reaction:

$$\text{rate} = k_2[HOCl][I^-]$$

Even though HOCl is a stable species because it is an intermediate in the reaction as written, technically we should not leave the rate law in this form. The concentration of HOCl can be expressed in terms of the reactants and products using the fast equilibrium approach:

$$K = \frac{k_1}{k_1{}'} = \frac{[HOCl][OH^-]}{[OCl^-]}$$

$$[HOCl] = \frac{k_1}{k_1{}'}\frac{[OCl^-]}{[OH^-]}$$

$$\text{rate} = \frac{k_2 k_1}{k_1{}'}\frac{[OCl^-][I^-]}{[OH^-]}$$

(c) An examination of the rate law shows that the rate is dependent upon the concentration of OH^-. As the pH increases, the concentration of OH^- increases and the rate decreases.

(d) If the reaction is carried out in an organic solvent, then H_2O is no longer the solvent and its concentration must be included in calculating the equilibrium concentration of HOCl:

$$K = \frac{k_1}{k_1{}'} = \frac{[HOCl][OH^-]}{[OCl^-][H_2O]}$$

$$[HOCl] = \frac{k_1}{k_1{}'}\frac{[OCl^-][H_2O]}{[OH^-]}$$

$$\text{rate} = \frac{k_2 k_1}{k_1'} \frac{[\text{OCl}^-][\text{I}^-][\text{H}_2\text{O}]}{[\text{OH}^-]}$$

The rate of reaction will then show a dependence upon the concentration of water, which will be obscured when the reaction is carried out with water as the solvent.

14.107 To get an expression for $t_{1/2}$ in terms of n, we need to evaluate an integral such as:

$$\int_{[A]_0}^{[A]} \frac{d[A]}{[A]^n} = -k \int_0^t dt = -kt$$

$$\frac{1}{n-1}\left(\frac{1}{[A]^{n-1}} - \frac{1}{[A]_0^{n-1}}\right) = kt$$

An expression for $t_{1/2}$ is then:

$$\frac{1}{n-1}\left(\frac{2^{n-1}}{[A]_0^{n-1}} - \frac{1}{[A]_0^{n-1}}\right) = kt_{1/2}$$

$$\frac{1}{n-1}\left(\frac{2^{n-1}-1}{[A]_0^{n-1}}\right) = kt_{1/2}$$

An expression for $t_{3/4}$ could be found by setting up $[A] = \frac{3}{4}[A]_0$

$$\frac{1}{n-1}\left(\frac{4^{n-1}}{3^{n-1}[A]_0^{n-1}} - \frac{1}{[A]_0^{n-1}}\right) = kt_{3/4}$$

$$\frac{1}{n-1}\left(\frac{\left(\frac{4}{3}\right)^{n-1}-1}{[A]_0^{n-1}}\right) = kt_{3/4}$$

The ratio $t_{1/2}/t_{3/4}$ is then

$$t_{1/2}/t_{3/4} = \left(\frac{2^{n-1}-1}{(4/3)^{n-1}-1}\right)$$

14.109

(a)

methyl isonitrile

cyanomethane

(b)   To convert from one isomer to the other, a single C-N bond is broken and a single C-C bond is formed, giving an approximate enthalpy of reaction of:

$305 \text{ kJ} \cdot \text{mol}^{-1} - 348 \text{ kJ} \cdot \text{mol}^{-1} = -43 \text{ kJ} \cdot \text{mol}^{-1}$.

Cyanomethane is the more stable isomer.

(c) The rate constant at 300 K is found using:

$\ln \dfrac{k'}{k} = \dfrac{E_a}{R}\left(\dfrac{1}{T} - \dfrac{1}{T'}\right)$. Substituting into this equation we find:

$\ln \dfrac{k'}{6.6 \times 10^{-4} \text{ s}^{-1}} = \dfrac{161000 \text{ J} \cdot \text{mol}^{-1}}{8.3145 \text{ J} \cdot \text{K}^{-1} \cdot \text{mol}^{-1}}\left(\dfrac{1}{500 \text{ K}} - \dfrac{1}{300 \text{ K}}\right)$

Solving for k':

$k' = 4.04 \times 10^{-15} \text{ s}^{-1}$

At this rate, the time required to reduce the initial concentration of the isomer by 75% is found by:

$\ln \dfrac{[CH_3NC]_t}{[CH_3NC]_0} = \ln(0.75) = -k \cdot t = -4.04 \times 10^{-15} \text{ s}^{-1} \cdot t$

$t = 7.12 \times 10^{13} \text{ s}$

(d) Since $\Delta H°$ for this isomerization reaction is exothermic (- 43 kJ.mol^{-1}), the reaction profile has the product lower than the reactant.

(e) We begin by finding the rate constant at which the concentration of the isomer will reduce by 75% in one hour:

$$\ln \frac{[CH_3NC]_t}{[CH_3NC]_0} = \ln(0.75) = -k \cdot t = -k \cdot 1\,h \left(\frac{60\,min}{h} \right) \left(\frac{60\,s}{min} \right)$$

Solving for k:

$$k = 7.99 \times 10^{-5}\,s^{-1}$$

We can now determine at what temperature this rate constant is reached using the equation from part (c) above:

$$\ln \frac{k'}{k} = \frac{E_a}{R} \left(\frac{1}{T} - \frac{1}{T'} \right).$$

$$\ln \frac{7.99 \times 10^{-5}\,s^{-1}}{6.6 \times 10^{-4}\,s^{-1}} = \frac{161000\,J \cdot mol^{-1}}{8.3145\,J \cdot K^{-1} \cdot mol^{-1}} \left(\frac{1}{500\,K} - \frac{1}{T} \right)$$

Solving for T:

$$T = 474\,K$$

(f) Argon serves as a collision partner. Collisions between the $CH_3NC(g)$ and argon atoms provide the energy needed to overcome the activation energy and allow the isomerization reaction to proceed toward products. The argon atoms also serve as an energy sink, accepting the energy released during the isomerization reaction.

(g) At high argon concentrations, collisions resulting in reaction are between the $CH_3NC(g)$ and argon atoms:

$$CH_3NC(g) + Ar(g) \rightarrow CH_3CN(g) + Ar(g).$$

Given this elementary reaction one would predict that the reaction was first order with respect to $CH_3NC(g)$ concentration and would appear first order overall if the concentration of $Ar(g)$ was large and unchanging. If, however, the concentration of $Ar(g)$ was greatly reduced, the isomerization reaction would also proceed through collisions between $CH_3NC(g)$ molecules, making the reaction second order with respect to $CH_3NC(g)$.

CHAPTER 15

THE ELEMENTS: THE MAIN GROUP ELEMENTS

15.1 (a) nitrogen; (b) potassium; (c) gallium; (d) iodine

15.3 (a) sulfur; (b) selenium; (c) sodium; (d) oxygen

15.5 Tellurium < selenium < oxygen

15.7 (a) antimony; (b) Antimony has higher effective nuclear charge.

15.9 (a) bromide ion; (b) Bromide has larger size.

15.11 $2 K(s) + H_2(g) \rightarrow 2 KH(s)$

15.13 (a) saline; (b) molecular; (c) molecular; (d) metallic

15.15 (a) acidic; (b) amphoteric; (c) acidic; (d) basic

15.17 (a) CO_2; (b) B_2O_3

15.19 (a) $C_2H_2(g) + H_2(g) \rightarrow H_2C{=}CH_2(g)$

Oxidation number of C in $C_2H_2 = -1$; of C in $H_2CRCH_2 = -2$, carbon has been reduced.

(b) $CO(g) + H_2O(g) \rightarrow CO_2(g) + H_2(g)$

(c) $BaH_2(s) + 2 H_2O(l) \rightarrow Ba(OH)_2 + 2 H_2(g)$

15.21 (a) $H_2(g) + Cl_2(g) \xrightarrow{\text{light}} 2\,HCl(g)$

(b) $H_2(g) + 2\,Na(l) \xrightarrow{\Delta} 2\,NaH(s)$

(c) $P_4(s) + 6\,H_2(g) \rightarrow 4\,PH_3(g)$

(d) $2\,Cu(s) + H_2(g) \rightarrow 2\,CuH(s)$

15.23 The reason for this trend is mainly due to the trend of the electronegativity of the central atom (N<O<F), which makes F more partially negative to attract H from another HF atom. Spatial hindrance may have contributions to this trend. There are three H atoms bonding to N, two to O, and there is only one H atom bond to F.

15.25 Lithium is the only Group 1 element that reacts directly with nitrogen to form lithium nitride:

$6\,Li(s) + N_2(g) \xrightarrow{\Delta} 2\,Li_3N(s)$

Lithium reacts with oxygen to form mainly the oxide:

$4\,Li(s) + O_2(g) \rightarrow 2\,Li_2O(s)$

The other members of the group form mainly the peroxide or superoxide. Lithium exhibits the diagonal relationship that is common to many first members of a group. Li is similar in many of its compounds to the compounds of Mg. This behavior is related to the small ionic radius of Li^+, 58 pm, which is closer to the ionic radius of Mg^{2+}, 72 pm, but substantially less than that of Na^+, 102 pm.

15.27 (a) $4\,Na(s) + O_2(g) \rightarrow 2\,Na_2O(s)$

(b) $6\,Li(s) + N_2(g) \xrightarrow{\Delta} 2\,Li_3N(s)$

(c) $2\,Na(s) + 2\,H_2O(l) \rightarrow 2\,NaOH(aq) + H_2(g)$

(d) $4\,KO_2(s) + 2\,H_2O(g) \rightarrow 4\,KOH(s) + 3\,O_2(g)$

15.29 $Mg(s) + 2 H_2O(l) \rightarrow Mg(OH)_2 + H_2(g)$

15.31 (a) $CaO(s) + H_2O(l) \rightarrow Ca(OH)_2(s)$

(b) $\Delta G^\circ_r = \Delta G^\circ_f(Ca(OH)_2, s) - [\Delta G^\circ_f(CaO, s) + \Delta G^\circ_f(H_2O, l)]$
$$= -898.49 \text{ kJ} \cdot \text{mol}^{-1} - [(-604.03 \text{ kJ} \cdot \text{mol}^{-1}) + (-237.13 \text{ kJ} \cdot \text{mol}^{-1})]$$
$$= -57.33 \text{ kJ} \cdot \text{mol}^{-1}$$

15.33 (a) $Mg(OH)_2(s) + 2 HCl(aq) \rightarrow MgCl_2(aq) + 2 H_2O(l)$

(b) $Ca(s) + 2 H_2O(l) \rightarrow Ca(OH)_2(aq) + H_2(g)$

(c) $BaCO_3(s) \xrightarrow{\Delta} BaO(s) + CO_2(g)$

15.35 (a)

$$:\ddot{C}l - Be - \ddot{C}l: \qquad Mg^{2+} \; 2\left[:\ddot{C}l:\right]^-$$

$MgCl_2$ is ionic; $BeCl_2$ is a molecular compound. (b) 180° (c) *sp*

15.37 The overall equation for the electrolytic reduction in the Hall process is
$$4 Al^{3+}(\text{melt}) + 6 O^{2-}(\text{melt}) + 3 C(s, gr) \rightarrow 4 Al(s) + 3 CO_2(g)$$

15.39 (a) $B_2O_3(s) + 3 Mg(l) \xrightarrow{\Delta} 2 B(s) + 3 MgO(s)$

(b) $2 Al(s) + 3 Cl_2(g) \rightarrow 2 AlCl_3(s)$

(c) $4 Al(s) + 3 O_2(g) \rightarrow 2 Al_2O_3(s)$

15.41 Since there are only 22 valence electrons, or 11 electron pairs, it is not possible to draw a good conventional Lewis structure for tetraborane, B_4H_{10}, that includes four B-H-B bridges. For the suggested structure given below, each bridging H and each four-coordinate B would have a formal charge of 1− while each of the six terminal H atoms and each three-coordinate B would have a formal charge of 0. The total formal charge adds up to 6– in this case even though the molecule is neutral. However, if the bridges are viewed as 3-center 2-electron bonds, then every atom can be assigned a formal charge of 0.

The second drawing shows the structure of tetraborane

See:

www.chem.leeds.ac.uk/boronweb/articles/incredible_boron/incredible.htm

15.43 The Lewis structure of $GaBr_4^-$ is:

The shape of the $GaBr_4^-$ is tetrahedral.

15.45 Silicon occurs widely in the Earth's crust in the form of silicates in rocks and as silicon dioxide in sand. It is obtained from quartzite, a form of quartz (SiO_2), by the following processes:

(1) reduction in an electric arc furnace

$$SiO_2(s) + 2\, C(s) \rightarrow Si(s,\, crude) + 2\, CO(g)$$

(2) purification of the crude product in two steps

$$Si(s,\, crude) + 2\, Cl_2(g) \rightarrow SiCl_4(l)$$

followed by reduction with hydrogen to the pure element

$$SiCl_4(l) + 2\, H_2(g) \rightarrow Si(s,\, pure) + 4\, HCl(g)$$

15.47 (a) +4; (b) +4; (c) +4

15.49 (a) $MgC_2(s)$ + 2 $H_2O(l)$ → $C_2H_2(g)$ + $Mg(OH)_2(s)$ (acid-base)

(b) 2 $Pb(NO_3)_2(s)$ → 2 $PbO(s)$ + 4 $NO_2(g)$ + $O_2(g)$ (Redox)

15.51 $SiO_2(s) + 2\ C(s) \rightarrow Si(s) + 2\ CO(g)$

$$\Delta H°_r = \Delta H°_f(\text{products}) - \Delta H°_f(\text{reactants})$$
$$= [(2)(-110.53\ \text{kJ} \cdot \text{mol}^{-1})] - [-910.94\ \text{kJ} \cdot \text{mol}^{-1}]$$
$$= +689.88\ \text{kJ} \cdot \text{mol}^{-1}$$

$$\Delta S°_r = S°(\text{products}) - S°(\text{reactants})$$
$$= [18.83\ \text{J} \cdot \text{K}^{-1} \cdot \text{mol}^{-1} + (2)(197.67\ \text{J} \cdot \text{K}^{-1} \cdot \text{mol}^{-1})]$$
$$- [41.84\ \text{J} \cdot \text{K}^{-1} \cdot \text{mol}^{-1} + (2)(5.740\ \text{J} \cdot \text{K}^{-1} \cdot \text{mol}^{-1})]$$
$$= +360.85\ \text{J} \cdot \text{K}^{-1} \cdot \text{mol}^{-1}$$

$$\Delta G°_r = \Delta H°_r - T\Delta S°_r$$
$$= 689.88\ \text{kJ} \cdot \text{mol}^{-1} - (298.15\ \text{K})(360.85\ \text{J} \cdot \text{K}^{-1} \cdot \text{mol}^{-1})/1000\ \text{J} \cdot \text{kJ}^{-1}$$
$$= +582.29\ \text{kJ} \cdot \text{mol}^{-1}$$

The temperature at which the equilibrium constant becomes greater than 1 is the temperature at which $\Delta G°_r = -RT \ln K = 0$ because $\ln 1 = 0$. Above this temperature, the equilibrium constant is greater than 1. $\Delta G°_r = 0$ when $T\Delta S°_r = \Delta H°_r$, or

$$T = \frac{\Delta H°_r}{\Delta S°_r} = \frac{+689.88 \times 10^3\ \text{J} \cdot \text{mol}^{-1}}{+360.85\ \text{J} \cdot \text{K}^{-1} \cdot \text{mol}^{-1}} = 1912\ \text{K}.$$

15.53 -3 NH_3, Li_3N, $LiNH_2$, NH_2^-

-2 H_2NNH_2

-1 N_2H_2, NH_2OH

$\ \ 0$ N_2

$+1$ N_2O, N_2F_2

+2	NO
+3	NF_3, NO_2^-, NO^+
+4	NO_2, N_2O_4
+5	HNO_3, NO_3^-, NO_2F

15.55 $CO(NH_2)_2(aq) + 2\,H_2O(l) \rightarrow (NH_4)_2CO_3(aq)$

$$\text{mass of } (NH_4)_2CO_3 = (4.0 \text{ kg urea})\left(\frac{10^3 \text{ g}}{1 \text{ kg}}\right)\left(\frac{1 \text{ mol}}{60.06 \,(NH_4)_2CO_3}\right)$$

$$\left(\frac{1 \text{ mol } (NH_4)_2CO_3}{1 \text{ mol urea}}\right)\left(\frac{96.09 \text{ g } (NH_4)_2CO_3}{1 \text{ mol } (NH_4)_2CO_3}\right)$$

$$= 6.4 \times 10^3 \text{ g (or 6.4 kg) } (NH_4)_2CO_3$$

15.57(a) 1 mole of $N_2(g)$ occupies 22.4 L at STP. For the reaction

$Pb(N_3)_2 \longrightarrow Pb + 3\,N_2$, the volume of $N_2(g)$ produced is

$$1.5 \text{ g } Pb(N_3)_2\left(\frac{1 \text{ mol } Pb(N_3)_2}{291.25 \text{ g } Pb(N_3)_2}\right)\left(\frac{3 \text{ mol } N_2}{1 \text{ mol } Pb(N_3)_2}\right)\left(\frac{22.4 \text{ L } N_2}{1 \text{ mol } N_2}\right)$$

$$= 0.35 \text{ L } N_2(g)$$

(b) $Hg(N_3)_2$ would produce a larger volume, because its molar mass is less. Note that molar mass occurs in the denominator in this calculation.
(c) Metal azides are good explosives because the azide ion is thermodynamically unstable with respect to the production of $N_2(g)$. This is because the N≡N triple bond is so strong and also because the production of a gas is favored entropically.

15.59 N_2O: $H_2N_2O_2$; $N_2O(g) + H_2O(l) \longrightarrow H_2N_2O_2(aq)$
N_2O_3: HNO_2; $N_2O_3(g) + H_2O(l) \longrightarrow 2\,HNO_2(aq)$
N_2O_5: HNO_3; $N_2O_5(g) + H_2O(l) \longrightarrow 2\,HNO_3(aq)$

15.61 Ammonia (NH_3) can undergo hydrogen bonding with itself while NF_3 can not.

15.63 (a) $4\,Li(s) + O_2(g) \xrightarrow{\Delta} 2\,Li_2O(s)$

(b) $2\,Na(s) + 2\,H_2O(l) \longrightarrow 2\,NaOH(aq) + H_2(g)$

(c) $2\,F_2(g) + 2\,H_2O(l) \longrightarrow 4\,HF(aq) + O_2(g)$

(d) $2\,H_2O(l) \longrightarrow O_2(g) + 4\,H^+(aq) + 4\,e^-$

15.65 (a) $2\,H_2S(g) + 3\,O_2(g) \xrightarrow{\Delta} 2\,SO_2(g) + 2\,H_2O(g)$

(b) $CaO(s) + H_2O(l) \longrightarrow Ca(OH)_2(aq)$

(c) $2\,H_2S(g) + SO_2(g) \xrightarrow{300°C,\,Al_2O_3} 3\,S(s) + 2\,H_2O(l)$

15.67 Both water and ammonia have four groups attached to their central atom and therefore both possess a tetrahedral electronic (or VSEPR) geometry. However, H_2O has two unshared electron pairs while NH_3 only has one, producing a larger dipole moment for H_2O.

15.69 (a)

$$\begin{array}{c} H \\ | \\ :\!\overset{\bullet\bullet}{O}\!-\!\overset{\bullet\bullet}{O}\!: \\ \overset{\bullet\bullet}{}\, | \\ H \end{array}$$

Each O in H_2O_2 is an AX_2E_2 structure; therefore, the bond angle is predicted to be $<109.5°$. In actuality, it is $97°$.

(b)–(e), The reduction potential of H_2O_2 is +1.78 V in acidic solution. It should, therefore, be able to oxidize any ion that has a reduction potential that is less than +1.78 V. For the ions listed, Cu^+ and Mn^{2+} will be oxidized. It would require an input of 1.98 V to oxidize Ag^+ to Ag^{2+} and 2.87 V to oxidize F^-.

15.71 $O_2^{2-} + H_2O \rightarrow HO_2^- + OH^-$ essentially complete

$$HO_2^- + H_2O \rightleftharpoons H_2O_2 + OH^- \quad K_b = \frac{K_w}{K_{a1}}$$

$$K_{a1} = 1.8 \times 10^{-12} \quad K_b = \frac{1.00 \times 10^{-14}}{1.8 \times 10^{-12}} = 5.6 \times 10^{-3}$$

Because this K_b is relatively small, we can assume that essentially all the OH^- is formed in the first ionization; therefore,

$$[OH^-] = \left(\frac{2.00 \text{ g Na}_2O_2}{0.200 \text{ L}} \right) \left(\frac{1 \text{ mol Na}_2O_2}{77.98 \text{ g Na}_2O_2} \right) \left(\frac{1 \text{ mol OH}^-}{1 \text{ mol Na}_2O_2} \right)$$

$$= 0.128 \text{ mol} \cdot \text{L}^{-1}$$

$$pOH = -\log(0.128) = 0.893 \quad pH = 14.00 - 0.893 = 13.11$$

If we do not ignore the second ionization, then the additional contribution to $[OH^-]$ can be approximately calculated as follows:

$$K_b = \frac{[H_2O_2][OH^-]}{[HO_2^-]} = \frac{x(0.128 + x)}{(0.128 - x)} = 5.6 \times 10^{-3}$$

To a first approximation, $x = 5.6 \times 10^{-3} \text{ mol} \cdot \text{L}^{-1}$

To a second approximation, $x = \dfrac{K_b(0.128 - 0.0056)}{(0.128 + 0.0056)} = 0.005$

Then $[OH^-] = 0.128 + 0.005 = 0.133$; $pOH = -\log(0.133) = 0.876$; and $pH = 13.12$. The difference between calculations is slight.

15.73 The weaker the H—X bond, the stronger the acid. H_2Te has the weakest bond; H_2O, the strongest. Therefore, the acid strengths are

$$H_2Te > H_2Se > H_2S > H_2O$$

15.75 (a) HIO(aq) $H = +1, O = -2$; therefore, $I = +1$

(b) ClO_2 $O = -2$; therefore, $Cl = +4$

(c) Cl_2O_7 $O = -2$; therefore, $Cl = +14/2 = +7$

(d) $NaIO_3$ $Na = +1, O = -2$; therefore, $I = +5$

15.77 (a) $4\,KClO_3(l) \xrightarrow{\Delta} 3\,KClO_4(s) + KCl(s)$

(b) $Br_2(l) + H_2O(l) \longrightarrow HBrO(aq) + HBr(aq)$

(c) $NaCl(s) + H_2SO_4(aq) \longrightarrow NaHSO_4(aq) + HCl(g)$

(d) (a) and (b) are redox reactions. In (a), Cl is both oxidized and reduced. In (b), Br is both oxidized and reduced. (c) is a Brønsted acid-base reaction; H_2SO_4 is the acid, and Cl^- the base.

15.79 (a) $HClO < HClO_2 < HClO_3 < HClO_4$ ($HClO_4$ is the strongest; HClO the weakest)

(b) The oxidation number of Cl increases from HClO to $HClO_4$. In $HClO_4$, chlorine has its highest oxidation number of +7, so $HClO_4$ will be the strongest oxidizing agent.

15.81 $:\!\ddot{Cl}\!-\!\ddot{O}\!-\!\ddot{Cl}\!:$, AX_2E_2, angular, about $109°$. The actual bond angle is $110.9°$.

15.83 (a) ICl_5 has too many highly electronegative chlorine atoms present on the central iodine. (b) IF_2 has an odd number of electrons and as a result is a radical and highly reactive. (c) $ClBr_3$ is sterically too crowded (too many large atoms on a small central atom).

15.85 (a) We can write the following general chemical equation for the process described:

$$x \text{ XeF}_2(\text{g}) + \text{I}_2(\text{g}) \longrightarrow 2 \text{ IF}_x(\text{s}) + x \text{ Xe(g)}$$

Since the pressure is directly proportional to the moles of each gas present and since we only need to find the combining ratio of iodine to fluorine to determine the chemical formula of IF_x, we can use pressure in lieu of moles for the purposes of this problem. Since all of the XeF_2 is gone at reactions end, and the reaction started with 3.6 atm of XeF_2 to start, there must be 3.6 atm of Xe present in the vessel at the end of the reaction. Therefore P (total at start) = 7.2 atm = P (XeF_2 start) + P (I_2 start); solving this gives us P (I_2 start) = 3.6 atm. At the reactions end, P (total at end) = 6.0 atm = P (Xe end) + P (I_2 end); solving this gives P (I_2 end) = 2.4 atm. This means that 1.2 atm of I_2 was used in the formation of IF_x. From this we can solve for x as follows:

$$1.2 \text{ atm I}_2 \left(\frac{x \text{ atm XeF}_2}{1 \text{ atm I}_2} \right) = 3.6 \text{ mol XeF}_2 \, ; x = 3$$

Therefore the chemical formula is IF_3.

(b) $3 \text{ XeF}_2(\text{g}) + \text{I}_2(\text{g}) \longrightarrow 2 \text{ IF}_3(\text{s}) + 3 \text{ Xe(g)}$

15.87 $\text{Cl}_2(\text{g}) + 2 \text{ e}^- \longrightarrow 2 \text{ Cl}^-(\text{aq}) \quad E^\circ = +1.36 \text{ V}$

$\text{MnO}_4^-(\text{aq}) + 8 \text{ H}^+(\text{aq}) + 5 \text{ e}^- \longrightarrow \text{Mn}^{2+}(\text{aq}) + 4 \text{ H}_2\text{O(l)} \quad E^\circ = +1.51 \text{ V}$

$E^\circ_{\text{cell}} = (1.36 - 1.51) \text{ V} = -0.15 \text{ V}$

Because E°_{cell} is negative, $\text{Cl}_2(\text{g})$ will not oxidize Mn^{2+} to form the permanganate ion in an acidic solution.

15.89 $\text{F}^-(\text{aq}) + \text{PbCl}_2(\text{s}) \longrightarrow \text{Cl}^-(\text{aq}) + \text{PbClF(s)}$

$$\text{molarity of F}^- \text{ ions} = \left(\frac{0.765 \text{ g PbClF}}{0.0250 \text{ L}} \right)\left(\frac{1 \text{ mol PbClF}}{261.64 \text{ g PbClF}} \right)\left(\frac{1 \text{ mol F}^-}{1 \text{ mol PbClF}} \right)$$

$$= 0.117 \text{ mol} \cdot \text{L}^{-1}$$

15.91 $3 \text{ NH}_4\text{ClO}_4(s) + 3 \text{ Al}(s) \longrightarrow \text{Al}_2\text{O}_3(s) + \text{AlCl}_3(s) + 6 \text{ H}_2\text{O}(g) + 3 \text{ NO}(g)$

The number of moles of NH_4ClO_4 is

$$\left(1.00 \text{ kg } \left(1000 \text{ g} \cdot \text{kg}^{-1} \right) \right) \div 117.49 \text{ g} \cdot \text{mol}^{-1} = 8.51 \text{ mol}$$

The number of moles of Al is

$$1.00 \text{ kg } \left(1000 \text{ g} \cdot \text{kg}^{-1} \right) \div 26.98 \text{ g} \cdot \text{mol}^{-1} = 37.06 \text{ mol}$$

The limiting reagent is the NH_4ClO_4.

The standard enthalpy for the reaction is given by

$$\Delta H°_r = (-1675.7 \text{ kJ} \cdot \text{mol}^{-1}) + (-704.2 \text{ kJ} \cdot \text{mol}^{-1}) + 6(-241.82 \text{ kJ} \cdot \text{mol}^{-1})$$
$$+3(90.25 \text{ kJ} \cdot \text{mol}^{-1}) - [3(-295.31 \text{ kJ} \cdot \text{mol}^{-1})]$$
$$= -2674.1 \text{ kJ} \cdot \text{mol}^{-1}$$

This value is the amount of heat released for 3 mol NH_4ClO_4. The amount released for 8.51 mol will be

$$8.51 \text{ mol NH}_4\text{ClO}_4 \left(\frac{-2674.1 \text{ kJ} \cdot \text{mol}^{-1}}{3 \text{ mol NH}_4\text{ClO}_4} \right) = -7.59 \times 10^3 \text{ kJ}$$

There will be $7.59 \times 10^3 \text{ kJ}$ of heat released.

15.93 Helium occurs as a component of natural gases found under rock formations in certain locations, especially some in Texas. Argon is obtained by distillation of liquid air.

15.95 (a) $\text{KrF}_2 : \text{F} = -1$; therefore, $\text{Kr} = +2$

(b) $\text{XeF}_6 : \text{F} = -1$; therefore, $\text{Xe} = +6$

(c) $\text{KrF}_4 : \text{F} = -1$; therefore, $\text{Kr} = +4$

(d) $XeO_4^{2-}: O = -2$, $N_{ox}(Xe) - 8 = -2$; therefore, $N_{ox}(Xe) = +6$

15.97 $XeF_4(aq) + 4\,H^+(aq) + 4\,e^- \rightarrow Xe(g) + 4\,HF(aq)$

15.99 Because H_4XeO_6 has more highly electronegative O atoms bonded to Xe, we predict that H_4XeO_6 is more acidic than H_2XeO_4.

15.101 (a)

(b)

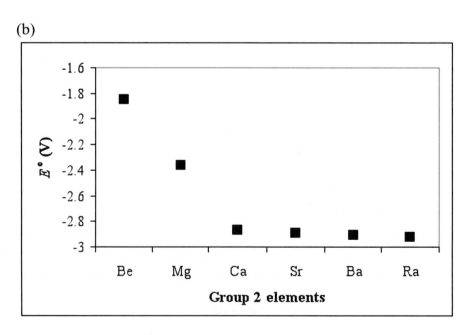

(b) For both groups, the trend in standard potentials with increasing atomic number is overall downward (they become more negative), but lithium is anomalous. This overall downward trend makes sense because we expect that it is easier to remove electrons that are farther away from the nuclei. However, because there are several factors that influence ease of removal, the trend is not smooth. The potentials are a net composite of the free energies of sublimation of solids, dissociation of gaseous molecules, ionization enthalpies, and enthalpies of hydration of gaseous ions. The origin of the anomalously strong reducing power of Li is the strongly exothermic energy of hydration of the very small Li^+ ion, which favors the ionization of the element in aqueous solution.

15.103 In the majority of its reactions, hydrogen acts as a reducing agent; that is, $H_2(g) \rightarrow 2\,H^+(aq) + 2\,e^-$, $E° = 0$ V. In these reactions, hydrogen resembles Group 1 elements, such as Na and K. However, as described in the text and in the answer to Exercise 14.91, it may also act as an oxidizing agent; that is, $H_2(g) + 2\,e^- \rightarrow 2\,H^-(aq)$, $E° = -2.25$ V. In these reactions, hydrogen resembles Group 17 elements, such as Cl and Br. Consequently, H_2 will oxidize elements with standard reduction potentials more negative than –2.25 V, such as the alkali and alkaline earth metals (except Be). The compounds formed are hydrides and contain the H^- ion; the singly charged negative ion is reminiscent of the halide ions. Hydrogen also forms diatomic molecules and covalent bonds like the halogens. The atomic radius of H is 78 pm, which compares rather well to that of F (64 pm) but not as well to that of Li (157 pm). The ionization energy of H is $1310\ kJ \cdot mol^{-1}$, which is similar to that of F ($1680\ kJ \cdot mol^{-1}$) but not similar to that of Li ($519\ kJ \cdot mol^{-1}$). The electron affinity of H is $+73\ kJ \cdot mol^{-1}$, that of F is $+328\ kJ \cdot mol^{-1}$, and that of Li is $60\ kJ \cdot mol^{-1}$. So in its atomic radius and ionization energy, H more closely resembles the Period 2 halogen, fluorine, in Group 17, than the Period 2 alkali metal, lithium, in Group 1; whereas in electron affinity, it more closely resembles lithium, Group 1. In electronegativity, H does not resemble elements in either Group 1 or Group 17, although its electronegativity is somewhat closer to those of Group 1. Consequently, hydrogen could be placed in either Group 1 or Group 17. But it is best to think of hydrogen as a unique element that has properties in common with both metals and nonmetals; therefore, it should probably be centered in the periodic table, as it is shown in the table in the text.

15.105 $H_2(g) + Br_2(l) \rightarrow 2\ HBr(g)$

$$\text{number of moles of } HBr = (0.135\ L\ H_2)\left(\frac{1\ mol\ H_2}{22.4\ L\ H_2}\right)\left(\frac{2\ mol\ HBr}{1\ mol\ H_2}\right)$$
$$= 0.0121\ mol$$

$$\text{molar concentration of } HBr = \frac{0.0121\ mol}{0.225\ L} = 0.0538\ mol \cdot L^{-1}$$

15.107 Species (a), (b), (c), and (d) can all function as greenhouse gases, while (e) cannot. Any molecule other than a homonuclear diatomic can exhibit a changing dipole moment as it vibrates with certain vibrational modes. Since argon is monoatomic it has no covalent bonds, no vibrational modes, and no dipole moment.

15.109 $CH_3OH(l) + \frac{3}{2}\ O_2(g) \rightarrow CO_2(g) + 2\ H_2O(l)$

$$\text{number of kg of } CO_2 = 1.00\ L\ CH_3OH\left(\frac{1000\ mL}{1\ L}\right)\left(\frac{0.791\ g\ CH_3OH}{1\ mL\ CH_3OH}\right)$$
$$\times\left(\frac{1\ mol\ CH_3OH}{32.04\ g\ CH_3OH}\right)\left(\frac{1\ mol\ CO_2}{1\ mol\ CH_3OH}\right)\left(\frac{44.02\ g\ CO_2}{1\ mol\ CO_2}\right)\left(\frac{1\ kg}{1000\ g}\right)$$
$$= 1.09\ kg\ CO_2$$

This mass of carbon dioxide is about half the amount generated by combusting an equivalent volume of octane (2.16 kg per liter).

However, we also need to consider how much energy is produced per liter of fuel and how the mass of carbon dioxide produced compares for a given amount of energy produced. Standard enthalpies of combustion given in Appendix 2 are $\Delta H_c^\circ = -5471 \text{ kJ} \cdot \text{mol}^{-1}$ for octane and $\Delta H_c^\circ = -726 \text{ kJ} \cdot \text{mol}^{-1}$ for methanol.

$$\text{energy per L methanol} = 1.00 \text{ L CH}_3\text{OH} \left(\frac{1000 \text{ mL}}{1 \text{ L}} \right) \left(\frac{0.791 \text{ g CH}_3\text{OH}}{1 \text{ mL CH}_3\text{OH}} \right)$$

$$\times \left(\frac{1 \text{ mole CH}_3\text{OH}}{32.04 \text{ g CH}_3\text{OH}} \right) \left(\frac{726 \text{ kJ}}{1 \text{ mole CH}_3\text{OH}} \right)$$

$$= 1.79 \times 10^4 \text{ kJ}$$

$$\text{energy per L octane} = 1.00 \text{ L C}_8\text{H}_{18} \left(\frac{1000 \text{ mL}}{1 \text{ L}} \right) \left(\frac{0.703 \text{ g C}_8\text{H}_{18}}{1 \text{ mL C}_8\text{H}_{18}} \right)$$

$$\times \left(\frac{1 \text{ mole C}_8\text{H}_{18}}{114.22 \text{ g C}_8\text{H}_{18}} \right) \left(\frac{5471 \text{ kJ}}{1 \text{ mole C}_8\text{H}_{18}} \right)$$

$$= 3.37 \times 10^4 \text{ kJ}$$

So the combustion of octane produces almost twice as much energy per liter as methanol (octane/methanol=1.88).

For an equivalent amount of combustion energy, methanol produces $1.88 \text{ L} \times 1.09 \text{ kg CO}_2 \cdot \text{L}^{-1} = 2.05 \text{ kg CO}_2$, which is still slightly less than octane. (However, it requires that the vehicle carry about 90% more fuel by volume, 1.9 vs. 1 L, and more than twice as much fuel by mass, 1.5 kg vs. 0.7 kg.)

15.111 (a) Diborane, B_2H_6, and Al_2Cl_6 (g) have the same basic structure in the way in which the atoms are arranged in space. (b) The bonding between the boron atoms and the bridging hydrogen atoms is electron deficient. There are three atoms and only two electrons to hold them together in a 3-center-2-electron bond. The bonding in Al_2Cl_6 is conventional in that all the bonds involve two atoms and two electrons. Here, the lone pair of a Cl atom is donated to an adjacent Al. (c) The hybridization is sp^3 at the B and Al atoms. (d) The molecules are not planar. The Group 13 element and the terminal atoms to which it is bound lie in a plane that is perpendicular to the plane that contains the main group element and the bridging atoms.

15.113 (a) The ionization energy of a molecule is the amount of energy required to strip one electron out of a gaseous molecule; it can be defined as follows:

$$\text{Molecule}\,(g) \longrightarrow \text{Molecule}^+(g) + 1e^-(g).$$

(b) SiI_4 has more electrons than $SiCl_4$; I is also less electronegative and much bigger (and therefore more polarizable) than Cl. As a result it should be easier to remove an electron from SiI_4 meaning that $SiCl_4$ should have the higher ionization energy.

15.115 (a) 10 electron species: NH_3 and H_3O^+; 15 electron species: NO and O_2^+; 22 electron species: N_2O^+ and NO_2^+. (b) Strongest Lewis acids:

H_3O^+, NO and NO_2^+; (c) Strongest oxidizing agents: NH_3, NO and N_2O^+

15.117 The structure of thiosulfuric acid and sulfuric acid are:

Due to the replacement of one of the $-OH$ groups in sulfuric acid with an $-SH$, it should be expected that an aqueous solution of thiosulfuric acid should be slightly less acidic; in addition, the boiling point should also be expected to be slightly lower (due to reduced hydrogen bonding).

15.119 The reaction of interest is: $XeF_4(s) + 2\,SF_4(s) \longrightarrow 2\,SF_6(s) + Xe(g)$

To determine how much product is formed, we must first identify the limiting reagent:

$$\text{moles of } XeF_4(s) = \frac{330.0 \text{ g}}{207.29 \text{ g} \cdot \text{mol}^{-1}} = 1.592 \text{ mol}$$

$$\text{moles of } SF_4(s) = \frac{250.0 \text{ g}}{108.07 \text{ g} \cdot \text{mol}^{-1}} = 2.313 \text{ mol}$$

Given the 1:2 stoichiometry of the reaction, $SF_4(s)$ is the limiting reagent and 2.313 mol of $SF_6(s)$ will be produced.

$(2.313 \text{ mol})(146.07 \text{ g} \cdot \text{mol}^{-1}) = 337.9$ g of $SF_6(s)$ produced.

15.121 Ion-ion forces are among the strongest intermolecular interactions. Therefore, the interactions in (a) are the strongest shown. Hydrogen bonding is stronger than a dipole-dipole interaction indicating that the interactions in (c) are stronger than those shown in (d).

15.123 (a)

$$\ddot{N}=\ddot{N}=\ddot{N}\Big]^{-}$$

AX$_2$, linear 180°

(b) F^-, 133 pm; N_3^-, 148 pm; Cl^-, 181 pm; therefore, between fluorine and chlorine.

(c) HCl, HBr, and HI are all strong acids. For HF, $K_a = 3.5 \times 10^{-4}$, so HF is slightly more acidic than HN_3. The small size of the azide ion suggests that the H—N bond in HN_3 is similar in strength to that of the H—F bond, so it is expected to be a weak acid.

(d) ionic: NaN_3, $Pb(N_3)_2$, AgN_3, etc.

covalent: HN_3, $B(N_3)_3$, FN_3, etc.

15.125 The solubility of the ionic halides is determined by a variety of factors, especially the lattice enthalpy and enthalpy of hydration. There is a delicate balance between the two factors, with the lattice enthalpy usually being the determining one. Lattice enthalpies decrease from chloride to iodide, so water molecules can more readily separate the ions in the latter. Less ionic halides, such as the silver halides, generally have a much lower solubility, and the trend in solubility is the reverse of the more ionic halides. For the less ionic halides, the covalent character of the bond allows the ion pairs to persist in water. The ions are not easily hydrated, making them less soluble. The polarizability of the halide ions, and thus, the covalency of their bonding, increases down the group.

15.127 (a) The molecular orbital diagram for NO^+ should have the oxygen orbitals slightly lower in energy than the nitrogen orbitals, because oxygen is more electronegative. This will cause the bonding to be more ionic than in either N_2 or O_2. There is an ambiguity, however, in that the MO diagram could be similar to either that of N_2 or that of O_2. Refer to Figures 3.31 and 3.32 where you will see that the σ_{2p} and π_{2p} have different relative energies. There are consequently two possibilities for the orbital energy diagram:

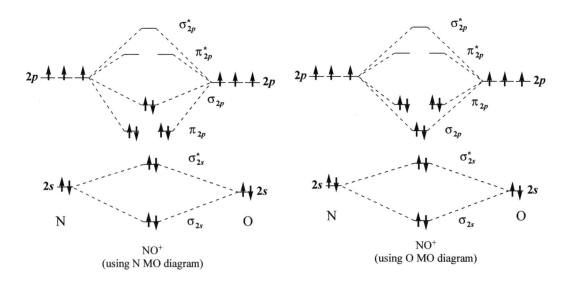

NO^+
(using N MO diagram)

NO^+
(using O MO diagram)

(b) The two orbital diagrams predict the same bond order (3) and same magnetic properties (diamagnetic), and so these properties cannot be used to determine which diagram is the correct one. That must be determined by more complex spectroscopic measurements.

15.129 (a) Cloud: [NO] = 860 ppt and $[NO_2]$ = 250 ppt; Clear air: [NO] = 480 ppt and $[NO_2]$ = 260 ppt;

(b) Clouds have the higher concentration of NO;

(c) in addition to jet engines and automobile emissions, NO is also formed when NO_2 dissolves in water:

$$3\,NO_2(g) + H_2O(l) \longrightarrow 2\,HNO_3(aq) + NO(g)$$

As a result one would expect to find more NO and less NO_2 in clouds, which also contain water vapor.

(d) the most likely product formed when NO (which is a radical) reacts with hydroxyl radical is nitrous acid:

$$\ddot{\underset{\cdot\cdot}{O}}=N-\ddot{\underset{\cdot\cdot}{O}}-H$$

(e) Nitrous acid would be expected to be more stable than either NO or hydroxyl radical since both of these species have unpaired electrons while nitrous acid does not.

(f) A possible mechanism that fits the observed rate law would be:

Step 1: $NO + O_3 \longrightarrow NO_2 + O_2$ $\left(slow\right)$

Step 2: $NO_2 + O \longrightarrow NO + O_2$ $\left(fast\right)$

(g) from part (f), rate = $k[O_3][NO]$; given that we know what the concentration of ozone is and the value of k, to determine the rate at which ozone molecules are being destroyed we need to determine the concentration of NO molecules present in the clear air sample. The total number of molecules in the sample per cm^3 can be calculated as follows:

$$220 \text{ mbar} \left(\frac{1 \text{ bar}}{1000 \text{ mbar}}\right)\left(\frac{1 \text{ kPa}}{1 \text{ bar}}\right)\left(\frac{1 \text{ atm}}{101.325 \text{ kPa}}\right) = 2.17 \times 10^{-3} \text{ atm}$$

$$\frac{n}{V} = \frac{P}{RT} = \frac{2.17 \times 10^{-3} \text{ atm}}{\left(0.08206 \text{ L} \cdot \text{atm} \cdot \text{mol}^{-1} \cdot \text{K}^{-1}\right)(243 \text{ K})} = 1.09 \times 10^{-4} \text{ mol of molecules} \cdot \text{L}^{-1}$$

$$= \left(1.09 \times 10^{-4} \text{ mol of molecules} \cdot \text{L}^{-1}\right)\left(6.022 \times 10^{23} \text{ molecules} \cdot \text{mol}^{-1}\right)\left(1 \times 10^{-3} \text{ L} \cdot \text{cm}^{-3}\right)$$

$$= 3.56 \times 10^{16} \text{ molecules} \cdot \text{cm}^{-3}$$

If the amount of NO present is 480 ppt, then the amount of NO in the sample in molecules per cm^3 is:

$$\left(6.56 \times 10^{16} \frac{\text{molecules}}{\text{cm}^3}\right)\left(\frac{480 \text{ NO molecules}}{1 \times 10^{12} \text{ molecules}}\right) = 3.15 \times 10^7 \text{ NO molecules} \cdot \text{cm}^{-3}$$

If $\left[O_3\right] = 4 \times 10^{-12} \text{ molec} \cdot \text{cm}^{-3}$ and $k = 6 \times 10^{-15} \text{ cm}^3 \cdot \text{molec}^{-1} \cdot \text{sec}^{-1}$; then

the rate at which ozone molecules is being destroyed in clear air is

$$\text{rate} = \left(6 \times 10^{-15} \text{ cm}^3 \cdot \text{molec}^{-1} \cdot \text{sec}^{-1}\right)\left(4 \times 10^{-12} \text{ molec } O_3 \cdot \text{cm}^{-3}\right)\left(3.15 \times 10^7 \text{ molec NO} \cdot \text{c}\right.$$

$$\text{rate} = 7.56 \times 10^{-19} \text{ molecules} \cdot \text{cm}^{-3} \cdot \text{sec}^{-1}$$

CHAPTER 16

THE ELEMENTS: THE *d* BLOCK

16.1 Elements at the left of the *d* block tend to have strongly negative standard potentials.

16.3 (a) Ti (b) One might expect gold to be larger than silver because it is a third row transition metal and silver is in row two; however, because of the lanthanide contraction, they are about the same size (Au, 144 pm; Ag, 144 pm) (c) Ta (d) Ir

16.5 (a) Fe; (b) Cu; (c) Pt; (d) Pd; (e) Ta

16.7 Hg is much more dense than Cd, because the shrinkage in atomic radius that occurs between $Z = 58$ and $Z = 71$ (the lanthanide contraction) causes the atoms following the rare earths to be smaller than might have been expected for their atomic masses and atomic numbers. Zn and Cd have densities that are not too dissimilar, because the radius of Cd is subject only to a smaller *d*-block contraction.

16.9 (a) Proceeding down a group in the *d* block (for example, from Cr to Mo to W), there is an increasing probability of finding the elements in higher oxidation states. That is, higher oxidation states become more stable on going down a group.

(b) The trend for the *p*-block elements is reversed. Because of the inert pair effect, the higher oxidation states tend to be less stable as one descends a group.

16.11 In MO_3, M has an oxidation number of +6. Of these three elements, the +6 oxidation state is most stable for Cr. See Fig. 16.6.

16.13 (a) $Ti(s)$, $MgCl_2(s)$

$$TiCl_4(g) + 2\ Mg(l) \xrightarrow{\Delta} Ti(s) + 2\ MgCl_2(s)$$

(b) $Co^{2+}(aq)$, $HCO_3^-(aq)$, $NO_3^-(aq)$

$$CoCO_3(s) + HNO_3\ (aq) \rightarrow Co^{2+}(aq) + HCO_3^-(aq) + NO_3^-(aq)$$

(c) $V(s)$, $CaO(s)$

$$V_2O_5(s) + 5\ Ca(l) \xrightarrow{\Delta} 2V(s)\quad 5\ CaO(s)$$

16.15 (a) titanium(IV) oxide, TiO_2

(b) iron(III) oxide, Fe_2O_3

(c) manganese(IV) oxide, MnO_2

(d) iron (II) chromite, $FeCr_2O_4$

16.17 (a) CO

(b) In Zones D & C,

$$3\ Fe_2O_3(s) + CO(g) \rightarrow 2\ Fe_3O_4(s) + CO_2(g)$$
$$Fe_3O_4(s) + CO(g) \rightarrow 3\ FeO(s) + CO_2(g)$$

These reactions combine to give

$$Fe_2O_3(s) + CO(g) \rightarrow 2\ FeO(s) + CO_2(g)$$

In Zone B,

$$Fe_2O_3(s) + 3\ CO(g) \rightarrow 2\ Fe(s) + 3\ CO_2(g)$$

$$FeO(s) + CO(g) \rightarrow Fe(s) + CO_2(g)$$

(c) carbon

16.19 (a) $Cr_2O_7^{2-} + 14\,H^+ + 6\,e^- \rightarrow 2\,Cr^{2+} + 7\,H_2O$ $E° = +1.33\ V$

$2\,Br^- \rightarrow Br_2 + 2\,e^-$ $E° = -1.09\ V$

$E°(\text{overall}) = +0.99\ V$

Therefore, Br^- will be oxidized to Br_2.

(b) $Cr_2O_7^{2-} + 14\,H^+ + 6\,e^- \rightarrow 2\,Cr^{2+} + 7\,H_2O$ $E° = +1.33\ V$

$Ag^+ \rightarrow Ag^{2+} + 1\,e^-$ $E° = -1.98\ V$

$E°(\text{overall}) = -0.65\ V$

Therefore, no reaction will occur.

16.21 (a) $V_2O_5(s) + 2\,H_3O^+(aq) \rightarrow 2\,VO_2^+(aq) + 3\,H_2O(l)$

(b) $V_2O_5(s) + 6\,OH^-(aq) \rightarrow 2\,VO_4^{3-}(aq) + 3\,H_2O(l)$

16.23 Even though all three Group 1B/11 metal atoms have the valence shell electron configuration $(n-1)d^{10}ns^1$, Cu is more reactive than Ag or Au. Metals ordinarily lose one or more electrons to form cations when they react with some other species. As the value of n increases, d and f electrons become less effective at shielding the outermost, highest energy electron(s) from the attractive charge of the nucleus. This higher effective nuclear charge makes it more difficult to oxidize the metal atom or ion. So, for example, Cu^{2+} exists in many common compounds (and can be formed by Cu^+ disproportionation in water) while Ag^{2+} does not. Furthermore, the valence electron orbital energies for most common Lewis bases match the orbitals of Cu and its cations more closely and would interact with them more favorably to form products.

16.25 (a) Cr^{3+} ions in water form the complex $[Cr(OH_2)_6]^{3+}(aq)$, which behaves as a Brønsted acid:

$[Cr(OH_2)_6]^{3+}(aq) + H_2O(l) \rightarrow [Cr(OH_2)_5OH]^{2+}(aq) + H_3O^+(aq)$

(b) The gelatinous precipitate is the hydroxide $Cr(OH)_3$. The precipitate

dissolves as the $Cr(OH)_4^-$ complex ion is formed:

$$Cr^{3+}(aq) + 3\,OH^-(aq) \rightarrow Cr(OH)_3(s)$$
$$Cr(OH)_3(s) + OH^-(aq) \rightarrow Cr(OH)_4^-(aq)$$

16.27 (a) hexacyanoferrate(II) ion

Let x = the oxidation number to be determined

$$x(Fe) + 6\,(-1) = -4$$
$$x(Fe) = -4 - (-6) = +2$$

(b) hexaamminecobalt(III) ion

$$x(Co) + 6(0) = +3$$
$$x(Co) = +3$$

(c) aquapentacyanocobaltate(III) ion

$$x(Co) + 5\,(-1) + 1(0) = -2$$
$$x(Co) = -2 - (-5) = +3$$

(d) pentaamminesulfatocobalt(III) ion

$$x(Co) + 1\,(-2) + 5\,(0) = +1$$
$$x(Co) = +1 - (-2) = +3$$

16.29 (a) $K_3[Cr(CN)_6]$

(b) $[Co(NH_3)_5(SO_4)]Cl$

(c) $[Co(NH_3)_4(H_2O)_2]Br_3$

(d) $Na[Fe(OH_2)_2(C_2O_4)_2]$

16.31 (a) The molecule $HN(CH_2CH_2NH_2)_2$ has three nitrogen atoms, each with

a lone pair of electrons that may be used for bonding to a metal center.

The molecule can thus function as a tridentate ligand.

(b) The CO_3^{2-} ion can bind to a metal ion through either one or two

oxygen atoms. It may, therefore, serve as a mono- or bidentate ligand.

(c) H_2O is always a monodenate ligand.

(d) The oxalate ion can bind through two oxygen atoms and is usually a bidentate ligand.

16.33 As shown below, only the molecule (b) can function as a chelating ligand. The two amine groups in (a) and (c) are arranged so that they would not be able to coordinate simultaneously to the same metal center. It is possible for each of the amine groups in (a) and (c) to coordinate to two different metal centers, however. This is not classified as chelating. When a single ligand binds to two different metal centers, it is known as a *bridging* ligand.

16.35 (a) 4 (b) 2 (c) 6 (en is bidentate) (d) 6 (EDTA is hexadentate)

16.37 (a) structural isomers, linkage isomers
(b) structural isomers, ionization isomers
(c) structural isomers, linkage isomers
(d) structural isomers, ionization isomers

16.39 (a) yes

Cl⁻ • H₂O

NH₃
Cl⁻ • H₂O

AND

trans-tetraamminedichloridocobalt (III)
chloride monohydrate

cis-tetraamminedichloridocobalt (III)
chloride monohydrate

(b) no

(c) yes

AND

cis-diamminedichloridoplatinum (II)

trans-diamminedichloridoplatinum (II)

16.41 Since nitro can bind as NO_2 or ONO, two linkage isomers are possible, as well as one ionization isomer:

16.43 (a)

complex

mirror image

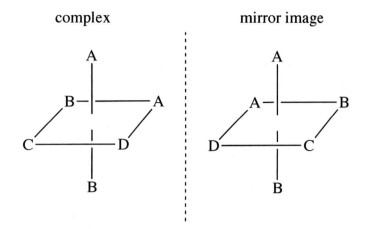

No rotation will make the complex and its mirror image match; therefore, it is chiral.

(b)

complex

mirror image

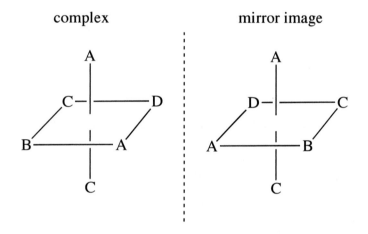

Two 90° rotations shows that the complex and its mirror image are superimposable:

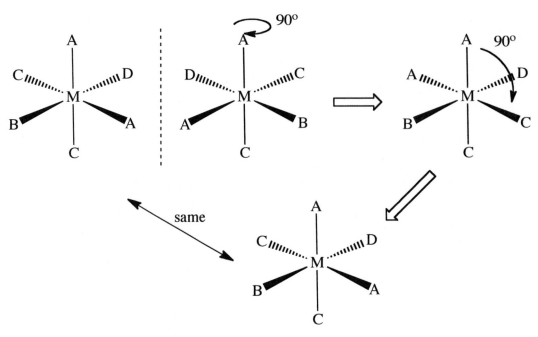

As a result, the complex is not chiral; in fact, the complex and its mirror image are identical (the same species) and are therefore not even isomers.

16.45 (a) 2; (b) 5; (c) 8; (d) 10; (e) 0 (or 8); (f) 10

16.47 (a) octahedral; strong-field ligand, 6 e⁻

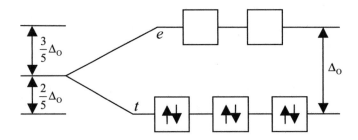

diamagnetic, no unpaired electrons

(b) tetrahedral: weak-field ligand, 8 e⁻

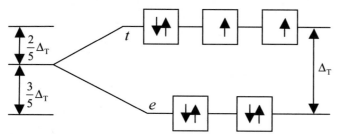

2 unpaired electrons

(c) octahedral: weak-field ligand, 5 e⁻

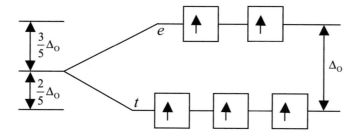

5 unpaired electrons

(d) octahedral: strong-field ligand, 5 e⁻

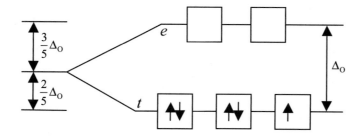

one unpaired electron

16.49 (a) $[Co(en)_3]^{3+}$ 6 e⁻

$[Co(en)_3]^{3+}$ has no unpaired electrons.

(b) $[Mn(CN)_6]^{3-}$ 4 e⁻

$[Mn(CN)_6]^{3-}$ has two unpaired electrons.

16.51 Weak-field ligands do not interact strongly with the *d*-electrons in the metal ion, so they produce only a small crystal field splitting of the *d*-electron energy states. The opposite is true of strong-field ligands. With weak-field ligands, unpaired electrons remain unpaired if there are unfilled orbitals; hence, a weak-field ligand is likely to lead to a high-spin complex. Strong-field ligands cause electrons to pair up with electrons in lower energy orbitals. A strong-field ligand is likely to lead to a low-spin complex. Ligands arranged in the spectrochemical series help to distinguish strong-field and weak-field ligands. Measurement of magnetic susceptibility (paramagnetism) can be used to determine the number of unpaired electrons, which, in turn, establishes whether the associated ligand is weak-field or strong-field in nature.

16.53 In both cases we are looking at the difference between an $Fe^{3+}(d^5)$ and an $Fe^{2+}(d^6)$. Since CN^- is a strong field ligand we expect that all the *d* electrons in the $[Fe(CN)_6]^{3-}$ complex are in the lower energy t_{2g} orbitals. Upon reduction, one more electron is added, resulting in all electrons being paired, giving a diamagnetic complex:

On the other hand, Cl^- is a weak field ligand; here we expect the *d* electrons to be distributed in both the t_{2g} and e_g the orbitals. Upon reduction, one more electron is added, resulting in only one electron being paired, leading to a complex that is still paramagnetic:

16.55 Because F^- is a weak-field ligand and en a strong-field ligand, the splitting between levels is less in (a) than in (b). Therefore, (a) will absorb light of longer wavelength than will (b) and consequently will display a shorter wavelength color. Blue light is shorter in wavelength than yellow light, so (a) $[CoF_6]^{3-}$ is blue and (b) $[Co(en)_3]^{3+}$ is yellow.

16.57 $E_{photon} = \left(\dfrac{209 \text{ kJ}}{\text{mol}} \right) \left(\dfrac{1 \text{ mol}}{6.02 \times 10^{23} \text{photons}} \right) = 3.47 \times 10^{-22} \text{ kJ} \cdot \text{photon}^{-1}$

$v = \dfrac{E}{h} = \dfrac{3.47 \times 10^{-19} \text{ J}}{6.626 \times 10^{-34} \text{ J} \cdot \text{s}} = 5.24 \times 10^{14} \text{ s}^{-1}$

$\lambda = \dfrac{c}{v} = \left(\dfrac{3.00 \times 10^8 \text{ m} \cdot \text{s}^{-1}}{5.24 \times 10^{14} \text{ s}^{-1}} \right) \left(\dfrac{10^9 \text{ nm}}{1 \text{ m}} \right) = 573 \text{ nm}$

This wavelength is in the yellow region of the visible spectrum. Since the complex absorbs yellow light, it transmits the complement, or purple (a.k.a. violet).

16.59 (a) There are no unpaired electrons in this complex (low spin, d^6).

(b) Both NH_3 and CN^- are strong field ligands, with cyanide being slightly stronger than ammonia (slightly larger splitting in $[Co(CN)_6]^{3-}$ than $[Co(NH_3)_6]^{3+}$). Exchange of the CN^- ligands for NH_3 should lead to the wavelength of the absorbed radiation to be slightly longer.

16.61 In Zn^{2+}, the 3d-orbitals are filled (d^{10}). Therefore, there can be no electronic transitions between the t and e levels; hence, no visible light is absorbed and the aqueous ion is colorless. The d^{10} configuration has no unpaired electrons, so Zn compounds would be diamagnetic, not paramagnetic.

16.63 (a) $\Delta_O = \dfrac{hc}{\lambda} = \dfrac{(6.63 \times 10^{-34}\ J \cdot s^{-1})(3.00 \times 10^8\ m \cdot s^{-1})}{740 \times 10^{-9}\ m} = 2.69 \times 10^{-19}\ J$

(b) $\Delta_O = \dfrac{hc}{\lambda} = \dfrac{(6.63 \times 10^{-34}\ J \cdot s^{-1})(3.00 \times 10^8\ m \cdot s^{-1})}{460 \times 10^{-9}\ m} = 4.32 \times 10^{-19}\ J$

(c) $\Delta_O = \dfrac{hc}{\lambda} = \dfrac{(6.63 \times 10^{-34}\ J \cdot s^{-1})(3.00 \times 10^8\ m \cdot s^{-1})}{575 \times 10^{-9}\ m} = 3.46 \times 10^{-19}\ J$

These numbers can be multiplied by 6.02×10^{23} to obtain $kJ \cdot mol^{-1}$.

(a) $2.69 \times 10^{-19}\ J\left(10^{-3}\ kJ \cdot J^{-1}\right)\left(6.02 \times 10^{23}\ mol^{-1}\right) = 162\ kJ \cdot mol^{-1}$

(b) $4.32 \times 10^{-19}\ J\left(10^{-3}\ kJ \cdot J^{-1}\right)\left(6.02 \times 10^{23}\ mol^{-1}\right) = 260\ kJ \cdot mol^{-1}$

(c) $3.46 \times 10^{-19}\ J\left(10^{-3}\ kJ \cdot J^{-1}\right)6.02 \times 10^{23}\ mol^{-1} = 208\ kJ \cdot mol^{-1}$

(d) $Cl^- < H_2O < NH_3$ (spectrochemical series)

16.65 The e_g set, which comprises the $d_{x^2-y^2}$ and d_{z^2} orbitals.

16.67 (a) The CN^- ion is a π-acid ligand accepting electrons into the empty π^* orbital created by the C—N multiple bond. (b) The Cl^- ion has extra lone pairs in addition to the one that is used to form the σ-bond to the metal, and so it can act as a π-base, donating electrons in a p-orbital to an empty d-orbital on the metal. (c) H_2O, like Cl^-, also has an "extra" lone pair of electrons that can be donated to a metal center, making it a weak π-base; (d) en is neither a π-acid nor a π-base, because it does not have any empty π-type antibonding orbitals nor does it have any extra lone pairs of electrons to donate. (e) $Cl^- < H_2O <$ en $< CN^-$. Note that the spectrochemical series orders the ligands as π-bases $< \sigma$-bond only ligands $< \pi$-acceptors.

16.69 Nonbonding or slightly antibonding. In a complex that forms only σ-bonds, the t_{2g} set of orbitals is nonbonding. If the ligands can function as weak π-donors (those close to the middle of the spectrochemical series, such as H_2O), the t_{2g} set becomes slightly antibonding by interacting with the filled p-orbitals on the ligands.

16.71 Antibonding. The e_g set of orbitals on an octahedral metal ion are always antibonding because of interactions with ligand orbitals that form the σ-bonds. This is true regardless of whether the ligands are π-acceptors, π-donors, or neither.

16.73 Water has two lone pairs of electrons. Once one of these is used to form the σ-bond to the metal ion, the second may be used to form a π-bond. This causes the t_{2g} set of orbitals to move up in energy, making Δ_O smaller; therefore, water is a weak-field ligand. Ammonia does not have this extra lone pair of electrons and consequently cannot function as a π-donor ligand.

16.75 The green color suggests chromium or copper but the blue solution formed when the mineral is dissolved in sulfuric acid points to copper as the metal ($CuSO_4$ is blue). The colorless gas that forms either upon heating or on treatment with acid is carbon dioxide; this is proven when the gas is bubbled into limewater (the cloudiness in the solution is due to calcium carbonate formation). CO_2 is formed when carbonates are either heated or treated with acid. The mineral is most likely basic copper carbonate ($Cu_2CO_3(OH)_2$).

16.77 (a) +3 is the most likely oxidation state of vanadium

(b) $2 V_2O_5(s) + 8 C(s) \rightarrow V_4C_3(s) + 5 CO_2(g)$

16.79 $Cr(OH)_3(s) + 3\,e^- \rightarrow Cr(s) + 3\,OH^-$ $E° = -1.34\ V$

$Cr(s) \rightarrow Cr^{3+} + 3\,e^-$ $E° = +0.74\ V$

$Cr(OH)_3(s) \rightarrow Cr^{3+} + 3\,OH^-$ $E° = -0.60\ V$

$\Delta G° = -nFE° = -RT\ln K$

$$\ln K = \frac{nFE°}{RT} = \frac{(3)(96\ 485\ \text{J}\cdot\text{V}^{-1}\cdot\text{mol}^{-1})(-0.60\ \text{V})}{(8.314\ \text{J}\cdot\text{K}^{-1}\cdot\text{mol}^{-1})(298\ \text{K})} = -70.1$$

$K_{sp} = e^{-70.1} = 3.6 \times 10^{-31}$

16.81   (a) $[PtBrCl(NH_3)_2]$

AND

cis-diamminebromidochloridoplatinum (II)   *trans*-diamminebromidochloridoplatinum (II)

(b)   If the compound were tetrahedral, there would be only one compound, not two.

16.83   (a)   The first, $[Ni(SO_4)(en)_2]Cl_2$, will give a precipitate of AgCl when $AgNO_3$ is added; the second will not.   (b)   The second, $[NiCl_2(en)_2]I_2$, will show free I_2 when mildly oxidized with, for example, Br_2, but the first will not.

16.85   (a)

$[MnCl_6]^{4-}$

5 e⁻, Cl⁻ is a weak-field ligand

$[Mn(CN)_6]^{4-}$

5 e⁻, CN⁻ is a strong-field ligand

(b) $[MnCl_6]^{4-}$: five; $[Mn(CN)_6]^{4-}$: one.

(c) Complexes with weak-field ligands absorb longer wavelength light, therefore, $[MnCl_6]^{4-}$ absorbs longer wavelengths.

16.87 If the prismatic structure were the true structural form of $\left[CoCl_2(NH_3)_4\right]$ one would expect to get four possible isomers, not the two seen $(X = NH_3)$:

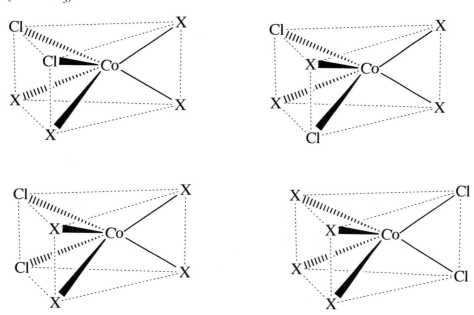

16.89 The correct structure for $[Co(NH_3)_6]Cl_3$ consists of four ions, $Co(NH_3)_6^{3+}$, and 3 Cl^- in aqueous solution. The chloride ions can be easily precipitated as AgCl. This would not be possible if they were bonded to the other (NH_3) ligands. If the structure were $Co(NH_3-NH_3-Cl)_3$, VSEPR theory would predict that the Co^{3+} ion would have a trigonal planar ligand arrangement. The splitting of the d-orbital energies would not be the same as the octahedral arrangement and would lead to different spectroscopic and magnetic properties inconsistent with the experimental evidence. In addition, neither optical nor geometrical isomers would be observed.

16.91 See Figure 16.26. Remove (mentally) the ligands along the $\pm z$ axes. From the shape of the atomic orbitals and their orientation with respect to the x, y, and z axes, it is clear that the $d_{x^2-y^2}$-orbital will have the greatest overlap, therefore repulsion, from the ligands in the xy plane. The d_{xy}-orbital will have the next strongest repulsion from these ligands, d_{z^2} next, and finally, the weakest repulsion will be with d_{xz} and d_{yz}. Therefore, the energy-level diagram will be (schematically):

The building-up principle is illustrated in the diagram with a d^8 ion (Pd^{2+}). The separation in energy between the $d_{x^2-y^2}$ and d_{xy} orbitals determines whether or not there will be unpaired electrons.

16.93 The electron configuration expected for Ni^{2+} is $\left[Ar\right]3d^8$. For this complex to have no unpaired electrons it would have to be square planar in its electronic geometry, as both the octahedral and tetrahedral geometries require a d^8 species to have two unpaired electrons. Square planar does not:

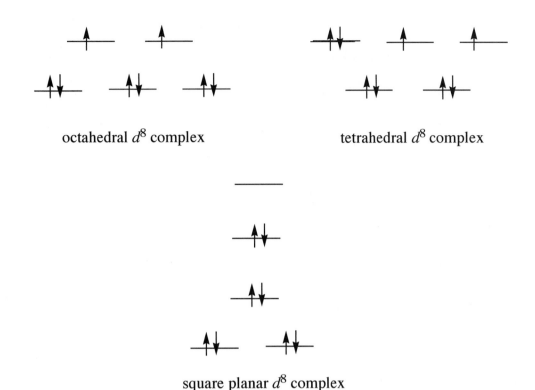

octahedral d^8 complex

tetrahedral d^8 complex

square planar d^8 complex

16.95 Since NH_3 is a strong-field ligand (large Δ_o) and water is a weak-field ligand (small Δ_o), of the octahedral cobalt (II) complexes, $\left[Co(NH_3)_6\right]^{2+}$ and $\left[Co(OH_2)_6\right]^{2+}$, it is expected that the hexaammine complex will absorb at a shorter wavelength and therefore transmit light of a higher wavelength than the hexaaquo complex. $\left[CoCl_4\right]^{2-}$ is a tetrahedral complex, and since the magnitude of Δ_t is much smaller than Δ_o it is expected this complex will absorb at the longest wavelength of the three and therefore transmit light of the shortest wavelength. Based on this $\left[Co(NH_3)_6\right]^{2+}$ is yellow, $\left[Co(OH_2)_6\right]^{2+}$ is pink and $\left[CoCl_4\right]^{2-}$ is blue.

16.97 AgCl will only form from Cl^- found outside the coordination sphere.

(a) the amount of Cl^- outside the coordination sphere in $CoCl_3 \cdot 5NH_3 \cdot H_2O$ is:

$$0.43 \text{ g AgCl} \left(\frac{1 \text{ mol AgCl}}{143.35 \text{ g AgCl}} \right)\left(\frac{1 \text{ mol Cl}^-}{1 \text{ mol AgCl}} \right) = 0.0030 \text{ mol Cl}^-\text{ ; since you}$$

started with 0.0010 mol of the salt the ratio of salt to Cl⁻ outside the coordination sphere is 1:3. The correct formula and name for this compound is $\left[Co(NH_3)_5 OH_2 \right]Cl_3$, pentaammineaquacobalt (III) chloride.

(b) the amount of Cl⁻ outside the coordination sphere in $CoCl_3 \cdot 5NH_3$ is:

$$0.29 \text{ g AgCl} \left(\frac{1 \text{ mol AgCl}}{143.35 \text{ g AgCl}} \right)\left(\frac{1 \text{ mol Cl}^-}{1 \text{ mol AgCl}} \right) = 0.0020 \text{ mol Cl}^-\text{ ; since you}$$

started with 0.0010 mol of the salt the ratio of salt to Cl⁻ outside the coordination sphere is 1:2. The correct formula and name for this compound is $\left[CoCl(NH_3)_5 \right]Cl_2$, pentaamminechloridocobalt (III) chloride.

16.99 (a) trigonal bipyramidal

(b) +1

(c) There are six possible isomers (including the one shown on the Web site).

nonsuperimposable mirror images
(*enantiomers*)

16.101 $Cu^+ + e^- \rightarrow Cu$ Eq 1. $E° = +0.52$ V

     $Cu^{2+} + e^- \rightarrow Cu^+$ Eq. 2 $E° = +0.15$ V

     Reversing Eq. 2 gives

     $Cu^+ \rightarrow Cu^{2+} + e^-$ Eq. 3

     The cathode process is given by Eq. 1; the anode process by Eq. 3.

Combining   them yields the overall (disproportionation) reaction:

     $2\,Cu^+ \rightarrow Cu^{2+} + Cu$

     The potential of this reaction is

     $E°_{cell} = E°_{cathode} - E°_{anode}$

     $E°_{cell} = +0.52$ V $- (+0.15$ V$) = +0.37$ V

     $\ln K = \dfrac{nFE}{RT} = \dfrac{2 \times (+0.37 \text{ V})}{0.025\ 693 \text{ V}} = +14.4$

     $K = 2 \times 10^6$

16.103 The equations for the reactions of Fe^{2+} with MnO_4^- or $Cr_2O_7^{2-}$ are:

     $MnO_4^-(aq) + 5\,Fe^{2+}(aq) + 8\,H^+(aq) \rightarrow 5\,Fe^{3+}(aq) + Mn^{2+}(aq) + 4\,H_2O(l)$

     $Cr_2O_7^{2-}(aq) + 6\,Fe^{2+}(aq) + 14\,H^+(aq) \rightarrow 6\,Fe^{3+}(aq) + 2\,Cr^{3+}(aq) + 7\,H_2O(l)$

     The amount of Fe^{2+} determined using 0.0210 M $Cr_2O_7^{2-}$ is:

     $25.20 \text{ mL} \left(\dfrac{1 \text{ L}}{1000 \text{ mL}} \right) \left(\dfrac{0.0210 \text{ mol } Cr_2O_7^{2-}}{1 \text{ L}} \right) \left(\dfrac{6 \text{ mol } Fe^{2+}}{1 \text{ mol } Cr_2O_7^{2-}} \right) =$

     3.175×10^{-3} mol Fe^{2+}.

     Based on this, the amount of 0.0420 M MnO_4^- needed to do this

     determination will be:

     3.175×10^{-3} mol $Fe^{2+} \left(\dfrac{1 \text{ mol } MnO_4^-}{5 \text{ mol } Fe^{2+}} \right) \left(\dfrac{1 \text{ L}}{0.0420 \text{ mol } MnO_4^-} \right) \left(\dfrac{1000 \text{ mL}}{1 \text{ L}} \right) =$

     15.12 mL.

CHAPTER 17

NUCLEAR CHEMISTRY

17.1 We assume that all the change in energy goes into the energy of the γ ray emitted. Then, in each case,

$$\nu = \frac{\Delta E}{h}, \qquad \lambda = \frac{c}{\nu}$$

$$\text{energy of 1 MeV} = \left(\frac{10^6 \text{ eV}}{1 \text{ MeV}}\right)\left(\frac{1.602 \times 10^{-19} \text{ J}}{1 \text{ eV}}\right)$$

$$= 1.602 \times 10^{-13} \text{ J} \cdot \text{MeV}^{-1}$$

(a) $\Delta E = (1.33 \text{ MeV})\left(\frac{1.602 \times 10^{-13} \text{ J}}{1 \text{ MeV}}\right) = 2.1307 \times 10^{-13} \text{ J}$

$$\nu = \frac{\Delta E}{h} = \frac{2.1307 \times 10^{-13} \text{ J}}{6.626 \times 10^{-34} \text{ J} \cdot \text{s}} = 3.22 \times 10^{20} \text{ s}^{-1} = 3.22 \times 10^{20} \text{ Hz}$$

$$\lambda = \frac{c}{\nu} = \frac{3.00 \times 10^8 \text{ m} \cdot \text{s}^{-1}}{3.22 \times 10^{20} \text{ s}^{-1}} = 9.32 \times 10^{-13} \text{ m}$$

(b) $\Delta E = (1.64 \text{ MeV})\left(\frac{1.602 \times 10^{-13} \text{ J}}{1 \text{ MeV}}\right) = 2.628 \times 10^{-13} \text{ J}$

$$\nu = \frac{\Delta E}{h} = \frac{2.628 \times 10^{-13} \text{ J}}{6.626 \times 10^{-34} \text{ J} \cdot \text{s}} = 3.97 \times 10^{20} \text{ s}^{-1} = 3.97 \times 10^{20} \text{ Hz}$$

$$\lambda = \frac{3.00 \times 10^8 \text{ m} \cdot \text{s}^{-1}}{3.97 \times 10^{20} \text{ s}^{-1}} = 7.56 \times 10^{-13} \text{ m}$$

(c) $\Delta E = (1.10 \text{ MeV})\left(\frac{1.602 \times 10^{-13} \text{ J}}{1 \text{ MeV}}\right) = 1.76 \times 10^{-13} \text{ J}$

$$\nu = \frac{\Delta E}{h} = \frac{1.76 \times 10^{-13} \text{ J}}{6.626 \times 10^{-34} \text{ J} \cdot \text{s}} = 2.66 \times 10^{20} \text{ s}^{-1} = 2.66 \times 10^{20} \text{ Hz}$$

$$\lambda = \frac{c}{\nu} = \frac{3.00 \times 10^8 \text{ m} \cdot \text{s}^{-1}}{2.65 \times 10^{20} \text{ s}^{-1}} = 1.13 \times 10^{-12} \text{ m}$$

17.3 Potassium-40 has 21 neutrons; $^{39}_{18}\text{Ar}$ and $^{41}_{20}\text{Ca}$ are isotones of K-40.

17.5 (a) $^{3}_{1}\text{T} \rightarrow {}^{0}_{-1}\text{e} + {}^{3}_{2}\text{He}$

(b) $^{83}_{39}\text{Y} \rightarrow {}^{0}_{1}\text{e} + {}^{83}_{38}\text{Sr}$

(c) $^{87}_{36}\text{Kr} \rightarrow {}^{0}_{-1}\text{e} + {}^{87}_{37}\text{Rb}$

(d) $^{225}_{91}\text{Pa} \rightarrow {}^{4}_{2}\alpha + {}^{221}_{89}\text{Ac}$

17.7 (a) $^{8}_{5}\text{B} \rightarrow {}^{0}_{1}\text{e} + {}^{A}_{Z}\text{E}$ $A = 8 - 0 = 8, Z = 5 - 1 = 4, \text{E} = \text{Be}$

so, $^{8}_{5}\text{B} \rightarrow {}^{0}_{1}\text{e} + {}^{8}_{4}\text{Be}$

(b) $^{63}_{28}\text{Ni} \rightarrow {}^{0}_{-1}\text{e} + {}^{A}_{Z}\text{E}$ $A = 63 - 0 = 63, \ Z = 28 - (-1) = 29, \text{E} = \text{Cu}$

so, $^{63}_{28}\text{Ni} \rightarrow {}^{0}_{-1}\text{e} + {}^{63}_{29}\text{Cu}$

(c) $^{185}_{79}\text{Au} \rightarrow {}^{4}_{2}\alpha + {}^{A}_{Z}\text{E}$ $A = 185 - 4 = 181, Z = 79 - 2 = 77, \text{E} = \text{Ir}$

so, $^{185}_{79}\text{Au} \rightarrow {}^{4}_{2}\alpha + {}^{181}_{77}\text{Ir}$

(d) $^{7}_{4}\text{Be} + {}^{0}_{-1}\text{e} \rightarrow {}^{A}_{Z}\text{E}$ $A = 7 + 0 = 7, \ Z = 4 - 1 = 3, \text{E} = \text{Li}$

so, $^{7}_{4}\text{Be} + {}^{0}_{-1}\text{e} \rightarrow {}^{7}_{3}\text{Li}$

17.9 (a) $^{24}_{11}\text{Na} \rightarrow {}^{24}_{12}\text{Mg} + {}^{0}_{-1}\text{e}$; a β particle is emitted.

(b) $^{128}_{50}\text{Sn} \rightarrow {}^{128}_{51}\text{Sb} + {}^{0}_{-1}\text{e}$; a β particle is emitted.

(c) $^{140}_{57}\text{La} \rightarrow {}^{140}_{56}\text{Ba} + {}^{0}_{1}\text{e}$; a positron ($\beta^{+}$) is emitted.

(d) $^{228}_{90}\text{Th} \rightarrow {}^{224}_{88}\text{Ra} + {}^{4}_{2}\alpha$; an α particle is emitted.

17.11 (a) $^{11}_{5}\text{B} + {}^{4}_{2}\alpha \rightarrow 2\,{}^{1}_{0}\text{n} + {}^{13}_{7}\text{N}$

(b) $^{35}_{17}\text{Cl} + {}^{2}_{1}\text{D} \rightarrow {}^{1}_{0}\text{n} + {}^{36}_{18}\text{Ar}$

(c) $^{96}_{42}\text{Mo} + {}^{2}_{1}\text{D} \rightarrow {}^{1}_{0}\text{n} + {}^{97}_{43}\text{Tc}$

(d) $^{45}_{21}\text{Sc} + {}^{1}_{0}\text{n} \rightarrow {}^{4}_{2}\alpha + {}^{42}_{19}\text{K}$

17.13 (a) $A/Z = 68/29 = 2.34 > (A/Z)_{based}$; hence, $_{29}^{68}\text{Cu}$

is neutron rich, and β decay is most likely. $_{29}^{68}\text{Cu} \rightarrow _{-1}^{0}\text{e} + _{30}^{68}\text{Zn}$

(b) $A/Z = 103/48 = 2.15 < (A/Z)_{based}$; therefore,

$_{48}^{103}\text{Cd}$ is proton rich, and β^+ decay is most likely. $_{48}^{103}\text{Cd} \rightarrow _{1}^{0}\text{e} + _{47}^{103}\text{Ag}$

(c) $_{97}^{243}\text{Bk}$ has $Z > 83$ and is proton rich; therefore, α decay is most likely.

$_{97}^{243}\text{Bk} \rightarrow _{2}^{4}\alpha + _{95}^{239}\text{Am}$

(d) $_{105}^{260}\text{Db}$ has $Z > 83$; therefore, α decay is most likely.

$_{105}^{260}\text{Db} \rightarrow _{2}^{4}\alpha + _{103}^{256}\text{Lr}$

17.15 α $\quad _{92}^{235}\text{U} \rightarrow _{2}^{4}\alpha + _{90}^{231}\text{Th}$

β $\quad _{90}^{231}\text{Th} \rightarrow _{-1}^{0}\text{e} + _{91}^{231}\text{Pa}$

α $\quad _{91}^{231}\text{Pa} \rightarrow _{2}^{4}\alpha + _{89}^{227}\text{Ac}$

β $\quad _{89}^{227}\text{Ac} \rightarrow _{-1}^{0}\text{e} + _{90}^{227}\text{Th}$

α $\quad _{90}^{227}\text{Th} \rightarrow _{2}^{4}\alpha + _{88}^{223}\text{Ra}$

α $\quad _{88}^{223}\text{Ra} \rightarrow _{2}^{4}\alpha + _{86}^{219}\text{Rn}$

α $\quad _{86}^{219}\text{Rn} \rightarrow _{2}^{4}\alpha + _{84}^{215}\text{Po}$

β $\quad _{84}^{215}\text{Po} \rightarrow _{-1}^{0}\text{e} + _{85}^{215}\text{At}$

α $\quad _{85}^{215}\text{At} \rightarrow _{2}^{4}\alpha + _{83}^{211}\text{Bi}$

β $\quad _{83}^{211}\text{Bi} \rightarrow _{-1}^{0}\text{e} + _{84}^{211}\text{Po}$

α $\quad _{84}^{211}\text{Po} \rightarrow _{2}^{4}\alpha + _{82}^{207}\text{Pb}$

17.17 To determine the charge and mass of the unknown particle, it helps to

write $_{1}^{1}\text{p}$ and $_{0}^{1}\text{n}$ for the proton and neutron, respectively; and $_{-1}^{0}\text{e}$ and $_{1}^{0}\text{e}$

for the β particle and positron, respectively.

(a) $_{7}^{14}\text{N} + _{2}^{4}\alpha \rightarrow _{8}^{17}\text{O} + _{1}^{1}\text{p}$

(b) $_{96}^{248}\text{Cm} + _{0}^{1}\text{n} \rightarrow _{97}^{249}\text{Bk} + _{-1}^{0}\text{e}$

(c) $^{243}_{95}\text{Am} + ^{1}_{0}\text{n} \rightarrow ^{244}_{96}\text{Cm} + ^{0}_{-1}\text{e} + \gamma$

(d) $^{13}_{6}\text{C} + ^{1}_{0}\text{n} \rightarrow ^{14}_{6}\text{C} + \gamma$

17.19 (a) $^{20}_{10}\text{Ne} + ^{4}_{2}\alpha \rightarrow ^{8}_{4}\text{Be} + ^{16}_{8}\text{O}$

(b) $^{20}_{10}\text{Ne} + ^{20}_{10}\text{Ne} \rightarrow ^{16}_{8}\text{O} + ^{24}_{12}\text{Mg}$

(c) $^{44}_{20}\text{Ca} + ^{4}_{2}\alpha \rightarrow \gamma + ^{48}_{22}\text{Ti}$

(d) $^{27}_{13}\text{Al} + ^{2}_{1}\text{H} \rightarrow ^{1}_{1}\text{p} + ^{28}_{13}\text{Al}$

17.21 For iron-56 we would expect the following nuclear reaction to occur:

$$^{56}_{26}\text{Fe} + 6\ ^{1}_{0}n \longrightarrow ^{62}_{26}\text{Fe} \longrightarrow ^{62}_{32}\text{Ge} + 6\ ^{0}_{-1}e$$

The isotope iron-56 will be converted to germanium-56.

17.23 In each case, identify the unknown particle by performing a mass and charge balance as you did in the solutions to Exercises 17.5 and 17.7. Then write the complete nuclear equation.

(a) $^{14}_{7}\text{N} + ^{4}_{2}\alpha \rightarrow ^{17}_{8}\text{O} + ^{1}_{1}\text{p}$

(b) $^{239}_{94}\text{Pu} + ^{1}_{0}\text{n} \rightarrow ^{240}_{95}\text{Am} + ^{0}_{-1}\text{e}$

17.25 (a) unbihexium, Ubh   (b) untrihexium, Uth   (c) binilnilium, Bnn

17.27 activity $= (5.3 \times 10^{5}\ \text{Bq})\left(\dfrac{1\,\text{Ci}}{3.7 \times 10^{10}\ \text{Bq}}\right) = 1.4 \times 10^{-5}\ \text{Ci}$

17.29 1 Bq $= 1$ disintegration per second (dps)

(a) $(2.5\ \mu\text{Ci})\left(\dfrac{10^{-6}\ \text{Ci}}{1\ \mu\text{Ci}}\right)\left(\dfrac{3.7 \times 10^{10}\ \text{dps}}{1\ \text{Ci}}\right) = 9.2 \times 10^{4}\ \text{dps}$

$$= 9.2 \times 10^{4}\ \text{Bq}$$

(b) $142\ \text{Ci} = (142)(3.7 \times 10^{10}\ \text{dps}) = 5.3 \times 10^{12}\ \text{Bq}$

(c) $(7.2 \text{ mCi})\left(\dfrac{10^{-3} \text{ Ci}}{1 \text{ mCi}}\right)\left(\dfrac{3.7 \times 10^{10} \text{ dps}}{1 \text{ Ci}}\right) = 2.7 \times 10^8 \text{ dps}$

$$= 2.7 \times 10^8 \text{ Bq}$$

17.31 $\text{dose} = \dfrac{15 \text{ J}}{2.0 \text{ kg}} = 0.75 \text{ J} \cdot \text{kg}^{-1}$

$\text{dose in rads} = 0.75 \text{ J} \cdot \text{kg}^{-1} \times \left(\dfrac{1 \text{ rad}}{10^{-2} \text{ J} \cdot \text{kg}^{-1}}\right) = 75 \text{ rad}$

$\text{dose equivalent in rems} = Q \times \text{dose in rads}$

$$= \left(\dfrac{1 \text{ rem}}{1 \text{ rad}}\right)(75 \text{ rad}) = 75 \text{ rem}$$

$75 \text{ rem} \div 100 \text{ rem/Sv} = 0.75 \text{ Sv}$

17.33 $1.0 \text{ rad} \cdot \text{day}^{-1} = (1.0 \text{ rad} \cdot \text{day}^{-1})\left(\dfrac{1 \text{ rem}}{1 \text{ rad}}\right) = 1 \text{ rem} \cdot \text{day}^{-1}$

$100 \text{ rem} = 1 \text{ rem} \cdot \text{day}^{-1} \times \text{time}$
$\text{time} = 100 \text{ day}$

17.35 $k = \dfrac{\ln 2}{t_{1/2}}$

(a) $k = \dfrac{\ln 2}{12.3 \text{ a}} = 5.64 \times 10^{-2} \text{ a}^{-1}$

(b) $k = \dfrac{\ln 2}{0.84 \text{ s}} = 0.83 \text{ s}^{-1}$

(c) $k = \dfrac{\ln 2}{10.0 \text{ min}} = 0.0693 \text{ min}^{-1}$

17.37 We know that initial activity $= N_0 = 2150 \text{ dpm}$ and final activity $= N = 1324 \text{ dpm}$ at $t = 6.0 \text{ h}$. Therefore,

$$\ln\left(\frac{N}{N_0}\right) = -kt \text{ and } k = \frac{\ln 2}{t_{1/2}}$$

$$\text{So } \ln\left(\frac{N}{N_0}\right) = -\frac{t\ln 2}{t_{1/2}}.$$

$$\ln\left(\frac{1342}{2150}\right) = -\frac{(6.0 \text{ h})\ln 2}{t_{1/2}} = -0.48$$

$$t_{1/2} = -\frac{(6.0 \text{ h})\ln 2}{-0.48}; \quad t_{1/2} = 8.8 \text{ h}.$$

17.39 In each case, $k = \dfrac{\ln 2}{t_{1/2}}$, $N = N_0 e^{-kt}$, $\dfrac{N}{N_0} = e^{-kt}$, and the percentage remaining

$$= 100\% \times (N/N_0)$$

(a) $k = \dfrac{\ln 2}{5.73 \times 10^3 \text{ a}} = 1.21 \times 10^{-4} \text{ a}^{-1}$

percentage remaining $= 100\% \times e^{-(1.21 \times 10^{-4} \text{ a}^{-1} \times 3000 \text{ a})} = 69.6\%$

(b) $k = \dfrac{\ln 2}{12.3 \text{ a}} = 0.0563 \text{ a}^{-1}$

percentage remaining $= 100\% \times e^{-(0.0563 \text{ a}^{-1} \times 12.0 \text{ a})} = 50.9\%$

17.41 Fraction of K-40 remaining $= \dfrac{N}{N_0} = \dfrac{3}{5}$;

$$t_{1/2} = 1.26 \times 10^9 \text{ a}, \quad k = \frac{\ln 2}{1.26 \times 10^9 \text{ a}} = 5.50 \times 10^{-10} \text{ a}^{-1}$$

$$\frac{3}{5} = e^{-kt}$$

$$\frac{3}{5} = e^{-(5.50 \times 10^{-10} \text{ a}^{-1} \times x)}$$

$$x = 9.29 \times 10^8 \text{ a}$$

17.43 Let dis = disintegrations

activity from "old" sample $= \dfrac{1500 \text{ dis}/0.250 \text{ g}}{10.0 \text{ h}} = 600 \text{ dis}\cdot\text{g}^{-1}\cdot\text{h}^{-1}$

activity from current sample $= 921 \text{ dis}\cdot\text{g}^{-1}\cdot\text{h}^{-1}$

$k = \dfrac{\ln 2}{t_{1/2}} = \dfrac{\ln 2}{5.73 \times 10^3 \text{ a}} = 1.21 \times 10^{-4} \text{ a}^{-1}$

old activity $\propto N$, current activity $\propto N_0$

$\dfrac{\text{old activity}}{\text{current activity}} = \dfrac{N}{N_0} = e^{-kt}, \dfrac{N_0}{N} = e^{kt}, \ln\left(\dfrac{N_0}{N}\right) = kt$

Solve for $t \ (= \text{age})$:

$t = \dfrac{\ln\left(\dfrac{N_0}{N}\right)}{k} = \dfrac{\ln\left(\dfrac{921}{600}\right)}{1.21 \times 10^{-4} \text{ a}^{-1}} = 3.54 \times 10^3 \text{ a}$

17.45 In each case, $k = \dfrac{\ln 2}{t_{1/2} \text{ (in s)}}$, activity in $Bq = k \times N$

activity in $Ci = \dfrac{\text{activity in } Bq}{3.7 \times 10^{10} \ Bq \cdot Ci^{-1}}$

Note: $Bq \ (= \text{disintegrating nuclei per second})$ has the units of $\text{nuclei}\cdot\text{s}^{-1}$

(a) $k = \left(\dfrac{\ln 2}{1.60 \times 10^3 \text{ a}}\right)\left(\dfrac{1 \text{ a}}{3.17 \times 10^7 \text{ s}}\right) = 1.37 \times 10^{-11} \text{ s}^{-1}$

$N = (1.0 \times 10^{-3} \text{ g})\left(\dfrac{1 \text{ mol}}{226 \text{ g}}\right)\left(\dfrac{6.02 \times 10^{23} \text{ nuclei}}{1 \text{ mol}}\right) = 2.7 \times 10^{18} \text{ nuclei}$

activity $= 1.37 \times 10^{-11} \text{ s}^{-1} \times 2.7 \times 10^{18} \text{ nuclei} \times \left(\dfrac{1 Ci}{3.7 \times 10^{10} \ Bq}\right)$

$= 9.9 \times 10^{-4} \text{ Ci}$

(b) $k = \left(\dfrac{\ln 2}{28.1 \text{ a}}\right)\left(\dfrac{1 \text{ a}}{3.17 \times 10^7 \text{ s}}\right) = 7.80 \times 10^{-10} \text{ s}^{-1}$

$N = (2.0 \times 10^{-3} \text{ g})\left(\dfrac{1 \text{ mol}}{90 \text{ g}}\right)\left(\dfrac{6.022 \times 10^{23} \text{ nuclei}}{1 \text{ mol}}\right) = 1.3 \times 10^{19} \text{ nuclei}$

$$\text{activity} = (7.80 \times 10^{-10} \text{ s}^{-1})(1.3 \times 10^{19} \text{ nuclei})\left(\frac{1\,\text{Ci}}{3.7 \times 10^{10}\,\text{Bq}}\right)$$

$$= 0.28 \text{ Ci}$$

(c) $\quad k = \left(\dfrac{\ln 2}{2.6\,\text{a}}\right)\left(\dfrac{1\,\text{a}}{3.17 \times 10^{7}\,\text{s}}\right) = 8.4 \times 10^{-9} \text{ s}^{-1}$

$$N = (0.43 \times 10^{-3} \text{ g})\left(\frac{1\,\text{mol}}{147\,\text{g}}\right)\left(\frac{6.022 \times 10^{23}\,\text{nuclei}}{1\,\text{mol}}\right) = 1.8 \times 10^{18} \text{ nuclei}$$

$$\text{activity} = (8.4 \times 10^{-9} \text{ s}^{-1})(1.8 \times 10^{18} \text{ nuclei})\left(\frac{1\,\text{Ci}}{3.7 \times 10^{10}\,\text{Bq}}\right) = 0.40 \text{ Ci}$$

17.47 $\quad k = \dfrac{\ln 2}{t_{1/2}} = \dfrac{\ln 2}{109 \text{ min}} = 6.36 \times 10^{-3} \text{ min}^{-1}$

$$N = N_0 e^{-kt} \quad \text{and} \quad \frac{N}{N_0} = e^{-kt}$$

Taking natural log of both sides gives

$$\ln\left(\frac{N}{N_0}\right) = -kt$$

Let $N_0 = 100\%$ and $N = 10\%$; we can then write

$$\ln\left(\frac{10\%}{100\%}\right) = -kt$$

Solving for t gives

$$t = -\left(\frac{1}{k}\right) \ln\left(\frac{10\%}{100\%}\right) = -\left(\frac{1}{6.36 \times 10^{-3} \text{ min}^{-1}}\right) \ln(0.10) = 400 \text{ min } (1 \text{ sf})$$

17.49 $\quad k = \dfrac{\ln 2}{t_{1/2}} = \dfrac{\ln 2}{5.27 \text{ a}} = 0.131 \text{ a}$

$$\frac{N}{N_0} = e^{-kt} = e^{-(0.131 \text{ a}^{-1})(2.50 \text{ a})} = \frac{0.266 \text{ g}}{N_0}$$

$$N_0 = 0.370 \text{ g}$$

$$\frac{0.370 \text{ g}}{1.40 \text{ g}} \times 100 = 26.4\%$$

17.51 Since radioactive decay follows first-order kinetics, the rate of loss of X is

$$\frac{d[X]}{dt} = -k_1[X]$$

(1) Y, which is an intermediate, is lost in the first reaction but formed in the second one, so its rate equation can be expressed as

$$\frac{d[Y]}{dt} = k_1[X] - k_2[Y]$$

(2) Z is the final product of the two consecutive reactions so its rate law is

$$\frac{d[Z]}{dt} = k_2[Y]$$

(3) As discussed in Chapter 13, the integrated form of equation (1) is

$$[X] = [X]_0 e^{-k_1 t}$$

(4) Substituting this expression into the rate law for Y and rearranging gives

$$\frac{d[Y]}{dt} + k_2[Y] = k_1[X]_0 e^{-k_1 t}$$

(5) This linear first-order differential equation has the solution

$$[Y] = \frac{k_1}{k_2 - k_1}(e^{-k_1 t} - e^{-k_2 t})[X]_0 \qquad \text{when } [Y]_0 = 0.$$

(6) Since $[X]+[Y]+[Z]=[X]_0$ at all times, $[Z]=[X]_0 -([X]+[Y])$, or

$$[Z] = [X]_0 - \left([X]_0 e^{-k_1 t} + \frac{k_1}{k_2 - k_1}(e^{-k_1 t} - e^{-k_2 t})[X]_0\right) = [X]_0\left(1 + \frac{k_1 e^{-k_2 t} - k_2 e^{-k_1 t}}{k_2 - k_1}\right)$$

(7) The values of the rate constants can be found from the half-lives:

$$k_1 = \frac{\ln 2}{27.4\ d} = 0.0253\ d^{-1} \qquad k_2 = \frac{\ln 2}{18.7\ d} = 0.0371\ d^{-1}$$

Using these constants and assuming $[X]_0 = 2.00$ g, equations (4), (6) and (7) are graphed below.

17.53 If isotopically enriched water, such as $H_2{}^{18}O$, is used in the reaction, the label can be followed. Once the products are separated, a suitable technique, such as vibrational spectroscopy or mass spectrometry, can be used to determine whether the product has incorporated the ^{18}O. For example, if the methanol ends up with the O atom from the water molecules, then its molar mass would be $34 \text{ g} \cdot \text{mol}^{-1}$, rather than $32 \text{ g} \cdot \text{mol}^{-1}$ found for methanol with elements present at their natural isotopic abundance.

17.55 The vibrational frequency is proportional to the reduced mass of the two atoms that form the bond according to the equation:

$$\nu = \frac{1}{2\pi}\sqrt{\frac{k}{\mu}}$$

where $\mu = \dfrac{m_A m_B}{m_A + m_B}$

Because we are not given v, it is easiest to make a relative comparison by taking the ratio of v for the C—D molecule versus v for the C—H molecule:

$$\frac{v_{C-D}}{v_{C-H}} = \frac{\frac{1}{2\pi}\sqrt{\frac{k}{\mu_{C-D}}}}{\frac{1}{2\pi}\sqrt{\frac{k}{\mu_{C-H}}}} = \sqrt{\frac{\mu_{C-H}}{\mu_{C-D}}} = \sqrt{\frac{\frac{m_C m_H}{m_C + m_H}}{\frac{m_C m_D}{m_C + m_D}}} = \sqrt{\frac{\frac{(12.011)(1.0078)}{12.011+1.0078}}{\frac{(12.011)(2.0140)}{12.011+2.0140}}}$$

$$= \sqrt{\frac{\left(\frac{12.105}{13.019}\right)}{\left(\frac{24.190}{14.025}\right)}} = 0.73422$$

We would thus expect the vibrational frequency for the C—D bond to be approximately 0.73 times the value for the C—H bond (lower in energy).

17.57 To determine the effective half-life we need to determine the effective rate constant, k_E. This constant is equal to the sum of the biological rate constant (k_B) and the radioactive decay rate constant (k_R), both of which can be obtained from the respective half-lives:

$$k_E = k_B + k_R = \frac{\ln 2}{90.0\ d} + \frac{\ln 2}{87.4\ d} = 1.56 \times 10^{-2}\ d^{-1}$$

$$t_{1/2}\ (\text{effective}) = \frac{\ln 2}{1.56 \times 10^{-2}\ d^{-1}} = 44.4\ d$$

17.59 Remember to convert g to kg.

(a) $E = mc^2 = (1.0 \times 10^{-3}\ kg)(3.00 \times 10^8\ m \cdot s^{-1})^2$
$$= 9.0 \times 10^{13}\ kg \cdot m^2 \cdot s^{-2} = 9.0 \times 10^{13}\ J$$

(b) $E = mc^2 = (9.109 \times 10^{-31}\ kg)(3.00 \times 10^8\ m \cdot s^{-1})^2$
$$= 8.20 \times 10^{-14}\ kg \cdot m^2 \cdot s^{-2} = 8.20 \times 10^{-14}\ J$$

(c) $E = mc^2$

$E = (1.0 \times 10^{-15}\ kg)(3.00 \times 10^8\ m \cdot s^{-1})^2 = 90.\ kg \cdot m^2 \cdot s^{-2} = 90.\ J$

(d) $E = mc^2$

$$E = (1.673 \times 10^{-27} \text{ kg})(3.00 \times 10^8 \text{ m} \cdot \text{s}^{-1})^2 = 1.51 \times 10^{-10} \text{ J}$$

17.61 $\Delta m = \dfrac{\Delta E}{c^2} = \dfrac{-3.9 \times 10^{26} \text{ J} \cdot \text{s}^{-1}}{(3.00 \times 10^8 \text{ m} \cdot \text{s}^{-1})^2} = -4.3 \times 10^9 \text{ kg} \cdot \text{s}^{-1}$

17.63 $1 \, m_u = 1.6605 \times 10^{-27} \text{ kg}$

In each case, calculate the difference in mass between the nucleus and the free particles from which it may be considered to have been formed. Then obtain the binding energy from the relation $E_{bind} = \Delta mc^2$.

(a) $^{62}_{28}\text{Ni}$: $28 \, ^1\text{H} + 34 \, \text{n} \rightarrow \, ^{62}_{28}\text{Ni}$

$\Delta m = 61.928346 \, m_u - (28 \times 1.0078 \, m_u + 34 \times 1.0087 \, m_u) = -0.586 \, m_u$

$\Delta m = (-0.586 \, m_u)\left(\dfrac{1.6605 \times 10^{-27} \text{ kg}}{1 \, m_u}\right) = -9.73 \times 10^{-28} \text{ kg}$

$E_{bind} = \left|-9.73 \times 10^{-28} \text{ kg}\right| \times (2.99792 \times 10^8 \text{ m} \cdot \text{s}^{-1})^2$

$\qquad = 8.74 \times 10^{-11} \text{ kg} \cdot \text{m}^2 \cdot \text{s}^{-2} = 8.74 \times 10^{-11} \text{ J}$

$E_{bind} / \text{nucleon} = \dfrac{8.74 \times 10^{-11} \text{ J}}{62 \text{ nucleons}} = 1.41 \times 10^{-12} \text{ J} \cdot \text{nucleon}^{-1}$

(b) $^{239}_{94}\text{Pu}$: $94 \, ^1\text{H} + 145 \, \text{n} \rightarrow \, ^{239}_{94}\text{Pu}$

$\Delta m = 239.0522 \, m_u - (94 \times 1.0078 \, m_u + 145 \times 1.0087 \, m_u) = -1.94 \, m_u$

$\Delta m = (-1.94 \, m_u)\left(\dfrac{1.6605 \times 10^{-27} \text{ kg}}{1 \, m_u}\right) = -3.23 \times 10^{-27} \text{ kg}$

$E_{bind} = \left|-3.23 \times 10^{-27} \text{ kg}\right| \times (2.99792 \times 10^8 \text{ m} \cdot \text{s}^{-1})^2$

$\qquad = 2.90 \times 10^{-10} \text{ kg} \cdot \text{m}^2 \cdot \text{s}^{-2} = 2.90 \times 10^{-10} \text{ J}$

$E_{bind} / \text{nucleon} = \dfrac{2.90 \times 10^{-10} \text{ J}}{239 \text{ nucleons}} = 1.21 \times 10^{-12} \text{ J} \cdot \text{nucleon}^{-1}$

(c) ^2_1H: $^1\text{H} + \text{n} \rightarrow \, ^2_1\text{H}$

$$\Delta m = 2.0141\, m_{\mathrm{u}} - (1.0078\, m_{\mathrm{u}} + 1.0087\, m_{\mathrm{u}}) = -0.0024\, m_{\mathrm{u}}$$

$$\Delta m = (-0.0024\, m_{\mathrm{u}})\left(\frac{1.6605\times10^{-27}\ kg}{1\, m_{\mathrm{u}}}\right) = -4.0\times10^{-30}\ \mathrm{kg}$$

$$E_{bind} = \left|-4.0\times10^{-30}\ \mathrm{kg}\right| \times (2.99792\times10^{8}\ \mathrm{m\cdot s^{-1}})^{2}$$

$$= 3.6\times10^{-13}\ \mathrm{kg\cdot m^{2}\cdot s^{-2}} = 3.6\times10^{-13}\ \mathrm{J}$$

$$E_{bind}/\mathrm{nucleon} = \frac{3.6\times10^{-13}\ \mathrm{J}}{2\ \mathrm{nucleons}} = 1.8\times10^{-13}\ \mathrm{J\cdot nucleon^{-1}}$$

(d) $_{1}^{3}\mathrm{H}$: $^{1}\mathrm{H} + 2\,\mathrm{n} \rightarrow\, _{1}^{3}\mathrm{H}$

$$\Delta m = 3.01605\, m_{\mathrm{u}} - (1.0078\, m_{\mathrm{u}} + 2\times1.0087\, m_{\mathrm{u}}) = -0.0092\, m_{\mathrm{u}}$$

$$\Delta m = (-0.0092\, m_{\mathrm{u}})\left(\frac{1.6605\times10^{-27}\ kg}{1\, m_{\mathrm{u}}}\right) = -1.5\times10^{-29}\ \mathrm{kg}$$

$$E_{bind} = \left|-1.5\times10^{-29}\ \mathrm{kg}\right| \times (2.99792\times10^{8}\ \mathrm{m\cdot s^{-1}})^{2}$$

$$= 1.3\times10^{-12}\ \mathrm{kg\cdot m^{2}\cdot s^{-2}} = 1.3\times10^{-12}\ \mathrm{J}$$

$$E_{bind}/\mathrm{nucleon} = \frac{1.3\times10^{-12}\ \mathrm{J}}{3\ \mathrm{nucleons}} = 4.3\times10^{-13}\ \mathrm{J\cdot nucleon^{-1}}$$

(e) $^{62}\mathrm{Ni}$ is the most stable, because it has the largest binding energy per nucleon.

17.65 In each case, we first determine the change in mass, Δm = (mass of products) × (mass of reactants). We then calculate the energy released from $\Delta E = (\Delta m)c^{2}$.

(a) $\mathrm{D} + \mathrm{D} \rightarrow\, ^{3}\mathrm{He} + \mathrm{n}$

$$2.0141\, m_{\mathrm{u}} + 2.0141\, m_{\mathrm{u}} \rightarrow 3.0160\, m_{\mathrm{u}} + 1.0087\, m_{\mathrm{u}}$$
$$4.0282\, m_{\mathrm{u}} \rightarrow 4.0247\, m_{\mathrm{u}}$$

$$\Delta m = (-0.0035\, m_{\mathrm{u}})\left(\frac{1.661\times10^{-27}\ kg}{1\, m_{\mathrm{u}}}\right) = -5.8\times10^{-30}\ \mathrm{kg}$$

$$\Delta E = \Delta mc^{2} = (-5.8\times10^{-30}\ \mathrm{kg})(3.00\times10^{8}\ \mathrm{m\cdot s^{-1}})^{2} = -5.2\times10^{-13}\ \mathrm{J}$$

$$\left(\frac{-5.2\times10^{-13}\ \mathrm{J}}{4.0282\, m_{\mathrm{u}}}\right)\left(\frac{1\, m_{\mathrm{u}}}{1.661\times10^{-24}\ \mathrm{g}}\right) = -7.8\times10^{10}\ \mathrm{J\cdot g^{-1}}$$

(b) $^{3}\mathrm{He} + \mathrm{D} \rightarrow\, ^{4}\mathrm{He} +\, _{1}^{1}\mathrm{H}$

$$3.0160\, m_u + 2.0141\, m_u \rightarrow 4.0026\, m_u + 1.0078\, m_u$$

$$5.0301\, m_u \rightarrow 5.0104\, m_u$$

$$\Delta m = (-0.0197\, m_u)\left(\frac{1.661\times10^{-27}\ \text{kg}}{1\, m_u}\right) = -3.27\times10^{-29}\,\text{kg}$$

$$\Delta E = \Delta m c^2 = -(3.27\times10^{-29}\ \text{kg})(3.00\times10^8\,\text{m}\cdot\text{s}^{-1})^2 = -2.94\times10^{-12}\ \text{J}$$

$$\left(\frac{-2.94\times10^{-12}\ \text{J}}{5.0301\, m_u}\right)\left(\frac{1\, m_u}{1.661\times10^{-24}\ \text{g}}\right) = -3.52\times10^{11}\ \text{J}\cdot\text{g}^{-1}$$

(c) $^7\text{Li} + {}^1_1\text{H} \rightarrow 2\ {}^4\text{He}$

$$7.0160\, m_u + 1.0078\, m_u \rightarrow 2(4.0026\, m_u)$$

$$8.0238\, m_u \rightarrow 8.0052\, m_u$$

$$\Delta m = (-0.0186\, m_u)\left(\frac{1.661\times10^{-27}\ \text{kg}}{1\, m_u}\right) = -3.09\times10^{-29}\ \text{kg}$$

$$\Delta E = \Delta m c^2 = (-3.09\times10^{-29}\ \text{kg})(3.00\times10^8\ \text{m}\cdot\text{s}^{-1})^2 = -2.78\times10^{-12}\ \text{J}$$

$$\left(\frac{-2.78\times10^{-12}\ \text{J}}{8.0238\, m_u}\right)\left(\frac{1\, m_u}{1.661\times10^{-24}\ \text{g}}\right) = -2.09\times10^{11}\ \text{J}\cdot\text{g}^{-1}$$

(d) $\text{D} + \text{T} \rightarrow {}^4\text{He} + \text{n}$

$$2.0141\, m_u + 3.0160\, m_u \rightarrow 4.0026\, m_u + 1.0087\, m_u$$

$$5.0301\, m_u \rightarrow 5.0113\, m_u$$

$$\Delta m = (-0.0188\, m_u)\left(\frac{1.661\times10^{-27}\ \text{kg}}{1\, m_u}\right) = -3.12\times10^{-29}\ \text{kg}$$

$$\Delta E = \Delta m c^2 = (-3.12\times10^{-29}\ \text{kg})(3.00\times10^8\ \text{m}\cdot\text{s}^{-1})^2 = -2.81\times10^{-12}\ \text{J}$$

$$\left(\frac{-2.81\times10^{-12}\ \text{J}}{5.0301\, m_u}\right)\left(\frac{1\, m_u}{1.661\times10^{-24}\ \text{g}}\right) = -3.36\times10^{11}\ \text{J}\cdot\text{g}^{-1}$$

17.67 (a) ${}^{24}_{11}\text{Na} \rightarrow {}^{24}_{12}\text{Mg} + {}^{\ 0}_{-1}\text{e}$

mass $({}^{24}_{11}\text{Na}) = 23.990\,96\ m_u$

mass $({}^{24}_{12}\text{Mg}) = 23.985\,04\ m_u$

The mass of the electron does not need to be explicitly included in the calculation because it is already included in the mass of Mg.

$$\Delta m = \text{mass} \left({}^{24}_{12}\text{Mg} \right) - \text{mass} \left({}^{24}_{11}\text{Na} \right) = 23.985\,04\ m_u - 23.990\,96\ m_u$$
$$= -5.92 \times 10^{-3}\ m_u$$

$$\Delta m \text{ (in kg)} = -5.92 \times 10^{-3}\ m_u \times 1.661 \times 10^{-27}\ \text{kg } m_u^{-1} = -9.83 \times 10^{-30}\ \text{kg}$$

(b) $\Delta E = \Delta mc^2 = -(9.83 \times 10^{-30}\ \text{kg})(3.00 \times 10^8\ \text{m} \cdot \text{s}^{-1})^2 = -8.85 \times 10^{-13}\ \text{J}$

(c) $\Delta E \text{ (per nucleon)} = \dfrac{-8.85 \times 10^{-13}\ \text{J}}{24 \text{ nucleons}} = -3.69 \times 10^{-14}\ \text{J} \cdot \text{nucleon}^{-1}$

This simple calculation works because the number of nucleons is the same on both sides of the equation.

17.69 (a) ${}^{244}_{95}\text{Am} \rightarrow {}^{134}_{53}\text{I} + {}^{107}_{42}\text{Mo} + 3\,{}^{1}_{0}\text{n}$

(b) ${}^{235}_{92}\text{U} + {}^{1}_{0}\text{n} \rightarrow {}^{96}_{40}\text{Zr} + {}^{138}_{52}\text{Te} + 2\,{}^{1}_{0}\text{n}$

(c) ${}^{235}_{92}\text{U} + {}^{1}_{0}\text{n} \rightarrow {}^{101}_{42}\text{Mo} + {}^{132}_{50}\text{Sn} + 3\,{}^{1}_{0}\text{n}$

17.71 (a) False; the dose equivalent is either equal to or larger than the actual dose, due to the Q factor.

(b) False; 1.8×10^8 Bq $= 0.003$ Ci which is much smaller than 10 Ci.

(c) True.

(d) True.

17.73 (a) $1\,\text{Ci} = 3.7 \times 10^{10}$ decays per second (dps)

decays per minute (dpm) for $4\,\text{pCi} = 4 \times 10^{-12}\ \text{Ci} \times 3.7 \times 10^{10}\ \text{dps} \times \left(\dfrac{60\ \text{s}}{1\ \text{min}} \right)$
$$= 9\ \text{dpm}$$

(b) $\text{volume(L)} = (2.0 \times 3.0 \times 2.5)\ \text{m}^3 \times \left(\dfrac{10^3\ \text{L}}{1\ \text{m}^3} \right) = 1.5 \times 10^4\ \text{L}$

$\text{number of decays} = (1.5 \times 10^4\ \text{L}) \left(\dfrac{4\ \text{pCi}}{1\ \text{L}} \right) \left(\dfrac{9\ \text{decays} \cdot \text{min}^{-1}}{4\ \text{pCi}} \right) (5.0\ \text{min})$
$$= 7 \times 10^5\ \text{decays}$$

17.75 N_0 = number of ^{222}Rn atoms = 2.0×10^{-5} mol $\times 6.0 \times 10^{23}$ atoms $\cdot$ mol^{-1}
$$= 1.2 \times 10^{19} \text{ atoms}$$

$$k = \frac{\ln 2}{t_{1/2}} = \frac{\ln 2}{3.82 \text{ d}} = 0.181 \text{ d}^{-1}$$

(a) rate of decay $= k \times N = \left(\frac{0.181}{\text{d}}\right)\left(\frac{1 \text{ d}}{8.64 \times 10^4 \text{ s}}\right)(1.2 \times 10^{19} \text{ atoms})$
$$= 2.52 \times 10^{13} \text{ atoms} \cdot \text{s}^{-1} \text{ (dps or Bq)}$$

initial activity $= (2.52 \times 10^{13} \text{ Bq})\left(\frac{1 \text{ Ci}}{3.7 \times 10^{10} \text{ Bq}}\right)\left(\frac{1 \text{ pCi}}{10^{-12} \text{ Ci}}\right)$
$$\times \left(\frac{1}{2000 \text{ m}^3}\right)\left(\frac{1 \text{ m}^3}{10^3 \text{ L}}\right)$$
$$= 3.4 \times 10^8 \text{ pCi} \cdot \text{L}^{-1}$$

(b) $N = N_0 e^{-kt} = 1.2 \times 10^{19} \text{ atoms} \times e^{-0.181 \text{ d}^{-1} \times 1 \text{ d}} = 1.0 \times 10^{19} \text{ atoms}$

(c) $\ln\left(\dfrac{\text{activity}}{\text{initial acitivity}}\right) = -kt$

$$t = -\left(\frac{1}{k}\right)\ln\left(\frac{\text{activity}}{\text{initial activity}}\right) = -\left(\frac{1}{0.181 \text{ d}^{-1}}\right)\ln\left(\frac{4}{3.4 \times 10^8}\right)$$
$$= 1 \times 10^2 \text{ days}$$

17.77 $k = \dfrac{\ln 2}{4.5 \times 10^9 \text{ a}} = 1.5 \times 10^{-10} \text{ a}^{-1}$

$$t(= \text{age}) = -\left(\frac{1}{k}\right)\ln\left(\frac{N}{N_0}\right)$$

$$\frac{N}{N_0} = \frac{\text{mass of } ^{238}\text{U}}{\text{initial mass of } ^{238}\text{U}} = \frac{1}{1 + \dfrac{\text{mass of } ^{206}\text{Pb}}{\text{mass of } ^{238}\text{U}}}$$

(a) $\dfrac{N}{N_0} = \dfrac{1}{1 + 1.00} = \dfrac{1}{2.00}$, therefore age $= t_{1/2} = 4.5$ Ga

(b) $\dfrac{N}{N_0} = \dfrac{1}{1 + \dfrac{1}{1.25}} = 0.556$

$$t(= \text{age}) = -\frac{\ln (0.556)}{\left(\dfrac{\ln 2}{4.5 \text{ Ga}}\right)} = 3.8 \text{ Ga}$$

17.79 (a) $\text{activity} = (17.3 \text{ Ci})\left(\dfrac{3.7 \times 10^{10} \text{ Bq}}{1 \text{ Ci}}\right)$

$$= 6.4 \times 10^{11} \text{ Bq} = 6.4 \times 10^{11} \text{ nuclei} \cdot \text{s}^{-1}$$

$$N = (2.0 \times 10^{-6} \text{ g})\left(\frac{1 \text{ u}}{1.661 \times 10^{-24} \text{ g}}\right)\left(\frac{1 \text{ nucleus}}{24 \text{ u}}\right) = 5.0 \times 10^{16} \text{ nuclei}$$

$$k = \frac{\text{activity}}{N} = \frac{6.4 \times 10^{11} \text{ nuclei} \cdot \text{s}^{-1}}{5.0 \times 10^{16} \text{ nuclei}} = 1.3 \times 10^{-5} \text{ s}^{-1} = 1.1 \text{ d}^{-1}$$

$$t_{1/2} = \frac{\ln 2}{k} = \frac{\ln 2}{1.3 \times 10^{-5} \text{ s}^{-1}} = 5.3 \times 10^4 \text{ s} = 15 \text{ h} = 0.63 \text{ d}$$

(b) $m = m_0 e^{-kt} = 2.0 \text{ mg} \times e^{-1.11 \text{ d}^{-1} \times 2.0 \text{ d}} = 0.22 \text{ mg}$

17.81 (a) Radioactive substances that emit γ radiation are most effective for diagnosis because they are the least destructive of the types of radiation listed. Additionally, γ rays pass easily through body tissues and can be counted, whereas α and β particles are stopped by the body tissues. (b) α particles tend to be best for this application because they cause the most destruction. (c) and (d) ^{131}I, 8d (used to image the thyroid); ^{67}Ga, 78 h (used most often as the citrate complex); ^{99m}Tc, 6 h (used for various body tissues by varying the ligands attached to the Tc atom).

17.83 (a) $^{98}_{42}\text{Mo} + ^{1}_{0}\text{n} \rightarrow ^{99}_{42}\text{Mo} \rightarrow ^{99m}_{43}\text{Tc} + ^{0}_{-1}\beta$

(b) Tc-99m ($N/Z = 1.30$ vs 1.36 for Mo-99)

17.85 You might say the following to a twelve year old: To make a small nuclear bomb you need what is called a "critical mass" of radioactive material; this is the minimum amount of material needed to achieve spontaneous fission, which is the process causes the bomb to explode. For plutonium, the amount of material needed is about 17 kg (or 33 pounds), the weight of an average three year old boy. Plutonium, however, is very dense (a large amount of it can fit in a very small space), so the actual SIZE of this amount of plutonium would only be the size of a grapefruit, or just a little bigger than a softball.

17.87 Radon-222 decays to polonium-218 by alpha emission with a half-life of 3.824 days.

$$^{222}_{86}Rn \rightarrow {}^{218}_{84}Po + {}^4_2\alpha \qquad\qquad t_{1/2} = 3.824 \text{ d}$$

An alpha particle is the nucleus of ^4He. Assuming that (1) the alpha particles are captured inside the container, (2) they behave as an ideal gas and (3) the temperature is constant at 298 K, we can find the volume of the container by calculating the number of moles of $^4_2\alpha$ formed in 15 days and then applying the ideal gas law.

$$n_{He} = n_{Po} = n_{initial\ Rn} - n_{final\ Rn} \qquad n_{final\ Rn} \propto N_{15\ days}$$

$$k = \frac{\ln 2}{t_{1/2}} = \frac{\ln 2}{3.824 \text{ d}} = 0.181 \text{ d}^{-1}$$

$$N_{15\ days} = N_0 e^{-kt} = (2.5 \text{ g})e^{-(0.181\ d^{-1})(15\ d)} = 0.165 \text{ g}$$

$$n_{final\ Rn} = (0.165 \text{ g }{}^{222}Rn)\left(\frac{1 \text{ mol }{}^{222}Rn}{222.0175 \text{ g }{}^{222}Rn}\right)$$
$$= 7.43 \times 10^{-4} \text{ mol }{}^{222}Rn$$

$$n_{initial\ Rn} = (2.5 \text{ g }{}^{222}Rn)\left(\frac{1 \text{ mol }{}^{222}Rn}{222.0175 \text{ g }{}^{222}Rn}\right)$$
$$= 1.13 \times 10^{-2} \text{ mol }{}^{222}Rn$$

$$n_{He} = n_{initial\ Rn} - n_{final\ Rn} = 1.13 \times 10^{-2} \text{ mol} - 7.43 \times 10^{-4} \text{ mol}$$
$$= 1.06 \times 10^{-2} \text{ mol}$$

$$PV = nRT$$

$$V = \frac{nRT}{P} = \frac{(1.06 \times 10^{-2} \text{ mol})(0.08206 \text{ L} \cdot \text{atm} \cdot \text{K}^{-1} \cdot \text{mol}^{-1})(298 \text{ K})}{1.00 \text{ atm}}$$
$$= 0.26 \text{ L}$$

17.89 $m_{tot} = m_{e^-} + m_{e^+}$

$$= 2m_{e^-} = 2(9.109\ 39 \times 10^{-31}\ \text{kg})$$

$$= 1.821\ 88 \times 10^{-30}\ \text{kg}$$

$E = mc^2$

$$= (1.821\ 88 \times 10^{-30}\text{kg})(2.997\ 92 \times 10^{8}\text{m} \cdot \text{s}^{-1})^2$$

$$= 1.637\ 42 \times 10^{-13}\text{J}$$

17.91   The average time to reach 8000 disintegrations or counts for the sample is 22.25 min; therefore the average disintegrations or counts per minute for the sample is 359.6 dpm.

Average time to reach 500 background disintegrations or counts = 5.11min; therefore the average disintegrations or counts per minute for the background is 97.8 dpm.

(a) Average level of radioactivity in the sample = 359.6 dpm – 97.8 dpm = 261.8 dpm = 262 dpm (3 sf).

(b) $(262\ \text{dpm})\left(\dfrac{1\ \text{dps}}{60\ \text{dpm}}\right)\left(\dfrac{1\ \text{Ci}}{3.7 \times 10^{10}\ \text{dps}}\right) = 1.18 \times 10^{-10}\ \text{Ci}$

$1.18 \times 10^{-10}\ \text{Ci} \times \left(\dfrac{1\ \mu\text{Ci}}{1 \times 10^{-6}\ \text{Ci}}\right) = 1.18 \times 10^{-4}\ \mu\text{Ci}$

17.93   Before beginning this problem it is important to recognize that essentially all of the radioactivity of the sample will still be present under the conditions employed for the Schilling test. This can be easily demonstrated using the half life of Co-58:

$$k = \frac{\ln 2}{t_{1/2}} = \frac{\ln 2}{72\ \text{d}} = 9.6 \times 10^{-3}\ \text{d}^{-1}$$

$$\text{fraction remaining} = \frac{N}{N_0} = e^{-kt}$$

$$= e^{-(9.6 \times 10^{-3}\text{d}^{-1} \times 1\ \text{d})} = 0.99$$

So 99% of the Co-58 remains after 24 h.

(a) For the standard: $3.0 \text{ mL} \times 0.4 \text{ nCi/mL} = 1.2 \text{ nCi}$; this produces 910 cpm; for the first sample obtained <u>without</u> intrinsic factor supplementation:

$$\frac{83 \text{ cpm}}{3.0 \text{ mL urine sample}} \times 1200 \text{ mL urine volume} = 3.3 \times 10^4 \text{ cpm for entire urine sample}$$

$$3.3 \times 10^4 \text{ cpm} \times \frac{1.2 \text{ nCi}}{910 \text{ cpm standard}} \times \frac{10^6 \ \mu\text{Ci}}{10^9 \text{ nCi}} = 0.044 \ \mu\text{Ci} \ ^{58}\text{Co excreted}$$

$$\frac{0.044 \ \mu\text{Ci} \ ^{58}\text{Co excreted}}{0.5 \ \mu\text{Ci} \ ^{58}\text{Co originally dosed}} \times 100 = 9\% \text{ excreted}$$

for the second sample obtained <u>with</u> intrinsic factor supplementation:

$$\frac{120 \text{ cpm}}{3.0 \text{ mL urine sample}} \times 1200 \text{ mL urine volume} = 4.8 \times 10^4 \text{ cpm}$$

$$4.8 \times 10^4 \text{ cpm} \times \frac{1.2 \text{ nCi}}{910 \text{ cpm standard}} \times \frac{10^6 \ \mu\text{Ci}}{10^9 \text{ nCi}} = 0.063 \ \mu\text{Ci} \ ^{58}\text{Co excreted}$$

$$\frac{0.063 \ \mu\text{Ci} \ ^{58}\text{Co excreted}}{0.5 \ \mu\text{Ci} \ ^{58}\text{Co originally dosed}} \times 100 = 13\% \text{ excreted}$$

(b) From before: $k = 9.6 \times 10^{-3} \text{ d}^{-1}$

So the fraction remaining $= \dfrac{N}{N_0} = e^{-kt}$

$$= e^{-(9.6 \times 10^{-3} \text{d}^{-1} \times 7 \text{ d})} = 0.94$$

94% of the activity remains after 7 days, which would be 0.47 μCi .

(c) $k_E = k_B + k_R = \dfrac{\ln 2}{180 \text{ d}} + \dfrac{\ln 2}{72 \text{ d}} = 1.35 \times 10^{-2} \text{ d}^{-1}$

$t_{1/2} \text{ (effective)} = \dfrac{\ln 2}{1.35 \times 10^{-2} \text{ d}^{-1}} = 51.4 \text{ d}$

(d) If the first dose was 0.5 μCi and 0.044 μCi was excreted (from part a), 0.456 μCi of ^{58}Co remained in the patient's system. One week (7 d) after the first dose, we can determine the amount of ^{58}Co remaining in the patients system using $k_E = 0.693/51.4\ \text{d} = 1.35\times10^{-2}\ \text{d}^{-1}$:

$$\frac{N}{N_0} = \frac{N}{0.456\ \mu\text{Ci}} = e^{-k_E t} = e^{-(1.35\times10^{-2}\ \text{d}^{-1}\times7\ \text{d})} = 0.91$$

$N = (0.91)0.456\ \mu\text{Ci} = 0.415\ \mu\text{Ci}$; this is the new N_0.

We can now determine how much of this is eliminated $(N_0 - N)$ in one day by using the biological half life:

$$\frac{N_0 - N}{N_0} = \frac{0.415\ \mu\text{Ci} - N}{0.415\ \mu\text{Ci}} = e^{-k_B t} = e^{-((0.693/180)\times7\ \text{d})} = 0.973$$

$$N = 0.415\ \mu\text{Ci} - \left[(0.973)0.415\ \mu\text{Ci}\right] = 0.011\ \mu\text{Ci}$$

From this we can determine the fraction of counts in the second sample due to the ^{58}Co left over from the first sample:

$$\frac{0.011\ \mu\text{Ci}}{0.063\ \mu\text{Ci}} = 0.17$$

(e) Only 1% of the ^{58}Co decays in the one day, meaning that only $0.005\ \mu\text{Ci}$ of ^{58}Co was converted to ^{59}Fe. Since each disintegration of a ^{58}Co atom yields a single ^{59}Fe atom, the mass of iron produced from this process can be determined as follows:

$$(5\times10^{-3}\ \mu\text{Ci})\times\left(\frac{1\ \text{Ci}}{10^6\ \mu\text{Ci}}\right)\times\left(\frac{3.7\times10^{10}\ \text{dps}}{1\ \text{Ci}}\right)\times\left(\frac{1\ \text{mol}\ ^{59}\text{Fe produced}}{6.022\times10^{23}\ \text{dps}}\right)\times$$

$$\left(\frac{59\ \text{g}\ ^{59}\text{Fe produced}}{1\ \text{mol}\ ^{59}\text{Fe produced}}\right) = 1.8\times10^{-20}\ \text{g of}\ ^{59}\text{Fe}.$$

Essentially none of the 2.5 g of iron in the patient will be due to ^{59}Fe.

CHAPTER 18

ORGANIC CHEMISTRY I:

THE HYDROCARBONS

18.1

(a)

H—C—C≡C—C—H

alkyne

(b)

H—C—C—C—C—H

alkane

(c)

H—C=C—C—C—H

alkene

(d)

H—C—C=C—C—C≡C—C—H

alkene and alkyne

(e)

C=C—C—C=C

alkene

SM-521

18.3 (a) $(CH_3)_3CH$ or C_4H_{10}, alkane; (b) $C_6H_7CH_3$ or C_7H_{10} alkene; (c) C_6H_{12}, alkane; (d) C_6H_{12} alkane

18.5 (a) $C_{12}H_{26}$, alkane; (b) $C_{13}H_{22}$, alkane; (c) C_7H_{14}, alkane; (d) $C_{14}H_8$ aromatic hydrocarbon

18.7 (a) propane; (b) butane; (c) heptane; (d) decane

18.9 (a) methyl; (b) pentyl; (c) propyl; (d) hexyl

18.11 (a) propane; (b) ethane; (c) pentane; (d) 2,3-dimethylbutane

18.13 (a) 4-methyl-2-pentene; (b) 2,3-dimethyl-2-phenylpentane

18.15 (a) $CH_2=CHCH(CH_3)CH_2CH_3$;
(b) $CH_3CH_2C(CH_3)_2CH(CH_2CH_3)(CH_2)_2CH_3$;
(c) $HC≡C(CH_2)_2C(CH_3)_3$; (d) $CH_3CH(CH_3)CH(CH_2CH_3)CH(CH_3)_2$

18.17

(a)

(b)

(c)

$$\underset{\underset{H}{|}}{\overset{\overset{H}{|}}{H-C}}-\underset{\underset{CH_3}{|}}{\overset{\overset{CH_3}{|}}{C}}-\underset{\underset{H}{|}}{\overset{\overset{H}{|}}{C}}-\underset{\underset{H}{|}}{\overset{\overset{CH_3}{|}}{C}}-\underset{\underset{H}{|}}{\overset{\overset{H}{|}}{C}}-H$$

(d)

$$H_3C-CH_2 \qquad H$$
$$\underset{H}{\overset{}{}}C=C\underset{CH_2-CH_3}{\overset{}{}}$$

18.19 (a)

(b) △   (c) ⬡

18.21 (a) hexenes:

$$H_3C\underset{CH_2}{\overset{CH_2}{}}\underset{CH}{\overset{CH_2}{}}CH_2$$

1-Hexene

$$\underset{H_3C}{\overset{CH=CH}{CH_2-CH_2}}CH_3$$

cis-2-Hexene

$$\underset{H_3C}{\overset{}{}}\underset{CH_2-CH_2}{\overset{CH=CH}{}}CH_3$$

trans-2-Hexene

$$H_3C\underset{CH_2}{\overset{CH=CH}{}}\underset{CH_2}{\overset{}{}}CH_3$$

cis-3-Hexene

$$H_3C\underset{CH}{\overset{CH_2}{}}\underset{CH_2}{\overset{CH}{}}CH_3$$

trans-3-Hexene

Pentenes:

$$H_3C\underset{CH}{\overset{}{}}\underset{CH}{\overset{CH_2}{}}CH_2$$
$$|$$
$$CH_3$$

4-Methyl-1-pentene

$$\underset{H_3C}{\overset{CH_3}{}}\underset{CH_2}{\overset{CH}{}}\underset{CH}{\overset{}{}}CH_2$$

3-Methyl-1-pentene

SM-523

H₃C—CH₂—C(CH₃)=CH₂

2-Methyl-1-pentene

H₃C—CH₂—CH=C(CH₃)(CH₃)

2-Methyl-2-pentene

H₃C—CH₂—C(CH₃)=CH—CH₃

cis-**3-Methyl-2-pentene**

(+ trans isomer)

H₃C—HC(CH₃)—CH=CH—CH₃

cis-**4-Methyl-2-pentene**

(+ trans isomer)

butanes:

(CH₃)₂C(CH₃)—CH=CH₂

3,3-Dimethyl-1-butene

(CH₃)₂CH—C(CH₃)=CH₂

2,3-Dimethyl-1-butene

(CH₃)₂C=C(CH₃)₂

2,3-Dimethyl-2-butene

(b) cyclic molecules:

Cyclohexane

Methylcyclopentane

ethylcyclobutane   1,1-dimethylcyclobutane

The following structures are drawn to emphasize the stereochemistry

cis-1,2-Dimethylcyclobutane

trans-1,2-Dimethylcyclobutane

(nonsuperimposable mirror images

trans-1,3-Dimethylcyclobutane   *cis*-1,3-Dimethylcyclobutane

Propylcyclopropane

Isoropylcyclopropane

or 2-cyclopropylpropane

1-Ethyl-1-methylcyclopropane

trans-1-Ethyl-2-methylcyclopropane

(nonsuperimposable mirror images)

1,1,2-Trimethylcyclopropane

(nonsuperimposable mirror images)

cis-1-Ethyl-2-methylcyclopropane

(nonsuperimposable mirror images)

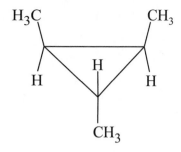

1,2,3-Trimethylcyclopropane   1,2,3-Trimethylcyclopropane

(all cis isomer)   (cis-trans isomer)

18.23 (a) Butane is C_4H_{10}, cyclobutane is C_4H_8. Because they have different formulas, they are not isomers.

(b) Same formula, but different structures; therefore, they are structural isomers.

(c) Same formula (C_5H_{10}), same structure (bonding arrangement is the same), but different geometry; therefore, they are geometrical isomers.

(d) Not isomers, because only their positions in space are different and these positions can be interchanged. Same molecule.

18.25 (a)

$$H_3C-\overset{\overset{\displaystyle H}{|}}{\underset{\underset{\displaystyle CH_3}{|}}{C}}-CH_3$$

(b) If only two isomeric products are formed and they are both branched, then the only possibilities are

$$H_3C-\overset{\overset{\displaystyle Cl}{|}}{\underset{\underset{\displaystyle CH_3}{|}}{C}}-CH_3 \qquad H_2C-\overset{\overset{\displaystyle Cl \quad H}{| \quad |}}{\underset{\underset{\displaystyle CH_3}{|}}{C}}-CH_3$$

18.27 An * designates a chiral carbon.

(a) optically active,

$$H_3C—\overset{\overset{\displaystyle H}{|}}{\underset{\underset{\displaystyle Br}{|}}{C}}\!\!{}^{*}—CH_2—CH_3$$

(b) not optically active,

$$Cl—\overset{\overset{\displaystyle H}{|}}{\underset{\underset{\displaystyle Cl}{|}}{C}}—\overset{\overset{\displaystyle H}{|}}{\underset{\underset{\displaystyle H}{|}}{C}}—CH_3$$

(c) optically active,

$$H—\overset{\overset{\displaystyle Br}{|}}{\underset{\underset{\displaystyle H}{|}}{C}}—\overset{\overset{\displaystyle Cl}{|}}{\underset{\underset{\displaystyle H}{|}}{C}}\!\!{}^{*}—CH_3$$

(d) optically active,

$$H_3C—\overset{\overset{\displaystyle Cl}{|}}{\underset{\underset{\displaystyle H}{|}}{C}}—\overset{\overset{\displaystyle Cl}{|}}{\underset{\underset{\displaystyle H}{|}}{C}}\!\!{}^{*}—CH_2—CH_2\text{-}CH_3$$

18.29 The difference can be traced to the weaker London forces that exist in branched molecules. Atoms in neighboring branched molecules cannot lie as close together as they can in unbranched isomers.

18.31 The balanced equations are

$$C_3H_8(g) + 5\,O_2(g) \longrightarrow 3\,CO_2(g) + 4\,H_2O(l)$$

$$C_4H_{10}(g) + 13/2\,O_2(g) \longrightarrow 4\,CO_2(g) + 5\,H_2O(l)$$

$$C_5H_{12}(g) + 8\,O_2(g) \longrightarrow 5\,CO_2(g) + 6\,H_2O(l)$$

The enthalpies of combustion that correspond to these reactions are listed in Appendix 2:

	(a) Enthalpy of combustion	(b) Heat released per g
Compound	kJ·mol^{-1}	kJ·g^{-1}
Propane	− 2220	50.3
Butane	− 2878	49.5
Pentane	− 3537	49.0

The molar enthalpy of combustion increases with molar mass as might be expected, because the number of moles of CO_2 and H_2O formed will increase as the number of carbon and hydrogen atoms in the compounds increases. The heat released per gram of these hydrocarbons is essentially the same because the H to C ratio is similar in the three hydrocarbons.

18.33 There are nine possible products:

one monochloro compound   two dichloro compounds

two trichloro compounds   two tetrachloro compounds

one pentachloro compound   one hexachloro compound

None of these form optical isomers.

18.35 (a) 4 σ-type single bonds

(b) 2 σ-type single bonds and 1 double bond with a σ- and a π-bond

(c) 1 σ-type single bond and 1 triple bond with a σ-bond and 2 π-bonds

18.37

cis-1,2-Dichloropropene   *trans*-1,2-Dichloropropene

cis-1,2-Dichloropropene is polar, although *trans*-1,2-Dichloropropene is slightly polar also.

18.39 (a)

```
        H    H    Br   H    H
        |    |    |    |    |
   H —— C —— C —— C —— C —— C —— H
        |    |    |    |    |
        H    H    H    H    H
```

3-Bromopentane

```
        H    Br   H    H    H
        |    |    |    |    |
   H —— C —— C —— C —— C —— C —— H
        |    |    |    |    |
        H    H    H    H    H
```

2-Bromopentane

(b) addition reaction

18.41 (a) $C_6H_{11}Br + NaOCH_2CH_3 \rightarrow C_6H_{10} + NaBr + HOCH_2CH_3$

(b)

+ NaOCH$_2$CH$_3$ $\longrightarrow$

+ NaBr + HOCH$_2$CH$_3$

(c) elimination reaction

18.43 $C_2H_4 + X_2 \longrightarrow C_2H_4X_2$

We will break one X—X bond and form two C—X bonds.

Using bond enthalpies:

Halogen	Cl	Br	I
X—X bond breakage (kJ$\cdot$mol^{-1})	+242	+193	+151
C—X bond formation (kJ$\cdot$mol^{-1})	-2(338)	-2(276)	-2(238)
Total (kJ$\cdot$mol^{-1})	-434	-359	-325

The reaction is less exothermic as the halogen becomes heavier. In general, the reactivity, and also the danger associated with use of the halogens in reactions, decreases as one descends the periodic table.

18.45 (a) 1-ethyl-3-methylbenzene; (b) pentamethylbenzene (1,2,3,4,5-pentamethylbenzene is also correct, but, because there is only one possible pentamethylbenzene, the use of the numbers is not necessary)

18.47 (a)

(b)

(c)

(d)

18.49

1,3-Dichloro-2-methylbenzene

1,4-Dichloro-2-methylbenzene

1,5-Dichloro-2-methylbenzene

1,3-Dichloro-5-methylbenzene

1,2-Dichloro-3-methylbenzene

1,2-Dichloro-4-methylbenzene

(b) All of these molecules are at least slightly polar.

18.51

Electrophiles tend to avoid the ortho and para positions that develop slight + charges in the resonance forms.

18.53 Two compounds can be produced. Resonance makes positions 1, 4, 6, and 9 equivalent. It also makes positions 2, 3, 7, and 8 equivalent. Positions 5 and 10 are equivalent but have no H atom.

18.55 These hydrocarbons are too volatile (they are all gases at room temperature) and would not remain in the liquid state.

18.57 $CH_2CHCH_3 + CH_3CH_2CH_2CH_3 \rightarrow CH_3(CH_2)_5CH_3$

18.59 (a) substitution, $CH_4 + Cl_2 \rightarrow CH_3Cl + HCl$
(b) addition $CH_2=CH_2 + Br_2 \rightarrow CH_2Br\text{-}CH_2Br$

18.61 Water is not used as the nonpolar reactants will not readily dissolve in a highly polar solvent like water. Also, the ethoxide ion reacts with water.

18.63 (a) $CH_3CH_2CH_3(g) + 5O_2(g) \rightarrow 3CO_2(g) + 4H_2O(g)$

CHCH + 2½O$_2$(g) → 2CO$_2$(g) + H$_2$O(g)

(b) propane, 44.10 g. mol^{-1}

$\Delta H_c° = \Sigma\Delta H_f°$(products) - $\Sigma\Delta H_f°$(reactants)

$= [3(-393.51) + 4(-241.82)] - [-103.85]$

$= -1180.53$ kJ. mol^{-1} $- 967.28$ kJ. mol^{-1} $+ 103.85$ kJ. mol^{-1}

$= -2043.96$ kJ. mol^{-1}

$$\frac{-2043.96 \text{ kJ. mol}^{-1}}{44.10 \text{ g. mol}^{-1}} = -46.35 \text{ kJ. g}^{-1}$$

ethyne, 26.04 g. mol^{-1}

$\Delta H_c° = [2(-393.51) + (-241.81)] - [+226.73]$

$= -787.02$ kJ. mol^{-1} $- 241.82$ kJ. mol^{-1} $- 226.73$ kJ. mol^{-1}

$= -1255.57$ kJ. mol^{-1}

$$\frac{-1255.57 \text{ kJ. mol}^{-1}}{26.04 \text{ g. mol}^{-1}} = -48.22 \text{ kJ. g}^{-1}$$

(c) More heat is released per gram of ethyne resulting in a hotter flame.

18.65 (a) 2-methyl-1-propene, no geometrical isomers; (b) *cis*-3-methyl-2-pentene, *trans*-3-methyl-2-pentene; (c) 1-hexyne, no geometrical isomers; (d) 3-hexyne, no geometrical isomers; (e) 2-hexyne, no geometrical isomers.

18.67 (a) C$_{10}$H$_{18}$; (b) naphthalene, , C$_{10}$H$_8$; (c) Yes. Cis and trans forms (relative to the C-C bond common to the two six-membered rings) are possible.

trans-Decalin   *cis*-Decalin

18.69 (a) 60°

(b) sever bond angel strain

(c) $C_3H_6(g) + 4\frac{1}{2}O_2(g) \rightarrow 3\ CO_2(g) + 3H_2O(g)$

using bond enthalpies: C – C 348 kJ. mol^{-1}

                                        C – H 412 kJ. mol^{-1}

                                        O = O 496 kJ. mol^{-1}

                                          C = O 743 kJ. mol^{-1}

                                          O – H 463 kJ. mol^{-1}

$$\Delta H_c^\circ = 3(348) + 6(412) + 4.5(496) - 6(743) - 6(463)$$
$$= 1044 + 2472 + 2232 - 4458 - 2778$$
$$= \text{-}1488 \text{ kJ. mol}^{-1}$$

(d) Cyclopropane releases more heat on combustion because it is not a stable molecule due to sever bond angle strain. The C – C and C – H bonds in cyclopropane are not as stable as regular C – C and C – H bonds and their true bond enthalpies are lower values.

18.71 number of moles of $H = \left(\dfrac{4.48 \text{ g H}_2\text{O}}{18.02 \text{ g H}_2\text{O/mol H}_2\text{O}} \right) \left(\dfrac{2 \text{ mol H}}{1 \text{ mol H}_2\text{O}} \right)$

                                    $= 0.497 \text{ mol H}$

number of moles of $C = \left(\dfrac{9.72 \text{ g CO}_2}{44.01 \text{ g CO}_2\text{/mol CO}_2} \right) \left(\dfrac{1 \text{ mol C}}{1 \text{ mol CO}_2} \right)$

                                    $= 0.221 \text{ mol C}$

$$\frac{0.497 \text{ mol H}}{0.221 \text{ mol C}} = \frac{9 \text{ mol H}}{4 \text{ mol C}}$$

The empirical formula is C_4H_9; the molecular formula might be C_8H_{18}, which matches the formula for alkanes (C_nH_{2n+2}). It is not likely an alkene or alkyne, because there is no reasonable Lewis structure for a compound having the empirical formula C_4H_9 and multiple bonds.

18.73 (a) 4-methyl-3-propylheptane

The longest chain has eight carbon atoms in it. The systematic name of the compound is 4-ethyl-5-methyloctane.

4-ethyl-5-methyloctane;

(b) 4,6-dimethyloctane

The compound name is almost correct, but the numbering scheme with the lowest numbers would be 3,5-dimethyloctane.

3,5-dimethyloctane;

(c) 2,2-dimethyl-4-propylhexane

The longest carbon chain in the molecule is seven carbon atoms long. The systematic name is 2,2-dimethyl-4-propylheptane.

2,2-dimethyl-4-ethylheptane;

(d) 2,2-dimethyl-3-ethylhexane.

The name is essentially correct except that ethyl should be listed first The systematic name is 3-ethyl-2,2-dimethylhexane

3-ethyl-2,2-dimethylhexane

18.75 Bromine is an electrophile which will undergo an addition reaction with alkenes in the dark. The lack of a reaction in the dark with Br_2 indicates that the molecule is not an alkene. In the presence of light, bromine will undergo a substitution reaction with alkanes. Therefore, the molecule is most likely an alkane and the only alkane with the molecular formula C_3H_6 is cyclopropane.

18.77 The NO_2 group is a meta-directing group and the Br atom is an ortho, para-directing group. Because the position para to Br is already substituted with the NO_2 group, further bromination will not occur there. The resonance forms show that the bromine atom will activate the position ortho to it as expected. The NO_2 group will deactivate the group ortho to itself, thus in essence enhancing the reactivity of the position meta to the NO_2 group. This position is ortho to the Br atom, so the effects of the Br and NO_2 groups reinforce each other. Bromination is thus expected to occur as shown:

18.79 (a) and (b)

(c) No, there are no cis/trans isomers for this molecule.

18.81 For a molecule such as 1,2-dichloro-4-diethylbenzene, $C_6H_3Cl_2(CH_2CH_3)$, 175.04 u, it is relatively easy to lose heavy atoms such as chlorine and groups of atoms such as methyl and ethyl fragments. Molecules can also lose hydrogen atoms. In mass spectrometry, P is used to represent the *parent ion*, which is the ion formed from the molecule without fragmentation. Fragments are then represented as $P - x$, where x is the particular fragment lost from the parent ion to give the observed mass.

Because the mass spectrum will measure the masses of individual molecules, the mass of carbon used will be 12.00 u (by definition) because the large majority of the molecules will have all ^{12}C. The mass of H is 1.0078 u. Some representative peaks that may be present are listed below.

Fragment formula	Relation to parent ion	Mass, u
$C_6H_3{}^{35}Cl_2(CH_2CH_3)$	P	174.00
$C_6H_3{}^{35}Cl^{37}Cl(CH_2CH_3)$	P	176.00
$C_6H_3{}^{37}Cl_2(CH_2CH_3)$	P	177.99
$C_6H_3{}^{35}Cl(CH_2CH_3)$	P-Cl	139.03
$C_6H_3{}^{37}Cl(CH_2CH_3)$	P-Cl	141.03
$C_6H_3{}^{35}Cl_2(CH_2)$	P-CH$_3$	158.98
$C_6H_3{}^{35}Cl^{37}Cl(CH_2)$	P-CH$_3$	160.97
$C_6H_3{}^{37}Cl_2(CH_2)$	P-CH$_3$	162.97
$C_6H_3{}^{35}Cl_2$	P-CH$_2$CH$_3$	144.96
$C_6H_3{}^{35}Cl^{37}Cl$	P-CH$_2$CH$_3$	146.96
$C_6H_3{}^{37}Cl_2$	P-CH$_2$CH$_3$	148.96
$C_6H_3{}^{35}Cl$	P-CH$_2$CH$_3$-Cl	109.99
$C_6H_3{}^{37}Cl$	P-CH$_2$CH$_3$-Cl	111.99
Etc.		

18.83 The presence of one bromine atom will produce in the ions that contain Br companion peaks that are separated by 2 u. Any fragment that contains Br will show this "doublet" in which the peaks are nearly but not exactly equal in intensity. Thus, seeing a mass spectrum of a compound that is known to have Br or that was involved in a reaction in which Br could have been added or substituted with such doublets, is almost a sure sign that Br is present in the compound. It is also fairly easy to detect Br atoms in the mass spectrum at 79 and 81 u, confirming their presence. If more than one Br atom is present, then a more complicated pattern is observed for the presence of the two isotopes. The possible combinations for a molecule of unknown formula with two Br atoms is $^{79}Br^{79}Br$, $^{79}Br^{81}Br$, $^{81}Br^{79}Br$, and $^{81}Br^{81}Br$. Thus, a set of three peaks (the two possibilities $^{79}Br^{81}Br$ and $^{81}Br^{79}Br$ have identical masses) will be generated that differ in mass by two units. The center peak, which is produced by the $^{79}Br^{81}Br$ and $^{81}Br^{79}Br$ combinations, will have twice the intensity of the outer two peaks, because statistically there are twice as many combinations that produce this mass. All modern mass spectrometers have spectral simulation programs that can readily calculate and print our the relative isotopic distribution pattern expected for any compound formulation, so that it is possible to easily match the expected pattern for a particular ion with the experimental result.

18.85 C_8H_{10} will have an absorption maximum at a longer wavelength. Molecular orbital theory predicts that in conjugated hydrocarbons (molecules which contain a chain of carbon atoms with alternating single and double bonds) electrons become delocalized and are free to move up and down the chain of carbon atoms. Such electrons may be described using the one dimensional "particle in a box" model introduced in Chapter 1. According to this model, as the box to which electrons are confined lengthens, the quantized energy states available to the electrons get closer together. As a result, the energy needed to excite an electron from the ground state to the next higher state is lower for electrons confined to longer boxes. Therefore, lower energy photons, i.e. photons with longer wavelengths, will be absorbed by the C_8H_{10} molecule because it provides a longer "box" than C_6H_8.

18.87 (a) Moles of C can be determined from moles of CO_2.

$$\frac{0.352\,g}{44.01\,g.mol^{-1}\,CO_2} = 0.008\,mol\,CO_2$$

$0.008\,mol\,C \times 12.01\,g.mol^{-1} = 0.096\,g\,C$

Moles of H can be determined from moles of H_2O.

$$\frac{0.109g}{18.01\,g.mol^{-1}\,H_2O} = 0.006\,mol\,H_2O$$

$0.012\,mol\,H \times 1.008\,g.mol^{-1} = 0.012\,g\,H$

Check if there is oxygen in the compound.

Total sample mass – (mass of C + mass of H)

0.108 g – (0.096 g + 0.012) = 0

Only C and H in the compound.

To find empirical formula divide by smallest number of moles:

C: $\dfrac{0.008\ \text{mole}}{0.008\ \text{mole}} = 1$

H: $\dfrac{0.012\ \text{mole}}{0.008\ \text{mole}} = 1.5$

Multiple by 2 to get the nearest integer. C_2H_3

(b)   molecular formula = (x)(empirical formula)

$$x = \frac{\text{mass of molecular formula}}{\text{mass of empirical formula}} = \frac{54.09\ \text{g.mol}^{-1}}{27.035\ \text{g.mol}^{-1}} = 2$$

molecular formula = $(2)(C_2H_3)$

molecular formula = (C_4H_6)

(c)

(d) Each of the first five compounds in (c) can add 4 H's which is what happens in a catalytic hydrogenation with a triple or double bond, and all five compounds would give butane (58.12 g.mol^{-1}). The sixth compound (cyclobutene) would add 2 H's to give cyclobutane (56.11 g.mol^{-1}) and this compound could therefore be eliminated.

(e) Remember that the total pressure of a mixture of gases is the sum of the partial pressures of its components. Partial pressure of the compound is 1.0 atm. Total pressure after adding HBr is 5.0 atm. Therefore partial pressure of HBr before the reaction is 4.0 atm, and after the reaction it is

2.0 atm. Since the initial pressure of the compound was 1.0 atm and the pressure of HBr decreased by 2.0 atm there is a 1 to 2 molar ratio between the reactants. We have determined that the product is the addition of 2 moles of HBr to the compound.

$$C_4H_6(g) + 2HBr(g) \rightarrow C_4H_8Br_2(g)$$

The molecular formula is $C_4H_8Br_2$.

(f) The two geometric isomers that the HBr is reacting with are the *trans*- and *cis*-1,3-butadiene.

Since addition of HBr occurs across each double bond there are three possible structural isomers.

1,4-dibromobutane   1,3-dibromobutane

2,3-dibromobutane

(g) All the carbons in the reactants are sp^2 hybridized.

(h) *trans*- and *cis*-1,3-butadiene.

CHAPTER 19

ORGANIC CHEMISTRY II: POLYMERS AND BIOLOGICAL COMPOUNDS

19.1 (a) RNH_2, R_2NH, R_3N (b) ROH (c) $RCOOH$ (d) $RCHO$

19.3 (a) ether; (b) ketone; (c) amine; (d) ester

19.5 (a) 2-iodo-2-butene; (b) 2,4-dichloro-4-methylhexane; (c) 1,1,1,-triiodoethane; (d) dichloromethane

19.7 (a)

$ClC_6H_4Cl(OH)$, phenol;

(b) $CH_3CH(CH_3)CH(OH)CH_2CH_3$, secondary alcohol; (c)

$CH_3CH_2CH(CH_3)CH_2CH(CH_3)CH_2OH$, primary alcohol;

(d) $CH_3C(CH_3)(OH)CH_2CH_3$, tertiary alcohol

19.9 (a) $CH_3CH_2OCH_3$; (b) $CH_3CH_2OCH_2CH_2CH_3$; (c) CH_3OCH_3

19.11 (a) butyl propyl ether; (b) methyl phenyl ether; (c) pentyl propyl ether

19.13 (a) aldehyde, ethanal; (b) ketone, propanone; (c) ketone, 3-pentanone

19.15 (a)

butanal

(b)

3-hexanone

(c)   2-heptanone

19.17 (a) ethanoic acid;   (b) butanoic acid;   (c) 2-aminoethanoic acid

19.19   (a)   (b)

   (c)   (d)

19.21 (a) methylamine;   (b) diethylamine;   (c) *o*-methylaniline, 2-methylaniline, *o*-methylphenylamine, or 1-amino-2-methylbenzene

19.23

   (a)   (b)

   (c)

19.25 Only (a) and (c) may function as nucleophiles, because they have lone pairs of electrons that will be attracted to a positively charged carbon center. CO_2 and SiH_4 have no lone pairs and cannot function as nucleophiles.

19.27 (a) ethanol; (b) 2-octanol; (c) 5-methyl-1-octanol. These reactions can be accomplished with an oxidizing agent such as acidified sodium dichromate, $Na_2Cr_2O_7$.

19.29

(a)

$$CH_3CH_2CH_2C \overset{\displaystyle O}{\underset{\displaystyle O-CH}{\|}} \overset{\displaystyle CH_3}{\underset{\displaystyle CH_3}{}}$$

(b)

$$CH_3C \overset{\displaystyle O}{\underset{\displaystyle O-CH_2CH_2CH_2CH_2CH_3}{\|}}$$

(c)

$$CH_3CH_2CH_2CH_2CH_2C \overset{\displaystyle O}{\underset{\displaystyle \underset{\displaystyle CH_3}{|}}{\underset{\displaystyle NCH_2CH_3}{\|}}}$$

(d)

$$CH_3C \overset{\displaystyle O}{\underset{\displaystyle NHCH_2CH_2CH_3}{\|}}$$

19.31 (a) addition; (b) condensation; (c) addition; (d) addition; (e) condensation

19.33 The following procedures can be used:

(1) Dissolve the compounds in water and use an acid-base indicator to look for a color change.

(2) $CH_3CH_2CHO \xrightarrow{\text{Tollens reagent}} CH_3CH_2COOH + Ag(s)$

(3) $CH_3COCH_3 \xrightarrow{\text{Tollens reagent}}$ no reaction

Procedure (1) will distinguish ethanoic acid; (2) and (3) will distinguish propanal from 2-propanone.

19.35 $CH_3CH_2COOH < CH_3COOH < ClCH_2COOH < Cl_3CCOOH$

The greater the electronegativities of the groups attached to the carboxyl group, the stronger the acid.

19.37 (a) $-CH_2 - C(CH_3)_2 - CH_2 - C(CH_3)_2 - CH_2 - C(CH_3)_2 -$

(b)

(c)

cis version

trans version

SM-546

19.39 $HOOCC_6H_{12}COOH$ $NH_2C_6H_4NH_2$

19.41   (a) $CHCl{=}CH_2$; (b) $CFCl{=}CF_2$

19.43   (a) $-OCCONH(CH_2)_4NHCOCONH(CH_2)_4NH-$;

(b) $-OC-CH(CH_3)-NH-OC-CH(CH_3)-NH-$

19.45   block copolymer

19.47   Polymers generally do not have definite molecular masses because there is no fixed point at which the chain-lengthening process will cease. The chain stops growing due to lack of nearby monomer units of the appropriate kind or a lack of properly oriented smaller polymeric aggregates. A polymer is, in a sense, not a pure compound, bur rather a mixture of similar compounds of different chain length. There is no fixed molar mass, only an average molar mass. Because there is no one unique compound, there is no one unique melting point, rather a range of melting points. Thus, there is no sharp transition between solid and liquid, and we say the solid softens rather than melts.

19.49   Larger average molar mass corresponds to longer average chain length. Longer chain length allows for greater intertwining of the chains, making them more difficult to pull apart. This twining results in (a) higher softening points,   (b) greater viscosity, and   (c) greater mechanical strength.

19.51   Highly linear, unbranched chains allow for maximum interaction between chains. The greater the intermolecular contact between chains, the stronger the forces between them, and the greater the strength of the material.

19.53 (a)

$$\overset{\displaystyle O}{\underset{\displaystyle \diagdown NH-}{\overset{\displaystyle \|}{-C}}}$$

(b) amide; (c) condensation

19.55 Side groups that contain hydroxyl, carbonyl, amino, and sulfide groups are all potentially capable of participating in hydrogen bonding that could contribute to the tertiary structure of the protein. Thus, serine, threonine, tyrosine, aspartic acid, glutamic acid, lysine, arginine, histidine, asparagine, and glutamine satisfy the criteria. Proline and tryptophan generally do not contribute through hydrogen bonding, because they are typically found in hydrophobic regions of proteins.

19.57

$$H-\overset{\displaystyle NH_2}{\underset{\displaystyle CH_2}{C}}-\overset{\displaystyle O}{\underset{\displaystyle NH-CH_2-COOH}{C}}$$

with a phenol (para-OH substituted benzene ring) attached to the CH$_2$ group.

19.59 (a) The functional groups are alcohols and aldehydes.

(b) The chiral carbon atoms are marked with asterisks (*).

$$OHC-\overset{\displaystyle H}{\underset{\displaystyle OH}{\overset{|}{C}*}}-\overset{\displaystyle H}{\underset{\displaystyle OH}{\overset{|}{C}*}}-\overset{\displaystyle OH}{\underset{\displaystyle H}{\overset{|}{C}*}}-\overset{\displaystyle OH}{\underset{\displaystyle H}{\overset{|}{C}*}}-CH_2OH$$

---

* An asterisk (*) denotes a chiral carbon atom.

19.61 (a) GTACTCAAT; (b) ACTTAACGT

19.63 (a) $C_5H_5N_5O$ (b) $C_6H_{12}O_6$ (c) $C_3H_7NO_2$

19.65 (a) alcohol, ether, aldehyde; (b) ketone, alkene; (c) amine, amide

19.67

(a)

(b)

19.69 (a)

(b)

19.71 (a)

| methylamine | dimethyl ether | acetic acid |

(b) methylamine, H_2NCH_3, 31.06 g. mol^{-1}

$2H_2NCH_3(g) + 4\tfrac{1}{2}O_2(g) \rightarrow 2CO_2(g) + 5H_2O_{(l)} + N_{2(g)}$

$\Delta H_c° = 4(388) + 2(305) + 6(412) + 4.5(496) - 4(743) - 10(463) - 944$

$= 1552 + 610 + 2472 + 2232 - 2972 - 4630 - 944$

$= -1680$ kJ. mol^{-1} reaction

Since the balanced reaction has 2 mol H_2NCH_3 the enthalpy per gram is:

$$\frac{-1680 \text{ kJ. mol}^{-1}}{2(31.06 \text{ g. mol}^{-1})} = -27.04 \text{ kJ. g}^{-1}$$

dimethyl ether, CH_3OCH_3, 46.07 g. mol^{-1}

$CH_3OCH_3(g) + 3\tfrac{1}{2}O_2(g) \qquad \rightarrow \qquad 2CO_2(g) + 3 H_2O_{(l)}$

$\Delta H_c° = 2(360) + 6(412) + 3.5(496) - 4(743) - 6(463)$

$= 720 + 2472 + 1736 - 2972 - 2778$

$= -822$ kJ. mol^{-1}

$$\frac{-822 \text{ kJ. mol}^{-1}}{46.07 \text{ g. mol}^{-1}} = -17.84 \text{ kJ. g}^{-1}$$

acetic acid, CH_3COOH, 60.05 g. mol^{-1}

$CH_3COOH_{(l)} + 2 O_2(g) \rightarrow 2CO_2(g) + 2 H_2O(g)$

$\Delta H_c° = 3(412) + 348 + 743 + 360 + 463 + 2(496) - 4(743) - 4(463)$

$= 3150 + 992 - 2972 - 1852$

$= -682$ kJ. mol^{-1}

$$\frac{-682 \text{ kJ. mol}^{-1}}{60.05 \text{ g. mol}^{-1}} = -11.36 \text{ kJ. g}^{-1}$$

(c) Methylamine is not suitable as a fuel as it is a controlled substance because it is a precursor to methamphetamine. Dimethyl ether is not toxic nor a controlled substance and is a promising clean-burning hydrocarbon fuel. Acetic acid is used in vinegar and is not toxic in dilute aqueous solutions. However in pure form it is highly corrosive which would be a problem. Also it has a low specific enthalpy of combustion and is therefore not suitable as a clean-burning hydrocarbon fuel.

19.73

(a)

Diethyl ether

(b)

1-butanol

1-Butanol can hydrogen bond with itself but diethyl ether cannot, so 1-butanol molecules are held together more strongly in the liquid and therefore 1-butanol has the higher boiling point. Both compounds can form hydrogen bonds with water and therefore have similar solubilities.

19.75

(a)

(b)

19.77 (a) $HOCH_2CH_2OH + 2\,CH_3(CH_2)_{16}COOH \rightarrow$

$$CH_3(CH_2)_{16}COOCH_2CH_2OOC(CH_2)_{16}CH_3 + 2\,H_2O$$

(b)

$$2\,CH_3CH_2OH + HOOCCOOH \rightarrow CH_3CH_2OOCCOOCH_2CH_3 + 2\,H_2O$$

(c) $CH_3CH_2CH_2CH_2OH + CH_3CH_2COOH \xrightarrow{\;\Delta\;}$

$$CH_3CH_2COOCH_2CH_2CH_2CH_3 + H_2O$$

19.79 (a) The polymer concentration is given by:

$$c = \frac{\Pi}{iRT} = \frac{0.325\ \text{atm}}{(1)\left(0.08206\ \tfrac{L\,atm}{K\,mol}\right)(298\ \text{K})} = 0.0133\ \text{M}$$

The moles of polymer in solution are:

$$n_{polymer} = (0.0133\ \text{M})(0.500\ \text{L}) = 0.00664\ \text{mol}$$

and the molar mass of the polymer is:

$$\text{M.M.} = \frac{47.7\ \text{g}}{0.00664\ \text{mol}} = 7180\ \tfrac{g}{mol}$$

(b) If a monomer in the polymer is $-CH_2CH(CN)-$, then the molar mass of a monomer is 53.06 g/mol and the average number of monomers in a polymer is: $(7180\text{g/mol})/(53.06\ \text{g/mol}) = 135$ monomers.

(c) The pressure of $H_2O(g)$ above the mixture is given by:

$$P_{H_2O} = X_{H_2O} \cdot P_{H_2O,\,pure}$$

in 100 mL, $n_{H_2O} = \dfrac{(100\ \text{mL})\left(1.00\ \tfrac{g}{mL}\right)}{18.02\ \tfrac{g}{mol}} = 5.55\ \text{mol}$

and, $n_{polymer} = (0.100\ \text{L})\left(0.01328\ \tfrac{mol}{L}\right) = 0.00133$ mol.

Therefore,

$$X_{H_2O} = \frac{5.55\ \text{mol}}{5.55\ \text{mol} + 0.00133\ \text{mol}} = 0.9998 \quad \text{and}$$

$$P_{H_2O} = (0.9998)(23.76\ \text{Torr}) = 23.75\ \text{Torr}$$

(d) Measuring the change in osmotic pressure proves to be a better method in this case. The osmotic pressure developed by the resulting polymer solution is readily measured while the change in partial pressure of $H_2O(g)$ changes by less than 0.1 % upon addition of the polymer.

19.81 (a) Primary structure is the sequence of amino acids along a protein chain. Secondary structure is the conformation of the protein, or the manner in which the chain is coiled or layered, as a result of interactions between amide and carboxyl groups. Tertiary structure is the shape into which sections of the proteins twist and intertwine, as a result of interactions between side groups of the amino acids in the protein. If the protein consists of several polypeptide units, then the manner in which the units stick together is the quaternary structure.

(b) The primary structure is held together by covalent bonds. Intermolecular forces provide the major stabilizing force of the secondary structure. The tertiary structure is maintained by a combination of London forces, hydrogen bonding, and sometimes ion-ion interactions. The same forces are responsible for the quaternary structure.

19.83 (a) $^+H_3NCH_2COOH(aq) + H_2O(l) \rightarrow {}^+H_3NCH_2COO^-(aq) + H_3O^+(aq)$

$^+H_3NCH_2COO^-(aq) + H_2O(l) \rightarrow H_2NCH_2COO^-(aq) + H_3O^+(aq)$

(b) $pK_{a1} = 2.35 \qquad pK_{a2} = 9.78$

$pH = 2,\ ^+H_3NCH_2COOH$

$pH = 5,\ ^+H_3NCH_2COO^-$

$pH = 12,\ H_2NCH_2COO^-$

19.85 Condensation polymerization involves the loss of a small molecule, often water or HCl, when monomers are combined. Dacron is more linear than the polymer obtained from benzene-1,2-dicarboxylic acid and ethylene glycol, so Dacron can be more readily spun into yarn.

19.87

$$H_2\ddot{N}-CH-\underset{\underset{\displaystyle CH_2}{|}}{\overset{\overset{\displaystyle :\ddot{O}:}{\|}}{C}}-H\ddot{N}-CH-\underset{\underset{\displaystyle CH_2}{|}}{\overset{\overset{\displaystyle :\ddot{O}:}{\|}}{C}}-H\ddot{N}-CH-\underset{\underset{\displaystyle CH_2}{|}}{\overset{\overset{\displaystyle :\ddot{O}:}{\|}}{C}}-\ddot{O}H$$

H₂N—CH-C—HN—CH-C—HN—CH-C—OH
with side chains:

- First residue: CH₂ — CH₂ — C=O — :OH
- Second residue: CH₂ — CH-CH₃ — CH₃
- Third residue: CH₂ — C=O — :OH

19.89

$$\left[-O-\underset{\underset{\displaystyle CH_3}{|}}{\overset{\overset{\displaystyle CH_3}{|}}{Si}}-O-\underset{\underset{\displaystyle CH_3}{|}}{\overset{\overset{\displaystyle CH_3}{|}}{Si}}-O-\underset{\underset{\displaystyle CH_3}{|}}{\overset{\overset{\displaystyle CH_3}{|}}{Si}}- \right]$$

19.91 (a)

NH₂ attached to a benzene ring

(b) sp^3 ; (c) sp^2 ; (d) each N atom carries one lone pair of electrons; (e) Yes, the N atoms help to carry the current because the unhybridized p orbital on each N atom is part of the extended π conjugation (delocalized π -bonds) that allows electrons to move freely along the polymer.

19.93 Two peaks are observed with relative overall intensities 3 : 1. The larger peak is due to the three methyl protons and is split into two lines with equal intensities. The smaller peak is due to the proton on the carbonyl carbon atom and is split into four lines with relative intensities 1 : 3 : 3 : 1.

19.95 The peaks in the spectrum can be assigned on the basis of the intensities and the coupling to other peaks. The hydrogen atoms of the CH_3 unit of the ethyl group will have an intensity of 3 and will be split into a triplet by the two protons on the CH_2 unit. This peak is found at $\delta \approx 1.2$. The CH_2 unit will have an intensity of 2 and will be split into a quartet by the three protons on the methyl group. This peak is found at $\delta = 4.1$.. The CH_2 group that is part of the butyl function will have an intensity of two but will appear as a singlet because there are no protons on adjacent carbon atoms. This is the signal found at $\delta = 2.1$. The remaining CH_3 groups are equivalent and also will not show coupling. They can be attributed to the signal at $\delta = 1.0$. Notice that the peak that is most downfield is the one for the CH_2 group attached directly to the electronegative oxygen atom, and that the second most downfield peak is the one attached to the carbonyl group.

19.97 If one considers the reaction, the products should be those arising from substitution of hydrogen atoms on the propane by chlorine atoms. We would expect then to form a chloropropane or, perhaps a dichloropropane. Remember that in the halogenation of alkanes, substitution becomes more difficult as more halogen atoms are introduced. If we then consider the NMR spectrum, we see that there is one large peak that sees a single proton, because it is split into a doublet. There is also a weaker feature, corresponding most likely to one proton, which is split into a septet. This indicates that the proton sees six equivalent protons. The structure that is consistent with this spectrum is 2-chloropropane.

19.99 (a) ^{13}C; (b) 1.11%; (c) No. The reason is that the probability of finding two ^{13}C nuclei next to each other is very low. A ^{12}C nucleus next to a ^{13}C nucleus will not interact with the ^{13}C nucleus because the ^{12}C nucleus has no spin. Because the natural abundance of ^{13}C is 1.11%, the probability of finding two ^{13}C nuclei next to each other in an organic molecule will be 0.0111×0.0111 or 1.23×10^{-4}. Although such coupling is possible, it is generally not observed because the signal is so much weaker than the signal due to the molecules with a single ^{13}C nucleus.

(d) Maybe. Because most of the carbon to which the protons are attached is ^{12}C, the bulk of the signal will not be split. The protons that are attached to the ^{13}C atoms will be split, but this amounts to only 1.11% of the sample, so the peaks are very small. Peaks that result from coupling to a small percentage of a magnetically-active isotope are referred to as satellites and may be observed if one has a very good spectrum.

(e) Yes. Although the splitting of protons by ^{13}C may not be observed because the amount of ^{13}C present is low, the opposite situation is not true. If a ^{13}C is attached to H atoms, the large majority of those H atoms will have a spin and so the ^{13}C will show fine structure due to splitting by the H atoms.

19.101 (a) 1; (b) 1; (c) 1; (d) 2; (e) Free rotation around the C-C and C-Cl bonds averages out the environments so that both CH_2 hydrogens are equivalent and all three CH_3 hydrogens are equivalent, resulting in just two different hydrogen signals.